Analysis of Queues
Methods and Applications

The Operations Research Series

Series Editor: A. Ravi Ravindran

Professor, Department of Industrial and Manufacturing Engineering
The Pennsylvania State University – University Park, PA

Published Titles:

Analysis of Queues: Methods and Applications
Natarajan Gautam

Integer Programming: Theory and Practice
John K. Karlof

Operations Research and Management Science Handbook
A. Ravi Ravindran

Operations Research Applications
A. Ravi Ravindran

Operations Research: A Practical Introduction
Michael W. Carter & Camille C. Price

Operations Research Calculations Handbook, Second Edition
Dennis Blumenfeld

Operations Research Methodologies
A. Ravi Ravindran

Probability Models in Operations Research
C. Richard Cassady & Joel A. Nachlas

Forthcoming Titles:

Introduction to Linear Optimization and Extensions with MATLAB®
Roy H. Kwon

Supply Chain Engineering: Models and Applications
A. Ravi Ravindran & Donald Paul Warsing

Analysis of Queues
Methods and Applications

Natarajan Gautam

CRC Press
Taylor & Francis Group
Boca Raton London New York

CRC Press is an imprint of the
Taylor & Francis Group, an **informa** business

CRC Press
Taylor & Francis Group
6000 Broken Sound Parkway NW, Suite 300
Boca Raton, FL 33487-2742

© 2012 by Taylor & Francis Group, LLC
CRC Press is an imprint of Taylor & Francis Group, an Informa business

No claim to original U.S. Government works

Printed in the United States of America on acid-free paper
Version Date: 20120312

International Standard Book Number: 978-1-4398-0658-6 (Hardback)

Library of Congress Cataloging-in-Publication Data

Gautam, Natarajan.
    Analysis of queues : methods and applications / Natarajan Gautam.
        p. cm. -- (Operations research series)
    Summary: "Analysis of queues is used in a variety of domains including call centers, web servers, internet routers, manufacturing and production, telecommunications, transportation, hospitals and clinics, restaurants, and theme parks. Combining elements of classical queueing theory with some of the recent advances in studying stochastic networks, this book covers a broad range of applications. It contains numerous real-world examples and industrial applications in all chapters. The text is suitable for graduate courses, as well as researchers, consultants and analysts that work on performance modeling or use queueing models as analysis tools"-- Provided by publisher.
    Includes bibliographical references and index.
    ISBN 978-1-4398-0658-6 (hardback)
    1. Queuing theory. I. Title.

T57.9.G38 2012
519.8'2--dc23
                                                                                    2012006570

Visit the Taylor & Francis Web site at
http://www.taylorandfrancis.com

and the CRC Press Web site at
http://www.crcpress.com

To my parents

Ramaa Natarajan and P.R. Natarajan

Contents

Preface.. xv
Acknowledgments xvii
Author... xix
List of Case Studies... xxi
List of Paradoxes xxiii

1. Introduction 1
   1.1 Analysis of Queues: Where, What, and How?.................... 2
        1.1.1 Where Is This Used? The Applications 2
        1.1.2 What Is Needed? The Art of Modeling.................... 5
        1.1.3 How Do We Plan to Proceed? Scope and Methods...... 7
   1.2 Systems Analysis: Key Results..................................... 8
        1.2.1 Stability and Flow Conservation 10
        1.2.2 Definitions Based on Limiting Averages................. 10
        1.2.3 Asymptotically Stationary and Ergodic Flow
               Systems.. 11
        1.2.4 Little's Law for Discrete Flow Systems.................. 12
        1.2.5 Observing a Flow System According to a Poisson
               Process.. 14
   1.3 Queueing Fundamentals and Notations 17
        1.3.1 Fundamental Queueing Terminology 21
        1.3.2 Modeling a Queueing System as a Flow System........ 26
        1.3.3 Relationship between System Metrics for $G/G/s$
               Queues 29
               1.3.3.1 $G/G/s/K$ Queue 34
        1.3.4 Special Case of $M/G/s$ Queue............................ 35
   1.4 Psychology in Queueing 36
   Reference Notes 41
   Exercises.. 42

2. Exponential Interarrival and Service Times: Closed-Form
   Expressions 45
   2.1 Solving Balance Equations via Arc Cuts......................... 46
        2.1.1 Multiserver and Finite Capacity Queue Model
               $(M/M/s/K)$... 48
        2.1.2 Steady-State Analysis 49
        2.1.3 Special Cases 55
   2.2 Solving Balance Equations Using Generating Functions 61
        2.2.1 Single Server and Infinite Capacity Queue
               $(M/M/1)$... 63

2.2.2   Retrial Queue with Application to Local Area
        Networks... 65
2.2.3   Bulk Arrival Queues ($M^{[X]}/M/1$) with a Service
        System Example 70
2.2.4   Catastrophic Breakdown of Servers in $M/M/1$
        Queues 74
2.3 Solving Balance Equations Using Reversibility 78
2.3.1   Reversible Processes.. 79
2.3.2   Properties of Reversible Processes 80
2.3.3   Example: Analysis of Bandwidth-Sensitive Traffic...... 81
Reference Notes 85
Exercises.. 85

3. **Exponential Interarrival and Service Times: Numerical
    Techniques and Approximations** 93
3.1 Multidimensional Birth and Death Chains....................... 93
3.1.1   Motivation: Threshold Policies in Optimal Control 94
3.1.2   Algorithm Using Recursively Tridiagonal Linear
        Equations 103
3.1.3   Example: Optimal Customer Routing 114
3.2 Multidimensional Markov Chains 117
3.2.1   Quasi-Birth-Death Processes.............................. 118
3.2.2   Matrix Geometric Method for QBD Analysis 122
3.2.3   Example: Variable Service Rate Queues in Computer
        Systems.. 124
3.3 Finite-State Markov Chains....................................... 132
3.3.1   Efficiently Computing Steady-State Probabilities 132
        3.3.1.1   Eigenvalues and Eigenvectors 133
        3.3.1.2   Direct Computation.............................. 134
        3.3.1.3   Using Transient Analysis 135
        3.3.1.4   Finite-State Approximation..................... 136
3.3.2   Example: Energy Conservation in Data Centers......... 136
Reference Notes 147
Exercises.. 147

4. **General Interarrival and/or Service Times: Closed-Form
    Expressions and Approximations** 151
4.1 Analyzing Queues Using Discrete Time Markov Chains....... 151
4.1.1 $M/G/1$ Queue 153
4.1.2 $G/M/1$ Queue 168
4.2 Mean Value Analysis 180
4.2.1   Explaining MVA Using an $M/G/1$ Queue 181
4.2.2   Approximations for Renewal Arrivals and General
        Service Queues.. 184
4.2.3   Departures from $G/G/1$ Queues........................... 187

4.3   Bounds and Approximations for General Queues 188
        4.3.1   General Single Server Queueing System ($G/G/1$) 189
        4.3.2   Multiserver Queues ($M/G/s$, $G/M/s$, and $G/G/s$) 194
        4.3.3   Case Study: Staffing and Work-Assignment in Call
                 Centers 196
                 4.3.3.1   TravHelp Calls for Help 197
                 4.3.3.2   Recommendation: Major Revamp 198
                 4.3.3.3   Findings and Adjustments...................... 200
4.4   Matrix Geometric Methods for $G/G/s$ Queues 201
        4.4.1   Phase-Type Processes: Description and Fitting 202
        4.4.2   Analysis of Aggregated Phase-Type Queue
                 ($\sum PH_i/PH/s$) 205
        4.4.3   Example: Application in Semiconductor
                 Wafer Fabs 209
4.5   Other General Queues but with Exact Results 213
        4.5.1 $M/G/\infty$ Queue: Modeling Systems with Ample
                 Servers... 213
        4.5.2 $M/G/1$ Queue with Processor Sharing:
                 Approximating CPUs 221
        4.5.3 $M/G/s/s$ Queue: Telephone Switch Application......... 227
Reference Notes 234
Exercises... 235

5. **Multiclass Queues under Various Service Disciplines** 241
5.1   Introduction.. 241
        5.1.1   Examples of Multiclass Systems........................... 242
        5.1.2   Preliminaries: Little's Law for the Multiclass
                 System... 244
        5.1.3   Work-Conserving Disciplines for Multiclass $G/G/1$
                 Queues 246
        5.1.4   Special Case: At Most One Partially Completed
                 Service 249
5.2   Evaluating Policies for Classification Based on Types:
        Priorities 253
        5.2.1   Multiclass $M/G/1$ with FCFS 254
        5.2.2 $M/G/1$ with Nonpreemptive Priority 256
        5.2.3 $M/G/1$ with Preemptive Resume Priority 262
        5.2.4   Case Study: Emergency Ward Planning 266
                 5.2.4.1   Service Received by Class-1 Patients........... 268
                 5.2.4.2   Experience for Class-2 Patients 269
                 5.2.4.3   Three-Class Emergency Ward Operation 271
5.3   Evaluating Policies for Classification Based on Location:
        Polling Models 272
        5.3.1   Exhaustive Polling... 274
        5.3.2   Gated Policy 277

5.3.3   Limited Service... 280
5.4   Evaluating Policies for Classification Based on Knowledge
      of Service Times 282
      5.4.1   Shortest Processing Time First............................ 284
      5.4.2   Preemptive Shortest Job First 285
      5.4.3   Shortest Remaining Processing Time 287
5.5   Optimal Service-Scheduling Policies............................ 293
      5.5.1   Setting and Classification................................. 293
      5.5.2   Optimal Scheduling Policies in Single Class
              Queues 296
      5.5.3   Optimal Scheduling Policies in Multiclass Queues...... 300
Reference Notes 303
Exercises... 304

6.   **Exact Results in Network of Queues:Product Form................... 311**
6.1   Acyclic Queueing Networks with Poisson Flows 312
      6.1.1   Departure Processes... 313
      6.1.2   Superpositioning and Splitting 315
      6.1.3   Case Study: Automobile Service Station 319
              6.1.3.1   System Description and Model 320
              6.1.3.2   Analysis and Recommendation................. 321
6.2   Open Jackson Networks 324
      6.2.1   Flow Conservation and Stability 325
      6.2.2   Product-Form Solution 326
      6.2.3   Examples.. 330
6.3   Closed Jackson Networks.. 339
      6.3.1   Product-Form Solution 340
      6.3.2   Arrivals See Time Averages (ASTA)...................... 346
      6.3.3   Single-Server Closed Jackson Networks 353
6.4   Other Product-Form Networks 357
      6.4.1   Open Jackson Networks with State-Dependent
              Arrivals and Service... 358
      6.4.2   Open Jackson–Like Networks with Deterministic
              Routing 360
      6.4.3   Multiclass Networks 364
      6.4.4   Loss Networks 367
Reference Notes 369
Exercises... 370

7.   **Approximations for General Queueing
      Networks... 377**
7.1   Single-Server and Single-Class General Queueing
      Networks 378
      7.1.1 $G/G/1$ Queue: Reflected Brownian Motion–Based
              Approximation... 379

7.1.2 Superpositioning, Splitting, and Flow through a
Queue... 390
    7.1.2.1 Superposition 390
    7.1.2.2 Flow through a Queue 392
    7.1.2.3 Bernoulli Splitting............................... 393
7.1.3 Decomposition Algorithm for Open Queueing
Networks... 394
7.1.4 Approximate Algorithms for Closed Queueing
Networks... 399
    7.1.4.1 Bottleneck Approximation for Large C 400
    7.1.4.2 MVA Approximation for Small C 402
7.2 Multiclass and Multiserver Open Queueing Networks
with FCFS 406
7.2.1 Preliminaries: Network Description 407
7.2.2 Extending $G/G/1$ Results to Multiserver, Multiclass
Networks... 409
    7.2.2.1 Flow through Multiple Servers 411
    7.2.2.2 Flow across Multiple Classes.................... 411
    7.2.2.3 Superposition and Splitting of Flows 413
7.2.3 QNA Algorithm.. 414
7.2.4 Case Study: Network Interface Card in Cluster
Computing... 421
7.3 Multiclass and Single-Server Open Queueing Networks
with Priorities... 424
7.3.1 Global Priorities: Exponential Case 425
7.3.2 Global Priorities: General Case 431
7.3.3 Local Priorities 435
    7.3.3.1 MVA-Based Algorithm........................... 436
    7.3.3.2 State-Space-Collapse-Based Algorithm........ 437
Reference Notes 441
Exercises... 442

8. **Fluid Models for Stability, Approximations, and Analysis of
Time-Varying Queues**... 447
8.1 Deterministic Fluid Queues: An Introduction 447
8.1.1 Single Queue with a Single Server 448
8.1.2 Functional Strong Law of Large Numbers and the
Fluid Limit.. 450
8.2 Fluid Models for Stability Analysis of Queueing
Networks 454
8.2.1 Special Multiclass Queueing Networks with Virtual
Stations 454
8.2.2 Stable Fluid Network Implies Stable Discrete
Network.. 467

8.2.3   Is the Fluid Model of a Given Queueing Network
        Stable? 475
8.3   Diffusion Approximations for Performance Analysis 483
      8.3.1   Diffusion Limit and Functional Central Limit
              Theorem 484
      8.3.2   Diffusion Approximation for Multiserver Queues 488
              8.3.2.1   Fix s, Increase λ 490
              8.3.2.2   Fix ρ, Increase λ and s 491
              8.3.2.3   Fix β, increase λ and s 492
      8.3.3   Efficiency-Driven Regime for Multiserver Queues
              with Abandonments 495
8.4   Fluid Models for Queues with Time-Varying Parameters 500
      8.4.1   Uniform Acceleration 502
      8.4.2   Diffusion Approximation 507
Reference Notes 511
Exercises 513

9. **Stochastic Fluid-Flow Queues: Characteristics and Exact**
    Analysis 515
9.1   Introduction 516
      9.1.1   Discrete versus Fluid Queues 516
      9.1.2   Applications 518
      9.1.3   Preliminaries for Performance Analysis 521
      9.1.4   Environment Process Characterization 522
              9.1.4.1   CTMC Environmental Processes 522
              9.1.4.2   Alternating Renewal Environmental
                        Processes 522
              9.1.4.3   SMP Environmental Processes 523
9.2   Single Buffer with Markov Modulated Fluid Source 525
      9.2.1   Terminology and Notation 525
      9.2.2   Buffer Content Analysis 530
      9.2.3   Steady-State Results and Performance Evaluation 534
      9.2.4   Examples 537
9.3   First Passage Times 550
      9.3.1   Partial Differential Equations and Boundary
              Conditions 551
      9.3.2   Examples 555
Reference Notes 584
Exercises 585

10. **Stochastic Fluid-Flow Queues: Bounds and Tail Asymptotics** 589
10.1 Introduction and Preliminaries 589
      10.1.1 Inflow Characteristics: Effective Bandwidths and
             ALMGF 590
      10.1.2 Computing Effective Bandwidth and ALMGF 593

        10.1.2.1 Effective Bandwidth of a CTMC Source 593

        10.1.2.2 Effective Bandwidth of a Semi-Markov
                Process (SMP) Source 598

        10.1.2.3 Effective Bandwidth of a General On/Off
                Source 601

     10.1.3 Two Extensions: Traffic Superposition and Flow
         through a Queue.. 605

        10.1.3.1 Superposition of Multiple Sources 605

        10.1.3.2 Effective Bandwidth of the Output from a
                Queue.. 607

10.2 Performance Analysis of a Single Queue......................... 609

     10.2.1 Effective Bandwidths for Tail Asymptotics.............. 611

     10.2.2 Chernoff Dominant Eigenvalue Approximation......... 622

     10.2.3 Bounds for Buffer Content Distribution.................. 627

10.3 Multiclass Fluid Queues 648

     10.3.1 Tackling Varying Output Capacity: Compensating
         Source 650

     10.3.2 Timed Round Robin (Polling) 651

     10.3.3 Static Priority Service Policy 662

     10.3.4 Other Policies... 674

Reference Notes 686

Exercises.. 687

Appendix A: Random Variables... 691

 A.1 Distribution and Moments 691

     A.1.1 Discrete Random Variables 693

     A.1.2 Continuous Random Variables 697

     A.1.3 Coefficient of Variation 702

 A.2 Generating Functions and Transforms 703

     A.2.1 Generating Functions 704

     A.2.2 Laplace–Stieltjes Transforms.............................. 705

     A.2.3 Laplace Transforms 707

 A.3 Conditional Random Variables 708

     A.3.1 Obtaining Probabilities 708

     A.3.2 Obtaining Expected Values 711

 A.4 Exponential Distribution... 714

     A.4.1 Characteristics... 714

     A.4.2 Properties 714

 A.5 Collection of IID Random Variables 716

     A.5.1 Poisson Process 717

     A.5.2 Renewal Process 720

Reference Notes 722

Exercises.. 722

Appendix B: Stochastic Processes **725**
  B.1   Discrete-Time Markov Chains 725
      B.1.1 Modeling a System as a DTMC 726
      B.1.2 Transient Analysis of DTMCs 729
      B.1.3 Steady-State Analysis of DTMCs........................... 730
  B.2   Continuous-Time Markov Chains................................... 732
      B.2.1 Modeling a System as a CTMC 733
      B.2.2 Transient Analysis of CTMCs.......................... 737
      B.2.3 Steady-State Analysis of CTMCs 738
  B.3   Semi-Markov Process and Markov Regenerative
      Processes 740
      B.3.1 Markov Renewal Sequence................................ 740
      B.3.2 Semi-Markov Process 742
      B.3.3 Markov Regenerative Processes 744
  B.4   Brownian Motion 746
      B.4.1 Definition of Brownian Motion 747
      B.4.2 Analysis of Brownian Motion.............................. 749
      B.4.3 Itô's Calculus 750
  Reference Notes 752
  Exercises... 753

References... **757**

Index **763**

Preface

The field of design, control, and performance analysis of queueing systems is extremely mature. The gap between the content of the texts used in undergraduate courses and what is published in leading journals like *Queueing Systems: Theory and Applications* is astounding. This gap can be bridged by offering a course or two covering a broad range of classical and contemporary topics in queueing. For this, one requires a text that not only deals with a wide range of concepts, but also has an exposition that would lend students to read up materials that are not covered in class on their own. This is precisely what motivated the content and presentation of this book. It has been written primarily keeping students in mind and has several solved examples, applications, case studies, and exercises for students to be able to read and understand the analytical nuances of various topics. Subsequently, they can dig deeper into specific concepts by taking specialized courses; working with their advisers; and/or reading research monographs, advanced texts, and journal articles to come up to speed with the state of the art.

Besides students, this book has also been written keeping instructors in mind. Having taught courses in queueing in two large industrial engineering departments, one of the major difficulties I have always faced is that the students are extremely heterogeneous, especially in terms of background. Some students are in the first year of their master's program, some are advanced master's students, many are early doctoral students, and a few are advanced doctoral students. Furthermore, students are from different departments, and they earn undergraduate degrees in several disciplines from various countries and institutions and at various points of time. One of the challenges to teach a course under such heterogeneous circumstances is to make sure that all students have the prerequisite material without having to teach that in class. This motivated the two appendix chapters, where concepts are explained in great detail with numerous examples and exercises. Students without the right prerequisite material can prepare themselves for the course by reading the appendices on their own, preferably before the course starts.

In summary, this book has been written to serve as a text in graduate courses on the topic of queueing models. Following are two alternative course structures based on the material presented in this book:

1. Teach one graduate course on queueing models that follows an introductory course on probability and random processes. For such a course on queueing models, since the contents of each chapter in this book are fairly modular, the instructor can easily select a subset of topics of his or her choice from each of the 10 main chapters. As a suggestion, instructors could cover the first half to two-thirds of

each chapter and leave out the remaining topics or assign them for independent reading.

2. Furthermore, this book would be perfect when used in two courses. In the first course, one could cover the appendices, followed by Chapters 1 through 4. And in the second (advanced) course, one could cover Chapters 5 through 10. In that case, it would be sufficient to require an undergraduate course on probability as a prerequisite to the first course in the sequence.

   The analytical methods presented in this book are substantiated using applications from a wide set of domains, including production, computer, communication, information, transportation, and service systems. This book could thus be used in courses in programs such as industrial engineering, systems engineering, operations research, statistics, management science, operations management, applied mathematics, electrical engineering, computer science, and transportation engineering. In addition, I sincerely hope that this book appeals to an audience beyond students and instructors. It would be appropriate for researchers, consultants, and analysts that work on performance modeling or use queueing models as analysis tools. This book has evolved based on my numerous offerings of entry-level to mid-level graduate courses on the theory and application of queueing systems. Those courses have been my favorite among graduate courses, and I am absolutely passionate about the subject area. I have truly enjoyed writing this book, and I sincerely hope you will enjoy reading it and getting value out of it.

Natarajan Gautam
College Station, Texas

For MATLAB® and Simulink® product information, please contact:

The MathWorks, Inc.
3 Apple Hill Drive
Natick, MA, 01760-2098 USA
Tel: 508-647-7000
Fax: 508-647-7001
E-mail: info@mathworks.com
Web: www.mathworks.com

Acknowledgments

I have come to realize after writing this book that just like it takes a village to raise a child, it does so to write a book as well. I would like to take this opportunity to express my gratitude to a small subset of that *village*.

I would like to begin by thanking my dissertation adviser Professor Vidyadhar G. Kulkarni for all the knowledge, guidance, and professional skills he has shared with me. His textbooks have been a source of inspiration and a wealth of information that have been instrumental in shaping this book. I would also like to acknowledge Professor Kulkarni's fabulous teaching style that I could only wish to emulate. Talking about excellent teachers, I would like to thank all the fantastic teachers I have had growing up. I was lucky to have fabulous mathematics teachers in high school—Mrs. Sarvamangala and Mr. Nainamalai. I am also grateful to my excellent instructors during my undergraduate program, including Professor G. Srinivasan for his course on operations research and Professor P.R. Parthasarathy for his course on probability and random processes. I would also like to thank Professor Shaler Stidham Jr., who taught me the only course on queueing that I have ever taken as a student.

Next, I would like to express my sincerest gratitude to some of my colleagues. In particular, I would like to thank Professor A. Ravindran for encouraging me to write this book and for all his tips for successfully completing it. I was also greatly motivated by the serendipitous conversation I had with Professor Sheldon Ross when he happened to sit by me during a bus ride at an INFORMS conference. In addition, I would also like to thank several colleagues that have helped me with this manuscript through numerous conversations, brainstorming sessions, and e-mail exchanges. They include Professors Karl Sigman, Ward Whitt, and David Yao from Columbia University; Dr. Mark Squillante from IBM; Professors Raj Acharya, Russell Barton, Jeya Chandra, Geroge Kesidis, Soundar Kumara, Anand Sivasubramaniam, Qian Wang, and Susan Xu from Penn State University; Professors J.-F. Chamberland, Guy Curry, Rich Feldman, Georgia-Ann Klutke, P.R. Kumar, Lewis Ntaimo, Henry Pfister, Don Phillips, Srinivas Shakkottai, Alex Sprintson, and Marty Wortman from Texas A&M University; Professor Rhonda Righter from the University of California at Berkeley; Professor Sunil Kumar from the University of Chicago; and Professors Anant Balakrishnan, John Hasenbein, David Morton, and Sridhar Seshadri from the University of Texas at Austin.

Some of the major contributions to the contents of this book are due to my former and current students that took my courses and collaborated on research with me. In particular, I would like to thank Vineet Aggarwal, Yiyu

Chen, Naveen Cherukuri, Prathi Chinnusami, Jenna Estep, Maria Emelia-nenko, Donna Ghosh, Nathan Gnanasambandham, Piyush Goel, Sai Rajesh Mahabhashyam, Cesar Rincon Mateus, Cenk Ozmutlu, Venky Sarangan, Mohamed Yacoubi, and Yanyong Zhang, as some of their research has appeared in example problems in this book. In addition, thanks also to the following students who not only did research that resulted in several examples in this book, but also took the time to read various chapters: Ezgi Eren, Jeff Kharoufeh, Young Myoung Ko, Arupa Mohapatra, Ronny Polansky, and Samyukta Sethuraman. Also, thanks to Youngchul Kim for some of the simulation results in this book.

Last but not the least, I would like to thank my friends and family, in starting with my friends from graduate school such as Suresh Acharya, Conrad Lautenbacher, Anu Narayanan, Srini Rajgopal, and Chris Rump who taught me things that I was reminded of while writing portions of this book. I am greatly indebted to my family members, especially Madhurika Arvind, Rohini Nath, and Arvind Ramakrishnan, who have continuously provided moral and emotional support during the entire book-writing process. I would like to dedicate this book to my parents for their endless contributions to my growth and development. They have kept such good tabs on my progress—every time I completed a chapter, I made sure I e-mailed it to them. A big thanks also goes to my brother Gokul Natarajan, who has been a pillar of support, and my other family members, including Srinithya Karthik, Karthik Muralidharan, Priya Ramakrishnan, as well as my parents-in-law for being there for me. I would like to conclude this acknowledgment by thanking my wife, Srividya Ramasubramanian, and my son, Sankalp Gautam, for their love, encouragement, and compassion, all of which were despite my absence, tantrums, and constant ramblings over the last few years. Also, thanks to Srividya for picking the artwork for the cover of this book.

Author

Natarajan Gautam is an associate professor in the Department of Industrial and Systems Engineering with a courtesy appointment in the Department of Electrical and Computer Engineering at Texas A&M University, College Station. Prior to joining Texas A&M University in 2005, he was on the industrial engineering faculty at Pennsylvania State University, University Park, for eight years. He received his PhD in operations research from the University of North Carolina at Chapel Hill. Gautam serves as an associate editor for *INFORMS Journal on Computing* and *Omega*. He has held officers positions in the INFORMS Applied Probability Society, INFORMS Telecommunication Section, and the Computer and Information Systems division of the IIE. He received the IIE Outstanding Young Industrial Engineer Award (education category) in 2006.

List of Case Studies

1. Staffing and work-assignment in call centers (Section 4.3.3)
2. Hospital emergency ward planning (Section 5.2.4)
3. Automobile service station (Section 6.1.3)
4. Network interface card in cluster computing (Section 7.2.4)

List of Paradoxes

1. $M/G/1$ queue busy period (Remark 8)
2. Braess' paradox (Problem 55)
3. Waiting time method (Problem 56)
4. Inspection paradox (Section A.5.2)

The author maintains a website of Paradoxes in queueing, refer to that for a more up to date list (http://ise.tamu.edu/people/faculty/Gautam/paradox.pdf). Also, please consider contributing to it. Further, there is a list of paradoxes in the following Wikipedia site: http://en.wikipedia.org/wiki/List_of_paradoxes. It has some good ones on probability (including the Monty Hall problem and the three cards problem). Interestingly, there is a slightly different list of paradoxes in another Wikipedia site (http://en.wikipedia.org/wiki/Category:Probability_theory_paradoxes).

1

Introduction

For a moment imagine being on an island where you do not have to *wait* for anything; you get everything the instant you want or need them! Sounds like a dream doesn't it? Well, let us not have any illusions about it and state upfront that this book is not about creating such an island, leave alone creating such a world. *Wait happens!* This book is about how to deal with it. In other words, how do you analyze systems to manage the waiting experienced by users of the system. Having said that waiting is inevitable, it is only fair to point out that in several systems waiting has been significantly reduced using modern technology. For example, at fast-food restaurants it is now possible to order online and your food is ready when you show up. At some amusement parks, especially for popular rides, you can pick up a ticket that gives you a time to show up at the ride to avoid long lines. With online services, waits at banks and post offices have reduced considerably. Most toll booths these days have automated readers that let you zip through without stopping. There are numerous such examples and it appears like there are only a few places like airport security where the wait has gotten longer over the years!

Before delving into managing waiting, here are some further comments to consider:

- There is no doubt that information technology is an enabler for most of the reduction in wait times discussed in the preceding text. However, this also means that the information infrastructure needs to be managed diligently. For this reason, a good number of examples in this book will be based on applications in computer-communication systems, an understanding of which is crucial for everyone in this hi-tech era.
- Many systems in this day and age exhibit tremendous amount of variability. The variability comes in many forms. It could be in terms of the types of different users such as those ordering online versus those ordering inside a fast-food restaurant, for example. The variability could also be in terms of the duration of processing times, which could have an extremely high coefficient-of-variation (i.e., ratio of standard deviation to mean). It is a challenge to manage waiting times under high variability.
- Reducing waiting time does come at a price. It certainly takes a lot of resources to drastically reduce waiting time. However, even

if the cost of the resources are outweighed by the revenue due to lower waiting times, there could be other undesired long-term implications. For example, sometimes, to maintain and use the information infrastructure results in expending tremendous amount of greenhouse gases for the electricity to cool the servers, which is an undesirable by-product of reducing wait times.

Having described the challenges and implications associated with waiting, we would next like to describe what role *analysis of queues* (part of the title of this book) plays in addressing them.

1.1 Analysis of Queues: Where, What, and How?

We would like to set the stage for this book by describing some introductory remarks with respect to analysis of queues. In particular, we aim to address questions such as the following: *Where* have analysis of queues been successfully used? *What* do we need as inputs to do the analysis and what can we expect as outputs? *How* do we plan to go about analyzing queues?

1.1.1 Where Is This Used? The Applications

Most of the applications we consider in this book would fall under one of the three domains described in the following text, although it is worthwhile pointing out that the results on queueing can be used in a much wider set of domains.

- *Computer, communication, and information systems*: There was a conference in Copenhagen (Denmark) recently to celebrate 100 years of queueing, a field that began with A.K. Erlang's paper on queueing models for managing telephone calls. As the field started to grow rapidly, Leonard Kleinrock, who has written one of the most widely used queueing theory books, laid the foundation for Internet communication. The irony is that both telephony and the Internet have come a long way in terms of high performance and fidelity that many perhaps do not even realize the role of queueing in them anymore! For that reason, it may be worthwhile to briefly mention about queues in computer, communication, and/or information systems. Say you are sitting with your laptop in a hotel by the beach typing up an email to your friend in another country. Before you even hit the send button, the information in the email has made it through several queues in your laptop. One of them would be that of the central processing unit (CPU) in your laptop, which is essentially a server that processes jobs from various applications including your email client. As soon as you click on *Send*, your email gets broken up into smaller pieces of information called *packets* that get forwarded through the Internet all the way from your laptop to your friend's email client. During this process, the packets typically are stored in several queues such as those in the Internet routers. Notice that every packet needs to be reliably transmitted from the source to the destination in a timely manner through multiple systems owned and operated by different organizations. This is a significant challenge and queueing theory has played a major role in its design and operation. Typically, while information flows from a source to a destination, a significant time is spent waiting in queues. But that wait time can be controlled, which is one of the main concerns for networking professionals. Having said that, it is important to say that the role of queueing goes way beyond *networking*. It is also used to analyze end systems such as servers, computers, and other components some of which we will see throughout this book.

- *Production systems*: From a historical standpoint, as queueing models were being used to analyze telephony, mass production started to emerge in a big way and presented a need for efficiency. That led to a flurry of research in production systems paving way for a plethora of results in queueing. In fact, a significant chunk of queueing theory was developed with production systems as the driving force. By production systems we mean manufacturing, logistics, and transportation involved during the life cycle of a product. Although it is just a matter of terminology, sometimes facilities such as repair shops are considered as service systems in the literature but in this book they would still fall under production systems. Further, from an analysis standpoint, it may not be relevant to model all the entities involved in a product's life cycle as the *system*. It is quite typical to consider a factory that manufactures products as a separate system, then the transportation network that distributes the product as a separate system, and warehouses involved in the process as separate systems, etc. One of the reasons for doing this is that usually a single company owns and operates such systems. Even if the same company does the logistics, transportation, warehousing, and retailing, it may be worthwhile to still consider those as separate subsystems and analyze them individually. With that thought consider a factory that transforms raw materials into finished products. Such a system typically consists of workstations (or work centers) and buffers. The workstations are machines that we would call "servers" generically and the buffers are queues where products wait to be processed. Inherent variability in processing times, unreliable machines as well as multiple product types forces the need to use buffers. Managing the flow of products and analyzing the buffers is key to successfully running a production facility. For that, queueing-based analysis come in handy as we will see in examples throughout this book.

- *Service systems*: Although it is not at all important with respect to the rest of this book to differentiate between a production system and a service system, here we make that distinction based on whether or not physical products are moved through the system. Some examples of service systems where queueing plays a major role are call centers, hospitals, and entertainment-related hospitality services such as theme parks. We explain them further in terms of the role of queueing by considering them one by one. An *inbound* call center consists of a collection of representatives and communication devices that receive calls from customers of one or more companies. When you call a toll-free help desk telephone number, chances are the phone call goes to an inbound call center. If there is a representative available to attend the call you get to talk to them immediately, otherwise you wait listening to music till a representative becomes available. Oftentimes, customers become impatient and abandon calls before they could talk to a representative and then try calling at a later time. Also, sometimes the solution provided by the help desk would not work and the customer makes a second call. The objective of the call center is to provide at the lowest possible cost excellent service levels that include short waiting times, low abandonment rates, and low probability of second calls. Queueing theory plays a monumental role in the design and operation of such call centers by developing algorithms implemented in decision-support systems for staffing, cross-training, and call handling. Having motivated the role of queueing in call centers, we move on to another service system where customers experience queueing and waiting, namely, hospitals and health clinics. Irrespective of whether it is a clinic or an emergency ward, they can be modeled as a multistation queueing system. For example, when patients check in, they wait in a queue for their turn, then they wait to be called inside, and once inside they wait one or more times to be treated. Here too, like call centers, queueing models can be used for staffing, cross-training, and developing information systems to handle patients. In fact, in the third service system example too, which is theme parks (or amusement parks), queueing models are used for staffing, training, and information systems. Here customers go from one station to another, wait in line, and get served (i.e., enjoy the ride).

In summary, queues or waiting lines are found everywhere and they need to be managed appropriately to balance costs, performance, and customer satisfaction. This book mainly deals with queues in applications or domains described earlier such as computer-communication networks, production systems, and service systems. However, it is crucial to point out that there are many other applications and domains that we have not talked about. Having said that, the presentation of this book would be made in a rather abstract manner so that the reader does not get the impression that the queueing model has been developed only with a particular application in mind. However, it is critical to realize that there is a modeling step that converts an application scenario into an analysis framework and vice versa. We address that next by presenting what is needed to start the analysis.

1.1.2 What Is Needed? The Art of Modeling

Modeling is the process of *abstracting* a real-life scenario in a manner that is conducive for analysis. This process is in itself an *art* because there is no single right way of going about it and the process is quite subjective. Although one could develop scientific principles as well as best practices for the modeling process, in the end it is still a piece of art that some like and some do not. One way that a lot of people rationalize is by saying "whatever works." In other words, a good model is one that produces results that are satisfactory for the intended purposes. Having said that, let us go over a model selection exercise using an example not related to queueing. Say you are interested in knowing whether or not a missile would strike a particular stationary target, given the setting from where it is launched. By "setting" we mean the speed, direction, weight of the missile, and location at the time of launch. For simplicity, assume that the missile has no control system to adjust on the fly (so it is more like a cannon ball from a traditional cannon). How would you go about determining whether or not the missile will strike the target?

There are many ways to answer this question. A rational approach is to abstract the scenario so that we can analyze it easily and inexpensively. We could abstract the missile as a point mass, use the formulas for projectile motion from basic physics, and determine whether or not the projectile would strike the target. But wait, how about the following: the projectile formula neglects air resistance; the acceleration due to gravity could be different from what is used and it could vary during flight; the initial conditions may not be totally accurate (e.g., if speed is set to 100 m/s, it may actually be somewhere between 95 and 105 m/s in practice); there could be an object in the path such as a bird flying; and many other such issues. So what do we do now? Should we worry about none, some, or all these issues? Of course we can never be sure that we have considered all the potential issues.

So whichever set of issues we use, we would end up making statements like: "under such and such assumptions we can state that the probability of striking the target is 0.953." Unfortunately, many a time we tend to ignore the "assumptions" and just make blank statements; one should be cognizant of that.

Further, is the preceding statement in quotes good enough to make decisions? That depends on what decisions are to be made and whether the assumptions appear reasonable. Knowing that is also an art and is quite subjective too. Hence we will not delve into it any further. However, there is another dimension to modeling. Notice that before modeling we do need to know the question being asked and whether our analysis can answer that question. For the preceding example, if the question is to determine whether the cannon can be on a boat to fire the missile, considering the boat's capability to sustain the recoil, perhaps a different model may be needed. Alternatively, if the missile is capable of adjusting its trajectory on the fly using its control systems, then the model would probably have to be accordingly modified. In summary, it is incredibly crucial to understand what assumptions go into a modeling-analysis framework and what the framework is capable of determining. Having said that, it is important to iterate that the power of queueing models is the ability to provide useful results under mild assumptions. Now, we next describe the framework and scope of this book.

As the title of this book suggests, we are mainly interested in systems that can be modeled as a set of queues or waiting lines. There are many phases or steps involved in solving problems experienced by these *real-life* systems. Three of those phases are modeling the system (which is to convert the real-life system into a simplified representation that is conducive for analysis); choosing an appropriate tool and analyzing the simplified system using existing methods or developing new ones; and using the output of the analysis for design, control, what-if analysis, decision making, etc., in the real-life system. This is represented in Figure 1.1. Essentially, the modeling process converts the real-life system into a simplified representation containing model description and analysis. Next the negotiation process kicks-in which matches the model with the analysis framework, if one exists. If there is no analysis framework, then the model description and assumptions are suitably modified. This process continues until the real-life system can be suitably represented as a queueing system. Then either using existing methods, tools, and algorithms, or developing new ones, the queueing system is analyzed and performance measures are obtained. Then a decision-making process uses the performance measures in optimization, control, or what-if analysis that can be implemented in the real-life system. Finally, the design and operation process is altered suitably to implement the changes suggested. This process would be repeated until a desired outcome is obtained.

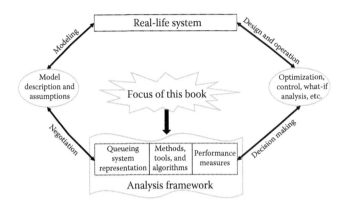

FIGURE 1.1
Framework and scope of this book.

1.1.3 How Do We Plan to Proceed? Scope and Methods

As described in Figure 1.1, the scope of this book is in the analysis framework. We will start with a queueing system representation and develop performance measures for that system. For that, we will consider several methods, tools, and algorithms, which would be the thrust of this book. Due to the subjectivity involved in modeling a real-life system as a queueing system we do not lay much emphasis on that. In addition, to develop a sense for how to model a system, it is critical to understand what goes into the analysis (so that the negotiation can be minimal and constructive). So in some sense, from a pedagogical standpoint, it is critical to present the analysis first and then move on to modeling. That way one knows what works and what does not, more importantly what systems can be analyzed and what cannot! Only if analysis of queues is well understood, do we have any hope of knowing how best to model a real-life system. Having described the scope of this book, next we describe the kind of methods used in this book.

The main focus of this book is to develop analytical methods for studying queues. The theoretical underpinnings of these analytical techniques can be categorized as *queueing theory*. An objective of queueing theory is to develop formulae, expressions, or algorithms for performance metrics such as average number of entities in a queue, mean time spent in the system, resource availability, and probability of rejection. The results from queueing theory can directly be used to solve design and capacity planning problems such as determining the number of servers, an optimum queueing discipline, schedule for service, number of queues, and system architecture. Besides making such strategic design decisions, queueing theory can also be used for tactical as well as operational decisions and controls. In summary, the scope of this book is essentially to develop performance metrics for queueing systems, which can subsequently be used in their design and operations.

In a nutshell, the methods in this book can be described as analytical approaches for system performance analysis. There are other approaches for system performance analysis such as simulations. It is critical to understand and appreciate situations when it is more appropriate to use queueing theory as well as situations where one is better off using simulations. Queueing theory is more appropriate when (a) several what-if situations need to be analyzed expeditiously, namely, what happens if the arrival rate doubles, triples, etc.; (b) insights into relationship between variables are required, namely, how is the waiting time related to service time; (c) to determine best course of action for any set of parameters, namely to answer the question of whether it is always better to have one queue with multiple servers than one queue for each server; (d) formulae are needed to plug into optimization routines, namely, to insert into a nonlinear program, the mean queue length must be written as a closed-form expression to optimize service speed. Simulations, on the other hand, are more appropriate when (a) system performance measures are required for a single set of numerical values, (b) performance of a set of given policies need to be evaluated numerically, and (c) assumptions needed for queueing models are unrealistic (which is arguably the most popular reason for using simulations). Having said that, in practice, it is not uncommon to use a simulation model to verify analytical results from queueing models or use analytical models for special cases to verify simulations. With that understanding, we would like to reiterate that for the purpose of this book we will mainly consider analytical models, and simulations will be used only to validate approximations. Next we proceed to describe some preliminary analytical results that are useful to a variety of systems, not just queueing systems.

1.2 Systems Analysis: Key Results

Consider Figure 1.2 that describes a flow system. Note that the flow system may or may not be a queueing system. Essentially, *entities* flow into the system, spend a finite time in the system and then leave the system. These entities could either be discrete (i.e., countable) or be continuous (i.e., uncountable). Examples of such flow systems and entities are as follows:

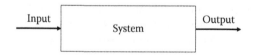

FIGURE 1.2
A flow system with inputs and outputs.

busses into which people enter and exit whenever the bus stops; cash register at a store into which money enters and exits; fuel reservoir in a gas station where gasoline enters when a fuel tanker fills it up and it exits when customers fill up their vehicle gas tanks; theme parks where riders arrive into the park, spend a good portion of their day going on rides and leave. There are many such examples of flow systems in everyday life. Not all such systems are necessarily best modeled as queueing systems. Nonetheless, there are a large number of fundamental results that we would like to present here with the understanding that although they are frequently used in the context of queueing, they can also be applied in wider domains such as inventory systems, for example.

We describe some notations that would be used in this chapter alone. The description is given as though the entities are discrete; however, by changing the word "number" to "amount," one can pretty much arrive at the same results if the entities were continuous. Let $\alpha(t)$ be the number of entities that flow into the system during the time interval $[0, t]$. Also, define $\gamma(t)$ as the number of entities in the system at time t with $\gamma(0) = 0$, that is, the system is empty initially. Finally, $\delta(t)$ denotes the number of entities that flow out of the system in time $[0, t]$. Due to *flow conservation*, we have

$$\alpha(t) = \gamma(t) + \delta(t) \tag{1.1}$$

which essentially states that all entities that arrived into the system during a time period of length t either left the system or are still in the system. In other words, entities are neither created nor destroyed. If one were careful, most flow systems can be modeled this way by appropriately choosing the entities.

For example, in systems like maternity wards in hospitals where it appears like the number of people checking in would be fewer than number of people checking out, by appropriately accounting for unborn children at the input itself, this balance can be attained. Although the previous example was said in jest, one must be careful especially in systems with losses. In our definition, entities that are lost are also included in the output (i.e., in the $\delta(t)$ definition) but one has to be very careful during analysis. To illustrate this point further, consider a system like a hotline where customers call for help. In such systems, some customers may wait and leave without being served, and some customers may leave without waiting (say due to a busy signal). One has to be very careful in classifying the customers and deriving performance measures accordingly for each class individually. The results presented in this section is by aggregating over all classes (unless accounted for explicitly). To clarify further, consider a production system where the raw material that flows in results in both defective and nondefective products. Clearly, when it comes to analysis, the emphasis we place on defective items may be significantly different than that for nondefective items, so it might be beneficial to derive individual performance measures. To model the production system as a whole, it may be beneficial to consider them as a single class. With this in mind, we next present a set of results that are asymptotic in time, that is, as $t \to \infty$.

1.2.1 Stability and Flow Conservation

The first thing that should come to mind when talking about asymptotic results of flow systems, is the notion of *stability* for that system. Although there is a mathematically rigorous way to describe the conditions for a system to be stable as well as the definition of stability, we would like to take a somewhat crude approach. Under rather mild conditions for the $\{\alpha(t), t \geq 0\}$ process (which by definition means the collection of $\alpha(t)$ values for every t from zero onward) we can state the condition of stability. We define the flow system to be stable if $\gamma(t)$ is finite for all t and in particular

$$\lim_{t \to \infty} \gamma(t) < \infty \tag{1.2}$$

almost surely. In words, the flow system is considered to be stable if the number of entities in the system at any instant (including after an infinite amount of time) would never blow off to infinity. The condition stated in Equation 1.2 would amount to $\gamma(t)/t \to 0$ as $t \to \infty$. Thus dividing Equation 1.1 by t and letting $t \to \infty$, we get

$$\lim_{t \to \infty} \frac{\alpha(t)}{t} = \lim_{t \to \infty} \frac{\delta(t)}{t}.$$

The preceding result is an important asymptotic result that is often misunderstood especially while applying to queueing networks. It states in words that the long-run average input rate (left side of the equation) equals the long-run average output rate (right side of the equation) if the flow system is stable. In that spirit, define for a stable flow system, the long-run average input rate Λ (assuming the limit exists) as

$$\Lambda = \lim_{t \to \infty} \frac{\alpha(t)}{t}. \tag{1.3}$$

Next we give a few more definitions.

1.2.2 Definitions Based on Limiting Averages

Consider a stable flow system and the notation described earlier. Let the long-run time-averaged number of entities in the system (assuming the limit exists) be defined as

$$H = \lim_{T \to \infty} \frac{\int_0^T \gamma(t)dt}{T}. \tag{1.4}$$

Although the preceding definition holds for discrete as well as continuous entities, all the definitions subsequently in this subsection are for discrete entities only. At first, we need to describe the indicator function $I(A)$, which is 1 if event A is true and 0 if event A is false. For example, for some j the indicator function $I(\gamma(t)=j)$ equals 1 or 0 if the number in the system at time t equals j or not j, respectively. In that light, define q_i (for $i=0,1,2,\ldots$) as the long-run fraction of time if there were exactly i number of entities in the system, that is,

$$q_i = \lim_{T\to\infty} \frac{\int_0^T I(\gamma(t) = i)dt}{T}. \tag{1.5}$$

Notice that the numerator of the term inside the limit essentially is the amount of time in the interval $[0,T]$ during which there were exactly i in the system. Verify that

$$H = \sum_{i=0}^{\infty} iq_i.$$

Let $\tau_1, \tau_2, \tau_3, \ldots$ be the time spent in the system by entity $1, 2, 3, \ldots$, assuming some arbitrary way of assigning numbers to entities (with the understanding that this assignment does not play any role in the result). Then, the long-run average time spent by an entity in the system, Θ, is

$$\Theta = \lim_{n\to\infty} \frac{\tau_1 + \tau_2 + \cdots + \tau_n}{n}. \tag{1.6}$$

We will subsequently establish a relationship between the various terms defined.

1.2.3 Asymptotically Stationary and Ergodic Flow Systems

We first define a stationary stochastic process $\{Z(t), t \geq 0\}$ for some arbitrary $Z(t)$ and then consider an *asymptotically stationary* processes subsequently. Although there is a rigorous definition for stationarity, for our purposes, at least for this introductory chapter, all we need is that

$$P\{z_l \leq Z(t) \leq z_u\} = P\{z_l \leq Z(t+s) \leq z_u\}$$

for any t, s, z_l, and z_u to call $\{Z(t), t \geq 0\}$ a stationary stochastic process. If the preceding result holds only as $t \to \infty$, then we call the stochastic process $\{Z(t), t \geq 0\}$ *asymptotically stationary*. For most results in this book, we only require asymptotic stationarity of stochastic processes unless explicitly stated otherwise. For example, the classical time-homogeneous, irreducible, and positive-recurrent continuous-time Markov chains (CTMCs) are asymptotically stationary stochastic processes. It is worthwhile to point out that sometimes in the literature, one defines stochastic processes as $\{Z(t), -\infty < t \leq \infty\}$ assuming that the process started at $t = -\infty$ and thereby at $t = 0$ the stochastic process is stationary. However, we will continue with the description provided earlier.

Having described the notion of asymptotically stationary stochastic processes, we now return to our flow system in Figure 1.2. Consider the number or amount of entities in the system at time t, that is, $\gamma(t)$. Notice that $\{\gamma(t), t \geq 0\}$ is a stochastic process that keeps track of the number or amount of entities in the system across time. Although we do not require $\{\gamma(t), t \geq 0\}$ to be a stationary stochastic process everywhere, we assume that it is asymptotically stationary. In general, $\{\gamma(t), t \geq 0\}$ will not be stationary for all t because $\gamma(0) = 0$ with probability 1.

Now, an asymptotically stationary and *ergodic* system is one where the long-run average quantities are equal to the steady-state expected values. The way that translates for the $\{\gamma(t), t \geq 0\}$ process is that if it is asymptotically stationary and ergodic, then

$$\lim_{t \to \infty} E[\gamma(t)] = H, \quad \lim_{t \to \infty} P\{\gamma(t) = i\} = q_i, \quad \text{and} \quad \lim_{n \to \infty} E[\tau_n] = \Theta,$$

where H, q_i, and Θ are defined in Equations 1.4 through 1.6. In simulations, one typically computes H, q_i, and Θ after considering a reasonable warm-up period to reduce the impact of the initial condition that $\gamma(0) = 0$.

Next, we present two results called Little's law and Poisson observations that can be derived for certain flow systems, especially when they are asymptotically stationary and ergodic. We will present some examples to illustrate the results. It may be worthwhile to pay attention to the assumptions (or conditions) stated to describe the results. Notice that Little's law is for discrete entity queues, not continuous (although one can derive the equivalent to Little's law for continuous).

1.2.4 Little's Law for Discrete Flow Systems

Consider a flow system that satisfies flow conservation. Further, assume that the flow system is stable and has *discrete* entities flowing through. Although we only require that the limits in Equations 1.3, 1.4, and 1.6 exist, we also assume that the flow system is asymptotically stationary and ergodic. According to Little's law

$$H = \Lambda\Theta \tag{1.7}$$

where Λ, H, and Θ are defined in Equations 1.3, 1.4, and 1.6, respectively. We do not provide a proof of the preceding result (for which it is more convenient if we have a stationary and ergodic system, although that is not a requirement). To illustrate Little's law, we first provide a brief example followed by a detailed numerical problem.

Consider a small company with 200 employees and an average attrition rate of 18.73 employees per year leaving the company. The policy of the company is not to grow beyond 200 employees and replace every vacant position. Therefore, the steady-state average recruiting rate is equal to the average attrition rate Λ. However, vacancies cannot be immediately filled and the time-averaged number of employees in the long run is calculated to be 187.3 (i.e., $H = 187.3$). Using Little's law, we have $\Theta = 10$ years, which is the average time an employee stays with the company. Next we consider a detailed numerical problem.

Problem 1

Couch-Potato is a high-end furniture store that carries a sofa set called Plush. Customers arrive into Couch-Potato requesting for a Plush sofa set according to a Poisson process at an average rate of 1 per week. Couch-Potato's policy is to not accept any back orders. So if there are no Plush sofa sets available in inventory, customers' requests are not fulfilled. It is also Couch-Potato's policy to place an order from the manufacturer for "five" Plush sofa sets as soon as the number of them in inventory goes down to "two". The manufacturer of Plush has an exponentially distributed delivery time with a mean of 1 week to deliver the set of "five" Plush sofa sets. Model the Plush sofa set system in Couch-Potato as a flow system. Is the system stable? Compute the average input rate Λ, the time-averaged number of Plush sofa sets in inventory (H), and the average number of weeks each Plush sofa set stays in Couch-Potato (Θ).

Solution

The Plush system in Couch-Potato is indeed a (discrete) flow system where with every delivery from the manufacturer, five sofa sets flow into the system. Also, with every fulfilled customer order, sofa sets exit the system. We let $\gamma(t)$ be the number of Plush sofa sets in the system at time t. Although we do not need $\gamma(0)$ to be zero for the analysis, assuming that would not be unreasonable. Also, notice that for all t, $\gamma(t)$ stays between "zero" and "seven". For example, if by the time the shipment arrived, two customers have already ordered Plushes, then the number in inventory would become zero. Likewise, a maximum of "seven" is because an order of "five" Plushes are placed when the inventory reaches "two", so if the shipment arrives before the next customer demand, there would be "seven" Plush sofa sets in the system. Notice that since $\gamma(t)$ never exceeds "seven", the system is stable.

To obtain the other performance measures, we model the stochastic process $\{\gamma(t), t \geq 0\}$ as a CTMC with state space $\{0, 1, 2, 3, 4, 5, 6, 7\}$ and corresponding infinitesimal generator matrix

$$Q = \begin{bmatrix} -1 & 0 & 0 & 0 & 0 & 1 & 0 & 0 \\ 1 & -2 & 0 & 0 & 0 & 0 & 1 & 0 \\ 0 & 1 & -2 & 0 & 0 & 0 & 0 & 1 \\ 0 & 0 & 1 & -1 & 0 & 0 & 0 & 0 \\ 0 & 0 & 0 & 1 & -1 & 0 & 0 & 0 \\ 0 & 0 & 0 & 0 & 1 & -1 & 0 & 0 \\ 0 & 0 & 0 & 0 & 0 & 1 & -1 & 0 \\ 0 & 0 & 0 & 0 & 0 & 0 & 1 & -1 \end{bmatrix}$$

in units of number of sofa sets per week. Details on modeling systems as CTMCs can be found in Section B.2. Let p_i be the probability that there are i Plush sofa sets at Couch-Potato in steady state. We can solve for the steady-state probabilities using

$$[p_0 \ p_1 \ p_2 \ p_3 \ p_4 \ p_5 \ p_6 \ p_7]Q = [0\,0\,0\,0\,0\,0\,0\,0]$$

and $p_0 + p_1 + \cdots + p_7 = 1$. We get

$$[p_0 \ p_1 \ p_2 \ p_3 \ p_4 \ p_5 \ p_6 \ p_7] = \frac{1}{21}[1\,1\,2\,4\,4\,4\,3\,2].$$

Note that an order for "five" Plush sofa sets is placed every time the inventory level reaches 2. So we pay attention to state 2 with corresponding steady-state probability $p_2 = 2/21$. In the long run, a fraction $2/21$ of time the system is in state 2 and on average state 2 lasts for half a week. Thus the average rate at which orders are placed is $2 \times 2/21$ per week. Hence the average input rate $\Lambda = 2 \times (2/21) \times 5 = 20/21$ Plush sofa sets per week. Also, the time-averaged number of Plush sofa sets in inventory (H) can be computed as

$$H = \sum_{i=0}^{7} ip_i = \frac{85}{21}.$$

Therefore, using Little's law, the average number of weeks each Plush sofa set stays in Couch-Potato (Θ) can be computed as

$$\Theta = \frac{H}{\Lambda} = 4.25 \text{ weeks.} \qquad \blacksquare$$

1.2.5 Observing a Flow System According to a Poisson Process

Consider the flow system in Figure 1.2 and notation described earlier, especially that $\gamma(t)$ is the number of entities in the system at time t with $\gamma(0) = 0$. Instead of observing the system continuously over time, say we observe it at times $t_1, t_2, t_3, \ldots$, according to a Poisson process (in other words, for any

$i > 0$, the inter-observation times $t_i - t_{i-1}$ are independent and identically distributed [IID] exponential random variables). Such discrete observations are common in many systems, especially when measurements need significant processing. For example, at traffic lights instead of making a video and processing it, one could just take stills at times $t_1, t_2, \ldots$, but according to a Poisson process. The question to ask is how the statistics taken continuously relate to those sampled according to a Poisson process.

The main result is that *the observations made according to a Poisson process yield the same results as those observed continuously*. To explain that further, recall the definition of H and q_i from Equations 1.4 and 1.5, respectively. Assuming the limits exist, we have

$$\lim_{n \to \infty} \frac{\sum_{j=1}^{n} \gamma(t_j)}{n} \to H.$$

Likewise for discrete entities, when $i = 0, 1, 2, \ldots$

$$\lim_{n \to \infty} \frac{\sum_{j=1}^{n} I(\gamma(t_j) = i)}{n} \to q_i.$$

Thus one can obtain time-averaged statistics by sampling the system at IID exponentially spaced intervals. This leads us to the following remark.

Remark 1

It is fairly common that systems are sampled at regularly spaced intervals (such as every 15 min, say) and then the data are averaged. That is not good practice because it could lead to unintended errors. For example, consider a traffic light that goes from red to green in 5-min cycles. If this light is observed at 15-min intervals and every time toward the end of a green period, then the observations would usually result in zero or a very small number of vehicles at the light. Although that is an extreme example, it is one where the averages based on discrete observations would not be equal to the time-averaged quantity. Now, instead, if observations are made according to a Poisson process with mean a inter-observation time of 15 min, then the average would indeed be similar to when the system is observed continuously. In fact, similar situations arise while monitoring computer networks. The probes sent to obtain statistics use what is called Poisson ping instead of equally spaced ping, again to avoid any biases induced in the averages. ∎

If the Poisson observations are made on an asymptotically stationary and ergodic system (namely, the $\{\gamma(t), t \geq 0\}$ process in steady state), then the Poisson observations will yield time averages. In particular,

$$\lim_{n\to\infty} E[\gamma(t_n)] \to H,$$

and

$$\lim_{n\to\infty} P\{\gamma(t_n) = i\} \to q_i,$$

where H and q_i are defined in Equations 1.4 and 1.5, respectively.

Problem 2

Consider a single-product inventory system with continuous review adopting what is known as the (K, R) policy, which we explain next. Demand of one unit arrives according to a Poisson process with parameter λ per week. Demand is satisfied using products stored in inventory, and no backorders allowed, that is, if a demand occurs when the inventory is empty, the demand is not satisfied. The policy adopted is called (K, R) policy wherein an order for K items is placed as soon as the inventory level reaches R. It takes a random time exponentially distributed with mean $1/\theta$ weeks for the order to be fulfilled (this is called lead time). Assume that $K > R$, but both R and K are fixed constants. Problem 1 is a special case of this single-product inventory system adopting the (K, R) policy with $K = 5$, $R = 2$, and $\theta = \lambda = 1$ per week. What would the distribution and expected value of the number of items in inventory be the instant a demand arrives? Also, determine the average product departure rates.

Solution

Let $X(t)$ be the number of products in inventory at time t. Clearly, $\{X(t), t \geq 0\}$ is a CTMC with state space $S = \{0, 1, \dots, R + K\}$ and rate diagram shown in Figure 1.3. Let p_i be the steady-state probability of i items in inventory. To obtain p_i for all $i \in [0, R + K]$, we use the balance equations

$$\theta \sum_{j=0}^{i-1} p_j = \lambda p_i, \ i = 1, \dots, R, \quad \theta \sum_{j=0}^{R} p_j = \lambda p_i, \ i = R+1, \dots, K,$$

$$\theta \sum_{j=i}^{R} p_j = \lambda p_{K+i}, \ i = 1 \dots, R.$$

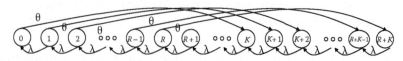

FIGURE 1.3
Rate diagram for (K, R) inventory system.

Then, p_0 can be obtained using $\sum_{i=0}^{K+R} p_i = 1 \implies$

$$p_0 \left[1 + \frac{\theta}{\lambda} \sum_{i=1}^{R} \phi^{i-1} + (K - R)\frac{\theta}{\lambda}\phi^R + \frac{\theta}{\lambda} \sum_{i=1}^{R}(\phi^R - \phi^{i-1}) \right] = 1,$$

where $\phi = 1 + (\theta/\lambda)$. Also, the steady-state distribution for $i > 0$ is

$$p_i = \begin{cases} \frac{\theta\phi^{i-1}}{\lambda+K\theta\phi^R} & 1 \le i \le R, \\[2ex] \frac{\theta\phi^R}{\lambda+K\theta\phi^R} & R < i \le K, \\[2ex] \frac{\theta(\phi^R-\phi^{i-k-1})}{\lambda+K\theta\phi^R} & K < i \le K+R. \end{cases}$$

We need to compute the distribution and expected value of the number of items in inventory the instant a demand arrives. However, the demands arrive according to a Poisson process and Poisson observations would match the long-run averages. Therefore, the distribution and expected value of the number of items in inventory the instant a demand arrives are p_i and $\sum_{i=0}^{R+K} i p_i$, respectively. Further, the average departure rate is $\lambda(1 - p_0)$. Verify that this is identical to the arrival rate derived for the special case in Problem 1. ∎

Having described some generic results for flow systems, we now delve into a special type of flow system called queueing systems.

1.3 Queueing Fundamentals and Notations

Although one could argue that any flow system such as the one depicted in Figure 1.2 is a queueing system, we typically consider a few minor distinguishing features for queueing-type flow systems. The potential arrivals (or inputs) into a flow system must take place on their own accord. For example, arrival of entities in an inventory system (which is a type of flow system) is due to the order placement and not on their own accord. In production systems literature inventories are known as pull systems while queues are known as push systems. Another aspect (although not necessary and perhaps not always true either) is that in a queueing system there is a notion of one or more servers (or processors or machines) that perform a service (or task or operation) for the entity. In an inventory system, there may be no true service rendered (or no task is performed) and the inventory is purely for storage purposes. With that in mind, we describe some features

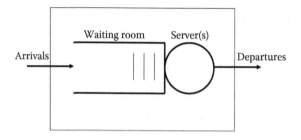

FIGURE 1.4
A single-station queueing system. (From Gautam, N., *Queueing Theory. Operations Research and Management Science Handbook*, A. Ravindran (ed.), CRC Press, Taylor & Francis Group, Boca Raton, FL, pp. 9.1–9.37, 2007. With permission.)

of canonical queueing systems. For the rest of this chapter, we only consider single-station queueing systems with one waiting line and one set of servers. Of course one could model a multistation queueing network as a single flow system, but in practice one typically models in a decomposed manner where each node in a network is a single-flow system. This justifies considering a single station. Also, at this stage, we do not make any distinctions between classes of entities. With that in mind, we present some details of queueing systems.

Consider a single-station queueing system as shown in Figure 1.4. This is also called a single-stage queue. There is a single waiting line and one or more servers (such as at a bank or post office). We will use the term "servers" generally but sometimes for specific systems we would call them processors or machines. We call the entities that arrive and flow through the queueing system as customers, jobs, products, parts, or just entities. Arriving customers enter the queueing system and wait in the waiting area if a server is not free (otherwise they go straight to a server). When a server becomes free, one customer is selected and service begins. Upon service completion, the customer departs the system. Usually, time between arrivals and time to serve customers are both random quantities. Therefore, to analyze queueing systems one needs to know something about the arrival process and the service times for customers. Other aspects that are relevant in terms of analysis include the number of servers, capacity of the system, and the policy used by the servers to determine the service order. Next we describe few key remarks that are needed to describe some generic, albeit basic, results for single-station queueing systems.

Remark 2

The entities that flow in the queueing system will be assumed to be discrete or countable. In fact, a bulk of this book is based on discrete queues with fluid queues considered in only two chapters toward the end. As described earlier, these entities would be called customers, jobs, products, parts, etc.   ■

Remark 3

Unless explicitly stated otherwise, the customer inter-arrival times, that is, the time between arrivals, are assumed to be IID. Thereby the arrival process is generally assumed to be what is called a renewal process. Some exceptions to that are when the arrival process is time varying or when it is correlated. But those exceptions will only be made in subsequent chapters. Further, all arriving customers enter the system if there is room to wait (that means unless stated otherwise, there is no balking). Also, all customers wait till their service is completed in order to depart (likewise, unless stated otherwise, there is no reneging).   ■

Remark 4

For the basic results some assumptions are made regarding the service process. In particular, we assume that the service times are IID random variables. Also, the servers are stochastically identical, that is, the service times are sampled from the same distribution for all servers. In addition, the servers adopt a work-conservation policy, that is, the server is never idle when there are customers in the system. The last, assumption means that as soon as a service is completed for a customer, the server starts serving the next customer instantaneously (if one is waiting for service). Thus while modeling one would have to appropriately define what all activities would be included in a service time.   ■

The assumptions made in the preceding remarks can and will certainly be relaxed as we go through the book. There are many instances in the book that do not require these assumptions. However, for the rest of this chapter, unless explicitly stated otherwise, we will assume that assumptions in Remarks 2, 3, and 4 hold. Next, using the assumptions, we will provide some generic results that will be useful to analyze queues. However, before we proceed to those results, recall that to analyze queueing systems one needs to know something about the arrival process, the service times for customers, the number of servers, capacity of the system, and the policy used by the servers to determine the service order. We will next describe queues using a compact nomenclature that takes all those into account.

In order to standardize description for queues we use a notation that is accepted worldwide called *Kendall notation* honoring the pioneering work by

D.G. Kendall. The notation has five fields:

$$AP/ST/NS/Cap/SD.$$

In the Kendall notation, AP denotes arrival process characterized by the inter-arrival distribution, ST denotes the service time distribution, NS is the number of servers in the system, Cap is the maximum number of customers in the whole system (with a default value of infinite), and SD denotes, service discipline that describes the service order such as first come first served (FCFS)—the default one, last come first served (LCFS), random order of service (ROS), and shortest processing time first (SPTF), etc. The fields AP and ST can be specific distributions such as exponential (denoted by M, which stands for memoryless or Markovian), Erlang with k phases (denoted by E_k), phase-type (PH), hyperexponential with k phases (H_k), and deterministic (D), etc. Sometimes instead of a specific distribution, AP and ST fields could be G or GI, which denote general distribution (although GI explicitly says "general independent," G also assumes independence considering the assumptions made in the remarks). Table 1.1 summarizes values that can be found in the five fields of Kendall notation.

For example, a queue that is $GI/H_2/4/6/LCFS$ implies that the arrivals are according to a renewal process with general distribution, service times are according to a two-phase hyperexponential distribution, there are four servers, a maximum of six customers are permitted in the system at a time (including four at the server), and the service discipline is LCFS. Also, $M/G/4/9$ implies that the inter-arrival times are exponentially distributed (thereby the arrivals are according to a Poisson process), service times are according to some general distribution, there are four servers, the system capacity is nine customers in total, and the customers are served according to FCFS. Since FCFS is the default scheme, it does not appear in the notation.

TABLE 1.1

Fields in the Kendall Notation

AP	$M, G, E_k, H_k, PH, D, GI$, etc.
ST	$M, G, E_k, H_k, PH, D, GI$, etc.
NS	denoted by s, typically $1, 2, \ldots, \infty$
Cap	denoted by K, typically $1, 2, \ldots, \infty$
	default: ∞
SD	FCFS, LCFS, ROS, SPTF, etc.
	default: FCFS

Source: Gautam, N., *Queueing Theory. Operations Research and Management Science Handbook,* A. Ravindran (ed.), CRC Press, Taylor & Francis Group, Boca Raton, FL, pp. 9.1–9.37, 2007. With permission.

In an $M/M/1$ queue, the arrivals are according to a Poisson process, service times exponentially distributed, there is one server, the waiting space is infinite, and the customers are served according to FCFS. Note that since both the system capacity and the service discipline take their default values, they do not appear in the notation. As a final example, consider the $E_4/M/3/\infty/LCFS$ queue. Of course the inter-arrival times are according to a four-phase Erlang distribution, service times are exponential, there are three servers, infinite capacity, and the discipline is LCFS. Note that in this case we do write down the default value for service capacity because the last field is not the default value.

These are only a few examples and it is not practical to list all the various cases possible for the various fields. However, a few are critical to be mentioned. Sometimes one sees discrete distributions like Geometric (represented as *Geo*) and that typically implies that the queue is observed at discrete times. At other times, one sees *MMPP* to mean Markov modulated Poisson process and *BMAP* to mean batch Markovian arrival process. It is also worthwhile noting that the Kendall notation is typically modified slightly for cases not described earlier. For example, if the system is time varying, then sometimes a subscript t is used in the notation (such as an $M_t/M/1$ queue, which is similar to an $M/M/1$ queue except the arrival rate is time varying, usually deterministically). Likewise, if customers arrive in batches or are served in batches, they are represented differently ($M^{[X]}/G/2$ implies that the arrival process is a compound Poisson process, that is, the inter-arrival time is exponential but with each arrival, a random number X entities enter the system). If we allow abandonment or reneging, then we can describe that too in the notation. We will describe such notations as and when they are needed in the book. Now, we will move to some generic analysis results that are true for any single-station queue.

1.3.1 Fundamental Queueing Terminology

Consider a single-station queueing system such as the one shown in Figure 1.4. Assume that this system can be described using Kendall notation. This means the inter-arrival time distribution, service time distribution, number of servers, system capacity, and service discipline are given. Recall that the assumptions in Remarks 2, 3, and 4 also hold. For such a system, we now describe some parameters and measures that we collectively call *queueing terminology*. As a convention, we assign numbers for customers with the nth arriving customer called customer-n. The only place where there may be some ambiguity is in the batch arrival case. For arrivals that occur simultaneously in a batch, we arbitrarily assign numbers for those customers as that would not affect our results. Most of the terminology and results presented in this section are also available in Kulkarni [67] with possibly different notations.

Define A_n as the time when the nth customer arrives, and thereby $A_n - A_{n-1}$ is the nth inter-arrival time if the arrivals are not in batches. Let S_n be the service time for the nth customer. Usually from the Kendall notations, especially when assumptions in Remarks 2, 3, and 4 hold, we typically know both $A_n - A_{n-1}$ and S_n stochastically for all n. In other words, we know the distributions of inter-arrival times and service times. In some sense they and the other Kendall notation terms form the "input." Next we describe some terms and performance measures that can be derived once we know the inputs.

Let D_n be the time when the nth customer departs. We denote $X(t)$ as the number of customers in the system at time t, X_n as the number of customers in the system just after the nth customer departs, and X_n^* as the number of customers in the system just before the nth customer arrives. Although in this chapter we would not go into details, it is worthwhile mentioning that $\{X(t), t \geq 0\}$, $\{X_n, n \geq 0\}$, and $\{X_n^*, n \geq 0\}$ are usually modeled as stochastic processes. We also define two other variables, which are usually not explicitly modeled. These are W_n, the waiting time of the nth customer, and $W(t)$, the total remaining workload at time t (this is the sum of the remaining service time for all the customers in the system at time t). The preceding variables are described in Table 1.2 for easy reference, where customer n denotes the nth arriving customer. Note that if we are given $A_1, A_2, \ldots$, as well as $S_1, S_2, \ldots$, we can obtain D_n, $X(t)$, X_n, X_n^* W_n, and $W(t)$. We describe that next for a special case (note that typically we do not know the explicit realizations of A_n and S_n for all n; we only know their distributions).

To illustrate the terms described in Table 1.2, consider a $G/G/1$ queue where the inter-arrival times are general and service times are general with a single server adopting FCFS and infinite space for customers to wait (refer

TABLE 1.2

Variables—Their Mathematical Notation as well as Meanings

Variable	Relation to Other Variables	Meaning
A_n		Arrival time of customer n
S_n		Service time of customer n
D_n		Departure time of customer n
$X(t)$		Number of customers in the system at time t
X_n	$X(D_n+)$	Number in system just after customer n's departure
X_n^*	$X(A_n-)$	Number in system just before customer n's arrival
W_n	$D_n - A_n$	Waiting time of customer n
$W(t)$		Total remaining workload at time t

Source: Gautam, N., *Queueing Theory. Operations Research and Management Science Handbook*, A. Ravindran (ed.), CRC Press, Taylor & Francis Group, Boca Raton, FL, pp. 9.1–9.37, 2007. With permission.

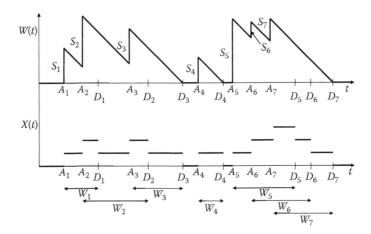

FIGURE 1.5
Sample path of workload and number in the system for a $G/G/1$ queue.

to Figure 1.5). Let $A_1, A_2, \ldots, A_7$ be the times that the first seven customers arrive to the queue. The customers require a service time of $S_1, S_2, \ldots, S_7$, respectively. Assume that the realizations of A_n and S_n are known (although in practice we only know them stochastically). The queue is initially empty. As soon as the first customer arrives (that happens at time A_1) the number in the queue jumps from 0 to 1 (note the jump in the $X[t]$ graph). Also, the workload in the system jumps up by S_1 because when the arrival occurs there is S_1 amount of work left to be done (note the jump in the $W[t]$ graph). Until the next arrival or service completion, the number in the system is going to remain a constant equal to 1. Hence the $X[t]$ graph stays flat at 1 till the next event. However, the workload keeps reducing because the server is working on the customer. Notice from the figure that before the first customer's service is completed, the second customer arrives. Hence the number in the system (the $X[t]$ graph) jumps up by 1 and the workload jumps up by S_2 (the $W[t]$ graph) at time A_2. Since there is only a single server and we use FCFS, the second customer waits while the first customer continues being served. Hence the number in the system (the $X[t]$ graph) stays flat at 2 and the workload reduces (the $W[t]$ graph) reduces continuously.

As soon as the server completes service of customer-1, that customer departs (this happens at time D_1). Note that the time spent in the system by customer-1 is $W_1 = D_1 - A_1 = S_1$. Immediately after customer-1 departs, the number in the system (the $X[t]$ graph) jumps down from 2 to 1. However, since the server adopts a work-conservation policy, it immediately starts working on the second customer. Hence the $W(t)$ graph has no jumps at time D_1. From the figure notice that the next event is arrival of customer-3, which happens while customer-2 is being served. Hence at time A_3 the

number in the system jumps from 1 to 2 and the workload jumps up by
S_3. The server completes serving customer-2 (at time D_2) and immediately
starts working on customer-3. Hence there is no jump in the workload process. However, the number in the system reduces by 1. Note that customer-2
spent $W_2 = D_2 - A_2$ amount of time in the system but it includes some waiting for service to begin plus S_2. Soon after time D_2 there is only one customer
in the system. Then at time D_3 customer-3 completes service and the system
becomes empty. This also means that the workload becomes zero and the
number in the system is also zero. Subsequently, customer-4 arrives into an
empty system at time A_4 and gets served before the fifth customer arrives.
So the system becomes empty again.

Then the fifth customer arrives at time A_5 and requires a large amount
of service S_5. By the time the server could complete service of the fifth customer, the sixth and seventh customers arrive. So the number in the system
goes from 0 to 1 at time A_5, then 1 to 2 at time A_6, and then 2 to 3 at time A_7.
The server completes serving the fifth, sixth, and seventh customer and the
system becomes empty once again. From the figure, notice that the server
is idle from time 0 to A_1, then becomes busy from A_1 to D_3, then becomes
idle from D_3 to A_4, and so on. The server essentially toggles between busy
and idle periods. The durations $D_3 - A_1$, $D_4 - A_4$, and $D_7 - A_5$ in the
figure are known as *busy periods*. The idle periods are continuous periods
of time when $W(t)$ is 0. Also, note that $X_1^* = 0$ since the number in the system
immediately before the first arrival is 0, whereas $X_1 = 1$ since the number
in the system immediately after customer-1 departs is 1. As another example note that $X_6^* = X(A_6-) = 1$ and also $X_5 = X(D_5+) = 2$. The terms X_n^* and
X_n are not depicted in the figure but can be easily obtained from the $X(t)$
graph since $X_n^* = X(A_n-)$ and $X_n = X(D_n+)$. To explain further, we consider
a problem next.

Problem 3

Consider the exact same arrival times of the first seven customers
$A_1, A_2, \ldots, A_7$ as well as the exactly same corresponding service time requirements of $S_1, S_2, \ldots, S_7$, respectively, as described earlier. However, the
system has two identical servers. Draw graphs of $W(t)$ and $X(t)$ across time
for the first seven customers assuming that the eighth customer arrives well
after all the seven previous customers are served. Compare and contrast the
graphs against those we saw earlier for the case of one server.

Solution

We assume that the system is empty at time $t = 0$. The graphs of $W(t)$ and
$X(t)$ versus t for this $G/G/2$ queue is depicted in Figure 1.6. The first customer arrives at time A_1 and one of the two servers processes this customer
and the workload process jumps up by S_1. While one server processes this
customer, another customer arrives and the workload jumps by S_2 at time

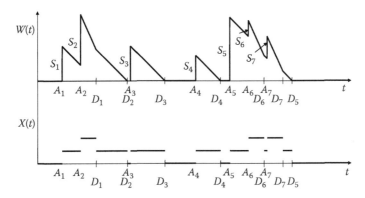

FIGURE 1.6
Sample path of workload and number in the system for a $G/G/2$ queue.

A_2. However, since there are two servers processing customers now, the workload can be reduced twice as fast (hence a different downward slope of $W[t]$ immediately after A_2). Then at time D_1, the first server completes serving the first customer and becomes idle. The second server subsequently completes serving customer-2 and the entire system becomes empty for a short period of time between D_2 and A_3 when the third customer arrives. Since the servers are identical, we do not specify which server processes the third customer but we do know that one of them is processing the customer. Then at time D_3, the system becomes empty. This process continues. Whenever there are two or more customers in the system the workload reduces at a faster rate than when there is one in the system. However, when there are no customers in the system $W(t)$ is 0.

Next we contrast the differences in Figures 1.5 and 1.6. The periods with the system empty have indeed grown, which is expected with having more servers. It is crucial to point out that the notion of busy period is unclear since a server could be idle but the system could have a customer served by the other server. However, the notion of the time when the system is empty is still consistent between the two figures (and that is when $W[t]$ and $X[t]$ are zero). Another difference between the figures that we described earlier is that in the $G/G/2$ case the downward slopes for the $W(t)$ process take on two different values depending on the number in the system. However, a crucial difference is that the customers do not necessarily depart in the order they arrived. For example, the seventh customer departs before the fifth customer (i.e., $D_7 < D_5$) in the $G/G/2$ figure. For this reason, we do not call this service discipline FIFO in this book (that means first-in-first-out) and instead stick to FCFS. However, the term "FIFO" does apply to the waiting area alone (not including the servers), and hence it is often found in the literature. However, to avoid any confusion we say FCFS. ∎

In a similar fashion, one could extend this to other queues and disciplines by drawing the $W(t)$ and $X(t)$ processes (see the exercises at the end of the chapter). However, typically we do not know realizations of A_n and S_n for all n but we only know the distributions of the inter-arrival times and service times. In that case can we say anything about $X(t)$, X_n, X_n^*, $W(t)$, D_n, and W_n? We will see that next.

1.3.2 Modeling a Queueing System as a Flow System

A natural question to ask is if we are given distributions for the inter-arrival times and service times, can we compute distributions for the quantities $X(t)$, $W(t)$, W_n, etc. It turns out that it is usually very difficult to obtain distributions of the random variables $X(t)$, X_n, X_n^*, $W(t)$, and W_n. However, some asymptotic results can be obtained by letting n and t go to infinite. But we just saw some asymptotic properties of flow systems in Section 1.2. Could we use them here? Of course, since the queueing system is indeed a flow system. But how do the notations relate to each other? The arrivals to the queueing system correspond to the inputs to the flow system. Note that $\alpha(t)$ described in Section 1.2 can be written as $\alpha(t) = \min\{n \geq 1 : A_n > t\} - 1$. In other words, $\alpha(t)$ for a queueing system counts the number of arrivals in time 0 to t. Also, $\gamma(t) = X(t)$, the number of customers in the queue at time t. Finally, $\delta(t)$ can be written as $\delta(t) = \min\{n \geq 1 : D_n > t\} - 1$. In other words, $\delta(t)$ for a queueing system counts the number of departures in time 0 to t. Thus, all the asymptotic properties that we described in Section 1.2 can be used here for queueing systems. We would henceforth not draw analogies between the notations used in Section 1.2 and here. Instead, we would just go ahead and use those results for flow systems here.

To describe the preceding asymptotic results, we consider the following performance measures. Let p_j be the long-run fraction of time that there are j customers in the system. Similarly, let π_j and π_j^* be the respective long-run fractions of departing and arriving customers that would see j other customers in the system. In addition, let $G(x)$ be the long-run fraction of time the workload is less than x. Likewise, let $F(x)$ be the long-run fraction of customers that spend less than x amount of time in the system. Finally, define L as the time-averaged number of customers in the system, and define W as the average waiting time (averaged across all customers). These performance metrics can be mathematically represented as follows (recall the indicator function $I[A]$ definition in Section 1.2.2):

$$p_j = \lim_{T \to \infty} \frac{\int_0^T I(X(t) = j)dt}{T},$$

$$\pi_j = \lim_{N \to \infty} \frac{\sum_{n=1}^{N} I(X_n = j)}{N},$$

$$\pi_j^* = \lim_{N \to \infty} \frac{\sum_{n=1}^{N} I(X_n^* = j)}{N},$$

$$G(x) = \lim_{T \to \infty} \frac{\int_0^T I(W(t) \le x)dt}{T},$$

$$F(x) = \lim_{N \to \infty} \frac{\sum_{n=1}^{N} I(W_n \le x)}{N},$$

$$L = \lim_{T \to \infty} \frac{\int_0^T X(t)dt}{T}$$

and

$$W = \lim_{N \to \infty} \frac{W_1 + W_2 + \cdots + W_N}{N}.$$

To illustrate the concept of time averages and indicator functions, as well as to get a better notion of the preceding terminologies we consider the following problem.

Problem 4

Consider the time between $t=0$ and $t=D_7$ for the $G/G/1$ queue in Figure 1.5. Assume that we have numerical values for all A_i and D_i for $i=1,\ldots,7$. What fraction of time between $t=0$ and $t=D_7$ were there for two customers in the system? What fraction of the seven arriving customers saw one customer in the system? What is time-averaged number of customers in the system between $t=0$ and $t=D_7$?

Solution

From Figure 1.5, note that there are two customers in the system between times A_2 and D_1, A_3 and D_2, A_6 and A_7, as well as D_5 and D_6. Thus the fraction of time between $t=0$ and $t=D_7$ that there were two customers in the system is $((D_1 - A_2) + (D_2 - A_3) + (A_7 - A_6) + (D_6 - D_5))/D_7$. Notice that the expression is identical to $\int_0^{D_7} I(X(t)=2)dt/D_7$.

From Figure 1.5, also note that customers 1, 4, and 5 saw zero customers in the system when they arrived; customers 2, 3, and 6 saw one in the system when they arrived; and customer 7 saw two customers in the system upon arrival. Thus a fraction 3/7 of the arriving customers saw one customer in the system. The fraction is indeed equal to $\sum_1^7 I(X_n^* = 1)/7$.

To obtain the time-averaged number of customers in the system between time 0 and D_7, we use the expression $\int_0^{D_7} X(t)dt/D_7$. Hence we have that value as

$$\frac{\begin{array}{l}(A_2 - A_1) + 2(D_1 - A_2) + (A_3 - D_1) + 2(D_2 - A_3) + (D_3 - D_2) + (D_4 - A_4) \\ + (A_6 - A_5) + 2(A_7 - A_6) + 3(D_5 - A_7) + 2(D_6 - D_5) + (D_7 - D_6)\end{array}}{D_7}$$

customers in the system by averaging over time from $t = 0$ to $t = D_7$. ∎

Although the preceding description looks cumbersome even for a small problem instance, it turns out that for many queueing systems it is possible to obtain expressions for p_j, π_j, π_j^*, $G(x)$, $F(x)$, L, and W. In particular, if we consider queues that are asymptotically stationary and ergodic flow systems, then it is possible to obtain some (if not all) of those expressions. It turns out that single-station queueing systems such as the one shown in Figure 1.4 that can be described using Kendall notation and that satisfy assumptions in Remarks 2, 3, and 4 are typically asymptotically stationary and ergodic. In that light, we redefine p_j, π_j, π_j^*, $G(x)$, $F(x)$, L, and W as the corresponding asymptotic measures as follows: p_j is also the probability that there would be j customers in the system in steady state; π_j and π_j^* would be the respective probabilities that in steady state a departing and, respectively, an arriving customer would see j other customers in the system; $G(x)$ and $F(x)$ would be the cumulative distribution functions of the workload in steady state and waiting time for an arrival into the system in steady state, respectively; L would be the expected number of customers in the system in steady state; and W would be the expected waiting time for a customer arriving at the system in steady state. Hence these performance metrics can also be represented as follows:

$$p_j = \lim_{t \to \infty} P\{X(t) = j\},$$

$$\pi_j = \lim_{n \to \infty} P\{X_n = j\},$$

$$\pi_j^* = \lim_{n \to \infty} P\{X_n^* = j\},$$

$$G(x) = \lim_{t \to \infty} P\{W(t) \le x\},$$

$$F(x) = \lim_{n \to \infty} P\{W_n \le x\},$$

$$L = \lim_{t \to \infty} E[X(t)]$$

and

$$W = \lim_{n \to \infty} E[W_n].$$

Since the system is asymptotically stationary and ergodic, the two definitions of p_j, π_j, π_j^*, $G(x)$, $F(x)$, L, and W would be equivalent. In fact, we would end up using the latter definition predominantly as we would be modeling the queueing system as stochastic processes and perform steady-state analysis.

One of the primary objectives of analysis of queues is to obtain closed-form expressions for the *performance metrics* p_j, π_j, π_j^*, $G(x)$, $F(x)$, L, and W given properties of the queueing system such as inter-arrival time distribution, service time distribution, number of servers, system capacity, and service discipline. Although we would derive the expressions for various settings in future chapters only, for the remainder of this section, we concentrate on describing the relationship between those measures.

1.3.3 Relationship between System Metrics for $G/G/s$ Queues

In this section, we describe some relationships between the metrics π_j, π_j^*, L, and W, as well as other new metrics that we will define subsequently. We consider the setting of a $G/G/s$ queue where the arrivals are according to a renewal process (such that the arrival times are at $A_1, A_2, \ldots$), the service times are according to some given distribution (with $S_1, S_2, \ldots$ being IID), there are s identical servers, there is an infinitely large waiting room, and the service is according to FCFS. The results can be generalized to other cases as well; in particular, we will consider a finite-sized waiting room at the end of this section. Also, we will describe whether the results can be used in a much broader setting whenever appropriate. In particular, the arrivals being according to a renewal process is not necessary for most results. Also, FCFS is not necessary for almost all the results (however, we do require for some results that the discipline be nonpreemptive and work conserving, which will be defined later). We begin by stating the relationship between π_j and π_j^*.

If the limits defining π_j and π_j^* exist, then

$$\pi_j = \pi_j^* \quad \text{for all } j \geq 0.$$

We explain this relation using an example illustration, but the rigorous proof uses what is known as a level-crossing argument. In a $G/G/s$ queue, note that the times customers arrive to an empty system are regenerative epochs. In other words, starting at a regenerative epoch, the future events are independent of the past. In Figure 1.5, times A_1, A_4, and A_5 are regenerative epochs. The process that counts the number of regenerative epoch is indeed a renewal process and the time between successive regenerative epochs are IID (although for this system the distribution is not easy to compute in general). We call this time between successive regenerative epochs as a *regenerative cycle*. For the regenerative process described previously in a $G/G/s$ system, we assume that the regenerative cycle times on average are finite and the system is stable (stability conditions will be explained later). It is crucial to note that within any regenerative cycle of such a $G/G/s$ queue, the number of arriving customers seeing j others in the system would be *exactly* equal to the number of departing customers seeing j others in the system.

For example, consider the regenerative cycle $[A_1, A_4)$ in Figure 1.5. There are three arrivals, two of which see one in the system (customers 2 and 3) and one sees zero in the system (customer 1). Observe that there must be exactly three departures (if there are three arrivals). Of the three departures, two see one in the system (customers 1 and 2) and one sees zero (customer 3). Similarly, in regenerative cycles $[A_4, A_5)$ and $[A_5, A_8)$ one can observe (pretending A_8 is somewhere beyond D_7) that the number of arriving customers that see j in the system (for any j) would be exactly equal to the number of departing customers that see j in the system. Since the entire time is composed of these regenerative cycles, by summing over infinitely large number of regenerative cycles we can see that the fraction of arriving customers that see j others in the system would be exactly equal to the fraction of departing customers that see j others in the system. Hence we get $\pi_j = \pi_j^*$. Before proceeding it is worthwhile to verify for the $G/G/2$ case in Figure 1.6 where the regenerative cycles are $[A_1, A_3)$, $[A_3, A_4)$, $[A_4, A_5)$, and $[A_5, A_8)$ assuming A_8 is somewhere beyond D_5. Also, this result can be generalized easily for a finite capacity queue and any service discipline as long as arrivals occur individually and service completions occur one by one.

Next we describe the relationship between L and W. For that we require some additional terminology. Define the following for a single stage $G/G/s$ queue (with characteristics described in the previous paragraph):

- λ: Average arrival rate into the system. By definition λ is the long-run average number of customers that arrive into $G/G/s$ queue per unit time. This definition holds even if the arrival process is not renewal. However, if the arrivals are indeed according to a renewal process then for any $n > 1$ (since the inter-arrival times are IID) we can define

$$\frac{1}{\lambda} = E[A_n - A_{n-1}].$$

  In other words, λ is the inverse of the expected inter-arrival time.
- μ: Average service rate of a server. By definition μ is the long-run average number of customers any one server can serve per unit time, if the server continuously served customers. Notice that we do

require the service times to be IID. Therefore, μ can easily be written for any arbitrary n as

$$\frac{1}{\mu} = E[S_n].$$

In other words, μ is the inverse of the average time needed to serve a customer. It is important to note that in our results we assume that the units for λ and μ are the same. In other words, both λ and μ should be expressed as per second or both should be per minute, etc.

- ρ: Traffic intensity of the system. By definition this is the load experienced by the system and is expressed as

$$\rho = \frac{\lambda}{s\mu}.$$

Note that ρ is a dimensionless quantity.

- L_q: Average number of customers waiting in the queue, not including ones in service. Note that the L we defined earlier includes customers at the servers.
- W_q: Average time spent by customers in the queue waiting, not including time spent in service. Note that the W we defined earlier includes customer service times.

Now, we are ready to present some relationships in terms of λ, μ, ρ, L_q, W_q, L, and W and then between p_j, π_j, and π_j^*. However, as a first step we describe the stability condition. The $G/G/s$ queue as described earlier is *stable* if

$$\rho < 1.$$

In other words, we require that $\lambda < s\mu$ for the system to be stable. This is intuitive because it says that there is enough capacity (service rate on average offered by all servers together is $s\mu$) to handle the arrivals. In the literature, $\rho > 1$ is called an overloaded system and $\rho = 1$ is called a critically loaded system. Next we present a remark for stable $G/G/s$ queues.

Remark 5

The average departure rate from a $G/G/s$ queue is defined as the long-run average number of customers that depart from the queue per unit time. If the $G/G/s$ queue is stable, then the average departure rate is λ. ∎

This remark is an artifact of the argument made in Section 1.2.1 that since this is a stable flow conserving system, the average input rate must be the average output rate. With that said, we describe the relationship between the preceding terms and then explain them subsequently.

A $G/G/s$ queue with notation described earlier in this section satisfies the following:

$$W = W_q + \frac{1}{\mu},\tag{1.8}$$

$$L = \lambda W \tag{1.9}$$

and

$$L_q = \lambda W_q. \tag{1.10}$$

We will now explain these equations. Equation 1.8 is directly from the definition; the total time in the system for any customer must be equal to the time spent waiting in the queue plus the service time; thus taking expectations we get Equation 1.8. Equations 1.9 and 1.10 are both due to Little's law described in Section 1.2.4. Essentially, if one considers the entire queueing system as a flow system and then suitably substitutes the terms in Equation 1.7, then Equation 1.9 can be obtained. However, if one considered just the waiting area as the flow system, then Equation 1.7 can be used once again to derive Equation 1.10. The preceding equations can be applied in more general settings than what we described. In particular, Equation 1.8 is applicable beyond the $G/G/s$ setting such as: it does not require renewal arrivals; if appropriately defined, it can be used for finite capacity queues as well as some non-FCFS disciplines. Likewise, Equations 1.9 and 1.10 are applicable in a much wider contexts since Little's law can be applied to any flow system, not just the $G/G/s$ queue setting. In particular, it can be extended to $G/G/s/K$ queues (as we will see at the end of this section) by appropriately picking λ values. Also, it is not required that the discipline be FCFS (even work-conservation is not necessary).

The key benefit of the three equations is that if we can compute one of L, L_q, W, or W_q, the other three can be obtained by solving the three equations for the three remaining unknowns. It is worthwhile pointing out that λ and μ are not unknowns. One can (usually) easily compute λ and μ from the $G/G/s$ description. We illustrate this using an example next.

Problem 5

A simulation of a $G/G/9$ queue yielded $W_q = 1.92$ min. The inputs to the simulation included a Pareto distribution (with mean 0.1235 min and coefficient of variation of 2) for the inter-arrival times and a gamma distribution (with mean 1 min and coefficient of variation of 1) for the service times. Compute L, L_q, and W_q.

Solution

Based on the problem description we have a $G/G/s$ queue with $s=9$, $\lambda=8.1$ per minute, and $\mu=1$ per minute. Also, $W_q=1.92$ min, which is the average time a customer waits to begin service. Using Equation 1.10 we can get $L_q=\lambda W_q=8.1\times1.92=15.552$ customers that can be seen waiting for service (on average) in the system. Also, using Equation 1.8 we have $W=W_q+1/\mu=2.92$ min (which is the mean time spent by each customer in the system). Finally, using Equation 1.9 we get $L=\lambda W=8.1\times2.92=23.652$ customers in the system on average in steady state. ∎

Next, we describe few more interesting results based on Equations 1.8 through 1.10. Multiplying Equation 1.8 by λ we get $\lambda W=\lambda W_q+\lambda/\mu$. However, since $L=\lambda W$ (Equation 1.9) and $L_q=\lambda W_q$ (Equation 1.10), we have

$$L = L_q + \frac{\lambda}{\mu}.$$

Also, since L is the expected number of customers in the system and L_q is the expected number of customers in the waiting, we have λ/μ as the expected number of customers at the servers. Therefore, the expected number of busy servers in steady state is $\lambda/\mu=s\rho$. Define random variable B_i as 1 if server i is busy in steady state and 0 otherwise (i.e., server i is idle). Since the servers are identical we define p_b as the probability that a particular server is busy, that is, $P(B_i=1)=p_b$. Also, $E[B_i]=p_b$ for all $i\in[1,2,\ldots,s]$. We saw earlier that $E[B_1+B_2+\cdots+B_s]=\lambda/\mu$ since the expected number of busy servers is λ/μ. But $E[B_1+B_2+\cdots+B_s]$ is also sp_b. Hence we have the probability that a particular server is busy p_b given by

$$p_b = \rho.$$

Also, for the special single server case of $s=1$, that is, $G/G/1$ queues, the probability that the system is empty, p_0 is

$$p_0 = 1 - \rho$$

since the probability that the server is busy in steady state is ρ. As we described earlier, most of the preceding results can be extended to more generalized settings (e.g., FCFS is not necessary). We next explain one such extension, namely, the $G/G/s/K$ queue for which some minor tweaks are needed.

1.3.3.1 G/G/s/K Queue

The description of a $G/G/s/K$ queue is identical to that of the $G/G/s$ queue given earlier with the only exception that the total number in the system cannot exceed K. We do assume that K is finite. The good news is that such a system is always stable. However, a complication arises because not all customers that "arrive" to the system actually "enter" the system. In other words, if an arriving customer finds K others in the system (i.e., a full system with no waiting room), that arriving customer leaves without service. It is crucial to notice that the Kendall notation is still for the *arrival* process and not the *entering* process. However, we must be careful while defining the performance measures by explicitly stating that they are for entering customers. In that light, the average entering rate $\bar{\lambda}$ can be defined as

$$\bar{\lambda} = \lambda \left(1 - \pi_K^*\right)$$

since a fraction π_K^* of arrivals would be turned away due to a full system. Also, note that the average rate of departure from the system after being served is also $\bar{\lambda}$. Using $\bar{\lambda}$, the average number of customers that enter the queueing system per unit time, we can write down Little's law as

$$L = \bar{\lambda}W. \tag{1.11}$$

It is important to note that W must be interpreted as the mean time in the system for customers that actually "enter" the system (and it does not include customers that were turned away). In other words W is a conditional expected value, conditioned on the arriving customer able to enter the system. However, if one were to consider all arriving customers, then the average time spent in the system would indeed be L/λ; however, this includes a fraction π_K^* that experienced zero time in the system.

Further, if we used the conditional definition of W and W_q (i.e., they are averaged only across customers that enter the system), then we can also show the following:

$$W = W_q + \frac{1}{\mu},$$

$$L_q = \bar{\lambda}W_q.$$

In addition, as we described earlier

$$\pi_j = \pi_j^*$$

for all j with them being equal to zero if $j > K$. Having said that, it is crucial to point out that for most of the book we will mainly concentrate on infinite capacity queues (with some exceptions especially in the very next chapter) due to issues of practicality and ease of analysis. From a practical standpoint, if a queue actually has finite capacity but the capacity is seldom reached, approximating the queue as an infinite capacity queue is reasonable.

1.3.4 Special Case of $M/G/s$ Queue

One special case of the $G/G/s$ queue is worth mentioning because of the additional results one can derive. That special case is when the arrival process is a Poisson process leading to an $M/G/s$ queue. Assume that the arrival process is a Poisson process with inter-arrival times according to an exponential distribution. The average inter-arrival time is still $1/\lambda$ as before and mean arrival rate is λ customers per unit time. All the other definitions are the same as that in the $G/G/s$ queue case.

With that said, the first result is based on the description in Section 1.2.5. There we stated that if a system is observed according to a Poisson process then the observations made according to a Poisson process yield the same results as those observed continuously. For the $M/G/s$ queue, we consider observing the number in the system. Clearly, p_j is the long-run fraction of time there are j in the system and it can be obtained by observing the system continuously. On the other hand, if the system is observed according to a Poisson process then a fraction p_j of those observations in the long run must result in the system being in state j. Now, if these Poisson observations are made by arriving customers (which is the case in $M/G/s$ queues), then a fraction p_j of arriving customers would observe the system being in state j in the long run. But the fraction of arriving customers that observe the system in state j by definition is π_j^*. Therefore,

$$p_j = \pi_j^*.$$

This result is called PASTA (Poisson arrivals see time averages) since the arrivals are seeing time-averaged quantities as described in Section 1.2.5.

The PASTA property can be used further to obtain relations between time averages and ensemble averages. Recall the definition of L, which can also be written as

$$L = \sum_{j=0}^{\infty} j p_j.$$

Hence L is the time-averaged number of customers in the system (as well as the expected number of customers in the system in steady state). In a similar manner, let $L_{(k)}$ be the kth factorial moment of the number of customers in the system in steady state, that is,

$$L_{(k)} = \sum_{j=k}^{\infty} k! \binom{j}{k} p_j.$$

Also, let $W^{(k)}$ be the kth moment of the waiting time in steady state, that is,

$$W^{(k)} = \lim_{n \to \infty} E[\{W_n\}^k].$$

Note that $W^{(1)} = W$ and $L_{(1)} = L$. Of course using Little's law we have $L = \lambda W$. But Little's law can be extended for the $M/G/s$ queue in the following manner. For an $M/G/s$ queue

$$L_{(k)} = \lambda^k W^{(k)}. \tag{1.12}$$

Therefore, all moments of the queue lengths are related to corresponding moments of waiting times for any $M/G/s$ queue. Note that from factorial moments it is easy to obtain actual moments.

In summary, we have shown the relationships between various performance measures (such as number in the system and their moments, time in the system and its moments) as well as steady-state distributions such as p_j, π_j, or π_j^*. However, we need to obtain some of the measures using the dynamics of the queueing system and then use the relationships derived earlier to obtain other metrics. To do that, that is, obtain some of the metrics by analyzing the dynamics of the system, would be the focus of most of the remaining chapters of the book. Such an approach of analysis is known as physics-based because the system dynamics and interactions are modeled. However, it is also possible to alleviate perceived waiting in systems, especially when humans are involved as entities. We address that next by studying psychological aspects.

1.4 Psychology in Queueing

The objective of many commercial systems is to design and control resources to maximize expected revenue (or minimize expected costs) in the long run. One of the key aspects for that objective is to ensure that the users or customers are satisfied. Customer satisfaction is a psychological concept and is often not linear, not rational, and not uniform across space (i.e., customers) as well as time. Also, customer satisfaction can sometimes be enhanced without improving the system performance. We present several examples to illustrate that.

The classic example presented in psychology of queues is what is known as the *elevator problem*. A hotel in a downtown of a big city was a high-rise building with multiple floors. Unfortunately, the hotel management got several complaints about how long it took from when a customer enters the hotel to go to his or her room. The basic problem was that the elevators did not have the capacity to handle the volume of customers especially during peak times. The hotel management considered several alternatives that were suggested. One of them was to replace the current elevators with faster ones. But that was soon dismissed because it was the stopping and letting customers out that was taking most of the time. Another suggestion was to try a different algorithm for the elevators to operate but simulations showed that did not dramatically improve the performance (time between showing up at the elevator and going to the room). A final suggestion that the management seriously considered is to build two new elevator shafts outside the building (which is the only space available).

Simulation studies showed that the final suggestion would certainly improve the system performance. But when the management did their cost analysis they found that it was exorbitant. The cost to build two new elevator shafts did not match up the benefits due to improved customer satisfaction. Frustrated, the hotel management approached a consultant. When the consultant went to the hotel to understand the problem, she immediately thought of a clever solution. Her suggestion was to invest a little to ensure that the perceived wait of customers are reduced. For that she suggested placing mirrors both in the elevator and in the waiting areas. That way customers would spend time grooming themselves instead of thinking they are waiting. She also suggested to play soothing music both in the waiting area and in the elevator and use fragrances. Sure enough, upon implementing, the complaints reduced dramatically. Although the actual time in the system did not improve, the perceived time surely did! The customers did not feel the agony of waiting.

There are several other systems where the operators have aimed at reducing the perceived waiting times. Doctors, offices and barber shops usually have magazines so that the patients and customers, respectively, can read them and not perceive the long wait. The waiting area in most entertainment facilities (such as theme parks, theaters, zoos, and aquariums) usually have videos or information about the exhibits on the walls for customers to enjoy while they are waiting. In fact, some dentist's offices even have a television mounted on the ceiling to distract customers that are truly experiencing agony! The grocery store checkout line has tabloids so that people are busy glancing through them not realizing they are waiting. That would certainly be better than other expensive solutions such as more checkout counters. In fact, these days most grocery stores have self-checkout lines too where customers are kept busy checking out (although usually the check out clerk is probably much faster, but since the customer is kept occupied it does not appear that way).

There are many such situations where humans perceive something as taking longer when it actually might be shorter. One such example is at fast-food restaurants. While designing the waiting area, one is faced with choosing whether to have one long serpentine waiting line or have one line in front of each server. We will see later in this book that one serpentine line is better than having one line in front of each server when it comes to minimizing the expected time spent in the system. But then why do we see fast-food restaurants having a line in front of each server? One reason is that most people feel happier to be on a shorter line than a longer line! Also, for example, if there are three servers in a fast-food restaurant taking orders, then most people feel better being second in line behind a server than being the fourth in a long serpentine line, although in terms of getting out of the restaurant it is better to be fourth in a serpentine line with three servers than second in a line with one server. Even though this appears irrational, the perception certainly matters while making design decisions.

Another aspect that is crucial for customer satisfaction is to reduce the anxiety involved in waiting. Providing information, estimates, and guarantees, as well as reducing uncertainties, goes a long way in terms of customer satisfaction. Getting an estimate of how long the wait is going to be can reduce the anxiety level. In many phone-based help desks one typically gets a message saying, "your call will be answered in approximately 5 minutes." By saying that one typically does not get impatient for that 5 min. In most restaurants when one waits for a table, the greeter usually gives an estimate of how long the wait is. In theme parks one is usually provided information such as "wait time is 45 minutes from this point." These days, to avoid road rage, on many city roads one sees estimated travel times displayed for various points of interest. It is also crucial to point out that in many instances providing this waiting information is not only useful in terms of reducing anxiety but it also enables the customer or user to make choices (such as considering alternative routes when a road is congested).

In many instances, providing information to customers so that they could make informed decisions actually improves the system performance. For example, a note on the highway (or on the radio) informing an accident would make some drivers take alternate routes. This certainly alleviates the congestion on the highway where the accident has taken place. Another example is at theme parks (such as at DisneyWorld) where as a guest you have the option of standing in a long line or picking up a token that gives you another window of time to show up for the ride (which also ensures short wait times). This is a win–win situation because it not only reduces the agony of waiting and improves the experience of the guests as now they can enjoy more things, but it also reduces bursts seen during certain times of the day by spreading them out and thereby running the systems more efficiently. This also allows for classifying customers and catering to their needs, as opposed to using FCFS for all customers. By not forcing all the customers to wait or to show up at times specified by the system, the system is able to satisfy customers that prefer one option versus the other.

The last comment made in the previous sentence essentially states that by putting the onus on the customers to choose the class they want to belong makes the system appear more fair and not skewed toward the preference of one type. The whole notion of fairness, especially in queues with human entities is a critical aspect. Customer satisfaction and frustration with waiting can get worse if there is a perception of the system being unfair. Consider a grocery store checkout line. Sometimes when a new counter opens up while everyone is waiting in long lines, the manager invites customers arbitrarily. A lot of customers consider that as unfair. In restaurants one tends to get annoyed when someone that arrived later gets seated first, although that might be for practical reasons such as a table of the right size becoming available. But usually when such an unfair thing occurs, the agony of waiting worsens. Therefore, most systems adopt FCFS as it is a common notion of being fair. But again that has been questioned by many researchers. Unfortunately, what is considered fair, is completely in the mind of the one experiencing the situation.

Talking about situations, usually the same duration of wait times could be tolerable in one situation and unbearable in another even though it is the same person. There are many reasons for that difference. One has to do with the customer's expectations. If one waits longer or shorter than expected, although in both cases the actual wait times are the same, the latter makes a customer more satisfied. In fact, that is why most of the times services overestimate the wait time while informing their customers. Also, whether a wait time is considered acceptable or not depends on what one is waiting for. There are many things that are considered "worth the wait," especially something precious. Of course as we described earlier, it also depends on what the customer is doing while waiting, an engaged customer would perceive the same wait time as shorter than when the same customer is idle. There are things that also appeal logically; for example, if it takes 5 min to download a very large file, that is alright but the same 5 min for a small web page is ridiculous. In summary, understanding human nature is important while making design and control decisions.

Human nature not only plays out while considering customer satisfaction but also in terms of behavior. Balking is a behavior when arriving customers decide not to join the queue. Although usually the longer the line, the greater the chance for balking. But that may not be true all the time as sometimes a longer line might imply better experience! In fact, people balk lines saying, "I wonder why no one is here, maybe it is awful." So it becomes important to understand the balking behavior and rationale. Same applies to reneging, which is abandoning a queue after waiting for some time and service does not begin. Again, understanding the behavior associated with reneging can help develop appropriate models. It was observed that the longer one waits, the lesser the propensity to renege. This is not intuitive because one expects customers to wait for some time and become impatient, so with time the reneging rate should have increased. In fact, systems like emergency management (such as 9-1-1 calls in the United States) use an LCFS policy because customers that are under true emergency situations tend to hang up and try again persistently with high rates of reneging. However, other callers behave in the opposite fashion, that is, wait patiently or renege and not retry. By understanding the behavior of true emergency callers, a system that prioritizes such callers without actually knowing their condition can be built.

Customer behavior and customer satisfaction go hand in hand. Satisfied customers behave in a certain way and unsatisfied customers behave in other ways. In other words, customer behavior is a reaction to customer satisfaction (or dissatisfaction). On the flip side, for organizations that provide service, understanding customer behavior and being able to model it goes a long way in providing customer satisfaction. There are three components of customer satisfaction: (a) quality of service (QoS), (b) availability, and (c) cost. A service system (with its limited resources) is depicted in Figure 1.7. Into such a system customers arrive, if resources are available they enter the system, obtain service for which the customers incur a cost, and then they leave the system. We define availability as the fraction of time arriving customers enter the system. Thereby QoS is provided only for customers that "entered" the system. From an individual customer's standpoint, the customer is satisfied if the customer's requirements over time on QoS, availability, and cost are satisfied.

The issue of QoS (sometimes called conditional QoS as the QoS is conditioned upon the ability to enter the system) versus availability needs further discussion. Consider the analogy of visiting a medical doctor. The ability to get an appointment translates to availability; however, once an appointment is obtained, QoS pertains to the service rendered at the clinic such as

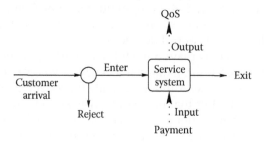

FIGURE 1.7
Customer satisfaction in a service system. (From Gautam, N., *Quality of Service Metrics. Frontiers in Distributed Sensor Networks*, S.S. Iyengar and R.R. Brooks (eds.), Chapman & Hall/CRC Press, Boca Raton, FL, pp. 613–628, 2004. With permission.)

waiting time and healing quality. Another analogy is airline travel. Getting a ticket on an airline at a desired time from a desired source to a desired destination is availability. QoS measures include delay, smoothness of flight, in-flight service, etc. One of the most critical business decisions is to find the right balance between availability and QoS. A service provider can increase availability by decreasing QoS and vice versa. A major factor that could affect QoS and availability is cost. As a final set of comments regarding customer satisfaction, consider the relationship between customer satisfaction and demand. If a service offers excellent customer satisfaction, very soon its demand would increase. However, if the demand increases, the service provider would no longer be able to provide high customer satisfaction, which eventually deteriorates, thereby decreasing demand. This is a cycle one has to plan carefully.

As a final comment, although this book mainly considers physics of queues, it is by no means the only way to design and operate systems. As we saw in the examples given earlier, by considering psychological issues it is certainly possible to alleviate anxiety and stress associated with waiting. In fact, a holistic approach would combine physics and psychology of queues to address design, control, and operational issues.

Reference Notes

There are several excellent books and papers on various aspects of theory and applications of queueing models. The list is continuously growing and something static such as this book may not be the best place for that list. However, there is an excellent site maintained by Hlynka [54], which is a phenomenal repository of queueing materials. The website includes a wonderful historical perspective of queues and cites several papers that touch upon the evolution of queueing theory. It also provides a large list of queueing researchers, books, course notes, and software among other things such as queueing humor.

This chapter as well as most of this book has been influenced by a subset of those fantastic books in Hlynka [54]. In particular, the general results based on queueing theory is out of texts such as Kleinrock [63], Wolff [108], Gross and Harris [49], Prabhu [89], and Medhi [80]. Then, the applications of queues to computer and communication systems have predominantly been influenced by Bolch et al. [12], Menasce and Almeida [81], and Gelenbe and Mitrani [45]. Likewise, applications to production systems are mostly due to Buzacott and Shanthikumar [15]. The theoretical underpinnings of this book in terms of stochastic processes are mainly from Kulkarni [67], which is also the source for many of the notations used in this chapter as well as others in this book.

Talking about theoretical underpinnings, it is important to note that they have not been treated in this chapter with the kind of rigor one normally expects in a queueing theory text. This is intentional, being an introductory chapter of this textbook. However, interested readers are strongly encouraged to consider books such as Baccelli and Bremaud [7] as well as Serfozo [96] for a discussion on Palm calculus, which is crucial to derive results such as Little's law and PASTA. Further generalizations of Little's law can also be found there. An addition, an excellent source for cutting-edge research on queueing is the journal *Queueing Systems: Theory and Applications* that is affectionately called QUESTA in the queueing community.

On a somewhat tangential note, the last section of this chapter on psychology of queues usually does not appear in most queueing theory texts. However, interested readers must follow the literature popularized greatly by the groundbreaking paper by Larson [72]. Also, readers interested in applications of queueing could visit the website (Hlynka [55]) for an up-to-date list of queueing software [57, 83]. On there, software that run on various other applications (such as MATLAB®, Mathematica, and Excel.) are explained and the most suitable one for the reader can be adopted.

Exercises

1.1 Consider a doctor's office where reps stop by according to a renewal process with an inter-renewal time according to a gamma distribution with mean 25 days and standard deviation 2 days. Whenever a rep stops, he or she drops off 10 samples of a medication. If a patient needs the medication, the doctor gives one of the free samples if available. Patients arrive at the doctor's office needing medication according to a Poisson process at an average rate of 1 per day. To make the model more tractable, assume that all samples given during a rep visit must be used or discarded before the next visit by the rep. Model the number of usable samples at the doctor's office as a flow system. Is it stable? Derive expressions for the average input rate Λ, the time-averaged number of free samples in the doctor's office H, and the average number of days each sample stays at the doctor's office Θ?

1.2 Consider a $G/D/1/2$ queue that is empty at time $t = 0$. The first eight arrivals occur at times $t = 1.5$, 2.3, 2.7, 3.4, 5.1, 5.2, 6.5, and 9.3. The service times are constant and equal to 1 time unit. Draw a graph of $W(t)$ and $X(t)$ from $t = 0$ to $t = 6.5$. Make sure to indicate arrival times and departure times. What is the time-averaged workload as well as number in the system from $t = 0$ to $t = 6.5$?

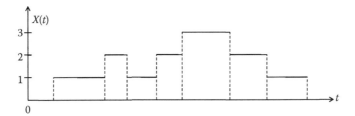

FIGURE 1.8
Plot of $X(t)$ for a $G/G/1$ queue.

1.3 Figure 1.8 represents the number in the system during the first few minutes of a $G/G/1$ queue that started empty. On the figure, mark A_3 and D_2. Also, what is $X(D_2+)$? Derive the time-averaged number in the system between $t=0$ and $t=D_4$ in terms of A_i and D_i for $i=1,2,3,4$.

1.4 An assembly line consists of three stages with a single machine at each stage. Jobs arrive to the first stage one by one and randomly with an average of one every 30 s. After processing is completed the jobs get processed at the second stage and then the third stage before departing. The first-stage machine can process 3 jobs per minute, the second-stage machine can process 2.5 jobs per minute, and the third-stage machine can process 2.25 jobs per minute. The average sojourn times (including waiting and service) at stages 1, 2, and 3 were 1, 2, and 4 min, respectively. What is the average number of jobs in the system for the entire assembly line and at each stage? What is the average time spent by each job in the system? What is the x-factor for the entire system that is defined as the ratio of the average time in the system to the average time spent processing for any job?

1.5 Consider a manual car wash station with three bays and no room for waiting (assume that cars that arrive when all the three bays are full leave without getting washed there). Cars arrive according to a Poisson process at an average rate of 1 per minute but not all cars enter. It is known that the long-run fraction of time there were 0, 1, 2, or 3 bays full are 6/16, 6/16, 3/16, and 1/16, respectively. What is L for this system? What about W and W_q? What is the average time to wash a car?

1.6 Consider a factory floor with two identical machines and jobs arrive externally at an average rate of one every 20 min. Each job is processed at the first available machine and it takes an average of 30 min to process a job. The jobs leave the factory as soon as processing is completed. The average work-in-process in the entire system is 24/7. Compute the steady-state average throughput (number of processed jobs exiting the system), cycle time (i.e., mean time in the system), and the long-run fraction of time each machine is utilized.

1.7 Consider a production system as a "black box." The system produces only one type of a customized item. The following information is known: the cycle time (i.e., average time between when an order is placed and a product is produced) for any item is 2 h and the throughput is one item every 15 min on average. It is also known that the average time spent on processing is only 40/3 min (the rest of the cycle time is spent waiting). In addition, the standard deviation of the processing times is also 40/3 min. What is the steady-state average number of products in the system? When the standard deviation of the processing time was reduced, it was observed that the average number in the system became 7, but the throughput remained the same. What is the new value for cycle time?

1.8 Two barbers own and operate a barber shop. They provide two chairs for customers who are waiting for a haircut, so the number of customers in the shop varies between 0 and 4. For $n = 0, 1, 2, 3, 4$, the probability p_n that exactly n customers are in the barber shop in steady-state is $p_0 = 1/16$, $p_1 = 4/16$, $p_2 = 6/16$, $p_3 = 4/16$, and $p_4 = 1/16$.

    (a) Calculate L and L_q.

    (b) Given that an average of four customers per hour arrive according to a Poisson process to receive a haircut, determine W and W_q.

    (c) Given that the two barbers are equally fast in giving haircuts, what is the average duration of a haircut?

1.9 Consider a discrete time queue where time is slotted in minutes. At the beginning of each minute, with probability p, a new customer arrives into the queue, and with probability $1 - p$, there are no arrivals at the beginning of that minute. At the end of a minute if there are any customers in the queue, one customer departs with probability q. Also, with probability $1-q$, no one departs a nonempty queue at the end of that minute. Let Z_n be the number of customers in the system at the beginning of the nth minute before any arrivals occur in that minute. Model $\{Z_n, n \geq 0\}$ as a discrete time Markov chain by drawing the transition diagram. What is the condition of stability? Assuming stability, obtain the steady-state probabilities for the number of customers in the system as seen by an arriving customer.

1.10 For an $M/M/2/4$ system with potential arrivals according to a Poisson process with mean rate 3 per minute and mean service time 0.25 min, it was found that $p_4 = 81/4457$ and $L = 0.8212$. What are the values of W and W_q for customers that enter the system?

2

Exponential Interarrival and Service Times: Closed-Form Expressions

---

The most fundamental queueing models, and perhaps the most researched as well, are those that can be modeled as continuous time Markov chains (CTMC). In this chapter, we are specifically interested in such queueing models for which we can obtain closed-form algebraic expressions for various performance measures. A common framework for all these models is that the potential customer arrivals occur according to a Poisson process with parameter λ, and the service time for each server is according to $\exp(\mu)$. Note that the inter-arrival times for potential customers are according to $\exp(\lambda)$ distribution, but all arriving customers may not enter the system.

The methods to analyze such queues to obtain closed-form expressions for performance measures are essentially by solving CTMC balance equations. The methods can be broadly classified into three categories. The first category is a network graph technique that uses flow balance across arcs called *arc cuts* on the CTMC rate diagram. Then, there are some instances where it is difficult to solve the balance equations using arc cuts for which *generating functions* would be more appropriate. Finally, in some instances where neither arc cuts nor generating functions work, it may be possible to take advantage of a property known as *reversibility* to obtain closed-form expressions for the performance measures.

In the next three sections, we describe those three methods with some examples. It is crucial to realize that the scenarios are indeed examples, but the methodologies can be used in many more instances. In fact, the focus of this chapter (and the entire book) is to explain the methodologies using examples and not showcase the results for various examples of queues. In other words, we would like to focus on the means and not the ends. There are several wonderful texts that provide results for every possible queueing system. Here we concentrate on the techniques used to get those results.

Since the methods in this chapter are based on CTMC analysis, we explain the basics of that first. Consider an arbitrary irreducible CTMC $\{Z(t), t \geq 0\}$ with state space S and infinitesimal generator matrix Q. Sometimes Q is also called rate matrix. In order to obtain the steady-state probability row vector $p = [p_i]$, we solve for $pQ = 0$ and normalize using $\sum_{i \in S} p_i = 1$. The set of equations for $pQ = 0$ is called balance equations and can be written for every $i \in S$ as

$$p_i q_{ii} + \sum_{j \neq i} p_j q_{ji} = 0.$$

If we draw the rate diagram, then what is immediately obvious is that since $q_{ii} = -\sum_{j \neq i} q_{ij}$, we have

$$\sum_{j \neq i} p_i q_{ij} = \sum_{j \neq i} p_j q_{ji}.$$

This means that across each node i, the flow out equals the flow in.

Many times it is not straightforward to solve the balance equations directly. The next three sections present various simplifying techniques to solve them and also use the results obtained for various other performance metrics besides the steady-state probability distribution.

2.1 Solving Balance Equations via Arc Cuts

Similar to the flow-in-equals-flow-out at each node (described above) is the notion of balance equations via arc cuts. If the sample space S is partitioned into two sets A and $S - A$ such that $A \subset S$, then across all the arcs between A and $S - A$ there is flow balance (flow from A to $S - A$ equals flow from $S - A$ to A). We have already seen a special case of that, which is the flow balance across a node, where A would just be node i and the rest of the nodes as $S - A$. So, in some sense, the node flow balance is the result of a special type of arc cut.

We further explain the arc cut by means of an example. See the rate diagram in Figure 2.1 with the arc cut depicted as a dashed line. Here $S = \{0,1,2,3,4,5\}$, $A = \{0,2\}$, and $S - A = \{1,3,4,5\}$. The flow balance across the arc cut becomes

$$\alpha p_0 + (\delta + \alpha + \gamma)p_2 = \beta(p_1 + p_3). \qquad (2.1)$$

As an exercise, the reader is encouraged to verify that the node balance equations

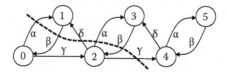

FIGURE 2.1
Arc cut example.

$$[p_0 \; p_1 \; p_2 \; p_3 \; p_4 \; p_5] \begin{bmatrix} -\alpha - \gamma & \alpha & \gamma & 0 & 0 & 0 \\ \beta & -\beta & 0 & 0 & 0 & 0 \\ 0 & \delta & -\alpha - \gamma - \delta & \alpha & \gamma & 0 \\ 0 & 0 & \beta & -\beta & 0 & 0 \\ 0 & 0 & 0 & \delta & -\alpha - \delta & \alpha \\ 0 & 0 & 0 & 0 & \beta & -\beta \end{bmatrix}$$

$$= [0 \, 0 \, 0 \, 0 \, 0 \, 0]$$

can be manipulated to get the flow balance across the arc cut resulting in Equation 2.1. It is crucial to understand that the arc cut made earlier is just to illustrate the theory. In practice, the cut chosen in the example would not be a good one to use. The objective of these cuts is to write down relationships between the unknowns so that the unknowns can be solved easily. Therefore, cuts that result in two or three p_i terms would be ideal to use. To illustrate that, we present the following example problem.

Problem 6

For the CTMC with rate diagram in Figure 2.1, compute the steady-state probabilities p_0, p_1, p_2, p_3, p_4, and p_5 by making appropriate arc cuts and solving the balance equations.

Solution

For this example, by making five suitable cuts (which ones?), we get the following relationships

$$(\alpha + \gamma)p_0 = \beta p_1$$

$$\gamma p_0 = \delta p_2$$

$$(\alpha + \gamma)p_2 = \beta p_3$$

$$\gamma p_2 = \delta p_4$$

$$\alpha p_4 = \beta p_5$$

from which it is relatively straightforward to write down p_1, p_2, p_3, p_4, and p_5 in terms of p_0 as follows:

$$p_1 = \frac{\alpha + \gamma}{\beta} p_0,$$

$$p_2 = \frac{\gamma}{\delta} p_0,$$

$$p_3 = \frac{(\alpha + \gamma)\gamma}{\delta\beta} p_0,$$

$$p_4 = \frac{\gamma^2}{\delta^2} p_0,$$

$$p_5 = \frac{\gamma^2 \alpha}{\delta^2 \beta} p_0.$$

Note how much simpler this is compared to solving the node balance equations. Once we know p_0, we have an expression for all p_i. Now, by solving for p_0 using the normalizing relation $p_0 + p_1 + p_2 + p_3 + p_4 + p_5 = 1$, we get

$$p_0 = \frac{1}{1 + (\alpha + \gamma)/\beta + \gamma/\delta + (\alpha + \gamma)\gamma/(\delta\beta) + \gamma^2/\delta^2 + \gamma^2\alpha/(\delta^2\beta)}. \quad \blacksquare$$

A word of caution is that when the cut set A is separated from the state space S, all the arcs going from A to $S - A$ must be considered. A good way to make sure of that is to clearly identify the cut set A as opposed to just performing a cut arbitrarily. Next, we use the arc cut method to obtain steady-state distributions of the number in the system for a specific class of queueing systems.

2.1.1 Multiserver and Finite Capacity Queue Model (*M/M/s/K*)

Consider a queueing system where potential arrivals occur one at a time according to a Poisson process at an average rate of λ per unit time. There are s stochastically identical servers and the capacity of the system is K where $K \geq s$. The service time at any server is exponentially distributed with mean $1/\mu$. An arriving customer that sees the system full is rejected without service. Such a system in Kendall notation is represented as an $M/M/s/K$ queue. Although Kendall notation uses a default first come first served (FCFS), it is not necessary for most expressions unless explicitly stated. We however do require that the service discipline be work conserving where a server would not be idle if there is work left to do. In this formulation, we also do not allow multiple servers to work on a single customer.

The resulting system can be modeled as a CTMC with $X(t)$ being the number in the system at time t. Note that $X(t)$ includes customers that are at the servers as well as those waiting to be served. The state space S is the set of all integers between 0 and K, that is, $S = \{0,1,2,\ldots,K\}$. Then, $\{X(t), t \geq 0\}$ is indeed a CTMC with generator matrix $Q = [q_{ij}]$ such that for $i \in S$ and $j \in S$, we have

$$q_{ij} = \begin{cases} \lambda & \text{if } j = i+1 \text{ and } i < K \\ i\mu & \text{if } j = i-1 \text{ and } 0 < i < s \\ s\mu & \text{if } j = i-1 \text{ and } s \le i \le K \\ -\min(1, K-i)\lambda - \min(i,s)\mu & \text{if } j = i \\ 0 & \text{otherwise.} \end{cases}$$

Although the methodological emphasis is on obtaining steady-state probabilities for the number in the system, it is worthwhile pointing out that transient analysis is also possible. When $K < \infty$, using transient CTMC analysis we have

$$P\{X(t+\tau) = j | X(\tau) = i\} = [\exp(Qt)]_{ij}$$

where $\exp(Qt) = I + Qt + Q^2 t^2 / 2 + Q^3 t^3 / 3! + Q^4 t^4 / 4! + \cdots$, and is called the exponential of matrix Qt. The previous equation states that if you are given that there were i in the system at time τ, then the probability that there would be j in the system after t time units is given by the ijth element of the $\exp(Qt)$ matrix. The exponential of a matrix is usually an in-built matrix operation in most mathematical software (e.g., in MATLAB® after defining Q and t, the command expm(Q*t) will return $\exp(Qt)$). For the case $K = \infty$, the reader is encouraged to refer to other techniques for CTMC transient analysis. Next, we proceed with the steady-state analysis.

2.1.2 Steady-State Analysis

Figure 2.2 is the rate diagram for an $M/M/s/K$ queue. By developing arc cuts between every pair of adjacent nodes, we obtain the following equations:

$$\lambda p_0 = \mu p_1$$

$$\lambda p_1 = 2\mu p_2$$

$$\lambda p_2 = 3\mu p_3$$

$$\vdots$$

$$\lambda p_{s-1} = s\mu p_s$$

$$\lambda p_s = s\mu p_{s+1}$$

$$\vdots$$

$$\lambda p_{K-1} = s\mu p_K$$

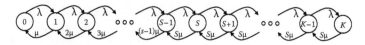

FIGURE 2.2
Rate diagram for $M/M/s/K$ queue.

where p_j for any $j \in S$ is the probability that there are j customers in the system in steady state, that is,

$$p_j = \lim_{t \to \infty} P\{X(t) = j\}.$$

In addition, since the CTMC is ergodic, p_j is also the long-run fraction of time the system has j customers, that is,

$$p_j = \lim_{T \to \infty} \frac{\int_0^T I_{\{X(t)=j\}} dt}{T}$$

where I_B is an indicator function such that it is 1 if event B occurs and 0 otherwise.

Next, solving for the previous arc cut equations in terms of p_0, and using the normalizing condition

$$\sum_{i=1}^{K} p_i = 1,$$

we get

$$p_0 = \left[\sum_{n=0}^{s} \left\{ \frac{1}{n!} \left(\frac{\lambda}{\mu} \right)^n \right\} + \frac{(\lambda/\mu)^s}{s!} \sum_{n=s+1}^{K} \rho^{n-s} \right]^{-1}$$

where

$$\rho = \frac{\lambda}{s\mu}.$$

Therefore, for all $j \in S$

$$p_j = \begin{cases} \frac{1}{j!} \left(\frac{\lambda}{\mu} \right)^j p_0 & \text{if } j < s, \\ \frac{1}{s! s^{j-s}} \left(\frac{\lambda}{\mu} \right)^j p_0 & \text{if } s \leq j \leq K. \end{cases}$$

Now that we have the steady-state probabilities, we can use them to obtain various performance measures for the system. The probability that an arriving customer in steady state is rejected is

$$p_K = \frac{(\lambda/\mu)^K}{s!s^{K-s}}p_0$$

since p_K is the probability that in steady state there are K customers in the system and that the arrivals are according to a Poisson process (due to PASTA property). Also note that the rate at which customers are rejected is λp_K on average and the mean queue *entering* rate is $\lambda(1 - p_K)$.

Another performance metric that we can quickly obtain is the long-run average number of customers in the system that are waiting for their service to begin (L_q). Using the definition of L_q and the p_j values, we can derive

$$L_q = \sum_{j=0}^{K} p_j \max(j - s, 0) = \frac{p_0(\lambda/\mu)^s \rho}{s!(1 - \rho)^2}[1 - \rho^{K-s} - (K-s)\rho^{K-s}(1 - \rho)].$$

Using L_q, it is relatively straightforward to obtain some related steady-state performance measures such as W_q (the average time spent by entering customers in the queue before service), W (the average sojourn time, i.e., time in the system, for entering customers), and L (the average number in the system) as follows:

$$W_q = \frac{L_q}{\lambda(1 - p_K)}$$

$$W = \frac{L_q}{\lambda(1 - p_K)} + \frac{1}{\mu}$$

$$L = L_q + \frac{\lambda(1 - p_K)}{\mu}$$

using the standard system analysis results via Little's law and the definitions (see Equations 1.8 through 1.10). Note that for Little's law, we use the entering rate $\lambda(1 - p_K)$ and not the arrival rate because not all arriving customers enter the system.

In a similar fashion, using p_j values, it would be possible to derive higher moments of the steady-state number in the system and number waiting to be served. See Exercise 2.1 at the end of this chapter for one such computation. However, computing higher moments of the time in the system and time in the queue in steady state for entering customers is a little more tricky. For that we first derive the steady-state distribution. Let Y_q and Y be random variables that respectively denote the time spent by an entering customer in the queue and in the system (including service). To obtain the cumulative distribution function (CDF) of Y_q and Y, we require that the *entities in the system are served according to FCFS*. The analysis until now did not require FCFS, and for any work-conserving discipline, the results would continue to hold. However, for the following analysis, we specifically take the default FCFS condition into account. Having said that, it is worthwhile to mention that other service disciplines can also be considered, but we will only consider FCFS here.

To obtain the CDF of Y_q that we denote as $F_{Y_q}(t)$, we first reiterate that Y_q is in fact a conditional random variable. In other words, it is the time spent waiting to begin service for a customer that entered the system in steady state, that is, *given* that there were less than K in the system upon arrival in steady state. Since the arrivals are according to a Poisson process, due to PASTA the probability that an arriving customer in steady state will see j in the system is p_j. Also, the probability that an *entering* customer in steady state would see j others in the system is the probability that an arriving customer would see j others in the system given that there are less than K in the system. Using a relatively straightforward conditional probability argument, one can show that the probability that an entering customer will see j in the system in steady state is $p_j/(1 - p_K)$ for $j = 0, 1, \ldots, K - 1$.

Also, if an entering customer sees j in the system, the time this customer spends before service begins is 0 if $0 \le j < s$ (why?) and an *Erlang*$(j - s + 1, s\mu)$ random variable if $s \le j < K$. The explanation for the Erlang part is that since all the s servers would be busy during the time, the entering customer waits before service, and this time corresponds to $j - s + 1$ service completions. However, since each service completion corresponds to the minimum of s random variables that are according to $\exp(\mu)$, service completions occur according to $\exp(s\mu)$ (due to *minimum of exponentials* property). Further, since the sum of $j - s + 1$ $\exp(s\mu)$ random variables is an *Erlang*$(j - s + 1, s\mu)$ random variable (due to the *sum of independently and identically distributed* [IID] *exponentials property*), we get the desired result. Thus, using the definition and CDF of the Erlang random variable, we can derive the following by conditioning on j customers in the system upon entering in steady state:

$$F_{Y_q}(t) = P\{Y_q \le t\}$$

$$= \sum_{j=0}^{s-1} 1 \frac{p_j}{1 - p_K} + \sum_{j=s}^{K-1} \left(1 - \sum_{r=0}^{j-s} e^{-s\mu t} \frac{(s\mu t)^r}{r!} \right) \frac{p_j}{1 - p_K}$$

$$= 1 - \sum_{j=s}^{K-1} \frac{p_j}{1 - p_K} \sum_{r=0}^{j-s} e^{-s\mu t} \frac{(s\mu t)^r}{r!}.$$

Once the CDF of Y_q is known, $F_Y(t)$, the CDF of Y can be obtained using the fact that $Y - Y_q$ is an $\exp(\mu)$ random variable, corresponding to the service time of this entering customer. Therefore, we have for $K > s > 1$ the CDF as

$$F_Y(t) = P\{Y \le t\}$$

$$= \int_0^t F_{Y_q}(t-u)\mu e^{-\mu u}\,du = \int_0^t \left(1 - e^{-\mu(t-u)}\right)dF_{Y_q}(u) + \left(1 - e^{-\mu t}\right)F_{Y_q}(0)$$

$$= \left(1 - e^{-\mu t}\right)\sum_{j=0}^{s-1}\frac{p_j}{1-p_K} + \sum_{j=s}^{K-1}\frac{p_j}{1-p_K} - \sum_{j=s}^{K-1}\frac{p_j}{1-p_K}\sum_{r=0}^{j-s}e^{-s\mu t}\frac{(s\mu t)^r}{r!}$$

$$- e^{-\mu t}\int_0^t e^{\mu u}\sum_{j=s}^{K-1}\frac{p_j}{1-p_K}\Bigg\{-s\mu e^{-s\mu u}$$

$$+ \sum_{r=1}^{j-s}e^{-s\mu u}s\mu\frac{(s\mu u)^{r-1}}{(r-1)!}\left(\frac{1-s\mu u}{r}\right)\Bigg\}du$$

$$= 1 - e^{-\mu t}\sum_{j=0}^{s-1}\frac{p_j}{1-p_K} - \sum_{j=s}^{K-1}\frac{p_j}{1-p_K}\sum_{r=0}^{j-s}e^{-s\mu t}\frac{(s\mu t)^r}{r!}$$

$$- e^{-\mu t}\Bigg[\sum_{j=s}^{K-1}\frac{p_j}{1-p_K}\Bigg\{-\frac{s}{s-1}\left(1 - e^{-(s-1)\mu t}\right)$$

$$+ \frac{s}{s-1}\left(1 - e^{-(s-1)\mu t} - (s-1)\mu t e^{-(s-1)\mu t}\right)$$

$$- \left(\frac{s}{s-1}\right)^{j-s+1}\left(1 - \sum_{i=0}^{j-s}e^{-(s-1)\mu t}\frac{((s-1)\mu t)^i}{i!}\right)$$

$$+ \frac{s}{s-1}\sum_{r=1}^{j-s}e^{-(s-1)\mu t}\frac{((s-1)\mu t)^{r+1}}{(r+1)!}\left(\frac{s}{s-1}\right)^r\Bigg\}\Bigg].$$

Note here that since W_q is a random variable with mass at 0, one has to be additionally careful with the convolution realizing that $F_{Y_q}(0)$ is nonzero. Also, when $K = s$, $F_Y(t) = 1 - e^{-\mu t}$ since the sojourn time equals service time for entering customers. The case $s = 1$ can be addressed by rederiving the integral using $s = 1$.

   Next, we present an alternative approach to obtain the distribution of Y since it is used for many of the special cases that we will see (such as $K = \infty$ and/or $s = 1$). It is based on the Laplace Stieltjes transform (LST) which is defined uniquely for all nonnegative random variables. The LST of $F_Y(\cdot)$ is defined as

$$\tilde{F}_Y(w) = E[e^{-wY}] = \int_0^\infty e^{-wu} dF_Y(u).$$

Once we know the LST, there are several techniques to invert it to obtain the CDF $F_Y(\cdot)$, for example, direct computation by looking up tables, converting to Laplace transform (LT) and inverting it, or by numerical transform inversion. However, moments of Y can easily be obtained by taking derivatives. With that understanding, the LST can be computed using the definition of Y (as opposed to taking the LST of $F_Y(\cdot)$) as

$$\tilde{F}_Y(w) = \sum_{j=0}^{s-1} \frac{p_j}{1 - p_K} \left(\frac{\mu}{\mu + w} \right) + \sum_{j=s}^{K-1} \frac{p_j}{1 - p_K} \left(\frac{\mu}{\mu + w} \right) \left(\frac{s\mu}{s\mu + w} \right)^{j-s+1}$$

$$= \sum_{j=0}^{s-1} \frac{p_j}{1 - p_K} \left(\frac{\mu}{\mu + w} \right) + \frac{p_0}{1 - p_K} \left(\frac{\mu}{\mu + w} \right) \frac{1}{s!} \left(\frac{s\mu}{s\mu + w} \right) \left(\frac{\lambda}{\mu} \right)^s$$

$$\times \left\{ \frac{1 - (\lambda/(s\mu + w))^{K-s}}{1 - \lambda/(s\mu + w)} \right\}.$$

   Having developed expressions for performance measures of the $M/M/s/K$ queue, a natural question to ask is what insights can we obtain from them? In other words, how do the performance measures vary with respect to the parameters they are expressed as (e.g., λ, μ, s, and K)? Although the sensitivity of the performance measures cannot easily be evaluated by just looking at the expressions, to determine that one can write a computer program (or use one of the several packages available on the web, see reference notes at the end of Chapter 1). The parameters μ, s, and K can be thought of as representing the system capacity and λ as the load on the system. Using the computer programs, one can do what-if analysis to quickly determine the performance under various scenarios by tweaking the input parameters (λ, μ, s, and K). As an interesting example, if one were to design for K keeping all other parameters a constant, then there is a trade-off in terms of W and p_K. This is because W increases with K but p_K decreases with K. It is worthwhile explaining that although W increasing with K appears counterintuitive as more capacity must improve performance, the reason for this is that W only accounts for customers that enter the queue (and larger K besides more capacity also corresponds to higher load). However, since we are unable to represent the performance measures as closed-form algebraic expressions of the input parameters, their relationships are not easily apparent. For this reason, we next present some special cases where the relationship between performance measures and input parameters is clearer.

2.1.3 Special Cases

There are several special cases of the $M/M/s/K$ queue that lend themselves nicely for closed-form algebraic expressions that are insightful. In fact, most queueing theory courses and textbooks would start with these special cases and subsequently move to the general case of $M/M/s/K$ queue. The special cases involve special values of one or both of s (1, s, or ∞) and K (s, K, or ∞). We present them as worked-out problems next.

Problem 7

Using the same notations as the $M/M/s/K$ queue earlier, derive the distribution for the number of entities in the system and also the sojourn time in the system for the classic $M/M/1$ queue. Do any conditions need to be satisfied? Obtain expressions for L, W, L_q, and W_q.

Solution

The $M/M/1$ queue is a special case of the $M/M/s/K$ queue with $s = 1$ and $K = \infty$. Most of the performance measures can be obtained by letting $s = 1$ and $K = \infty$ in the $M/M/s/K$ analysis. Hence, unless necessary, the results will not be derived, but the reader is encouraged to verify them. However, there is one issue. While letting $K = \infty$, we need to ensure that the CTMC that models the number of customers in the system for the $M/M/1$ queue is positive recurrent. The condition for positive recurrence (and hence stability of the system) is

$$\rho = \frac{\lambda}{\mu} < 1.$$

In other words, the average arrival rate (λ) must be smaller than the average service rate (μ) to ensure stability. This is intuitive because in order for the system to be stable, the server should be able to remove customers on average faster than the speed at which they enter. Note that when K is finite, instability is not an issue.

The long-run probability that the number of customers in the system is j (for all $j \geq 0$) is given by $p_j = \rho^j p_0$, which can be obtained by writing the balance equations for p_j in terms of p_0. Now the normalizing equation $\sum_j p_j = 1$

can be solved only when $\rho < 1$, and hence this is called the condition for positive recurrence. Therefore, when $\rho < 1$, we have

$$p_0 = 1 - \rho$$

and

$$p_j = (1 - \rho)\rho^j \quad \text{for all } j \geq 0.$$

Therefore, the number of customers in the system in steady state follows a modified geometric distribution. Further, the long-run probability that there are more than n customers in the system is ρ^n (why?).

Either using p_j from earlier equations (and using Little's law wherever needed) or by letting $s = 1$ and $K = \infty$ in the $M/M/s/K$ results, we have

$$L = \frac{\lambda}{\mu - \lambda},$$

$$L_q = \frac{\lambda^2}{\mu(\mu - \lambda)},$$

$$W = \frac{1}{\mu - \lambda},$$

$$W_q = \frac{\lambda}{\mu(\mu - \lambda)}.$$

It is crucial to realize that all these results require that $\rho < 1$. Also note that while using Little's law, one can use λ as the entering rate as no customers are going to be turned away. Besides these performance metrics, one can also derive higher moments of the steady-state number in the system. However, to obtain the higher moments of the time spent in the system by a customer arriving at steady state, one technique is to use the distribution of the time in the system Y. By letting $s = 1$, $K = \infty$, and using $p_j = (1 - \lambda/\mu)(\lambda/\mu)^j$ in the $M/M/s/K$ results, we get the LST after some algebraic manipulation as

$$\tilde{F}_Y(w) = \frac{\mu - \lambda}{\mu - \lambda + w}.$$

Inverting the LST, we get $F_Y(y) = P\{Y \leq y\} = 1 - e^{(\lambda - \mu)y}$ for $y \geq 0$. Therefore, the sojourn time for a customer arriving to the system in steady state is according to an exponential distribution with mean $1/(\mu - \lambda)$, that is,

$$Y \sim \exp(\mu - \lambda).$$

Verify that $E[Y] = W = 1/(\mu - \lambda)$.   ∎

Problem 8

Using the same notations as the $M/M/s/K$ described earlier in this section, derive the distribution for the number of entities in the system and also the sojourn time in the system for the multiserver $M/M/s$ queue. What is the stability condition that needs to be satisfied? Obtain expressions for L, W, L_q, and W_q.

Solution

The $M/M/s$ queue is a special case of the $M/M/s/K$ queue with $K=\infty$. Most of the performance measures can be obtained by letting $K=\infty$ in the $M/M/s/K$ analysis; hence unless necessary the results will not be derived but the reader is encouraged to verify them. Similar to the $M/M/1$ queue, here too we need to be concerned about stability while letting $K=\infty$. The condition for stability is

$$\rho = \frac{\lambda}{s\mu} < 1.$$

In other words, the average arrival rate (λ) must be smaller than the average service capacity ($s\mu$) across all servers to ensure stability.

By writing down the balance equations for the CTMC or by letting $K=\infty$ in the $M/M/s/K$ results, we can derive the long-run probability that the number of customers in the system is j (when $\rho < 1$) as

$$p_j = \begin{cases} \frac{1}{j!}\left(\frac{\lambda}{\mu}\right)^j p_0 & \text{if } 0 \leq j \leq s-1 \\ \frac{1}{s!\, s^{j-s}}\left(\frac{\lambda}{\mu}\right)^j p_0 & \text{if } j \geq s \end{cases}$$

where

$$p_0 = \left[\sum_{n=0}^{s-1}\left\{\frac{1}{n!}\left(\frac{\lambda}{\mu}\right)^n\right\} + \frac{(\lambda/\mu)^s}{s!}\frac{1}{1-\lambda/(s\mu)}\right]^{-1}.$$

Either using p_j from the previous equation (and using Little's law wherever needed) or by letting $K=\infty$ in the $M/M/s/K$ results, we have

$$L_q = \frac{p_0(\lambda/\mu)^s\lambda}{s!s\mu[1-\lambda/(s\mu)]^2},$$

$$L = \frac{\lambda}{\mu} + \frac{p_0(\lambda/\mu)^s\lambda}{s!s\mu[1-\lambda/(s\mu)]^2},$$

$$W = \frac{1}{\mu} + \frac{p_0(\lambda/\mu)^s}{s!s\mu[1 - \lambda/(s\mu)]^2},$$

$$W_q = \frac{p_0(\lambda/\mu)^s}{s!s\mu[1 - \lambda/(s\mu)]^2}.$$

It is a worthwhile exercise to verify that by letting $s = 1$, we get the $M/M/1$ results. Also note that all the previous results require that $\rho = \lambda/(s\mu) < 1$. Other than the previous performance metrics, we can also derive higher moments of the steady-state number in the system. In order to obtain the higher moments of the time spent in the system by a customer arriving at steady state, we use the distribution of the time in the system, Y. By taking the limit $K \to \infty$ in the $M/M/s/K$ results with the understanding that $\lambda < s\mu$, we get the LST as

$$\tilde{F}_Y(w) = \sum_{j=0}^{s-1} p_j \left(\frac{\mu}{\mu + w} \right) + p_0 \left(\frac{\mu}{\mu + w} \right) \frac{1}{s!} \left(\frac{s\mu}{s\mu + w - \lambda} \right) \left(\frac{\lambda}{\mu} \right)^s.$$

Using partial fractions we can write

$$\frac{1}{(\mu + w)(s\mu + w - \lambda)} = \frac{1}{(\mu + w)(s\mu - \mu - \lambda)} - \frac{1}{(s\mu - \lambda - \mu)(s\mu + w - \lambda)}.$$

Then by inverting the LST, we get

$$F_Y(y) = \sum_{j=0}^{s-1} p_j(1 - e^{-\mu y}) + p_0 \frac{(\lambda/\mu)^s}{s!} \left\{ \frac{s\mu}{(s-1)\mu - \lambda}(1 - e^{-\mu y}) \right.$$

$$\left. - \frac{s\mu^2}{(s\mu - \lambda)[(s-1)\mu - \lambda]}(1 - e^{-(s\mu - \lambda)y}) \right\}$$

for $y \geq 0$. The reader is encouraged to verify that $E[Y]$ results in the expression for W, and that by letting $s = 1$ we get $Y \sim \exp(\mu - \lambda)$, the $M/M/1$ sojourn time expression. ∎

Problem 9

Using the same notations as the $M/M/s/K$ described earlier in this section, derive the distribution for the number of entities in the system and also the sojourn time in the system for the single-server finite capacity $M/M/1/K$ queue. What is the rate at which customers enter the system on an average? Obtain expressions for L, W, L_q, and W_q.

Solution

The $M/M/1/K$ queue is a special case of the $M/M/s/K$ system with $s=1$. All the results presented here are obtained by letting $s=1$ for the corresponding $M/M/s/K$ results. We define $\rho=\lambda/\mu$; however, since K is finite ρ can be greater than one and the system would still be stable. The steady-state probability that there are j customers in the system (for $0 \leq j \leq K$) is given by

$$p_j = \frac{\rho^j[1-\rho]}{1-\rho^{K+1}}.$$

In particular, the fraction of arrivals that are turned away due to a full system is

$$p_K = \frac{\rho^K[1-\rho]}{1-\rho^{K+1}}.$$

Note that p_K is also the probability that a potential arrival is rejected. Therefore, the effective entering rate into the system is $\lambda(1-p_K)$.

Using the previous results or by letting $s=1$ in the $M/M/s/K$ results, we can derive the following:

$$L_q = \frac{\rho}{1-\rho} - \frac{\rho\left(K\rho^K + 1\right)}{1-\rho^{K+1}},$$

$$W_q = \frac{1}{\mu}\left[\frac{(K-1)\rho^{K+1}+\rho-K\rho^K}{(1-\rho)(1-\rho^K)}\right],$$

$$W = \frac{1}{\mu}\left[\frac{K\rho^{K+1}-(K+1)\rho^K+1}{(1-\rho)(1-\rho^K)}\right],$$

$$L = \frac{K\rho^{K+2}-(K+1)\rho^{K+1}+\rho}{(1-\rho)(1-\rho^{K+1})}.$$

Recall that $L_q = \lambda(1-p_K)W_q$ since a fraction p_K of arriving customers are turned away. The only situation where the previous results are problematic is when $\rho=1$. In that case, one would need to carefully take limits for the earlier expressions as $\rho \to 1$. In addition, using p_j one can obtain higher moments of the number in the system in steady state.

For customers that enter the system, using the $M/M/s/K$ results with $s=1$, the LST of the CDF of the steady-state sojourn time in the system Y is given by

$$\tilde{F}_Y(w) = \sum_{j=0}^{K-1} \frac{p_j}{1 - p_K} \left(\frac{\mu}{\mu + w}\right)^{j+1}.$$

This LST can be inverted using the fact that $(\mu/\mu + w)^{j+1}$ is the LST of an *Erlang*$(j + 1, \mu)$ distribution to get

$$F_Y(y) = P\{Y \le y\} = \sum_{j=0}^{K-1} \frac{p_j}{1 - p_K} \left(1 - e^{-\mu y} \sum_{r=0}^{j} \frac{(\mu y)^r}{r!}\right)$$

for $y \ge 0$. The reader is encouraged to cross-check all the results with the $M/M/1$ queue by letting $K \to \infty$ and assuming $\rho < 1$. ∎

Problem 10

Using the same notations as the $M/M/s/K$ described earlier in this section, derive the distribution for the number of entities in the queue-less $M/M/s/s$ system. What if $s = \infty$? Is it trivial to obtain the sojourn time distributions for customers that enter the system? Obtain expressions for L, W, L_q, and W_q.

Solution

The last of the special cases of the $M/M/s/K$ system is the case when $s = K$, which gives rise to the $M/M/s/s$ system. Note that no customers wait for service. If there is a server available, an arriving customer enters the system otherwise the customer is turned away. This is also known as the *Erlang loss system*. Similar to the previous special cases, here too one can either work with the $M/M/s/K$ system letting $s = K$ or start with the CTMC.

The probability that there are j (for $j = 0, \dots, s$) customers in the system in steady state is

$$p_j = \frac{(\lambda/\mu)^j/j!}{\sum_{i=0}^{s} (\lambda/\mu)^i/i!}.$$

Therefore, the *Erlang loss formula* is the probability that a customer arriving to the system in steady state is rejected (or the fraction of arriving customers that are lost in the long run) and is given by

$$p_s = \frac{(\lambda/\mu)^s/s!}{\sum_{i=0}^{s} (\lambda/\mu)^i/i!}.$$

Although we do not derive the result here (see Section 4.5.3), it is worthwhile to point out the remarkable fact that the previous formulae hold even for the

$M/G/s/s$ system with mean service time $1/\mu$. In other words, the steady-state distribution of the number in the system for an $M/G/s/s$ queue does not depend on the distribution of the service time.

Using the steady-state number in the system, we can derive

$$L = \frac{\lambda}{\mu}(1 - p_s).$$

Since the effective entering rate into the system is $\lambda(1 - p_s)$, we get $W = 1/\mu$. This is intuitive because for customers that enter the system, since there is no waiting for service, the average sojourn time is indeed the average service time. For the same reason the sojourn time distribution for customers that enter the system is same as that of the service time, that is, $Y \sim \exp(\mu)$. In addition, since there is no waiting for service, $L_q = 0$ and $W_q = 0$.

It is customary to consider a special case of the $M/M/s/s$ system which is when $s = \infty$. We call that $M/M/\infty$ queue since K takes the default value of infinite. For the $M/M/\infty$ system, the probability that there are j customers in the system in the long run is

$$p_j = (\lambda/\mu)^j \frac{1}{j!} e^{-\lambda/\mu}.$$

In other words, the steady-state number in the system is according to a Poisson distribution. This result also holds good for the $M/G/\infty$ system (see Section 4.5.1). Also, $L = \lambda/\mu$ and $W = 1/\mu$. ∎

Having described the $M/M/s/K$ queue and its special cases in detail, next we move to other CTMC-based queueing systems for which arc cuts are inadequate and we demonstrate other methodological tools.

2.2 Solving Balance Equations Using Generating Functions

There are many queueing systems where the resulting CTMC balance equations cannot be solved in closed form using arc cuts (or for that matter directly using the node balance equations). Under such cases, especially when the CTMC has a state space with infinite elements, the next option to try is using generating functions. The reader is encouraged to read generating function of a random variable in Section A.2.1, if required, before proceeding further. The best way to explain the use of generating functions is by means of examples. We first present a short example (although it can be easily solved using arc cuts) for illustration purposes, followed by some detailed ones.

Problem 11

Consider a CTMC with $S = \{0, 1, 2, 3, \ldots\}$ for which we are interested in obtaining the steady-state probabilities $p_0, p_1, \ldots$ represented using a generating function. For all $i \in S$ and $j \in S$, let the elements of the infinitesimal generator matrix Q be

$$
q_{ij} = \begin{cases}
\alpha & \text{if } j = i + 1, \\
-\alpha - i\beta & \text{if } j = i, \\
i\beta & \text{if } j = i - 1 \text{ and } i > 0, \\
0 & \text{otherwise.}
\end{cases}
$$

Using the balance equations to obtain the generating function

$$
\Psi(z) = \sum_{i=0}^{\infty} p_i z^i.
$$

Solution

The balance equations can be written for $i > 0$ as

$$
-p_i(\alpha + i\beta) + p_{i-1}\alpha + p_{i+1}(i + 1)\beta = 0, \tag{2.2}
$$

and for $i = 0$ as

$$
-p_0\alpha + p_1\beta = 0. \tag{2.3}
$$

Now by multiplying Equation 2.2 by z^i, Equation 2.3 by z^0, and summing over all $i \geq 0$ we get

$$
\sum_{i=0}^{\infty} p_i z^i(\alpha + i\beta) = \sum_{i=0}^{\infty} \left(p_i z^i \alpha z + i p_i z^{i-1} \beta \right).
$$

If this derivation is not clear, it may be better to write down the balance equations for $i = 0, 1, 2, 3, \ldots$, multiply them by $1, z, z^2, z^3, \ldots$, and then see how that results in the previous equation. We can rewrite the previous equation in terms of $\Psi(z)$ as

$$
\alpha\Psi(z) + \beta z\Psi'(z) = \alpha z\Psi(z) + \beta\Psi'(z),
$$

where $\Psi'(z) = d\Psi(z)/dz$. Upon rearranging terms, we get the differential equation

$$\Psi'(z) = \frac{\alpha}{\beta}\Psi(z)$$

which can be solved by dividing the equation by $\Psi(z)$ and integrating both sides with respect to z to get

$$\log(\Psi(z)) = \frac{\alpha}{\beta}z + c.$$

Using the condition $\Psi(1) = 1$ (why?), the constant c can be obtained as equal to $-\alpha/\beta$. Thus, we obtain the generating function $\Psi(z)$ as

$$\Psi(z) = e^{-\alpha/\beta}e^{(\alpha/\beta)z}. \qquad\blacksquare$$

Note that in the general case we typically do not compute p_i values explicitly, but can obtain them if necessary taking the ith derivative of $\Psi(z)$ with respect to z, letting $z = 0$ and dividing by $i!$. Although that is the general technique, in Problem 11 we could just write down the infinite sum to get p_i directly. Also, in the general case several performance measures such as the average number in the system can be computed directly with $\Psi(z)$. We will see that in the following examples as well as a breadth of problems where generating functions can be used.

2.2.1 Single Server and Infinite Capacity Queue (M/M/1)

Although we have already obtained the steady-state distribution for the number in the system for the $M/M/1$ queue in Problem 7, we once again consider it here to further illustrate the generating function technique as well as cross-check our results.

Problem 12

Consider an $M/M/1$ queue with arrival rate λ per hour and service rate μ per hour. For $j = 0, 1, 2, \ldots$, let p_j be the steady-state probability that there are j in the system. Using the balance equations, derive an expression for the generating function

$$\Psi(z) = \sum_{i=0}^{\infty} p_i z^i.$$

Using that, compute L, L_q, W, and W_q.

Solution

The node balance equations are

$$p_0\lambda = p_1\mu$$
$$p_1(\lambda + \mu) = p_0\lambda + p_2\mu$$
$$p_2(\lambda + \mu) = p_1\lambda + p_3\mu$$
$$p_3(\lambda + \mu) = p_2\lambda + p_4\mu$$

$$\vdots \quad \vdots \quad \vdots$$

and multiply the first equation by z^0, the second by z^1, the third by z^2, the fourth by z^3, and so on. Then, upon adding we get

$$(p_0z^0 + p_1z^1 + p_2z^2 + p_3z^3 + \cdots)(\lambda + \mu) - p_0\mu$$

$$= (p_0z^0 + p_1z^1 + p_2z^2 + p_3z^3 + \cdots)\lambda z$$
$$+ (p_0z^0 + p_1z^1 + p_2z^2 + p_3z^3 + \cdots)\frac{\mu}{z} - \frac{p_0\mu}{z}.$$

Multiplying the previous equation by z and using the relation

$$\Psi(z) = \sum_{i=0}^{\infty} p_i z^i$$

we get

$$(\lambda + \mu)z\Psi(z) - p_0\mu z = \lambda z^2\Psi(z) + \mu\Psi(z) - p_0\mu.$$

Therefore, we can write

$$\Psi(z) = \frac{p_0\mu(1-z)}{\mu - (\lambda + \mu)z + \lambda z^2} = \frac{p_0\mu}{(\mu - \lambda z)}.$$

The only unknown in this equation is p_0. Using $\Psi(0) = p_0$ would not result in anything meaningful. However, using $\Psi(1) = 1$ would imply $p_0\mu/(\mu - \lambda) = 1$, and hence $p_0 = 1 - \lambda/\mu$. It is crucial to note that for p_0 to be a probability, we require that $\lambda < \mu$, which is the condition for stability. Hence, when $\lambda < \mu$ we can write

$$\Psi(z) = \frac{\mu - \lambda}{\mu - \lambda z} = \frac{1 - \rho}{1 - \rho z}$$

where $\rho = \lambda/\mu$ is the traffic intensity. We can write $\Psi(z) = (1 - \rho)(1 + \rho z + (\rho z)^2 + (\rho z)^3 + \cdots)$ and hence $p_i = (1 - \rho)\rho^i$ for all $i \geq 0$, which is consistent with the results in Problem 7.

To obtain the performance metrics using $\Psi(z)$, we first compute L. By definition we have

$$L = 0 p_0 + 1 p_1 + 2 p_2 + 3 p_3 + \cdots$$

$$= \Psi'(1)$$

where $\Psi'(z) = d\Psi(z)/dz$. Using the expression for $\Psi(z)$, we have

$$\Psi'(z) = \frac{(1 - \rho)\rho}{(1 - \rho z)^2}.$$

Therefore, we have

$$L = \Psi'(1) = \frac{\rho}{1 - \rho} = \frac{\lambda}{\mu - \lambda}.$$

Finally, based on the relations $W = L/\lambda$, $W_q = W - 1/\mu$, and $L_q = \lambda W_q$, we have (when $\lambda < \mu$) the following results:

$$L_q = \frac{\lambda^2}{\mu(\mu - \lambda)},$$

$$W = \frac{1}{\mu - \lambda},$$

$$W_q = \frac{\lambda}{\mu(\mu - \lambda)}. \qquad \blacksquare$$

Although we have seen these results before in Problem 7, the main reason they are presented here is to get an appreciation of the generating function as an alternate technique. In the next few examples, the flow balance equations may not be easily solved (also the arc cuts would not be useful) and the power of generating functions will become clearly evident.

2.2.2 Retrial Queue with Application to Local Area Networks

In the $M/M/s/s$ queue that we studied in Problem 10, if an arriving customer finds no available server, that customer is dropped or rejected. Here we consider a modification of that system where the customer that finds no server upon arrival, instead of getting dropped, waits for a random time exponentially distributed with mean $1/\theta$ and retries. At this time of retrial, if a server is available the customer's service begins, otherwise the customer retries after another exp(θ) time. This process continues until the customer is served. This system is called a retrial queue. A popular application of this is the telephone switch. If there are s lines in a telephone switch and all of them are busy, the customer making a call gets a message "all lines are busy please try your call later" and the customer retries after a random time. In the following example, we consider another application (albeit a much simplified model than what is observed in practice) where $s = 1$, which is used in modeling Ethernets with exponential back-off.

Problem 13

Consider a simplified model of the Ethernet (an example of a local area network). Requests arrive according to a Poisson process with rate λ per unit time on average to be transmitted on the Ethernet cable. If the Ethernet cable is free, the request is immediately transmitted and the transmission time is exponentially distributed with mean $1/\mu$. However, if the cable is busy transmitting another request, this request waits for an exp(θ) time and retries (this is called exponential back-off in the networking literature). Note that every time a retrial occurs, if the Ethernet cable is busy the request gets backed off for a further exp(θ) time. Model the system using a CTMC and write down the balance equations. Then obtain the following steady-state performance measures: probability that the system is empty (i.e., no transmissions and no backlogs), fraction of time the Ethernet cable is busy (i.e., utilization), mean number of requests in the system, and cycle time (i.e., average time between when a request is made and its transmission is completed).

Solution

The state of the system at time t can be modeled using two variables: $X(t)$ denoting the number of backlogged requests and $Y(t)$ the number of requests being transmitted on the Ethernet cable. The resulting bivariate stochastic process $\{(X(t), Y(t)), t \geq 0\}$ is a CTMC with rate diagram given in Figure 2.3. To explain this rate diagram, consider node (3,0). This state represents three messages that have been backed off and each of them would retry after

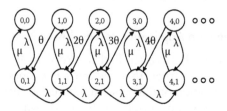

FIGURE 2.3
Rate diagram for the retrial queue.

exp(θ) time. Hence, the first of them would retry after exp(3θ) time resulting in state (2,1). Also, a new request could arrive when the system is in state (3,0) at rate λ which would result in a new state (3,1). Note that in state (3,0), there are no requests being transmitted. Now consider state (3,1). A new request arrival at rate λ would be backed off resulting in the new state (4,1); however, a transmission completion at rate μ would result in the new state (3,0). Note that a retrial in state (3,1) would not change the state of the system and is not included. However, even if one were to consider the event of retrial, it would get canceled in the balance equations and hence we need not include it.

Although one could write down the node balance equations, it is much simpler when we consider arc cuts. Specifically, cuts around nodes $(i, 0)$ for all i would result in the following balance equations:

$$p_{0,0}\lambda = p_{0,1}\mu$$

$$p_{1,0}(\lambda + \theta) = p_{1,1}\mu$$

$$p_{2,0}(\lambda + 2\theta) = p_{2,1}\mu$$

$$p_{3,0}(\lambda + 3\theta) = p_{3,1}\mu$$

$$\vdots \ \vdots \ \vdots$$

and vertical cuts on the rate diagram would result in the following balance equations:

$$p_{0,1}\lambda = \theta p_{1,0}$$

$$p_{1,1}\lambda = 2\theta p_{2,0}$$

$$p_{2,1}\lambda = 3\theta p_{3,0}$$

$$p_{3,1}\lambda = 4\theta p_{4,0}$$

$$\vdots \ \vdots \ \vdots$$

and we will leave as an exercise for the reader to solve for $p_{0,0}$ using the previous set of equations. We consider using generating functions. Let $\Psi_0(z)$ and $\Psi_1(z)$ be defined as follows:

$$\Psi_0(z) = \sum_{i=0}^{\infty} p_{i,0}z^i$$

and

$$\Psi_1(z) = \sum_{i=0}^{\infty} p_{i,1} z^i.$$

Also, define the derivative

$$\Psi_0'(z) = \frac{d\Psi_0(z)}{dz} = \sum_{i=0}^{\infty} i p_{i,0} z^{i-1}.$$

For the first set of balance equations, if we multiply the first equation by z^0, the second by z^1, the third by z^2, the fourth by z^3, and so on, then upon adding we get

$$\lambda\Psi_0(z) + z\theta\Psi_0'(z) = \mu\Psi_1(z). \tag{2.4}$$

Likewise, if we multiply the first equation in the second set of balance equations by z^0, the second equation by z^1, the third by z^2, the fourth by z^3, and so on, then upon adding we get

$$\lambda\Psi_1(z) = \theta\Psi_0'(z). \tag{2.5}$$

Rearranging Equations 2.4 and 2.5, we get

$$\Psi_1(z) = \frac{\lambda}{\mu - \lambda z}\Psi_0(z), \tag{2.6}$$

$$\frac{\lambda}{\mu}\Psi_0(z) + \frac{\theta z}{\mu}\Psi_0'(z) = \frac{\theta}{\lambda}\Psi_0'(z). \tag{2.7}$$

Using Equation 2.6 and the fact that $\Psi_0(1) + \Psi_1(1) = 1$, we get

$$\Psi_1(1) = \frac{\lambda}{\mu} = 1 - \Psi_0(1)$$

provided $\lambda < \mu$ which is the condition for stability. Therefore, if $\lambda < \mu$, the utilization or fraction of time Ethernet cable is busy is $\Psi_1(1) = \lambda/\mu$. Hence, the fraction of time the Ethernet cable is idle in steady state is $1 - \lambda/\mu$. Also, if we can solve for $\Psi_0(z)$ in the (differential) Equation 2.7, then we can immediately compute $\Psi_1(z)$ using Equation 2.6. This is precisely what we do next.

Letting $y = \Psi_0(z)$, we can rewrite Equation 2.7 as

$$\frac{1}{y}dy = \frac{\lambda/\mu}{(\theta/\lambda) - (\theta/\mu)z}dz.$$

Integrating both sides of this equation, we get

$$\log(y) + k = -\frac{\lambda}{\theta} \log\left(\frac{\theta}{\lambda} - \frac{\theta z}{\mu}\right),$$

where k is a constant that needs to be determined. But since $\Psi_0(1) = 1 - \lambda/\mu$ (i.e., if $z = 1$ then $y = 1 - \lambda/\mu$), we have $k = -\log(1 - \lambda/\mu) - (\lambda/\theta) \log(\theta/\lambda - \theta/\mu)$. Hence, we have

$$\Psi_0(z) = \left(1 - \frac{\lambda}{\mu}\right) \left(\frac{1}{\lambda} - \frac{z}{\mu}\right)^{-\lambda/\theta} \left(\frac{1}{\lambda} - \frac{1}{\mu}\right)^{\lambda/\theta}$$

$$= \left(1 - \frac{\lambda}{\mu}\right)^{(\lambda/\theta)+1} \left(\frac{\mu}{\mu - \lambda z}\right)^{\lambda/\theta}.$$

Also using Equation 2.6, we have

$$\Psi_1(z) = \frac{\lambda}{\mu - \lambda z} \left(1 - \frac{\lambda}{\mu}\right)^{(\lambda/\theta)+1} \left(\frac{\mu}{\mu - \lambda z}\right)^{\lambda/\theta}.$$

Next, we obtain the performance metrics using $\Psi_0(z)$ and $\Psi_1(z)$. We have already computed the utilization of the Ethernet cable. The probability that the system is empty with no transmissions or backlogs is $p_{0,0}$, which is equal to $\Psi_0(0)$, and hence

$$p_{0,0} = \left(1 - \frac{\lambda}{\mu}\right)^{(\lambda/\theta)+1}.$$

The mean number of backlogged requests in the system is

$$\Psi_0'(1) + \Psi_1'(1) = \frac{\lambda^2}{\theta\mu} \left[\frac{\mu + \theta}{\mu - \lambda}\right]$$

and the mean number of requests being transmitted in the system is λ/μ (i.e., the cable utilization). Therefore, the mean number of requests in the system (L) is

$$L = \frac{\lambda}{\mu} + \frac{\lambda^2}{\theta\mu} \left[\frac{\mu + \theta}{\mu - \lambda}\right] = \frac{\lambda(\lambda + \theta)}{\theta(\mu - \lambda)}.$$

Hence, using Little's law, the cycle time (W) which is the average time between when a request is made and its transmission is completed, is

$$W = \frac{\lambda + \theta}{\theta(\mu - \lambda)}. \qquad \blacksquare$$

2.2.3 Bulk Arrival Queues ($M^{[X]}/M/1$) with a Service System Example

So far we have only considered the case of individual arrivals. However, in practice it is not uncommon to see bulk arrivals into a system. For example, arrivals into theme parks is usually in groups and arrivals as well as service in restaurants is in groups. We do not consider bulk service in this text; the reader is referred to other books in the queueing literature on that subject. We only discuss the single server bulk arrival queue here.

Consider an infinite-sized queue with a single server (with service times $\exp(\mu)$). Arrivals occur according to a Poisson process with average rate λ per unit time. Each arrival brings a random number X customers into the queue. The server processes the customers one by one taking an independent $\exp(\mu)$ time for each. This system is called an $M^{[X]}/M/1$ queue. Let a_i be the probability that arriving batch is of size i, that is, $a_i = P\{X=i\}$ for $i > 0$. The generating function of the probability mass function (PMF) of X is $\phi(z)$, which is given by

$$\phi(z) = E[z^X] = \sum_{i=1}^{\infty} P(X = i)z^i = \sum_{i=1}^{\infty} a_i z^i.$$

Note that $\phi(z)$ is either given or can be computed since a_i is known. In addition, we can compute $E[X] = \phi'(1)$ (i.e., the derivative of $\phi(z)$ with respect to z at $z = 1$) and $E[X(X - 1)] = \phi''(1)$ (i.e., the second derivative of $\phi(z)$ with respect to z at $z = 1$) from which we can derive $E[X^2]$ and thereby $Var[X]$. With that we present Problem 14.

Problem 14

Consider a single server fast-food restaurant where customers arrive in groups according to a Poisson process with rate λ per unit time on average. The size of each group is independent and identically distributed with a probability a_i of having a batch of size i (with generating function $\phi(z)$ described earlier). Customers are served one by one, even though they may have arrived in batches, and it takes the server an $\exp(\mu)$ time to serve each customer. Model the system using a CTMC and write down the balance equations. Define $\Psi(z)$ as the generating function

$$\Psi(z) = \sum_{j=0}^{\infty} p_j z^j,$$

where p_j is the steady-state probability of j customers being in the system. Derive an expression for $\Psi(z)$. Then obtain L (the long-run average number of customers in the system) and W (the average time spent by each customer in the system in steady state).

Solution

Let $X(t)$ denote the number of customers in the system at time t. The stochastic process $\{X(t), t \geq 0\}$ is a CTMC with rate diagram given in Figure 2.4. To explain this rate diagram, consider node 1. This state represents one customer in the system that finishes service in $\exp(\mu)$ time resulting in state 0. However, if a batch of size i arrives the resulting state is $i + 1$, and this happens at rate λa_i.

Using the rate diagram, we write down the node balance equations as

$$p_0 \lambda = \mu p_1$$

$$p_1(\lambda + \mu) = \mu p_2 + \lambda a_1 p_0$$

$$p_2(\lambda + \mu) = \mu p_3 + \lambda a_1 p_1 + \lambda a_2 p_0$$

$$p_3(\lambda + \mu) = \mu p_4 + \lambda a_1 p_2 + \lambda a_2 p_1 + \lambda a_3 p_0$$

$$p_4(\lambda + \mu) = \mu p_5 + \lambda a_1 p_3 + \lambda a_2 p_2 + \lambda a_3 p_1 + \lambda a_4 p_0$$

$$\vdots \quad \vdots \quad \vdots$$

and multiply the first equation by z^0, the second by z^1, the third by z^2, the fourth by z^3, and so on. Then, upon adding we get

$$\lambda \Psi(z) + \mu \Psi(z) - \mu p_0 = \frac{\mu}{z} \Psi(z) - \frac{\mu}{z} p_0 + \lambda a_1 z \Psi(z)$$
$$+ \lambda a_2 z^2 \Psi(z) + \lambda a_3 z^3 \Psi(z) + \cdots .$$

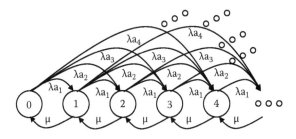

FIGURE 2.4
Rate diagram for the $M^{[X]}/M/1$ queue.

Since $\phi(z) = a_1 z + a_2 z^2 + a_3 z^3 + \cdots$, we can write down the previous equation as

$$\lambda z \Psi(z) + \mu z \Psi(z) - \mu z p_0 = \mu \Psi(z) - \mu p_0 + \lambda z \Psi(z) \phi(z)$$

which can be rewritten as

$$\Psi(z) = \frac{\mu p_0 (1 - z)}{\mu(1 - z) + \lambda z(\phi(z) - 1)}.$$

The only unknown in the previous equation is p_0. Similar to the previous examples, here too $\Psi(0) = p_0$ does not yield anything useful in terms of solving for p_0. But we can use the normalizing equation which corresponds to $\Psi(1) = 1$. However, since $\phi(1) = 1$, we would get a 0/0 type expression by substituting $z = 1$. Therefore, we first consider $A(z)$ defined as

$$A(z) = \frac{\lambda z \phi(z) - \lambda z}{1 - z}$$

so that

$$\Psi(z) = \frac{\mu p_0}{\mu + A(z)}.$$

Therefore, to compute $\Psi(1)$ it would suffice if we obtained $A(1)$, which we do as follows:

$$A(1) = \lim_{z \to 1} \frac{\lambda z \phi(z) - \lambda z}{1 - z} = -\lambda \phi'(1)$$

where the last equality is using L'Hospital's rule since $A(z)$ would also result in a 0/0 form by substituting $z = 1$. However, we showed earlier that $\phi'(1) = E[X]$ and hence $A(1) = -\lambda E[X]$. Thus, $\Psi(1) = 1$ implies that

$$p_0 = \frac{1 - \lambda E[X]}{\mu}$$

provided $\lambda E[X] < \mu$. The condition $\lambda E[X] < \mu$ is necessary for stability, and it is intuitive since $\lambda E[X]$ is the effective average arrival rate of customers. Thus, we have

$$\Psi(z) = \frac{\mu(1 - \lambda E[X]/\mu)(1 - z)}{\mu(1 - z) + \lambda z(\phi(z) - 1)}$$

when $\lambda E[X]/\mu < 1$.

From the previous expression, L can be computed as

$$L = \Psi'(1)$$

$$= \lim_{z \to 1} \frac{d\Psi(z)}{dz}$$

$$= \mu p_0 \lim_{z \to 1} \left(\frac{-(\mu(1-z)+\lambda z(\phi(z)-1))-(1-z)(-\mu+\lambda(\phi(z)-1)+\lambda z\phi'(z))}{\{\mu(1-z)+\lambda z(\phi(z)-1)\}^2} \right)$$

$$= \mu p_0 \lim_{z \to 1} \left(\frac{-\mu(1-z)-\lambda z\phi(z)+\lambda z+(1-z)\mu-\lambda\phi(z)+\lambda+\lambda z\phi(z)-\lambda z-\lambda z\phi'(z)+\lambda z^2\phi'(z)}{\{\mu(1-z)+\lambda z(\phi(z)-1)\}^2} \right)$$

$$= \mu p_0 \lim_{z \to 1} \left(\frac{-\lambda\phi(z)+\lambda-\lambda z\phi'(z)+\lambda z^2\phi'(z)}{\{\mu(1-z)+\lambda z(\phi(z)-1)\}^2} \right).$$

However, taking the limit results in a 0/0 format, we use L'Hospital's rule and continue as follows:

$$L = \mu p_0 \lim_{z \to 1} \left(\frac{-\lambda\phi'(z) - \lambda\phi'(z) - \lambda z\phi''(z) + 2\lambda z\phi'(z) + \lambda z^2\phi''(z)}{2\{\mu(1-z) + \lambda z(\phi(z)-1)\}\{-\mu + \lambda\phi(z) - \lambda + \lambda z\phi'(z)\}} \right)$$

$$= \mu p_0 \lim_{z \to 1} \left(\frac{2\lambda(z-1)\phi'(z) + \lambda z(z-1)\phi''(z)}{2\{\mu(1-z) + \lambda z(\phi(z)-1)\}\{-\mu + \lambda\phi(z) - \lambda + \lambda z\phi'(z)\}} \right)$$

$$= \mu p_0 \lim_{z \to 1} \left(\frac{2\lambda\phi'(z) + \lambda z\phi''(z)}{2\left\{-\mu + \lambda z\left(\frac{\phi(z)-1}{z-1}\right)\right\}\{-\mu + \lambda\phi(z) - \lambda + \lambda z\phi'(z)\}} \right)$$

$$= \mu p_0 \left(\frac{2\lambda\phi'(1) + \lambda\phi''(1)}{2\{-\mu + \lambda E[X]\}\{-\mu + \lambda\phi(1) - \lambda + \lambda\phi'(1)\}} \right).$$

The last equation uses the result we earlier derived namely $\lim_{z \to 1}(\phi(z) - 1)/(z - 1) = -A(1) = E[X]$. Also since $E[X] = \phi'(1)$, $E[X^2] - E[X] = \phi''(1)$, and realizing $\mu p_0 = \mu - \lambda E[X]$, we can rewrite L as

$$L = (\mu - \lambda E[X]) \left(\frac{2\lambda E[X] + \lambda E[X^2] - \lambda E[X]}{2\{-\mu + \lambda E[X]\}\{-\mu + \lambda E[X]\}} \right)$$

$$= \frac{\lambda E[X] + \lambda E[X^2]}{2\{\mu - \lambda E[X]\}}.$$

To compute W, we use Little's law (note that the effective customer entering rate is $\lambda E[X]$) to obtain

$$W = \frac{\lambda E[X] + \lambda E[X^2]}{2\lambda E[X]\{\mu - \lambda E[X]\}}. \qquad \blacksquare$$

Note that although $\Psi(z)$ requires the entire distribution of X, in order to compute L and W all we need are $E[X]$ and $E[X^2]$. So if one is only interested in L and W, instead of estimating a_i for all i, one could just compute the mean and variance of the batch size (besides λ and μ).

2.2.4 Catastrophic Breakdown of Servers in $M/M/1$ Queues

Server breakdown and repair issues have been well studied in the queueing literature by modeling the server being in two states, working or under repair. However, in those models the effect of machines breaking down is only felt in terms of customers getting delayed. Here we consider a situation where when a server breaks down, all the customers are ejected from the system and while the machine is down no new customers enter the system. For that reason, this is termed as catastrophic breakdown. Such catastrophic breakdowns are especially typical in computer systems like web servers. Consider a web server with a single queue where the entities are requests for various files that are stored in the server. These requests are from browsers and when the requests are satisfied, the web page gets displayed at the browser. From a queueing standpoint, the request arrivals to the web server queue correspond to customer arrivals. And the process of finding or creating a file as well as transmitting it back to the browser corresponds to a service.

We have all experienced the inability to reach a web server when we type a URL on our browsers. If the reason is due to a problem at the server's end, then it is perhaps due to a catastrophic breakdown. They could be due to the server shutting down when the room is over-heated, a serious power outage, a denial-of-service attack, or something more benign like losing connectivity to the network, or a congestion in the network. When one of this happens, the web server loses all its pending requests in the system and also new requests cannot enter. As described in the previous paragraph, this is termed as catastrophic breakdown. Next, we model such a system as a queue with catastrophic breakdowns and repairs of servers by stating the problem and providing a solution. This is based on Gautam [42] which also provides additional details in terms of pricing, performance, and availability.

Problem 15

As an abstract model, we have a single server queue where customers arrive according to a Poisson process with mean arrival rate λ. Service times are exponentially distributed with mean $1/\mu$. There is infinite room for requests to wait. The server stays "on" for a random time distributed exponentially with mean $1/\alpha$ after which a catastrophic breakdown occurs. When the server turns off (i.e., breaks down), all customers in the system are ejected. Note that the server can break down when there are no customers in the system. The server stays off for a random time distributed exponentially with mean $1/\beta$. No requests can enter the system when the server is off (typically after a time-out the client browser would display something to the effect of *unable to reach server*). Model the system as a CTMC, obtain steady-state probabilities, and performance measures such as average number in the system, fraction of requests lost, and sojourn time.

Solution

Note that the system behaves as an $M/M/1$ queue when the server is on, and the system is empty when the server is off. The server toggles between on and off states irrespective of the queueing process. We model the system as a CTMC. Let $X(t) = i$ (for $i = 0, 1, 2, 3, \ldots$) imply that there are i requests in the system and the server is on at time t. In addition, let $X(t) = D$ denote that the server is down (and there are no customers) at time t. Clearly, $\{X(t), t \geq 0\}$ is a CTMC with rate diagram shown in Figure 2.5. The CTMC is ergodic, and for $j = D, 0, 1, 2, \ldots$, let

$$p_j = \lim_{t \to \infty} P\{X(t) = j\}.$$

To obtain the steady-state probabilities p_j, consider the following balance equations:

$$\alpha(p_0 + p_1 + \cdots) = \beta p_D$$

$$\beta p_D + \mu p_1 = (\lambda + \alpha)p_0$$

$$\mu p_2 + \lambda p_0 = (\lambda + \alpha + \mu)p_1$$

$$\mu p_3 + \lambda p_1 = (\lambda + \alpha + \mu)p_2$$

$$\mu p_4 + \lambda p_2 = (\lambda + \alpha + \mu)p_3$$

$$\vdots \quad \vdots \quad \vdots$$

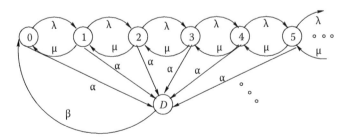

FIGURE 2.5
Rate diagram of the CTMC. (From Gautam, N., *J. Revenue Pricing Manag.*, 4(1), 7, 2005.)

From the first equation, we have $p_D = \alpha/(\alpha + \beta)$ since $p_0 + p_1 + \cdots = 1 - p_D$. Multiplying the second equation by 1, third equation by z, fourth equation by z^2, and so on, and summing up we get

$$\beta p_D + \frac{\mu(\psi(z) - p_0)}{z} + \lambda z \psi(z) = (\lambda + \alpha + \mu)\psi(z) - \mu p_0$$

where $\psi(z) = p_0 + p_1 z + p_2 z^2 + p_3 z^3 + p_4 z^4 + \cdots$ (note that unlike typical moment generation functions, here $\psi(1) = 1 - p_D$). We can rearrange the terms and write down $\psi(z)$ as

$$\psi(z) = \frac{\mu p_0 - z\beta p_D - p_0 \mu z}{\mu + \lambda z^2 - \lambda z - \alpha z - \mu z}. \tag{2.8}$$

Since we already know $p_D = \alpha/(\alpha + \beta)$, the only unknown in Equation 2.8 is p_0. However, standard techniques such as $\psi(0) = p_0$ and $\psi(1) = \beta/(\alpha + \beta)$ do not yield a solution for p_0. Hence, we need to do something different to determine p_0 and thereby $\psi(z)$.

Note that since $\psi(z)$ is $p_0 + p_1 z + p_2 z^2 + p_3 z^3 + p_4 z^4 + \cdots$, it is a continuous, differentiable, bounded, and increasing function over $z \in [0, 1]$. However, from Equation 2.8, $\psi(z)$ is of the form $\phi(z) = A(z)/B(z)$, where $A(z)$ and $B(z)$ are polynomials corresponding to the numerator and denominator of the equation. If there exists a $z^* \in [0, 1]$ such that $B(z^*) = 0$, then $A(z^*) = 0$ (otherwise it violates the condition that $\psi(z)$ is a bounded and increasing function over $z \in [0, 1]$). We now use the previous realization to derive a closed-form algebraic expression for $\psi(z)$.

By setting the denominator of $\psi(z)$ in Equation 2.8 to zero, we get

$$z^* = \frac{(\lambda + \mu + \alpha) - \sqrt{(\lambda + \mu + \alpha)^2 - 4\lambda\mu}}{2\lambda},$$

as the unique solution such that $z^* \in [0, 1]$. Setting the numerator of $\psi(z)$ in Equation 2.8 to zero at $z = z^*$, we get

$$p_0 = \frac{\alpha\beta z^*}{(\alpha + \beta)\mu(1 - z^*)}.$$

By substituting for z^*, we get p_0 as

$$p_0 = \frac{\alpha\beta}{\mu(\alpha + \beta)} \left[\frac{\lambda + \mu + \alpha - \sqrt{(\lambda + \mu + \alpha)^2 - 4\lambda\mu}}{\lambda - \mu - \alpha + \sqrt{(\lambda + \mu + \alpha)^2 - 4\lambda\mu}} \right]. \tag{2.9}$$

Also, by rearranging terms in Equation 2.8, we get the function $\psi(z)$ as

$$\psi(z) = \frac{\mu p_0(1-z) - z\alpha\beta/(\alpha+\beta)}{\lambda z^2 - (\lambda + \mu + \alpha)z + \mu}. \tag{2.10}$$

Based on this equation, we can also derive an asymptotic result. We let the server up- and downtimes to be extremely large (especially in comparison to the arrival and service rates). Then we can remark that if $\lambda < \mu$, $\alpha \to 0$ and $\beta \to 0$ such that $\alpha/\beta \to r$, then $p_D = r/(1+r)$ and for $i = 0, 1, 2, \ldots$, $p_i = (1-p_D)(1-\lambda/\mu)(\lambda/\mu)^i$. This statement confirms our intuition that since the system reaches steady state when the server is up, the probability that there are i requests in steady state is the product of the probability that the server is up and the probability there are i requests in steady state given the server is up.

Next, we derive few steady-state performance measures, namely, average number in the system, fraction of requests lost, and sojourn time (also known as response time) for the nonasymptotic case. Let P_ℓ be the probability that a request is lost and let W be the average sojourn time (or response time) at the server as experienced by users that receive a response. Note that the latter is a conditional expected value, conditioned on receiving a response (this is analogous to the $M/M/s/K$ queue case where W was computed only for those customers that entered). Note that measures P_ℓ and W are such that the lower their values, the better the quality of service experienced by the users. Both measures are in terms of L, the time-averaged number of requests in the system in the long run (note that it includes the downtimes when there are no requests in the system) which we derive first.

By definition

$$L = 0p_D + 0p_0 + 1p_1 + 2p_2 + 3p_3 + \cdots,$$

and clearly that can be written as $L = \psi'(1)$. By taking the derivative of $\psi(z)$ in Equation 2.10, and then letting $z = 1$, we get the average number of requests in the web server system as

$$L = \frac{1}{\alpha}\left[\frac{\lambda\beta - \mu\beta + p_0\mu(\alpha+\beta)}{\alpha+\beta}\right],$$

where p_0 is described in Equation 2.9.

The number of requests that are dropped per unit time in steady state is $\alpha(1p_1 + 2p_2 + 3p_3 + \cdots) = \alpha L$. Therefore, the fraction of requests that entered the queue and were dropped when the server switched from on to off is $\alpha L/\lambda(1-p_D)$. The probability that an arriving request will complete service, given that it arrived when the server was up, is given by (conditioning on the number of requests seen upon arrival)

$$\sum_{j=0}^{\infty} \left(\frac{p_j}{1 - p_D} \right) \left(\frac{\mu}{\mu + \alpha} \right)^{j+1} = \frac{\mu}{\mu + \alpha} \frac{1}{1 - p_D} \psi \left(\frac{\mu}{\mu + \alpha} \right)$$

$$= \frac{\mu}{1 - p_D} \frac{\beta - p_0(\alpha + \beta)}{\lambda(\alpha + \beta)}.$$

Therefore, the rate at which requests exit the queue is $\mu(\beta)/(\alpha + \beta) - \mu p_0$, which makes sense as whenever there are one or more requests in the system, the exit rate is μ. In addition, since the drop rate (derived earlier) is αL, we can write $\mu((\beta)/(\alpha + \beta)) - \mu p_0 = \lambda(1 - p_D) - \alpha L$, which again makes sense and the total arrival rate when web server is on is $\lambda(1 - p_D)$.

We also have a fraction p_D requests that are rejected when the server is down. Therefore, the loss probability is $(\lambda p_D + \alpha L)/\lambda$, and by substituting for p_D we obtain P_ℓ in terms of L as

$$P_\ell = \frac{\alpha L(\alpha + \beta) + \lambda \alpha}{\lambda(\alpha + \beta)}.$$

Using Little's law, we can derive W in the following manner. The expected number of requests in the system when the server is on is $L/(1 - p_D)$. In steady state, of these requests a fraction $(\lambda(1 - p_D) - \alpha L)/\lambda(1 - p_D)$ only will receive service. Therefore, the average sojourn time (or response time) at the server as experienced by users that receive a response is given by $L/\lambda(1 - p_d)^2$, which yields W in terms of L as

$$W = \frac{L(\alpha + \beta)^2}{\lambda \beta^2}. \qquad \blacksquare$$

2.3 Solving Balance Equations Using Reversibility

There are some CTMCs, especially those with multidimensional and finite state space, where balance equations cannot be solved using either arc cuts or generating functions. Yet in some instances, the solutions to the balance equations have a tractable closed-form structure. One such class of problems that lend to elegant algebraic expressions is when the CTMC is reversible. We first explain what a reversible process is, then go over some properties, and finally using an example application illustrate the methodology of solving balance equations using reversibility.

2.3.1 Reversible Processes

We first explain a reversible process in words. Consider observing the states of a system continuously over time and videotaping it. Now, if one were unable to statistically distinguish between running the videotape forward versus backward, then the states of the observed system form a reversible process. This is explained mathematically next.

At time t, let $X(t)$ be the state of the system described earlier that is continuously observed, for all $t \in (-\infty, \infty)$. If stochastic process $\{X(t), -\infty < t < \infty\}$ is stochastically identical to the process $\{X(\tau - t), -\infty < t < \infty\}$ for all $\tau \in (-\infty, \infty)$, then $\{X(t), -\infty < t < \infty\}$ is a *reversible process*. The process $\{X(\tau - t), -\infty < t < \infty\}$ for any $\tau \in (-\infty, \infty)$ is known as the reversed process at τ. Although this is the definition of a reversed process, it is usually hard to show a CTMC is reversible based on that directly. Instead we resort to one of the properties of reversible processes that are especially applied to CTMCs.

The first property is an easy check to see if a CTMC is not reversible. In the rate diagram if there is an arc from node i to node j of the CTMC, then there must be an arc from j to i as well for the CTMC to be reversible. This is straightforward because only if you can go from j to i in the forward video, you can go from i to j in the reversed video. Note that this is necessary but not sufficient. For example, the CTMC corresponding to the rate diagram in Figure 2.1 is not reversible because it is possible to only go from 0 to 2 but not 2 to 0 in one transition. However, the CTMC corresponding to the $M/M/s/K$ queue depicted in Figure 2.2 is such that this necessary condition is satisfied. The next result would help us decide if it is indeed reversible.

An ergodic CTMC $\{X(t), -\infty < t < \infty\}$ with state space S, infinitesimal generator matrix Q (with q_{ij} being the ijth element of Q), and limiting distribution $\{p_i, i \in S\}$ is *reversible* if and only if for all $i \in S$ and $j \in S$

$$p_i q_{ij} = p_j q_{ji}.$$

Note how this also requires the necessary condition which can also be mathematically shown that if q_{ij} is zero or nonzero, then q_{ji} is also respectively zero or nonzero (since p_i and p_j are nonzero). It is not necessary for i to be a scalar; it just represents a possible value that $X(t)$ can take.

It is worthwhile to note that the CTMC corresponding to the $M/M/s/K$ queue for any s and K (as long as the queue is stable) is reversible. In essence, the condition $p_i q_{ij} = p_j q_{ji}$ is identical to either the balance equation corresponding to arc cuts between successive nodes or corresponding to the case $q_{ij} = q_{ji} = 0$. In essence, any one-dimensional birth and death process that is ergodic would be reversible (for the same reason as the $M/M/s/K$ queue). However, it is a little tricky to check if other CTMCs (that satisfy the necessary condition) are reversible. To address this shortcoming, we next explain some properties of reversible processes that can be used to determine if a given CTMC is reversible and how to compute its steady-state probabilities.

2.3.2 Properties of Reversible Processes

We now describe two properties of reversible processes that are useful in checking if a process is reversible. Here we just describe the properties, but in the next section we will illustrate using an example the power and benefit of reversible processes which would not be apparent here. The first of the two properties essentially extends one-dimensional reversible processes to multiple dimensions.

Remark 6

Joint processes of independent reversible processes are reversible.   ∎

Mathematically the previous property can be interpreted as follows. Let $\{X_1(t), -\infty < t < \infty\}$, $\{X_2(t), -\infty < t < \infty\}$, $\ldots$, $\{X_n(t), -\infty < t < \infty\}$ be n independent *reversible* process. Then the joint process

$$\{(X_1(t), X_2(t), \ldots, X_n(t)), -\infty < t < \infty\}$$

is *reversible*. In fact, the steady-state probabilities would also just be the product of those of the corresponding states of the individual reversible processes. Although the previous result is fairly general, from a practical standpoint, if $\{X_i(t), -\infty < t < \infty\}$ for all i are one-dimensional birth and death processes (which is usually easy to verify as being reversible), then $\{(X_1(t), X_2(t), \ldots, X_n(t)), -\infty < t < \infty\}$ is an n-dimensional birth and death process which is also reversible. It is easy to check that the resulting steady-state probabilities satisfy the condition $p_i q_{ij} = p_j q_{ji}$ (where i and j are n-dimensional). As described earlier, we will illustrate an application of this in the following section. Now we move on to the second property.

Consider a reversible CTMC $\{X(t), -\infty < t < \infty\}$ with infinitesimal generator $Q = [q_{ij}]$ defined on state space S and steady-state probabilities p_j that the CTMC is in state j for all $j \in S$. Now consider another CTMC $\{Y(t), -\infty < t < \infty\}$ which is a truncated version of $\{X(t), -\infty < t < \infty\}$ defined on state space A such that $A \subset S$. By truncated we mean that we first perform an arc cut described in Section 2.1 to the rate diagram of $\{X(t), -\infty < t < \infty\}$ to separate the state space S into A and $S - A$. Then, if we remove all arcs going from A to $S - A$, the resulting connected graph on set A would be the rate diagram of the truncated CTMC $\{Y(t), -\infty < t < \infty\}$. In other words, for all i and j such that $i \neq j$ and $i \in A$ as well as $j \in A$, the rate of going from state i to j is q_{ij}. Only the diagonal elements would change in the CTMC $\{Y(t), -\infty < t < \infty\}$. We next describe a result characterizing the truncated CTMC $\{Y(t), -\infty < t < \infty\}$.

Remark 7

Truncated processes of reversible processes are reversible. ■

This remark essentially says that $\{Y(t), -\infty < t < \infty\}$ described earlier is reversible. Next is to obtain the steady-state distribution of. $\{Y(t), -\infty < t < \infty\}$. For the reversible process $\{X(t), -\infty < t < \infty\}$, we have $p_i q_{ij} = p_j q_{ji}$ for all $i \in \mathcal{A}$ and $j \in \mathcal{A}$. Also, since q_{ij} for $\{X(t), -\infty < t < \infty\}$ and $\{Y(t), -\infty < t < \infty\}$ are identical, the steady-state probability that the CTMC $\{Y(t), -\infty < t < \infty\}$ is in state j is proportional to p_j. However, using the normalizing condition that all the steady-state probabilities must add to 1, we have the steady-state probability that the CTMC $\{Y(t), -\infty < t < \infty\}$ is in state j as

$$\frac{p_j}{\sum_{k \in \mathcal{A}} p_k}$$

for all $j \in \mathcal{A}$.

As an illustration of this, consider the $M/M/s$ queue. Let $\lambda < s\mu$ where λ and μ are, respectively, the arrival and service rates. If $X(t)$ is the number of customers in the system at time t, then $\{X(t), -\infty < t < \infty\}$ is a reversible CTMC on state space $\mathcal{S} = \{0, 1, 2, 3, \ldots\}$ with steady-state probabilities $p_j^{M/M/s}$ given in Section 2.1. Now the truncated process $\{Y(t), -\infty < t < \infty\}$, where $Y(t)$ is the number of customers in the system in an $M/M/s/K$ queue, is also reversible and defined on state space $\mathcal{A} = \{0, 1, \ldots, K\}$. Verify from the results in Section 2.1 that the steady-state probabilities $p_j^{M/M/s/K}$ satisfy

$$p_j^{M/M/s/K} = \frac{p_j^{M/M/s}}{\sum_{i=0}^{K} p_i^{M/M/s}}$$

for $j = 0, 1, \ldots, K$.

2.3.3 Example: Analysis of Bandwidth-Sensitive Traffic

Consider the access link between a web server farm and the Internet. This link is usually a bottleneck considering the amount of bandwidth-sensitive traffic it transmits onto the Internet. Let the capacity of the link be C kbps. We consider N classes of connections. Class i connection requests arrive according to a Poisson process at rate λ_i. The holding time for class i requests is according to $\exp(\mu_i)$. During the entire duration of the connection, each class i request uses b_i kbps of bandwidth. Note that these applications are usually real-time multimedia traffic and we assume no buffering takes place. In the traditional telephone network, we have $N = 1$ class, and each call uses 60 kbps with C/b_1 being the number of lines a telephone switch could handle. This problem is just a multiclass version of that.

Let $X_i(t)$ be the number of ongoing class i connections at time t across the bottleneck link under consideration. Clearly, there is a constraint at all times t:

$$b_1 X_1(t) + b_2 X_2(t) + \cdots + b_N X_N(t) \leq C.$$

In this example, we use a complete sharing admission policy; by this we mean that if a request of class i seeks admission, we admit the request if the available capacity at the time of request exceeds b_i kbps. In other words, if a class i request arrives at time t and $b_i + b_1 X_1(t) + b_2 X_2(t) + \cdots + b_N X_N(t) > C$, then we reject the request. However, if $b_i + b_1 X_1(t) + b_2 X_2(t) + \cdots + b_N X_N(t) \leq C$, we accept the request. A more general version of this problem is to devise an admission control policy to decide based on the state of the system and the class of the request whether or not to admit it (subject to satisfying the bandwidth constraint described earlier). This is known as the *stochastic knapsack problem* analogous to the deterministic equivalent $\max\{\sum_i a_i y_i\}$ subject to $\sum_i b_i y_i < C$, where a_i is the reward for accepting a class i entity into the knapsack.

In this example, our objective is not to derive the optimal policy, but given a policy we would like to evaluate its performance. In other words, given a policy we could compute the average number of class i requests in the system in steady state, the fraction of class i requests rejected in the long run, time-averaged fraction of capacity C used, etc. For the performance analysis, we only consider the complete sharing policy, that is, admit a request of any class as long as there is bandwidth available for that class at the time of arrival. The stochastic process $\{(X_1(t), X_2(t), \ldots, X_N(t)), -\infty < t < \infty\}$ is a CTMC with an N-dimensional state space constrained by $b_1 X_1(t) + b_2 X_2(t) + \cdots + b_N X_N(t) \leq C$ at all t. Let us call this constrained state space $\mathcal{A}$. We next describe a methodology based on the reversibility properties to derive the steady-state distribution for the stochastic process.

Let $C = \infty$, then the resulting N-dimensional CTMC would be unconstrained with state space $\mathcal{S} = \mathbb{Z}_+^N$, essentially the N-dimensional grid of nonnegative integers. Clearly, our original CTMC $\{(X_1(t), X_2(t), \ldots, X_N(t)), -\infty < t < \infty\}$ on $\mathcal{A}$ is a truncated process of this process with state space $\mathcal{S}$. Further, this N-dimensional CTMC on $\mathcal{S}$ is just a joint process of N-independent processes (corresponding to each of the N classes). Therefore, when $C = \infty$, let $X_i^\infty(t)$ be the number of class i connections in the system at time t. Then, the stochastic process $\{X_i^\infty(t), -\infty < t < \infty\}$ is a reversible

CTMC which is independent of other CTMCs $\left\{X_j^\infty(t), -\infty < t < \infty\right\}$ for $i \neq j$. In addition, $\left\{X_i^\infty(t), -\infty < t < \infty\right\}$ is the queue length process of an $M/M/\infty$ queue with arrival rate λ_i and service rate μ_i for each server. The steady-state probabilities for this queue are for $i = 1, 2, \ldots, N$ and $j = 0, 1, \ldots$

$$p_j^i(\infty) = e^{-\lambda_i/\mu_i} \left(\frac{\lambda_i}{\mu_i}\right)^j \frac{1}{j!}$$

where ∞ in $p_j^i(\infty)$ denotes $C = \infty$.

Working our way backward, the joint process $\{(X_1^\infty(t), X_2^\infty(t), \ldots, X_N^\infty(t)),$ $-\infty < t < \infty\}$ of independent reversible processes $\left\{X_i^\infty(t), -\infty < t < \infty\right\}$ is a reversible process with joint distribution

$$P\left\{X_1^\infty(t) = x_1, X_2^\infty(t) = x_2, \ldots, X_N^\infty(t) = x_N\right\}$$

$$= p_{x_1}^1(\infty) p_{x_2}^2(\infty) \ldots p_{x_N}^N(\infty) = \left(e^{-\sum_{i=1}^N \lambda_i/\mu_i}\right) \prod_{i=1}^N \left(\frac{\lambda_i}{\mu_i}\right)^{x_i} \frac{1}{x_i!}.$$

Since the joint process $\left\{(X_1^\infty(t), X_2^\infty(t), \ldots, X_N^\infty(t)), -\infty < t < \infty\right\}$ is a reversible process, its truncated process $\{(X_1(t), X_2(t), \ldots, X_N(t)), -\infty < t < \infty\}$ is also reversible with the steady-state probability of having x_1 class-1 connections, x_2 class-2 connections, $\ldots$, x_N class-N connections as

$$p_{x_1, x_2, \ldots, x_N} \propto e^{-\sum_{i=1}^N \lambda_i/\mu_i} \prod_{i=1}^N \left(\frac{\lambda_i}{\mu_i}\right)^{x_i} \frac{1}{x_i!}$$

such that $b_1 x_1 + b_2 x_2 + \cdots + b_N x_N \leq C$. Hence, we can rewrite the steady-state probability as

$$p_{x_1, x_2, \ldots, x_N} = R \prod_{i=1}^N \left(\frac{\lambda_i}{\mu_i}\right)^{x_i} \frac{1}{x_i!} \tag{2.11}$$

subject to $b_1 x_1 + b_2 x_2 + \cdots + b_N x_N \leq C$, where R is the normalizing constant which can be obtained by solving

$$\sum_{x_1, x_2, \ldots x_N : b_1 x_1 + b_2 x_2 + \cdots b_N x_N \leq C} p_{x_1, x_2, \ldots, x_N} = 1.$$

In other words

$$R = \left[\sum_{x_1, x_2, \dots x_N : b_1 x_1 + b_2 x_2 + \cdots b_N x_N \leq C} \prod_{i=1}^{N} \left(\frac{\lambda_i}{\mu_i} \right)^{x_i} \frac{1}{x_i!} \right]^{-1}. \qquad (2.12)$$

It is important to realize that although we have a closed-form solution for the steady-state probabilities, there is a significant computational overhead in obtaining R. There are several researchers who have described ways to efficiently obtain R. One such technique is to iteratively compute R for $C = 0, 1, \dots, C$ assuming C is an integer. However, for the purpose of this book we do not go into the details.

Next, we present a numerical example to illustrate the methodology and showcase its capability.

Problem 16

Consider a channel with capacity 700 kbps on which two classes of bandwidth-sensitive traffic can be transmitted. Class-1 uses 200 kbps bandwidth and class-2 uses 300 kbps bandwidth. Let λ_1 and λ_2 be the parameters of the Poisson processes corresponding to the arrivals of the two class. Also, let each admitted request spend $\exp(\mu_i)$ time holding onto the bandwidth they require for $i = 1, 2$. Let $X_1(t)$ and $X_2(t)$ be the number of ongoing class-1 and class-2 requests at time t. Model the CTMC $\{(X_1(t), X_2(t)), t \geq 0\}$ and obtain its steady-state probabilities.

Solution

Note that this is a special case when $N = 2$, $C = 700$, $b_1 = 200$, and $b_2 = 300$. The CTMC $\{(X_1(t), X_2(t)), t \geq 0\}$ can be modeled as the rate diagram in Figure 2.6. Since we have the constraint $200 X_1(t) + 300 X_2(t) \leq 700$, the

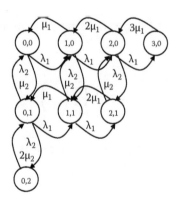

FIGURE 2.6
Arc cut example.

state space is {(0,0), (0,1), (0,2), (1,0), (1,1), (2,0), (2,1), (3,0)}. In order to fully appreciate the power of using reversibility results, the reader is encouraged to solve for $p_{0,0}$, $p_{0,1}$, $p_{0,2}$, $p_{1,0}$, $p_{1,1}$, $p_{2,0}$, $p_{2,1}$, and $p_{3,0}$ using the node balance equations.

Now, using Equation 2.12, we can obtain the normalizing constant as

$$R = \frac{1}{1 + \frac{\lambda_1}{\mu_1} + \frac{\lambda_1^2}{2\mu_1^2} + \frac{\lambda_2}{\mu_2} + \frac{\lambda_1^3}{6\mu_1^3} + \frac{\lambda_2^2}{2\mu_2^2} + \frac{\lambda_1\lambda_2}{\mu_1\mu_2} + \frac{\lambda_1^2\lambda_2}{2\mu_1^2\mu_2}}.$$

Therefore, we can derive the steady-state probabilities as

$$p_{0,0} = R, \quad p_{0,1} = \frac{\lambda_2}{\mu_2}R, \quad p_{0,2} = \frac{\lambda_2^2}{2\mu_2^2}R, \quad p_{1,0} = \frac{\lambda_1}{\mu_1}R,$$

$$p_{1,1} = \frac{\lambda_1\lambda_2}{\mu_1\mu_2}R, \quad p_{2,0} = \frac{\lambda_1^2}{2\mu_1^2}R, \quad p_{2,1} = \frac{\lambda_1^2\lambda_2}{2\mu_1^2\mu_2}R, \quad p_{3,0} = \frac{\lambda_1^3}{6\mu_1^3}R. \qquad \blacksquare$$

Reference Notes

The underlying theme of this chapter, as the title suggests, is queues that can be modeled and analyzed using CTMCs. The sources for the three main thrusts, namely, arc cuts for birth and death processes, generating functions, and reversibility, have been different. In particular, the $M/M/s/K$ queue and special cases are mainly from Gross and Harris [49]. The section on using generating functions is largely influenced by Kulkarni [67], Medhi [80], and Prabhu [89]. The reversibility material is presented predominantly from Kelly [59]. The exercise problems are essentially a compilation of homework and exam questions in courses taught by the author over the last several years. However, a large number of the exercise problems were indeed adapted from some of the books described earlier.

Exercises

2.1 Compute the variance of the number of customers in the system in steady state for an $M/M/s/K$ queue.

2.2 Consider an $M/M/s/K$ queue with $\lambda = 10$ and $\mu = 1.25$. Write a computer program to plot p_K and W for various values of K from s to $s + 19$. Consider two cases (i) $s = 10$ and (ii) $s = 5$.

2.3 The second factorial moment of an integer-valued random variable
Z is defined as $E[Z(Z-1)]$ and the second moment of a continuous
random variable Y is defined as $E[Y^2]$. Consider an $M/M/1$ queue.
Let $L_{(2)}$ be the second factorial moment of the number in the system
and let W_2 be the second moment of the time in the system experienced by a customer in steady state (assume $\rho < 1$). Verify based on
the steady-state distributions that

$$L_{(2)} = \lambda^2 W_2.$$

2.4 Let U be a random variable that denotes the time between successive
departures (in the long run) from the system in an $M/M/1$ queueing system. Assume that $\lambda < \mu$. Show by conditioning whether or
not a departure has left the system empty that U is an exponentially
distributed random variable with mean $1/\lambda$.

2.5 *Feedback queue.* In the $M/M/1$ system suppose that with probability q,
a customer who completes their service rejoins the queue for further
service. What is the stability condition for this queue? Assuming
conditions for stability hold, derive expressions for L and W.

2.6 *Static control.* Consider an $M/M/1$ queue where the objective is to
pick a service rate μ in an optimal fashion. There are two types of
costs associated: (i) a service-cost rate, c (cost per unit time per unit
of service rate) and (ii) a holding-cost rate h (cost per unit time per
customer in the system). In other words, (i) if we choose a service
rate μ, then we pay a service cost $c\,\mu$ per unit time; (ii) the system
incurs $h\,i$ monetary units of holding cost per unit time while i customers are present. Let $C(\mu)$ be the expected steady-state cost per
unit time, when service rate μ is chosen, that is,

$$C(\mu) = c\mu + h \lim_{t \to \infty} E[X(t)].$$

The objective is to choose a μ that minimizes $C(\mu)$. What is the
optimal value μ^* for the service rate and what is the optimal cost
$C(\mu^*)$?

2.7 Consider a queueing system with two servers. You have to decide
if it is better to have two queues, one in front of each server or
just one queue. Assume that the service times are exponentially distributed with mean service rate μ for each server and the servers
follow an FCFS discipline. Assume that the system has infinite waiting room. The arrival process to the system is $PP(\lambda)$. In the system
with two queues, each arriving customer chooses either queue with
equal probability (assume that customers are not able to join the
shortest queue or jockey between queues at any time). By comparing

the average waiting time W, decide which system you will go with, single-queue or two-queue system?

2.8 Consider a single-server queue with two classes. Class i customers arrive according to $PP(\lambda_i)$ for $i=1,2$. For both classes, the service times are according to $\exp(\mu)$. If the total number of customers (of both classes) in the system is greater than or equal to K, class-1 customers do not join the system, whereas class-2 customers always join the system. Model this system as a CTMC. When is this system stable? Under stability, what is the steady-state distribution of the number of customers in the system? Compute the expected sojourn time for each type of customer in steady state, assuming they exist. Note that for type 1 customers, you are only required to obtain the mean waiting time for those customers that join the system.

2.9 There is a single line to order drinks at a local coffee shop. When the number of customers in the line is three or less, only one person does the check out as well as making beverages. This takes $\exp(\mu_1)$ time. When there are more than three persons in the line, the store manager comes in to help. In this case, the service rate increases to $\mu_2 > \mu_1$ (i.e., the reduced service times now become $\exp(\mu_2)$). Assume that the arrival process is $PP(\lambda)$. Model this queue as a CTMC.

2.10 Consider a standard $M/M/1$ queue with arrival rate λ and service rate μ. The server toggles between being busy and idle. Let B and I denote random times the server is busy and idle, respectively, in a cycle. Obtain an expression for the ratio $E(B)/E(I)$. Using that relation, show that

$$E(B) = \frac{1}{\mu - \lambda}.$$

Assume that stability condition holds.

2.11 In the usual $M/M/2$ model, we assume that both servers are identical. Now consider the following modification, although with respect to the customers things are the same (i.e., arrivals according to $PP(\lambda)$ and $\exp(\mu)$ amount of work required). Speed of server i is c_i, that is, if there is x amount of work it takes x/c_i time to complete ($i=1,2$). Let $c_1 > c_2$. The customers join a common queue. Arriving customers go to server 1 if available; otherwise they go to server 2 if available; else they wait. Model this as a CTMC. Derive the condition of stability and the steady-state distribution.

2.12 Consider two infinite-capacity queueing systems: system 1 has s servers, each serving at rate μ; system 2 has a single server, serving at rate $s\mu$. Both systems are subject to $PP(\lambda)$ arrivals and exponentially distributed service times. Show that in steady state, assuming the systems are stable, the expected number of customers in system 2 is less than that in system 1. Do the same analysis for the expected number of customers in the queue L_q.

2.13 Consider a machine shop with two machines. The lifetime of a machine is an $\exp(\lambda)$ random variable. When a machine fails, it joins a queue of failed machines. There is one repair person who repairs these machines according to FCFS. The repair times are $\exp(\mu)$ random variables. Assume that up- and downtimes are independent. Suppose that every working machine produces revenues at a rate of r dollars per unit time. The cost of repairing the machine is C dollars per repair. Compute the long-run net rate at which the system earns profits (revenue rate – cost rate).

2.14 Consider the bulk arrival $M^{[X]}/M/1$ queue. For the following PMFs for X (i.e., the random number of customers in a batch), obtain the long-run average number in the system (note that a_i is the probability that a batch of size i arrives to the queue):

(a) $a_i = (1 - q)q^{i-1}$ for $i \geq 1$ and $0 < q < 1$ (i.e., the geometric distribution)

(b) $a_i = 1$ for $i = K$ and for all other i, $a_i = 0$ (i.e., the constant size batch case)

(c) $a_i = 1/K$ for $i = 1, \ldots, K$ and for all other i, $a_i = 0$ (i.e., the discrete uniform distribution)

(d) $a_i = \left(e^{-\theta}/(1 - e^{-\theta})\right)\theta^i/i!$ for $i \geq 1$ (i.e., the modified Poisson distribution)

2.15 Consider the infinite server bulk arrival $M^{[X]}/M/\infty$ queue. As usual, batches arrive according to $PP(\lambda)$ and each customer's service takes $\exp(\mu)$ time. The batch sizes are geometrically distributed such that probability that a batch of size i arrives to the queue is $a_i = (1 - q)q^{i-1}$ for $i \geq 1$ and $0 < q < 1$. Model the number of customers in the system as a CTMC and write down the balance equations. Using the balance equations, derive the following differential equation for the generating function $P(z)$:

$$P'(z) = \frac{\lambda}{\mu(1 - qz)}P(z),$$

where $P(z) = \sum_{i=0}^{\infty} p_i z^i$. Then, show that the following is a solution to this differential equation:

$$P(z) = \left[\frac{1 - q}{1 - qz}\right]^{\lambda/(\mu q)}.$$

2.16 Customers arrive into a single-server queue according to $PP(\lambda)$. An arriving customer belongs to class-1 with probability α and class-2 with probability β (note that $\beta = 1 - \alpha$). The service times of class i customers are IID $\exp(\mu_i)$ for $i = 1,2$ such that $\mu_1 \neq \mu_2$. Customers form a single queue and are served according to FCFS. Let $X(t)$ be the number of customers in the system at time t and $Y(t)$ be the class of the customer in service at time t (assume $Y(t) = 0$ if $X(t) = 0$). Model the process $\{(X(t), Y(t)), t \geq 0\}$ as a CTMC. Let p_{ik} be the long-run probability that there are i customers in the system and the one in service is of class k. Define the function $\psi_k(z) = \sum_{i=1}^{\infty} p_{ik} z^i$, for $k = 1, 2$. Derive the following expression for $\psi_1(z)$:

$$\psi_1(z) = \frac{\alpha\lambda(\lambda(1-z) + \mu_2)}{\mu_1\mu_2/z - \lambda\mu_1(1 - \alpha/z) - \lambda\mu_2(1 - \beta/z) - \lambda^2(1-z)} p_{00}.$$

The expression for $\psi_2(z)$ is obtained by swapping α with β and μ_1 with μ_2. Compute p_{00} and show that the condition of stability is $\lambda(\alpha/\mu_1 + \beta/\mu_2) < 1$.

2.17 For a stable $M/M/2$ queue with $PP(\lambda)$ arrivals and service rate μ for each server, the following are the balance equations:

$$\lambda p_0 = \mu p_1$$

$$\lambda p_1 = 2\mu p_2$$

$$\lambda p_2 = 2\mu p_3$$

$$\lambda p_3 = 2\mu p_4$$

$$\lambda p_4 = 2\mu p_5$$

$$\vdots\ \vdots\ \vdots$$

Define $\Psi(z)$ as the generating function

$$\Psi(z) = p_0 + p_1 z + p_2 z^2 + \cdots .$$

By multiplying the ith balance equation by z^i for $i = 0, 1, \ldots$, and summing, show that

$$\Psi(z) = \frac{2\mu p_0 + \mu z p_1}{2\mu - \lambda z}.$$

Solve for the unknowns p_0 and p_1 (for this do not use the results from $M/M/s$ queue but feel free to verify).

2.18 Solve the retrial queue steady-state equations in Section 2.2.2 and compute p_{00} using the arc cut method.

2.19 Consider a post office with two queues: queue 1 for customers without any financial transactions (such as waiting to pick up mail) and queue 2 for customers requiring financial transactions (such as mailing a parcel). For $i = 1,2$, queue i gets arrivals according to a Poisson process with parameter λ_i, service time for each customer is according to $\exp(\mu_i)$, and has i servers. Due to safety reasons, a maximum of four customers are allowed inside the post office at a time. Model the system as a reversible CTMC and derive the steady-state probabilities.

2.20 Consider a queueing system with two parallel queues and two servers, one for each queue. Customers arrive to the system according to $PP(\lambda)$ and each arriving customer joins the queue with the fewer number of customers. If both queues have the same number of customers, then the arriving customer picks either with equal probability. The service times are exponentially distributed with mean $1/\mu$ at either server. When a service is completed at one queue and the other queue has two more customers than this queue, then the customer at the end of the line instantaneously switches to the shorter queue to balance the queues. Let $X_1(t)$ and $X_2(t)$ be the number of customers in queues 1 and 2, respectively, at time t. Assuming that $X_1(0) = X_2(0) = 0$, we have $|X_1(t) - X_2(t)| \leq 1$ for all t. Model the bivariate stochastic process $\{(X_1(t), X_2(t)), t \geq 0\}$ as a CTMC by writing down the state space and drawing the rate diagram. Assuming stability, let

$$p_{ij} = \lim_{t \to \infty} P\{X_1(t) = i, X_2(t) = j\}.$$

Compute the steady-state probabilities (p_{ij}) for the CTMC either using the rate diagram or otherwise.

2.21 Consider a queueing system with infinite waiting room where customers arrive in batches. Batch arrivals occur according to $PP(\lambda)$ and each batch has either one or two customers, with equal probability. Shuttle cars arrive according to $PP(\mu)$. Each shuttle car can carry a maximum of two customers. When a shuttle car arrives, it instantaneously picks up as many customers as possible from the queueing system and leaves immediately. If the system is empty when a shuttle car arrives, it leaves empty. Likewise, if the system has one customer when a shuttle car arrives, it leaves with this one customer. Otherwise the shuttle car leaves with two customers (i.e., when there are two or more customers when a shuttle arrives). Model the number of customers in the queueing system at time t as a CTMC and write down the balance equations. Obtain the generating function of the number of customers in the system in steady state for the special case $\lambda = \mu$. Compute L and W for this queueing system.

2.22 Consider an $M/M/1$ queue where customers in queue (but not the one in service) may get discouraged and leave without receiving service. Each customer who joins the queue will leave after an $\exp(\gamma)$ time, if the customer does not enter service by that time. Assume FCFS.

(a) What fraction of arrivals are served? Hence, what are the average departure rates both after service and without service.

(b) Suppose an arrival finds one customer in the system. What is the probability that this customer is served?

(c) On an average, how long do customers that get served wait in the queue before beginning service?

2.23 For an $M^{[X]}/M/2$ queue with batch arrival rate λ, constant batch size 4, $\exp(\mu)$ service time, and traffic intensity $\rho = 2\lambda/\mu < 1$, show that the generating function $P(z) = \sum_{n=0}^{\infty} p_n z^n$ for the distribution of the number of customers in the system is

$$P(z) = \frac{(1-z)(1-\rho)(16+4\rho z)}{\left[\rho z^5 - (\rho+4)z + 4\right](4+\rho)}.$$

2.24 Justify using a brief reasoning whether each of the following is TRUE or FALSE.

(a) Consider two $M/M/1$ queues: one has arrival rate λ and service rate μ, while the other has arrival and service rates as 2λ and 2μ, respectively. Is the following statement TRUE or FALSE? On an average, both queues have the same number of customers in the system in steady state.

(b) Consider two stable queues: one is an $M/M/1$ queue with arrival rate λ and service rate μ, while the other is an $M/M/2$ queue with arrival rate λ and service rate μ for EACH server. Is the following statement TRUE or FALSE? On an average, twice as many entities depart from the $M/M/2$ queue as compared to the $M/M/1$ queue in steady state.

(c) Consider an $M/M/1$ queue with reneging. The arrival rate is λ, the service rate is μ, but the reneging rate is also equal to μ (i.e., $\theta = \mu$). Note that the birth and death process is identical to that of an $M/M/\infty$ queue. Is the following statement TRUE or FALSE? For this $M/M/1$ queue with reneging, we have $L_q = 0$.

(d) Consider two stable queues: one is an $M^{[X]}/M/1$ queue with batch arrival rate λ, constant batch size N (i.e., $P(X = N) = 1$), and service rate μ, while the other is an $M/M/1$ queue with arrival rate $N\lambda$ and service rate μ. Note that both queues have the same effective entity-arrival rate. Is the following statement TRUE or FALSE? On an average, entities spend more time in the system in the $M^{[X]}/M/1$ queue as compared to the $M/M/1$ queue in steady state.

(e) Consider a stable $M/M/1$ queue that uses processor sharing discipline (see Section 4.5.2). Arrivals are according to $PP(\lambda)$, and it would take $\exp(\mu)$ time to process an entity if it were the only one in the system. Is the following statement TRUE or FALSE? The average workload in the system at an arbitrary point in steady state is $\lambda/(\mu(\mu - \lambda))$.

3

Exponential Interarrival and Service Times: Numerical Techniques and Approximations

In the previous chapter, we considered queueing systems with exponential interarrival and service times where we were able to compute the steady-state distribution of performance measures such as the number of entities in the system as closed-form algebraic expressions. However, there are several instances where it is not possible to obtain closed-form algebraic expressions and we resort to numerical techniques or approximations. The word *resort* should not be taken in a negative sense, especially since most of these techniques produce extraordinarily accurate results extremely quickly. In fact, in some cases, these techniques are as good. For example, Feldman and Valdez-Flores [35] recommend using numerical techniques to solve the steady-state probabilities for the $M^{[X]}/M/1$ queue for which we used the generating function approach in the previous chapter. In the following sections, we will first explain some numerical techniques (mostly based on matrix analysis) followed by some approximations.

3.1 Multidimensional Birth and Death Chains

The continuous time Markov chain (CTMC) corresponding to the $M/M/s/K$ queue in the previous chapter is a special case of a one-dimensional birth and death process (or chain). A one-dimensional birth and death chain is a CTMC $\{X(t), t \geq 0\}$ on $S = \{0, 1, 2, \ldots\}$ with generator matrix Q, which is of the form

$$
q_{ij} = \begin{cases} \lambda_i & \text{if } j = i+1 \\ \mu_i & \text{if } j = i-1 \text{ and } i > 0 \\ 0 & \text{otherwise} \end{cases}
$$

for all $i \in S$ and $j \in S$. Essentially with every transition, the stochastic process jumps to a state one value higher (*birth*) or one value lower (*death*). These are also called *skip-free* CTMCs because it is not possible to go from state i to state j without going through every state in between (i.e., no skipping of states is allowed). The rates $\lambda_0, \lambda_1, \ldots$ are known as birth rates and the rates

$\mu_1, \mu_2, \ldots$ are known as death rates. For the $M/M/1$ queue, all birth rates are equal to λ and all death rates are equal to μ. The steady-state distribution of the one-dimensional birth and death process (or chain) is easy to compute using arc cuts.

However, in the multidimensional case, it is not as easy unless one can use reversibility argument discussed toward the end of the previous chapter. In this chapter (particularly in this section), we will show numerical techniques to obtain the steady-state distribution of the multidimensional birth and death chains that are not reversible. For that, we first define a multidimensional birth and death chain. An n-dimensional CTMC $\{(X_1(t), X_2(t), \ldots, X_n(t)), t \geq 0\}$ is multidimensional birth and death chain if with every transition the CTMC jumps to a state one value higher or lower in exactly one of the dimensions. In other words, if $X(t)$ is an n-dimensional row vector $[X_1(t) \, X_2(t) \ldots X_n(t)]$ and e_i is an n-dimensional unit (row) vector (i.e., one in the ith dimension and zero everywhere else), then the next state the CTMC $\{X(t), t \geq 0\}$ goes to from $X(t)$ is either $X(t) + e_i$ or $X(t) - e_i$ for some $i \in [1, 2, \ldots, n]$. It is worthwhile to point out that the discrete time version of this is called a multidimensional random walk, although sometimes the terms "random walk" and "birth–death" are used interchangeably.

In the remainder of this section, we first motivate the need to study multidimensional birth–death chains using an example in optimal control. Then, we provide an efficient algorithm to obtain the steady-state probabilities. Finally, we provide an example based on energy conservation in data centers where this approach comes in handy.

3.1.1 Motivation: Threshold Policies in Optimal Control

This book mainly focuses on performance analysis of queues, that is, obtaining analytical expressions for various measures of performance. However, there is a vast literature on control of queues that we do not address in this book, except in a few minor places including here. Clearly, the two areas are extremely interrelated, albeit using potentially different methodologies to address them. In particular, the emphasis in the field of control of queues is on optimization and methods that address it. A simple example that we have all seen in control of queues is in grocery stores. The store manager typically decides to open up new checkout counters when the existing counters start experiencing long lines. In terms of terminology, adding a new counter is a *control action*, and deciding when to add or reduce counters is a *control policy*. Note that looking at the number of customers in line and adding a server when this crosses a "threshold" is called a *threshold policy*. Such kinds of threshold policies are very common in the control of queues. We next present an example of a queueing system where the optimal policy is of threshold type.

Problem 17

Consider two single server queues that work in parallel. Both queues have finite waiting rooms of size B_i, and the service times are exponentially distributed with mean $1/\mu_i$ in queue i (for $i=1,2$). Arrivals occur into this two-queue system according to a Poisson process with mean rate λ. At every arrival, a scheduler observes the number in the system in each queue and decides to take one of three control actions: reject the arrival, send the arrival to queue 1, or send the arrival to queue 2. Assume that the control actions happen instantaneously and customers cannot jump from one queue to the other or leave the system before their service is completed. The system earns a reward r dollars for every accepted customer and incurs a holding cost h_i dollars per unit time per customer held in queue i (for $i=1,2$). Assume that the reward and holding cost values are such that the scheduler rejects an arrival only if both queues are full. Describe the structure of the scheduler's optimal policy.

Solution

The system is depicted in Figure 3.1. For $i=1,2$, let $X_i(t)$ be the number of customers in the system in queue i at time t (including any customers at the servers). If an arrival occurs at time t, the scheduler looks at $X_1(t)$ and $X_2(t)$ to decide whether the arrival should be rejected or sent to queue 1 or queue 2. Note that because of the assumption that the scheduler rejects an arrival only if both queues are full, the scheduler's action in terms of whether to accept or reject a customer is already made. Also, if only one of the queues is full, then the assumption requires sending the arrival to the nonfull queue. Therefore, the problem is simplified so that the control action is essentially which queue to send an arrival to when there is space in both (we also call this *routing policy*, i.e., decision to send to queue 1 or 2 depending on the number in each queue).

Intuitively, the optimal policy when there is space in both queues is to send an arriving request to queue i if it is "shorter" than queue $3-i$ for $i=1,2$. If $\mu_1=\mu_2$, $B_1=B_2$, and $h_1=h_2$, then it can be shown that routing

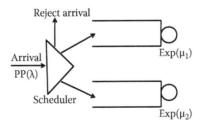

FIGURE 3.1
Schematic for scheduler's options at arrivals.

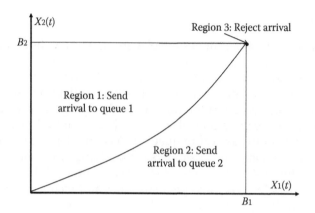

FIGURE 3.2
Structure of the optimal policy given arrival at time t.

to the shorter queue is optimal. However, in the generic case, we expect a threshold policy or switching curve as the optimal policy (with joining the shorter queue being a special case of that). Before deriving this result, we first illustrate the structure of the optimal policy in Figure 3.2. As shown in the figure, there are three regions. If an arrival occurs at time t when there are $X_1(t)$ in queue 1 and $X_2(t)$ in queue 2, then depending on the coordinates of $(X_1(t), X_2(t))$, the optimal action is taken. In particular, if $(X_1(t), X_2(t))$ is in region 1, 2, or 3, then the optimal action is to send the arrival at time t to queue 1, queue 2, or reject, respectively. Note that region 3 is the single point (B_1, B_2).

Although the threshold policy or switching curve in Figure 3.2 is intuitively appealing as the optimal policy, we still need to show mathematically that it is indeed the case. We do this next. First, we need some notation. Let x denote a vector $x = (x_1, x_2)$ such that x_1 and x_2 are the number of customers in queue 1 and queue 2, respectively, when a new customer arrives. In essence, suppose x denotes the state of the system as seen by an arrival. Let e_i denote a unit vector in the ith dimension, for $i = 1, 2$. Clearly, $e_1 = (1, 0)$ and $e_2 = (0, 1)$. If the arriving customer is routed to queue 1, then the new state becomes $x + e_1$ and if the arriving customer is routed to queue 2, the new state becomes $x + e_2$. To show the *monotonic* switching curve in Figure 3.2 is optimal, all we need to show are the following:

- If the optimal policy in state x is to route to queue 1, then the optimal policy in states $x + e_2$ and $x - e_1$ would also be to route to queue 1.
- If the optimal policy in state x is to route to queue 2, then the optimal policy in states $x + e_1$ and $x - e_2$ would also be to route to queue 2.

Before proceeding ahead, it is critical to convince oneself of that.

To show the previous set of results, we need to first formulate the problem as a semi-Markov decision process (SMDP) and then investigate the optimal policy in various states. The reader is encouraged to read any standard text on stochastic dynamic programming or Markov decision processes for a thorough understanding of this material. We first define the value function $V(x)$, which is the maximal expected total discounted net benefit over an infinite horizon, starting from state x, that is, (x_1, x_2). Note that although x is a vector, $V(x)$ is a scalar. We also use the term "discounted" because we consider a discount factor α and $V(x)$ denotes the expected present value. It is customary in the SMDP literature to pick appropriate time units so that

$$\alpha + \lambda + \mu_1 + \mu_2 = 1,$$

and therefore in our analysis to compute $V(x)$, we do not have to be concerned about α. To obtain $V(x)$, we first describe some notation. Let $h(x)$ be the holding cost incurred per unit time in state x, that is,

$$h(x) = h_1 x_1 + h_2 x_2.$$

Let a^+ denote $\max\{a, 0\}$, if a is a scalar and $x^+ = (x_1^+, x_2^+)$. Now, we are in a position to write down the optimality or Bellman equation.

The value function $V(x)$ satisfies the following optimality equation: for $x_1 \in [0, B_1)$ and $x_2 \in [0, B_2)$,

$$V(x) = -h(x) + \lambda \max\{r + V(x + e_1), r + V(x + e_2)\} + \mu_1 V((x - e_1)^+)$$

$$+ \mu_2 V((x - e_2)^+). \tag{3.1}$$

We will not derive this optimality equation (the reader is encouraged to refer to any standard text on stochastic dynamic programming or Markov decision processes). However, there is merit in going over the equation itself. When the system is in state x, a holding cost of $h(x)$ is incurred per unit time (the negative sign in front of $h(x)$ is because it is a cost as opposed to a benefit). If an arrival occurs (at rate λ), a revenue of r is obtained and depending on whether the arrival is sent to queue 1 or 2, the new state becomes $x + e_1$ or $x + e_2$, respectively. From the new state $x + e_i$ for $i = 1, 2$, the net benefit is $V(x + e_i)$. It is quite natural to select queue 1 or 2 depending on which has a higher net benefit, hence the maximization. Instead of the arrival, if the next event is a service completion at queue i (for $i = 1, 2$), then the new state becomes $(x - e_i)^+$ at rate μ_i and the value function is $V((x - e_i)^+)$. In summary, the left-hand side (LHS) $V(x)$ equals the (negative of) holding cost incurred in state x, plus the net benefit that depends on the next event (arrival routed to queue 1 or 2, service completion at queue 1, and service completion at queue 2), which would lead to a new state. The reason it appears as if the units do not match in Equation 3.1 is that in reality, the entire right-hand side (RHS) should be multiplied by $\frac{1}{\alpha+\lambda+\mu_1+\mu_2}$, and in our case, that is equal to 1. As a matter of fact, if $x_i = 0$ for $i = 1$ or 2, then the actual equation is $(\alpha+\lambda+\mu_{3-i})V(x) = -h(x)+\lambda\max\{r+V(x+e_1), r+V(x+e_2)\}+\mu_{3-i}V((x-e_{3-i})^+)$ since server i cannot complete service as there are no customers. However, we add $\mu_i V(x)$ (since it is equal to $\mu_i V((x - e_i)^+)$) to both sides to get $V(x)(\alpha + \lambda + \mu_1 + \mu_2) = -h(x) + \lambda\max\{r + V(x + e_1), r + V(x + e_2)\} + \mu_1 V((x - e_1)^+) + \mu_2 V((x - e_2)^+)$, which is identical to Equation 3.1 since $\alpha+\lambda+\mu_1+\mu_2 = 1$. A similar argument can be made for $V(x)$ when $x_1 = x_2 = 0$, and hence we do not have to be concerned about that either.

Also, by looking at the states for which Equation 3.1 holds, we still need the value function at the boundaries, that is, $x_1 = B_1$ or $x_2 = B_2$, which we present next. For $x_1 = B_1$ and $x_2 < B_2$ where arrivals are routed to queue 2

$$V(x) = -h(x) + \lambda(r + V(x + e_2)) + \mu_1 V((x - e_1)^+) + \mu_2 V((x - e_2)^+).$$

For $x_2 = B_2$ and $x_1 < B_1$ where arrivals are routed to queue 1

$$V(x) = -h(x) + \lambda(r + V(x + e_1)) + \mu_1 V((x - e_1)^+) + \mu_2 V((x - e_2)^+).$$

For $x_1 = B_1$ and $x_2 = B_2$ where arrivals are rejected

$$V(x) = -h(x) + \lambda V(x) + \mu_1 V((x - e_1)^+) + \mu_2 V((x - e_2)^+).$$

Since no arrivals enter when $x_1 = B_1$ and $x_2 = B_2$, this equation is derived from $(\alpha + \mu_1 + \mu_2)V(x) = -h(x) + \mu_1 V((x-e_1)^+) + \mu_2 V((x-e_2)^+)$ by adding $\lambda V(x)$ to both sides. For the remainder, we will just use Equation 3.1 with the understanding that at the boundary, the appropriate one among the previous three equations would be used instead.

The following are sufficient conditions that $V(x)$ must satisfy for all x to guarantee that the *monotonic* switching curve in Figure 3.2 is optimal (i.e., the bulleted items described near the figure are met):

$$V(x + e_2 + e_1) - V(x + e_2 + e_2) \geq V(x + e_1) - V(x + e_2), \qquad (3.2)$$

$$V(x + e_1 + e_2) - V(x + e_1 + e_1) \geq V(x + e_2) - V(x + e_1), \qquad (3.3)$$

$$V(x + e_2) - V(x + e_1 + e_2) \geq V(x) - V(x + e_1). \qquad (3.4)$$

It is relatively straightforward to check (and the reader is encouraged to do so) that if conditions (3.2) and (3.3) are met, then: (a) if the optimal policy in state x is to route to queue 1, then the optimal policy in states $x + e_2$ and $x - e_1$ would also be to route to queue 1; (b) if the optimal policy in state x is to route to queue 2, then the optimal policy in states $x + e_1$ and $x - e_2$ would also be to route to queue 2. For example, if condition (3.2) is met for all x, then it can be shown that: if the optimal policy in state x is to route to queue 1, then the optimal policy in state $x + e_2$ would also be to route to queue 1; if the optimal policy in state x is to route to queue 2, then the optimal policy in state $x - e_2$ would also be to route to queue 2. Likewise for condition (3.3). However, we are yet to show why the inequality (3.4) is needed; we will address that subsequently.

At this time, all we need to show is that for all x, $V(x)$ satisfies conditions (3.2), (3.3), and (3.4). For that we use the principle of mathematical induction. Let $V_0(x) = 0$ for all x. Define the iterative relation for $n = 0, 1, 2, \ldots$.

$$V_{n+1}(x) = -h(x) + \lambda \max\{r + V_n(x + e_1), r + V_n(x + e_2)\} + \mu_1 V_n((x - e_1)^+)$$

$$+ \mu_2 V_n((x - e_2)^+). \tag{3.5}$$

Incidentally, this is known as value iteration and is typically used to compute $V(x)$ for all x by computing $V_1(x)$ using $V_0(x) = 0$, then $V_2(x)$ using $V_1(x)$, and so on so that $V_k(x)$ converges to $V(x)$ as $k \to \infty$. Trivially, $V_0(x)$ satisfies conditions (3.2), (3.3), and (3.4). Next, we assume that $V_n(x)$ satisfies conditions (3.2), (3.3), and (3.4). Using that we need to show that $V_{n+1}(x)$ satisfies conditions (3.2), (3.3), and (3.4). Thus by mathematical induction, $V_k(x)$ satisfies those conditions for all k and in the limit so will $V(x)$. Of course, for this to make sense, we do mean to replace $V(x)$ by $V_0(x)$, $V_n(x)$, $V_{n+1}(x)$, and $V_k(x)$, wherever appropriate to check conditions (3.2), (3.3), and (3.4). We use that slight abuse of notation in the arguments that follow as well.

Recall that, all that is left to show is if $V_n(x)$ satisfies conditions (3.2), (3.3), and (3.4), then $V_{n+1}(x)$ also satisfies those three conditions. To do that, consider the value iteration Equation 3.5. Assuming that $V_n(x)$ satisfies the three conditions, we need to show that (a) $-h(x)$; (b) $\max\{r + V_n(x + e_1), r + V_n(x + e_2)\}$; and (c) $V_n((x - e_i)^+)$ for $i = 1, 2$ all satisfy the conditions (3.2), (3.3), and (3.4) as well. Then we are done because from Equation 3.5, $V_{n+1}(x)$ will also satisfy those three conditions. It is relatively straightforward to show (a) and (c), and hence they are left as an exercise for the reader (see the end of the chapter). The only extra result we need is that $V_n(x)$ is nonincreasing for all n and this can be shown once again by induction ($V_0(x) = 0$ is nonincreasing, $-h(x)$ is nonincreasing, and if $V_n(x)$ is nonincreasing, then from Equation 3.5, $V_{n+1}(x)$ is also nonincreasing). Thus, we are just left to show that $\max\{r + V_n(x + e_1), r + V_n(x + e_2)\}$ satisfies the conditions (3.2), (3.3), and (3.4). Let us denote $g(x)$ by $g(x) = \max\{r + V_n(x + e_1), r + V_n(x + e_2)\}$.

To show that $g(x) = \max\{r + V_n(x + e_1), r + V_n(x + e_2)\}$ satisfies the conditions (3.2), (3.3), and (3.4) when $V_n(x)$ satisfies them for all x, we first start with condition (3.4) and consider four cases representing the actions in states x and $x + e_1 + e_2$:

1. *Action in x is route to 1 and in $x + e_1 + e_2$ is route to 1.*
   This implies $g(x) = r + V_n(x + e_1)$ and $g(x + e_1 + e_2) = r + V_n(x + 2e_1 + e_2)$.

Also, $g(x + e_1) \geq r + V_n(x + 2e_1)$ (since $g(x + e_1) = \max\{r + V_n(x + 2e_1), r + V_n(x + e_1 + e_2)\}$) and $g(x + e_2) \geq r + V_n(x + e_1 + e_2)$ (since $g(x + e_2) = \max\{r + V_n(x + e_1 + e_2), r + V_n(x + 2e_2)\}$). Using these relations, we show that $g(x)$ satisfies (3.4), that is, $g(x + e_2) - g(x + e_1 + e_2) - g(x) + g(x + e_1) \geq 0$. For this, we begin with the LHS and show it is ≥ 0:

$$\text{LHS} = g(x + e_2) - g(x + e_1 + e_2) - g(x) + g(x + e_1)$$

$$= g(x + e_2) - r - V_n(x + 2e_1 + e_2) - r - V_n(x + e_1) + g(x + e_1)$$

$$\geq r + V_n(x + e_1 + e_2) - r - V_n(x + 2e_1 + e_2) - r - V_n(x + e_1) + r$$

$$+ V_n(x + 2e_1)$$

$$= V_n(x + e_1 + e_2) - V_n(x + 2e_1 + e_2) - V_n(x + e_1) + V_n(x + 2e_1) \geq 0.$$

The last inequality is because $V_n(x + e_1)$ satisfies condition (3.4). Hence, for this case, $g(x)$ does satisfy (3.4).

2. *Action in x is route to 1 and in $x + e_1 + e_2$ is route to 2.*
   This implies $g(x) = r + V_n(x + e_1)$ and $g(x + e_1 + e_2) = r + V_n(x + e_1 + 2e_2)$. Also, $g(x + e_1) \geq r + V_n(x + e_1 + e_2)$ (since $g(x + e_1) = \max\{r + V_n(x + 2e_1), r + V_n(x + e_1 + e_2)\}$) and $g(x + e_2) \geq r + V_n(x + e_1 + e_2)$ (since $g(x + e_2) = \max\{r + V_n(x + e_1 + e_2), r + V_n(x + 2e_2)\}$). Using these relations, we show that $g(x)$ satisfies (3.4), that is, $g(x + e_2) - g(x + e_1 + e_2) - g(x) + g(x + e_1) \geq 0$. For this we begin with the LHS and show it is ≥ 0:

$$\text{LHS} = g(x + e_2) - g(x + e_1 + e_2) - g(x) + g(x + e_1)$$

$$= g(x + e_2) - r - V_n(x + e_1 + 2e_2) - r - V_n(x + e_1) + g(x + e_1)$$

$$\geq r + V_n(x + e_1 + e_2) - r - V_n(x + e_1 + 2e_2) - r - V_n(x + e_1) + r$$

$$+ V_n(x + e_1 + e_2)$$

$$= V_n(x + e_1 + e_2) - V_n(x + e_1 + 2e_2) - V_n(x + e_1)$$

$$+ V_n(x + e_1 + e_2) \geq 0.$$

Since $V_n(x + e_1)$ satisfies conditions (3.2) and (3.4), by adding them up we get the last inequality (which also shows that $V_n(x + e_1)$ is concave in x_2 for a fixed $x_1 + 1$). Hence, for this case, $g(x)$ does satisfy (3.4).

3. *Action in x is route to 2 and in $x + e_1 + e_2$ is route to 1.*
   This implies $g(x) = r + V_n(x + e_2)$ and $g(x + e_1 + e_2) = r + V_n(x + 2e_1 + e_2)$.

The remaining arguments are symmetric to case 2 (i.e., action in x is route to 1 and in $x + e_1 + e_2$ is route to 2). Hence, in a symmetric manner, we can show that for this case, $g(x)$ does satisfy (3.4).

4. *Action in x is route to 2 and in $x + e_1 + e_2$ is route to 2.*
   This implies $g(x) = r + V_n(x + e_2)$ and $g(x + e_1 + e_2) = r + V_n(x + 2e_2 + e_1)$. The remaining arguments are symmetric to case 1 (i.e., action in x is route to 1 and in $x + e_1 + e_2$ is route to 1). Hence, in a symmetric manner, we can show that for this case, $g(x)$ does satisfy (3.4).

Therefore, $g(x) = \max\{r + V_n(x + e_1), r + V_n(x + e_2)\}$ satisfies condition (3.4) when $V_n(x)$ satisfies conditions (3.2), (3.3), and (3.4).

To show that $g(x) = \max\{r + V_n(x + e_1), r + V_n(x + e_2)\}$ satisfies condition (3.3), when $V_n(x)$ satisfies (3.2), (3.3), and (3.4) for all x, we once again consider four cases but representing the actions in states $x + e_2$ and $x + 2e_1$:

1. *Action in $x + e_2$ is route to 1 and in $x + 2e_1$ is route to 1.*
   This implies $g(x+e_2) = r + V_n(x + e_1 + e_2)$ and $g(x + 2e_1) = r + V_n(x + 3e_1)$. Also, $g(x + e_1 + e_2) \geq r + V_n(x + 2e_1 + e_2)$ (since $g(x + e_1 + e_2) = \max\{r + V_n(x + 2e_1 + e_2), r + V_n(x + e_1 + 2e_2)\}$) and $g(x + e_1) \geq r + V_n(x + 2e_1)$ (since $g(x + e_1) = \max\{r + V_n(x + 2e_1), r + V_n(x + e_1 + e_2)\}$). Using these relations, we show that $g(x)$ satisfies (3.3), that is, $g(x + e_1 + e_2) - g(x + e_1 + e_1) - g(x + e_2) + g(x + e_1) \geq 0$. For this, we begin with the LHS and show it is ≥ 0:

   $$\text{LHS} = g(x + e_1 + e_2) - g(x + e_1 + e_1) - g(x + e_2) + g(x + e_1)$$

   $$= g(x + e_1 + e_2) - r - V_n(x + 3e_1) - r - V_n(x + e_1 + e_2) + g(x + e_1)$$

   $$\geq r + V_n(x + 2e_1 + e_2) - r - V_n(x + 3e_1) - r - V_n(x + e_1 + e_2) + r$$

   $$+ V_n(x + 2e_1)$$

   $$= V_n(x + 2e_1 + e_2) - V_n(x + 3e_1) - V_n(x + e_1 + e_2) + V_n(x + 2e_1) \geq 0.$$

   The last inequality is because $V_n(x + e_1)$ satisfies condition (3.3). Hence, for this case, $g(x)$ does satisfy (3.3).

2. *Action in $x + e_2$ is route to 1 and in $x + 2e_1$ is route to 2.*
   This implies $g(x + e_2) = r + V_n(x + e_1 + e_2)$ and $g(x + 2e_1) = r + V_n(x + 2e_1 + e_2)$. Also, $g(x + e_1 + e_2) \geq r + V_n(x + 2e_1 + e_2)$ (since $g(x + e_1 + e_2) = \max\{r + V_n(x + 2e_1 + e_2), r + V_n(x + e_1 + 2e_2)\}$) and $g(x + e_1) \geq r + V_n(x + e_1 + e_2)$ (since $g(x + e_1) = \max\{r + V_n(x + 2e_1), r + V_n(x + e_1 + e_2)\}$). Using these relations, we show that $g(x)$ satisfies (3.3), that is,

$g(x + e_1 + e_2) - g(x + e_1 + e_1) - g(x + e_2) + g(x + e_1) \geq 0$. For this, we begin with the LHS and show it is ≥ 0:

$$\text{LHS} = g(x + e_1 + e_2) - g(x + e_1 + e_1) - g(x + e_2) + g(x + e_1)$$

$$= g(x + e_1 + e_2) - r - V_n(x + 2e_1 + e_2) - r - V_n(x + e_1 + e_2)$$

$$+ g(x + e_1)$$

$$\geq r + V_n(x + 2e_1 + e_2) - r - V_n(x + 2e_1 + e_2) - r - V_n(x + e_1 + e_2)$$

$$+ r + V_n(x + e_1 + e_2)$$

$$\geq 0.$$

Hence, for this case, $g(x)$ does satisfy (3.3).

3. *Action in $x + e_2$ is route to 2 and in $x + 2e_1$ is route to 1.*
   This implies $g(x+e_2) = r + V_n(x+2e_2)$ and $g(x+2e_1) = r + V_n(x+3e_1)$. The remaining arguments are symmetric to case 2 (i.e., action in $x+e_2$ is route to 1 and in $x + 2e_1$ is route to 2). Hence, in a symmetric manner, we can show that for this case, $g(x)$ does satisfy (3.2).

4. *Action in $x + e_2$ is route to 2 and in $x + 2e_1$ is route to 2.*
   This implies $g(x+e_2) = r + V_n(x+2e_2)$ and $g(x+2e_1) = r + V_n(x+2e_1+e_2)$. The remaining arguments are symmetric to case 1 (i.e., action in $x + e_2$ is route to 1 and in $x + 2e_1$ is route to 1). Hence, in a symmetric manner, we can show that for this case, $g(x)$ does satisfy (3.2).

Hence $g(x) = \max\{r + V_n(x+e_1), r + V_n(x+e_2)\}$ satisfies condition (3.3) when $V_n(x)$ satisfies conditions (3.2), (3.3), and (3.4). To show that $g(x) = \max\{r + V_n(x + e_1), r + V_n(x + e_2)\}$ satisfies condition (3.2), when $V_n(x)$ satisfies (3.2), (3.3), and (3.4) for all x, one needs to follow the argument symmetric to what we showed for condition (3.3). However, because of the symmetry the solution steps are identical and we do not present that here.

Hence $g(x) = \max\{r + V_n(x+e_1), r + V_n(x+e_2)\}$ satisfies the conditions (3.2), (3.3), and (3.4) when $V_n(x)$ satisfies them for all x, and therefore $V_{n+1}(x)$ also satisfies all those conditions implying by induction that $V(x)$ satisfies them as well. Therefore, the *monotonic* switching curve in Figure 3.2 is optimal. It is reasonable to wonder why condition (3.4) is necessary since even without it, $g(x)$ would satisfy conditions (3.2) and (3.3) which are the ones we really need. As it turns out (see the exercises at the end of this chapter), condition (3.4) would also need to be satisfied for $V_n(x)$, to show that conditions (3.2) and (3.3) are satisfied for $V_n((x-e_i)^+)$. In fact, besides (3.2), $V_n(x)$ would also need to be nonincreasing. ∎

Similarly, there are several situations in control of queues where the optimal policy is threshold type (or a switching curve). However, this only explains the structure of the optimal policy but not the optimal policy itself. For example, if numerical values for the preceding problem are given, where does the optimal line that separates region 1 from 2 lie? In other words, can we draw Figure 3.2 precisely for a given set of numerical values? The answer is yes. For every candidate switching curve, we can model the resulting system as a CTMC and evaluate its performance using the steady-state probabilities. For example, one algorithm would be to start with the switching curve being the straight line from $(0,0)$ to (B_1, B_2) and evaluate the expected discounted net benefit (via the steady-state probabilities). Then try all possible neighbors to determine the optimal switching curve that would maximize the expected discounted net benefit. We explain the algorithm in detail and provide a numerical example at the end of the next subsection. However, we first need a method to quickly compute steady-state probabilities of such CTMCs so that we can efficiently search through the space of switching curves swiftly, which is the objective of the next subsection.

3.1.2 Algorithm Using Recursively Tridiagonal Linear Equations

Matrix operations using standard matrix structures and algorithms are slow and consume large amounts of memory when applied to large sparse matrices, which are typical in multidimensional birth and death processes. Therefore, solving the CTMC balance equations for steady-state probabilities of multidimensional birth and death processes is computationally intensive using standard techniques, and to do it quickly one needs efficient algorithms that are specialized. One such algorithm uses matrix geometric technique and is explained in the following section. The technique requires a repetitive structure but is generalizable to multidimensional CTMCs that are not birth and death type. However, here we explain another algorithm for multidimensional birth and death processes, which is generic enough and does not require any repetitive structure. The algorithm is based on Servi [97] and will be termed as the *Servi algorithm*. We present only the two-dimensional and nonsingular version here and refer the reader to Servi [97] for variants. We explain the algorithm for two-dimensional birth and death processes using an example problem.

Problem 18

Two classes of requests arrive to a computer system according to a Poisson process with rate λ_i per second for class i (for $i = 1, 2$). The number of bytes of processing required for class i requests are exponentially distributed with mean $1/\mu_i$ MB. Assume that there is a 1 MB/s single processor that simultaneously processes all the requests using a full processor-sharing regime. In other words, if there are two class-1 requests and three class-2 requests currently running, then each of the five requests get 200 kB/s (assuming there are 1000 kB in 1 MB, which is not technically correct as there ought to be

1024 kB in 1 MB). However, in practice, the processor cycles through the five requests processing each for a tiny amount of time called time-quantum, and this is approximated as a full processor-sharing discipline. Further, there is a restriction that a maximum of four requests of class1 and three of class2 can be simultaneously served at any given time. Model the system as a two-dimensional birth and death process and obtain the steady-state distribution of the number of customers of each class in the system.

Solution

Let $X_i(t)$ be the number of class i customers in the system at time t (for $i = 1, 2$). Then the stochastic process $\{(X_1(t), X_2(t)), t \geq 0\}$ is a CTMC on state space $S = \{(0,0), (0,1), (0,2), (0,3), (1,0), \ldots, (4,2), (4,3)\}$ and infinitesimal generator matrix Q such that

$$Q - Diag(Q) =$$

$$
\begin{bmatrix}
0 & \lambda_2 & 0 & 0 & \lambda_1 & 0 & 0 & 0 & 0 & 0 & 0 & 0 & 0 & 0 & 0 & 0 & 0 & 0 & 0 & 0 \\
\mu_2 & 0 & \lambda_2 & 0 & 0 & \lambda_1 & 0 & 0 & 0 & 0 & 0 & 0 & 0 & 0 & 0 & 0 & 0 & 0 & 0 & 0 \\
0 & \mu_2 & 0 & \lambda_2 & 0 & 0 & \lambda_1 & 0 & 0 & 0 & 0 & 0 & 0 & 0 & 0 & 0 & 0 & 0 & 0 & 0 \\
0 & 0 & \mu_2 & 0 & 0 & 0 & 0 & \lambda_1 & 0 & 0 & 0 & 0 & 0 & 0 & 0 & 0 & 0 & 0 & 0 & 0 \\
\mu_1 & 0 & 0 & 0 & 0 & \lambda_2 & 0 & 0 & \lambda_1 & 0 & 0 & 0 & 0 & 0 & 0 & 0 & 0 & 0 & 0 & 0 \\
0 & \frac{\mu_1}{2} & 0 & 0 & \frac{\mu_2}{2} & 0 & \lambda_2 & 0 & 0 & \lambda_1 & 0 & 0 & 0 & 0 & 0 & 0 & 0 & 0 & 0 & 0 \\
0 & 0 & \frac{\mu_1}{3} & 0 & 0 & \frac{2\mu_2}{3} & 0 & \lambda_2 & 0 & 0 & \lambda_1 & 0 & 0 & 0 & 0 & 0 & 0 & 0 & 0 & 0 \\
0 & 0 & 0 & \frac{\mu_1}{4} & 0 & 0 & \frac{3\mu_2}{4} & 0 & 0 & 0 & 0 & \lambda_1 & 0 & 0 & 0 & 0 & 0 & 0 & 0 & 0 \\
0 & 0 & 0 & 0 & \mu_1 & 0 & 0 & 0 & 0 & \lambda_2 & 0 & 0 & \lambda_1 & 0 & 0 & 0 & 0 & 0 & 0 & 0 \\
0 & 0 & 0 & 0 & 0 & \frac{2\mu_1}{3} & 0 & 0 & \frac{\mu_2}{3} & 0 & \lambda_2 & 0 & 0 & \lambda_1 & 0 & 0 & 0 & 0 & 0 & 0 \\
0 & 0 & 0 & 0 & 0 & 0 & \frac{2\mu_1}{4} & 0 & 0 & \frac{2\mu_2}{4} & 0 & \lambda_2 & 0 & 0 & \lambda_1 & 0 & 0 & 0 & 0 & 0 \\
0 & 0 & 0 & 0 & 0 & 0 & 0 & \frac{2\mu_1}{5} & 0 & 0 & \frac{3\mu_2}{5} & 0 & 0 & 0 & 0 & \lambda_1 & 0 & 0 & 0 & 0 \\
0 & 0 & 0 & 0 & 0 & 0 & 0 & 0 & \mu_1 & 0 & 0 & 0 & 0 & \lambda_2 & 0 & 0 & \lambda_1 & 0 & 0 & 0 \\
0 & 0 & 0 & 0 & 0 & 0 & 0 & 0 & 0 & \frac{3\mu_1}{4} & 0 & 0 & \frac{\mu_2}{4} & 0 & \lambda_2 & 0 & 0 & \lambda_1 & 0 & 0 \\
0 & 0 & 0 & 0 & 0 & 0 & 0 & 0 & 0 & 0 & \frac{3\mu_1}{5} & 0 & 0 & \frac{2\mu_2}{5} & 0 & \lambda_2 & 0 & 0 & \lambda_1 & 0 \\
0 & 0 & 0 & 0 & 0 & 0 & 0 & 0 & 0 & 0 & 0 & \frac{3\mu_1}{6} & 0 & 0 & \frac{3\mu_2}{6} & 0 & 0 & 0 & 0 & \lambda_1 \\
0 & 0 & 0 & 0 & 0 & 0 & 0 & 0 & 0 & 0 & 0 & 0 & \mu_1 & 0 & 0 & 0 & 0 & \lambda_2 & 0 & 0 \\
0 & 0 & 0 & 0 & 0 & 0 & 0 & 0 & 0 & 0 & 0 & 0 & 0 & \frac{4\mu_1}{5} & 0 & 0 & \frac{\mu_2}{5} & 0 & \lambda_2 & 0 \\
0 & 0 & 0 & 0 & 0 & 0 & 0 & 0 & 0 & 0 & 0 & 0 & 0 & 0 & \frac{4\mu_1}{6} & 0 & 0 & \frac{2\mu_2}{6} & 0 & \lambda_2 \\
0 & 0 & 0 & 0 & 0 & 0 & 0 & 0 & 0 & 0 & 0 & 0 & 0 & 0 & 0 & \frac{4\mu_1}{7} & 0 & 0 & \frac{3\mu_2}{7} & 0
\end{bmatrix}
$$

where $Diag(Q)$ represents the diagonal of Q replacing the diagonal of an identity matrix. This is purely because of space restrictions, and one can easily obtain Q by summing up the rows and adding the negative of that to the diagonal.

By drawing the rate diagram (left as an exercise to the reader), one can see that the system can be modeled as a two-dimensional birth and death process with state space and Q matrix given earlier. Next, we obtain the steady-state distribution of the number of customers of each class in the system. For that we first write down Q as

$$Q = \begin{bmatrix} Q_{00} & Q_{01} & \bar{0} & \bar{0} & \bar{0} \\ Q_{10} & Q_{11} & Q_{12} & \bar{0} & \bar{0} \\ \bar{0} & Q_{21} & Q_{22} & Q_{23} & \bar{0} \\ \bar{0} & \bar{0} & Q_{32} & Q_{33} & Q_{34} \\ \bar{0} & \bar{0} & \bar{0} & Q_{43} & Q_{44} \end{bmatrix}$$

where

$$\bar{0} = \begin{bmatrix} 0 & 0 & 0 & 0 \\ 0 & 0 & 0 & 0 \\ 0 & 0 & 0 & 0 \\ 0 & 0 & 0 & 0 \end{bmatrix},$$

$$Q_{00} = \begin{bmatrix} -\lambda_1 - \lambda_2 & \lambda_2 & 0 & 0 \\ \mu_2 & -\lambda_1 - \lambda_2 - \mu_2 & \lambda_2 & 0 \\ 0 & \mu_2 & -\lambda_1 - \lambda_2 - \mu_2 & \lambda_2 \\ 0 & 0 & \mu_2 & -\lambda_1 - \mu_2 \end{bmatrix},$$

$$Q_{01} = \begin{bmatrix} \lambda_1 & 0 & 0 & 0 \\ 0 & \lambda_1 & 0 & 0 \\ 0 & 0 & \lambda_1 & 0 \\ 0 & 0 & 0 & \lambda_1 \end{bmatrix},$$

$$Q_{10} = \begin{bmatrix} \mu_1 & 0 & 0 & 0 \\ 0 & \frac{\mu_1}{2} & 0 & 0 \\ 0 & 0 & \frac{\mu_1}{3} & 0 \\ 0 & 0 & 0 & \frac{\mu_1}{4} \end{bmatrix},$$

$$Q_{11} = \begin{bmatrix} -\lambda_1 - \lambda_2 - \mu_1 & \lambda_2 & 0 \\ \frac{\mu_2}{2} & -\lambda_1 - \lambda_2 - \frac{\mu_1}{2} - \frac{\mu_2}{2} & \lambda_2 \\ 0 & \frac{2\mu_2}{3} & -\lambda_1 - \lambda_2 - \frac{\mu_1}{3} - \frac{2\mu_2}{3} \\ 0 & 0 & \frac{3\mu_2}{4} \end{bmatrix}$$
$$\begin{matrix} 0 \\ 0 \\ \lambda_2 \\ -\lambda_1 - \frac{3\mu_2}{4} - \frac{\mu_1}{4} \end{matrix} \Bigg],$$

$$Q_{12} = \begin{bmatrix} \lambda_1 & 0 & 0 & 0 \\ 0 & \lambda_1 & 0 & 0 \\ 0 & 0 & \lambda_1 & 0 \\ 0 & 0 & 0 & \lambda_1 \end{bmatrix},$$

$$
Q_{21} = \begin{bmatrix} \mu_1 & 0 & 0 & 0 \\ 0 & \frac{2\mu_1}{3} & 0 & 0 \\ 0 & 0 & \frac{2\mu_1}{4} & 0 \\ 0 & 0 & 0 & \frac{2\mu_1}{5} \end{bmatrix},
$$

$$
Q_{22} = \begin{bmatrix} -\lambda_1 - \lambda_2 - \mu_1 & \lambda_2 & 0 \\ \frac{\mu_2}{3} & -\lambda_1 - \lambda_2 - \frac{2\mu_1}{3} - \frac{\mu_2}{3} & \lambda_2 \\ 0 & \frac{2\mu_2}{4} & -\lambda_1 - \lambda_2 - \frac{2\mu_1}{4} - \frac{2\mu_2}{4} \\ 0 & 0 & \frac{3\mu_2}{5} \end{bmatrix}
$$

$$
\begin{bmatrix} 0 \\ 0 \\ \lambda_2 \\ -\lambda_1 - \frac{3\mu_2}{5} - \frac{2\mu_1}{5} \end{bmatrix},
$$

$$
Q_{23} = \begin{bmatrix} \lambda_1 & 0 & 0 & 0 \\ 0 & \lambda_1 & 0 & 0 \\ 0 & 0 & \lambda_1 & 0 \\ 0 & 0 & 0 & \lambda_1 \end{bmatrix},
$$

$$
Q_{32} = \begin{bmatrix} \mu_1 & 0 & 0 & 0 \\ 0 & \frac{3\mu_1}{4} & 0 & 0 \\ 0 & 0 & \frac{3\mu_1}{5} & 0 \\ 0 & 0 & 0 & \frac{3\mu_1}{6} \end{bmatrix},
$$

$$
Q_{33} = \begin{bmatrix} -\lambda_1 - \lambda_2 - \mu_1 & \lambda_2 & 0 \\ \frac{\mu_2}{4} & -\lambda_1 - \lambda_2 - \frac{3\mu_1}{4} - \frac{\mu_2}{4} & \lambda_2 \\ 0 & \frac{2\mu_2}{5} & -\lambda_1 - \lambda_2 - \frac{3\mu_1}{5} - \frac{2\mu_2}{5} \\ 0 & 0 & \frac{3\mu_2}{6} \end{bmatrix}
$$

$$
\begin{bmatrix} 0 \\ 0 \\ \lambda_2 \\ -\lambda_1 - \frac{3\mu_2}{6} - \frac{3\mu_1}{6} \end{bmatrix},
$$

$$
Q_{34} = \begin{bmatrix} \lambda_1 & 0 & 0 & 0 \\ 0 & \lambda_1 & 0 & 0 \\ 0 & 0 & \lambda_1 & 0 \\ 0 & 0 & 0 & \lambda_1 \end{bmatrix},
$$

$$Q_{43} = \begin{bmatrix} \mu_1 & 0 & 0 & 0 \\ 0 & \frac{4\mu_1}{5} & 0 & 0 \\ 0 & 0 & \frac{4\mu_1}{6} & 0 \\ 0 & 0 & 0 & \frac{4\mu_1}{7} \end{bmatrix}$$

and $Q_{44} = \begin{bmatrix} -\lambda_2 - \mu_1 & \lambda_2 & 0 & 0 \\ \frac{\mu_2}{5} & -\lambda_2 - \frac{4\mu_1}{5} - \frac{\mu_2}{5} & \lambda_2 & 0 \\ 0 & \frac{2\mu_2}{6} & -\lambda_2 - \frac{4\mu_1}{6} - \frac{2\mu_2}{6} & \lambda_2 \\ 0 & 0 & \frac{3\mu_2}{7} & -\frac{3\mu_2}{7} - \frac{4\mu_1}{7} \end{bmatrix}.$

Now, to obtain $p = [p_{0,0} \quad p_{0,1} \quad \cdots \quad p_{4,2} \quad p_{4,3}]$, we typically solve $pQ = [0 \ 0 \ \cdots \ 0 \ 0]$ and normalize using

$$\sum_{i=0}^{4} \sum_{j=0}^{3} p_{i,j} = 1.$$

However, that process gets computationally intensive for large state spaces. Therefore, we describe an alternate procedure to obtain p, which is essentially the Servi algorithm that we would subsequently describe for a general two-dimensional birth and death process.

Instead of obtaining the 1×20 row vector p directly, we write $p = a[R_0 \ R_1 \ R_2 \ R_3 \ R_4]$ where a is a 1×4 row vector and R_i is a 4×4 matrix (for $i = 0, 1, 2, 3, 4$). The vector a and matrix R_i (for $i = 0, \ldots, 4$) are unknown and need to be obtained recursively. Since $pQ = [0 \ \cdots \ 0]$, we have

$$a[R_0 \ R_1 \ R_2 \ R_3 \ R_4]Q = [0 \ \cdots \ 0]. \tag{3.6}$$

Using Q as

$$Q = \begin{bmatrix} Q_{00} & Q_{01} & \bar{0} & \bar{0} & \bar{0} \\ Q_{10} & Q_{11} & Q_{12} & \bar{0} & \bar{0} \\ \bar{0} & Q_{21} & Q_{22} & Q_{23} & \bar{0} \\ \bar{0} & \bar{0} & Q_{32} & Q_{33} & Q_{34} \\ \bar{0} & \bar{0} & \bar{0} & Q_{43} & Q_{44} \end{bmatrix}$$

we can rewrite Equation 3.6 as the following set:

$$a[R_0 Q_{00} + R_1 Q_{10}] = [0\ 0\ 0\ 0],$$

$$a[R_0 Q_{01} + R_1 Q_{11} + R_2 Q_{21}] = [0\ 0\ 0\ 0],$$

$$a[R_1 Q_{12} + R_2 Q_{22} + R_3 Q_{32}] = [0\ 0\ 0\ 0],$$

$$a[R_2 Q_{23} + R_3 Q_{33} + R_4 Q_{43}] = [0\ 0\ 0\ 0],$$

$$a[R_3 Q_{34} + R_4 Q_{44}] = [0\ 0\ 0\ 0].$$

The following set of R_i (for $i = 0, \ldots, 4$) and a values would ensure that the previous set of equations are satisfied:

$$R_0 = I \quad \text{the identity matrix,}$$

$$R_1 = -R_0 Q_{00} Q_{10}^{-1},$$

$$R_2 = [-R_0 Q_{01} - R_1 Q_{11}] Q_{21}^{-1},$$

$$R_3 = [-R_1 Q_{12} - R_2 Q_{22}] Q_{32}^{-1},$$

$$R_4 = [-R_2 Q_{23} - R_3 Q_{33}] Q_{43}^{-1}$$

and a is computed as a solution to $a[R_3 Q_{34} + R_4 Q_{44}] = [0\ 0\ 0\ 0]$. Next, we let $\theta = a[R_0\ R_1\ R_2\ R_3\ R_4]$ and obtain $p = \theta/(\theta \bar{1})$, where $\bar{1}$ is a 20 × 1 column vector of ones. Therefore, note that $p = [p_{0,0}\ p_{0,1}\ \cdots\ p_{4,2}\ p_{4,3}]$ also satisfies

$$\sum_{i=0}^{4} \sum_{j=0}^{3} p_{i,j} = 1. \qquad \blacksquare$$

Initially, it is natural for one to wonder why it would be a better idea to solve for a, R_0, ..., R_4 in the previous example as opposed to computing 20 unknowns of the p vector directly. The main reason is that Q is a sparse matrix and solving for p directly would result in more computational time. For example, consider numerical values $\lambda_1 = 1$, $\lambda_2 = 1.5$, $\mu_1 = 2$, and $\mu_2 = 4$. For this example, solving p directly took about 0.9 ms on average, whereas solving using a, R_0, ..., R_4 took about 0.4 ms on average.

One of the main benefits is that inverses are taken of smaller matrices, which are also tridiagonal (hence much easier to invert). In fact, this is the key concept explored in the Servi algorithm. The percentage gains in terms of computations increase as the size of Q increases. Although the algorithm is only presented for two-dimensional birth and death processes, the recursive tridiagonal structure is also leveraged upon in the higher dimensional algorithm. The only concern is when the inverses do not exist due to singularities, and Servi [97] explains how to get around that. With that motivation, we present the Servi algorithm.

Let $\{(X_1(t), X_2(t)), t \geq 0\}$ be a finite-state CTMC that can be modeled as a two-dimensional birth and death process. For given finite integers b_1 and b_2, $0 \leq X_1(t) \leq b_1$ and $0 \leq X_2(t) \leq b_2$ for all t. Arrange the states of the CTMC in the following order: $(0,0)$, $(0,1)$, $(0,2)$, ..., $(0,b_2)$, $(1,0)$, $(1,1)$, $(1,2)$, ..., $(1,b_2)$, ..., $(b_1,0)$, $(b_1,1)$, $(b_1,2)$, ..., (b_1,b_2). Then, since the CTMC is a two-dimensional birth and death process, the Q matrix takes the form (see exercise problems)

$$Q = \begin{bmatrix} Q_{0,0} & Q_{0,1} & \bar{0} & \bar{0} & \cdots & \bar{0} & \bar{0} \\ Q_{1,0} & Q_{1,1} & Q_{1,2} & \bar{0} & \cdots & \bar{0} & \bar{0} \\ \bar{0} & Q_{2,1} & Q_{2,2} & Q_{2,3} & \cdots & \bar{0} & \bar{0} \\ \bar{0} & \bar{0} & Q_{3,2} & Q_{3,3} & \cdots & \bar{0} & \bar{0} \\ \vdots & \vdots & \vdots & \ddots & \ddots & \vdots & \vdots \\ \bar{0} & \bar{0} & \bar{0} & \bar{0} & \cdots & Q_{b_1-1,b_1-1} & Q_{b_1-1,b_1} \\ \bar{0} & \bar{0} & \bar{0} & \bar{0} & \cdots & Q_{b_1,b_1-1} & Q_{b_1,b_1} \end{bmatrix}$$

where $\bar{0}$ is a $(b_2+1) \times (b_2+1)$ matrix of zeros and for all i, j, $Q_{i,j}$ are $(b_2+1) \times (b_2+1)$ matrices. Assuming that the CTMC is irreducible (i.e., it is possible to go from every state to every other state in one or more transitions), our objective is to determine the steady-state probabilities $p_{i,j}$ for $0 \leq i \leq b_1$ and $0 \leq j \leq b_2$ where

$$p_{i,j} = \lim_{t \to \infty} P\{X_1(t) = i, X_2(t) = j\}.$$

Of course one approach is to solve $pQ = 0$ (where p is a $1 \times (b_1+1)(b_2+1)$ row vector of $p_{i,j}$ values) and $\sum_i \sum_j p_{i,j} = 1$. We next present an alternative approach that we call Servi algorithm, which is computationally much more efficient.

Servi algorithm

1. Let $R_0 = I$, that is, an identity matrix of size $(b_2+1) \times (b_2+1)$.

2. Obtain R_1 as $R_1 = -R_0 Q_{0,0} Q_{1,0}^{-1}$.

3. For $i = 1, \ldots, b_1 - 1$, obtain R_{i+1} as $R_{i+1} = [-R_{i-1}Q_{i-1,i} - R_i Q_{i,i}]Q_{i+1,i}^{-1}$.

4. Find a $1 \times (b_2+1)$ row vector a that satisfies $a[R_{b_1-1}Q_{b_1-1,b_1} + R_{b_1}Q_{b_1,b_1}] = [0 \cdots 0]$.

5. Compute the $1 \times (b_1+1)(b_2+1)$ row vector θ as $\theta = a[R_0\ R_1\ \ldots\ R_{b_1}]$.

6. The steady-state probability vector p is $p = \theta/(\theta\bar{1})$, where $\bar{1}$ is a $(b_1+1)(b_2+1) \times 1$ column vector of ones.

Next, we present a numerical example to illustrate the algorithm.

Problem 19

Consider a bilingual customer service center where there are two finite capacity queues: one for English-speaking customers and other for Spanish-speaking customers. A maximum of three Spanish-speaking customers can be in the system at any time. Likewise, a maximum of four English-speaking customers can be in the system simultaneously. Spanish-speaking and English-speaking customers arrive into their respective queues according to a Poisson process with respective rates 4, and 6 per hour. There is a Spanish-speaking server that takes on an average 12 min to serve each of his customers and there is an English-speaking server that takes on an average 7.5 min to serve each of her customers. Assume that none of the customers are bilingual; however, the manager who oversees the two servers can speak English and Spanish. Whenever the number of customers in one of the queues has two or more customers than the other queue, the manager helps out the server with the longer queue, thereby increasing the service rate by 2 per hour. Assume that all service times are exponentially distributed.

Model the bilingual customer service center system using a two-dimensional birth and death process. Then use the Servi algorithm to obtain the steady-state probabilities of the number of customers in the system speaking Spanish and English. What fraction of customers of each type is rejected without service? Determine the average time spent by each type of accepted customer in the system.

Solution

Let $X_1(t)$ be the number of Spanish-speaking customers in the system at time t and $X_2(t)$ be the number of English-speaking customers in the system at time t. From the problem description, $b_1 = 3$ and $b_2 = 4$. Clearly, $\{(X_1(t), X_2(t)), t \geq 0\}$ is a finite-state CTMC that can be modeled as a two-dimensional birth and death process with $0 \leq X_1(t) \leq 3$ and $0 \leq X_2(t) \leq 4$ for all t. The state space is $S = \{\ (0,0),\ (0,1),\ (0,2),\ (0,3),\ (0,4),\ (1,0),\ (1,1),\ (1,2),\ (1,3),\ (1,4),\ (2,0),\ (2,1),\ (2,2),\ (2,3),\ (2,4),\ (3,0),\ (3,1),\ (3,2),\ (3,3),\ (3,4)\ \}$. Note that when the Spanish-speaking server is by himself, the service rate is 5 per hour, and when the English-speaking server is by herself, the service rate is 8 per hour. However, when the manager comes to assist, the service rate of the Spanish-speaking server becomes 7 per hour and that of the English-speaking server becomes 10 per hour.

Using that and the arrival rates, the Q matrix (in the order of states in S) is given as

$$\left[\begin{array}{cccccccccccccccccccc}
-10 & 6 & 0 & 0 & 0 & 4 & 0 & 0 & 0 & 0 & 0 & 0 & 0 & 0 & 0 & 0 & 0 & 0 & 0 & 0\\
8 & -18 & 6 & 0 & 0 & 0 & 4 & 0 & 0 & 0 & 0 & 0 & 0 & 0 & 0 & 0 & 0 & 0 & 0 & 0\\
0 & 10 & -20 & 6 & 0 & 0 & 0 & 4 & 0 & 0 & 0 & 0 & 0 & 0 & 0 & 0 & 0 & 0 & 0 & 0\\
0 & 0 & 10 & -20 & 6 & 0 & 0 & 0 & 4 & 0 & 0 & 0 & 0 & 0 & 0 & 0 & 0 & 0 & 0 & 0\\
0 & 0 & 0 & 10 & -14 & 0 & 0 & 0 & 0 & 4 & 0 & 0 & 0 & 0 & 0 & 0 & 0 & 0 & 0 & 0\\
5 & 0 & 0 & 0 & 0 & -15 & 6 & 0 & 0 & 0 & 4 & 0 & 0 & 0 & 0 & 0 & 0 & 0 & 0 & 0\\
0 & 5 & 0 & 0 & 0 & 8 & -23 & 6 & 0 & 0 & 0 & 4 & 0 & 0 & 0 & 0 & 0 & 0 & 0 & 0\\
0 & 0 & 5 & 0 & 0 & 0 & 8 & -23 & 6 & 0 & 0 & 0 & 4 & 0 & 0 & 0 & 0 & 0 & 0 & 0\\
0 & 0 & 0 & 5 & 0 & 0 & 0 & 10 & -25 & 6 & 0 & 0 & 0 & 4 & 0 & 0 & 0 & 0 & 0 & 0\\
0 & 0 & 0 & 0 & 5 & 0 & 0 & 0 & 10 & -19 & 0 & 0 & 0 & 0 & 4 & 0 & 0 & 0 & 0 & 0\\
0 & 0 & 0 & 0 & 0 & 7 & 0 & 0 & 0 & 0 & -17 & 6 & 0 & 0 & 0 & 4 & 0 & 0 & 0 & 0\\
0 & 0 & 0 & 0 & 0 & 0 & 5 & 0 & 0 & 0 & 8 & -23 & 6 & 0 & 0 & 0 & 4 & 0 & 0 & 0\\
0 & 0 & 0 & 0 & 0 & 0 & 0 & 5 & 0 & 0 & 0 & 8 & -23 & 6 & 0 & 0 & 0 & 4 & 0 & 0\\
0 & 0 & 0 & 0 & 0 & 0 & 0 & 0 & 5 & 0 & 0 & 0 & 8 & -23 & 6 & 0 & 0 & 0 & 4 & 0\\
0 & 0 & 0 & 0 & 0 & 0 & 0 & 0 & 0 & 5 & 0 & 0 & 0 & 10 & -19 & 0 & 0 & 0 & 0 & 4\\
0 & 0 & 0 & 0 & 0 & 0 & 0 & 0 & 0 & 0 & 7 & 0 & 0 & 0 & 0 & -13 & 6 & 0 & 0 & 0\\
0 & 0 & 0 & 0 & 0 & 0 & 0 & 0 & 0 & 0 & 0 & 7 & 0 & 0 & 0 & 8 & -21 & 6 & 0 & 0\\
0 & 0 & 0 & 0 & 0 & 0 & 0 & 0 & 0 & 0 & 0 & 0 & 5 & 0 & 0 & 0 & 8 & -19 & 6 & 0\\
0 & 0 & 0 & 0 & 0 & 0 & 0 & 0 & 0 & 0 & 0 & 0 & 0 & 5 & 0 & 0 & 0 & 8 & -19 & 6\\
0 & 0 & 0 & 0 & 0 & 0 & 0 & 0 & 0 & 0 & 0 & 0 & 0 & 0 & 5 & 0 & 0 & 0 & 8 & -13
\end{array}\right]$$

Now, to use the Servi algorithm, we write down

$$Q = \begin{bmatrix} Q_{0,0} & Q_{0,1} & \bar{0} & \bar{0} \\ Q_{1,0} & Q_{1,1} & Q_{1,2} & \bar{0} \\ \bar{0} & Q_{2,1} & Q_{2,2} & Q_{2,3} \\ \bar{0} & \bar{0} & Q_{3,2} & Q_{3,3} \end{bmatrix}$$

where each $Q_{i,j}$ is a 5×5 matrix for all i, j and $\bar{0}$ is a 5×5 matrix of zeros. It is important to point out here that it would have been wiser to have switched $X_1(t)$ and $X_2(t)$ since they would have resulted in 4×4 matrices, which would have been computationally simpler to manipulate, especially inverses. We now use the Servi algorithm to determine the steady-state probabilities $p_{i,j}$ for all $(i, j) \in S$.

1. Let $R_0 = I$, and hence

$$R_0 = \begin{bmatrix} 1 & 0 & 0 & 0 & 0 \\ 0 & 1 & 0 & 0 & 0 \\ 0 & 0 & 1 & 0 & 0 \\ 0 & 0 & 0 & 1 & 0 \\ 0 & 0 & 0 & 0 & 1 \end{bmatrix}.$$

2. Obtain R_1 as $R_1 = -R_0 Q_{0,0} Q_{1,0}^{-1}$, which implies

$$R_1 = \begin{bmatrix} 2 & -1.2 & 0 & 0 & 0 \\ -1.6 & 3.6 & -1.2 & 0 & 0 \\ 0 & -2 & 4 & -1.2 & 0 \\ 0 & 0 & -2 & 4 & -1.2 \\ 0 & 0 & 0 & -2 & 2.8 \end{bmatrix}.$$

3. For $i = 1, \ldots, 2$, obtain R_{i+1} as $R_{i+1} = [-R_{i-1}Q_{i-1,i} - R_i Q_{i,i}]Q_{i+1,i}^{-1}$. Hence

$$R_2 = \begin{bmatrix} 5.0857 & -7.9200 & 1.4400 & 0 & 0 \\ -7.5429 & 19.6000 & -9.8400 & 1.4400 & 0 \\ 2.2857 & -15.6000 & 22.4000 & -10.8000 & 1.4400 \\ 0 & 3.2000 & -17.2000 & 24.0000 & -9.3600 \\ 0 & 0 & 4.0000 & -15.6000 & 12.2400 \end{bmatrix},$$

$$R_3 = \begin{bmatrix} 20.2596 & -31.3420 & 16.1280 & -1.7280 & 0 \\ -39.8041 & 80.0539 & -70.1280 & 18.4320 & -1.7280 \\ 23.3796 & -77.6735 & 135.8400 & -78.4800 & 18.4320 \\ -3.6571 & 30.1714 & -119.7600 & 146.5600 & -63.4080 \\ 0 & -4.5714 & 43.3600 & -99.4400 & 62.9920 \end{bmatrix}.$$

4. Find a 1×5 row vector a that satisfies $a[R_2 Q_{2,3} + R_3 Q_{3,3}] = [0 \ \cdots \ 0]$, which results in

$$a = [0.7269 \ 0.5447 \ 0.3395 \ 0.2087 \ 0.1269].$$

5. The 1×20 row vector θ is computed as $\theta = a[R_0 \ R_1 \ R_2 \ R_3]$. Therefore, $\theta = [0.7269 \ 0.5447 \ 0.3395 \ 0.2087 \ 0.1269 \ 0.5823 \ 0.4097 \ 0.2867 \ 0.1736 \ 0.1050 \ 0.3641 \ 0.2914 \ 0.2085 \ 0.1471 \ 0.0891 \ 0.2185 \ 0.1730 \ 0.1446 \ 0.1094 \ 0.0779]$.

6. The steady-state probability vector p is $p = \theta/(\theta\bar{1})$, where $\bar{1}$ is a 20×1 column vector of ones. Then, we get $p = [p_{00} \ p_{01} \ p_{02} \ p_{03} \ p_{04} \ p_{10} \ p_{11} \ p_{12} \ p_{13} \ p_{14} \ p_{20} \ p_{21} \ p_{22} \ p_{23} \ p_{24} \ p_{30} \ p_{31} \ p_{32} \ p_{33} \ p_{34}] = [0.1364 \ 0.1022 \ 0.0637 \ 0.0392 \ 0.0238 \ 0.1093 \ 0.0769 \ 0.0538 \ 0.0326 \ 0.0197 \ 0.0683 \ 0.0547 \ 0.0391 \ 0.0276 \ 0.0167 \ 0.0410 \ 0.0325 \ 0.0271 \ 0.0205 \ 0.0146]$.

Having obtained the steady-state probabilities, the next step is to answer the system performance questions. The fraction of Spanish-speaking customers that are rejected is $p_{30} + p_{31} + p_{32} + p_{33} + p_{34} = 0.1357$. This is due to Poisson arrivals see time process (PASTA) since every time a potential Spanish-speaking customer arrives, he/she will see there are three other Spanish-speaking customers in the system with probability $p_{30} + p_{31} + p_{32} + p_{33} + p_{34}$. Likewise, the fraction of English-speaking customers that are rejected is $p_{04} + p_{14} + p_{24} + p_{34} = 0.0748$.

Therefore, the rejection and entering rates of Spanish-speaking customers are $4 \times 0.1357 = 0.5428$ and $4 \times 0.8643 = 3.4572$ per hour, respectively. Likewise, the rejection and entering rates of English-speaking customers are $6 \times 0.0748 = 0.4488$ per hour and $6 \times 0.9252 = 5.5512$ per hour, respectively. The average number of Spanish-speaking customers in the system is

$$\sum_{i=0}^{3} \sum_{j=0}^{4} i p_{ij} = 1.1122$$

and the average number of English-speaking customers in the system is

$$\sum_{i=0}^{3} \sum_{j=0}^{4} j p_{ij} = 1.2926.$$

Hence, from Little's law, the average time spent by accepted Spanish- and English-speaking customers is $1.1122/3.4572 = 0.3217\,\text{h}$ and $1.2926/5.5512 = 0.2329\,\text{h}$, respectively.   ∎

3.1.3 Example: Optimal Customer Routing

In Section 3.1.1, we saw that threshold policy or switching curve policy was optimal in many queueing control problems. Then, in Section 3.1.2, we introduced an effective algorithm to expeditiously compute steady-state probabilities of multidimensional birth and death processes. Here, we seek to close the loop by considering a scenario where the objective is to obtain the switching curve based on the fact that it is optimal from Section 3.1.1. We use the algorithm in Section 3.1.2 to quickly compare candidate switching curves by computing the steady-state system performance measures. By searching through the solution space of switching curves, we obtain the optimal one numerically. We illustrate the approach using an example of optimal customer routing.

Problem 20

Laurie's Truck Repair offers emergency services for which they have two facilities: one in the north end of town and the other in the south end. All trucks that require an emergency repair call a single number to schedule their repair. When a call is received, the operator must determine whether to accept the repair request. If a repair is accepted, the operator must also determine whether to send it to the north or south facility. The company has installed a sophisticated system that can track the status of all the repairs (this means the operator can know how many outstanding repairs are in progress at each facility). Due to space restrictions to park the trucks, the north-side facility can handle at most three requests at a time, whereas the south-side facility can only handle two simultaneous requests. Use the following information to determine the routing strategy: calls for repair arrive according to a Poisson process with mean rate of four per day; the service times are exponentially distributed at both facilities; however, north-side facility can repair three trucks per day on average, whereas the south-side facility can repair two per day on average. The average revenue earned per truck repair is $100. The holding cost per truck per day is $20 in the north side and $10 in the south side (the difference is partly due to the cost of insurance in the two neighborhoods). Assume the following: the time to take a truck to a repair facility is negligible compared to the service time; decisions to accept/reject and route are made instantaneously and are based only on the number of committed outstanding repairs at each facility; once accepted at a facility, the truck does not leave it until the repair is complete; the operator would never reject a call if there is space in at least one of the facilities to repair; at either location, trucks are repaired one at a time; the revenue earned for a truck repair is independent of the time to repair, location, and type of repair.

Solution

For notational convenience, we use subscript "1" to denote the north side and subscript "2" for south side. Note that the problem description is almost identical to that of Problem 17 with $B_1 = 3$, $B_2 = 2$, $r = \$100$, $\mu_1 = 3$ per day, $\mu_2 = 2$ per day, $\lambda = 4$ per day, $h_1 = \$20$ per truck per day, and $h_2 = \$10$ per truck per day. However, one key difference between Problem 17 and this one is that here the solution needs to be the actual policy (not the structure as required in Problem 17). In other words, if a request for service is made when there are i trucks in the north side and j in the south side, should we accept the request, and if we do accept, should it be sent to the north or south side?

Let $X_1(t)$ and $X_2(t)$ denote the number of trucks under repair in the north- and south-side facilities, respectively. Ignoring the time to schedule as well as to drive to the appropriate facility, the system can be modeled as a two-dimensional birth and death process $\{(X_1(t), X_2(t)), t \geq 0\}$ with state space $S = \{(0,0), (0,1), (0,2), (1,0), (1,1), (1,2), (2,0), (2,1), (2,2), (3,0), (3,1), (3,2)\}$. The action in state $(3,2)$ is to reject requests for repairs. When the system is in states $(0,2)$, $(1,2)$, and $(2,2)$, the optimal action is to route to facility 1 (i.e., north) since there is no space in facility 2. Likewise, when the system is in states $(3,0)$ and $(3,1)$, the optimal action is to route to facility 2 (i.e., south) since there is no space in facility 1. Thus, we only need to determine the actions in the six states $(0,0)$, $(0,1)$, $(1,0)$, $(1,1)$, $(2,0)$, and $(2,1)$.

Although there are $2^6 = 64$ possible actions in the six states together where we need to determine the optimal action (route to 1 or 2), from the solution to Problem 17, we know the optimal solution is a monotonic switching curve. Therefore, we are reduced to only 10 different sets of actions that we need to consider, which are summarized in Table 3.1. Let A_{ij} be the action in state (i, j) such that $A_{ij} = 1$ implies routing to 1 and $A_{ij} = 2$ implies routing to 2. Therefore, there is considerable computation and time savings from 64 possible alternatives to 10. The only possible concern is that in Problem 17, the objective is in terms of long-run average discounted cost, whereas here it is the long-run average cost per unit time. As it turns out, the average cost per unit time case also yields the same structure of the optimal policy.

Each one of the 10 alternative actions in Table 3.1 yields a two-dimensional birth and death process. As described earlier, for all the 10 alternatives, $X_1(t)$ and $X_2(t)$ denote the number of trucks under repair in the north- and south-side facilities, respectively, and $\{(X_1(t), X_2(t)), t \geq 0\}$ would be a two-dimensional birth and death process with state space $S = \{(0,0), (0,1), (0,2), (1,0), (1,1), (1,2), (2,0), (2,1), (2,2), (3,0), (3,1), (3,2)\}$. The key difference among the 10 alternatives would be the Q matrix. Therefore, although notationally we have the same set of $p_{i,j}$ values, they would depend on the Q matrix. The objective is to determine the optimal one among the 10 alternatives that would maximize the expected net revenue per unit

TABLE 3.1

Alternatives for Actions in the 6 States
Where They Are to Be Determined

A_{00}	A_{01}	A_{10}	A_{11}	A_{20}	A_{21}
1	1	1	1	1	1
1	1	1	1	2	1
1	1	1	1	2	2
1	1	2	1	2	1
1	1	2	1	2	2
2	1	2	1	2	1
1	1	2	2	2	2
2	1	2	1	2	2
2	1	2	2	2	2
2	2	2	2	2	2

time. For a given alternative, if the steady-state probabilities $p_{i,j}$ can be computed for all i and j such that $(i,j) \in S$, the steady-state expected net revenue per unit time is

$$r\lambda(1 - p_{3,2}) - \sum_{i=0}^{3}\sum_{j=0}^{2}(ih_1 + jh_2)p_{i,j}$$

dollars per day. This is due to the fact that a fraction $(1 - p_{3,2})$ of requests are accepted (which arrive at rate λ on average) and every request on average brings a revenue of r dollars; hence, the average revenue is $r\lambda(1 - p_{3,2})$. The remaining term is the average holding cost expenses that need to be subtracted from the revenue. Since at any given time there are i trucks in location 1 and j trucks in location 2 with probability $p_{i,j}$, by conditioning the number of trucks in each location, we can obtain the holding cost per unit time as $ih_1 + jh_2$ if there are i trucks in location 1 and j in location 2.

For each of the candidate alternate actions, to evaluate the objective function, that is, the steady-state expected net revenue per unit time, we solve for $p_{i,j}$ and compute the objective function. To speed up the process to obtain $p_{i,j}$, we use Servi algorithm in Section 3.1.2 but do not present the details here. Using the numerical values for r, h_1, h_2, λ, μ_1, and μ_2, we obtain the *optimal* set of actions as $A_{00} = 1$, $A_{01} = 1$, $A_{10} = 2$, $A_{11} = 1$, $A_{20} = 2$, and $A_{21} = 2$ with an objective function value of $320.8905 per day. This optimal action set yields a two-dimensional birth and death process with rate diagram depicted in Figure 3.3. For this system, obtaining the steady-state probability using Servi algorithm is left as an exercise. ∎

There are many such queueing control problems where the objective is to determine the optimal control actions in each state. In a majority of

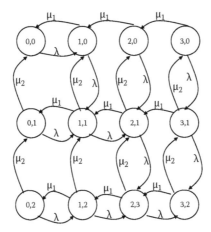

FIGURE 3.3
Two-dimensional birth and death process corresponding to optimal action.

those problems, the structure of the optimal policy is a switching curve or a threshold policy. To determine the optimal switching curve or threshold from a set of candidate values, if the resulting CTMC is a multidimensional birth and death process, we can use Servi algorithm to quickly compute the steady-state probabilities, which are used in the objective function to determine the optimal actions (especially when other techniques such as using the value iteration directly, or using a linear programming formulation of such Markov decision processes, are more time consuming). The next natural question is what if we need the steady-state probabilities of a multidimensional CTMC that is not necessarily a birth and death process. We address such multidimensional stochastic processes in the following section.

3.2 Multidimensional Markov Chains

There are several queueing systems with Poisson arrivals and exponentially distributed service times that can be modeled as multidimensional CTMCs. One such example is the multidimensional birth and death process explained in the previous section. Although the methodologies are related, some of the key differences between the multidimensional birth and death chains in the previous section and the multidimensional CTMCs in this section are as follows: here we consider infinite number of states in one of the dimensions, whereas in the previous section, all dimensions were finite; here we require a repetitive structure (we will see this subsequently), whereas in the previous section, the structure could be arbitrary; and of course, here we do not require a birth and death structure unlike the previous section. Having made that last comment, interestingly, although our multidimensional CTMCs will not be birth and death processes, the resulting Q matrix will have a birth-and-death-like structure and, hence, we call them quasi-birth-and-death processes. This will be the focus of the following section.

3.2.1 Quasi-Birth-Death Processes

A quasi-birth-death (QBD) process is a CTMC whose infinitesimal generator matrix Q is of the following block diagonal form:

$$Q = \begin{pmatrix} B_1 & B_0 & 0 & 0 & 0 & 0 & \cdots \\ B_2 & A_1 & A_0 & 0 & 0 & 0 & \cdots \\ 0 & A_2 & A_1 & A_0 & 0 & 0 & \cdots \\ 0 & 0 & A_2 & A_1 & A_0 & 0 & \cdots \\ 0 & 0 & 0 & A_2 & A_1 & A_0 & \cdots \\ \vdots & \vdots & \vdots & \vdots & \vdots & \vdots & \ddots \end{pmatrix}$$

where A_0, A_1, A_2, B_0, B_1, and B_2 are finite-sized matrices (instead of scalars, which is the case in one-dimensional birth and death processes; hence the name quasi-birth–death process). Usually, A_0, A_1, and A_2 are square matrices of size $m \times m$, whereas B_0, B_1, and B_2 are of sizes $\ell \times m$, $\ell \times \ell$, and $m \times \ell$, respectively. It is important to note that $m < \infty$ and $\ell < \infty$. The zeros in the Q matrix are also matrices of appropriate sizes. The terminology used for each block in Q is called *level*. Each level consists of multiple phases of the original CTMC. In particular the first level has ℓ phases and the remaining levels have m phases, the states can jump down one level, stay in the same level, or jump up one level. Hence, the CTMC is said to be skip-free between levels.

An intuitive reason for observing QBD processes in queues is that if one considers the system as a whole, customers are arriving into the system one by one according to a Poisson process (unless they are in batches) and get served one after the other (hence depart from the system one by one). Therefore, it is natural that the total number in the system most likely goes up or down by one and thereby in that dimension one does have a birth-and-death-like structure. Hence, overall one sees birth and death processes. Next, we present a few examples of queueing systems where the multidimensional CTMCs can be modeled as QBDs.

Problem 21

Consider a queueing system with two parallel queues and two servers, one for each queue. Customers arrive to the system according to $PP(\lambda)$, and each arriving customer joins the queue with fewer number of customers. If both queues have the same number of customers, then the arriving customer picks either with equal probability. The service times are exponentially distributed with mean μ_i at server i for $i = 1, 2$. When a service is completed at one queue and the other queue has two more customers than this queue, then the customer at the end of the line instantaneously switches to the shorter queue (this is called jockeying) to balance the queues. Let $X_1(t)$ and $X_2(t)$ be the number of customers in queues 1 and 2, respectively, at time t. Assuming that $X_1(0) = X_2(0) = 0$, we have $|X_1(t) - X_2(t)| \leq 1$ for all t. Model the bivariate stochastic process $\{(X_1(t), X_2(t)), t \geq 0\}$ as a QBD by obtaining A_0, A_1, A_2, B_0, B_1, and B_2.

Solution

The bivariate CTMC $\{(X_1(t), X_2(t)), t \geq 0\}$ has a state space

$$S = \{(0,0), (1,0), (0,1), (1,1), (2,1), (1,2), (2,2), (3,2), (2,3), \ldots\}.$$

By writing down the Q matrix of the CTMC and considering sets of three states as a "level" with three "phases," it is easy to verify that Q has a QBD structure with

$$A_0 = B_0 = \begin{pmatrix} 0 & 0 & 0 \\ \lambda & 0 & 0 \\ \lambda & 0 & 0 \end{pmatrix}$$

$$A_2 = B_2 = \begin{pmatrix} 0 & \mu_2 & \mu_1 \\ 0 & 0 & 0 \\ 0 & 0 & 0 \end{pmatrix}$$

$$A_1 = \begin{pmatrix} -\lambda - \mu_1 - \mu_2 & \lambda/2 & \lambda/2 \\ \mu_1 + \mu_2 & -\lambda - \mu_1 - \mu_2 & 0 \\ \mu_1 + \mu_2 & 0 & -\lambda - \mu_1 - \mu_2 \end{pmatrix} \text{ and}$$

$$B_1 = \begin{pmatrix} -\lambda & \lambda/2 & \lambda/2 \\ \mu_2 & -\lambda - \mu_2 & 0 \\ \mu_1 & 0 & -\lambda - \mu_1 \end{pmatrix}.$$

Care should be taken to write the states in the order (i, i), then $(i + 1, i)$, followed by $(i, i + 1)$ for all $i = 0, 1, 2, \ldots$ and use those three states as part of a level. Note that the previous CTMC does not resemble a birth and death process at all. ∎

Problem 22

Consider a single server queue with infinite waiting room where the service times are $\exp(\mu)$. Into this queue, arrivals occur according to a Poisson process with one of three possible rates λ_0, λ_1, or λ_2. The rates are governed by a CTMC $\{Z(t), t \geq 0\}$ called the environment process on states $\{0, 1, 2\}$ and infinitesimal generator matrix

$$\begin{bmatrix} -\alpha_1 - \alpha_2 & \alpha_1 & \alpha_2 \\ \beta_0 & -\beta_0 - \beta_2 & \beta_2 \\ \gamma_0 & \gamma_1 & -\gamma_0 - \gamma_1 \end{bmatrix}.$$

For $i = 0, 1, 2$ and any $t \in [0, \infty]$, if the CTMC is in state i at time t (i.e., $Z(t) = i$), then arrivals into the queue occur according to a $PP(\lambda_i)$. Let $X(t)$ be the number of entities in the system at time t. Model the bivariate stochastic process $\{(X(t), Z(t)), t \geq 0\}$ as a QBD by obtaining $A_0, A_1, A_2, B_0, B_1,$ and B_2.

Solution

The bivariate CTMC $\{(X(t), Z(t)), t \geq 0\}$ has a state space

$$S = \{(0,0), (0,1), (0,2), (1,0), (1,1), (1,2), (2,0), (2,1), (2,2), \ldots\}.$$

By writing down the Q matrix of the CTMC and considering sets of three states as a "level" with three "phases," it is easy to verify that Q has a QBD structure with

$$A_0 = B_0 = \begin{pmatrix} \lambda_0 & 0 & 0 \\ 0 & \lambda_1 & 0 \\ 0 & 0 & \lambda_2 \end{pmatrix}$$

$$A_2 = B_2 = \begin{pmatrix} \mu & 0 & 0 \\ 0 & \mu & 0 \\ 0 & 0 & \mu \end{pmatrix}$$

$$A_1 = \begin{pmatrix} -\lambda_0 - \mu - \alpha_1 - \alpha_2 & \alpha_1 & \alpha_2 \\ \beta_0 & -\lambda_1 - \mu - \beta_0 - \beta_2 & \beta_2 \\ \gamma_0 & \gamma_1 & -\lambda_2 - \mu - \gamma_0 - \gamma_1 \end{pmatrix} \text{ and}$$

$$B_1 = \begin{pmatrix} -\lambda_0 - \alpha_1 - \alpha_2 & \alpha_1 & \alpha_2 \\ \beta_0 & -\lambda_1 - \beta_0 - \beta_2 & \beta_2 \\ \gamma_0 & \gamma_1 & -\lambda_2 - \gamma_0 - \gamma_1 \end{pmatrix}.$$

Note that the levels correspond to the number in the system and phase corresponds to the state of the environment CTMC $\{Z(t), t \geq 0\}$. In this example, the phases and levels have a true meaning unlike the previous example. In fact, historically these types of queues were first studied as QBDs and hence some of that terminology remain. Further, such types of time-varying arrival processes are called Markov-modulated Poisson processes.

Although the arrival process is indeed a piecewise constant nonhomogeneous Poisson process (NPP), since the rates change randomly (as opposed to deterministically), the literature on NPP usually does not consider this. In practice, typically the $X(t)$ changes much more frequently than $Z(t)$, and while $Z(t)$ is a constant, the queue does behave like an $M/M/1$ queue. Also, unlike the previous example, this CTMC does resemble a birth and death process. ∎

Problem 23

Customers arrive at a single server facility according to $PP(\lambda)$. An arriving customer, independent of everything else, belongs to class-1 with probability α and class-2 with probability $\beta = 1 - \alpha$. The service time of class i (for $i = 1, 2$) customers are IID $\exp(\mu_i)$ such that $\mu_1 \neq \mu_2$. The customers form a single line and are served according to first come first served (FCFS). Let $X(t)$ be the total number of customers in the system at time t, and $Y(t)$ be the class of the customer in service (with $Y(t) = 0$ if there are no customers in the system). Model the bivariate stochastic process $\{(X(t), Y(t)), t \geq 0\}$ as a QBD by obtaining A_0, A_1, A_2, B_0, B_1, and B_2.

Solution

The bivariate CTMC $\{(X(t), Y(t)), t \geq 0\}$ has a state space

$$S = \{(0,0), (1,1), (1,2), (2,1), (2,2), (3,1), (3,2), \ldots\}.$$

By writing down the Q matrix of the CTMC and considering sets of two states as a "level" with two "phases," note that Q has a QBD structure with $\ell = 3, m = 2$

$$A_0 = \begin{pmatrix} \lambda & 0 \\ 0 & \lambda \end{pmatrix}$$

$$B_0 = \begin{pmatrix} 0 & 0 \\ \lambda & 0 \\ 0 & \lambda \end{pmatrix}$$

$$A_2 = \begin{pmatrix} \alpha\mu_1 & \beta\mu_1 \\ \alpha\mu_2 & \beta\mu_2 \end{pmatrix}$$

$$B_2 = \begin{pmatrix} 0 & \alpha\mu_1 & \beta\mu_1 \\ 0 & \alpha\mu_2 & \beta\mu_2 \end{pmatrix}$$

$$A_1 = \begin{pmatrix} -\lambda - \mu_1 & 0 \\ 0 & -\lambda - \mu_2 \end{pmatrix} \quad \text{and}$$

$$B_1 = \begin{pmatrix} -\lambda & \alpha\lambda & \beta\lambda \\ \mu_1 & -\lambda - \mu_1 & 0 \\ \mu_2 & 0 & -\lambda - \mu_2 \end{pmatrix}. \qquad \blacksquare$$

3.2.2 Matrix Geometric Method for QBD Analysis

Matrix geometric method (MGM) is widely used for exact analysis of frequently encountered type of queueing systems that can be modeled as QBDs. It is worthwhile to point out that similar techniques can be used in other contexts that are not QBDs, which are called the more generic matrix analytic methods. However, the MGM can only be applied if the system can be decomposed into two parts: the initial and the repetitive portions which are typically found in all QBDs. To perform MGM-based analysis, we assume that the CTMC with generator Q is irreducible. Further, we assume that $A_0 + A_1 + A_2$ is an irreducible infinitesimal generator with stationary probability π (a $1 \times m$ row vector) such that $\pi(A_0 + A_1 + A_2) = \bar{0}'$ and $\pi\bar{1} = 1$, where for $i = 0, 1$, $\bar{i}$ is an $m \times 1$ column vector of i. The meaning of π is very subtle; given any level greater than zero, the elements of π denote the probabilities of being at the respective phase. If one were to draw the rate diagram and perform arc cuts between levels i and $i+1$ for any $i \geq 0$, then the net flow from left to right is $\pi A_0 \bar{1}$ and likewise the net flow from right to left is $\pi A_2 \bar{1}$ (note that the transitions in A_1 correspond to those within a level; however, A_0 are those that go up a level and A_2 those that go down a level). For the system to be stable (much like the $M/M/1$ queue result), the net flow to the right ought to be smaller than the net flow to the left. Thus, once π is obtained, the condition that the CTMC is stable can be stated as

$$\pi A_2 \bar{1} > \pi A_0 \bar{1}.$$

If the CTMC is stable based on this condition, the steady-state probabilities exist; let us call them p as usual. We write p as $[p_0 \ p_1 \ p_2 \ \ldots]$, where p_0 is an $1 \times \ell$ row vector and $p_1, p_2, \ldots$, are $1 \times m$ row vectors. Since $pQ = 0$, we have the following set of equations:

$$p_0 B_1 + p_1 B_2 = 0$$

$$p_0 B_0 + p_1 A_1 + p_2 A_2 = 0$$

$$p_1 A_0 + p_2 A_1 + p_3 A_2 = 0$$

$$p_2 A_0 + p_3 A_1 + p_4 A_2 = 0$$

$$p_3 A_0 + p_4 A_1 + p_5 A_2 = 0$$

$$\vdots \quad \vdots \quad \vdots$$

for which we try the solution for all $i \geq 1$

$$p_i = p_1 R^{i-1}.$$

This is known as the matrix geometric solution due to the matrix geometric relation between the stationary probabilities. Clearly, for that to be a solution, we need $A_0 + R A_1 + R^2 A_2 = 0$, where R is an unknown. In fact, the crux of the MGM is in computing the R that satisfies $A_0 + R A_1 + R^2 A_2 = 0$. The matrix R is known as the auxiliary matrix. Then, once R is known, we can obtain p_0 in terms of p_1 so that it satisfies both $p_0 B_1 + p_1 B_2 = 0$ and $p_0 B_0 + p_1 A_1 + p_1 R A_2 = 0$. Note that the result in $\ell + m$ sets of equations with ℓ unknowns for p_0 and m unknowns for p_1. However, as with any CTMC, the $\ell + m$ equations are not linearly independent. Thus, we would have to drop one of them and use the normalizing condition that the elements of p sum to one. For this, it is convenient to write down all the p_i terms are in terms of p_1. The normalizing condition can be written as

$$p_0 \underline{1} + (p_1 + p_2 + p_3 + \cdots) \overline{1} = 1,$$

$$p_0 \underline{1} + (p_1 + p_1 R + p_1 R^2 + \cdots) \overline{1} = 1,$$

$$p_0 \underline{1} + p_1 (I + R + R^2 + \cdots) \overline{1} = 1,$$

$$p_0 \underline{1} + p_1 (I - R)^{-1} \overline{1} = 1$$

provided that all eigenvalues of R are in the open interval between -1 and 1. This is also sometimes written as the spectral radius of R should be less than 1. An intuitive way to think about that result is that if x is an eigenvector and k an eigenvalue, then $Rx = kx$ and $R^i x = k^i x$. Thus, the sum $(I + R + R^2 + \cdots) x$ can be written as $(1 + k + k^2 + \cdots) x$ which converges provided $|k| < 1$. Thus, the sum $(I + R + R^2 + \cdots)$ converges to $I - R$ if $|k| < 1$. Since the spectral radius of R is the largest $|k|$, by ensuring it is less than 1, all eigenvalues are between -1 and 1.

Summarizing, we obtain the steady-state probabilities $p = [p_0 \ p_1 \ p_2 \ \cdots]$ by solving for p_0 and p_1 in the following set of equations (after dropping one of the $\ell + m$ equations from the first two sets):

$$p_0 B_1 + p_1 B_2 = 0$$

$$p_0 B_0 + p_1 A_1 + p_1 R A_2 = 0$$

$$p_0 \underline{1} + p_1 (I - R)^{-1} \bar{1} = 1$$

where
$\underline{1}$ is an $\ell \times 1$ column vector
R is the minimal nonnegative solution to the equation

$$R^2 A_2 + R A_1 + A_0 = 0.$$

Then, using $p_i = p_1 R^{i-1}$ for all $i \geq 2$, we obtain p_2, p_3, p_4, etc. Sometimes R can be obtained analytically in closed form by solving $A_0 + R A_1 + R^2 A_2 = 0$, especially when A_0, A_1, and A_2 are simple. Otherwise, R can be obtained numerically, especially recursively using $A_0 + R A_1 + R^2 A_2 = 0$. We will not delve into solving for R using $A_0 + R A_1 + R^2 A_2 = 0$ and assume that is something that can be done. Although it is worthwhile to point out that there have been several articles in the literature that focus on efficiently computing R recursively, we would like to reiterate that once R is computed, p can be obtained which could be used to determine steady-state performance measures such as queue length and waiting times, among others. Next, we illustrate the MGM technique for a specific example.

3.2.3 Example: Variable Service Rate Queues in Computer Systems

Resource sharing, especially the processor, is fairly common in computer systems that run a variety of applications. From a single application's point of view, the processing rate (say in terms of number of bytes per second) can vary with time during the course of processing its tasks. At this time, we will describe the system in an abstract fashion; however, at the end of this section, there are some concrete examples. Consider a single server queue whose service capacity varies over time. That is, the speed with which it serves a customer is determined by an external environment process. In particular, we assume that the server speed changes according to a CTMC that is independent of the arrival process and service requirements of the customer. Each customer brings a certain random amount of work; however, the rate at which this work is completed is time-varying. It is important to clarify that during the service of a request, the processing rate of the server could change. Other than that, the queue is a standard system. We assume that the customers in the queue are served in an FCFS manner. For this model, we are interested in obtaining the distribution of the steady-state number in system.

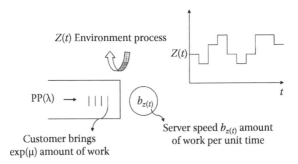

FIGURE 3.4
Schematic representation. (From Mahabhashyam, S., and Gautam, N., *Queueing Syst. Theory Appl.*, 51(1–2), 89, 2005. With permission.)

The system is represented schematically in Figure 3.4. Customers arrive into the single server queue according to a Poisson process with mean rate λ customers per unit time. Each arriving customer brings a certain amount of work distributed exponentially with mean $1/\mu$ bytes. Let $X(t)$ be the number of customers in queue at time t. Let $Z(t)$ be the state of the environment process which governs the server speed at time t such that $\{Z(t), t \geq 0\}$ is an irreducible finite-state CTMC with m states and infinitesimal generator matrix $Q_z = [q_{i,j}]$ for $i \in \{1, 2, \ldots, m\}$ and $j \in \{1, 2, \ldots, m\}$ (we use Q_z to distinguish it from Q that is reserved for the MGM analysis). When the state of the environment process $Z(t) = i$, the service speed available is b_i bytes per unit time. That is, the server can process b_i bytes per unit time. Let θ_i be the instantaneous service completion rate when the environment is in state i. Typically, $\theta_{Z(t)} = \mu b_{Z(t)}$. One can show that the bivariate stochastic process $\{(Z(t), X(t)), t \geq 0\}$ is a CTMC. Next, we use MGM to obtain the steady-state probabilities of this CTMC since none of the other techniques introduced earlier would be feasible.

The two-dimensional CTMC $\{(Z(t), X(t)), t \geq 0\}$ is a QBD process on state space $\{(1, 0), (2, 0), \ldots, (m, 0), (1, 1), (2, 1), \ldots, (m, 1), (1, 2), (2, 2), \ldots, (m, 2), \ldots\}$ with infinitesimal generator matrix Q of form

$$
Q = \begin{pmatrix}
B_1 & B_0 & 0 & 0 & 0 & 0 & \cdots \\
B_2 & A_1 & A_0 & 0 & 0 & 0 & \cdots \\
0 & A_2 & A_1 & A_0 & 0 & 0 & \cdots \\
0 & 0 & A_2 & A_1 & A_0 & 0 & \cdots \\
0 & 0 & 0 & A_2 & A_1 & A_0 & \cdots \\
\vdots & \vdots & \vdots & \vdots & \vdots & \vdots & \ddots
\end{pmatrix}
$$

where

$$A_0 = B_0 = \begin{pmatrix} \lambda & 0 & 0 & \cdots & 0 \\ 0 & \lambda & 0 & \cdots & 0 \\ 0 & 0 & \lambda & \cdots & 0 \\ \vdots & \vdots & \vdots & \vdots & \vdots \\ 0 & 0 & \cdots & 0 & \lambda \end{pmatrix}$$

$$A_2 = B_2 = \begin{pmatrix} \theta_1 & 0 & 0 & \cdots & 0 \\ 0 & \theta_2 & 0 & \cdots & 0 \\ 0 & 0 & \theta_3 & \cdots & 0 \\ \vdots & \vdots & \vdots & \vdots & \vdots \\ 0 & 0 & \cdots & 0 & \theta_m \end{pmatrix}$$

$$A_1 = \begin{pmatrix} s(1) & q_{1,2} & q_{1,3} & \cdots & q_{1,m} \\ q_{2,1} & s(2) & q_{2,3} & \cdots & q_{2,m} \\ q_{3,1} & q_{3,2} & s(3) & q_{3,4} & \cdots \\ \vdots & \vdots & \vdots & \vdots & \vdots \\ q_{m,1} & q_{m,2} & \cdots & q_{m,m-1} & s(m) \end{pmatrix}$$

such that for $i = 1, \ldots, m$, $s(i) = q_{i,i} - \lambda - \theta_i$ and

$$B_1 = \begin{pmatrix} u(1) & q_{1,2} & q_{1,3} & \cdots & q_{1,m} \\ q_{2,1} & u(2) & q_{2,3} & \cdots & q_{2,m} \\ q_{3,1} & q_{3,2} & u(3) & q_{3,4} & \cdots \\ \vdots & \vdots & \vdots & \vdots & \vdots \\ q_{m,1} & q_{m,2} & \cdots & q_{m,m} & u(m) \end{pmatrix}$$

such that for $i = 1, \ldots, m$, $u(i) = q_{i,i} - \lambda$. Recall that $q_{i,j}$ values in these matrices are from Q_z. Note that A_0, A_1, and A_2 are square matrices of size $m \times m$. In addition, B_0, B_1, and B_2 are also of sizes $m \times m$; in essence, the ℓ value corresponding to the B_i matrices for $i = 0, 1, 2$ equals m. The zeros in the Q matrix are also of size $m \times m$.

In summary, $\{(Z(t), X(t)), t \geq 0\}$ is a level-independent infinite-level QBD process. Next, we use MGM to obtain the steady-state probabilities. Since $Q_z = A_0 + A_1 + A_2$ and $\{Z(t), t \geq 0\}$ is an irreducible CTMC, we satisfy the requirement that $A_0 + A_1 + A_2$ is an irreducible. Let π be the stationary probability for the CTMC $\{Z(t), t \geq 0\}$. The $1 \times m$ row vector $\pi = [\pi_1 \ \ldots \ \pi_m]$ can be obtained by solving $\pi(A_0 + A_1 + A_2) = [0 \ 0 \ \ldots \ 0]$ and $\pi \bar{1} = 1$, where $\bar{1}$ is an $m \times 1$ column vector. Since the condition for the CTMC to be stable is

$$\pi A_2 \bar{1} > \pi A_0 \bar{1},$$

we have the necessary and sufficient condition for stability as

$$\mu \sum_{i=1}^{m} \pi_i b_i > \lambda.$$

Note that this condition implies that the average arrival rate must be smaller than the average service rate for stability. However, an interesting observation is that the average service time experienced by a customer is not the reciprocal of the time-averaged service rate (described earlier).

Having described the stability condition, assuming it is met, the next step is to obtain the steady-state probabilities p of the QBD process with rate matrix Q. As described in the MGM analysis, we write p as $[p_0 \ p_1 \ p_2 \ \dots]$, where $p_0, p_1, p_2, \dots$, are $1 \times m$ row vectors. We obtain the steady-state probabilities $p = [p_0 \ p_1 \ p_2 \ \dots]$ by solving for p_0 and p_1 in the following set of equations:

$$p_0 B_1 + p_1 B_2 = 0$$

$$p_0 B_0 + p_1 A_1 + p_1 R A_2 = 0$$

$$p_0 \bar{1} + p_1 (I - R)^{-1} \bar{1} = 1$$

where R is the minimal nonnegative solution to the equation

$$R^2 A_2 + R A_1 + A_0 = 0.$$

Then, using $p_i = p_1 R^{i-1}$ for all $i \geq 2$, we obtain p_2, p_3, p_4, etc. In this case, R can only be obtained numerically using a recursive computation once Q_z, λ, and θ_i for $i = 1, \dots, m$ are known.

Once p_0 and p_1 are obtained, the average number of customers in the system (L) is

$$L = \sum_{i=1}^{\infty} i p_i \bar{1} = \sum_{i=1}^{\infty} i p_1 R^{i-1} \bar{1} = p_1 (I - R)^{-2} \bar{1}.$$

Therefore, the expected waiting (including service) time of a job in the system can be calculated using Little's law as follows:

$$W = \lambda^{-1} p_1 (I - R)^{-2} \bar{1} \tag{3.7}$$

where W is the average sojourn time in the system for an arbitrary arrival in steady state.

Next, to illustrate the previous results, we present a numerical example on analyzing multimedia traffic at web servers. Web servers often transmit multimedia information that can typically be classified into two types: bandwidth sensitive and loss sensitive. Bandwidth-sensitive traffic usually requires information to be streamed at a certain bandwidth; however, when the available capacity is reached, requests for bandwidth-sensitive information are dropped (or lost). On the other hand, loss-sensitive traffic uses up available capacity not used by the bandwidth-sensitive traffic. Hence, it is called elastic traffic. Bandwidth-sensitive traffic is usually audio or video, whereas loss-sensitive traffic is usually data. In fact, data traffic (loss-sensitive) requests are always accepted; however, their transmission rates can vary depending on the available capacity.

From the data traffic standpoint, this system is similar to a single server with variable service rate. Consider a channel with finite capacity through which this bandwidth-sensitive and loss-sensitive traffic are transmitted. Analyzing the bandwidth-sensitive traffic is straightforward as it can be analyzed separately considering the loss-sensitive traffic has no effect on its dynamics. However, the queueing process for the loss-sensitive traffic is extremely dependent on the bandwidth-sensitive traffic. Hence, we analyze the queueing process of the elastic traffic (loss-sensitive) such that the available processor capacity varies with time and governed by an environment process, which essentially is the dynamics of the bandwidth-sensitive traffic. To illustrate this analysis, we present the following numerical example.

Problem 24

Consider a web server that streams video traffic at different bandwidths. This is very common in websites that broadcast sports over the Internet. The users are given an option to select one of the two bandwidths offered depending on their connection speed. Let us denote the two bandwidths by $r_1 = 0.265$ Mbps (low bandwidth) and $r_2 = 0.350$ Mbps (high bandwidth). Let the processing capacity of the web server be $C = 0.650$ Mbps. The arrival rates of requests for the two bandwidths (low and high, respectively) are exponentially distributed with parameters $\lambda_1 = 1$ per second and $\lambda_2 = 2$ per second. The service rates are exponentially distributed with parameters $\mu_1 = 2$ per second and $\mu_2 = 3$ per second for the two bandwidths, respectively. Note that service rate corresponds to the holding time that the streaming request stays connected streaming traffic at its bandwidth. Besides the streaming traffic, there is also elastic traffic, which is usually data. The arrival rate and file size of elastic traffic are exponentially distributed with parameter $\lambda = 3$ per second and $\mu = 8$ per MB (note the unit of the file size parameter). The elastic traffic uses whatever remaining capacity (out of C) the processor has. Compute mean number of elastic traffic requests in the system in steady state as well as the steady-state response time they experience.

Solution

Let the state of the environment be a two-dimensional vector denoting the number of low- and high-bandwidth streaming requests in the system at time t. Note that the low- and high-bandwidth requirements are $r_1 = 0.265$ Mbps and $r_2 = 0.350$ Mbps, respectively, with total capacity $C = 0.650$ Mbps. Hence, the possible states for the environment are $(0,0)$, $(1,0)$, $(2,0)$, $(0,1)$, $(1,1)$, where the first tuple represents the number of ongoing low-bandwidth requests and the second one represents the number of ongoing high-bandwidth requests. Without loss of generality, we map the states $(0,0)$, $(1,0)$, $(2,0)$, $(0,1)$, and $(1,1)$ to 1, 2, 3, 4, and 5, respectively. Therefore, the environment process $\{Z(t), t \geq 0\}$ is a CTMC on state space $\{1,2,3,4,5\}$. The corresponding available bandwidths (in Mbps) for the elastic traffic in those five states are $b_1 = 0.650$, $b_2 = 0.385$, $b_3 = 0.120$, $b_4 = 0.300$, and $b_5 = 0.035$. Further, the infinitesimal generator matrix Q_z for the irreducible CTMC $\{Z(t), t \geq 0\}$ is given by

$$
Q_z = \begin{bmatrix}
-\lambda_1 - \lambda_2 & \lambda_1 & 0 & \lambda_2 & 0 \\
\mu_1 & -\mu_1 - \lambda_1 - \lambda_2 & \lambda_1 & 0 & \lambda_2 \\
0 & 2\mu_1 & -2\mu_1 & 0 & 0 \\
\mu_2 & 0 & 0 & -\mu_2 - \lambda_1 & \lambda_1 \\
0 & \mu_2 & 0 & \mu_1 & -\mu_1 - \mu_2
\end{bmatrix}
$$

$$
= \begin{bmatrix}
-3 & 1 & 0 & 2 & 0 \\
2 & -5 & 1 & 0 & 2 \\
0 & 4 & -4 & 0 & 0 \\
3 & 0 & 0 & -4 & 1 \\
0 & 3 & 0 & 2 & -5
\end{bmatrix}.
$$

Now, consider the elastic traffic. Elastic traffic requests arrive into a single server queue according to a Poisson process with mean rate $\lambda = 3$ per second. Each arriving request brings a certain amount of work distributed exponentially with mean $1/\mu = 0.125$ Mb that is processed at varying rates, depending on the available capacity left over by the streaming traffic. In particular, if the environment process is in state i, the request (if any) in process is served at rate $\theta_i = \mu b_i$ for $i = 1, 2, 3, 4, 5$. Let $X(t)$ be the number of elastic requests in queue at time t. Let $Z(t)$ be the state of the environment process, which governs the server speed at time t and $\{Z(t), t \geq 0\}$ is an irreducible finite-state CTMC with $m = 5$ states and infinitesimal generator matrix $Q_z = [q_{i,j}]$ for $i \in \{1, 2, \ldots, m\}$ and $j \in \{1, 2, \ldots, m\}$ described earlier. When the state of the environment process $Z(t) = i$, the service speed available is b_i bytes per unit time, which is also given earlier for $i = 1, \ldots, 5$. Clearly, the bivariate stochastic process $\{(Z(t), X(t)), t \geq 0\}$ is a two-dimensional CTMC on state space $\{(1,0), (2,0), \ldots, (5,0), (1,1), (2,1), \ldots, (5,1), (1,2), (2,2), \ldots, (5,2), \ldots\}$

with infinitesimal generator matrix Q of form

$$
Q = \begin{pmatrix}
B_1 & B_0 & 0 & 0 & 0 & 0 & \cdots \\
B_2 & A_1 & A_0 & 0 & 0 & 0 & \cdots \\
0 & A_2 & A_1 & A_0 & 0 & 0 & \cdots \\
0 & 0 & A_2 & A_1 & A_0 & 0 & \cdots \\
0 & 0 & 0 & A_2 & A_1 & A_0 & \cdots \\
\vdots & \vdots & \vdots & \vdots & \vdots & \vdots & \ddots
\end{pmatrix}
$$

where

$$
A_0 = B_0 = \begin{pmatrix}
\lambda & 0 & 0 & 0 & 0 \\
0 & \lambda & 0 & 0 & 0 \\
0 & 0 & \lambda & 0 & 0 \\
0 & 0 & 0 & \lambda & 0 \\
0 & 0 & 0 & 0 & \lambda
\end{pmatrix}, \quad
A_2 = B_2 = \begin{pmatrix}
\theta_1 & 0 & 0 & 0 & 0 \\
0 & \theta_2 & 0 & 0 & 0 \\
0 & 0 & \theta_3 & 0 & 0 \\
0 & 0 & 0 & \theta_4 & 0 \\
0 & 0 & 0 & 0 & \theta_5
\end{pmatrix},
$$

$$
A_1 = \begin{pmatrix}
-\lambda_1 - \lambda_2 - \lambda - \theta_1 & \lambda_1 & 0 \\
\mu_1 & -\mu_1 - \lambda_1 - \lambda_2 - \lambda - \theta_2 & \lambda_1 \\
0 & 2\mu_1 & -2\mu_1 - \lambda - \theta_3 \\
\mu_2 & 0 & 0 \\
0 & \mu_2 & 0
\end{pmatrix}
$$

$$
\begin{pmatrix}
\lambda_2 & 0 \\
0 & \lambda_2 \\
0 & 0 \\
-\mu_2 - \lambda_1 - \lambda - \theta_4 & \lambda_1 \\
\mu_1 & -\mu_1 - \mu_2 - \lambda - \theta_5
\end{pmatrix}
$$

where $\lambda = 3$, $\lambda_1 = 1$, $\lambda_2 = 2$, $\mu_1 = 2$, $\mu_2 = 3$, $\theta_1 = 5.20$, $\theta_2 = 3.08$, $\theta_3 = 0.96$, $\theta_4 = 2.40$, and $\theta_5 = 0.28$ (all in per second), and

$$
B_1 = \begin{pmatrix}
-\lambda_1 - \lambda_2 - \lambda & \lambda_1 & 0 \\
\mu_1 & -\mu_1 - \lambda_1 - \lambda_2 - \lambda & \lambda_1 \\
0 & 2\mu_1 & -2\mu_1 - \lambda \\
\mu_2 & 0 & 0 \\
0 & \mu_2 & 0
\end{pmatrix}
$$

$$
\begin{pmatrix}
\lambda_2 & 0 \\
0 & \lambda_2 \\
0 & 0 \\
-\mu_2 - \lambda_1 - \lambda & \lambda_1 \\
\mu_1 & -\mu_1 - \mu_2 - \lambda
\end{pmatrix}.
$$

Now, to obtain the steady-state probabilities for the bivariate CTMC $\{(Z(t), X(t)), t \geq 0\}$, we use MGM for the analysis. For that, we begin by letting π be the stationary probability for the CTMC $\{Z(t), t \geq 0\}$. By solving $\pi Q_z = [0\ 0\ 0\ 0\ 0]$ and $\pi \bar{1} = 1$, we get $\pi = [0.3810\ 0.1905\ 0.0476\ 0.2540\ 0.1270]$. Next, we check if the condition for stability

$$\mu \sum_{i=1}^{5} \pi_i b_i > \lambda$$

is satisfied. Since

$$\mu \sum_{i=1}^{5} \pi_i b_i = 3.2585$$

and $\lambda = 3$, clearly the stability condition is satisfied.

The next step is to describe the steady-state probabilities p of the QBD process with rate matrix Q. As described in the MGM analysis, we write p as $[p_0\ p_1\ p_2\ \ldots]$, where $p_0, p_1, p_2, \ldots$ are 1×5 row vectors. We first obtain R as the minimal nonnegative solution to the equation

$$R^2 A_2 + R A_1 + A_0 = 0.$$

For this, we initialize R to be a 5×5 matrix of zeros and iterate writing $R = -(R^2 A_2 + A_0) A_1^{-1}$ so that the RHS is the old R and the LHS is the new one. This iteration continues until R converges. The methodology to obtain R as the minimal nonnegative solution to $R^2 A_2 + R A_1 + A_0 = 0$ is a greatly researched topic and the reader is encouraged to investigate that as a poor choice of initial R (such as an identity matrix) using the previous iteration would result in a wrong R. A wise choice would be to select R so that the spectral radius is less than 1 (for the identity matrix it is one). Also, the reader must be warned that for large matrices, the convergence of R is not fast. For this, we get R as

$$R = \begin{bmatrix} 0.4551 & 0.0845 & 0.0131 & 0.1456 & 0.0400 \\ 0.2526 & 0.4187 & 0.0600 & 0.1273 & 0.1210 \\ 0.2585 & 0.2917 & 0.4400 & 0.1291 & 0.0907 \\ 0.2959 & 0.0905 & 0.0145 & 0.4774 & 0.0830 \\ 0.2958 & 0.2373 & 0.0363 & 0.2367 & 0.4574 \end{bmatrix}.$$

Then, by solving for p_0 and p_1 in the following set of equations:

$$p_0 B_1 + p_1 B_2 = 0$$

$$p_0 B_0 + p_1 A_1 + p_1 R A_2 = 0$$

$$p_0 \bar{1} + p_1 (I - R)^{-1} \bar{1} = 1$$

we get $p_0 = [0.0345\ 0.0115\ 0.0020\ 0.0168\ 0.0052]$ and $p_1 = [0.0256\ 0.0111\ 0.0024\ 0.0160\ 0.0067]$. Note that there are 11 equations and 10 unknowns in the previous set. However, the 10 equations corresponding to $p_0 B_1 + p_1 B_2 = 0$ and $p_0 B_0 + p_1 A_1 + p_1 R A_2 = 0$ are linearly independent. Hence, the best approach is to use 9 of those 10 equations and use the normalizing equation $p_0 \bar{1} + p_1 (I - R)^{-1} \bar{1} = 1$ as the 10th equation. The average number of customers in the system is $L = p_1 (I - R)^{-2} \bar{1} = 14.4064$ and the expected response time for a job in steady state is $W = \lambda^{-1} L = 4.8021$ s. ∎

3.3 Finite-State Markov Chains

We have seen so far only CTMCs that have a nice structure, such as (a) a multidimensional birth and death process, (b) a repeating block structure where one can use MGMs, or (c) the ability to obtain closed-form expressions such as in Chapter 1. In this section, we will consider CTMCs that do not have a nice structure and we would have to solve for the steady-state probability by direct matrix manipulations. Typically, this is computationally intensive and would not be appropriate for very large matrices that have to be solved repeatedly (such as to obtain the boundaries of control regions optimally). However, software packages such as MATLAB®, Mathematica, and Maple have extremely good inbuilt functions to perform matrix manipulations efficiently. Next, we provide some techniques to further speed up the matrix manipulations for finite-state CTMCs.

3.3.1 Efficiently Computing Steady-State Probabilities

We begin by considering approaches to solve for the steady-state probabilities of finite-state irreducible CTMCs using software packages. As an example, consider a CTMC on state space $S = \{1, 2, 3, 4\}$ with infinitesimal generator matrix

$$Q = \begin{bmatrix} -4 & 1 & 2 & 1 \\ 1 & -3 & 2 & 0 \\ 2 & 2 & -6 & 2 \\ 3 & 0 & 0 & -3 \end{bmatrix}. \tag{3.8}$$

The steady-state probabilities $p = [p_1 \ p_2 \ p_3 \ p_4]$ can be obtained by solving for $pQ = [0 \ 0 \ 0 \ 0]$ and $p_1 + p_2 + p_3 + p_4 = 1$. This example can be solved in many ways; however, we present a few generic techniques that can be used for any Q matrix.

3.3.1.1 Eigenvalues and Eigenvectors

Note that one of the eigenvalues of Q is 0 and p is in fact a left eigenvector. Therefore, one approach to solve for p is to compute the left eigenvalues of Q and pick the one corresponding to eigenvalue of 0. Since eigenvectors are not unique, each software package uses its own way to represent eigenvectors. However, since the sum of the p_i values must be 1, a quick way to obtain p is to take the appropriate left eigenvector and divide each element by the sum of the elements of that eigenvector. It is also worthwhile to point out that some software packages only compute right eigenvectors, so one would have to use the transpose of Q to obtain the left eigenvectors.

Problem 25

Obtain the left eigenvector of the Q matrix in Equation 3.8 corresponding to eigenvalue 0 and normalize it so that it adds to 1 to obtain p.

Solution

The left eigenvectors of Q are $[-0.6528 \ -0.4663 \ -0.3730 \ -0.4663]$, $[-0.5 \ 0.5 \ -0.5 \ 0.5]$, $[-0.0000 \ 0.4082 \ -0.8165 \ 0.4082]$, and $[0.4126 \ -0.7220 \ -0.2063 \ 0.5157]$. They correspond to eigenvalues 0, -6, -7, and -3, respectively. The left eigenvector corresponding to eigenvalue of 0 is $[-0.6528 \ -0.4663 \ -0.3730 \ -0.4663]$. Normalizing by dividing each element by the sum of the elements of the previous eigenvector, we get

$$p = [0.3333 \ 0.2381 \ 0.1905 \ 0.2381]. \qquad \blacksquare$$

It is a good idea to verify this result by performing $pQ = 0$. A word of caution is that obtaining eigenvalues and eigenvectors is extremely time-consuming. However, most software packages have efficient routines to solve for them, even for sparse matrices.

3.3.1.2 Direct Computation

The most common approach to solve for $pQ = 0$ and $\sum_{i \in S} p_i = 1$, where Q is an $m \times m$ infinitesimal generator matrix, is to compute it directly. There are $m + 1$ equations (m equations for $pQ = 0$ and 1 for $\sum p_i = 1$) and m unknowns. Since Q has linearly dependent columns (because the rows of Q add to 1), drop one of the columns of Q and use the normalizing equation $\sum p_i = 1$ instead. We illustrate this with an example.

Problem 26

For the Q matrix in Equation 3.8, write down p in matrix form using Q so that it can be used to solve for p.

Solution

Since the steady-state probabilities $p = [p_1 \ p_2 \ p_3 \ p_4]$ can be obtained by solving for $pQ = [0 \ 0 \ 0 \ 0]$ and $p_1 + p_2 + p_3 + p_4 = 1$, we have

$$[p_1 \ p_2 \ p_3 \ p_4] \begin{bmatrix} -4 & 1 & 2 & 1 \\ 1 & -3 & 2 & 0 \\ 2 & 2 & -6 & 2 \\ 3 & 0 & 0 & -3 \end{bmatrix} = [0\ 0\ 0\ 0]$$

and

$$[p_1 \ p_2 \ p_3 \ p_4] \begin{bmatrix} 1 \\ 1 \\ 1 \\ 1 \end{bmatrix} = 1.$$

By knocking off the last column of $pQ = 0$ and replacing it with the matrix equation corresponding to $\sum p_i = 1$, we get

$$[p_1 \ p_2 \ p_3 \ p_4] \begin{bmatrix} -4 & 1 & 2 & 1 \\ 1 & -3 & 2 & 1 \\ 2 & 2 & -6 & 1 \\ 3 & 0 & 0 & 1 \end{bmatrix} = [0\ 0\ 0\ 1]$$

and that can be used to get p as

$$[p_1 \ p_2 \ p_3 \ p_4] = [0\ 0\ 0\ 1] \begin{bmatrix} -4 & 1 & 2 & 1 \\ 1 & -3 & 2 & 1 \\ 2 & 2 & -6 & 1 \\ 3 & 0 & 0 & 1 \end{bmatrix}^{-1}$$

$$= [0.3333 \ 0.2381 \ 0.1905 \ 0.2381]. \qquad \blacksquare$$

It is a good idea to verify this result by performing $pQ = 0$. A word of caution is that obtaining inverses could be time-consuming and some software packages prefer to solve $xA = b$ using $x = b/A$ as opposed to $x = bA^{-1}$.

3.3.1.3 Using Transient Analysis

When the CTMC is irreducible with not a very large number of states, one of the commonly used procedures is to just use transient analysis taking the limit. The attractiveness of this approach is that it takes one line of code to write as opposed to several lines for the other techniques. Be warned that this method is neither accurate nor computationally efficient. Based on transient analysis of finite-state irreducible CTMCs where Q is an $m \times m$ infinitesimal generator matrix, we know that

$$p = [1/m\ 1/m\ \ldots 1/m] \lim_{t \to \infty} \exp(Qt)$$

where

$$\exp(Qt) = I + \sum_{j=1}^{\infty} \frac{(Qt)^j}{j!}.$$

Therefore, by using a large enough value of t, one could obtain p using the previous equations (the reason for it being a one-line code is that in most mathematical software packages, exponential of a matrix is a built-in function, namely, in MATLAB, the command is expm(Q*t)). We show this via an example.

Problem 27

For the Q matrix in Equation 3.8, obtain p as the limit of a transient analysis.

Solution

Since the steady-state probabilities $p = [p_1\ p_2\ p_3\ p_4]$ can be obtained by

$$p = [1/4\ 1/4\ 1/4\ 1/4] \lim_{t \to \infty} \exp(Qt),$$

we substitute $t = 100,000$ and

$$Q = \begin{bmatrix} -4 & 1 & 2 & 1 \\ 1 & -3 & 2 & 0 \\ 2 & 2 & -6 & 2 \\ 3 & 0 & 0 & -3 \end{bmatrix}$$

to approximate p as $[1/4 \ 1/4 \ 1/4 \ 1/4] \exp(Qt)$. For that, we get

$$[p_1 \ p_2 \ p_3 \ p_4] = [0.3333 \ 0.2381 \ 0.1905 \ 0.2381]. \qquad \blacksquare$$

It is a good idea to verify this result by performing $pQ = 0$. A word of caution is that the choice of t should be large enough for p to be accurate. This is sometimes hard to check if the CTMC is quite large. Therefore, this method would not be recommended unless m is really small, such as less than 10.

3.3.1.4 Finite-State Approximation

A natural question to ask is that if there is a CTMC with infinite states for which an analytical solution is impossible, how does one obtain steady-state performance measures? The most intuitive approximation under such circumstances is to truncate the state space to some finite amount and do the analysis as though it is a finite-state CTMC. There are no rules of thumb to determine where the cutoff needs to be made; it depends on the parameters of the Q matrix but can be selected using a trial and error procedure. Next, we present an example where the state space of an infinite-state CTMC is truncated and the performance measures are obtained repeatedly to determine the optimal control actions. However, we first describe the setting as it is a timely topic, then the control policy, and finally compute the optimal control in various states.

3.3.2 Example: Energy Conservation in Data Centers

Data centers are facilities that comprise hundreds to thousands of computer systems, servers, and associated components, such as telecommunications and storage systems. The main purpose of a data center is to run applications that handle the core business and operational data of one or more organizations. Practically every single organization uses the services of a data center (in-house or outsourced, usually the latter) for their day-to-day operations. In particular, the data centers acquire, analyze, process, store, retrieve, and disseminate different types of information for their clients. Large data centers are frequently referred to as Internet hotels because of the way the thousands of servers are arranged in the facility in racks along aisles.

Data centers consume a phenomenal amount of energy and emit a staggering amount of greenhouse gases. We realize that laptops and desktops themselves generate quite a bit of heat, and those of us that have been to a room with servers know how hot that can get. Now, imagine a data center with thousands of servers. Since so many servers are packed into a small area, data centers have energy densities 10 times greater than that of commercial office buildings, and their energy use is doubling every 4 years (as of year 2006). In a study conducted in January 2006, almost half the fortune 500 IT executives identified power and cooling as problems in their data centers. A study identified that $100,000$ ft^2 data center would cost about \$44 million per year just to power the servers and \$18 million per year to power the cooling infrastructure.

One of the biggest concerns for data centers is to find a way to significantly reduce the energy consumed. Although there are several strategic initiatives to design *green* data centers, by controlling their operations energy consumption in data centers can be significantly conserved. For example, instead of running one application per server, collect a set of applications and run them on multiple servers. If the load for a collection of applications is low, then one could turn off a few servers. For example, if we have 8 applications $a_1, a_2, \ldots, a_8$ and 10 servers $s_1, s_2, \ldots, s_{10}$, then a possible assignment is as follows: the set of applications $\{a_1, a_3, a_6\}$ are assigned to each of servers s_3, s_5, s_7, and s_{10}; applications $\{a_2, a_5, a_7, a_8\}$ are assigned to each of servers s_1, s_2, s_8, and s_9; and application $\{a_4\}$ is assigned to each of servers s_4 and s_6. Then, if the load for the set of applications $\{a_1, a_3, a_6\}$ is low, medium, or high, then one could perhaps turn off two, one, or zero servers, respectively, from the set of servers assigned to this application $\{s_3, s_5, s_7, s_{10}\}$.

Two of the difficulties for turning off servers are that (a) turning servers on and off frequently causes wear and tear and reduces their lifetime in addition to spending energy for powering on and off; and (b) it takes few minutes for a server to be powered on, and therefore any unforeseen spikes in load would cause degradation in service to the clients. To address concern (a), in the model that we develop there would be a cost to boot a server (also popularly known as switching cost in the queueing control literature). Further, a strategy for concern (b) is to perform what is known as dynamic voltage/frequency scaling. In essence, the speed at which the server processes information (which is related to the CPU frequency) can be reduced by scaling down the voltage dynamically. By doing so, the CPU not only consumes lesser energy (which is proportional to the cube of the frequency), but also can switch instantaneously to a higher voltage when a spike occurs in the load. However, even at the lowest frequency (similar to hibernation on a laptop), the server consumes energy and therefore the best option is still to turn servers off.

Next, we describe a simple model to develop a strategy for controlling the processing speed of the servers as well as powering servers on and off. We assume that applications are already assigned to servers and consider a single collection of applications, all of which are loaded on K identical servers. At any time, a subset of the K servers are on (the rest are off) and all on servers run at the same frequency. There is a single queue for the set of K servers and requests arrive according to a Poisson process with mean rate λ. The number of bytes to process for each request is assumed to be IID with an exponential distribution. There are ℓ possible frequencies to run the servers. Therefore, we assume that at any time, the service times are exponentially distributed with mean rate μ_1, or μ_2, or ..., μ_ℓ. Without loss of generality, we assume that

$$\mu_1 < \mu_2 < \cdots < \mu_\ell.$$

At every arrival and departure, an action is taken to determine the number of "on" servers and their service rate. The number of requests at any time in the system can go up to infinite. However, for stability, we assume that $\lambda < K\mu_\ell$ (if all the servers are on and running at full speed, they system must be stable). Note that the system is a "piecewise" $M/M/s$ queue where the number of servers (i.e., s such that $s \leq K$) and service rate remain a constant between events (entering arrivals or departure).

The objective is to determine the optimal control action in each state (i.e., how many servers should be "on" and at what rate?) so that the long-run average cost per unit time is minimized. There are three types of costs: power cost, booting cost, and holding cost. We explain them next. At an arbitrary time in steady state, let there be s servers running and processing jobs at rate μ_i (for an arbitrary s such that $1 \leq s \leq K$ and any $\mu_i \in \{\mu_1, \mu_2, \ldots, \mu_\ell\}$). The power cost (in dollars) incurred per unit time is $s\left(C_0 + C\mu_i^3\right)$ for operating the s servers at rate μ_i. Note that the power cost is a cubic equation in terms of the request processing rates of the servers and linear in terms of the number of servers. A booting cost of B dollars is incurred every time a server is powered on. There is no cost for turning a server off and there is no cost for switching the service rate. Also, service rates can be switched almost instantaneously and that is realistic. However, we make an assumption for modeling convenience that servers can also be powered on and off instantaneously. Finally, a holding cost of h dollars is incurred per unit time for every customer (i.e., request) in the system. Actually, the holding cost does not have to be linear and can be modeled based on expectations of users that send requests.

Similar to the optimal control problems described earlier in this chapter, this problem can also be formulated as an SMDP to determine the optimal policy in various states. However, as described earlier, the objective of this book is not on control of queues but rather on performance analysis; hence, we simply state the structure of the optimal policy without deriving it. Note that there are two actions to be performed in each state, determining the number of servers to be on and also their speed. Since there is a cost to power up a server, it would not make sense to have a pure threshold policy on the number of servers for the following reason. For example, a server was switched on upon an arrival, then if the policy is threshold, when a service is completed (and no new arrivals occur), the server must be switched off and this process would continue as customers arrive and depart. To avoid frequently paying a switching cost (booting cost to be precise), a hysteretic policy is considered to be more optimal than threshold. In that case, there

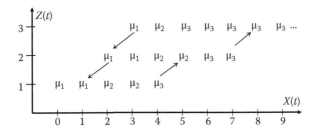

FIGURE 3.5
Schematic for hysteretic policy.

are two thresholds (let us call them θ_1 and θ_2 such that $\theta_1 \le \theta_2$), and the hysteretic policy suggests that if the queue length is larger than θ_2, then switch on a server, but do not switch it off until the queue length goes below θ_1. However, for the frequencies, the optimal policy is threshold type.

Next, we illustrate the hysteretic policy for the number of servers and threshold policy for the service rate using Figure 3.5. Let $X(t)$ denote the number of customers in the system at time t and let $Z(t)$ be the number of servers on at time t. For the purpose of illustration, let there be three servers (i.e., $K=3$) and the number of service rates possible is three (i.e., $\ell=3$ and hence three rates μ_1, μ_2, and μ_3). We represent the state of the system at time t using the two-tuple $(X(t), Z(t))$. Since there are two actions in each state, the actions depicted in Figure 3.5 for each state need some explanation. When the system is empty and one server is running (this corresponds to state $(0,1)$ in the figure), the server runs at rate μ_1. Now, if a new customer arrives in this state (which is the only possible event), the action based on the policy is to continue with one server and run the server in this new state $(1,1)$ at rate μ_1. In state $(1,1)$ if an arrival occurs, the policy is to continue with one server; however, the server will run at rate μ_2 in the new state $(2,1)$. Whereas, if a service is completed in state $(1,1)$, the policy is to stick with one server and run the server at rate μ_1 in the new state $(0,1)$. As long as there are less than or equal to four requests in the system, only one server will run at rate μ_i, which will depend on the number in the system (i.e., μ_1 for 0 or 1 in the system, μ_2 for 2 or 3 in the system, and μ_3 for 4 in the system).

Now, in state $(4,1)$, if an arrival occurs, from the policy in Figure 3.5, a new server is added to the system as the first action. For the second action, we observe the new state after arrival as $(5,2)$, where the action is to run both servers at rate μ_2 (note that prior to this in state $(4,1)$, the single server was running at μ_3 and that is slowed down). Note that in state $(5,2)$, if a service is completed, we do not immediately go back to 1 server but wait till the number of customers goes below 2. In other words, once there are two servers running and the number of customers are between 2 and 7, the action with respect to number of servers is to do nothing (i.e., no addition or subtraction). The action in terms of service rates of both servers is to use μ_1 for 2 or 3 in the system, μ_2 for 4 or 5 in the system, and μ_3 for 6 or 7 in the system. In state (2,2), if a service is completed, the action is to turn off a server (of course the natural choice is to select the server, which completed the service). Also, in the new state (1,1), the single server would run at rate μ_1. Likewise, if an arrival occurs in state (7,2), then the action is to power on a new server and in the new state (8,3), all three servers would run at μ_3. As long as there are three or more customers in the system, all the three servers would continue running at rates μ_1 for 3 in the system, μ_2 for 4 in the system, and μ_3 for 5 or more in the system. When the number in the system reaches 3 with three servers running and one service completes, one of the servers are turned off and the remaining two servers in the new state (2,2) run at rate μ_1. All this is depicted in the policy schematic in Figure 3.5. The policy is also described in Table 3.2.

Although the optimal policy has a structure as described earlier (hysteretic for number of servers and threshold for service rates), it is not clear what the exact optimal policy is. For that, we need to search across all such policies and determine the one that results in the minimal long-run average cost per unit time. For that, we explain how to compute the long-run average cost per unit time for a given policy. For any given policy, we can describe a CTMC $\{(X(t), Z(t)), t \geq 0\}$ with state space S. As an example, consider the policy in Table 3.2 for a system with $K = 3$ and $\ell = 3$. The rate diagram of the corresponding CTMC is depicted in Figure 3.6. Assume that for this CTMC, we can obtain the steady-state probabilities $p_{i,j}$ for all $(i,j) \in S$. Then, the long-run average cost per unit time is

$$B\lambda(p_{4,1} + p_{7,2}) + \sum_{(i,j) \in S} j(C_0 + C\mu(i,j)^3)p_{i,j} + \sum_{(i,j) \in S} h i p_{i,j}.$$

This expression deserves an explanation. Whenever the system is in state (4,1) or (7,2) and an arrival occurs, a booting cost of B dollars is incurred for powering on a server. Hence, the average booting cost incurred per unit time is $B\lambda(p_{4,1} + p_{7,2})$. The notation $\mu(i,j)$ denotes the server speed in state (i,j) according to given policy. For example, $\mu(3,1) = \mu_2$, $\mu(6,2) = \mu_3$, and $\mu(3,3) = \mu_1$. Therefore, the second expression states that if the system is in state (i,j) (that happens with probability $p_{i,j}$), a power cost of $j(C_0 + C\mu(i,j)^3)$ is incurred per unit time for running j servers at speed $\mu(i,j)$. In a similar manner, the last expression is the holding cost per unit time, which can be computed as the holding cost in state (i,j) is hi per unit time times the probability of being in state (i,j) summed over all $(i,j) \in S$. Likewise, the long-run average cost per unit time for each policy can be computed and the best policy can be selected. To navigate through policies and quickly compute $p_{i,j}$ for all $(i,j) \in S$ for such CTMCs that do not have a nice structure, we resort to the numerical techniques mentioned earlier in this section. Although not necessary, one technique is to truncate the state space to create a finite-state

TABLE 3.2

Hysteretic Policy in Tabular Form

Current State $(X(t), Z(t))$	Event	Server Action	New State	Rate Action
(0,1)	Arrival	Do nothing	(1,1)	μ_1
(1,1)	Arrival	Do nothing	(2,1)	μ_2
	Departure	Do nothing	(0,1)	μ_1
(2,1)	Arrival	Do nothing	(3,1)	μ_2
	Departure	Do nothing	(1,1)	μ_1
(3,1)	Arrival	Do nothing	(4,1)	μ_3
	Departure	Do nothing	(2,1)	μ_2
(4,1)	Arrival	Add 1 server	(5,2)	μ_2
	Departure	Do nothing	(3,1)	μ_2
(2,2)	Arrival	Do nothing	(3,2)	μ_1
	Departure	Remove 1	(1,1)	μ_1
(3,2)	Arrival	Do nothing	(4,2)	μ_2
	Departure	Do nothing	(2,2)	μ_1
(4,2)	Arrival	Do nothing	(5,2)	μ_2
	Departure	Do nothing	(3,2)	μ_1
(5,2)	Arrival	Do nothing	(6,2)	μ_3
	Departure	Do nothing	(4,2)	μ_2
(6,2)	Arrival	Do nothing	(7,2)	μ_3
	Departure	Do nothing	(5,2)	μ_2
(7,2)	Arrival	Add 1 server	(8,3)	μ_3
	Departure	Do nothing	(6,2)	μ_3
(3,3)	Arrival	Do nothing	(4,3)	μ_2
	Departure	Remove 1	(2,2)	μ_1
(4,3)	Arrival	Do nothing	(5,3)	μ_3
	Departure	Do nothing	(3,3)	μ_1
(5,3)	Arrival	Do nothing	(6,3)	μ_3
	Departure	Do nothing	(4,3)	μ_2
(6,3)	Arrival	Do nothing	(7,3)	μ_3
	Departure	Do nothing	(5,3)	μ_3
(7,3)	Arrival	Do nothing	(8,3)	μ_3
	Departure	Do nothing	(6,3)	μ_3
(8,3)	Arrival	Do nothing	(9,3)	μ_3
	Departure	Do nothing	(7,3)	μ_3
(9,3)	Arrival	Do nothing	(10,3)	μ_3
	Departure	Do nothing	(8,3)	μ_3
$\vdots$	Arrival	Do nothing	$\vdots$	μ_3
$\vdots$	Departure	Do nothing	$\vdots$	μ_3

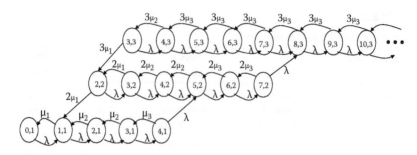

FIGURE 3.6
Rate diagram corresponding to hysteretic policy.

CTMC. Next, we present a small numerical example to completely solve and obtain the optimal policy.

For that, we require the following notation. For $i=1,2,\ldots,K$, let l_i be the smallest number of customers for which there would be i servers running and u_i be the largest number of customers for which there would be i servers running. In other words, if there are l_i customers in the system with i servers running and a service is completed, one server would be turned off. Likewise, a server would be powered on if there are l_i in the system with i servers running and a new arrival occurs. Both l_1 and u_K can be predetermined, since clearly we need $l_1=0$ and $u_K=\infty$. Further, for $j=2,\ldots,\ell$, we let m_{ij} to be the smallest number of customers in the system when there are i servers running to be processing at rate μ_j. In other words, if there are m_{ij} customers in the system and i servers are running, a service completion would result in switching to service rate μ_{j-1} for all i servers. The objective of the optimization problem is to select l_i, u_i, and m_{ij} values for all $i=1,2,\ldots,K$ and $j=2,\ldots,\ell$ so that the long-run average cost per unit time is minimized. We illustrate this computation next for a small numerical example.

Problem 28

Using the notation described so far in this section, consider a small example of $K=3$ servers and $\ell=2$ different service rates (such that $\mu_1=10$ and $\mu_2=15$). For this system, obtain optimal values of l_2, l_3, u_1, u_2, m_{12}, m_{22}, and m_{32} so that the long-run average cost per unit time is minimized subject to the following constraints:

$$2 \le m_{12} \le u_1,$$
$$2 \le l_2 \le m_{22} \le u_2,$$
$$l_2 \le u_1+1 \le u_2,$$
$$l_2+1 \le l_3 \le m_{32},$$
$$l_3 \le u_2+1.$$

In addition, use the following numerical values: $\lambda = 30$, $C_0 = 6500$, $C = 1$, $B = 5000$, and $h = 1000$.

Solution

We formulate the problem description as a mathematical program where the integer-valued decision variables are l_2, l_3, u_1, u_2, m_{12}, m_{22}, and m_{32}, which need to be obtained optimally. The objective function is to minimize the long-run average cost per unit time subject to satisfying the set of constraints described earlier. It is crucial to point out that not all of these constraints are due to the hysteretic/threshold structure of the optimal policy, although most are. Also, we do not check if the value function corresponding to the SMDP is submodular. The main aim is to illustrate the use of numerical techniques to solve for the stationary probabilities expeditiously and use it in the optimization setting. Since the problem size is small, we use a complete enumeration scheme, searching through the entire space of possible solutions to obtain the optimal one. We describe our procedure stepwise next.

1. For each l_2, l_3, u_1, u_2, m_{12}, m_{22}, and m_{32} that satisfies the following constraints

$$2 \leq m_{12} \leq u_1$$

$$2 \leq l_2 \leq m_{22} \leq u_2$$

$$l_2 \leq u_1 + 1 \leq u_2$$

$$l_2 + 1 \leq l_3 \leq m_{32}$$

$$l_3 \leq u_2 + 1$$

we determine the objective function value by going through the remaining steps.

2. Define $X(t)$ as the number of requests in the system and $Z(t)$ the number of servers powered on at time t. Given l_2, l_3, u_1, u_2, m_{12}, m_{22}, and m_{32}, the CTMC $\{(X(t), Z(t)), t \geq 0\}$ has a state space S given by

$$S = \{(0,1), (1,1), \ldots, (u_1 - 1, 1), (u_1, 1), (l_2, 2), (l_2 + 1, 2), \ldots, (u_2 - 1, 2),$$

$$(u_2, 2), (l_3, 3), (l_3 + 1, 3), \ldots\}.$$

We use $l_1 = 0$ and $u_3 = \infty$ for the remainder of this exercise. The infinitesimal generator Q can be obtained for all $(i, j) \in S$ and $(k, l) \in S$ using $q_{(i,j),(k,l)}$, which is the rate of going from state (i, j)

to (k, l) and is given by

$$
q_{(i,j),(k,l)} = \begin{cases}
\lambda & \text{if } k = i+1, l = j, i < u_j \\
\lambda & \text{if } k = i+1, l = j+1, i = u_j \\
j\mu_1 & \text{if } k = i-1, l = j, i < m_{j2}, i > l_j \\
j\mu_1 & \text{if } k = i-1, l = j-1, i < m_{j2}, i = l_j \\
j\mu_2 & \text{if } k = i-1, l = j, i \geq m_{j2}, i > l_j \\
j\mu_2 & \text{if } k = i-1, l = j-1, i \geq m_{j2}, i = l_j \\
-\lambda - j\mu_1 & \text{if } k = i, l = j, 0 < i < m_{j2} \\
-\lambda - j\mu_2 & \text{if } k = i, l = j, 0 < i \geq m_{j2} \\
-\lambda & \text{if } k = i, l = j, i = 0, i < m_{j2} \\
-\lambda & \text{if } k = i, l = j, i = 0, i \geq m_{j2} \\
0 & \text{otherwise.}
\end{cases}
$$

The rate diagram for a special case of $l_2 = 2$, $l_3 = 4$, $u_1 = 6$, $u_2 = 15$, $m_{12} = 2$, $m_{22} = 5$, and $m_{32} = 7$ is depicted in Figure 3.7. We will see later that this special case is indeed the optimal solution.

3. Given l_2, l_3, u_1, u_2, m_{12}, m_{22}, and m_{32}, as well as the corresponding CTMC $\{(X(t), Z(t)), t \geq 0\}$ as described in the previous step, next we obtain the steady-state probabilities $p_{i,j}$ for all $(i, j) \in S$ such that

$$
p_{i,j} = \lim_{t \to \infty} P\{X(t) = i, Z(t) = j\}.
$$

Although there are many ways to numerically obtain $p_{i,j}$ for all $(i, j) \in S$ (especially by writing all $p_{i,j}$ values in terms of $p_{u_2+1,3}$), we use a finite state approximation by truncating the state space. In particular, we pick a large M such that M is much larger than m_{32}. Since $\lambda = 30$ and $3\mu_2 = 45$, the terms in the state space beyond

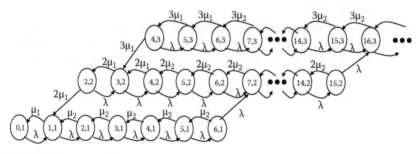

FIGURE 3.7
Rate diagram corresponding to optimal solution.

$(M, 3)$ would contribute negligibly to the objective function or for that matter to the $p_{i,j}$ values. Now, we define the truncated state space $\overline{S}$ as $\overline{S} = \{(0, 1), (1, 1), \ldots, (u_1 - 1, 1), (u_1, 1), (l_2, 2), (l_2 + 1, 2), \ldots, (u_2 - 1, 2), (u_2, 2), (l_3, 3), (l_3 + 1, 3), \ldots, (M, 3)\}$.

The truncated infinitesimal generator matrix $\overline{Q}$ is such that for all $(i, j) \in \overline{S}$ and for all $(k, l) \in \overline{S}$, $\overline{q}_{(i,j),(k,l)} = q_{(i,j),(k,l)}$ with the exception of $q_{(M,3),(M,3)} = -3\mu_2$. Let $\overline{p}$ be a row vector that can be found by solving for $\overline{p}\overline{Q} = 0$ and $\overline{p}\overline{1} = 1$, where $\overline{1}$ is a column vector of ones. Therefore, for the original CTMC (not the truncated), we use the approximation: for all $(i, j) \in S$

$$
p_{i,j} = \begin{cases} \overline{p}_{i,j} & \text{if } (i, j) \in \overline{S} \\ 0 & \text{otherwise.} \end{cases}
$$

4. Now that we have an expression for $p_{i,j}$ for all $(i, j) \in S$ given $l_2, l_3, u_1, u_2, m_{12}, m_{22}$, and m_{32}, the next step is to obtain the objective function. Let us denote the objective function as $f(l_2, l_3, u_1, u_2, m_{12}, m_{22}, m_{32})$. Using the expression described prior to the problem statement for the long-run average cost per unit time, the objective function is

$$
f(l_2, l_3, u_1, u_2, m_{12}, m_{22}, m_{32})
$$

$$
= B\lambda(p_{u_1,1} + p_{u_2,2}) + \sum_{j=1}^{3} \sum_{i=l_j}^{m_{j2}-1} j\left(C_0 + C\mu_1^3\right) p_{i,j}
$$

$$
+ \sum_{j=1}^{3} \sum_{i=m_{j2}}^{u_j} j\left(C_0 + C\mu_2^3\right) p_{i,j} + \sum_{j=1}^{3} \sum_{i=l_j}^{u_j} h i p_{i,j}.
$$

It is worthwhile to point out that the various $p_{i,j}$ values are themselves functions of $l_2, l_3, u_1, u_2, m_{12}, m_{22}$, and m_{32}. In particular, we write $p_{i,j} = g_{ij}(l_2, l_3, u_1, u_2, m_{12}, m_{22}, m_{32})$ although we do not know $g_{ij}(\cdot)$ explicitly, and this is done purely to write down a mathematical program. Therefore, the relationship between the decision variables and the objective function does not exist in closed form. The mathematical programming formulation to optimally select *integers* $l_2, l_3, u_1, u_2, m_{12}, m_{22}$, and m_{32} is

Minimize $f(l_2, l_3, u_1, u_2, m_{12}, m_{22}, m_{32})$

Subject to

$$2 \leq m_{12} \leq u_1$$

$$2 \leq l_2 \leq m_{22} \leq u_2$$

$$l_2 \leq u_1 + 1 \leq u_2$$

$$l_2 + 1 \leq l_3 \leq m_{32}$$

$$l_3 \leq u_2 + 1$$

$$f(l_2, l_3, u_1, u_2, m_{12}, m_{22}, m_{32}) = B\lambda(p_{u_1,1} + p_{u_2,2})$$

$$+ \sum_{j=1}^{3} \sum_{i=l_j}^{m_{j2}-1} j\left(C_0 + C\mu_1^3\right) p_{i,j}$$

$$+ \sum_{j=1}^{3} \sum_{i=m_{j2}}^{u_j} j\left(C_0 + C\mu_2^3\right) p_{i,j} + \sum_{j=1}^{3} \sum_{i=l_j}^{u_j} h i p_{i,j}$$

$$p_{i,j} = g_{ij}(l_2, l_3, u_1, u_2, m_{12}, m_{22}, m_{32})$$

$$\forall \ (i,j) : j = 1, 2, 3; l_j \leq i \leq u_j.$$

Although there may be several ways to solve the mathematical program, we just use a brute force approach due to the small size. We completely enumerate all possible solutions and select the minimum from that. From the complete set of l_2, l_3, u_1, u_2, m_{12}, m_{22}, and m_{32} values, select the one with the lowest long-run average cost per unit time as the set of optimal values l_2^*, l_3^*, u_1^*, u_2^*, m_{12}^*, m_{22}^*, and m_{32}^*.

For our example with numerical values $\lambda = 30$, $\mu_1 = 10$, $\mu_2 = 15$, $C_0 = 6500$, $C = 1$, $B = 5000$, and $h = 1000$, the optimal solution is $l_2^* = 2$, $l_3^* = 4$, $u_1^* = 6$, $u_2^* = 15$, $m_{12}^* = 2$, $m_{22}^* = 5$, and $m_{32}^* = 7$. The CTMC corresponding to this optimal solution is depicted in Figure 3.7 as the appropriate rate diagram. ■

Reference Notes

The material presented in this chapter is rather unusual in a queueing theory text, and in fact in most universities this material is not part of a graduate level course on queueing. However, there are two topics in this chapter that have received a tremendous amount of attention in the literature: matrix analytical methods and control of queues. There are several excellent books on matrix analytical methods, and one of the pioneering works is by Neuts [85]. Other books include Stewart [99], and Latouche and Ramaswami [73]. The essence of matrix analytical methods is to use numerical and iterative methods for Markov chains, and it is general enough to be used in a variety of settings beyond what is considered in this chapter. We will visit this technique in the next chapter as well to get a full appreciation.

The topic of control of queues is also widespread. However, the literature on performance analysis and control is quite distinct. There is a clear optimization flavor in control of queues and the use of stochastic dynamic programming. A recent book by Stidham [100] meticulously covers the topic of optimization in queues by design and control. The crucial point made in this chapter is that stochastic dynamic programming only provides the structure of the optimal policy (such as threshold, switching curve, and hysteretic). Whereas to obtain the exact optimal policy, one needs to explore the space of solutions that satisfy the given policy. This chapter, especially the first and third parts, explicitly explores that.

Exercises

3.1. Refer to the notation in Problem 17 in Section 3.1.1. First show that $-h(x)$ satisfies conditions (3.2), (3.3), and (3.4). Then, assuming that $V_n(x)$ is nonincreasing and satisfies the conditions (3.2), (3.3), and (3.4), show that $V_n((x - e_i)^+)$ for $i = 1, 2$ also satisfy the conditions (3.2), (3.3), and (3.4).

3.2. A finite two-dimensional birth and death process on a rectangular lattice for some given a and b has a state space given by $S = \{(i,j) : 0 \leq i \leq a, 0 \leq j \leq b\}$, that is, all integer points on the XY plane such that the x points are between 0 and a and the y points are between 0 and b. The rate of going from state (i,j) to $(i+1,j)$ is α_i for $i < a$ and the rate of going from state (i,j) to $(i,j+1)$ is γ_j for $j < b$ such that $(i,j) \in S$. The rates α_i and γ_j are known as birth rates. Likewise, define death rates β_i and δ_j such that they are probabilities of going from (i,j) to $(i-1,j)$ and $(i,j-1)$, respectively, when $i > 0$ and $j > 0$. Show that this generic two-dimensional birth and death process has

Q matrix of the block tridiagonal form (as described before the Servi algorithm in Section 3.1.2) by writing it down in that form.

3.3. Consider Problem 20 in Section 3.1.3. The two-dimensional birth and death process corresponding to the optimal action is described in Figure 3.3. Using the numerical values stated in that problem, compute the steady-state probabilities for that two-dimensional birth and death process using the Servi algorithm. Also, compute the long-run average cost per unit time.

3.4. For an $M^{[X]}/M/1$ batch arrival queue with individual service where batches arrive according to $PP(\lambda)$ and each batch is of size 1, 2, 3, or 4 with probability 0.4, 0.3, 0.2, and 0.1, respectively. If the service rate is μ, model the number of customers in the system as a QBD process. Obtain the condition for stability and the steady-state probabilities using MGM. For the special case of $\lambda = 1$ and $\mu = 2.5$, obtain numerical values for the steady-state probabilities and the average number in the system in steady state.

3.5. Consider a CPU of a computer that processes tasks from a software agent as well as other tasks on the computer in parallel by sharing the computer's processor. The software agent submits tasks according to a Poisson process with parameter λ and each task has $\exp(\mu)$ work (in terms of kilobytes) in it that the CPU has to perform. If the only process running on the CPU is that of the agent, it receives all the CPU speed. However, if there are few other processes running on the CPU, only a fraction of the CPU speed is available depending on how many processes, running at the same time. Model the system as a queue with time-varying service rates according to an external environment process (the other processes that run on the CPU). For this, let the available capacity vary according to a CTMC $\{Z(t),\ t \geq 0\}$ with generator matrix Q_z such that at time t the available processing speed for the agent tasks is $b_{Z(t)}$ (kilobytes per second). There are five possible server speeds, that is, $Z(t)$ takes values 1–5. They are $b_1 = 1$, $b_2 = 2$, $b_3 = 3$, $b_4 = 4$, and $b_5 = 5$. The infinitesimal generator matrix Q_z is a 5×5 matrix given by

$$
Q_z =
\begin{bmatrix}
-6 & 2 & 1 & 2 & 1 \\
1 & -7 & 3 & 2 & 1 \\
3 & 2 & -8 & 2 & 1 \\
2 & 1 & 1 & -5 & 1 \\
3 & 4 & 1 & 2 & -10
\end{bmatrix}.
$$

The mean arrival rate $\lambda = 2.5$ and the mean task size $1/\mu = 1$. Compute the mean response time for jobs posted by the software agent to the CPU. Use a similar framework as Problem 24 in Section 3.2.3.

3.6. Consider CTMCs $\{U(t), t \geq 0\}$ and $\{Y(t), t \geq 0\}$ with respective state spaces $S_U = \{0, 1, 2\}$ and $S_Y = \{0, 1\}$ and infinitesimal generator matrices

$$Q_U = \begin{bmatrix} -\alpha & \alpha & 0 \\ \delta & -\beta - \delta & \beta \\ 0 & \gamma & -\gamma \end{bmatrix}$$

and

$$Q_Y = \begin{bmatrix} -\theta & \theta \\ \nu & -\nu \end{bmatrix}.$$

For $i = 0, 1, 2$, whenever the CTMC $\{U(t), t \geq 0\}$ is in state i, arrivals to a single server queue with exponentially distributed service times and infinite waiting room occur according to a Poisson process with rate λ_i. Likewise, for $j = 0, 1$, whenever the CTMC $\{Y(t), t \geq 0\}$ is in state j, the service rate is μ_j. Wireless channels are sometimes characterized this way where the service rates change depending on signal strength and the arrival rates vary depending on the type of information flow. Model this system as a QBD process by representing the Q matrix in block diagonal form and stating A_0, A_1, A_2, B_0, B_1, and B_2.

3.7. Customers arrive into a queue with infinite waiting room according to $PP(\lambda)$. The service times are exponentially distributed with mean $1/\mu$. There is a single permanent server in the system and a maximum of K temporary servers at any given time. Whenever a customer finishes service, with probability p, the customer becomes a temporary server if there are less than K of them in the system. This is the only way temporary servers are created. Each temporary server acts as a server for an $\exp(\theta)$ time and leaves the system. Peer-to-peer systems are sometimes characterized this way. Model the system as a QBD process. Obtain the average number of customers waiting to begin service in the system in steady state using the following numerical values: $\lambda = 10$, $\mu = 6$, $K = 2$, $p = 0.5$, and $\theta = 1$.

3.8. Grab-a-grub is a tiny outfit at an airport that makes a variety of sandwiches and burritos. Because of space restrictions, at any given time, only sandwiches or burritos are made and there is a setup time involved in switching from making one to the other. However, once a sandwich or burrito is made, it is inspected, carefully packaged, and labeled before it is put on display. It takes an $\exp(\lambda_s)$ time to make a sandwich and an $\exp(\lambda_b)$ time to make a burrito. Grab-a-grub makes N_s sandwiches, then spends $\exp(\theta)$ time setting up to make burritos and makes N_b burritos. After that, another $\exp(\theta)$

time is incurred to switch back to making sandwiches. This cycle continues and after making every sandwich or burrito, it enters a single server queue for inspection, packaging, and labeling, which together takes exp(μ) time per sandwich or burrito. Let $X(t)$ be the number of food products (sandwiches and burritos together) in this queue including any in the process of being inspected, packaged, or labeled. Using a suitable CTMC $\{Z(t), t \geq 0\}$ to characterize the sandwich- and burrito-making process, model $\{(Z(t), X(t)), t \geq 0\}$ as a QBD process.

3.9. Consider an FCFS single server queue where the arrivals are according to $PP(\lambda)$ and service times are IID exp(μ) random variables. Upon arrival, a customer enters the system with probability r^i if there are i other customers in the system. Also, each customer that entered the system has a patience time distributed exponentially with mean $1/\beta$ so that if the patience time elapses before service begins, the customer abandons the queue. Model the system as a one-dimensional birth and death process. Obtain an expression using infinite sums and products for the average number of customers in the system (L) and the average departure rate from the server after service in steady state (throughput). Using those, compute numerical values of L and throughput for $\lambda = 10$, $r = 0.99$, $\mu = 12$, and $\beta = 1$. Now, using a finite-state approximation of the birth and death process, compute L and throughput for the same numerical values.

3.10. Using an appropriate matrix software package, compare the solution times for the three techniques in Section 3.3.1 (i.e., eigenvectors, direct computation, and transient analysis) to numerically solve for the steady-state probabilities. For that, randomly generate a 100×100 Q matrix where it is possible to go from every state to every other state in one step.

4

General Interarrival and/or Service Times: Closed-Form Expressions and Approximations

In Chapters 2 and 3, we studied queues for which both the interarrival times and service times were exponentially distributed. This enabled us to use CTMCs to model and analyze such queues. In this chapter, we study how to analyze queues with other distributions for interarrival times or service times or both. Of course, if the distributions are mixtures of exponentials (such as exponential, Erlang, hyperexponential, hypoexponential, and phase-type), we could still analyze using CTMCs. Otherwise, we would need other techniques and approximations to model and analyze these queues. The objective of this chapter is to present a breadth of methods to analyze queues with general distributions for either interarrival times or service times or both. We present techniques that would result in closed-form algebraic expressions, numerical values, or approximations for various measures of performance such as distributions and moments of queue length as well as sojourn times. We begin by considering discrete time Markov chains to model some queueing systems to obtain closed-form algebraic expressions with the understanding that such Markov models can be widely used for many other settings. We also end the chapter with closed-form algebraic expressions; however, the techniques presented are somewhat specialized for those settings. The middle of the chapter describes techniques that can be used for obtaining performance measures numerically or through approximations.

4.1 Analyzing Queues Using Discrete Time Markov Chains

In this section, we will consider a few queueing systems where we would use discrete time Markov chains for their analysis. In that light, we first describe the fundamentals involved in DTMC analysis and then use it in two systems, the $M/G/1$ queue and the $G/M/1$ queue. Unlike the CTMC models where in most cases all we needed was the solution to the balance equations, here in the DTMC models we need more analysis after the balance equations are solved in order to obtain the required performance metrics. They are fairly specialized and so we will consider them individually for the $M/G/1$ queue and the $G/M/1$ queue. But first we take a look at the steady-state probabilities in a DTMC.

Consider a DTMC $\{Z_n, n \geq 0\}$ where Z_n is the state of the system at the nth observation. Although in many DTMC examples, one considers these observations being made at equally spaced intervals, that is not necessary. In fact, in both analyses of this section, the inter-observation times will not be equal. However, as we will see subsequently, it is crucial to point out that the choices of when to observe the system and what to capture in Z_n (the state) must be made carefully so that the resulting process $\{Z_n, n \geq 0\}$ is indeed a DTMC. Assuming that is done, consider a DTMC with state space S and one-step transition probability matrix P. Denote p_{ij} as an element of P corresponding to row i and column j. By definition

$$p_{ij} = P\{Z_{n+1} = j | Z_n = i\}$$

and p_{ij} is not a function of n, that is, time-homogeneous.

In this chapter we are only interested in irreducible DTMCs, that is, DTMCs where it is possible to go from every state in S to every other state in S in one or more steps. In order to obtain the steady-state probability vector $\pi = [\pi_i]$, we solve for $\pi P = \pi$ and normalize using $\sum_{i \in S} \pi_i = 1$. The set of equations for $\pi P = \pi$ is also called balance equations and can be written for every $i \in S$ and $j \in S$ as

$$\pi_i p_{ii} + \sum_{j \neq i} \pi_j p_{ji} = \pi_i.$$

If we draw the transition diagram, then what is immediately obvious is that since $p_{ii} = 1 - \sum_{j \neq i} p_{ij}$ we have

$$\sum_{j \neq i} \pi_i p_{ij} = \sum_{j \neq i} \pi_j p_{ji}.$$

This means that across each node i, the flow out equals the flow in just like in the CTMCs. Notice that the preceding expression does not include p_{ii}; however, if that is preferred, we could add $\pi_i p_{ii}$ to both sides of the equation to get

$$\sum_{j \in S} \pi_i p_{ij} = \sum_{j \in S} \pi_j p_{ji}.$$

Since the balance equations along with the normalizing conditions are exactly analogous to those of the CTMC (essentially replacing q_{ij} by p_{ij}

and p_i by π_i), the methodologies to solve for them to get the steady-state probabilities are also identical. All the techniques explained in Chapters 2 and 3 can be applied to DTMCs as well and there would be no point in repeating those here. In particular, we will use generating functions in the first example ($M/G/1$ queue) to follow and that is identical to that explained in Chapter 2. However, for the second DTMC ($G/M/1$ queue) we will use a trial approach that can potentially also be used for CTMCs although we do not explicitly address it anywhere in Chapters 2 or 3.

4.1.1 The $M/G/1$ Queue

We begin by describing the $M/G/1$ queue and setting the notation. Consider a single server queue with infinite waiting room. Arrivals to this queue occur according to a Poisson process with mean rate λ arrivals per unit time. The service times are independent and identically distributed random variables with common CDF $G(t)$ such that if S_i is the service time of the ith customer, then

$$G(t) = P\{S_i \leq t\}$$

for all i. Service times are nonnegative and hence $G(t) = 0$ for all $t < 0$. However, there are no other restrictions on the random variables S_i; in fact, they could be discrete, continuous, or a mixture of discrete and continuous. For the sake of notational convenience, we let the mean and variance of service times to be $1/\mu$ and σ^2, respectively, such that for all i,

$$E[S_i] = \frac{1}{\mu}$$

and

$$Var[S_i] = \sigma^2.$$

Since the mean and variance of the service times can easily be derived from the CDF $G(t)$, sometimes μ and σ are not specified while describing the service times.

We have almost everything in place to call the preceding system an $M/G/1$ queue. The only aspect that remains is the service discipline. Although Kendall notation specifies that the default discipline is FCFS and we will derive all our results assuming FCFS, it is worthwhile to discuss the generality of the analysis. For most of the analysis, we only require that if there is an entity in the system, useful work is always performed and at most one entity in the queue can have incomplete service. This deserves some explanation. Firstly, the server cannot be idle when there are customers in the system. That also means that if there are customers waiting, as soon as service completes for a customer, the service for the next customer starts instantaneously. This is an aspect that is frequently overlooked while collecting service time data. The simplest way to fix the problem is to add any time spent between service of customers to the customers' service time. Secondly, since at most one customer can have incomplete service, this precludes disciplines involving preemption or processor sharing. However, schemes such as LCFS without preemption and random order of service can be analyzed.

Having described the setting for the $M/G/1$ queue, we next model and analyze the system. Notice that unless the service times are according to some mixture of exponential distributions, we cannot model the system using a CTMC. In fact, we will see that even for the mixture of exponential case, modeling as a DTMC provides some excellent closed-form algebraic expressions that the CTMC model may fail to provide. With that said, we begin modeling the system. The first question that comes to mind is when to observe the system so that the resulting process is a DTMC. We can immediately rule out observing the system any time in the middle of a service since the remaining service time would now depend on history and Markov property would not be satisfied. Therefore, the only options are to observe at the beginning and/or end of service times. Although it may indeed be possible to model the system as a DTMC by observing both at the beginning and at the end of a service, we will see that it is sufficient if we observed the system at the end of a service. In other words, we will observe the system whenever a customer departure occurs. The next question is that during these departure epochs, the number in the system goes down by one—so should we observe before or after a departure? Although either case would work, we will observe immediately after the departure so that the departing customer is not included in the state.

With that in place, we let X_n be the number of customers in the system immediately after the nth departure. The state space, that is, set of all possible values of X_n, for all n is $\{0, 1, 2, 3, \ldots\}$. For some arbitrary n, let $X_n = i$ such that $i > 0$. We now derive a distribution for X_{n+1}, given $X_n = i$. If $X_n = i$ immediately after the nth departure, we will have one customer at the server and $i - 1$ waiting. When this customer at the service completes service, we observe the system next, that is, the $n+1$st departure. So X_{n+1} would be equal to $i - 1$ plus all the customers that arrived during the service time that just completed. In other words, X_{n+1} would be $i - 1 + j$ with probability a_j, where a_j is the probability that j customers arrive during a service (for $j = 0, 1, 2, \ldots$). Hence, we write mathematically

$$P\{X_{n+1} = i - 1 + j | X_n = i\} = a_j$$

for all $i > 0$ and $j \geq 0$.

The case $X_n = 0$ is a little trickier. When the nth customer departs, the system is empty. The next event is necessarily an arrival. After the arrival occurs, during this customer's service if j arrivals occur, then when this $n+1$st customer departs, there would be j in the system. In other words, if $X_n = 0$, X_{n+1} would be j with probability a_j, where a_j once again is the probability that j customers arrive during a service (for $j = 0, 1, 2, \ldots$). This we write mathematically as

$$P\{X_{n+1} = j | X_n = 0\} = a_j$$

for all $j \geq 0$. This deserves a little explanation as it is a little different from the case where $X_n > 0$ when the time between observations was equal to a service time. Notice that when $X_n = 0$, the next observation is after two events, one arrival and one service, in that order. Clearly, X_{n+1} would be equal to the number of customers that arrive during the second phase, that is, a service. Thus, we are able to use the same notation a_j. Of course, we do not have an expression for a_j and would need to derive it. We will do that after modeling the system as a DTMC.

From the earlier description, to determine X_{n+1} we only need to be given X_n and not the history. Also the probability of transitioning from X_n to X_{n+1} does not depend on n. Therefore, we can model $\{X_n, n \geq 0\}$ as a DTMC with state space $\{0, 1, 2, \ldots\}$ and transition probability matrix

$$P = \begin{bmatrix} a_0 & a_1 & a_2 & a_3 & \cdots \\ a_0 & a_1 & a_2 & a_3 & \cdots \\ 0 & a_0 & a_1 & a_2 & \cdots \\ 0 & 0 & a_0 & a_1 & \cdots \\ 0 & 0 & 0 & a_0 & \cdots \\ \vdots & \vdots & \vdots & \vdots & \vdots \end{bmatrix}$$

which is an example of an upper Hessenberg matrix (which is an upper triangular matrix plus an additional off-diagonal entry of a_0 values). The only parameters that need to be determined are a_j values for all $j \geq 0$ (where a_j is the probability that j arrivals occur during one service completion). To obtain a_j, we condition on the service time being t. Therefore, given the service time equals t, the probability that j arrivals occur during that service time is $e^{-\lambda t}(\lambda t)^j / j!$ because the number of arrivals in time t is according to a Poisson distribution with parameter (λt). By unconditioning on the service time, we get for any $j = 0, 1, 2, \ldots$,

$$a_j = \int_0^\infty e^{-\lambda t} \frac{(\lambda t)^j}{j!} dG(t).$$

A minor technicality is that Riemann integral should be replaced by Lebesgue integral if S_i is a mixture random variable with point masses at discrete values. For the remainder of this chapter, we continue to use

Riemann integral pretending that S_i is purely continuous and unless stated otherwise, it can easily be generalized to the mixture case. With that we have the P matrix completely characterized in terms of the "inputs," namely, λ and $G(t)$.

Since $a_j > 0$ for all finite j, the DTMC $\{X_n, n \geq 0\}$ is irreducible as it is possible to go from every state to every other state in one or more transitions. Next, we obtain the steady-state probabilities for the DTMC and the condition for stability. Assuming that the $M/G/1$ queue is stable (we will later derive the condition for stability), let π_j be the limiting probability that in steady state a departing customer sees j others in the system, that is,

$$\pi_j = \lim_{n \to \infty} P\{X_n = j\}.$$

The limiting distribution $\pi = (\pi_0 \ \pi_1 \dots)$ can be obtained by solving $\pi = \pi P$ and $\sum \pi_j = 1$. To solve the balance equations (i.e., the equations that correspond to $\pi = \pi P$ for each node), we use the generating function approach seen in Chapter 2. The balance equations are

$$\pi_0 = a_0 \pi_0 + a_0 \pi_1$$

$$\pi_1 = a_1 \pi_0 + a_1 \pi_1 + a_0 \pi_2$$

$$\pi_2 = a_2 \pi_0 + a_2 \pi_1 + a_1 \pi_2 + a_0 \pi_3$$

$$\pi_3 = a_3 \pi_0 + a_3 \pi_1 + a_2 \pi_2 + a_1 \pi_3 + a_0 \pi_4$$

$$\vdots \vdots \vdots$$

We multiply the first equation by z^0, the next by z^1, the next by z^2, and so on. Upon summing we get

$$\pi_0 z^0 + \pi_1 z^1 + \pi_2 z^2 + \pi_3 z^3 + \cdots = \pi_0(a_0 z^0 + a_1 z^1 + a_2 z^2 + \cdots)$$

$$+ \pi_1(a_0 z^0 + a_1 z^1 + a_2 z^2 + \cdots)$$

$$+ \pi_2(a_0 z^1 + a_1 z^2 + a_2 z^3 + \cdots)$$

$$+ \pi_3(a_0 z^2 + a_1 z^3 + a_2 z^4 + \cdots) + \cdots$$

which we can rewrite as

$$\pi_0 z^0 + \pi_1 z^1 + \pi_2 z^2 + \pi_3 z^3 + \cdots = \pi_0(a_0 z^0 + a_1 z^1 + a_2 z^2 + \cdots)$$
$$+ \pi_1(a_0 z^0 + a_1 z^1 + a_2 z^2 + \cdots)$$
$$+ \pi_2 z(a_0 z^0 + a_1 z^1 + a_2 z^2 + \cdots)$$
$$+ \pi_3 z^2(a_0 z^0 + a_1 z^1 + a_2 z^2 + \cdots) + \cdots.$$

By collecting common terms we get

$$\pi_0 z^0 + \pi_1 z^1 + \pi_2 z^2 + \pi_3 z^3 + \cdots$$
$$= (a_0 z^0 + a_1 z^1 + a_2 z^2 + \cdots)(\pi_0 + \pi_1 + \pi_2 z + \pi_3 z^2 + \cdots).$$

Since we are going to use generating functions, we can rewrite the preceding equation as

$$\pi_0 z^0 + \pi_1 z^1 + \pi_2 z^2 + \pi_3 z^3 + \cdots = (a_0 z^0 + a_1 z^1 + a_2 z^2 + \cdots)$$
$$\left[\pi_0 + \frac{1}{z}(-\pi_0 + \pi_0 z^0 + \pi_1 z^1 + \pi_2 z^2 + \pi_3 z^3 + \cdots) \right]$$

which using summations becomes

$$\sum_{i=0}^{\infty} \pi_i z^i = \left(\sum_{j=1}^{\infty} a_j z^j \right) \left[\pi_0 + \frac{1}{z} \left(-\pi_0 + \sum_{i=0}^{\infty} \pi_i z^i \right) \right]. \tag{4.1}$$

Clearly, there are two generating functions to consider: one corresponding to the π_i values and the other corresponding to the a_j terms. For that, consider generating functions $\phi(z)$ and $A(z)$ such that

$$\phi(z) = \sum_{i=0}^{\infty} \pi_i z^i,$$

$$A(z) = \sum_{j=0}^{\infty} a_j z^j.$$

Therefore, we can rewrite Equation 4.1 in terms of the two generating functions as

$$\phi(z) = A(z) \left[\pi_0 + \frac{1}{z}(-\pi_0 + \phi(z)) \right].$$

Note that $A(z)$ can be computed based on the inputs λ and $G(t)$. Hence, by rewriting the preceding equation, we get $\phi(z)$ as

$$\phi(z) = \frac{\pi_0 A(z)(1-z)}{A(z) - z}. \tag{4.2}$$

The only unknown on the RHS of Equation 4.2 is π_0. To obtain that, we first need to write down some properties for $A(z)$.

From the definition of $A(z)$, we have $A(1) = 1$ since $a_0 + a_1 + \cdots = 1$. However, to get other properties, we first write $A(z)$ in the simplest possible form in terms of the parameters in the problem definition, namely, λ and $G(t)$. By using the definition of a_j, we get

$$A(z) = \sum_{j=0}^{\infty} a_j z^j = \sum_{j=0}^{\infty} \int_0^{\infty} e^{-\lambda t} z^j \frac{(\lambda t)^j}{j!} dG(t) = \int_0^{\infty} e^{-\lambda t} \left(\sum_{j=0}^{\infty} z^j \frac{(\lambda t)^j}{j!} \right) dG(t)$$

$$= \int_0^{\infty} e^{-\lambda t} \left(\sum_{j=0}^{\infty} \frac{(z\lambda t)^j}{j!} \right) dG(t) = \int_0^{\infty} e^{-\lambda t} e^{z\lambda t} dG(t)$$

$$= \int_0^{\infty} e^{-(1-z)\lambda t} dG(t) = \tilde{G}((1-z)\lambda)$$

where the last expression $\tilde{G}((1-z)\lambda)$ is the LST of $G(t)$ at $(1-z)\lambda$. By definition, if S is a random variable corresponding to the service times, then the LST of $G(t)$ at u is

$$E\left[e^{-uS} \right] = \int_{u=0}^{\infty} e^{-ut} dG(t) = \tilde{G}(u).$$

Also, using the properties of LSTs, $\tilde{G}(0) = 1$, $\tilde{G}'(0) = -E[S] = -1/\mu$, and $\tilde{G}''(0) = E[S^2] = 1/\mu^2 + \sigma^2$. Therefore, from the earlier results and the relation $A(z) = \tilde{G}((1-z)\lambda)$, we get

$$A'(1) = -\lambda \tilde{G}'(0) = \frac{\lambda}{\mu}, \tag{4.3}$$

$$A''(1) = \lambda^2 \tilde{G}''(0) = \frac{\lambda^2}{\mu^2} + \lambda^2 \sigma^2. \tag{4.4}$$

Now we get back to Equation 4.2. To obtain π_0 we first try $\phi(0)$ and that gives $\phi(0) = \pi_0$, which is true but does not help us to get π_0. Next we try

$\phi(1) = 1$. To get $\phi(1)$, we do the following:

$$\phi(1) = \lim_{z \to 1} \frac{\pi_0 A(z)(1-z)}{A(z) - z}$$

$$= \pi_0 A(1) \lim_{z \to 1} \frac{(1-z)}{A(z) - z}.$$

Using $A(1) = 1$ and realizing that the limit is of the type $0/0$, we use L'Hospital's rule to get

$$\phi(1) = \pi_0 \lim_{z \to 1} \frac{(-1)}{A'(z) - 1}$$

$$= \pi_0 \frac{1}{1 - A'(1)}.$$

Using $\phi(1) = 1$ and $A'(1) = \lambda/\mu$ from Equation 4.3, we get

$$\pi_0 = 1 - \frac{\lambda}{\mu}.$$

Clearly, for a meaningful solution to π_0 and thereby $\phi(z)$, we require that $0 < \pi_0 < 1$, and since λ as well as μ are positive, we necessarily require that $\lambda < \mu$. Therefore, the stability condition is

$$\lambda < \mu.$$

This is similar to the stability condition of the $M/M/1$ queue in Problem 7 in Section 2.1.3. It states that on average the arrival rate must be smaller than the service rate for the system to be stable, which is quite intuitive. We use the term ρ as the traffic intensity such that

$$\rho = \frac{\lambda}{\mu}.$$

Therefore, if $\rho < 1$, we can write down the generating function $\phi(z)$ as

$$\phi(z) = \frac{(1-\rho)(1-z)\tilde{G}(\lambda - \lambda z)}{\tilde{G}(\lambda - \lambda z) - z}. \tag{4.5}$$

Since $\phi(z) = \pi_0 + \pi_1 z + \pi_2 z^2 + \pi_3 z^3 + \cdots$, in theory we can obtain $\pi_1, \pi_2, \ldots$ from $\phi(z)$ by either writing down the RHS in Equation 4.5 as a polynomial and matching terms corresponding to $z^1, z^2, \ldots$, or taking the first, second, $\ldots$, derivatives of $\phi(z)$ and letting $z \to 0$. However, it may be possible to obtain certain performance measures without actually obtaining π_j values and that is what we will consider in the next few problems.

Problem 29

Consider a stable $M/G/1$ queue with $PP(\lambda)$ arrivals, mean service time $1/\mu$, and variance of service time σ^2. Compute L, the long-run time-averaged number of entities in the system. Using that obtain W the average sojourn time spent by an entity in the system in steady state.

Solution

Recall that π_j is the long-run fraction of time a departing customer sees j others in the system. It is also known as departure-point steady-state probability. However, to compute L, we need the time-averaged (and not as seen by departing customers) fraction of time spent in state j, which we call p_j. For that, we know from PASTA (Poisson arrivals see time averages) described in Section 1.3.4 that p_j must be equal to the long-run fraction of time an arriving customer sees j others in the system, that is, π_j^*. In other words, $p_j = \pi_j^*$. But we also know from Section 1.3.3 that $\pi_j^* = \pi_j$ for any $G/G/s$ queue and hence $p_j = \pi_j$ for all j.

Using that logic, the average number of customers in the system is

$$L = \sum_{j=0}^{\infty} j p_j = \sum_{j=0}^{\infty} j \pi_j = \phi'(1).$$

Therefore, to get L we need to compute $\phi'(1)$, which involves several steps that we address next. Rewriting Equation 4.5 as

$$\phi(z)\left(\tilde{G}(\lambda - \lambda z) - z\right) = (1 - \rho)(1 - z)\tilde{G}(\lambda - \lambda z)$$

and differentiating both sides, we get

$$\phi'(z)\left(\tilde{G}(\lambda - \lambda z) - z\right) + \phi(z)\left(-\lambda \tilde{G}'(\lambda - \lambda z) - 1\right)$$
$$= -(1 - \rho)\tilde{G}(\lambda - \lambda z) - (1 - \rho)(1 - z)\lambda \tilde{G}'(\lambda - \lambda z).$$

We can rewrite the preceding equation as

$$\phi'(z) =$$
$$\frac{(1 - \rho)\tilde{G}(\lambda - \lambda z) + (1 - \rho)(1 - z)\lambda \tilde{G}'(\lambda - \lambda z) - (1 + \lambda \tilde{G}'(\lambda - \lambda z))\phi(z)}{z - \tilde{G}(\lambda - \lambda z)}.$$

Earlier in this section, we saw that $\tilde{G}(0) = 1$ and $\tilde{G}'(0) = -1/\mu$. Using those (and also realizing $\phi(1) = 1$) to compute $\phi'(1)$ by taking the limit as z approaches one, we get a $0/0$ expression. Therefore, using L'Hospital's rule, we get

$$\phi'(1) = \lim_{z \to 1} \frac{-2(1-\rho)\lambda\tilde{G}'(\lambda-\lambda z) - ((1-\rho)(1-z) - \phi(z))\lambda^2\tilde{G}''(\lambda-\lambda z) - (1+\lambda\tilde{G}'(\lambda-\lambda z))\phi'(z)}{1 + \lambda\tilde{G}'(\lambda-\lambda z)}.$$

Notice that both sides of the preceding equation has $\phi'(z)$. Now by taking the limits as $z \to 1$ using $\tilde{G}(0) = 1$, $\tilde{G}'(0) = -1/\mu$, $\tilde{G}''(0) = 1/\mu^2 + \sigma^2$, and $\phi(1) = 1$, we get

$$\phi'(1) = \rho + \frac{\lambda^2}{2}\frac{(\sigma^2 + 1/\mu^2)}{1 - \rho}$$

where $\rho = \lambda/\mu$. Since $L = \phi'(1)$, we have

$$L = \rho + \frac{\lambda^2}{2}\frac{(\sigma^2 + 1/\mu^2)}{1 - \rho}. \tag{4.6}$$

The preceding equation is known as the Pollaczek–Khintchine equation. Using Little's law, we can write $W = L/\lambda$ where W is the average steady-state sojourn time for an entity in the system as

$$W = \frac{1}{\mu} + \frac{\lambda}{2}\frac{(\sigma^2 + 1/\mu^2)}{1 - \rho}. \qquad \blacksquare$$

Notice that the preceding equation for W as well as Equation 4.6 for L are in terms of λ, μ, and σ only. The entire service time distribution $G(t)$ is not necessary for those expressions. From a practical standpoint, this is very useful because μ and σ can be estimated more robustly compared to $G(t)$.

Next, having discussed the distribution of the queue length, it is quite natural to consider the distribution of the sojourn time in the system that we call waiting time. For this, we require FCFS and it is the first time we truly require FCFS discipline. Up until now, all the results can be derived for any work conserving discipline with a maximum of one customer having incomplete service at any given time. We describe the next result on the sojourn time distribution as a problem, by recognizing that we already know the mean sojourn time from the previous problem.

Problem 30

Let Y be the sojourn time in the system for a customer arriving into a stable $M/G/1$ queue in steady state. If the service is FCFS, then show that the LST of the CDF of Y is

$$E[e^{-sY}] = \frac{(1-\rho)s\tilde{G}(s)}{s - \lambda(1-\tilde{G}(s))}.$$

Solution

As before, X_n denotes the number of customers in the $M/G/1$ queueing system as seen by the nth departing customer. Let B_n be the number of customers that arrive during the nth customer's sojourn, which includes time spent waiting (if any) and time for service. Since the service discipline is FCFS, we have

$$X_n = B_n.$$

That directly implies that for any z

$$E[z^{X_n}] = E[z^{B_n}].$$

However, since $\phi(z)$ is defined as $\phi(z) = \pi_0 + \pi_1 z^1 + \cdots$, we can rewrite it as

$$\phi(z) = \lim_{n\to\infty} \sum_{i=0}^{\infty} P\{X_n = i\}z^i = \lim_{n\to\infty} E[z^{X_n}].$$

Therefore, from the equality $E[z^{X_n}] = E[z^{B_n}]$ and the earlier expression, we have

$$\lim_{n\to\infty} E[z^{B_n}] = \phi(z). \qquad (4.7)$$

Next, we develop an expression for $E[z^{B_n}]$ using an entirely different approach (and it will give an idea of the connection to Y defined in the problem).

Let W_n be the sojourn time (or waiting time) of customer n with CDF $H_n(w)$ such that $H_n(w) = P\{W_n \leq w\}$. We seek to obtain an expression for $H_n(w)$ as $n \to \infty$ as that would be the sojourn time CDF in steady state with $W_n \to Y$ as $n \to \infty$. However, since B_n is indeed the number of customers that arrived during the random time W_n, we can write down $E[z^{B_n}]$

by conditioning on W_n as follows:

$$E[z^{B_n}] = \int_0^\infty E[z^{B_n}|W_n = w]dH_n(w)$$

$$= \int_0^\infty \sum_{i=0}^\infty e^{-\lambda w} \frac{(\lambda w)^i}{i!} z^i dH_n(w)$$

where the last equation uses the fact that the arrivals are $PP(\lambda)$ and the probability of getting i arrivals in time w is $e^{-\lambda w}(\lambda w)^i/i!$. Therefore, we have

$$E\left[z^{B_n}\right] = \int_0^\infty e^{-\lambda w} \left(\sum_{i=0}^\infty \frac{(\lambda w)^i}{i!} z^i \right) dH_n(w)$$

$$= \int_0^\infty e^{-\lambda w} e^{\lambda w z} dH_n(w) = \int_0^\infty e^{-(1-z)\lambda w} dH_n(w)$$

$$= \tilde{H}_n(\lambda(1-z)) \tag{4.8}$$

where $\tilde{H}_n(s)$ is the LST of $H_n(w)$ such that

$$E[e^{-sW_n}] = \tilde{H}_n(s).$$

Taking the limits as $n \to \infty$ of the preceding expression, and using the fact that $W_n \to Y$ as $n \to \infty$, we get

$$E[e^{-sY}] = \tilde{H}(s).$$

By substituting $z = 1 - s/\lambda$ in Equation 4.8 and letting $n \to \infty$, we have

$$E[e^{-sY}] = \tilde{H}(s) = \lim_{n \to \infty} E\left[\left(\frac{1-s}{\lambda} \right)^{B_n} \right].$$

However, from Equation 4.7, we have

$$\lim_{n \to \infty} E\left[\left(\frac{1-s}{\lambda}\right)^{B_n}\right] = \phi\left(\frac{1-s}{\lambda}\right).$$

Now from the previous two equations, we have

$$E[e^{-sY}] = \tilde{H}(s) = \phi\left(\frac{1-s}{\lambda}\right). \qquad (4.9)$$

Using the expression for $\phi(z)$ in Equation 4.5 in terms of z, λ, $\tilde{G}(\cdot)$, and μ, we have by letting $z = 1 - s/\lambda$

$$E[e^{-sY}] = \frac{(1-\rho)s\tilde{G}(s)}{s - \lambda(1 - \tilde{G}(s))}$$

where $\rho = \lambda/\mu$. ∎

Before proceeding, it is worthwhile to verify if the earlier result is accurate for the M/M/1 queue. Consider $G(t) = 1 - e^{-\mu t}$ for $t \geq 0$, that is, the service times are $\exp(\mu)$ random variables. We also know from Problem 7 in Section 2.1.3 that $Y \sim \exp(\mu - \lambda)$ and $E[e^{-sY}] = (\mu - \lambda)/(s + \mu - \lambda)$ when the queue is stable. Therefore, we need to check if the RHS of the previous equation $E[e^{-sY}] = (1 - \rho)s\tilde{G}(s)/[s - \lambda(1 - \tilde{G}(s))]$ is equal to $(\mu - \lambda)/(s + \mu - \lambda)$. Since $\tilde{G}(s) = \mu/(s + \mu)$, by substituting in the RHS of the previous equation, we get $E[e^{-sY}] = (\mu - \lambda)/(s + \mu - \lambda)$.

Reverting to the more general case of $G(t)$ in the M/G/1 queue, notice that the average sojourn time W can be computed from $E[e^{-sY}]$ by taking the derivative with respect to s, multiplying by (-1), and letting $s \to 0$. We leave this as an exercise problem for the reader (see the last section of this chapter). That brings about an interesting comment about Little's law and its extension. Note that for an FCFS M/G/1 queue, we can actually prove Little's law using Equation 4.9. Since $\tilde{H}(s) = \phi(1 - s/\lambda)$, taking the derivative with respect to s on both sides, we get $\tilde{H}'(s) = -(1/\lambda)\phi'(1 - s/\lambda)$. Letting $s \to 0$, we get $\tilde{H}'(0) = -(1/\lambda)\phi'(1)$. But we know that $\tilde{H}'(0) = -W$ and $\phi'(1) = L$ and hence we have $L = \lambda W$. However, notice that $\tilde{H}(s) = \phi(1 - s/\lambda)$ is essentially a relationship between the LST of the sojourn time and the generating function of the queue length, perhaps more can be said in terms of relating higher moments of the respective quantities. We do that next through a problem.

Problem 31

Derive the relationship between higher moments of the steady-state sojourn time against those of the number in the system for an $M/G/1$ queue with FCFS service.

Solution

Let $E[Y^r]$ be the rth moment of the steady-state sojourn time, for $r = 1, 2, 3, \ldots,$ which can be computed as

$$E[Y^r] = \lim_{s \to 0} (-1)^r \frac{d^r \tilde{H}(s)}{ds^r}.$$

Likewise, let $L^{(r)}$ be the rth factorial moment of the steady-state number in the system. Note that for a discrete random variable X on $0, 1, 2, \ldots$, the rth factorial moment is defined as $E[X(X-1)(X-2) \ldots (X - r + 1)]$. Therefore, $L^{(1)}$ is L itself. Then, $L^{(2)}$ can be used to compute the variance of the number in the system as $L^{(2)} + L - L^2$. Likewise, higher moments of the number in the system can be obtained. However, notice that

$$E[L^{(r)}] = \lim_{z \to 1} \frac{d^r \tilde{\phi}(z)}{ds^r}.$$

Making a change of variable $z = 1 - s/\lambda$, we get by taking the rth derivative of $\tilde{H}(s) = \phi(1 - s/\lambda)$ and letting $s \to 0$

$$E[L^{(r)}] = \lambda^r E[Y^r]. \tag{4.10}$$

■

There are many results like the one in the preceding text that can be easily derived for the $M/G/1$ queue that sometimes we do not even mention while talking about the special case $M/M/1$, although methodologically they would require quite different approaches. Having said that, the next result is one that would typically be analyzed using identical methods for $M/G/1$ and $M/M/1$, followed by a curious paradox. That result is presented as a problem.

Problem 32

In a single-server queue, a busy period is defined as a consecutive stretch of time when the server is busy serving customers. A busy period starts when a customer arrives into an empty single-server queue and ends when the system becomes empty for the first time after that. With that definition, obtain the LST of the busy period distribution of an $M/G/1$ queue.

Solution

Let Z be a random variable denoting the busy period initiated by an arrival into an empty queue in steady state. Also, let S be the service time of this customer that just arrived. Remember that we are only going to consider nonpreemptive schemes, although this result would not alter if we considered preemption, as long as it is work conserving. But the proof would have to be altered, hence the assumption. Let N be the number of customers that arrive during the service of this "first" customer, that is, in time S. Of course, if N is zero, then the busy period is S itself. Let us remember this case but for now assume $N > 0$. We keep these N customers aside in what we call *initial pool*. Take one from the initial pool and serve that customer and in the mean time if any new customers arrive serve them one by one till there are no customers in the system except the $N - 1$ in the initial pool. It is critical to realize that the time to serve the first customer in the initial pool and all the customers that came subsequently till the queue did not have any customers that are not part of the initial pool is stochastically equal to a busy period. We call this time Z_1. Next, pick the second customer (if any) from the initial pool and spend a busy period (of length Z_2) serving that customer and all that arrive until the queue only has customers from the initial pool. Repeat the process until there are no customers left in the initial pool. We use this to write down the conditional relation for some $u \geq 0$:

$$E[e^{-uZ} | S = t, N = n] = E[e^{-u(t + Z_1 + Z_2 + \cdots + Z_n)}].$$

But this conditional relation also works when $N = 0$. So from now on, we remove restriction on N and say that the preceding is true for all $N \geq 0$. Unconditioning the earlier equation using $P\{N = n | S = t\} = e^{-\lambda t}(\lambda t)^n / n!$, we get

$$E[e^{-uZ}] = \int_0^\infty \sum_{n=0}^\infty e^{-ut} E[e^{-uZ}]^n e^{-\lambda t} \frac{(\lambda t)^n}{n!} dG(t), \qquad (4.11)$$

since $Z, Z_1, Z_2, \ldots$, are IID random variables. We use the notation $\tilde{F}_Z(u)$ as the LST of the CDF of Z that is defined mathematically as

$$\tilde{F}_Z(u) = E[e^{-uZ}] = \int_0^\infty e^{-uz} dF_Z(z)$$

where $F_Z(z) = P\{Z \leq z\}$. Notice that we do not know $F_Z(z)$ and are trying to obtain it via the LST $\tilde{F}_Z(u)$. Rewriting Equation 4.11 in terms of the LST of the busy period distribution, we get

$$\tilde{F}_Z(u) = \int_0^\infty \sum_{n=0}^\infty e^{-ut} \frac{[\tilde{F}_Z(u)\lambda t]^n}{n!} e^{-\lambda t} dG(t).$$

By summing over, we get

$$\tilde{F}_Z(u) = \int_0^\infty e^{-ut} e^{[\tilde{F}_Z(u)\lambda t]} e^{-\lambda t} dG(t) = \tilde{G}(u + \lambda - \lambda\tilde{F}_Z(u)).$$

Although we have the LST of the busy period distribution as an embedded relation $\tilde{F}_Z(u) = \tilde{G}(u + \lambda - \lambda\tilde{F}_Z(u))$, it is possible to obtain $E[Z] = 1/(\mu - \lambda)$ (see Exercises) as well as higher moments $E[Z^r]$. ∎

Although we do not have a closed-form algebraic expression for the LST of the busy period distribution, we have it as a solution to an equation $\tilde{F}_Z(u) = \tilde{G}(u + \lambda - \lambda\tilde{F}_Z(u))$. However, for the $M/M/1$ queue, we can obtain a closed-form expression. Notice that for the $M/M/1$ as well, the mean busy period is $1/(\mu - \lambda)$. But wait, that looks like the average sojourn time of an $M/M/1$ queue. Is that strange? We state a paradox next as a remark.

Remark 8

For an $M/G/1$ queue with $\sigma > 1/\mu$ since the mean busy period $E[Z] = 1/(\mu - \lambda)$ and the mean sojourn time $W = 1/\mu + \lambda/2(\sigma^2 + 1/\mu^2)/(1 - \rho)$, we get $E[Z] < W$. In other words, the mean busy period is smaller than the mean waiting time when $\sigma > 1/\mu$. However, this appears like a paradox because if you take any busy period, the waiting time of a customer that entered and left during this busy period is always smaller than the busy period itself. But is the expected waiting time greater than the expected busy period? How could that be? ∎

Although we are not providing a rigorous argument to support the remark, it is worthwhile describing the intuition. One of the characteristics of heavy-tailed random variables (i.e., ones with larger standard deviation than mean) is that the realizations are usually small values with occasional large ones. Therefore, when the service times are heavy-tailed, the busy period equals a single service time very often (that is because the interarrival time is usually much larger than the service time). Occasionally, a very large service time would result in a long busy period. That would develop into a cycle alternating between several short busy periods (single service) followed by one long busy period (with several customers). Thus, the average busy period would be fairly short. But typically there may be several customers stuck in the long busy period and, averaging over customers, the sojourn times would end up being long. A simulation might help with the intuition and the reader is encouraged to try one out. Having described this, we wrap up the topic $M/G/1$ queue and move on to its counterpart, the $G/M/1$ queue.

4.1.2 $G/M/1$ Queue

Consider a single server queue with infinite waiting room. Arrivals to this queue occur according to a renewal process with interrenewal time CDF $G(t)$ such that if T_i is the ith interarrival time, then for all $t \geq 0$

$$G(t) = P\{T_i \leq t\}.$$

Assume that $G(0) = 0$ to represent that interarrival times cannot be of length 0, that is, no batch arrivals. However, there are no other restrictions on the random variables; in fact, they could be discrete, continuous, or a mixture of discrete and continuous. Notice that we are using the same letter $G(\cdot)$ for the interarrival time CDF that we used for the service times of the $M/G/1$ queue, a choice that has pros and cons but is commonly done in the literature. However, unlike the $M/G/1$ case, where for L and W we only really need the moments of the service time, here for the $G/M/1$ queue's L and W we need the entire interarrival distribution $G(t)$. Further, the average interarrival time is $1/\lambda$, that is, $E[T_i] = 1/\lambda$ for all i. The service times are independent and identically distributed exponential random variables with mean $1/\mu$. Although the default service discipline is FCFS, most of the results hold for any work conserving discipline, including processor sharing and preemptive schemes (although they were not allowed for $M/G/1$). The memoryless property of the exponential service time distribution primarily causes the extra flexibility.

Having looked at modeling an $M/G/1$ queue (and the title of this section), it should not be surprising that we would use DTMC to model a $G/M/1$ queue. However, the similarities end right there after modeling. The analysis would be significantly different and so would the properties. We first explain the modeling. We can immediately rule out observing the system any time in the middle of two arrivals, since the remaining time for the next arrival would now depend on the history and the Markov property would not be satisfied. Therefore, the only options are to observe at arrival time points. The next question is that during these arrival epochs, the number in the system goes up by one, so should we observe before or after an arrival? Although either case would work, we will observe immediately before the arrival so that the arriving customer is not included in the state.

With that justification in place, we let X_n^* be the number of customers in the $G/M/1$ system just before the nth arrival. Then the stochastic process

$\{X_n^*, n \geq 0\}$ is a DTMC with state space $S = \{0, 1, 2, \ldots\}$, and

$$P = \begin{bmatrix} b_0 & a_0 & 0 & 0 & \cdots \\ b_1 & a_1 & a_0 & 0 & \cdots \\ b_2 & a_2 & a_1 & a_0 & \cdots \\ b_3 & a_3 & a_2 & a_1 & \cdots \\ b_4 & a_4 & a_3 & a_2 & \cdots \\ \vdots & \vdots & \vdots & \vdots & \vdots \end{bmatrix}$$

where a_j is the probability that j departures occur between two consecutive arrivals and b_j is given by

$$b_j = \sum_{i=j+1}^{\infty} a_i \text{ for all } j \geq 0$$

The transition probability matrix P needs some explanation. Consider an arriving customer that sees there are i (for $i = 0, 1, 2, \ldots$) customers already in the system (this corresponds to the ith row of P which is $[b_i \; a_i \; \cdots \; a_0 \; 0 \; 0 \; \cdots]$). Notice that as soon as this customer arrives, there would be $i + 1$ in the system. When the next customer arrives, if there are k departures (among the $i + 1$) between this arrival and the previous one (for $k \leq i$), then this arriving customer sees $i + 1 - k$ in the system. That happens with probability a_k and the new state is $i + 1 - k$. For example, if $i = 4$ (i.e., the current arrival sees 4 in the system) and $k = 2$ with probability a_2 (i.e., two service completions occur before next arrival), then the next arrival sees $i + 1 - k = 3$ in the system upon arrival. Further, a_j for $j = 0, 1, 2, \ldots$, can be computed by conditioning on the interarrival time to be t as

$$a_j = \int_0^{\infty} e^{-\mu t} \frac{(\mu t)^j}{j!} dG(t).$$

Notice that the preceding is derived using exactly the same argument as that in the $M/G/1$ queue (the reader is encouraged to that verify). Also, the preceding equation assumes the interarrival times as being purely continuous. We will continue to treat the interarrival times that way with the understanding that if there were discrete-valued point masses, then the Riemann integral would be replaced by the Lebesgue integral.

The case not considered in the preceding text is when there are actually $k = i + 1$ departures, where i is the number of customers in the system when the previous arrival occurred. Then, when the next customer arrives, there would be no other customer in the system. However, the probability of going from i to 0 in the DTMC is not a_{i+1}. This is because a_{i+1} denotes the probability there are exactly $i+1$ departures during one interarrival time interval. But, the $i+1$ departures would have occurred before the interarrival time period ended and if there were more in the system, perhaps there could have been more departures. Hence, if there were an abundant number of customers in the system, then during the interarrival time interval, there would be $i+1$ or more departures. Hence, the probability of transitioning from state i to 0 in the DTMC is $a_{i+1} + a_{i+2} + a_{i+3} + \cdots$, which we call b_i as defined earlier. Notice that the rows add to 1 in the P matrix and this is a lower Hessenberg matrix.

Having modeled the $G/M/1$ queue as a DTMC, next we analyze the steady-state behavior and derive performance measures. Let π_j^* be the limiting probability that in the long run an arriving customer sees j in the system, that is,

$$\pi_j^* = \lim_{n \to \infty} P\left\{X_n^* = j\right\}.$$

The limiting distribution $\pi^* = \left(\pi_0^* \; \pi_1^* \; \ldots\right)$, if it exists, can be obtained by solving $\pi^* = \pi^* P$ and $\sum_j \pi_j^* = 1$. The balance equations that arise out of solving for $\pi^* = \pi^* P$ are

$$\pi_0^* = b_0 \pi_0^* + b_1 \pi_1^* + b_2 \pi_2^* + b_3 \pi_3^* + \cdots$$

$$\pi_1^* = a_0 \pi_0^* + a_1 \pi_1^* + a_2 \pi_2^* + a_3 \pi_3^* + \cdots$$

$$\pi_2^* = a_0 \pi_1^* + a_1 \pi_2^* + a_2 \pi_3^* + a_3 \pi_4^* + \cdots$$

$$\pi_3^* = a_0 \pi_2^* + a_1 \pi_3^* + a_2 \pi_4^* + a_3 \pi_5^* + \cdots$$

$$\vdots \quad \vdots$$

and we solve it using a technique we have not used before in this text. Since there is a unique solution to the balance equations (if it exists), we try some common forms for the steady state-probabilities. In particular, we try the form $\pi_i^* = (1 - \alpha)\alpha^i$ for $i = 0, 1, 2, \ldots$, where α is to be determined. The justification for that choice is that for the $M/M/1$ system, π_i^* is of that form. The very first equation from the earlier set $\pi_0^* = b_0 \pi_0^* + b_1 \pi_1^* + b_2 \pi_2^* + b_3 \pi_3^* + \cdots$, is a little tricky, but all others are straightforward. Plugging in $\pi_i^* = (1 - \alpha)\alpha^i$ and $b_i = a_{i+1} + a_{i+2} + a_{i+3}$ for $i = 0, 1, 2, \ldots$, we get

$$\pi_0^* = b_0 \pi_0^* + b_1 \pi_1^* + b_2 \pi_2^* + b_3 \pi_3^* + \cdots$$

$$(1 - \alpha) = (a_1 + a_2 + \cdots)(1 - \alpha) + (a_2 + a_3 + \cdots)(1 - \alpha)\alpha$$

$$+ (a_3 + a_4 + \cdots)(1 - \alpha)\alpha^2 + (a_4 + a_5 + \cdots)(1 - \alpha)\alpha^3 + \cdots$$

$$= (1 - \alpha)[(a_1) + (a_2 + \alpha a_2) + (a_3 + \alpha a_3 + \alpha^2 a_3) + \cdots]$$

$$= (1 - \alpha)\left[a_1 \frac{1 - \alpha}{1 - \alpha} + a_2 \frac{1 - \alpha^2}{1 - \alpha} + a_3 \frac{1 - \alpha^3}{1 - \alpha} + \cdots \right]$$

$$= a_0(1 - \alpha^0) + a_1(1 - \alpha^1) + a_2(1 - \alpha^2) + a_3(1 - \alpha^3) + \cdots$$

$$= (a_0 + a_1 + a_2 + a_3 + \cdots) - \sum_{i=0}^{\infty} a_i \alpha^i$$

and since $a_0 + a_1 + a_2 + a_3 + \cdots = 1$, we need α to satisfy

$$\alpha = \sum_{i=0}^{\infty} a_i \alpha^i.$$

Next we check the remaining balance equations by plugging in $\pi_i^* = (1 - \alpha)\alpha^i$ for $i = 0, 1, 2, \ldots$, to convert

$$\pi_1^* = a_0 \pi_0^* + a_1 \pi_1^* + a_2 \pi_2^* + a_3 \pi_3^* + \cdots$$

$$\pi_2^* = a_0 \pi_1^* + a_1 \pi_2^* + a_2 \pi_3^* + a_3 \pi_4^* + \cdots$$

$$\pi_3^* = a_0 \pi_2^* + a_1 \pi_3^* + a_2 \pi_4^* + a_3 \pi_5^* + \cdots$$

$$\vdots \quad \vdots$$

respectively, to

$$(1 - \alpha)\alpha = a_0(1 - \alpha) + a_1(1 - \alpha)\alpha + a_2(1 - \alpha)\alpha^2 + a_3(1 - \alpha)\alpha^3 + \cdots$$

$$(1 - \alpha)\alpha^2 = a_0(1 - \alpha)\alpha + a_1(1 - \alpha)\alpha^2 + a_2(1 - \alpha)\alpha^3 + a_3(1 - \alpha)\alpha^4 + \cdots$$

$$(1 - \alpha)\alpha^3 = a_0(1 - \alpha)\alpha^2 + a_1(1 - \alpha)\alpha^3 + a_2(1 - \alpha)\alpha^4 + a_3(1 - \alpha)\alpha^5 + \cdots$$

$$\vdots \quad \vdots$$

which are all satisfied if α is the solution to $\alpha = \sum_{i=0}^{\infty} a_i \alpha^i$. Let us first write down the condition $\alpha = \sum_{i=0}^{\infty} a_i \alpha^i$ in terms of the variables in the G/M/1 system. Using the expression for a_i, we get

$$\alpha = \sum_{i=0}^{\infty} a_i \alpha^i = \sum_{i=0}^{\infty} \alpha^i \int_0^{\infty} e^{-\mu t} \frac{(\mu t)^i}{i!} dG(t) = \int_0^{\infty} e^{-\mu t} \sum_{i=0}^{\infty} \frac{(\alpha \mu t)^i}{i!} dG(t)$$

$$= \int_0^{\infty} e^{-(1-\alpha)\mu t} dG(t) = \tilde{G}((1-\alpha)\mu) = E\left[e^{-(1-\alpha)\mu T_j} \right]$$

where $\tilde{G}(s)$ is the LST of $G(t)$ at some $s \geq 0$. In summary, the limiting proba-bility π_i^* exists for $i = 0, 1, 2, \ldots$, and is equal to $\pi_i^* = (1-\alpha)\alpha^i$ if there exists a solution to $\alpha = \tilde{G}((1-\alpha)\mu)$ such that $\alpha \in (0, 1)$. Next, we check the condition when $\alpha = \tilde{G}((1-\alpha)\mu)$ has a solution such that $\alpha \in (0, 1)$. As it turns out, that would be the stability condition for the DTMC $\{X_n^*, n \geq 0\}$.

We use a graphical method to describe the condition for stability for the G/M/1 queue, which is the same as the condition for positive recurrence for the DTMC $\{X_n^*, n \geq 0\}$. We write $\tilde{G}((1-\alpha)\mu)$ as $\tilde{G}(\mu - \alpha\mu)$. Note that from the definition, $\tilde{G}(\mu - \alpha\mu) = \int_0^{\infty} e^{-(1-\alpha)\mu t} dG(t)$, where α only appears on the exponent, and $\tilde{G}(\mu - \alpha\mu)$ is a nondecreasing convex function of α. Also, $\tilde{G}(0) = 1$ and hence one solution to $\alpha = \tilde{G}((1-\alpha)\mu)$ is indeed $\alpha = 1$. With these properties of $\tilde{G}(\mu - \alpha\mu)$ in mind, refer to Figure 4.1. We plot $\tilde{G}(\mu - \alpha\mu)$ versus α as well as the 45° line, that is, the function $f(\alpha) = \alpha$. Since $\tilde{G}(\mu - \alpha\mu)$ is nondecreasing and convex it, would intersect the 45° line at two points. If the slope of $\tilde{G}(\mu - \alpha\mu)$ at $\alpha = 1$ is greater than 1, then $\tilde{G}(\mu - \alpha\mu)$ would intersect the 45° line once at some $\alpha \in (0, 1)$. This is depicted in the LHS of Figure 4.1. However, if the slope is less than 1, then the intersection occurs at some $\alpha \geq 1$ depicted in the RHS of Figure 4.1. In fact, if the slope is exactly 1, then the two points on intersection converge to one point and the 45° line just becomes a tangent (we do not show this in Figure 4.1). Therefore,

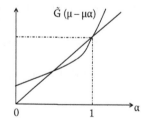 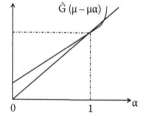

FIGURE 4.1
Two possibilities for $\tilde{G}(\mu - \mu\alpha)$ vs. α.

the condition for stability is that the slope of $\tilde{G}(\mu - \alpha\mu)$ at $\alpha=1$ must be greater than 1.

For this, we let $\alpha=1$ after computing $d\tilde{G}(\mu - \alpha\mu)/d\alpha$ and require that $-\mu\tilde{G}'(0) > 1$. However, $\tilde{G}'(0) = -1/\lambda$ since the first moment or mean interarrival time is $1/\lambda$. Therefore, the condition for stability of the $G/M/1$ queue or the condition for positive recurrence of the irreducible DTMC $\{X_n^*, n \geq 0\}$ is

$$\rho = \frac{\lambda}{\mu} < 1$$

which is the necessary and sufficient condition to find a solution to $\alpha = \tilde{G}((1 - \alpha)\mu)$ such that $\alpha \in (0, 1)$. In summary, if $\rho < 1$, we can show that the probability an arriving customer sees j others in the system in steady state is

$$\pi_j^* = (1 - \alpha)\alpha^j$$

which is the unique solution to the DTMC balance equations, where α is the solution in $(0, 1)$ to

$$\alpha = \tilde{G}(\mu - \mu\alpha).$$

Problem 33

For a stable $G/M/1$ queue with FCFS service policy, using the preceding results derive the distribution for the sojourn time spent by a customer in the system.

Solution

Let Y be the sojourn time experienced by an arbitrary arrival into the system in steady state. This is also referred to as the waiting time or time in the system. Since we know that in steady state the probability an arriving customer sees j in the system is π_j^*, we can obtain the LST of the distribution of Y by conditioning on the number in the system as seen by an arrival. Therefore, we have the LST as

$$E\left[e^{-sY}\right] = \sum_{j=0}^{\infty} E\left[e^{-sY} \,|\, X_\infty^* = j\right]\pi_j^*$$

where X_∞^* is the number in the system as seen by an arriving customer in steady state with $P\{X_\infty^* = j\} = \pi_j^*$. Note that if there are j customers in the system upon arrival, the sojourn time for this customer is the sum of service times of the j customers ahead as well as the customer's own service times. Hence, the conditional sojourn time is the sum of $j + 1$ exponentials with parameter μ (i.e., according to *Erlang*$(j + 1, \mu)$). Thus, we have

$$
E[e^{-sY}] = \sum_{j=0}^{\infty} E\left[e^{-sY} \,|\, X_{\infty}^* = j\right] \pi_j^*
$$

$$
= \sum_{j=0}^{\infty} \left(\frac{\mu}{\mu + s}\right)^{j+1} \pi_j^* = \sum_{j=0}^{\infty} \left(\frac{\mu}{\mu + s}\right)^{j+1} (1 - \alpha)\alpha^j
$$

$$
= \frac{\mu}{\mu + s}(1 - \alpha) \sum_{j=0}^{\infty} \left(\frac{\mu\alpha}{\mu + s}\right)^j = \frac{\mu(1 - \alpha)}{s + \mu(1 - \alpha)}
$$

where the last equation uses the fact that since $0 < \alpha < 1$ and $0 < \mu/(\mu + s) < 1$, the infinite geometric sum converges. Therefore, $Y \sim \exp(\mu(1 - \alpha))$, that is, the sojourn time is exponentially distributed with parameter $\mu(1 - \alpha)$. ∎

Using the preceding results, we can immediately see that the average time in the system (sojourn time or waiting time) is

$$
W = E[Y] = \frac{1}{\mu(1 - \alpha)}.
$$

Then, from Little's law we have the average number of customers in the system as

$$
L = \frac{\lambda}{\mu(1 - \alpha)}.
$$

Although we do have the average number of customers in the system, we do not yet have the distribution of the number in the system in steady state. Notice that so far all the techniques obtained that first, but in this $G/M/1$ queue analysis, it is not particularly straightforward. In fact, unlike the $M/G/1$ queue where we could invoke PASTA property, that would not be possible here since arrivals are not necessarily Poisson. We see next in a problem setting how to compute the distribution of the number in the system in steady state.

Problem 34

Consider a stable $G/M/1$ queue and let $X(t)$ be the number of customers in the system at time t. Define p_j as the probability that there are j in the system in steady state, that is,

$$p_j = \lim_{t \to \infty} P\{X(t) = j\}.$$

Show that p_j can be solved as

$$p_0 = 1 - \rho$$
$$p_j = \rho \pi^*_{j-1} \quad \text{for } j > 0$$

where $\rho = \lambda/\mu$.

Solution

Let A_n be the time of the nth arrival in the system with $A_0 = 0$. The bivariate stochastic process $\{(X^*_n, A_n), n \geq 0\}$ is a Markov renewal sequence with Kernel $K(x) = [K_{ij}(x)]$ such that $K_{ij}(x) = P\{X^*_{n+1} = j, A_{n+1} - A_n \leq x \mid X^*_n = i\}$. Actually the expression for the Kernel is not necessary; all we need is to realize that $\pi^* = [\pi^*_j]$ satisfies $\pi^* = \pi^* K(\infty)$ and we already have π^*_j. Also, since the arrivals are renewal, we have the conditional expected sojourn times $\beta_j = E[A_{n+1} - A_n \mid X^*_n = j] = 1/\lambda$ for all j. Now, the stochastic process $\{X(t), t \geq 0\}$ is a Markov regenerative process (MRGP) with embedded Markov renewal sequence $\{(X^*_n, A_n), n \geq 0\}$. Therefore, from MRGP theory (see Section B.3.3), we have

$$p_j = \frac{\sum_{i=0}^{\infty} \pi^*_i \gamma_{ij}}{\sum_{i=0}^{\infty} \pi^*_i \beta_i}$$

where γ_{ij} is the expected time spent in state j between time A_n and A_{n+1} given that $X^*_n = i$. We first compute γ_{ij} and then derive p_j. For that, we use the indicator function I_A, which is one if A is true and zero if A is false. Using the definition of γ_{ij}, the indicator function, and properties of Poisson processes, we can derive the following:

$$\gamma_{ij} = E\left[\int_{A_n}^{A_{n+1}} I_{\{X(t)=j\}} dt \mid X^*_n = i \right]$$

$$= \int_0^{\infty} E\left[\int_0^u I_{\{X(t)=j\}} dt \mid X^*_0 = i, A_1 = u \right] dG(u)$$

$$= \int_0^\infty \int_0^u P\{X(t) = j | X(0) = i\} dt \, dG(u)$$

$$= \int_0^\infty \int_0^u e^{-\mu t} \frac{(\mu t)^{i+1-j}}{(i+1-j)!} dt \, dG(u)$$

for $i + 1 \geq j$ and $j \geq 1$. For a given j such that $j \geq 1$, from MRGP theory (and using the fact that $\beta_i = 1/\lambda$ and $\gamma_{ij} = 0$ if $i + 1 < j$), we have

$$p_j = \frac{\sum_{i=0}^\infty \pi_i^* \gamma_{ij}}{\sum_{i=0}^\infty \pi_i^* \beta_i}$$

$$= \frac{\sum_{i=j-1}^\infty \pi_i^* \int_0^\infty \int_0^u e^{-\mu t} \frac{(\mu t)^{i+1-j}}{(i+1-j)!} dt \, dG(u)}{\sum_{i=0}^\infty \pi_i^* \frac{1}{\lambda}}$$

$$= \lambda \int_0^\infty \int_0^u e^{-\mu t} \sum_{i=j-1}^\infty (1-\alpha)\alpha^i \frac{(\mu t)^{i+1-j}}{(i+1-j)!} dt \, dG(u)$$

$$= \lambda \int_{u=0}^\infty \int_{t=0}^u e^{-\mu t} (1-\alpha)\alpha^{j-1} e^{\alpha \mu t} dt \, dG(u)$$

$$= \frac{\lambda}{\mu} \int_{u=0}^\infty \int_{t=0}^u e^{-(1-\alpha)\mu t} (1-\alpha)\mu \alpha^{j-1} dt \, dG(u)$$

$$= \frac{\lambda}{\mu} \int_{u=0}^\infty \left(1 - e^{-(1-\alpha)\mu t}\right) \alpha^{j-1} dG(u)$$

$$= \frac{\lambda}{\mu} \alpha^{j-1} (1 - \tilde{G}(\mu(1-\alpha)))$$

$$= \frac{\lambda}{\mu} \alpha^{j-1} (1 - \alpha) = \frac{\lambda}{\mu} \pi_{j-1}^*$$

where the last line uses the fact that $\alpha = \tilde{G}(\mu(1-\alpha))$. Therefore, for $j \geq 1$, $p_j = \rho \pi_{j-1}^*$ where $\rho = \lambda/\mu$. From this, p_0 can be easily computed as $p_0 = 1 - \sum_{j \geq 1} p_j = 1 - \rho$. ∎

Therefore, there is a relationship between the arrival point probabilities π_j^* and the steady-state probabilities p_j, but since the arrivals are not Poisson they are not equal. A natural question to ask is if the arrivals were indeed Poisson for the $G/M/1$ queue (which would become $M/M/1$), would the arrival point probabilities be the same as the steady-state probabilities? In fact, it would be a good idea to verify all the results we have derived thus far by assuming that the arrivals are Poisson, which is the theme of our next problem.

Problem 35

Using the results derived for the $G/M/1$ queue, obtain α, π_j^*, p_j, $P\{Y \leq y\}$, L, and W when $G(t) = 1 - e^{-\lambda t}$ for $t \geq 0$ when the queue is stable.

Solution

Note that the interarrival times are exponentially distributed; therefore, some of our results can be verified using those of the $M/M/1$ queue and the reader is encouraged to do that. The LST of the interarrival time is $\tilde{G}(s) = \lambda/(\lambda + s)$. Therefore, we solve for α in $\alpha = \tilde{G}(\mu - \alpha\mu)$, that is, $\alpha = \lambda/(\lambda + (1 - \alpha)\mu)$. We get two solutions to that quadratic equation. Since we require $\alpha \in (0, 1)$, we do not consider the solution $\alpha = 1$. However, we know the queue is stable (hence $\lambda/\mu < 1$) and thus we have

$$\alpha = \frac{\lambda}{\mu}.$$

Using that we get for $j = 0, 1, 2, \ldots,$

$$\pi_j^* = (1 - \alpha)\,\alpha^j = \left(1 - \frac{\lambda}{\mu}\right)\left(\frac{\lambda}{\mu}\right)^j.$$

Using $\rho = \lambda/\mu$ we get

$$p_0 = 1 - \rho = \pi_0^*$$
$$p_j = \rho\pi_{j-1}^* = (1 - \rho)\rho^j = \pi_j^* \quad \text{for } j > 0.$$

Therefore, $p_j = \pi_j^*$ for all j, which is not surprising due to PASTA. Also notice that p_j is identical to what was derived in the $M/M/1$ queue analysis. Further,

$$P\{Y \leq y\} = 1 - e^{-(1-\alpha)\mu y} = 1 - e^{-(\mu - \lambda)y}.$$

Since $W = 1/\mu(1 - \alpha)$ we get

$$W = \frac{1}{\mu - \lambda}$$

and using $L = \lambda W$ we have

$$L = \frac{\lambda}{\mu - \lambda}.$$

Notice that L and W as well as the distribution of Y are identical to those of the $M/M/1$ queue. ∎

In a similar manner, one can obtain the preceding expressions for other interarrival time distributions as well, some of which are given in the exercises at the end of the chapter. Before wrapping up this section on DTMC-based analysis, it is worthwhile to describe one more example. This is the $G/M/2$ queue. It is crucial to point out that the generic $G/M/s$ queue can hence be analyzed in a similar fashion. However, analyzing the $M/G/s$ queue using DTMC is quite intractable for $s \geq 2$. The reason for that is the $G/M/s$ queue, if observed at arrivals, is Markovian, whereas the $M/G/s$ queue observed at departures is not Markovian. Now to the $G/M/2$ queue.

Problem 36

Consider a $G/M/2$ queue. Let X_n^* be the number of customers just before the nth arrival. Show that $\{X_n^*, n \geq 0\}$ is a DTMC by computing the transition probability matrix. Derive the condition for stability and the limiting distribution for X_n^*.

Solution

We begin with some notation. Let a_j be the probability that j departures occur between two arrivals when both servers are working throughout the time between the two arrivals. Then

$$a_j = \int_0^\infty e^{-2\mu t} \frac{(2\mu t)^j}{j!} dG(t).$$

Let c_j be the probability that j departures occur between two arrivals where both servers are working until the jth departure, after which only one server is working but does not complete service. Then

$$c_0 = \int_0^\infty e^{-\mu t} dG(t) = \tilde{G}(\mu)$$

for $j > 0$,

$$c_j = \int_0^\infty \int_0^t e^{-\mu(t-s)} e^{-2\mu s} \frac{(2\mu s)^{j-1}}{(j-1)!} 2\mu ds \, dG(t)$$

$$= \int_0^\infty 2^j e^{-\mu t} \left\{ 1 - \sum_{i=0}^{j-1} e^{-\mu t} \frac{(\mu t)^i}{i!} \right\} dG(t)$$

and b_j is given by

$$b_j = 1 - c_j - \sum_{i=1}^j a_{i-1}.$$

Let X_n^* be the number of customers in the system just before the nth arrival. Then $\{X_n^*, n \geq 0\}$ is a DTMC with

$$P = \begin{bmatrix} b_0 & c_0 & 0 & 0 & \cdots \\ b_1 & c_1 & a_0 & 0 & \cdots \\ b_2 & c_2 & a_1 & a_0 & \cdots \\ b_3 & c_3 & a_2 & a_1 & \cdots \\ b_4 & c_4 & a_3 & a_2 & \cdots \\ \vdots & \vdots & \vdots & \vdots & \cdots \end{bmatrix}.$$

Let π_j^* be the limiting probability that in steady state an arriving customer sees j in the system, that is,

$$\pi_j^* = \lim_{n \to \infty} P\{X_n^* = j\}.$$

By trying the solution $\pi_j^* = \beta \alpha^j$, we can show the following:

1. The solution $\pi_j^* = \beta \alpha^j$ works for $j > 0$ if there is a unique solution to $\alpha = \tilde{G}(2\mu(1 - \alpha))$ such that $\alpha \in (0,1)$, which is the condition of stability and can be written as $\lambda/(2\mu) < 1$.
2. Also, $\pi_0^* = \beta \alpha[1 - 2\tilde{G}(\mu)]/[(1 - 2\alpha)\tilde{G}(\mu)]$.
3. Therefore, using $\sum_{j=0}^\infty \pi_j^* = 1$, we can derive that $\beta = (1 - \alpha)$ $(1 - 2\alpha)\tilde{G}(\mu)/[\alpha(1 - \alpha) - \alpha\tilde{G}(\mu)]$. This can be used to compute $\pi^* = (\pi_0^* \ \pi_1^* \ \cdots)$.

It is worthwhile to point out that one can obtain the distribution of the sojourn time in the system using an analysis similar to that in Problem 33. That is left as an exercise for the reader.   ■

4.2 Mean Value Analysis

So far in this chapter, we have focused on queues that can be modeled as appropriate stochastic processes so that distributions of performance measures such as time and number in the system can be obtained. However, if one was not interested in the distribution but just the expected value of the performance measure, then a technique aptly known as mean value analysis (MVA) can be adopted. In fact, MVA can also be used to obtain excellent approximations when systems cannot easily be modeled as suitable stochastic processes. MVA leverages upon two properties of random variables:

1. Expected value of a linear combination of random variables is equal to the linear combination of the expected values, that is, if Y_1, Y_2, ..., Y_k are random variables with finite means and $a_1, a_2, \ldots, a_k$ are known finite constants, then

$$E[a_1 Y_1 + a_2 Y_2 + \cdots + a_k Y_k] = a_1 E[Y_1] + a_2 E[Y_2] + \cdots + a_k E[Y_k].$$

   The preceding result holds even if the random variables Y_1, Y_2, ..., Y_k are dependent on each other.

2. Expected value of a product of independent random variables is equal to the product of the expected values of those random variables, that is, if Y_1, $Y_2, \ldots$, Y_k are independent random variables, then

$$E[Y_1 Y_2 \ldots Y_k] = E[Y_1] E[Y_2] \ldots E[Y_k].$$

   The preceding result requires that the random variables Y_1, $Y_2, \ldots$, Y_k be independent.

The easiest way to explain MVA is using an example. We specifically consider an example that we have seen so far. This way, it is possible to compare the results obtained and contrast the approaches. In that light, we first consider the $M/G/1$ queue. Subsequently, we will illustrate the use of MVA to analyze $G/G/1$ queues.

4.2.1 Explaining MVA Using an $M/G/1$ Queue

Consider an $M/G/1$ queue where the arrivals are according to a Poisson process with mean λ per unit time. The service times are IID random variables sampled from a distribution with mean $1/\mu$ and variance σ^2. There is a single server and infinite room for customers to wait. Customers are served according to FCFS. Let X_n be the number of customers in the system immediately after the nth departure (this is identical to the definition in the DTMC analysis of $M/G/1$ queues described in Section 4.1.1). Let U_n be the number of arrivals that occurred during the service of the nth customer. For the MVA, we also require the following asymptotic values based on X_n and U_n defined earlier:

$$\pi_0 = \lim_{n \to \infty} P\{X_n = 0\},$$

$$L = \lim_{n \to \infty} E[X_n],$$

$$U = \lim_{n \to \infty} E[U_n].$$

Having described the notation, now we are ready for MVA. We first write down a relation between X_n and X_{n+1}. If $X_n > 0$, then $X_{n+1} = X_n - 1 + U_{n+1}$ because the number in the system as seen by the $n + 1$st departure is equal to what the nth departure sees (which is X_n and also includes the $n + 1$st customer since $X_n > 0$) plus all the customers that arrived during the service of the $n + 1$st customer minus one (since only the number remaining in the system is described in X_{n+1}). However, if $X_n = 0$, then $X_{n+1} = U_{n+1}$ since when the nth customer departs, the system becomes empty, and then the $n + 1$st customer arrives and starts getting served immediately, and all the customers that showed up during that service would remain when the $n+1$st customer departs. Thus, we can write down the following relation:

$$X_{n+1} = \max\{X_n - 1, 0\} + U_{n+1} = (X_n - 1)^+ + U_{n+1}. \tag{4.12}$$

Taking expected values on both sides of the preceding relation, by suitably conditioning on whether $X_n = 0$ or $X_n > 0$, we can derive the following:

$$E[X_{n+1}] = E[(X_n - 1)^+] + E[U_{n+1}]$$

$$= E[(X_n - 1)^+ | X_n > 0]P(X_n > 0) + E[(X_n - 1)^+ | X_n = 0]P(X_n = 0)$$

$$+ E[U_{n+1}]$$

$$= E[X_n | X_n > 0]P(X_n > 0) - P(X_n > 0) + E[U_{n+1}]$$

$$= E[X_n|X_n > 0]P(X_n > 0) + E[X_n|X_n = 0]P(X_n = 0)$$

$$- P(X_n > 0) + E[U_{n+1}]$$

$$= E[X_n] - P(X_n > 0) + E[U_{n+1}].$$

Taking the limit as $n \to \infty$, we get $L = L - (1 - \pi_0) + U$. Canceling L's (assuming $L < \infty$) we get

$$U = 1 - \pi_0.$$

Now, to derive an expression for U, notice that if we condition on the service times (say S_n is the service time of the nth customer), then

$$U = \lim_{n \to \infty} E[U_n] = \lim_{n \to \infty} E[E[U_n|S_n]] = \lim_{n \to \infty} E[\lambda S_n] = \frac{\lambda}{\mu}.$$

Since $\rho = \lambda/\mu$, using $U = 1 - \pi_0$ we have

$$\pi_0 = 1 - \rho.$$

Notice that the preceding equation was derived in the $M/G/1$ analysis using DTMCs. Also, the condition $0 < \pi_0 < 1$ implies $\rho < 1$, which is the stability condition, and thus $L < \infty$.

Continuing with the MVA by squaring Equation 4.12, we get

$$X_{n+1}^2 = (X_n - 1)^2 I_{\{X_n > 0\}} + 2U_{n+1}(X_n - 1)I_{\{X_n > 0\}} + U_{n+1}^2$$

where $I_{\{X_n > 0\}}$ is an indicator function that is one if $X_n > 0$ and zero otherwise. Taking the expected value of the preceding equation, we get

$$E\left[X_{n+1}^2\right] = E\left[\left(X_n^2 - 2X_n + 1\right)I_{\{X_n > 0\}}\right]$$

$$+ 2E[U_{n+1}]E[(X_n - 1)I_{\{X_n > 0\}}] + E\left[U_{n+1}^2\right] \tag{4.13}$$

since U_{n+1} is independent of $(X_n - 1)I_{\{X_n > 0\}}$. We derive each term of the RHS of Equation 4.13 separately starting from the right extreme. Conditioning on the service time of the $n + 1$st customer, we get $E\left[U_{n+1}^2\right] = E\left[E\left[U_{n+1}^2|S_{n+1}\right]\right]$ $= E[Var[U_{n+1}|S_{n+1}] + \{E[U_{n+1}|S_{n+1}]\}^2] = E\left[\lambda S_{n+1} + \lambda^2 S_{n+1}^2\right] = \lambda E[S_{n+1}] + \lambda^2 E\left[S_{n+1}^2\right] = \rho + \lambda^2 \sigma^2 + \rho^2$. Using an identical argument described earlier to compute $E[(X_n - 1)^+]$ (see expressions following Equation 4.12), we have the middle term $E[(X_n - 1)I_{\{X_n > 0\}}] = E[X_n] - P(X_n > 0)$. Of course, we also saw earlier that $E[U_{n+1}] = \rho$, which leaves us with the first expression that can be derived as follows:

$$E\left[\left(X_n^2 - 2X_n + 1\right) I_{\{X_n > 0\}}\right] = E\left[\left(X_n^2 - 2X_n + 1\right) I_{\{X_n > 0\}} | X_n > 0\right] P(X_n > 0)$$

$$+ E\left[\left(X_n^2 - 2X_n + 1\right) I_{\{X_n > 0\}} | X_n = 0\right] P(X_n = 0)$$

$$= E\left[\left(X_n^2 - 2X_n + 1\right) | X_n > 0\right] P(X_n > 0)$$

$$= E\left[\left(X_n^2 - 2X_n\right) | X_n > 0\right] P(X_n > 0) + P(X_n > 0)$$

$$= E\left[\left(X_n^2 - 2X_n\right) | X_n > 0\right] P(X_n > 0)$$

$$+ E\left[\left(X_n^2 - 2X_n\right) | X_n = 0\right] P(X_n = 0)$$

$$+ P(X_n > 0)$$

$$= E\left[X_n^2 - 2X_n\right] + P(X_n > 0).$$

Therefore, we can rewrite Equation 4.13 as

$$E\left[X_{n+1}^2\right] = E\left[X_n^2\right] - 2E[X_n] + P(X_n > 0) + 2\rho E[X_n]$$

$$- 2\rho P(X_n > 0) + \rho + \lambda^2 \sigma^2 + \rho^2.$$

Taking the limit as $n \to \infty$, canceling the LHS with the first term in the RHS, and rearranging the terms we get

$$L = \rho + \frac{\lambda^2 \sigma^2 + \rho^2}{2(1 - \rho)}.$$

Verify that the preceding expression is identical to the Pollaczek–Khintchine formula given in Equation 4.6.

Notice that we did not use a DTMC to model $\{X_n, n \geq 0\}$ for MVA. All we did was write down a recursive relation and computed the expected value. Of course, the trick is in writing down a suitable relation and carefully rearranging terms. This would be more pronounced in the next example. However, the overall method is fairly consistent over all examples. Next, we consider a $G/G/1$ queue and use MVA to develop some approximations.

4.2.2 Approximations for Renewal Arrivals and General Service Queues

Consider a $G/G/1$ queue where the arrivals are according to a renewal process with known interarrival time distribution from which the mean and variance of interarrival times can be obtained. In addition, the service times are IID and sampled from a known distribution from which the mean and variance can be computed. There is a single server that uses FCFS discipline and there is an infinitely big waiting room. The objective is to obtain the mean number in the system in steady state and the mean waiting time. Notice that we have not so far obtained any performance measures for the $G/G/1$ queue. All we have done so far is develop relations such as Little's law. Since a generic $G/G/1$ queue cannot be modeled as a suitable stochastic process, it is difficult to obtain analytical closed-form expressions for the steady-state performance metrics. However, there are numerous approximation techniques, some of which we will present in this chapter. The first of those approximations use MVA that we describe next.

Similar to the MVA for $M/G/1$ queue, we describe some notations (see Table 4.1). The six variables defined can be divided into three categories: T_n and S_n are known in distribution (i.e., they form an IID sequence each sampled from given distributions); W_n and I_n are performance measures not known a priori and need to be computed using MVA; and A_n and D_n are mainly for convenience to derive the approximation. We are going to use MVA to approximately obtain $E[W_n]$ and $E[I_n]$ as $n \to \infty$ in terms of the first two moments of the interarrival and service times. In particular, define the mean arrival rate λ, mean service rate μ, squared coefficient of variation (SCOV) of arrivals C_a^2, and SCOV of service C_s^2 as follows:

$$\lambda = \frac{1}{E[T_n]},$$

$$\mu = \frac{1}{E[S_n]},$$

TABLE 4.1

Notation for the $G/G/1$ MVA

A_n	Time of nth arrival
$T_{n+1} = A_{n+1} - A_n$	The $n + 1$st interarrival time
S_n	Service time of the nth customer
W_n	Time spent (sojourn) in the system by the nth customer
$D_n = A_n + W_n$	Time of nth departure
$I_{n+1} = (A_{n+1} - A_n - W_n)^+$	Idle time between nth and $n + 1$st service

$$C_a^2 = \frac{Var[T_n]}{(E[T_n])^2} = \lambda^2 Var[T_n],$$

$$C_s^2 = \frac{Var[S_n]}{(E[S_n])^2} = \mu^2 Var[S_n].$$

Also define the performance measures W, I_d, and $I^{(2)}$ as

$$W = \lim_{n \to \infty} E[W_n],$$

$$I_d = \lim_{n \to \infty} E[I_n],$$

$$I^{(2)} = \lim_{n \to \infty} E[I_n^2].$$

Using all the preceding definitions, we carry out MVA by first writing down the sojourn time of the $n + 1$st customer, W_{n+1} being equal to that customer's service time plus any time the customer spent waiting for service to begin (this happens if the customer arrived before the previous one departed). In other words,

$$W_{n+1} = S_{n+1} + \max\{D_n - A_{n+1}, 0\}.$$

Using the definitions of D_n, T_n, and I_n in Table 4.1, we can write down the following sets of equations:

$$W_{n+1} = S_{n+1} + \max\{D_n - A_{n+1}, 0\}$$

$$= S_{n+1} + \max\{W_n + A_n - A_{n+1}, 0\}$$

$$= S_{n+1} + W_n + A_n - A_{n+1} + \max\{A_{n+1} - A_n - W_n, 0\}$$

$$= S_{n+1} + W_n - T_{n+1} + I_{n+1}.$$

Taking expectations, we get

$$E[W_{n+1}] = E[S_{n+1}] + E[W_n] - E[T_{n+1}] + E[I_{n+1}].$$

Now, by letting $n \to \infty$ and using the notation defined earlier, we see that

$$W = \frac{1}{\mu} + W - \frac{1}{\lambda} + I_d.$$

Thus, if $W < \infty$, that is, a stable $G/G/1$ queue, we have

$$I_d = \frac{1}{\lambda} - \frac{1}{\mu} = \frac{(1-\rho)}{\lambda}$$

where $\rho = \lambda/\mu$. From the definition of I_d, we have $I_d > 0$ requiring $\rho < 1$, which is the stability condition.

Then continuing with the MVA, recall that $W_{n+1} = S_{n+1} + W_n - T_{n+1} + I_{n+1}$. Similar to what we did for $M/G/1$ queue, here too we square the terms. However, before doing that, we want to carefully rearrange terms by noticing the following three things:

1. Recall the definitions of W_{n+1}, S_{n+1}, and I_{n+1}. Notice that $W_{n+1} - S_{n+1}$ is the time the $n + 1$st customer waits for service to begin and I_{n+1} is the idle time between serving the nth and $n + 1$st customers. Based on that, we have $(W_{n+1} - S_{n+1})I_{n+1} = 0$, since when there is a nonzero idle time, the $n + 1$st customer does not wait for service and vice versa.

2. The time a customer waits for service to begin is independent of the service time of that customer, hence $(W_{n+1} - S_{n+1})$ is independent of S_{n+1}.

3. Also, the sojourn time of the nth customer is independent of the time between the nth and $n + 1$st arrivals. Hence, we have W_n independent of T_{n+1}.

The preceding three results prompt us to rearrange $W_{n+1} = S_{n+1} + W_n - T_{n+1} + I_{n+1}$ as $W_{n+1} - S_{n+1} - I_{n+1} = W_n - T_{n+1}$. Squaring and rearranging terms leads to the following results:

$$W_{n+1} - S_{n+1} - I_{n+1} = W_n - T_{n+1},$$

$$(W_{n+1} - S_{n+1} - I_{n+1})^2 = (W_n - T_{n+1})^2,$$

$$W_{n+1}^2 - 2(W_{n+1} - S_{n+1})S_{n+1} - S_{n+1}^2 + I_{n+1}^2$$

$$-2(W_{n+1} - S_{n+1})I_{n+1} = W_n^2 - 2W_n T_{n+1} + T_{n+1}^2.$$

Now using the facts that $(W_{n+1} - S_{n+1})I_{n+1} = 0$, $(W_{n+1} - S_{n+1})$ and S_{n+1} are independent, as well as W_n is independent of T_{n+1}, and taking the expected value of the preceding expression, we get

$$E\left[W_{n+1}^2\right] - 2E[(W_{n+1} - S_{n+1})]E[S_{n+1}] - E\left[S_{n+1}^2\right] + E\left[I_{n+1}^2\right]$$

$$= E\left[W_n^2\right] - 2E[W_n]E[T_{n+1}] + E\left[T_{n+1}^2\right].$$

Notice that $E[T_{n+1}] = 1/\lambda$, $E[S_{n+1}] = 1/\mu$, $E\left[T_{n+1}^2\right] = \left(C_a^2 + 1\right)/\lambda^2$, and $E\left[S_{n+1}^2\right] = \left(C_s^2 + 1\right)/\mu^2$. Making those substitutions and taking the limit as $n \to \infty$, we get

$$W = \frac{1}{\mu} + \frac{\rho^2 C_s^2 + C_a^2 + (1 - \rho)^2 - \lambda^2 I^{(2)}}{2\lambda\{1 - \rho\}} \tag{4.14}$$

where $\rho = \lambda/\mu$.

The only unknown quantity in the preceding expression for W is $I^{(2)}$. Therefore, suitable bounds and approximations for W can be obtained by cleverly bounding and approximating $I^{(2)}$. Section 4.3 is devoted to bounds and approximations for queues, and to obtain some of those bounds, we will use Equation 4.14. However, for the sake of completing this analysis, we present a simple upper bound for W. Since the variance of the idle time for a server between customers must be positive, we have $I^{(2)} \geq (I_d)^2 = (1 - \rho)^2/\lambda^2$. Thus, we have $-\lambda^2 I^{(2)} \leq -(1 - \rho)^2$ and plugging into Equation 4.14, we get

$$W \leq \frac{1}{\mu} + \frac{\rho^2 C_s^2 + C_a^2}{2\lambda\{1 - \rho\}}.$$

A key point to notice is that the preceding bound only depends on the mean and variance of the interarrival times and service times. Therefore, we really do not need the entire distribution information. Of course, the preceding bound for W is quite weak and one can obtain much better bounds and approximations that we would describe in Section 4.3, which would also use only λ, μ, C_a^2, and C_s^2. However, before that we present another result for $G/G/1$ queue using the MVA results.

4.2.3 Departures from $G/G/1$ Queues

The MVA for the $G/G/1$ system can be immediately extended to characterizing the output from a $G/G/1$ queue. In a queueing network setting (used in Chapter 7), output from one queue may act as an input to another queue. In that light, we strive to obtain the mean and SCOV of the interdeparture times. There are some notations involved, but we will describe them as we go along with the explanation. Let V_{n+1} be the interdeparture time between the nth and $n + 1$st departure. By definition, we have

$$V_{n+1} = D_{n+1} - D_n = \max(A_{n+1} - D_n, 0) + S_{n+1}$$

using the substitution $D_{n+1} = \max(A_{n+1}, D_n) + S_{n+1}$. Based on the definition of I_{n+1}, we can rewrite V_{n+1} as

$$V_{n+1} = I_{n+1} + S_{n+1}. \tag{4.15}$$

Taking the expected value of the preceding equation, we get $E[V_{n+1}] = E[I_{n+1}] + E[S_{n+1}]$ and letting $n \to \infty$ we obtain

$$\lim_{n \to \infty} E(V_{n+1}) = I_d + \frac{1}{\mu} = \frac{1}{\lambda} - \frac{1}{\mu} + \frac{1}{\mu} = \frac{1}{\lambda}.$$

This is not a surprising result, as when the queue is stable the average departure rate is the same as the average arrival rate as no customers are created or destroyed in the queue (see conservation law in Section 1.2.1).

The SCOV of the interdeparture times C_d^2 is a little more involved. For that, we go back to Equation 4.15. Since I_{n+1} is independent of S_{n+1}, taking variance on both sides of Equation 4.15 we get $Var(V_{n+1}) = Var(I_{n+1}) + Var(S_{n+1})$. By letting $n \to \infty$ we obtain

$$\lim_{n \to \infty} Var(V_{n+1}) = I^{(2)} - I_d^2 + \frac{C_s^2}{\mu^2}.$$

However, using the definition of C_d^2 and substituting for $I^{(2)}$ from Equation 4.14, we get

$$C_d^2 = \lim_{n \to \infty} \frac{Var(V_{n+1})}{(E[V_{n+1}])^2} = \lambda^2 \left(I^{(2)} - I_d^2 + \frac{C_s^2}{\mu^2} \right)$$

$$= C_a^2 + 2\rho^2 C_s^2 + 2\rho(1 - \rho) - 2\lambda W(1 - \rho). \qquad (4.16)$$

The reason this is written in terms of W is that now we only need a good approximation or bound for W. Once we have that, we get a good bound or approximation for C_d^2 automatically. Hence, in the next section we mainly focus on obtaining bounds and approximations for only W.

4.3 Bounds and Approximations for General Queues

In this section, we consider $G/G/s$-type queues that are hard to model using Markov chains. For such queues using MVA and other techniques, we obtain bounds as well as approximations for average performance measures such as L or W. There are several applications such as flexible manufacturing systems for which these results are extremely useful and there is almost no other way of analytically obtaining performance measures. The bounds and approximations presented here are based on one or more of the following analysis techniques:

1. Using MVA along with additional properties of random variables
2. Using $M/M/s$ results and adjusting for $G/G/s$
3. Using heavy-traffic approximations
4. Using empirically derived results

We begin with the single server $G/G/1$ queue and continue from where we left off in the previous section. Then we show bounds and approximations for multiserver queues for the remainder of this section.

4.3.1 General Single Server Queueing System $(G/G/1)$

Here we describe some bounds and approximations for W (the sojourn time in the system) in a $G/G/1$ queue in terms of λ the arrival rate, μ the service rate, C_a^2 the SCOV for interarrival times, and C_s^2 the SCOV for service times. Wherever convenient, we will describe the results in terms of $\rho = \lambda/\mu$, the traffic intensity. Other performance metrics such as L, C_d^2, L_q, and W_q can be obtained using their relationship to W. In particular, recall that

$$L = \lambda W,$$

$$C_d^2 = C_a^2 + 2\rho^2 C_s^2 + 2\rho(1 - \rho) - 2\lambda W(1 - \rho),$$

$$L_q = \lambda W - \rho,$$

$$W_q = W - \frac{1}{\mu}.$$

Thus, if we obtain bounds and approximations for W, they can easily be translated to the preceding performance metrics.

In that light, first we describe some bounds on W based on MVA and properties of random variables. Recall from Equation 4.14 that the MVA results for W in terms of the unknown $I^{(2)}$ are

$$W = \frac{1}{\mu} + \frac{\rho^2 C_s^2 + C_a^2 + (1 - \rho)^2 - \lambda^2 I^{(2)}}{2\lambda\{1 - \rho\}}.$$

For any random variable Y, we have $E[Y^2] \geq \{E[Y]\}^2$. Therefore, restating the results in the previous section, the idle time between the nth and $n +$ 1st services denoted as I_{n+1} also obeys $E\left[I_{n+1}^2\right] \geq \{E[I_{n+1}]\}^2$, and by letting $n \to \infty$ we get $I^{(2)} = \lim_{n\to\infty} E\left[I_{n+1}^2\right] \geq \lim_{n\to\infty}\{E[I_{n+1}]\}^2 = (I_d)^2$. But since $(I_d)^2 = (1 - \rho)^2/\lambda^2$, we get

$$W \leq \frac{1}{\mu} + \frac{\rho^2 C_s^2 + C_a^2}{2\lambda\{1 - \rho\}}.$$

However, better bounds can be obtained using more properties of random variables. We describe two here, both from Buzacott and Shanthikumar [15].

1. For any two positive-valued random variables Y and Z, we have

$$E[\{(Y-Z)^+\}^2] \geq \frac{E[Y^2]}{(E[Y])^2} \{E[(Y-Z)^+]\}^2.$$

Using $Y = T_{n+1} = A_{n+1} - A_n$ the $n+1$st interarrival time and $Z = W_n$ the nth customer's sojourn in the system, we get from the earlier relation

$$E\left[I_{n+1}^2\right] \geq \frac{E\left[T_{n+1}^2\right]}{(E[T_{n+1}])^2} \{E[I_{n+1}]\}^2 = \left(C_a^2 + 1\right)\{E[I_{n+1}]\}^2.$$

Letting $n \to \infty$ in the preceding expression we get $I^{(2)} \geq (C_a^2 + 1)(I_d)^2$ and plugging into Equation 4.14, we get

$$W \leq \frac{\rho(2-\rho)C_a^2 + \rho^2 C_s^2}{2\lambda(1-\rho)} + \frac{1}{\mu}.$$

2. For any positive-valued continuous random variable X, the hazard rate function $h(x)$ is defined as

$$h(x) = \frac{F'(x)}{1 - F(x)}$$

where $F(x) = P\{X \leq x\}$, the CDF and $F'(x)$ its derivative. Some random variables are such that $h(x)$ increases with x and they are called IFR (increasing failure rate) random variables. There are also some random variables such that $h(x)$ decreases with x and they are called DFR (decreasing failure rate) random variables. Of course, there are many random variables that are neither IFR or DFR and the following result cannot be used for those. For two positive-valued random variables Y and Z, if Y is IFR, we have

$$E[\{(Y-Z)^+\}^2] \leq \frac{E[Y^2]}{E[Y]} E[(Y-Z)^+],$$

whereas if Y is DFR, we have

$$E[\{(Y-Z)^+\}^2] \geq \frac{E[Y^2]}{E[Y]} E[(Y-Z)^+].$$

Using $Y = T_{n+1} = A_{n+1} - A_n$ the $n + 1$st interarrival time and $Z = W_n$ the nth customer's sojourn in the system, we get from the earlier relation if T_{n+1} is IFR

$$E\left[I_{n+1}^2\right] \leq \frac{E\left[T_{n+1}^2\right]}{E[T_{n+1}]} E[I_{n+1}] = \frac{(C_a^2 + 1)}{\lambda} E[I_{n+1}],$$

whereas if Y is DFR, we have

$$E\left[I_{n+1}^2\right] \geq \frac{E\left[T_{n+1}^2\right]}{E[T_{n+1}]} E[I_{n+1}] = \frac{(C_a^2 + 1)}{\lambda} E[I_{n+1}].$$

Actually, the preceding results do not require IFR or DFR but a much weaker condition (that they be decreasing or increasing mean residual life, respectively). However, we use the stronger requirement of IFR or DFR to obtain the following bound by letting $n \to \infty$ in the preceding expressions to get $I^{(2)} \leq (C_a^2 + 1)/\lambda I_d$ if interarrival times are IFR and $I^{(2)} \geq (C_a^2 + 1)/\lambda I_d$ if interarrival times are DFR. Plugging into Equation 4.14, we get

$$W \geq \frac{\rho\left(C_a^2 - 1 + \rho\right) + \rho^2 C_s^2}{2\lambda(1 - \rho)} + \frac{1}{\mu} \quad \text{if interarrival time is IFR,}$$

$$W \leq \frac{\rho\left(C_a^2 - 1 + \rho\right) + \rho^2 C_s^2}{2\lambda(1 - \rho)} + \frac{1}{\mu} \quad \text{if interarrival time is DFR.}$$

Having presented some bounds, next we consider some approximations for W that are special for the $G/G/1$ queue. These are all from Buzacott and Shanthikumar [15]. The key idea is that we have a closed-form expression for W from the Pollaczek–Khintchine formula for $M/G/1$ queues in Equation 4.6. Therefore, the first two approximations are just appropriate factors multiplying the bounds for the $G/G/1$ queue so that when $C_a^2 = 1$, the results are identical to that of the Pollaczek–Khintchine formula. The third approximation is based on a heuristic that results in fairly accurate results when $C_a^2 \leq 1$.

Approximation	W
1	$\left(\frac{\rho^2(1+C_s^2)}{1+\rho^2 C_s^2}\right)\left(\frac{C_a^2+\rho^2 C_s^2}{2\lambda(1-\rho)}\right) + \frac{1}{\mu}$
2	$\left(\frac{\rho(1+C_s^2)}{2-\rho+\rho C_s^2}\right)\left(\frac{\rho(2-\rho)C_a^2+\rho^2 C_s^2}{2\lambda(1-\rho)}\right) + \frac{1}{\mu}$
3	$\frac{\rho^2(C_a^2+C_s^2)}{2\lambda(1-\rho)} + \frac{(1-C_a^2)C_a^2\rho}{2\lambda} + \frac{1}{\mu}$

There are other approximations for heavy-traffic queues that we will see in the $G/G/s$ setting where one can use $s = 1$ and get $G/G/1$ approximations. The reader is encouraged to review those approximations as well. The literature also has several empirical approximations. Care must be taken to ensure that the test cases that were used to obtain the empirical approximations and conclusions are identical to those considered by the reader. It is worthwhile to point out that the steady-state mean waiting time and number in the system can also be obtained using simulations that we use for testing our approximations. In fact, one does not even need sophisticated simulation software for that, we explain it next using an example.

Problem 37

For a $G/G/1$ queue, develop an algorithm to simulate and obtain the mean number in the system in steady-state, given the CDF of interarrival times $F(t)$ and service times $G(t)$.

Solution

Clearly from the problem description, for all $n \geq 0$, $F(t) = P\{T_n \leq t\}$ and $G(t) = P\{S_n \leq t\}$. Using U_n and V_n as uniform $(0, 1)$ random variables that come fairly standard in any computational package, we can obtain samples of T_n and S_n as $F^{-1}(U_n)$ and $G^{-1}(V_n)$, respectively. Notice that $F^{-1}(\cdot)$ is the inverse of the function $F(\cdot)$, for example, if $F(t) = 1 - e^{-\lambda t}$, then $T_n = F^{-1}(U_n) = (-1/\lambda)\log_e(1 - U_n)$. Now we describe the following algorithm using T_n and S_n for all n:

1. Initialize $A_0 = 0$ and assume the system is empty, thus $D_0 = S_0$.
2. For $n \geq 0$, $A_{n+1} = A_n + T_n$, $D_{n+1} = S_{n+1} + \max(A_{n+1}, D_n)$, and $W_n = D_n - A_n$. Notice that the expression for D_{n+1} essentially states that the $n+1$st departure would occur a service time S_{n+1} after either the nth departure or the $n + 1$st arrival, whichever happens later (which is the time when this service begins).
3. Iteratively compute the previous step a large number of times, delete the first several values (warm-up period), and take the sample mean of remaining W_n values as $\hat{W}$. This would be a good estimate of W, the mean time in the system. Using Little's law, the mean number in the system L can be estimated as $\hat{W}/E[T_n]$. ∎

A major concern is that almost all the approximations in the literature are based on the first two moments of the interarrival time and service time distributions which could lead the readers to misunderstand that only the first two moments matter. We illustrate that in the next example.

Problem 38

Consider a $G/G/1$ queue where the service times are IID uniform random variables between 0 and $2/\mu$. Obtain the mean waiting time using simulations when interarrival times (T_n) are according to a Pareto distribution whose CDF is

$$P(T_n \le x) = \begin{cases} 1 - (K/x)^\beta & \text{if } x \ge K \\ 0 & \text{otherwise} \end{cases}$$

with parameters $\beta = 1 + \sqrt{2}$ and $K = (\beta - 1)/(\beta\lambda)$. Show that $C_a^2 = 1$ and then compare the mean waiting time with that of the $M/G/1$ queue. Use $\lambda = 10$ and $\mu = 15$.

Solution

First let us analyze the arrival process. Using the CDF we can compute

$$E[T_n] = \frac{K\beta}{\beta - 1} = \frac{1}{\lambda}$$

when $\beta > 1$ (which is needed for the first equality and it is the case here) and

$$Var[T_n] = \frac{K^2\beta}{(\beta - 1)^2(\beta - 2)} = \frac{1}{\lambda^2}$$

when $\beta > 2$ (for the first equality, which is also the case in this problem) and the second equality is because $\beta = 1 + \sqrt{2}$. Thus, the SCOV of the arrival process $C_a^2 = 1$. Since the service times are according to a uniform distribution, the mean service time is $1/\mu$, the variance is $1/(3\mu^2)$, and SCOV $C_s^2 = 1/3$.

Now to obtain the mean waiting time W for this $G/G/1$ queue via simulation, we use the solution to Problem 37. Since $F(t) = P\{T_n \le t\} = 1 - (K/t)^\beta$, samples of T_n can be obtained as $F^{-1}(U_n) = K(1 - U_n)^{-1/\beta}$. The service times S_n can easily be obtained by multiplying the uniform random variables between 0 and 1 by $2/\mu$.

Now using the algorithm given in Problem 37, we obtain $W = 0.094$ time units for $\lambda = 10$ and $\mu = 15$. Now for the $M/G/1$ queue we get $W = 0.156$ time units, which is significantly different from the simulation result. In other words, if we used just λ, μ, C_a^2, and C_s^2 we could get a fairly inaccurate result for W. That is, we need the entire distribution, not just the first two moments. ∎

4.3.2 Multiserver Queues (*M/G/s*, *G/M/s*, and *G/G/s*)

Consider the most general of the three cases, namely, the *G/G/s* queue, where the arrivals are according to a renewal process with known interarrival time distribution from which the mean $(1/\lambda)$ and SCOV (C_a^2) can be obtained. In addition, the service times are IID and sampled from a known distribution from which the mean $(1/\mu)$ and SCOV (C_s^2) can be computed. There are *s* identical servers that use FCFS discipline and there is an infinite waiting room. The objective is to obtain approximations and bounds for *W*, the mean sojourn time (i.e., time spent) in the system when $s \geq 2$. Having said that, there are several empirical approximations that would work for $s = 1$ and, hence, not described in the previous section for *G/G/1* queues.

If the service times are indeed according to an exponential distribution, then it is possible to obtain *W* exactly (see Gross and Harris [49]) for the *G/M/s* queue. The analysis is an extension of the *G/M/2* system described in Problem 36, but we do not present it here. We only describe approximations and bounds for that system. Our goal is to obtain the approximations and bounds in terms of λ, μ, C_a^2, and C_s^2. Oftentimes, the results are written in terms of the traffic intensity ρ given by $\rho = \lambda/(s\mu)$. In addition, the results are mostly based on either heavy-traffic approximations using fluid and diffusion models or empirical techniques. By heavy traffic we mean that ρ is close to 1, however ρ has to be less than 1 for stability.

We do not explicitly present the fluid and diffusion approximation, since we would do so in a much broader setting of a network of queues in Chapter 7. However, we present a simple argument of how heavy-traffic approximation can be used in the *G/G/1* case that we saw in the previous section using MVA. Consider Equation 4.14 based on MVA that describes *W* in terms of the unknown $I^{(2)}$ as

$$W = \frac{1}{\mu} + \frac{\rho^2 C_s^2 + C_a^2 + (1-\rho)^2 - \lambda^2 I^{(2)}}{2\lambda\{1-\rho\}}.$$

Notice that $I^{(2)}$ is the second moment of the server idle time between successive arrivals. However, under heavy traffic, only for a small fraction of time $(1 - \rho)$ would the server experience nonzero idle time. Therefore, a reasonable heavy-traffic approximation as ρ approaches one for the *G/G/1* queue is

$$W \approx \frac{1}{\mu} + \frac{\rho^2 C_s^2 + C_a^2}{2\lambda\{1-\rho\}}.$$

Extending this heavy-traffic approximation for *G/G/s* queues, we get

$$W \approx \frac{1}{\mu} + \frac{\rho^2 C_s^2 + C_a^2}{2\lambda(1-\rho)}$$

where $\rho = \lambda/(s\mu)$. This approximation is especially accurate for the $G/M/s$ systems and can be proved using fluid and diffusion scaling. Likewise, an empirical approximation for $G/G/s$ queues (originally developed for $M/G/s$ queues) is

$$W \approx \frac{1}{\mu} + \frac{\alpha_s}{\mu}\left(\frac{1}{1-\rho}\right)\left(\frac{C_a^2 + C_s^2}{2s}\right),$$

where one should choose α_s such that

$$\alpha_s = \begin{cases} \frac{\rho^s + \rho}{2} & \text{if } \rho > 0.7, \\ \rho^{\frac{s+1}{2}} & \text{if } \rho < 0.7. \end{cases}$$

Another empirical approximation for $G/G/s$ queues is

$$W_q \approx \frac{W_{q,M/M/s}}{W_{q,M/M/1}} W_{q,G/G/1}$$

where $W_{q,M/M/s}$ is the expected time before service begins in an $M/M/s$ queue with the same λ and μ as the $G/G/s$ queue, likewise $W_{q,M/M/1}$ and $W_{q,G/G/1}$ are the wait times before service for the corresponding queues. However, notice that $W_{q,G/G/1}$ still relies on approximations, and further both $W_{q,G/G/1}$ and $W_{q,M/M/1}$ use the same ρ as the $G/G/s$.

Bounds for the $G/G/s$ system are described in Bolch et al. [12] using Kingman's upper bound and Brumelle/Marchal's lower bound as

$$\frac{1}{\mu} + \frac{\rho^2 C_s^2 - \rho(2-\rho)}{2\lambda(1-\rho)} - \frac{s-1}{s}\frac{\left(C_s^2+1\right)}{2\mu} \leq W \leq \frac{1}{\mu} + \frac{C_a^2 + s\rho^2 C_s^2 + (s-1)\rho^2}{2\lambda(1-\rho)}.$$

Of course, by suitably substituting for appropriate terms, these bounds could naturally be used for the $M/G/s$ and the $G/M/s$ queues. The bounds are reasonable when ρ is small or when s is small. Fortunately, there are heavy-traffic approximations and other asymptotic results that can be used when ρ approaches 1 and s is large. Hence in some sense, based on the situation, an appropriate method can be used.

Having described the $G/G/s$ bounds and approximations, we next describe a case study of a call center, which is one domain where the results have been extensively used for planning and operations. The main objective of the case study is to illustrate the process of making decisions for design and operation of a system. The case study has been inspired by "Interstate Rail and Trucking Company: A Case Study in Applying Queueing Theory" (at the time of writing this book, the exact source is unknown). Upon completing the case study, in the next section we will move to other methods to analyze $G/G/s$ queues.

4.3.3 Case Study: Staffing and Work-Assignment in Call Centers

TravHelp, a mid-size in-bound call center, is located in a suburb of Philadelphia. As the name suggests, TravHelp provides customer support for clients for a variety of travel-related companies that are web-based. Although the user-interface and options for online reservations for travel have come a long way, there still are many instances where individuals feel the need to pick up the phone and talk to a human so that their needs are addressed. For example, the call center might get a call from a user who entered all information including her credit card details, and before a confirmation number was given, her computer quit on her. So she is calling the number listed for customer service to see if her airline ticket is confirmed, obtain a confirmation number, and have an e-ticket emailed to her. For something like this, TravHelp is equipped with the latest computer telephony integration technology that enables their agents to not only speak to the customers needing help but also view the data they have entered.

In essence, there are three parties involved in the system. One is TravHelp, which is a call center providing customer support. The second party is the set of web-based travel-related companies or travel agencies that outsource customer support to TravHelp so that their customers can call TravHelp for their difficulties experienced while making reservations. These travel agencies are of various sizes, from large well-known companies to small mom-and-pop travel agents. And finally, the third party is the customer. Customer calls come in many flavors, some for business, some for pleasure and others for both. Sometimes customers are travelers themselves, other times they are office staff of the travelers. Under this three-way relationship, the objective of TravHelp is to provide customer-support for their travel-agent clients' web-based systems at the highest quality while incurring the minimum cost. However, since the client (i.e., travel agency) is removed from the customer-support operations, sometimes it becomes necessary for TravHelp representatives (or reps) to call the client for authorization.

To achieve the objective described in the preceding text, TravHelp organized its reps into *clusters*. In each cluster, there are about four reps. Recall the variability in terms of size of travel-agency clients. Besides that, the customer base is all over the United States and Canada, and a variety of services are being catered to (e.g., airline tickets, hotels, car rental, and cruise). Thus, the clusters were formed, some based on the major clients, some based on the service offered, and some by geographic location. TravHelp made sure that the workloads were fairly divided among the clusters and incoming calls were appropriately routed. In the beginning of 2011, TravHelp had grown so much that they had 122 reps during the *peak shift* of a work day. The 122 reps were divided into 32 clusters. However, even with all these reps, customers sometimes had to wait over 5 min to speak with a rep. That seemed unacceptably high to the owner of TravHelp and felt something needed to be done. So the owner of TravHelp decided to call one of his former classmates from Wharton who runs a professional consulting firm ProCon.

4.3.3.1 TravHelp Calls for Help

Jacob, who works for ProCon, just finished a long consulting project for a large airline company and was about to leave for the day when his boss called him. Jacob was briefed about the customer service problems TravHelp was facing and was assigned to work on that as his next project. Jacob did not know much about call centers. His main exposure was through the TV series *Outsourced* and a movie by the same name. Based on those, it did not occur to him that there could be issues in call centers, for those actors seemed to be so satisfied and stress-free at their jobs. So Jacob did his homework and showed up the next morning at TravHelp. After understanding the issues TravHelp was facing, Jacob decided to dig deeper and get more data. Fortunately, TravHelp had maintained electronic logs of customer call arrival times, call completion times, reps they spoke to, duration of each call, wait times to talk to reps, and details about the callers (company, geographic location, and travel-request type).

Jacob burned the data on a CD and took it back to ProCon to analyze it as well as to brainstorm with his colleagues. As a result, some interesting findings emerged. The average call-arrival rate was 2030.3 per hour during the peak shift. At that time, 122 reps were employed and there were 32 clusters of reps. A histogram of interarrival times of calls is depicted in Figure 4.2 and an exponential distribution seemed to be a good fit. In terms of service times, the results showed that there was no statistically significant difference in the mean and variance of the service times across client-company size, location, or travel-request type. Thus, the aggregated service times across all customers in the sample data was used. It had a mean of 3 min 20 s and a standard deviation of about 2 min 30 s. Further, almost 40% of the customers waited at least 10 s to speak to a representative and the overall average wait time was close to 2 min (this includes the 60% customers that waited less than 10 s, most of which were practically 0 s).

Jacob wanted to dig deeper and build a computer simulation of the system. He used a gamma distribution for the service times as they fit the data best among the distributions he tested. He was able to calibrate his simulation results against the waiting time data available for all customers. The waiting time histogram using the simulations matched closely that of the data collected. Jacob was thrilled but he was not sure of what to do. He twiddled his thumbs for a while and played with the simulation features a little. Then suddenly, Jacob noticed something when he was running his simulations in animation mode. He saw that sometimes there were long lines at certain clusters of reps with many customers on hold, whereas at the same time there were other clusters where the reps were idle. Although the load was balanced across time, at any given time the system was inefficient. Jacob

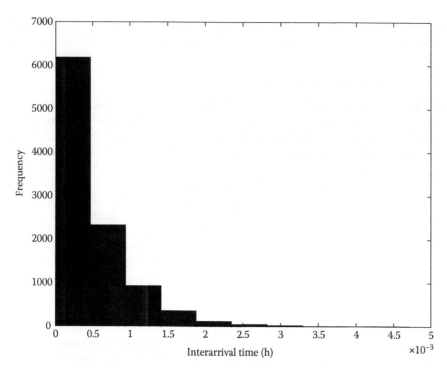

FIGURE 4.2
Histogram of interarrival times of customer calls.

immediately realized that this was because of the 32 separate queueing systems. He soon recalled a homework problem from his queueing theory class, where it was shown that a single queue would be more efficient than having multiple parallel lines.

4.3.3.2 Recommendation: Major Revamp

Jacob was elated because he knew all that was needed was to get rid of the clusters and throw all the reps into a single large pool of servers with a single queue. Of course, he still has to calculate the average time each customer would wait before talking to a rep so that a meaningful comparison could be made with the current system. However, he realized there was only a little time for the next brainstorming meeting with his colleagues and boss. Clearly, he was not going to be able to run a full-blown simulation study with many replications. But asking himself how hard would it be to analyze it as an $M/G/s$ queue since the arrivals were according to a Poisson process (with $\lambda = 2030.3$ per hour and $C_a^2 = 1$), the service times generally distributed (with $\mu = 18$ per hour and $C_s^2 = 0.5625$) and $s = 122$ servers. He quickly pulled out the text he used for his queueing theory course and found two formulae for $G/G/s$ queues in there that would be appropriate to use. The first was a heavy-traffic approximation

$$W_q \approx \frac{\rho^2 C_s^2 + C_a^2}{2\lambda(1 - \rho)}$$

where $\rho = \lambda/(s\mu) = 0.956$, which appeared to satisfy the heavy-traffic condition. The second approximation was also appealing because it was originally developed for $M/G/s$ queues:

$$W_q \approx \frac{\alpha_s}{\mu} \left(\frac{1}{1 - \rho} \right) \left(\frac{C_a^2 + C_s^2}{2s} \right),$$

where $\alpha_s = (\rho^s + \rho)/2$ when $\rho > 0.7$ (in this case, ρ is well above 0.7). Plugging in the numbers for W_q, he got the average waiting time for each customer to speak to a rep as 0.29 min and 0.13 min using the first and second formula, respectively.

Although Jacob realized these were approximations, he felt that they were clearly lower than the 2 min wait that the current system customers experience on average. Jacob wondered what if there were a fewer number of reps. He checked for $s = 116$ reps (a reduction of 6 reps) and the average wait times for customers to speak to a rep became 0.82 and 0.41 min using the first and second formula, respectively. Jacob was thrilled, he looked at his watch and there was enough time to grab a quick latte from a nearby coffee shop before his meeting. At the brainstorming meeting, Jacob presented his recommendation, which is to consolidate the 32 clusters into a single large cluster. When a customer service call arrives, it would go to a free rep if one is available, otherwise the call would be put on hold until a rep became free. Jacob also suggested reducing the number of reps to 116. His colleagues liked the idea. One of them also added another recommendation: to use a monitor to display the number of customers on hold that all the reps can see. That way, if the number of customers on hold is zero, then the reps who are busy can spend more time talking to their customers projecting greater concern. Whereas if there is a large number of calls on hold, reps can quickly wrap up their calls. Jacob liked the idea, and when he saw his latte he realized his coffee shop also adopts a similar notion where if there is a long line, the orders are taken quickly and if it is empty, the workers spend time chatting with customers.

Jacob rechecked all his calculations to make sure everything was alright. Then he proceeded to TravHelp. He made the recommendations that were discussed. TravHelp decided to adopt them but continue with the 122 reps. It was an easy redesign for them and they assigned calls to available reps in some fair round-robin fashion. TravHelp also decided to use monitors that were spread throughout the call center and reps knew how long the lines were at all times. TravHelp monitored their system as usual and also collected data electronically as they had done before. Jacob told TravHelp he would return in a week to see how things were going and analyze the data to see the actual improvements.

4.3.3.3 Findings and Adjustments

Four days later, when Jacob checked his voicemail, there was a frantic call from TravHelp. They wanted him there right away. Jacob raced over and found that TravHelp was inundated with complaints both from customers and from clients. Customers who have used TravHelp before found it was taking a lot longer for reps to address their concerns. Clients felt that reps were calling them for approval for the same things over and over. Jacob was told that in the previous system with clusters, since the reps were organized based on clients, if reps checked something with the clients, they would immediately share that with the other reps in their cluster. Hence, repeated calls to clients about the same issue were not made. Also, when Jacob talked to the reps he found that their morale was low. That is attributed to not only due to customers and clients complaining to them, but also because they felt pressurized when they saw a large number of calls on hold. This resulted in the reps' loss of personal satisfaction.

Jacob quickly collected the four days' worth data on a CD and began analyzing it. He found three things that he did not anticipate. First off, the average waiting time to speak to a rep was over 90 s (which although is still better than almost 2 min in the original system, it is still quite unacceptable and not what he had calculated). Next, the aggregate service times had a mean of 3.5 min and a standard deviation of 4.75 min. These were significantly higher than the corresponding values in the original system. Finally, a more curious result was that several customers called multiple times. Upon interviewing the reps, Jacob found out that in the original system since the reps were clustered according to the major clients, services offered, or geographic locations, they were extremely familiar with their tasks. Now they had to learn more with the new system. They also had to call clients, talk to another fellow rep or open up several computer screens, or perform detailed searches. These clearly increased the mean service times. In particular, the difficult-to-answer questions took much longer and hence the standard deviation of service times increased.

In addition, as described earlier, clients were frustrated because different reps would call to ask the same questions, which did not happen in the original system. The reps were constantly getting complaints not only from clients but also customers. It pressurized the reps so much and their morale went down. Jacob immediately realized he needed to develop a better staffing and work-assignment scheme for the call center. He figured it was crucial to retain the clusters with major clients. He also felt there was merit in clustering according to some but not all services. He found out from the managers in TravHelp that the geographic clusters were mainly for personnel reasons (reps started at different times and to accommodate that they used different time zones). So this time Jacob carefully redesigned the system with two layers of reps. In the first layer, he recommended a set of 20 reps that made the initial contact to determine the appropriate cluster to forward the call. These calls lasted less than 30 s each. At the second layer, there were 23 specialized clusters each with 3–5 reps, as well as a large pool of 40 reps. The specialized small clusters were for the large volume of quick calls that were either client specific or for a single service type. The remaining calls were being handled by the large pool.

To develop this design, Jacob had to perform several what-if analyses and used queueing approximations to obtain quick results. He also worked closely with the managers and reps at TravHelp to understand the implications and estimate quantities for service times. Finally, before implementing the solution, Jacob developed a simulation of the system. It revealed that the average wait time of customers (not including the time they spend speaking to reps) was less than a minute. However, if one were to classify customers into groups, those that require longer service times and those that require short ones, then the wait times were larger for the former set of customers. But this was in line with customers' expectations. Thus, this differentiated service was palatable to customers as well as clients and the reps' morale was restored. TravHelp implemented the new system and the results matched those predicted by Jacob's simulations. Jacob was delighted about that. He was also appreciative of the use of queueing approximations for doing quick analysis. And last but not the least, he realized the importance of considering behavioral aspects while making decisions, the criticality to talk to the individuals involved to understand the situation better, and finally that perceptions of customers is a crucial thing to consider.

4.4 Matrix Geometric Methods for $G/G/s$ Queues

In the previous section, we saw a set of techniques that use the exact model-input (such as interarrival times and service times) to get approximate performance measures. Another approach that would be the focus of this section is to approximate the model-input itself and then do an exact analysis for the performance measures. In particular, we approximate the arrival process and service process as "mixtures" of exponentials called *phase-type* distributions. Then, it would be possible to model the resulting system as a multidimensional CTMC. The added benefit beyond being able to analyze exactly is that LSTs can also be easily obtained. In fact, performance measures such as distribution of sojourn times and number in the system (under moderate traffic, not light, or heavy traffic) can only be obtained this way. However, for a simple system like a $G/G/s$ queue, it is fairly computationally intensive. But this is a powerful technique that can be effectively used in a wide variety of applications beyond queueing.

4.4.1 Phase-Type Processes: Description and Fitting

The key requirement before using matrix geometric methods is to model the arrival processes and service times using phase-type distributions. In theory, it is possible to find a phase-type distribution to approximate the CDF of any positive-valued random variable as closely as desired. In other words, if one wishes to approximate a CDF $G(t)$ by a phase-type distribution, then given an $\epsilon > 0$, there exists a phase-type CDF $F(t)$ such that $|G(t) - F(t)| < \epsilon$ for all $t \geq 0$. It is true even for subexponential distributions such as Pareto with several moments equaling infinite. However, it may not be the easiest thing in practice to find such an $F(t)$. There are several articles that we would refer to later in this section on fitting distributions. Before that we first explain the phase-type distribution.

A phase-type distribution is essentially a mixture of exponential distributions. In particular, we consider m exponential distributions mixed in various forms to form a phase-type distribution. Typically, the phase-type random variable can be mapped into a stochastic process that starts in state i (for $i = 1, \ldots, m$), spends an exponentially distributed random time, then perhaps jumps to state j (for $j = 1, \ldots, m$), spends an exponentially distributed random time, and so on. Finally, the transitions stop and the process ends. The time spent in the process is according to a phase-type distribution and it converges in distribution to the original random variable. The stochastic process can be nicely modeled as a CTMC with $m + 1$ states including an absorption state to indicate the end of transitions.

Consider a CTMC $\{Z(t), t \geq 0\}$ defined on $m + 1$ states $1, 2, \ldots, m, m + 1$, and an infinitesimal generator matrix

$$Q = \begin{bmatrix} T & T^* \\ \overline{0} & 0 \end{bmatrix}$$

where
  T is an $m \times m$ matrix,
  T^* an $m \times 1$ vector,
  $\overline{0}$ a $1 \times m$ vector of zeros.

We explicitly assume that $m + 1$ is the absorption state. Also notice that the rows add up to zero, hence $T^* = -T\overline{1}$ where $\overline{1}$ is an $m \times 1$ vector of ones. Let Y be a random variable that denotes the first passage time to state $m + 1$ given an $1 \times (m + 1)$ initial probability vector $[\overline{p} \ \ p_{m+1}]$, with $\overline{p}$ a $1 \times m$ vector. In other words,

$$Y = \min\{t \geq 0 : Z(t) = m + 1\}.$$

The random variable Y is according to a phase-type distribution with m phases. Also note that the stochastic process $\{Z(t), t \geq 0\}$ could transition to any phase zero or more times. Unless stated otherwise in this book, we will always assume $p_{m+1} = 0$.

For all $y \geq 0$, the CDF of the phase-type random variable is

$$F(y) = P\{Y \leq y\} = 1 - \bar{p}\exp(Ty)\bar{1},$$

so that the PDF is

$$f(y) = \frac{dF(y)}{dy} = \bar{p}\exp(Ty)T^*.$$

Notice the use of exponential of a matrix that is defined for a square matrix M as

$$\exp(M) = I + M + \frac{M^2}{2!} + \frac{M^3}{3!} + \cdots$$

and when M is a scalar, it reduces to the regular e^M expression. We can also compute the moments of Y as

$$E[Y^k] = (-1)^k k! \bar{p} T^{-k} \bar{1}.$$

The key idea is that if a positive-valued random variable X with CDF $G(\cdot)$ needs to be approximated as a phase-type distribution Y with CDF $F(\cdot)$, then there exists at least one m, $\bar{p}$, and Q that would ensure that $F(y)$ is arbitrarily close to $G(y)$ for all $y \geq 0$. However, in practice, choosing or finding the appropriate m, $\bar{p}$, and Q is nontrivial. To alleviate that concern of over-parameterization, one typically considers the following special types of phase-type distributions (with much fewer parameters to estimate):

1. *Hypoexponential distribution*: A phase-type distribution with $\bar{p} = [1\ 0\ \ldots\ 0]$, $p_{m+1} = 0$, and $m \times m$ matrix

$$T = \begin{bmatrix} -\lambda_1 & \lambda_1 & 0 & \ldots & 0 \\ 0 & -\lambda_2 & \lambda_2 & \ldots & 0 \\ \vdots & \vdots & \ddots & \ldots & \vdots \\ 0 & 0 & 0 & \ldots & \lambda_{m-1} \\ 0 & 0 & 0 & \ldots & -\lambda_m \end{bmatrix}.$$

In this case, m needs to be determined (usually by trial and error), then the only parameters to be estimated are $\lambda_1, \lambda_2, \ldots, \lambda_m$. If the random variable to be approximated has SCOV < 1, then the hypoexponential distribution would be ideal. A special case of the hypoexponential distribution is the Erlang distribution, when $\lambda_1 = \lambda_2 = \cdots = \lambda_m$.

2. *Hyperexponential distribution*: A phase-type distribution with $\bar{p} = [p_1 \ p_2 \ \cdots \ p_m]$, $p_{m+1} = 0$ (such that $p_1 + p_2 + \cdots + p_m = 1$) and $m \times m$ diagonal matrix

$$
T = \begin{bmatrix}
-\lambda_1 & 0 & 0 & \cdots & 0 \\
0 & -\lambda_2 & 0 & \cdots & 0 \\
\vdots & \vdots & \ddots & \cdots & \vdots \\
0 & 0 & 0 & \cdots & 0 \\
0 & 0 & 0 & \cdots & -\lambda_m
\end{bmatrix}.
$$

In this case, m needs to be determined (usually by trial and error), then the parameters to be estimated are $\lambda_1, \lambda_2, \ldots, \lambda_m$ as well as p_1, $p_2, \ldots, p_m$. If the random variable to be approximated has SCOV > 1, then a hyperexponential distribution can be found that is extremely accurate.

3. *Generalized Coxian distribution*: A phase-type distribution with $\bar{p} = [p_1 \ p_2 \ \cdots \ p_m]$, $p_{m+1} = 0$ (such that $p_1 + p_2 + \cdots + p_m = 1$), and $m \times m$ matrix

$$
T = \begin{bmatrix}
-\lambda_1 & \alpha_1\lambda_1 & 0 & \cdots & 0 \\
0 & -\lambda_2 & \alpha_2\lambda_2 & \cdots & 0 \\
\vdots & \vdots & \ddots & \cdots & \vdots \\
0 & 0 & 0 & \cdots & \alpha_{m-1}\lambda_{m-1} \\
0 & 0 & 0 & \cdots & -\lambda_m
\end{bmatrix}.
$$

In this case, m needs to be determined (usually by trial and error), then the parameters to be estimated are $\lambda_1, \lambda_2, \ldots, \lambda_m$, $p_1, p_2, \ldots,$ p_m, as well as $\alpha_1, \alpha_2, \ldots, \alpha_{m-1}$. Notice how the hypoexponential and hyperexponential distributions are special cases of the generalized Coxian distribution. In practice, usually m is selected as a small value such as 2 or 3. A more commonly used approximation is the regular Coxian distribution that requires $\bar{p} = [1 \ 0 \ \cdots \ 0]$, that is, one starts in the first phase.

Before concluding this section and analyzing phase-type queues, here are a few words in terms of *fitting* a positive-valued random variable X with CDF $G(\cdot)$ as a phase-type distribution Y with CDF $F(\cdot)$. There are several papers that discuss the selection of m, $\bar{p}$, and Q so that the resulting phase-type distribution fits well. A recent paper (Fackrell [34]) nicely summarizes various estimation techniques (such as maximum likelihood, moment matching, least squares, and distance minimization) and suggests ways to cope with over-parameterization. One method to estimate r parameters, say, is to solve for equations of the type $F(x_1) = G(x_1)$, $F(x_2) = G(x_2)$, ..., $F(x_{r-2}) = F(x_{r-2})$ and also match the first two moments. Of course, for this the choice of x_i for $i = 1, \ldots, r - 2$ are important. Feldmann and Whitt [36] discuss some drawbacks of moment-matching methods and suggest alternative methods of fitting hyperexponential distributions for heavy-tailed distributions (such as Weibull and Pareto). A recent paper by Osogami and Harchol-Balter [86] describes mapping a general distribution into a regular Coxian distribution. Since hypoexponential and Erlang distributions are special cases of the regular Coxian distribution, the mapping can directly be used for those too. Note that the hyperexponential distribution is only a special case of generalized Coxian but not the regular Coxian distribution. Further, the references in Osogami and Harchol-Balter [86] also describe other techniques for fitting phase-type distributions.

4.4.2 Analysis of Aggregated Phase-Type Queue ($\sum PH_i/PH/s$)

We originally only described how to convert the $G/G/s$ queue to the $PH/PH/s$ queue (PH is *phase-type* in the Kendall notation) with phase-type interarrival time and IID service times with s servers. The analysis can be generalized with a very similar notation in two ways: (1) by considering a superposition of many phase-type arrivals and (2) by considering heterogeneous servers. However, (2) would result in an explosion of the state-space unless s is small such as less than 3. Therefore, we *only* consider (1) here and call such a system as a $\sum PH_i/PH/s$ queue. The results are from Curry and Gautam [21] and they have been derived by extending the single server $\sum PH_i/PH/1$ system considered in Bitran and Dasu [11] to multiple servers. What enables us to develop a more general system beyond just the $PH/PH/s$ queue is the ability to use Kronecker products ($\otimes$) and sums ($\oplus$) to write down expressions more elegantly. It may be worthwhile to explain Kronecker sums and products first.

Consider a matrix $\overline{L}$ that is made up of elements l_{ij} and another matrix $\overline{M}$ that is made up of elements m_{ij} so that we write $\overline{L} = [l_{ij}]$ and $\overline{M} = [m_{ij}]$. Then the Kronecker product of $\overline{L}$ and $\overline{M}$ is given by

$$\overline{L} \otimes \overline{M} = [l_{ij}\overline{M}].$$

For example, if we choose $\overline{L}$ and $\overline{M}$ such that

$$\overline{L} = \begin{bmatrix} l_{11} & l_{12} & l_{13} \\ l_{21} & l_{22} & l_{23} \\ l_{31} & l_{32} & l_{33} \end{bmatrix} \quad \text{and} \quad \overline{M} = \begin{bmatrix} m_{11} & m_{12} \\ m_{21} & m_{22} \end{bmatrix}$$

then we have

$$\bar{L} \otimes \bar{M} = \begin{bmatrix} l_{11}m_{11} & l_{11}m_{12} & l_{12}m_{11} & l_{12}m_{12} & l_{13}m_{11} & l_{13}m_{12} \\ l_{11}m_{21} & l_{11}m_{22} & l_{12}m_{21} & l_{12}m_{22} & l_{13}m_{21} & l_{13}m_{22} \\ l_{21}m_{11} & l_{21}m_{12} & l_{22}m_{11} & l_{22}m_{12} & l_{23}m_{11} & l_{23}m_{12} \\ l_{21}m_{21} & l_{21}m_{22} & l_{22}m_{21} & l_{22}m_{22} & l_{23}m_{21} & l_{23}m_{22} \\ l_{31}m_{11} & l_{31}m_{12} & l_{32}m_{11} & l_{32}m_{12} & l_{33}m_{11} & l_{33}m_{12} \\ l_{31}m_{21} & l_{31}m_{22} & l_{32}m_{21} & l_{32}m_{22} & l_{33}m_{21} & l_{33}m_{22} \end{bmatrix}.$$

To develop the Kronecker sum, we define identity matrices I_L and I_M corresponding to the sizes of $\bar{L}$ and $\bar{M}$, respectively. In the previous example,

$$I_L = \begin{bmatrix} 1 & 0 & 0 \\ 0 & 1 & 0 \\ 0 & 0 & 1 \end{bmatrix} \quad \text{and} \quad I_M = \begin{bmatrix} 1 & 0 \\ 0 & 1 \end{bmatrix}.$$

The Kronecker sum of any two matrices $\bar{L}$ and $\bar{M}$ is given by

$$\bar{L} \oplus \bar{M} = \bar{L} \otimes I_M + I_L \otimes \bar{M}.$$

An interesting property of the Kronecker sums and products when applied to the exponential of a matrix is that

$$\exp(\bar{L}) \otimes \exp(\bar{M}) = \exp(\bar{L} \oplus \bar{M})$$

for square matrices $\bar{L}$ and $\bar{M}$.

Now, using the Kronecker product and sum characterization, we analyze the $\sum PH_i/PH/s$ system. Consider a queue with infinite waiting room and s servers that are identical. Arrivals occur into this queue as a superposition of K streams (or sources). For $i = 1, \ldots, K$ let the interarrival times for the ith arrival stream be according to a phase-type distribution with parameters $m_{A,i}$, $\bar{p}_{A,i}$, and $T_{A,i}$ representing the number of phases, the initial probability vector, and the T matrix, respectively. For the sake of notation, we also define $T_{A,i}^*$ that can easily be derived from $T_{A,i}$ as described earlier. Let $Z_i(t)$ be the state (corresponding to the phase) of the ith arrival process at time t. In a similar manner, let the service times at the server be according to a phase-type distribution with parameters m_S, $\bar{p}_S$, and T_S. Once again we also define T_S^*, which can easily be derived from T_S. Let $U_j(t)$ be the state (corresponding to the phase) of the jth working server at time t, for $j = 1, \ldots, s$.

Define $X(t)$ as the number of entities (or customers) in the system at time t. Then the multidimensional stochastic process

$$\{(X(t), Z_1(t), \ldots, Z_K(t), U_1(t), \ldots, U_{\min[X(t),s]}(t)), t \geq 0\}$$

(although if $X(t) = 0$, then the state is just $\{(X(t), Z_1(t), \ldots, Z_K(t)\})$ is a CTMC. The CTMC has lexicographically ordered states with infinitesimal generator (i.e., Q) matrix of the QBD block diagonal form:

$$
\begin{pmatrix}
B_{0,0} & B_{0,1} & 0 & 0 & \cdots & 0 & 0 & 0 & 0 & 0 & 0 & 0 & \cdots \\
B_{1,0} & B_{1,1} & B_{1,2} & 0 & \cdots & 0 & 0 & 0 & 0 & 0 & 0 & 0 & \cdots \\
0 & B_{2,1} & B_{2,2} & B_{2,3} & \cdots & 0 & 0 & 0 & 0 & 0 & 0 & 0 & \cdots \\
\vdots & \ddots & \ddots & \ddots & \cdots & \vdots & \vdots & \vdots & \vdots & \vdots & \vdots & \vdots & \cdots \\
0 & 0 & 0 & 0 & \cdots & B_{s-1,s-2} & B_{s-1,s-1} & B_{s-1,s} & 0 & 0 & 0 & 0 & \cdots \\
0 & 0 & 0 & 0 & \cdots & 0 & B_{s,s-1} & A_1 & A_0 & 0 & 0 & 0 & \cdots \\
0 & 0 & 0 & 0 & \cdots & 0 & 0 & A_2 & A_1 & A_0 & 0 & 0 & \cdots \\
0 & 0 & 0 & 0 & \cdots & 0 & 0 & 0 & A_2 & A_1 & A_0 & 0 & \cdots \\
0 & 0 & 0 & 0 & \cdots & 0 & 0 & 0 & 0 & A_2 & A_1 & A_0 & \cdots \\
\vdots & \vdots & \vdots & \vdots & \cdots & \vdots & \vdots & \vdots & \vdots & \ddots & \ddots & \ddots & \ddots
\end{pmatrix}.
$$

The matrices, $B_{i,j}$, correspond to transition rates from states where the number in the system is i to states where the number in the system is j for $i, j \leq s$. Also, A_0, A_1, and A_2 are identical to those in the QBD description and valid when there are more than s customers in the system. We determine the matrices A_0, A_1, A_2, and $B_{i,j}$ using Kronecker sums and products as follows (for $i = 1, \ldots, s$):

$$
A_2 = I_{\prod_{i=1}^K m_{A,i}} \otimes \left[\bigoplus_s (T_S^* \otimes \overline{p_S}) \right]
$$

$$
A_1 = T_{A,1} \oplus T_{A,2} \oplus \cdots \oplus T_{A,K} \oplus \left[\bigoplus_s (T_S) \right]
$$

$$
A_0 = \left[\left(T_{A,1}^* \otimes \overline{p_{A,1}} \right) \oplus \left(T_{A,2}^* \otimes \overline{p_{A,2}} \right) \oplus \cdots \oplus \left(T_{A,K}^* \otimes \overline{p_{A,K}} \right) \right] \otimes I_{(m_S)^s}
$$

$$
B_{0,0} = T_{A,1} \oplus T_{A,2} \oplus \cdots \oplus T_{A,K}
$$

$$
B_{i,i-1} = I_{\prod_{i=1}^K m_{A,i}} \otimes \left[\bigoplus_i (T_S^*) \right]
$$

$$
B_{i,i} = T_{A,1} \oplus T_{A,2} \oplus \cdots \oplus T_{A,K} \oplus \left[\bigoplus_i (T_S) \right]
$$

$$
B_{i-1,i} = \left[\left(T_{A,1}^* \otimes \overline{p_{A,1}} \right) \oplus \left(T_{A,2}^* \otimes \overline{p_{A,2}} \right) \oplus \cdots \oplus \left(T_{A,K}^* \otimes \overline{p_{A,K}} \right) \right] \otimes I_{(m_S)^{i-1}} \otimes \overline{p_S}
$$

where $\bigoplus_j(M)$ is the Kronecker sum of matrix M with itself j times, that is, $\bigoplus_2(M) = M \oplus M$ and $\bigoplus_3(M) = M \oplus M \oplus M$.

Having modeled the system as a QBD, the next step is to calculate the steady-state probabilities using MGM. Notice that this is just a minor extension of the MGM described in Chapter 3 and we will just use the same analysis here. The reader is encouraged to read Section 3.2.2 before proceeding. First, we assume that $A_0 + A_1 + A_2$ is an irreducible infinitesimal generator with stationary probability π (a $1 \times m$ row vector) such that $\pi(A_0 + A_1 + A_2) = \bar{0}'$ and $\pi\bar{1} = 1$, where for $i = 0, 1, \bar{i}$ is a column vector of i's. The irreducibility assumption is automatically satisfied for phase-type distributions such as hypoexponential, hyperexponential, and Coxian. Once π is obtained, the condition that the $\sum PH/PH/s$ queue is stable is

$$\pi A_2 \bar{1} > \pi A_0 \bar{1}$$

and this usually corresponds to the total mean service rate due to all s servers being larger that the arrival rate on average.

If the queue is stable, then the next step is to find R that is the minimal nonnegative solution to the equation

$$A_0 + RA_1 + R^2 A_2 = 0.$$

Now let p_i be a row-vector of steady-state probabilities, where the number in the system is i for $i = 0, 1, 2, \ldots$ and satisfies the flow balance equations. Then using MGM results, we know that p_i for $i = 0, 1, \ldots, s$ can be computed by solving (with $p_{s-2} = 0$ if $s = 1$):

$$p_0 B_{0,0} + p_1 B_{1,0} = 0$$

$$p_0 D_{0,1} + p_1 B_{1,1} + p_2 B_{2,1} = \bar{0}$$

$$p_1 B_{1,2} + p_2 B_{2,2} + p_3 B_{3,2} = 0$$

$$p_2 B_{2,3} + p_3 B_{3,3} + p_4 B_{4,3} = 0$$

$$\vdots \quad \vdots \quad \vdots$$

$$p_{s-2} B_{s-2,s-1} + p_{s-1} B_{s-1,s-1} + p_s B_{s,s-1} = 0$$

$$p_{s-1} B_{s-1,s} + p_s A_1 + p_s R A_2 = 0$$

$$p_0 \bar{1} + p_1 \bar{1} + \cdots + p_{s-1} \bar{1} + p_s (I - R)^{-1} \bar{1} = 1$$

where $\bar{1}$ is an appropriately sized column vector of ones. Then, using $p_i = p_s R^{i-s}$ for all $i \geq s+1$, we obtain $p_{s+1}, p_{s+2}, p_{s+3}$, etc. Thus, the steady-state distribution of the number of entities in the system can be evaluated as

$$\lim_{t\to\infty} P\{X(t) = j\} = p_j\overline{1}.$$

Using this, it is rather straightforward to obtain performance measures such as L, L_q, W, and W_q. In particular,

$$L_q = \sum_{i=1}^{\infty} i p_{s+i}\overline{1} = \sum_{i=1}^{\infty} i p_s R^i\overline{1} = p_s R(I - R)^{-2}\overline{1}.$$

Using λ as the mean arrival rate and μ as the mean service rate for each server (both of which can be computed from the phase-type distributions), we can write down $W_q = L_q/\lambda$, $W = W_q + 1/\mu$, and $L = \lambda W$. However, what is not particularly straightforward is the sojourn time distribution. For that we need to know the arrival point probabilities in steady state, that is, the distribution of the state of the system when an entity arrives into the system. Once that is known by conditioning on the arrival point probabilities, then computing the LST of the conditional sojourn time for the arriving customer, and then by unconditioning, we can obtain the sojourn time distribution. Even for a simple example, this computation is fairly tedious and hence not presented here. In the next section, we consider an example application to illustrate the results seen in this section.

4.4.3 Example: Application in Semiconductor Wafer Fabs

The semiconductor industry consists of extremely long and complex manufacturing processes for various products. These products are manufactured via multiple layered production processes where they make many passes through a sequence of similar processing steps. With the high cost and high value of the resulting products, it is important to carefully analyze these production systems. It is often fruitful to develop analytical models of these systems and subprocesses. One of the most widely utilized analytical tool in semiconductor fab systems analysis is based on queueing. To effectively develop an analytical model of a complex network of production processes such as those encountered in the semiconductor industry, it is necessary to utilize a modeling approach of decomposing the complex interconnected system of processing workstations. The decomposition approach treats each workstation individually by characterizing the inflow and outflow streams. Therefore, it is paramount that we have accurate characterizations of these product flow streams. In this section, we obtain the performance of a single workstation assuming that we know the characteristics of the superimposed arrival streams and service times.

We consider a simple system with two distinct arrival streams ($K = 2$) and two identical servers ($s = 2$). The two independent arrival streams are modeled as two-phase (thus $m_{A,1} = m_{A,2} = 2$) exponential processes with a given probability of using the second phase (therefore, a Coxian distribution). Even though the individual arrival processes are independent, the time between arrivals in the superpositioned process are correlated. In a similar fashion, the service processes are independent (but identical) Coxian processes with $m_S = 2$. We now describe the parameters of the arrival and service processes. The first arrival stream is such that the interarrival time is equal to either an $\exp(\lambda_1)$ distribution with probability $1 - \alpha$ or a sum of two exponentials (with parameters λ_1 and λ_2) with probability α. Hence, the resultant process is a Coxian distribution, which is a special type of phase-type distribution with $m_{A,1} = 1, \overline{p_{A,1}} = [1 \ 0]$:

$$T_{A,1} = \begin{bmatrix} -\lambda_1 & \alpha\lambda_1 \\ 0 & -\lambda_2 \end{bmatrix} \quad \text{and} \quad T^*_{A,1} = \begin{bmatrix} (1-\alpha)\lambda_1 \\ \lambda_2 \end{bmatrix}.$$

Likewise, the second arrival stream as well as service times are two-phase Coxian distributions. In particular, the second arrival stream has parameters $m_{A,2} = 1, \overline{p_{A,2}} = [1 \ 0]$,

$$T_{A,2} = \begin{bmatrix} -\gamma_1 & \beta\gamma_1 \\ 0 & -\gamma_2 \end{bmatrix} \quad \text{and} \quad T^*_{A,1} = \begin{bmatrix} (1-\beta)\gamma_1 \\ \gamma_2 \end{bmatrix}.$$

Each entity requires a service at one of the two identical servers so that the service time is according to a Coxian distribution with parameters $m_S = 1$, $\overline{p_S} = [1 \ 0]$,

$$T_S = \begin{bmatrix} -\mu_1 & \delta\mu_1 \\ 0 & -\mu_2 \end{bmatrix} \quad \text{and} \quad T^*_{A,1} = \begin{bmatrix} (1-\delta)\mu_1 \\ \mu_2 \end{bmatrix}.$$

Next we consider a numerical example to illustrate the methodology described in the previous section.

Problem 39

A semiconductor wafer fab has a bottleneck workstation with two identical machines. Products arrive into the workstation from two sources and they wait in a line to be processed by one of the two identical machines. Data suggests that the arrival streams as well as service times can be modeled as two-phase Coxian distributions described earlier. In particular, for the first arrival stream $(\lambda_1, \alpha, \lambda_2) = (20, 0.25, 5)$, for the second arrival stream $(\gamma_1, \beta, \gamma_2) = (9.091, 0.9, 10)$, and for the service times $(\mu_1, \delta, \mu_2) = (10, 0.3333, 20)$. Model the system as a CTMC by writing down the infinitesimal generator in QBD form using Kronecker sums and products.

Then solve for the steady-state probabilities to obtain the average number of products in the system in the long run.

Solution

As defined earlier, $X(t)$ is the number of products in the system, $Z_i(t)$ is the phase of the ith arrival process, and $U_i(t)$ is the phase ith service process if there is a product in service, all at time t and for $i = 1, 2$. Then the multidimensional stochastic process

$$\{(X(t), Z_1(t), Z_2(t), U_1(t), U_{\min[X(t),s]}(t)), t \geq 0\},$$

with lexicographically ordered states, is a CTMC. The preceding is with the understanding that if $X(t) = 0$, then the state is just $\{(X(t), Z_1(t), Z_2(t))\}$, if $X(t) = 1$, then the state is just $\{(X(t), Z_1(t), Z_2(t), U_1(t))\}$, else when $X(t) > 1$, the state $\{(X(t), Z_1(t), Z_2(t), U_1(t), U_2(t))\}$. The infinitesimal generator matrix of the CTMC is of block diagonal form (hence QBD):

$$\begin{pmatrix} B_{0,0} & B_{0,1} & 0 & 0 & 0 & \cdots \\ B_{1,0} & B_{1,1} & B_{1,2} & 0 & 0 & \cdots \\ 0 & B_{2,1} & A_1 & A_0 & 0 & \cdots \\ 0 & A_2 & A_1 & A_0 & \cdots \\ \vdots & \vdots & \ddots & \ddots & \ddots & \ddots \end{pmatrix}$$

where

$$A_2 = I_4 \otimes \left[(T_S^* \otimes \overline{p_S}) \oplus (T_S^* \otimes \overline{p_S}) \right]$$

$$A_1 = T_{A,1} \oplus T_{A,2} \oplus T_S \oplus T_S$$

$$A_0 = \left[\left(T_{A,1}^* \otimes \overline{p_{A,1}} \right) \oplus \left(T_{A,2}^* \otimes \overline{p_{A,2}} \right) \right] \otimes I_4$$

$$B_{0,0} = T_{A,1} \oplus T_{A,2}$$

$$B_{0,1} = \left[\left(T_{A,1}^* \otimes \overline{p_{A,1}} \right) \oplus \left(T_{A,2}^* \otimes \overline{p_{A,2}} \right) \right] \otimes \overline{p_S}$$

$$B_{1,0} = I_4 \otimes T_S^*$$

$$B_{1,1} = T_{A,1} \oplus T_{A,2} \oplus T_S$$

$$B_{1,2} = \left[\left(T_{A,1}^* \otimes \overline{p_{A,1}} \right) \oplus \left(T_{A,2}^* \otimes \overline{p_{A,2}} \right) \right] \otimes I_2 \otimes \overline{p_S}$$

$$B_{2,1} = I_4 \otimes \left[T_S^* \oplus T_S^* \right].$$

Notice that we have numerical values for $T_{A,i}$ and $\overline{p_{A,i}}$ for $i = 1,2$ as well as T_S and $\overline{p_S}$. Therefore, the preceding matrices can be computed. However, since some of the matrices are too huge to be displayed in the following text (e.g., A_0, A_1, and A_2 are 16×16 matrices), we only show a few computations to illustrate the Kronecker product and sum calculations. In particular, verify that

$$B_{0,0} = T_{A,1} \oplus T_{A,2} = \begin{bmatrix} -29.091 & 8.182 & 5 & 0 \\ 0 & -30 & 0 & 5 \\ 0 & 0 & -14.091 & 8.182 \\ 0 & 0 & 0 & -15 \end{bmatrix},$$

also to obtain the 4×8 matrix $B_{0,1}$ we need

$$\left(T_{A,1}^* \otimes \overline{p_{A,1}}\right) \oplus \left(T_{A,2}^* \otimes \overline{p_{A,2}}\right) = \begin{bmatrix} 15.909 & 0 & 0 & 0 \\ 10 & 15 & 0 & 0 \\ 5 & 0 & 0.909 & 0 \\ 0 & 5 & 10 & 0 \end{bmatrix},$$

for the 16×16 A_1 matrix we use

$$T_S \oplus T_S = \begin{bmatrix} -20 & 10/3 & 10/3 & 0 \\ 0 & -30 & 0 & 10/3 \\ 0 & 0 & -30 & 10/3 \\ 0 & 0 & 0 & -20 \end{bmatrix},$$

and the 16×8 matrix $B_{2,1}$ requires

$$T_S^* \oplus T_S^* = \begin{bmatrix} 40/3 & 0 \\ 20 & 20/3 \\ 20 & 20/3 \\ 0 & 40 \end{bmatrix}.$$

Next, assuming that the infinitesimal generator matrix can be obtained numerically (notice that most mathematical software such as MATLAB® and Mathematica have inbuilt programs for Kronecker sums and products) we are ready to perform the MGM analysis. We first check if the condition for stability is met. Notice that the aggregate mean interarrival time can be computed as 0.0667 and the mean service time is 0.01167, resulting in a traffic intensity (arrival rate over twice the service rate) of 87.5%. Hence, the queue is indeed stable.

We find the 16×16 matrix R by obtaining the minimal nonnegative solution to the equation:

$$A_0 + RA_1 + R^2A_2 = 0.$$

Now let p_i be a row-vector of steady-state probabilities where the number in the system is i for $i = 0, 1, 2, \ldots$ and satisfies the flow balance equations. Notice that p_0, p_1, and p_i for $i > 1$ are of lengths 4, 8, and 16, respectively. They can be computed by solving

$$p_0 B_{0,0} + p_1 B_{1,0} = 0$$

$$p_0 B_{0,1} + p_1 B_{1,1} + p_2 B_{2,1} = 0$$

$$p_1 B_{1,2} + p_2 A_1 + p_2 R A_2 = 0$$

$$p_0 \bar{1} + p_1 \bar{1} + p_2 (I - R)^{-1} \bar{1} = 1$$

where $\bar{1}$ is an appropriately sized column vector of ones. Then, using $p_i = p_2 R^{i-2}$ for all $i \geq 3$, we obtain p_3, p_4, p_5, etc. Thus, the steady-state distribution of the number of entities in the system can be evaluated as

$$\lim_{t \to \infty} P\{X(t) = j\} = p_j \bar{1}.$$

Using this we can obtain

$$L_q = \sum_{i=1}^{\infty} i p_{2+i} \bar{1} = \sum_{i=1}^{\infty} i p_2 R^i \bar{1} = p_2 R (I - R)^{-2} \bar{1}.$$

Based on that, we can compute L as 8.5588 customers on average in steady state. ∎

4.5 Other General Queues but with Exact Results

In this section, we present results for few special queues with generally distributed service times. For the analysis of these queues, we will see methods that we have not previously used for the other systems.

4.5.1 M/G/∞ Queue: Modeling Systems with Ample Servers

An $M/G/\infty$ queue is one where the arrivals are according to a Poisson process and the service time for each arriving entity is generally distributed with cumulative distribution function (CDF) $G(\cdot)$. Further, there are an infinite number of servers in the system and, therefore, the sojourn time in the system equals the service time. Let λ be the average arrival rate. Also, let τ denote the mean service time (with respect to the previous notation,

$\tau = 1/\mu$). We will typically consider the mean service time to be finite, that is, $\tau < \infty$. Also, for nontriviality we assume $\tau > 0$. It is crucial to note that there is no restriction in terms of the service time random variable; it could be discrete, continuous, or a mixture of discrete and continuous (however, our analysis is in terms of continuous, for the others the Riemann integral must be replaced by the Lebesgue integral).

Some of the performance measures are relatively straightforward, such as the sojourn time distribution is identical to the service time distribution. Therefore, the key analysis is to obtain the distribution of the number in the system. Define $X(t)$ as the number (of entities) in the system at time t. The objective of transient and steady-state analysis is to obtain a probability distribution of $X(t)$ for finite t and as $t \to \infty$, respectively. We would first present the transient analysis and then take the limit for steady-state analysis.

Transient analysis typically depends on the initial state of the queue. To obtain simple closed-form expressions, we need to make one of the following three assumptions: (i) the queue is empty at $t = 0$, that is, $X(0) = 0$; (ii) the queue started empty in the distant past, that is, $X(-\infty) = 0$, and hence the stochastic process $\{X(t), t \geq 0\}$ is stationary; and (iii) if $X(0) > 0$, then the service for all the $X(0)$ entities *begin* at $t = 0$. If one of the preceding three assumptions are not satisfied, we will have to know the times when service began for each of the $X(0)$ customers in order to obtain a distribution for $X(t)$.

For the transient analysis here, we make the first assumption, that is, $X(0) = 0$. Hence, we are interested in computing $p_j(t)$ defined as

$$p_j(t) = P\{X(t) = j | X(0) = 0\}.$$

Note that if we made the second assumption, then $X(t)$ would be according to the steady state distribution for all $t \geq 0$. However, if we made the third assumption, then

$$P\{X(t) = j | X(0) = i, B_i\} = \sum_{k=0}^{\min(i,j)} \binom{i}{k} [1 - G(t)]^k [G(t)]^{i-k} p_{j-k}(t),$$

with the event B_i denoting that service for all i initial customers begins at $t = 0$. Since this is straightforward once $p_j(t)$ is known, we continue with obtaining $p_j(t)$ by making the first assumption.

Notice that $\{X(t), t \geq 0\}$ is a regenerative process with regeneration epochs corresponding to when the queueing system becomes empty. To obtain $p_j(t)$, consider an arbitrary arrival at time x such that $x \leq t$. Let q_x be the probability that this arriving entity is in the system at time t. Clearly,

$$q_x = 1 - G(t - x)$$

since it is the probability that this entity's service time is larger than $t - x$. Let $\{N(t), t \geq 0\}$ be a Poisson process with parameter λ that counts the number of arrivals in time $(0, t]$ for all $t \geq 0$. Now consider a nonhomogeneous Bernoulli splitting of the Poisson (arrival) process $\{N(t), t \geq 0\}$ such that with probability q_x an entity arriving at time x will be included in the split process. Since the split process counts the number in the $M/G/\infty$ system at time t, we have

$$p_j(t) = \exp\left\{-\lambda \int_0^t q_x \, dx\right\} \frac{\left\{\lambda \int_0^t q_x \, dx\right\}^j}{j!}.$$

The proof is described in Kulkarni [67], Gross and Harris [49], and Wolff [108]. It is based on conditioning the number of arrivals in time t and using the fact that each of the given arrivals occur uniformly in $[0, t]$. The argument is similar to the derivation of the number of departures from the $M/G/\infty$ queue in time t in Problem 41.

Next we write down $p_j(t)$ in terms of entities given in the model, namely, λ and $G(t)$. As a result of using $q_x = 1 - G(t - x)$ and change of variables, we get

$$p_j(t) = \exp\left\{-\lambda \int_0^t q_x \, dx\right\} \frac{\left\{\lambda \int_0^t q_x \, dx\right\}^j}{j!}$$

$$= \exp\left\{-\lambda \int_0^t [1 - G(t - x)] dx\right\} \frac{\left\{\lambda \int_0^t [1 - G(t - x)] dx\right\}^j}{j!}$$

$$= \exp\left\{-\lambda \int_0^t [1 - G(u)] du\right\} \frac{\left\{\lambda \int_0^t [1 - G(u)] du\right\}^j}{j!}.$$

Using $p_j(t)$ we can also obtain

$$E[X(t)] = \lambda \int_0^t [1 - G(u)] du$$

and

$$Var[X(t)] = \lambda \int_0^t [1 - G(u)] du$$

by observing that for a given t, $X(t)$ is a Poisson random variable with parameter $\lambda \int_0^t [1 - G(u)]du$.

For the steady-state analysis, we let $t \to \infty$ in the transient analysis to derive all results. Let $X(t)$ converge in distribution to X as $t \to \infty$ such that $p_j(t) \to p_j$, where $p_j = \lim_{t \to \infty} P\{X(t) = j\}$. Taking the limit $t \to \infty$ for $p_j(t)$ we find

$$p_j = e^{-\lambda \tau} \frac{(\lambda \tau)^j}{j!},$$

where τ is the average service time. Then, with an integration-by-parts step, it can be shown that

$$\tau = \int_0^\infty [1 - G(u)]du.$$

In addition, the *mean and variance of the number in the system in steady state* are both $\lambda \tau$ since X is a Poisson random variable with parameter $\lambda \tau$. Note that the mean and variance of the sojourn times correspond, respectively, to the mean and variance of the service times since there is no waiting for service to begin. Before concluding this section, it is worthwhile observing that the steady-state probability p_j is identical to those of the $M/M/\infty$ system, and thus p_j does not depend on the CDF $G(\cdot)$ but just the mean service time.

Problem 40

Obtain the distribution of the busy period, that is, the continuous stretch of time when there are one or more entities in the system beginning with the arrival of an entity into an empty system.

Solution

As we described earlier, $\{X(t), t \geq 0\}$ is a regenerative process with regeneration epochs corresponding to times when the number in the system goes from 1 to 0. Each regeneration time corresponds to one idle period followed by one *busy period* (time when there are one or more entities in the $M/G/\infty$ system). Let U be the regeneration time and $U = I + B$, where the idle time $I \sim \exp(\lambda)$ (i.e., time for next arrival in a Poisson process), and the busy period B has a CDF $H(\cdot)$. We need to determine $H(\cdot)$. For that, we develop the following explanation based on Example 8.17 in Kulkarni [67].

Let $F(\cdot)$ be the CDF of U. Using a renewal argument by conditioning on $U = u$, we can write down a renewal-type equation for $p_0(t) = P\{X(t) = 0 | X(0) = 0\}$ as

$$p_0(t) = \int_0^t p_0(t-u)dF(u) + \int_t^\infty P\{I > t | U = u\}dF(u).$$

Since $U = I + B$ and $P\{I > t | U = u\} = 0$ if $u \le t$, we can rewrite

$$p_0(t) = \int_0^t p_0(t-u)dF(u) + \int_0^\infty P\{I > t | U = u\}dF(u)$$

$$= \int_0^t p_0(t-u)dF(u) + P\{I > t\}$$

$$= p_0 * F(t) + e^{-\lambda t}$$

with $p_0 * F(t)$ is in the convolution notation.

We already derived an expression for $p_0(t)$ as

$$p_0(t) = \exp\left\{-\lambda \int_0^t [1 - G(u)]du\right\},$$

where $G(\cdot)$ is the CDF of the service times. One way to obtain the unknown $F(t)$ function is to numerically solve for $F(t)$ in $p_0(t) = p_0 * F(t) + e^{-\lambda t}$ and similarly solve another convolution equation to get $H(t)$. An alternate approach, typically standard when there are convolutions, is to use transforms. Therefore, taking the LST on both sides of the equation $p_0(t) = p_0 * F(t) + e^{-\lambda t}$, we get

$$\tilde{p}_0(s) = \tilde{p}_0(s)\tilde{F}(s) + \frac{s}{s + \lambda}.$$

However, since we are ultimately interested in $H(t)$, we use the relation

$$\tilde{F}(s) = \left(\frac{\lambda}{\lambda + s}\right)\tilde{H}(s)$$

to obtain

$$\tilde{H}(s) = 1 + \frac{s}{\lambda} - \frac{s}{\lambda \tilde{p}_0(s)}.$$

In the most general case, the preceding LST is not easy to invert and obtain $H(t)$. Another challenge is to obtain $\tilde{p}_0(s)$ from $p_0(t)$, which is not trivial. However, there are several software packages available (such as MATLAB and Mathematica) that can be used to numerically compute as well as invert LSTs.

Having said that, it is relatively straightforward to obtain $E[B]$, the mean busy period. Using the fact that $\tilde{p}_0(0) = p_0(\infty) = e^{-\lambda\tau}$, we can obtain

$$E[B] = -\tilde{H}'(0) = \frac{e^{\lambda\tau} - 1}{\lambda}.$$

Another way to obtain it is to use regenerative process results and solve for $E[B]$ in $1 - p_0 = E[B]/(E[B] + 1/\lambda)$. ∎

Problem 41

Compute the distribution of the interdeparture times both in the transient case and in the steady-state case.

Solution

Since the output from an $M/G/\infty$ may flow into some other queue, it is critical to analyze the departure process. Let $D(t)$ be the number of departures from the $M/G/\infty$ system in time $[0, t]$ given that $X(0) = 0$. For any arbitrary t, we seek to obtain the distribution of the random variable $D(t)$ and thereby characterize the stochastic process $\{D(t), t \geq 0\}$. Similar to the analysis for the number in the system, here too we first consider transient and then describe steady-state results. The results follow the analysis in Gross and Harris [49]. However, it is crucial to point out that there are many other elegant ways of analyzing departures from $M/G/\infty$ queues and extending them, some of which we will see toward the end of this section.

If we are given that n arrivals occurred in time $[0, t]$, then using standard Poisson process results we know that the time of arrival of any of the n arrivals is uniformly distributed over $[0, t]$ and it is independent of the time of other arrival times. Therefore, consider one of the n arrivals that occurred in time $[0, t]$. The probability $\theta(t)$ that this entity would have departed before time t can be obtained by conditioning on the time of arrival and unconditioning as

$$\theta(t) = \frac{1}{t}\int_0^t G(t - x)dx = \frac{1}{t}\int_0^t G(u)du.$$

In addition, the probability that out of the n arrivals in time $[0, t]$, exactly i of those departed before time t is $\binom{n}{i}[\theta(t)]^i[1 - \theta(t)]^{n-i}$.

Now, to compute the distribution of $D(t)$, we condition on $N(t)$, which is the number of arrivals in time $[0, t]$. In order to remind us that we do make the assumption that $X(0) = 0$, we include this condition in the expressions. Therefore, we have

$$P\{D(t) = i|X(0) = 0\} = \sum_{n=i}^{\infty} P\{D(t) = i|N(t) = n, X(0) = 0\}e^{-\lambda t}\frac{(\lambda t)^n}{n!}$$

$$= \sum_{n=i}^{\infty} \binom{n}{i}[\theta(t)]^i[1 - \theta(t)]^{n-i}e^{-\lambda t}\frac{(\lambda t)^n}{n!}$$

$$= \frac{[\theta(t)]^i}{i!}e^{-\lambda t}(\lambda t)^i \sum_{n=i}^{\infty}[1 - \theta(t)]^{n-i}\frac{(\lambda t)^{n-i}}{(n-i)!}$$

$$= \frac{[\lambda t\theta(t)]^i}{i!}e^{-\lambda t\theta(t)}.$$

Therefore, for a given t, $D(t)$ is a Poisson random variable with parameter $\lambda t\theta(t)$. In other words, the transient distribution of the number of departure in time t is Poisson with parameter $\lambda t\theta(t)$. Having described the transient analysis, we just let $t \to \infty$ for steady-state analysis.

It can be shown from the definition of $\theta(t)$ that for very large t, $t\theta(t)$ approaches $t - \tau$. However, since we assume that τ is finite, as we let $t \to \infty$, the departure becomes a Poisson process with parameter $\lambda(1 - \tau/t) \to \lambda$. Of course, as we have not shown stationary and independent properties of the departure process, the reader is encouraged to refer to Serfozo [96] Section 9.6 for a derivation based on stationary marked point processes. Using the earlier results as well as stationary and independence properties, we can conclude that the departure process from an $M/G/\infty$ queue in steady-state is the same as the arrival process, that is, a Poisson process with parameter λ. Further, similar to the steady-state probability p_j being identical to that of the $M/M/\infty$ system and being independent of the CDF $G(\cdot)$, the departure process in steady state is also identical to that of the $M/M/\infty$ system and depends only on the mean service time (τ) and not its distribution. ■

Problem 42

Consider an extension to the $M/G/\infty$ queue. The arrival process is Poisson, however, the parameter of the Poisson process is time varying. The average arrival rate at time t (for all t in $(-\infty, \infty)$) is a deterministic function of t represented as $\lambda(t)$. Hence, the arrival process is defined as a nonhomogeneous Poisson process. Everything else is the same as the regular $M/G/\infty$ queue. We call such a system an $M_t/G/\infty$ queue. Perform transient analysis for this system.

Solution

This summary of results for the $M_t/G/\infty$ queue is based on Eick, Massey, and Whitt [27]. Recall that we need to make one of the three assumptions for initial condition, otherwise we would need to know when service started for each of the customers at a reference time, say $t = 0$. We make the second assumption, although making one of the other two would not significantly alter the analysis but the results would not be as neat. In other words, assume that the system was empty at $t = -\infty$. Assume that for all finite t, $\lambda(t)$ is nonnegative, measurable, and integrable. With these assumptions in mind, we state the results.

The results are in terms of the equilibrium random variable of the service times. In particular, let S be a random variable corresponding to a service time. We have already defined the CDF $G(x) = P\{S \leq x\}$ and mean $\tau = E[S]$. The equilibrium random variable S_e corresponding to the service times has a CDF $G_e(x)$ defined as

$$G_e(x) = P\{S_e \leq x\} = \frac{1}{\tau} \int_0^x [1 - G(u)]du.$$

One encounters equilibrium random variables in renewal theory. If a renewal process with interrenewal random variable Y starts at $t = -\infty$, then at time $t = 0$, the remaining time for a renewal is according to Y_e.

Recall the notation $X(t)$. Here it denotes the number (of entities) in the $M_t/G/\infty$ system at time t. Then for any t, $X(t)$ has a Poisson distribution with parameter $\mu(t)$ given by

$$\mu(t) = E[\lambda(t - S_e)]E[S]$$

with $E[X(t)] = Var[X(t)] = \mu(t)$. It is important to note that $\lambda(\cdot)$ is the function defined in this section and not make the mistake of thinking that at $t = 0$ the average number in the system is negative! In fact, observe that the number in the system at any time depends on the arrival rate S_e time units ago. Further, the departure process also has a similar time lag effect where the average departure rate at time t is $E[\lambda(t - S)]$ and the resulting process is nonhomogeneous Poisson.

For $u > 0$ and any t,

$$Cov[X(t), X(t + u)] = E\left[\lambda\left(t - (S - u)_e^+\right)\right]E[(S - u)^+],$$

where the notation $(y)^+$ is $\max(y, 0)$. In fact, this result can be derived for the homogeneous case (which we have not done earlier but is extremely useful especially in computer-communication traffic with long-range dependence). For an $M/G/\infty$ queue where $\lambda(t) = \lambda$,

$$Cov[X(t), X(t + u)] = \lambda P\{S_e > u\}E[S].$$

There are several results for networks of $M_t/G/\infty$ queues. Since the entities do not interact and departure processes are Poisson, the analysis is fairly convenient. The reader is referred to Eick et al. [27] as well as the references therein for further results. ∎

Notice that the methods used to analyze the $M/G/\infty$ system and its extensions are significantly different from the others in this book. In fact, even the related system, the $M/G/s/s$ queue would be analyzed in Section 4.5.3 differently using a multidimensional continuous-state Markov process. To describe that method, we first explain the $M/G/1$ queue with a special discipline called processor sharing, then use the same technique for the $M/G/s/s$ queue subsequently.

4.5.2 $M/G/1$ Queue with Processor Sharing: Approximating CPUs

Processor sharing is a work-conserving service discipline like FCFS, LCFS, random order of service, etc. Essentially all the jobs or entities in the system equally share the processor at any given time, hence the name processor sharing. For example, if there is one entity in the system and during its service another entity arrives, then the old job and the new job would both be processed, each at half the rate that the old job was being processed. It is fairly common to model computer systems (especially CPUs) as ones with processor sharing. For example, if the processor speed is c bytes per second and there are n jobs, then each job gets c/n bytes per second of processor capacity. In practice, what actually happens is that the processor spends a *time-quantum* of ϵ micro seconds to process one job and *context-switches* to the next job and spends another ϵ micro second. If the processor went in a round-robin fashion across existing jobs, then as $\epsilon \to 0$, the scheduling mechanism converges to processor sharing if the context switching time is zero. It is also fair to mention that in practice there is a multiprogramming limit (MPL), which is the maximum number of simultaneous jobs a processor can handle. Here we assume that MPL is infinite.

Interestingly, this special type of service discipline does produce closed-form results. The analysis uses multidimensional Markov processes with continuous state space similar to the $M/G/s/s$ system that we will see in Section 4.5.3. We first explain the system. Arrivals into the system occur according to a Poisson process with mean rate λ per second. Let S_i be the amount of work (say in kilo bytes) customer i brings, and $S_1, S_2, \ldots$ are IID random variables. There is a single server with processing capacity of 1 (say kilo bytes per second). If there are n entities in the system at time t, then each entity is processed at rate $1/n$ (say kilo bytes per second). However, since the processing capacity is 1, if S_i is the amount of work for the ith job or customer, it would take S_i seconds if it were the only job on the processor during its sojourn. In that light, we say that S_i is the service time requirement for the ith customer so that the CDF is $G(t) = P\{S_i \leq t\}$. We assume that the system is stable, that is, $\rho < 1$, where $\rho = \lambda E[S_i]$. Notice that the stability condition is the same as that of the $M/G/1$ queue with FCFS service discipline since both queues are work conserving.

Now we model the $M/G/1$ processor sharing queue to obtain performance measures such as distribution of the number in the system and mean sojourn time. Let $X(t)$ be the number of customers in the system at time t and $R_i(t)$ be the remaining service time for the ith customer in the system. The multidimensional stochastic process $\{(X(t), R_1(t), R_2(t), \ldots, R_{X(t)}(t)), t \geq 0\}$ satisfies the Markov property (since to predict the future states we only need the present state and nothing from the past) and hence it is a Markov process. However, notice that most of the elements in the state space are continuous, unlike the discrete ones we have seen before. Typically such Markov processes are difficult to analyze unless they have a special structure like this one (and the $M/G/s/s$ queue we will see in Section 4.5.3).

Define $F_n(t, y_1, y_2, \ldots, y_n)$ as the following joint probability

$$F_n(t, y_1, y_2, \ldots, y_n) = P\{X(t) = n, R_1(t) \leq y_1, R_2(t) \leq y_2, \ldots, R_n \leq y_n\}.$$

Thereby, the density function $f_n(t, y_1, y_2, \ldots, y_n)$ is defined as

$$f_n(t, y_1, y_2, \ldots, y_n) = \frac{\partial^n F_n(t, y_1, y_2, \ldots, y_n)}{\partial y_1 \, \partial y_2 \, \ldots \, \partial y_n}.$$

Our objective is to write down a partial differential equation for $f_n(t, y_1, y_2, \ldots, y_n)$ that captures the queue dynamics. For that we consider the system at an infinitesimal time h after t and write down an expression for $f_n(t + h, y_1, y_2, \ldots, y_n)$. For that we need to realize that if $X(t + h) = n$, then at time t one of three things would have happened: $X(t) = n + 1$ and one of the customers departed between t and $t + h$; $X(t) = n - 1$ and one new customer arrived between t and $t + h$; or $X(t) = n$ with no new arrivals or departures between t and $t + h$. Of course, there are other possibilities but they would disappear as we take the limit $h \to 0$. We write this down more formally and in the process we use the notation $o(h)$, which is a collection of terms of order higher than h such that

$$\lim_{h \to 0} \frac{o(h)}{h} = 0.$$

It is important to realize that $o(h)$ is not a specific function of h, but a convenient way of collecting higher-order terms but not state them explicitly since anyway they would go to zero when the limit is taken.

Now we are ready to write down $f_n(t + h, y_1, y_2, \ldots, y_n)$ in terms of the states at time t. Based on the discussion in the previous paragraph, we have

$$f_n(t+h, y_1, y_2, \ldots, y_n)$$

$$= (1 - \lambda h) f_n(t, y_1 + h/n, y_2 + h/n, \ldots, y_n + h/n)$$

$$+ (1 - \lambda h) \sum_{i=0}^{n} \int_0^{\frac{h}{n+1}} f_{n+1}\left(t, y_1 + \frac{h}{n+1}, \ldots, y_{i-1} + \frac{h}{n+1}, y, y_i\right.$$

$$+ \frac{h}{n+1}, \ldots, y_n + \frac{h}{n+1}\right) dy$$

$$+ \lambda h \sum_{i=1}^{n} \frac{G'(y_i)}{n} f_{n-1}\left(t, y_1 + \frac{h}{n-1}, \ldots, y_{i-1}\right.$$

$$+ \frac{h}{n-1}, y_{i+1} + \frac{h}{n-1}, \ldots, y_n + \frac{h}{n-1}\right) + o(h). \tag{4.17}$$

The preceding equation perhaps deserves some explanation. Since the service discipline is processor sharing, if there is y_i amount of service remaining at time $t + h$, then at time t there would have been $y_i + h/n$ service remaining when there are n customers in the system during time t to $t+h$. The probability that there are no arrivals in a time-interval h units long is $(1 - \lambda h) + o(h)$ and the probability of exactly one arrival is $\lambda h + o(h)$. First consider the case that there are no new arrivals in time t to $t + h$, then one of two things could have happened: no service completions during that time interval (first expression in the preceding equation) or one service completion such that at time t there are $n + 1$ customers and the one with less than $h/(n + 1)$ service remaining would complete (second expression in the preceding equation). Therefore, the first term is pretty straightforward and the second term incorporates via the integral, the probability of having less than $h/(n+1)$ service in any of the $(n + 1)$ spots around the n customers in time $t + h$. The third term considers the case of exactly one arrival. This arrival could have been customer i with workload y_i. Notice that the $G'(y_i)$ is the PDF of the service times at y_i, however this could be any of the n customers with probability $1/n$ and hence the summation.

To simplify Equation 4.17, we use the following Taylor-series expansion

$$f_n\left(t, y_1 + \frac{h}{n}, y_2 + \frac{h}{n}, \ldots, y_n + \frac{h}{n}\right) = f_n(t, y_1, y_2, \ldots, y_n)$$

$$+ \frac{h}{n} \sum_{i=1}^{n} \frac{\partial f_n(t, y_1, y_2, \ldots, y_n)}{\partial y_i} + o(h),$$

as well as the fundamental calculus area-under-the-curve result

$$
\int_{0}^{h/(n+1)} f_{n+1}\left(t, y_1 + \frac{h}{n+1}, \ldots, y_{i-1} + \frac{h}{n+1}, y, y_i + \frac{h}{n+1}, \ldots, y_n + \frac{h}{n+1}\right) dy
$$

$$
= f_{n+1}\left(t, y_1, \ldots, y_{i-1}, 0, y_i, \ldots, y_n\right) \frac{h}{n+1} + o(h).
$$

Using the preceding two results, we can rewrite Equation 4.17 as

$$
f_n(t + h, y_1, y_2, \ldots, y_n)
$$

$$
= (1 - \lambda h) f_n(t, y_1, y_2, \ldots, y_n) + (1 - \lambda h) \frac{h}{n} \sum_{i=1}^{n} \frac{\partial f_n(t, y_1, y_2, \ldots, y_n)}{\partial y_i}
$$

$$
+ (1 - \lambda h) \sum_{i=0}^{n} f_{n+1}(t, y_1, \ldots, y_{i-1}, 0, y_i, \ldots, y_n) \frac{h}{n+1}
$$

$$
+ \lambda h \sum_{i=1}^{n} \frac{G'(y_i)}{n} f_{n-1}\left(t, y_1 + \frac{h}{n-1}, \ldots, y_{i-1} + \frac{h}{n-1},\right.
$$

$$
\left. y_{i+1} + \frac{h}{n-1}, \ldots, y_n + \frac{h}{n-1}\right) + o(h).
$$

Subtracting $f_n(t, y_1, y_2, \ldots, y_n)$ on both sides and dividing by h in the preceding equation, we get

$$
\frac{f_n(t + h, y_1, y_2, \ldots, y_n) - f_n(t, y_1, y_2, \ldots, y_n)}{h}
$$

$$
= -\lambda f_n(t, y_1, y_2, \ldots, y_n)
$$

$$
+ (1 - \lambda h) \frac{1}{n} \sum_{i=1}^{n} \frac{\partial f_n(t, y_1, y_2, \ldots, y_n)}{\partial y_i}
$$

$$
+ (1 - \lambda h) \sum_{i=0}^{n} f_{n+1}(t, y_1, \ldots, y_{i-1}, 0, y_i, \ldots, y_n) \frac{1}{n+1}
$$

$$+ \lambda \sum_{i=1}^{n} \frac{G'(y_i)}{n} f_{n-1}\left(t, y_1 + \frac{h}{n-1}, \dots, y_{i-1} + \frac{h}{n-1}, \right.$$

$$\left. y_{i+1} + \frac{h}{n-1}, \dots, y_n + \frac{h}{n-1}\right) + \frac{o(h)}{h}.$$

Taking the limits as $h \to 0$ in the preceding equation, we get

$$\frac{\partial f_n(t, y_1, y_2, \dots, y_n)}{\partial t} = -\lambda f_n(t, y_1, y_2, \dots, y_n) + \frac{1}{n} \sum_{i=1}^{n} \frac{\partial f_n(t, y_1, y_2, \dots, y_n)}{\partial y_i}$$

$$+ \sum_{i=0}^{n} f_{n+1}(t, y_1, \dots, y_{i-1}, 0, y_i, \dots, y_n) \frac{1}{n+1}$$

$$+ \lambda \sum_{i=1}^{n} \frac{G'(y_i)}{n} f_{n-1}(t, y_1, \dots, y_{i-1}, y_{i+1}, \dots, y_n).$$

The preceding equation is similar to the Chapman–Kolmogorov equation and we will see next how to derive the steady-state balance equation from that. Since the stochastic process $\{X(t), R_1(t), R_2(t), \dots, R_{X(t)}(t)\}$ is a Markov process and the stability condition $\rho < 1$ is satisfied, in steady-state, the stochastic process converges to a stationary process. In other words as $t \to \infty$, we have $\partial f_n(t, y_1, y_2, \dots, y_n)/\partial t = 0$ and $f_n(t, y_1, y_2, \dots, y_n)$ converges to the stationary distribution $f_n(y_1, y_2, \dots, y_n)$, that is, $f_n(t, y_1, y_2, \dots, y_n) \to f_n(y_1, y_2, \dots, y_n)$. Therefore, from the preceding equation as we let $t \to \infty$, we get the following balance equation:

$$0 = -\lambda f_n(y_1, y_2, \dots, y_n) + \frac{1}{n} \sum_{i=1}^{n} \frac{\partial f_n(y_1, y_2, \dots, y_n)}{\partial y_i}$$

$$+ \sum_{i=0}^{n} f_{n+1}(y_1, \dots, y_{i-1}, 0, y_i, \dots, y_n) \frac{1}{n+1}$$

$$+ \lambda \sum_{i=1}^{n} \frac{G'(y_i)}{n} f_{n-1}(y_1, \dots, y_{i-1}, y_{i+1}, \dots, y_n).$$

One way is to solve the balance equations by trying various n values starting from 0. Another way is to find a candidate solution and check if it satisfies the balance equation. We try the second approach realizing that if we have a solution it is the unique solution. In particular, we consider the $M/M/1$ queue with processor sharing. There we can show (left as an exercise for the reader) that

$$f_n(y_1, y_2, \ldots, y_n) = (1 - \rho)\lambda^n \prod_{i=1}^{n} [1 - G(y_i)].$$

As a first step we check if the earlier solution satisfies the balance equations for the $M/G/1$ with processor sharing case.

In fact, when $f_n(y_1, y_2, \ldots, y_n) = (1 - \rho)\lambda^n \prod_{i=1}^{n} [1 - G(y_i)]$, it would imply that

$$\sum_{i=0}^{n} f_{n+1}(y_1, \ldots, y_{i-1}, 0, y_i, \ldots, y_n) \frac{1}{n+1} = \lambda f_n(y_1, y_2, \ldots, y_n)$$

since $f_{n+1}(y_1, \ldots, y_{i-1}, 0, y_i, \ldots, y_n) = (1 - \rho)\lambda^{n+1}[1 - G(0)] \prod_{i=1}^{n}[1 - G(y_i)] = \lambda f_n(y_1, y_2, \ldots, y_n)$ since $G(0) = 0$. In addition, if $f_n(y_1, y_2, \ldots, y_n) = (1 - \rho)\lambda^n \prod_{i=1}^{n}[1 - G(y_i)]$, then

$$\frac{\partial f_n(y_1, y_2, \ldots, y_n)}{\partial y_i} = -\lambda G'(y_i) f_{n-1}(y_1, \ldots, y_{i-1}, y_{i+1}, \ldots, y_n)$$

since

$$\frac{\partial f_n(y_1, y_2, \ldots, y_n)}{\partial y_i} = (1 - \rho)\lambda^n(-G'(y_i)) \prod_{j=1}^{i-1}[1 - G(y_j)] \prod_{j=i+1}^{n} [1 - G(y_j)]$$

$$= -\lambda(1 - \rho)G'(y_i)\lambda^{n-1} \prod_{j=1}^{i-1}[1 - G(y_j)] \prod_{j=i+1}^{n} [1 - G(y_j)]$$

$$= -\lambda G'(y_i) f_{n-1}(y_1, \ldots, y_{i-1}, y_{i+1}, \ldots, y_n).$$

Thus, $f_n(y_1, y_2, \ldots, y_n) = (1 - \rho)\lambda^n \prod_{i=1}^{n}[1 - G(y_i)]$ satisfies the balance equation

$$0 = -\lambda f_n(y_1, y_2, \ldots, y_n) + \frac{1}{n} \sum_{i=1}^{n} \frac{\partial f_n(y_1, y_2, \ldots, y_n)}{\partial y_i}$$

$$+ \sum_{i=0}^{n} f_{n+1}(y_1, \ldots, y_{i-1}, 0, y_i, \ldots, y_n) \frac{1}{n+1}$$

$$+ \lambda \sum_{i=1}^{n} \frac{G'(y_i)}{n} f_{n-1}(y_1, \ldots, y_{i-1}, y_{i+1}, \ldots, y_n).$$

Also, it is straightforward to check that

$$\sum_{n=0}^{\infty} \int_{y_1=0}^{\infty} \int_{y_2=0}^{\infty} \cdots \int_{y_n=0}^{\infty} f_n(y_1, y_2, \ldots, y_n) dy_n \ldots dy_2 dy_1 = 1$$

for $f_n(y_1, y_2, \ldots, y_n) = (1 - \rho)\lambda^n \prod_{i=1}^{n}[1 - G(y_i)]$ and hence is the steady-state solution.

Now, to obtain the performance measures, let p_i be the steady-state probability and there are i in the system, that is,

$$p_i = \lim_{t \to \infty} P\{X(t) = i\}.$$

Using the fact that (via integration by parts)

$$\int_{y_j=0}^{\infty} [1 - G(y_j)] dy_j = \int_{y_j=0}^{\infty} y_j G'(y_j) dy_j = \frac{1}{\mu},$$

we have

$$p_i = \int_{y_1=0}^{\infty} \int_{y_2=0}^{\infty} \cdots \int_{y_i=0}^{\infty} f_i(y_1, y_2, \ldots, y_i) dy_i \ldots dy_2 dy_1 = (1 - \rho)\rho^i.$$

Notice that this is identical to the number in the system for an $M/M/1$ queue with FCFS discipline. Thus, $L = \rho/(1 - \rho)$ and $W = 1/(\mu - \lambda)$. It is also possible to obtain the expected conditional sojourn time for a customer arriving in steady state with a workload S as $S/(1 - \rho)$. It uses the fact that the expected number of customers in the system throughout the sojourn time (due to stationarity of the stochastic process) is one plus the average number, that is, $1 + \rho/(1 - \rho) = 1/(1 - \rho)$. Hence, a workload of S would take $S/(1 - \rho)$ time to complete processing at a processing rate of 1.

4.5.3 *M/G/s/s* Queue: Telephone Switch Application

Consider an $M/G/s/s$ queue where the arrivals are according to a Poisson process with mean rate λ per unit time. There are s servers but no waiting room. If all s servers are busy, arriving customers are immediately rejected. However, if there is at least one available server, an arriving customer gets served in a random time S with CDF $G(t) = P\{S \leq t\}$ such that $E[S] = 1/\mu$. We assume that all servers are identical and the service times are IID random variables. There are several applications for such queues, one of them is telephone switches that in fact started the field of queueing theory by A. K. Erlang. When a landline caller picks up his or her telephone to make a call, this amounts to an arrival to the switch. If the caller hears a dial tone, it means a line is available and the caller punches the number he or she wishes to call. If a line is not available, the caller would get a tone stating all lines are busy (these are also quite common in cellular phones where messages such as "the network is busy" are received). The telephone switch has s lines and each line is held for a random time S by a caller and this time is also frequently known as holding times. The pioneering work by A. K. Erlang resulted in the computation of the blocking probability (or the probability a potential caller is rejected).

For this, let $X(t)$ be the number of customers in the system at time t and $R_i(t)$ be the remaining service time at the ith busy server. The multidimensional stochastic process $\{(X(t), R_1(t), R_2(t), \ldots, R_{X(t)}(t)), t \geq 0\}$ satisfies the Markov property (since to predict the future states we only need the present state and nothing from the past) and hence it is a Markov process. It is worthwhile to make two observations here. First of all, this analysis is almost identical to that of the $M/G/1$ processor sharing queue seen in the previous section. Some of the terms used here such as $o(h)$ have been defined in that section and the reader is encouraged to go over that. Second, it is possible to model the system as the remaining service time in each of the s servers. However, additional constraints on whether or not the server is busy imposes more bookkeeping. Hence, we just stick to the $X(t)$ busy servers at time t with the understanding that the alternating formulation could also be used.

Define $F_n(t, y_1, y_2, \ldots, y_n)$ as the following joint probability

$$F_n(t, y_1, y_2, \ldots, y_n) = P\{X(t) = n, R_1(t) \leq y_1, R_2(t) \leq y_2, \ldots, R_n \leq y_n\}.$$

Thereby, the density function $f_n(t, y_1, y_2, \ldots, y_n)$ is defined as

$$f_n(t, y_1, y_2, \ldots, y_n) = \frac{\partial^n F_n(t, y_1, y_2, \ldots, y_n)}{\partial y_1 \, \partial y_2 \, \ldots \, \partial y_n}.$$

Our objective is to write down a partial differential equation for $f_n(t, y_1, y_2, \ldots, y_n)$ that captures the queue dynamics. For that, we consider the system at an infinitesimal time h after t and write down an expression for $f_n(t + h, y_1, y_2, \ldots, y_n)$. For that, we need to realize that if $X(t + h) = n$, then at time t one of three things would have happened: $X(t) = n + 1$ and one of the customers departed between t and $t + h$, $X(t) = n - 1$ and one new customer arrived between t and $t + h$, or $X(t) = n$ with no new arrivals or departures between t and $t + h$. Of course, there are other possibilities but they would disappear as we take the limit $h \to 0$. We do the analysis for the case $0 < n < s$, however, for $n = 0$ and $n = s$, it is just a matter of not counting events that cannot occur (such as a departure when $n = 0$ or arrival when $n = s$).

Now we are ready to write down $f_n(t + h, y_1, y_2, \ldots, y_n)$ in terms of the states at time t. Based on the discussion in the previous paragraph, we have

$$f_n(t + h, y_1, y_2, \ldots, y_n)$$

$$= (1 - \lambda h) f_n(t, y_1 + h, y_2 + h, \ldots, y_n + h)$$

$$+ (1 - \lambda h) \sum_{i=0}^{n} \int_0^h f_{n+1}(t, y_1 + h, \ldots, y_{i-1} + h, y, y_i + h, \ldots, y_n + h) dy$$

$$+ \lambda h \sum_{i=1}^{n} \frac{G'(y_i)}{n} f_{n-1}(t, y_1 + h, \ldots, y_{i-1} + h, y_{i+1} + h, \ldots, y_n + h) + o(h).$$

$$(4.18)$$

The preceding equation perhaps deserves some explanation. If there is y_i amount of service remaining at time $t + h$, then at time t there would have been $y_i + h$ service remaining. The probability that there are no arrivals in a time-interval h units long is $(1 - \lambda h) + o(h)$ and the probability of exactly one arrival is $\lambda h + o(h)$. First consider the case that there are no new arrivals in time t to $t + h$, then one of two things could have happened: no service completions during that time interval (first expression in the preceding equation) or one service completion such that at time t there are $n + 1$ customers and the one with less than h amount of service remaining would complete (second expression in the preceding equation). Therefore, the first term is pretty straightforward and the second term incorporates via the integral, the probability of having less than h service in any of the $(n + 1)$ spots around the n customers in time $t + h$. The third term considers the case of exactly one arrival. This arrival could have been customer i with workload y_i. Notice that the $G'(y_i)$ is the PDF of the service times at y_i, however, this could be any of the n customers with probability $1/n$ and hence the summation.

To simplify Equation 4.18, we use the following Taylor-series expansion:

$$f_n(t, y_1 + h, y_2 + h, \ldots, y_n + h) = f_n(t, y_1, y_2, \ldots, y_n)$$

$$+ h \sum_{i=1}^{n} \frac{\partial f_n(t, y_1, y_2, \ldots, y_n)}{\partial y_i} + o(h),$$

as well as the fundamental calculus area-under-the-curve result:

$$\int_0^h f_{n+1}(t, y_1 + h, \ldots, y_{i-1} + h, y, y_i + h, \ldots, y_n + h) dy$$

$$= f_{n+1}(t, y_1, \ldots, y_{i-1}, 0, y_i, \ldots, y_n) h + o(h).$$

Using the earlier two results, we can rewrite Equation 4.18 as

$$f_n(t+h, y_1, y_2, \ldots, y_n)$$

$$= (1 - \lambda h) f_n(t, y_1, y_2, \ldots, y_n) + (1 - \lambda h) h \sum_{i=1}^{n} \frac{\partial f_n(t, y_1, y_2, \ldots, y_n)}{\partial y_i}$$

$$+ (1 - \lambda h) \sum_{i=0}^{n} f_{n+1}(t, y_1, \ldots, y_{i-1}, 0, y_i, \ldots, y_n) h$$

$$+ \lambda h \sum_{i=1}^{n} \frac{G'(y_i)}{n} f_{n-1}(t, y_1 + h, \ldots, y_{i-1} + h, y_{i+1}$$

$$+ h, \ldots, y_n + h) + o(h).$$

Subtracting $f_n(t, y_1, y_2, \ldots, y_n)$ on both sides and dividing by h in the preceding equation, we get

$$\frac{f_n(t+h, y_1, y_2, \ldots, y_n) - f_n(t, y_1, y_2, \ldots, y_n)}{h}$$

$$= -\lambda f_n(t, y_1, y_2, \ldots, y_n) + (1 - \lambda h) \sum_{i=1}^{n} \frac{\partial f_n(t, y_1, y_2, \ldots, y_n)}{\partial y_i}$$

$$+ (1 - \lambda h) \sum_{i=0}^{n} f_{n+1}(t, y_1, \ldots, y_{i-1}, 0, y_i, \ldots, y_n)$$

$$+ \lambda \sum_{i=1}^{n} \frac{G'(y_i)}{n} f_{n-1}(t, y_1 + h, \ldots, y_{i-1} + h, y_{i+1} + h, \ldots, y_n + h) + \frac{o(h)}{h}.$$

Taking the limits as $h \to 0$ in the preceding equation, we get

$$\frac{\partial f_n(t, y_1, y_2, \ldots, y_n)}{\partial t} = -\lambda f_n(t, y_1, y_2, \ldots, y_n) + \frac{1}{n} \sum_{i=1}^{n} \frac{\partial f_n(t, y_1, y_2, \ldots, y_n)}{\partial y_i}$$

$$+ \sum_{i=0}^{n} f_{n+1}(t, y_1, \ldots, y_{i-1}, 0, y_i, \ldots, y_n) \frac{1}{n+1}$$

$$+ \lambda \sum_{i=1}^{n} \frac{G'(y_i)}{n} f_{n-1}(t, y_1, \ldots, y_{i-1}, y_{i+1}, \ldots, y_n).$$

The preceding equation is similar to the Chapman–Kolmogorov equation and we will see next how to derive the steady-state balance equation from that. Since the stochastic process $\{X(t), R_1(t), R_2(t), \ldots, R_{X(t)}(t)\}$ is a stable Markov process, in steady-state the stochastic process converges to a stationary process. In other words as $t \to \infty$, we have $\partial f_n(t, y_1, y_2, \ldots, y_n)/\partial t = 0$ and $f_n(t, y_1, y_2, \ldots, y_n)$ converges to the stationary distribution $f_n(y_1, y_2, \ldots, y_n)$, that is, $f_n(t, y_1, y_2, \ldots, y_n) \to f_n(y_1, y_2, \ldots, y_n)$. Therefore, from the preceding equation as we let $t \to \infty$, we get the following balance equation:

$$0 = -\lambda f_n(y_1, y_2, \ldots, y_n) + \sum_{i=1}^{n} \frac{\partial f_n(y_1, y_2, \ldots, y_n)}{\partial y_i}$$

$$+ \sum_{i=0}^{n} f_{n+1}(y_1, \ldots, y_{i-1}, 0, y_i, \ldots, y_n)$$

$$+ \lambda \sum_{i=1}^{n} \frac{G'(y_i)}{n} f_{n-1}(y_1, \ldots, y_{i-1}, y_{i+1}, \ldots, y_n).$$

One way is to solve the balance equations by trying various n values starting from 0. Another way is to find a candidate solution and check if it satisfies the balance equation. We try the second approach realizing that if we have a solution, it is the unique solution. In particular, we consider the $M/M/s/s$ queue. There we can show that

$$f_n(y_1, y_2, \ldots, y_n) = K \frac{\lambda^n}{n!} \prod_{i=1}^{n} [1 - G(y_i)],$$

where K is a constant. As a first step, we check if the earlier solution satisfies the balance equations for the $M/G/s/s$ queue.

In fact, when $f_n(y_1, y_2, \ldots, y_n) = K(\lambda^n/n!) \prod_{i=1}^{n} [1 - G(y_i)]$, it would imply that

$$\sum_{i=0}^{n} f_{n+1}(y_1, \ldots, y_{i-1}, 0, y_i, \ldots, y_n) = \lambda f_n(y_1, y_2, \ldots, y_n)$$

since $f_{n+1}(y_1, \ldots, y_{i-1}, 0, y_i, \ldots, y_n) = (1 - \rho)[\lambda^{n+1}/(n+1)!][1 - G(0)] \prod_{i=1}^{n} [1 - G(y_i)] = \lambda/(n+1) f_n(y_1, y_2, \ldots, y_n)$ with $G(0) = 0$. In addition, if $f_n(y_1, y_2, \ldots, y_n) = K(\lambda^n/n!) \prod_{i=1}^{n} [1 - G(y_i)]$, then

$$\frac{\partial f_n(y_1, y_2, \ldots, y_n)}{\partial y_i} = -\frac{\lambda}{n} G'(y_i) f_{n-1}(y_1, \ldots, y_{i-1}, y_{i+1}, \ldots, y_n)$$

since

$$\frac{\partial f_n(y_1, y_2, \ldots, y_n)}{\partial y_i} = K\frac{\lambda^n}{n!}(-G'(y_i)) \prod_{j=1}^{i-1}[1 - G(y_j)] \prod_{j=i+1}^{n}[1 - G(y_j)]$$

$$= -K\frac{\lambda}{n}G'(y_i)\frac{\lambda^{n-1}}{(n-1)!} \prod_{j=1}^{i-1}[1 - G(y_j)] \prod_{j=i+1}^{n}[1 - G(y_j)]$$

$$= -\frac{\lambda}{n}G'(y_i)f_{n-1}(y_1, \ldots, y_{i-1}, y_{i+1}, \ldots, y_n).$$

Thus, $f_n(y_1, y_2, \ldots, y_n) = K(\lambda^n/n!) \prod_{i=1}^{n}[1 - G(y_i)]$ satisfies the balance equation

$$0 = -\lambda f_n(y_1, y_2, \ldots, y_n) + \sum_{i=1}^{n} \frac{\partial f_n(y_1, y_2, \ldots, y_n)}{\partial y_i}$$

$$+ \sum_{i=0}^{n} f_{n+1}(y_1, \ldots, y_{i-1}, 0, y_i, \ldots, y_n)$$

$$+ \lambda \sum_{i=1}^{n} \frac{G'(y_i)}{n} f_{n-1}(y_1, \ldots, y_{i-1}, y_{i+1}, \ldots, y_n).$$

Also, to obtain the constant K, we use

$$\sum_{n=0}^{s} \int_{y_1=0}^{\infty} \int_{y_2=0}^{\infty} \cdots \int_{y_n=0}^{\infty} f_n(y_1, y_2, \ldots, y_n) dy_n \ldots dy_2 dy_1 = 1$$

for $f_n(y_1, y_2, \ldots, y_n) = K(\lambda^n/n!) \prod_{i=1}^{n}[1 - G(y_i)]$ to get

$$K = \frac{1}{\sum_{j=0}^{s} \frac{1}{j!}(\lambda/\mu)^j}.$$

Now, to obtain the performance measures, let p_i be the steady-state probability there are i in the system, that is,

$$p_i = \lim_{t \to \infty} P\{X(t) = i\}.$$

Using the fact that (via integration by parts)

$$\int_{y_j=0}^{\infty} [1 - G(y_j)]dy_j = \int_{y_j=0}^{\infty} y_j G'(y_j)dy_j = \frac{1}{\mu},$$

we have

$$p_i = \int_{y_1=0}^{\infty} \int_{y_2=0}^{\infty} \cdots \int_{y_i=0}^{\infty} f_i(y_1, y_2, \ldots, y_i)dy_i \ldots dy_2 dy_1 = K\frac{\lambda^i}{i!\mu^i}.$$

Therefore, for $0 \leq i \leq s$,

$$p_i = \frac{(\lambda/\mu)^i/i!}{\sum_{k=0}^{s} (\lambda/\mu)^k/k!}.$$

The *Erlang loss formula* due to A. K. Erlang is the probability that an arriving customer is rejected (or the fraction of arriving customers that are lost in steady state) and is given by

$$p_s = \frac{(\lambda/\mu)^s/s!}{\sum_{i=0}^{s} (\lambda/\mu)^i/i!}.$$

Notice that the distribution of the number in the system in steady state for an $M/G/s/s$ queue does not depend on the distribution of the service time. Using the steady-state number in the system, we can derive

$$L = \frac{\lambda}{\mu}(1 - p_s).$$

Since the effective entering rate into the system is $\lambda(1 - p_s)$, we get $W = 1/\mu$. This is intuitive since there is no waiting for service for customers that enter the system, the average sojourn time is indeed the average service time. For the same reason, the sojourn time distribution for customers that enter the system is same as that of the service time. In addition, since there is no waiting for service, $L_q = 0$ and $W_q = 0$. We conclude by making a remark without proof.

Remark 9

The Markov process defined here

$$\{(X(t), R_1(t), R_2(t), \ldots, R_{X(t)}(t)), t \geq 0\}$$

in steady state is in fact a *reversible* process. In simple terms, that means if the process is recorded and viewed backward, it would be stochastically identical to running it forward. One of the artifacts of reversibility is the existence of *product-form* solutions such as the expression for $f_n(y_1, y_2, \ldots, y_n)$. Further, because of reversibility, the departures from the original system correspond to arrivals in the reversed system. Therefore, the departure process from the $M/G/s/s$ queue is a Poisson process with rate $(1 - p_s)\lambda$ departures per unit time on average. ∎

Reference Notes

Unlike most of the other chapters in this book, this chapter is a hodgepodge of techniques applied to a somewhat common theme of nonexponential interarrival and/or service times. We start with DTMC methods, then gravitate toward MVA, develop bounds and approximations, then present CTMC models, and finally, some special-purpose models. For that reason, it has been difficult to present the complete details of all the methods. In that light, we have provided references along with the description so that the readers can immediately get to the source to find out the missing steps. These include topics such as $G/G/s$ and $\sum PH/PH/s$ queues, phase type distributions and fitting, $M/G/\infty$ queue, $M/G/s/s$ queue, and $M/G/1$ with processor sharing. Leaving out some of the details was a difficult decision to make considering that most textbooks on queues also typically leave those out. But perhaps there is a good reason for doing so. Nevertheless, thanks to Prof. Don Towsley's class notes for all the details on $M/G/1$ processor sharing queues that was immensely useful here.

The approximations and bounds presented in this chapter using MVA are largely due to Buzacott and Shanthikumar [15]. All the empirical approximations are from Bolch et al. [12]. However, topics such as $M/G/1$ and $G/M/1$ queues have been treated in a similar vein as Gross and Harris [49]. Many of the results presented on those topics have also been heavily influenced by Kulkarni [67]. A lot of the results presented here on those topics have been explained in a lot more crisp and succinct fashion in Wolff [108]. Further, there is a rich literature on using fluid and diffusion approximations as well as methodologies to obtain tail distributions. The main reason for leaving them out in this chapter is that those techniques lend themselves nicely to analyze a network of queues, which, is the focus of the other chapters of this book.

Exercises

4.1 For a stable $M/G/1$ with FCFS service, derive the average sojourn time in the system

$$W = \frac{1}{\mu} + \frac{\lambda}{2} \frac{(\sigma^2 + 1/\mu^2)}{1 - \rho}$$

from $E[e^{-sY}]$ by taking the derivative with respect to s, multiplying by (-1) and letting $s \to 0$.

4.2 Using the LST of the busy period distribution of an $M/G/1$ queue that is a solution to $\tilde{F}_Z(u) = \tilde{G}(u + \lambda - \lambda \tilde{F}_Z(u))$, obtain the average busy period length as $E[Z] = 1/(\mu - \lambda)$ as well as the second moment of the busy period length as $E[Z^2] = (\sigma^2 + 1/\mu^2)/(1 - \lambda/\mu)^3$.

4.3 For a stable $G/M/1$ queue, obtain the second moment of the sojourn time and the second factorial moment of the number in the system in steady state. Is there a relation between the two terms like we saw for the $M/G/1$ queue?

4.4 Compute the expected queue length (L_q) in an $M/G/1$ queue with the following service time distributions (all with mean $= 1/\mu$):
(a) Exponential with parameter μ
(b) Uniform over $[0, 2/\mu]$
(c) Deterministic with mean $1/\mu$
(d) Erlang with parameters $(k, k\mu)$
Which service time produces the largest congestion? Which one produces the lowest?

4.5 Consider a single-server queue that gets customers from k independent sources. Customers from source i arrive according to $PP(\lambda_i)$ and demand $\exp(\mu_i)$ of service time. All customers form a single queue and served according to FCFS. Let $X(t)$ be the number of customers in the system at time t. Model $\{X(t), t \geq 0\}$ as an $M/G/1$ queue and compute its limiting expected value when it exists.

4.6 *An $M/G/1$ queue with server vacations.* Consider an $M/G/1$ queue where the server goes on vacation if the system is empty after a service completion. If the system is empty upon the return of the server from vacation, the server goes on another vacation; otherwise, the server begins service. Successive server vacations are IID random variables. Let $\psi(z)$ be the generating function of the number of arrivals during a vacation (the vacation length may depend upon the arrival process during the vacation). Let X_n be the number of customers in the system after the nth service completion. Show that $\{X_n, n \geq 0\}$ is a DTMC by describing the transition probability matrix. Then:

(a) Show that the system is stable if $\rho = \lambda/\mu < 1$.

(b) Assuming that $\rho < 1$, show that the generating function $\phi(z)$ of the steady-state distribution of X_n is given by

$$\phi(z) = \left(\frac{1-\rho}{m}\right) \frac{\tilde{G}(\lambda - \lambda z)}{z - \tilde{G}(\lambda - \lambda z)}(\psi(z) - 1),$$

where m is the average number of arrivals during a vacation.

(c) Show that

$$L = \rho + \frac{1}{2}\frac{\rho^2}{1-\rho}(1 + \sigma^2\mu^2) + \frac{m^{(2)}}{2m},$$

where $m^{(2)} = \psi''(1)$ is the second factorial moment of the number of arrivals during a vacation.

4.7 Consider a $G/M/1$ queue with interarrival distribution

$$G(x) = r(1 - e^{-\lambda_1 x}) + (1 - r)(1 - e^{-\lambda_2 x}),$$

where $0 < r < 1$ and $\lambda_i > 0$ for $i = 1, 2$. Find the stability condition and derive an expression for p_j, the steady-state probability that there are j customers in the system.

4.8 A service station is staffed with two identical servers. Customers arrive according to a $PP(\lambda)$. The service times are IID $\exp(\mu)$. Consider the following two operational policies used to maintain two separate queues:

(a) Every customer is randomly assigned to one of the two servers with equal probability.

(b) Customers are alternately assigned to the two servers.
Compute the expected number of customers in the system in steady state for both cases. Which operating policy is better?

4.9 Requests arrive to a web server according to a renewal process. The interarrival times are according to an Erlang distribution with mean $10\,$s and standard deviation $\sqrt{50}\,$s. Assume that there is infinite waiting room available for the requests to wait before being processed by the server. The processing time (i.e., service time) for the server is according to a Pareto distribution with CDF

$$G(x) = 1 - \left(\frac{K}{x}\right)^{\beta}, \quad \text{if } x \geq K.$$

The mean service time is $K\beta/(\beta - 1)$, if $\beta > 1$, and the variance of the service time is $K^2\beta/[(\beta - 1)^2(\beta - 2)]$ if $\beta > 2$. Use $K = 5$ and $\beta = 2.25$ so that the mean and standard deviation of the service times are 9 and 12 s, respectively. Using the results for the $G/G/1$ queue, obtain bounds as well as approximations for the average response time (i.e., waiting time in the system including service) for an arbitrary request in the long run. Pick any bound or approximation from the ones given in this chapter.

4.10 Consider a stable $M/G/1$ queue with $PP(\lambda)$ arrivals and $\tilde{G}(w)$ as the LST of the CDF of the service times such that $\tilde{G}'(0) = -1/\mu$. Write down the LST of the interdeparture times (between two successive departures picked arbitrarily in steady state) in terms of λ, μ, and $\tilde{G}(w)$. (Note that w is used instead of s in the LST to avoid confusion with S, the service time.)

4.11 Consider a stable $G/M/1$ queue with traffic intensity ρ and parameter α, which is a solution to $\alpha = \tilde{G}(\mu - \mu\alpha)$, where $G(t)$ is the CDF of the interarrival time and μ is the mean service rate. Derive an expression for the generating function $\Psi(z)$ of the number of entities in the system in steady state as a closed-form expression in terms of ρ, α, and z.

4.12 Answer the following multiple choice questions:

(i) For a stable $G/M/1$ queue with $\tilde{G}(s)$ being the LST of the interarrival time CDF and service rate μ, which of the following statements are not true?

(a) There is a unique solution for α in $(0,1)$ to the equation $\alpha = \tilde{G}(\mu - \mu\alpha)$.

(b) In steady state, the time spent by an arbitrary arrival in the system before service begins is exponentially distributed.

(c) If $\tilde{G}(s) = \lambda/(\lambda + s)$, then the average total time in the system (i.e., W) is $1/(\mu - \lambda)$.

(d) The fraction of time the server is busy in the long run is $-1/[\tilde{G}'(0)\mu]$.

(ii) Consider a stable $M/G/1$ queue and the notation given in this chapter. Which of the following statements are not true?

(a) For $j = 1, 2, \ldots$, we have $\pi_j = \pi_j^* = p_j$.

(b) Average time between departures in steady state is $1/\lambda$.

(c) Server idle times are according to $\exp(\lambda)$.

(d) Average workload in the system seen by an arriving customer in steady state is W.

(iii) For a stable $M/G/1$ queue with traffic intensity ρ and squared coefficient of variation of service times C_s^2,

    (a) L increases with ρ but decreases with C_s^2.

    (b) L decreases with both ρ and C_s^2.

    (c) L increases with both ρ and C_s^2.

    (d) L decreases with ρ but increases with C_s^2.

4.13 For the following TRUE or FALSE questions, give a brief reason why you picked the statements to be true or false.

(a) Consider a stable $G/D/1$ queue where the interarrival times are independent and identically distributed with mean a seconds and standard deviation σ_A seconds. The service time is constant and equal to τ seconds. Is the following statement TRUE or FALSE? The average time in the system (W) is given by

$$W = \tau + \frac{\sigma_A^2}{2\{a - \tau\}}.$$

(b) Consider a stable $M/M/1$ queue that uses processor sharing discipline. Arrivals are according to $PP(\lambda)$ and it would take $\exp(\mu)$ time to process an entity if it were the only one in the system. Is the following statement TRUE or FALSE? The average workload in the system at an arbitrary point in steady state is $\lambda/[\mu(\mu - \lambda)]$.

(c) The Pollaczek–Khintchine formula to compute L in $M/G/1$ queues requires the service discipline to be FCFS. TRUE or FALSE?

4.14 Compare an $M/E_2/1$ and $E_2/M/1$ queue's L values for the case when both queues have the same ρ. The term E_2 denotes an Erlang-2 distribution.

4.15 Which is better: an $M/G/1$ queue with $PP(\lambda)$ and processing speed of 1 or one with $PP(2\lambda)$ and processing speed 2? By processing speed, we mean the amount of work the server can process per unit time, so if there is x amount of work brought by an arrival and the processing speed is c, then the service time would be x/c. Assume that the amount of work has a finite mean and finite variance. Also, use mean sojourn time to compare the two systems.

4.16 Consider an $M/G/1$ queue with mean service time $1/\mu$ and variance $1/(3\mu^2)$. The interarrival times are exponentially distributed with mean $1/\lambda$ and the service times are according to an Erlang distribution. Let W_n and S_n, respectively, be the time in the system and service time for the nth customer. Define a random variable called *slowdown* for customer n as W_n/S_n. Compute the mean and variance of slowdown for a customer arriving in steady state, that is, compute $E[W_n/S_n]$ and $Var[W_n/S_n]$ as $n \to \infty$. Assume stability. Note that the term x-factor defined in the exercises of Chapter 1 is $E[W_n]/E[S_n]$, however the means low down is $E[W_n/S_n]$.

4.17 Consider a $G/M/2$ queue which is stable. Obtain the cumulative distribution function (CDF) of the time in the system for an arbitrary customer in steady state by first deriving its Laplace Stieltjes transform (LST) and then inverting it. Also, based on it, derive expressions for W and thereby L.

4.18 Consider an $M/G/1$ queue with mean service time $1/\mu$ and second moment of service time $E[S^2]$. In addition, after each service completion, the server takes a vacation of random length V with probability q or continues to serve other units in the queue with probability p (clearly $p = 1 - q$). However, the server always takes a vacation of length V as soon as the system is empty; at the end of it, the server starts service, if there is any unit waiting, and otherwise it waits for units to arrive. Let $\tilde{V}(s) = E[e^{-sV}]$ be the LST of the vacation time distribution. Use MVA to derive the following results for p_0, the long-run probability the system is empty, and L, the long-run average number of units in the system:

$$p_0 = \frac{1 - (\lambda/\mu) - qE(V)\lambda}{\tilde{V}(\lambda) + pE(V)\lambda},$$

$$L = \frac{\lambda}{\mu} + \frac{\lambda^2}{2\left(1 - (\lambda/\mu) - \lambda qE(V)\right)} \left\{ E[S^2] + qE(V^2) + \frac{2q}{\mu}E(V) \right\}$$

$$+ \frac{\lambda^2 pE(V^2)}{2\left[\tilde{V}(\lambda) + pE(V)\lambda\right]}.$$

Assume stability.

4.19 Let $X(t)$ be the number of entities in the system in an $M/M/1$ queue with processor sharing. The arrival rate is λ and the amount of service requested is according to $\exp(\mu)$ so that the traffic intensity is $\rho = \lambda/\mu$. Model $\{X(t), t \geq 0\}$ as a birth and death process and obtain the steady-state distribution. Using that and the properties of the exponential distribution, derive an expression for

$$\lim_{t\to\infty} F_n(t, y_1, y_2, \ldots, y_n),$$

where

$$F_n(t, y_1, y_2, \ldots, y_n) = P\{X(t) = n, R_1(t) \le y_1, R_2(t) \le y_2, \ldots, R_n \le y_n\}$$

and $R_i(t)$ is the remaining service for the ith customer in the system. From that result, show that

$$f_n(y_1, y_2, \ldots, y_n) = (1 - \rho)\lambda^n \prod_{i=1}^{n} [1 - G(y_i)],$$

where $G(t) = 1 - e^{-\mu t}$.

5

Multiclass Queues under Various Service Disciplines

In the models considered in the previous chapters, note that there was only a single class of customers in the system. However, there are several applications where customers can be differentiated into classes and each class has its own characteristics. For example, consider a hospital emergency ward. The patients can be classified into emergency, urgent, and regular cases with varying arrival rates and service requirements. The question to ask is: In what order should the emergency ward serve its patients? It is natural to give highest priority to critical cases and serve them before others. But how does that impact the quality of service for each class? We seek to address such questions in this chapter. We begin by describing some introductory remarks, then evaluate performance measures of various service disciplines, and finally touch upon the notion of *optimal* policies to decide the order of service.

5.1 Introduction

The scenario considered in this chapter is an abstract system into which multiple classes of customers enter, get "served," and depart. Why are there multiple classes? First, the system might be naturally classified into various classes because there are inherently different items requiring their own performance measures (e.g., in a flexible manufacturing system, if a machine produces three different types of parts, and it is important to measure the in-process inventory of each of them individually, then it makes sense to model the system using three classes). Second, when the service times are significantly different for different customers, then it might be beneficial to classify the customers based on service times (e.g., in most grocery stores there are special checkout lines for customers that have fewer items). Third, due to physical reasons of where the customers arrive and wait, it might be practical to classify customers (e.g., at fast-food restaurants customers can be classified as drive-through and in-store depending on where they arrive).

Next, given that there are multiple classes, how should the customers be served? In particular, we need to determine a service discipline to serve the customers. For that one typically considers measures such as cost, fairness, performance, physical constraints, goodwill, customer satisfaction, etc. Although we will touch upon the notion of optimal service disciplines toward the end of this chapter, we will assume until that point that we have a system with a *given* service discipline and evaluate the performance experienced by each class in that system. This is with the understanding that in many systems one may be restricted to using a particular type of service policy. To get a better feel for that, in the next section we present example of several systems with multiple classes. Before proceeding with that, as a final comment, it is worth mentioning that this work typically falls in the literature under the umbrella of *stochastic scheduling*.

5.1.1 Examples of Multiclass Systems

The objective of this section is to give the reader a better understanding of multiclass systems using real-world examples, as well as an appreciation for various factors considered to determine service disciplines. The examples are grouped together based on the application domain.

1. *Road transportation*: One of the classic examples of classifying customers due to physical reasons is at a traffic light. As a simple case, consider a four-way intersection where each flow is a single lane. Then the system can be modeled as one with four classes. A service-scheduling policy is to offer a green to one of the four flows in a round-robin fashion with a fixed green time for each flow. The notion of dollar cost is quite nebulous, so the timing of the lights is usually a combination of fairness, performance, and physical constraints. Note how a manual police-operated traffic intersection usually works a little differently. Another road transportation example is a toll booth with multiple servers (or lanes). Some lanes are reserved for electronic toll collection, some for manual with exact changes, some for manual with a human that doles out change, and perhaps some for larger vehicles. Note that this is not a single-server system, customers have the choice to select their lane (with the appropriate exceptions); however, there isn't a notion of service discipline besides first come first served (FCFS) within a lane. Note the similarity between this and the grocery checkout line with special lanes for customers having fewer items. There are many other examples of road transportation, including high occupancy vehicle lanes, traffic intersection with yield sign (i.e., priority queue), etc. The key point is that in roadway transportation systems alone, there are a variety of service disciplines implemented; however, the unique feature is that the entities in the queue are

humans (this is also common to the hospitality industry, especially restaurants).

2. *Production*: Although road transportation focused on human entities, the other end of the spectrum is in production, where the entities are products for which fairness is not all that critical but cost and performance are crucial aspects. In the 1980s, a large volume of literature surfaced when researchers became interested in scheduling jobs on flexible manufacturing systems. Many of the results presented in this chapter are based on this. There is a plethora of applications in the production domain and we mention only a few here. In particular, we alluded to an example earlier where a flexible machine (essentially a machine capable of doing different things) produces multiple parts. The machine processes one part at a time. However, because of the setup time involved in changing the configuration from one type to another, it makes sense to process several parts of one class, then switch over to another class, and continue in a round-robin fashion. There is a trade-off in minimizing the time wasted in switching classes against the time that products wait to be processed. The issues are essentially cost, performance, and physical constraints for the most part that drive the service disciplines. Another example is an automobile repair shop (although some would argue that this is not a production but a service system). Vehicles are dropped off, and the supervisor gets to decide the order in which vehicles are to be repaired. Knowing the approximate duration for the repair times based on what the vehicle owner reported, the supervisor can use the shortest expected repair time policy to, for example, schedule repairs.

3. *Computer communications*: The topic of stochastic scheduling resurfaced recently with an explosion of applications and new needs in computer-communication systems. We describe a very small number of examples where entities can be classified based on individual types, service times, or physical location. In the networking arena, it is common to classify packets into TCP or UDP types, and, perhaps in the future, into classes of service (such as gold, silver, bronze, and best effort). At a higher level, the traffic can be classified as file transfer, streaming video, hypertext transfer, etc. All these are based on types. Entities are also classified based on service times such as web server requests (large files are also called *elephants* and small ones are called *mice*). In addition, there is a large literature on service disciplines in the router design area (generalized processor sharing, weighted fair queueing, weighted round-robin, random early detection, etc.) that we do not cover in this chapter, considering that it is too specialized for a single application. However, it is worth mentioning that the queues in the routers are typical examples of

physical-location-based classification. Having said that, one of the major applications of polling models for which we have allocated an entire section in this chapter is in computer systems. For example, a CPU would process jobs in a round-robin fashion context switching from time to time; the CPU also polls various queues, say in network interface cards. The token-ring local area network protocol and most metropolitan area networks like FDDI are also polling based. On a completely different note, most of the acronyms in this paragraph are not expanded because the acronyms are generally more popular than their expansions.

These examples are meant to motivate the reader but by no means an indication of the variety of circumstances in which multiclass queueing systems occur. While presenting various service disciplines, we will describe some more examples to put things in better perspective. In fact, many of our following examples would fall under service systems (such as hospitals), which is another application domain that has received a lot of attention recently. Next, we present some results that are applicable to any multiclass queueing system with "almost" any reasonable service discipline.

5.1.2 Preliminaries: Little's Law for the Multiclass System

Consider a generic flow-conserving system (does not have to be a queueing system) where K classes of customers arrive, spend some time in the system, and then depart. Customers belonging to class i ($i \in \{1, 2, \ldots, K\}$) arrive into the system according to a renewal process with mean rate λ_i per unit time. Assume that the arrival process is independent of other classes and class switching is not allowed inside the system. Let L_i be the long-run average number of i customers in the system. Also, let W_i be the mean sojourn time in the system for class-i customers. Irrespective of how the customers interact with each other, Little' law holds. In particular, we have for every $i \in \{1, 2, \ldots, K\}$

$$L_i = \lambda_i W_i.$$

Of course we can aggregate across all K classes of customers and state the usual Little's law as

$$L = \lambda W$$

where
  L is the total number of customers in the system on average in steady state across all classes
  W is the sojourn time averaged over all customers (of all classes)
  λ is the aggregate arrival rate

Since the system is flow conserving and class switching is not permitted, we have

$$\lambda = \lambda_1 + \lambda_2 + \cdots + \lambda_K,$$
$$L = L_1 + L_2 + \cdots + L_K.$$

Next we consider a special case of the generic flow-conserving system without class switching, namely the $G/G/s$ queue with multiple classes. It is required that customers enter the system at mean rate λ_i for class i ($i \in \{1, 2, \ldots, K\}$), get served once, and leave the system immediately after service. Customers belonging to class i (which henceforth would be called class-i customers) require an average service time of $1/\mu_i$ (in same time units as $1/\lambda_j$ for all i and j). The service times are independent and identically distributed (IID) across all the customers of a particular class. We assume that the distribution of interarrival times and service times are known for each class (although not required for the results in this section). Unless otherwise stated, we also assume that arrival times and service times are not known until arrivals occur and service is completed, respectively. This is sometimes also known as nonanticipative, although that term does have other related meanings in the literature. There are systems such as web servers where the service time is declared upon arrival (for example when file size to be downloaded is known). Although that would technically not be nonanticipative, we could get around it by considering each file as a separate class.

Note that so far we have not specified the scheduling policy (i.e., service discipline) for the multiclass system. We now describe some results that are invariant across scheduling policies (or at least a subset of policies). Recall that L_i and W_i are the mean queue length and mean sojourn time, respectively, for class-i customers. Irrespective of the scheduling policy, Little's law holds for each class, so for all $i \in [1, K]$ with $[1, K]$ being a compact representation of the set $\{1, 2, \ldots, K\}$

$$L_i = \lambda_i W_i.$$

Also, similar results can be derived for L_{iq} and W_{iq} which, respectively, denote the average number waiting in queue (not including customers at servers) and average time spent waiting before service. In particular, for all $i \in [1, K]$

$$L_{iq} = \lambda_i W_{iq}$$
$$W_i = W_{iq} + \frac{1}{\mu_i}$$
$$L_i = L_{iq} + \rho_i$$

where

$$\rho_i = \frac{\lambda_i}{\mu_i}.$$

Note that ρ_i here is **not** the traffic intensity offered by class i when $s > 1$. But we will mainly consider the case $s = 1$ for most of this chapter which would result in ρ_i being the traffic intensity and will use the previous results. In addition, L and W are the overall mean number of customers and mean sojourn time averaged over all classes. Recall that $L = L_1 + L_2 + \cdots + L_K$ and if $\lambda = \lambda_1 + \lambda_2 + \cdots + \lambda_K$, the net arrival rate, then $W = L/\lambda$. For the $G/G/1$ case with multiple classes, more results can be derived for a special class of scheduling policies called work-conserving disciplines which we describe next.

5.1.3 Work-Conserving Disciplines for Multiclass $G/G/1$ Queues

We would first like to state explicitly that we restrict ourselves to multiclass queues with a single server. In other words, we let $s = 1$ for the system in the previous section and only study $G/G/1$ queues with multiple classes. There are a few reasons for that. First of all, note that except for the very last section of this chapter, we are typically interested in obtaining performance measures for a given policy. However, recall from Chapter 4 that even the single class $M/G/s$ queue was intractable to analyze. Another reason for using single-server queue is that we will see subsequently that multi-server queues are generally not work conserving. That would become more apparent once we define what work conserving is. Having said that, we would like to add that we will visit multiple servers when we discuss optimal policies in Section 5.5. Until then we will assume there is only one server.

It is worthwhile to describe the notation used here. We consider a multiclass $G/G/1$ queue. There are a total of K classes. For $i = 1, \ldots, K$, class-i customers arrive into queue according to an independent renewal process at average rate of λ_i per unit time. Further, class-i customers require a random amount of service which is denoted generically as S_i for all $i \in [1, K]$. The mean service time is $1/\mu_i = E(S_i)$. Also, $E(S_i^2)$ is the second moment of the service time. The traffic intensity for class i is λ_i/μ_i. In this section, we concentrate on a subset of service-scheduling policies (i.e., service disciplines) seen before. In particular, we consider *work-conserving disciplines* where more results for the $G/G/1$ queue can be obtained. In fact, many of these results have not been explained in the single class case in the previous chapters but by letting $K = 1$ (i.e., only one class), they can easily be accomplished.

Recall that whenever a customer arrives into a $G/G/1$ queue, the workload in the system jumps by a quantity equal to the service time of that customer. Also, whenever there is work to be performed, the workload

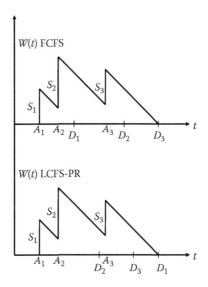

FIGURE 5.1
$W(t)$ vs. t for FCFS and LCFS-PR.

decreases at unit rate. A sample path of the workload in the system at time t, $W(t)$, is described in Figure 5.1 with A_n denoting the time of the nth arrival and S_n its service time requirement for $n = 1, 2, 3$. The figure gives departure times D_n for the nth arriving customer using two service disciplines, FCFS and LCFS-PR (with preemptive resume) across all classes. Note that although the departure times are different under the two service disciplines, the workload $W(t)$ is identical for all t. In other words, the workload $W(t)$ at time t is conserved. But that does not always happen. If the server were to idle when $W(t) > 0$ or if we considered LCFS with preemptive repeat (and S_n is not exponentially distributed), the workload would not have been conserved.

In this and the next section, we only consider the class of service disciplines that result in the workload being conserved. The essence of *work-conserving disciplines* is that the system workload at every instant of time remains unchanged over all work-conserving service-scheduling disciplines. Intuitively this means that the server never idles whenever there is work to do, and the server does not do any wasteful work. The server continuously serves customers if there are any in the system. For example, FCFS, LCFS, and ROS are work conserving. Certain priority policies that we will see later such as nonpreemptive and preemptive resume policies are also work conserving. Further, disciplines such as processor sharing, shortest expected processing time first, and round-robin policies are also work conserving when the switch-over times are zero. There are policies that are nonwork conserving such as preemptive repeat (unless the service times are exponential) and preemptive identical (i.e., the exact service time is repeated as opposed to preemptive repeat where the service time is resampled). Usually when the server takes a vacation from service or if there is a switch-over time (or setup time) during moving from classes, unless those can be explicitly accounted for in the service times, those servers are nonwork conserving.

Having described the concept of work conservation, next we present some results for queues with such service disciplines. Note that across all work-conserving service-scheduling disciplines, not only is $W(t)$ identical for all t, but also the busy period and idle time sample paths are identical. Therefore, all the results that depend only on the busy period distribution can be derived for all work-conserving disciplines. We present some of those next. Consider the notation used earlier in this section as well as those in Section 5.1.2. Define ρ, the overall traffic intensity, as

$$\rho = \sum_{i=1}^{K} \rho_i.$$

A K-class $G/G/1$ queue with a work-conserving scheduling discipline is stable if

$$\rho < 1.$$

This result can be derived using the fact that ρ is just the overall arrival rate times the average service time across all classes. It is also equal to the ratio of the mean busy period to the mean busy period plus idle period. The busy period and idle period are identical across all work-conserving disciplines, and we know the previous result works for the single class $G/G/1$ queue with FCFS. Therefore, the FCFS across all classes is essentially a single class FCFS with traffic intensity ρ. This result works for all work-conserving service-scheduling disciplines. In a similar manner, we can show that when a $G/G/1$ system is work conserving, the probability that the system is empty is $1 - \rho$.

In the next section, we describe a few more results for multi-class $G/G/1$ queues with work-conserving scheduling disciplines. However, we require an additional condition that rules out some of the work-conserving schemes. For that we also need additional notation. Let S_i be the random variable denoting the service time of a class-i customer (this is different from S_n we defined earlier which is the service time realization of the nth arriving customer). It is also crucial to point out that since the service times for all customers of particular class are IID, we use a generic random variable, such as S_i for class i. Using that notation, the second moment of the overall service time is

$$E[S^2] = \frac{1}{\lambda} \sum_{i=1}^{K} \lambda_i E[S_i^2].$$

5.1.4 Special Case: At Most One Partially Completed Service

In this section, we consider a special case of work-conserving disciplines where at any time there can be at most one customer in the system that has had any service performed. Policies such as FCFS, nonpreemptive LCFS, ROS, nonpreemptive priority, and round-robin with complete service (without switch-over times) fall under this category where once the service begins for a customer, the service is not interrupted and service rate remains unchanged until the service is complete. Typically policies that allow preemption, round-robin with incomplete service, and processor sharing do not fall under this category because there could be more than one customer whose service is incomplete. The only exception is when the service times are exponentially distributed. Because of memoryless property, when a service is only partially completed, the remaining service time is still exponentially distributed with the same service time parameter. Therefore, it is equivalent to the case of at most one partially completed service even under preemption, round-robin with incomplete service, and processor sharing. With that we make a quick remark before proceeding.

Remark 10

What we consider here is a special case of the head-of-the-line (HL) discipline considered in Dai [25]. Under HL, there could be at most one partially completed service within each class. However, across classes preemption, round-robin with incomplete service, and processor sharing are allowed. ∎

We now explain the method used here. For that the first step is to ensure that the stochastic system is stable. That can be checked fairly quickly; in fact for the multiclass $G/G/1$ queue, all we need to check is if $\rho < 1$. If the system is stable, the key idea is to observe the system at an arbitrary time in steady state. Then the observation probabilities correspond to the steady-state probabilities. For example, in a stable $G/G/1$ queue with multiple classes, the probability that an arbitrary observation in steady state would result in an empty system is $1 - \rho$, the steady-state probability of having no customers in the system. Further, if the system is stationary and ergodic, then the observation just needs to be made at an arbitrary time and the results would also indicate time-averaged behavior. Thus, the expected number in the system during this arbitrary observation in steady state is L (the steady-state mean number in the system). This should not be confused with customer-stationary process results such as Poisson arrivals see time averages (PASTA). Note that here we are considering just one observation and the observation does not correspond to a customer arrival or departure.

For the next step of the analysis, recall that our system is a $G/G/1$ queue with K classes and a service discipline that is work conserving with at most one partially completed service allowed. We can divide this system into two subsystems, one the waiting area and the other the service area. The workload in the system at any time is the sum of the workload in the service area and that in the waiting area. Note that all arriving customers go to the waiting area (although they may spend zero time there), then go to the service area and exit the system (without going back to the waiting area). With that in mind, we present two results that are central to such special cases of work-conserving disciplines. These results were not presented for the single class case (but they are easily doable by letting the number of classes $K = 1$). We present these results as two problems.

Problem 43

Consider a $G/G/1$ queue with K classes and a service discipline that is work conserving with at most one partially completed service allowed. Assume that the queue is stable, that is, $\rho < 1$. If the system is observed at an arbitrary time in steady state, then show that for that observation, the expected workload at the server (i.e., expected remaining service time) is $\lambda E[S^2]/2$.

Solution

Let C be the class of the customer in service when the observation is made at an arbitrary time in steady state with $C = 0$ implying there is no customer in service. Also, let R be the remaining service time when this observation is made (again R is indeed the workload at the server). We need to compute $E[R]$.

Note that $P\{C = 0\} = 1 - \rho$ and for $i = 1, \ldots, K$, $P\{C = i\} = \rho_i$. Although $P\{C = 0\}$ has been mentioned earlier in this section, $P\{C = i\}$ deserves some explanation. First consider $P\{C = i|C > 0\}$ which would be the probability that if all the service times were arranged back to back at the server and an arbitrary time point was picked. For this consider a long stick made up of n_i small sticks of random lengths sampled from S_i (the service times of class i) for $i = 1, \ldots, K$. If a point is selected uniformly on this stick, then the point would be on a class i small stick with probability $n_i E[S_i]/(\sum_j n_j E[S_j])$. The number of sticks n_i correspond to the number of class-i customers sampled. As the sample size grows, $n_i/\sum_j n_j \to \lambda_i/\sum_j \lambda_j$

because the fraction of class-i customers is proportional to their arrival rates. Therefore, $P\{C = i|C > 0\} = \lambda_i E[S_i]/(\sum_j \lambda_j E[S_j]) = \rho_i/\rho$. Since $P\{C > 0\}\rho$, unconditioning we get for $i = 1, \ldots, K$, $P\{C = i\} = \rho_i$.

To compute $E[R]$, we now condition on C. Clearly $E[R|C = 0] = 0$ since the remaining service time is zero if there is no customer at the server. For the case $C > 0$, recall the result from renewal theory (see Section A.5.2) that the remaining time for a renewal to occur at an arbitrary time in steady state is according to the equilibrium distribution. Further, the expected remaining time for the renewal is half the ratio of the second moment to the mean interrenewal time. Therefore, we have for $i = 1, \ldots, K$

$$E[R|C = i] = \frac{E[S_i^2]}{2E[S_i]}.$$

Unconditioning, we get

$$E[R] = \sum_{i=0}^{K} E[R|C = i]P\{C = i\}$$

$$= 0 + \sum_{i=1}^{K} \frac{E[S_i^2]\rho_i}{2E[S_i]}$$

$$= \sum_{i=1}^{K} \frac{E[S_i^2]\lambda_i E[S_i]}{2E[S_i]}$$

$$= \sum_{i=1}^{K} \frac{E[S_i^2]\lambda_i}{2}$$

$$= \frac{\lambda E[S^2]}{2}$$

using the notation $\mu_i = 1/E[S_i]$ and

$$E[S^2] = \frac{1}{\lambda} \sum_{i=1}^{K} \lambda_i E[S_i^2].$$

Hence, we have shown that the expected workload at the server is $\lambda E[S^2]/2$ for the system under consideration. Although we did not explicitly state it, note that for this result we do require that the service discipline is work conserving with at most one partially completed service. ∎

In the previous problem, we did not use all the power of work conservation. In particular, the workload at some time t, $W(t)$ is conserved over all work-conserving disciplines. The next problem discusses the second subsystem, namely the waiting area, and uses the fact that $W(t)$ does not depend on the service discipline.

Problem 44

Consider a stable $G/G/1$ queue with K classes and a service discipline that is work conserving with at most one partially completed service allowed. Let W_{iq} be the average waiting time in the queue (not including service) for a class-i customer. Let the system be observed at an arbitrary time in steady state. Show that the expected workload in the waiting area (not including any at the server) at that observation is

$$\sum_{i=1}^{K} \rho_i W_{iq}.$$

Further, show that this expression is a **constant** over all work-conserving disciplines.

Solution

The expected workload in the waiting area is the sum of the expected workloads of all classes. Let the expected number of class-i customers in the waiting area in steady state be L_{iq}. Therefore, the expected workload due to class-i customers in the waiting area is $L_{iq}E[S_i]$ since each class-i customer brings in an average workload of $E[S_i]$. However, from Little's law we have $L_{iq} = \lambda_i W_{iq}$, and hence $L_{iq}E[S_i] = \rho_i W_{iq}$. Thus, the expected workload in the waiting area (not including any at the server) at an observation made at an arbitrary time in steady state is

$$\sum_{i=1}^{K} \rho_i W_{iq}.$$

This result only tells us the average workload in the waiting area subsystem for any given work-conserving discipline where at most one customer can be at the server. However, we are yet to show that this quantity when computed for any such service discipline would be a constant. For that consider the basic work conservation result that states that the amount of work in the system at a given time point (here we consider an arbitrary time point in steady state) is the same irrespective of the service discipline, as long as it is work conserving. Naturally the expected value of the amount of work is also conserved. However, the expected value of the amount of work at an arbitrary point in steady state is

$$\frac{\lambda E[S^2]}{2} + \sum_{i=1}^{K} \rho_i W_{iq} \tag{5.1}$$

and is conserved across disciplines. However, the term $\lambda E[S^2]/2$ remains a constant across disciplines. Therefore

$$\sum_{i=1}^{K} \rho_i W_{iq}$$

is a constant over all work-conserving disciplines. ∎

It is worthwhile to point out that in this expression typically W_{iq} *does* change with service discipline (ρ_i does not). One of the biggest benefits of this result is to verify expressions derived for W_{iq}. Also, the concept of work conservation and dividing the system into the waiting area plus server area subsystem would be exploited in the analysis to follow. Talking about analysis, note that so far we have written down several expressions in this chapter relating quantities such as L, W, L_i, and W_i (and the respective quantities with the q subscript). However, we have not yet provided a way to compute any of them. That would be the objective of the remainder of this chapter (except the last section). In particular, we will compute L, W, L_i, and W_i (and the respective quantities with the q subscript) for a few service disciplines. Note that unless $K = 1$, the expressions would typically vary based on the service discipline used. The most general system for which it is possible to derive the previous expressions in closed form is the multiclass $M/G/1$ queue which we will use in the next few sections.

5.2 Evaluating Policies for Classification Based on Types: Priorities

In this section, we consider a set of service disciplines and evaluate performance measures such as sojourn time and number in the system for each class of customers in a single-server multiclass queue. One way to think about the set of service disciplines addressed in this section is that the customers belong to different *types* and each type has its own requirement. In the sections to follow, we consider classification based on physical location

(Section 5.3) and knowledge of service times (Section 5.4). In this section where we consider classification based on customer type, we also assume that there is no switch-over time from customer to customer or class to class. In addition, we assume that the arrival and service times are nonanticipative, specifically the realized service time is known only upon service completion.

Before describing the model and analysis technique for such service disciplines, we describe some physical examples of multiclass queues with classification based on types. The emergency ward situation that we described earlier and a case study we will address later is a canonical example of such a system. In particular, depending on the urgency of the patient, they can be classified as critical, urgent, or regular. Their needs in terms of queue performance are significantly different. Although in this example the service performed could be different, next is the one where they are similar. Many fast-food restaurants nowadays accept online orders. Typically those customers are classified differently compared to the ones that stand in line and order physically. The needs of the online orders in terms of sojourn times are certainly different but the service times are no different from the other class. In addition there are many examples in computer systems, networking, transportation, and manufacturing where the analysis to follow can be applied.

Although there are several applications, we present a fairly generic model for the system. Consider a special case of the $G/G/1$ queue with K classes where the arrival process is $PP(\lambda_i)$ for class i ($i = 1, 2, \ldots, K$). The service times are IID with mean $E[S_i] = 1/\mu_i$, second moment $E[S_i^2]$, CDF $G_i(\cdot)$, and $\rho_i = \lambda_i/\mu_i$ for class i (for $i = 1, 2, \ldots, K$). There is a single server with infinite waiting room. From an analysis standpoint, it does not matter whether there is a queue for each class or whether all customers are clubbed into one class as long as the class of each customer and the order of arrival are known to the server. We present results for three work-conserving disciplines: FCFS, nonpreemptive priority, and preemptive resume priority. Note that the case of preemptive repeat (identical or random) is not considered but is available in the literature. Other schemes such as round-robin will be dealt with in subsequent sections.

5.2.1 Multiclass $M/G/1$ with FCFS

In this service-scheduling scheme, the customers are served according to FCFS. None of the classes receives any preferential treatment. In other words, although the server knows the arrival time and class of each customer in the system, the server ignores the class and serves only according to the arrivals. One purpose to study this queue is to benchmark against priority policy, and more importantly some of the priority policy results are actually based on this one. In some sense, this is the easiest work-conserving discipline we can consider for analysis. Therefore, we first derive performance measures for this system. It also reinforces some of the results we derived earlier in this chapter.

The analysis of a multiclass $M/G/1$ queue with FCFS is done by aggregating all K classes into a single class. This is because the server does not differentiate between classes and serves in the order the customers arrive. First consider the arrival process. The superposed net customer arrival process is $PP(\lambda)$ with $\lambda = \lambda_1 + \lambda_2 + \cdots + \lambda_K$. Let S be a random variable denoting the "effective" service time for an arbitrary customer. In essence, S can be thought of as the service time incurred by the server aggregated over all customers. Then we can derive the following results by conditioning on the class of customer being served (the probability that a customer that enters the server is of class i is λ_i/λ):

$$G(t) = P(S \le t) = \frac{1}{\lambda}\sum_{i=1}^{K}\lambda_i G_i(t),$$

$$E[S] = \frac{1}{\mu} = \frac{1}{\lambda}\sum_{i=1}^{K}\lambda_i E[S_i],$$

$$E[S^2] = \sigma^2 + \frac{1}{\mu^2} = \frac{1}{\lambda}\sum_{i=1}^{K}\lambda_i E[S_i^2],$$

$$\rho = \lambda E[S].$$

Note that the average aggregate service rate is μ and the variance of the aggregate service time is σ^2. Also, $\rho = \rho_1 + \cdots + \rho_K$.

Assume that the system is stable, that is, $\rho < 1$. Note that the system is identical to that of a single class $M/G/1$ queue with $PP(\lambda)$ arrivals and service times with CDF $G(t)$, mean $1/\mu$, and variance σ^2. Then using Pollaczek–Khintchine formula (see Equation 4.6) for a single class $M/G/1$ queue, we get the following results:

$$L = \rho + \frac{1}{2}\frac{\lambda^2 E[S^2]}{1-\rho},$$

$$W = \frac{L}{\lambda},$$

$$W_q = W - 1/\mu,$$

$$L_q = \frac{1}{2}\frac{\lambda^2 E[S^2]}{1-\rho}.$$

Now we need to derive the performance measures for each class i for $i = 1, \ldots, K$. The key result that enables us to derive performance measures for the individual classes based on aggregate measures is $W_{iq} = W_q$ for all i. This result essentially states that the expected time spent waiting for service is identical for all classes and is equal to that of the aggregate W_q. This deserves an explanation. Note that the aggregate arrivals and class i arrivals are according to Poisson processes with parameters λ and λ_i, respectively. Due to PASTA, the average workload in the system in steady state is as seen by the aggregate arrivals, and it is also the same as that seen by class-i arrivals. Further, the steady-state average workload also corresponds to the expected remaining work before service begins for that arrival. Based on all that, $W_{iq} = W_q$ for all i.

Using that result we can show that the expected number of class-i customers in the system (L_i) as well as those waiting in the queue (L_{iq}) and the expected sojourn time in the system for class i (W_i) as well as waiting in the queue (L_{iq}) are given by

$$W_{iq} = W_q = \frac{1}{2}\frac{\lambda E[S^2]}{1 - \rho},$$

$$L_{iq} = \lambda_i W_{iq},$$

$$L_i = \rho_i + L_{iq},$$

$$W_i = W_{iq} + \frac{1}{\mu_i}.$$

Often one is tempted to take intuitive guesses at the relationship between the individual class performance and that of the aggregate class. For example, one can ask whether $L_i = \lambda_i L/\lambda$, or still better, is $L_i = \rho_i L/(1 - \rho)$? Although they are intuitively appealing, they are both not true. In fact from the previous results we have

$$L_i = \frac{\lambda_i L}{\lambda} - \frac{\lambda_i}{\mu} + \rho_i.$$

However, note that the result $L_1 + L_2 + \cdots + L_K = L$ holds.

Thus, we have analyzed the performance of each class of a multiclass $M/G/1$ queue with FCFS service discipline. Next we move on to systems with priorities; in other words, the classes correspond to priority order of service. We describe them first and then analyze.

5.2.2 M/G/1 with Nonpreemptive Priority

In this service-scheduling scheme, the customers are classified and served according to a priority rule. We assume that the server knows the order of arrivals and class of each customer in the system. To keep the mapping between classes and priorities simple, we let class-1 to be the highest priority and class K the lowest (we will see later how to do this optimally). Also, service discipline within a class is FCFS. Therefore, the server upon a service completion always starts serving a customer of the highest class among those waiting for service, and the first customer that arrived within that class. Of course, if there are no customers waiting, the server selects the first customer that arrives subsequently. However, it is important to clarify that the server completes serving a customer before considering whom to serve next. The meaning of *nonpreemptive priority* is that a customer in service does not get preempted (or interrupted) while in service by another customer of high priority (essentially it is only in the waiting room where there is priority).

Recall that we are considering an $M/G/1$ queue with K classes where the arrival process is $PP(\lambda_i)$ for class i ($i = 1, 2, \ldots, K$); the service times are IID with mean $E[S_i] = 1/\mu_i$, second moment $E[S_i^2]$, CDF $G_i(\cdot)$, and $\rho_i = \lambda_i/\mu_i$ for class i (for $i = 1, 2, \ldots, K$). There is a single server with infinite waiting room. From an analysis standpoint, it does not matter whether there is a queue for each class or whether all customers are clubbed into one class as long as the class of each customer and the order of arrival are known to the server. However, from a practical standpoint it is easiest to create at least a "virtual" queue for each class and pick from the head of the nonempty line with highest priority. Further, we assume that the system is stable, that is, $\rho < 1$. For the rest of this section, we will use the terms class and priority interchangeably. With that said, we are ready to analyze the system and obtain steady-state performance measures.

To analyze the system, we consider a class-i customer that arrives into the system in steady state for some $i \in \{1, \ldots, K\}$. We reset the clock and call that time as 0. Another way to do this is to assume the system is stationary at time 0 and a customer of class i arrives. Consider the *random variables* defined in Table 5.1 (albeit with some abuse of notation). It is crucial to note that the terms in Table 5.1 would perhaps have other meanings in the rest of this book.

When a customer of class i arrives in the stationary queue at time 0, this customer first waits for anyone at the server to be served (i.e., for a random time U). Then the customer also waits for all customers that are in the system

TABLE 5.1

Random Variables Used for Nonpreemptive and Preemptive Resume Cases

W_i^q	Waiting time in the queue (not including service) for customer of class i (note that this is a random variable and not the expected value)
U	Remaining service time of the customer in service (this is zero if the server is idle)
R_j	Time to serve all customers of type j who are waiting in the queue at time 0 (for $1 \leq j \leq i$)
T_j	Time to serve all customers of type j who arrive during the interval $[0, W_i^q]$ (for $1 \leq j < i$)

of equal or higher priority at time 0 (i.e., for a random time $\sum_{j=1}^{i} R_j$). Note that during the time this customer waits in the system to begin service (i.e., W_i^q), there could be other customers of higher priority that may have arrived and served before this customer. Thus, this customer waits a further $\sum_{j=1}^{i-1} T_j$ (with the understanding that the term is zero if $i = 1$) before service begins. Therefore, we have

$$W_i^q = U + \sum_{j=1}^{i} R_j + \sum_{j=1}^{i-1} T_j.$$

Taking expectations we have

$$E[W_i^q] = E[U] + \sum_{j=1}^{i} E[R_j] + \sum_{j=1}^{i-1} E[T_j]. \tag{5.2}$$

We need to derive expressions for $E[U]$, $E[R_j]$, and $E[T_j]$ which we do next.

Problem 45

Derive the following results (for the notations described in Table 5.1):

$$E[U] = \frac{\lambda}{2} E[S^2]$$

$$E[R_j] = \rho_j E[W_j^q]$$

$$E[T_j] = \rho_j E[W_i^q]$$

where $\rho_i = \lambda_i E[S_i]$.

Solution

Recall from Problem 43 that if a stable $G/G/1$ queue with K classes is observed at an arbitrary time in steady state, then the expected remaining service time is $\lambda E[S^2]/2$. Of course this requires a service discipline that is work conserving with at most one partially completed service allowed, which is true here. However, since the arrivals are Poisson, due to PASTA and $M/G/1$ system being ergodic, an arriving class-i customer in steady state would observe an expected remaining service time of $\lambda E[S^2]/2$. In other words, the arrival-point probability is the same as the steady-state probability. Thus, from the definition of U we have

$$E[U] = \frac{\lambda}{2} E[S^2].$$

Once again because of PASTA this arriving customer at time 0 will see L_{jq} customers of class j waiting for service to begin. Therefore, by the definition of R_j (time to serve all customers of type j waiting in the queue at time 0), we have $E[R_j] = E[E[R_j|N_j]] = E[N_j/\mu_j] = L_{jq}/\mu_j$, where N_j is a random variable denoting the number of class-j customers in steady state waiting for service to begin. Next, using Little's law $L_{jq} = \lambda_j W_{jq}$, we have

$$E[R_j] = \lambda_j W_{jq} \frac{1}{\mu_j} = \rho_j E[W_j^q].$$

Now, to compute $E[T_j]$ note the definition that it is the time to serve all customers of type j who arrive during the interval $[0, W_i^q]$ for any $j < i$. Clearly, the expected number of type j arrivals in time t is $\lambda_j t$ because the arrivals are according to a Poisson process and each of those arrivals require $1/\mu_j$ service time. Hence, we have $E[T_j|W_i^q] = \lambda_j W_i^q/\mu_j$. Taking expectations we get

$$E[T_j] = \rho_j E[W_i^q]. \qquad \blacksquare$$

Plugging these results into Equation 5.2, we have

$$E[W_i^q] = \frac{\lambda}{2} E[S^2] + \sum_{j=1}^{i} \rho_j E[W_j^q] + E[W_i^q] \sum_{j=1}^{i-1} \rho_j.$$

We rewrite this equation using the notation that we have used earlier for the average waiting time before service, that is, $W_{iq} = E[W_i^q]$. Thus, we have for all $i \in \{1, \ldots, K\}$

$$W_{iq} = \frac{\lambda}{2} E[S^2] + \sum_{j=1}^{i} \rho_j W_{jq} + W_{iq} \sum_{j=1}^{i-1} \rho_j.$$

Note that we have K equations with K unknowns W_{iq}. However, for $i = 1$, there is only one unknown W_{1q} which we can solve. Then we consider $i = 2$. Since we already know W_{1q}, we can write W_{2q} in terms of W_{1q}. In that manner we can solve recursively for W_{iq} starting with $i = 1$ all the way to $i = K$. For that we let $\alpha_i = \rho_1 + \rho_2 + \cdots + \rho_i$ with $\alpha_0 = 0$. Then it is possible to show that for all $i \in \{1, \ldots, K\}$

$$W_{iq} = \frac{\frac{1}{2} \sum_{j=1}^{K} \lambda_j E[S_j^2]}{(1 - \alpha_i)(1 - \alpha_{i-1})}.$$

Now, using W_{iq} we can derive the other performance measures as follows for all $i \in \{1, \ldots, K\}$:

$$L_{iq} = \lambda_i W_{iq},$$

$$W_i = W_{iq} + E[S_i],$$

$$L_i = L_{iq} + \rho_i.$$

Now, using the performance measures for individual classes, we can easily obtain aggregate performance measures across all classes as follows:

$$L = L_1 + L_2 + \cdots + L_K,$$

$$W = \frac{L}{\lambda},$$

$$W_q = W - \frac{1}{\mu},$$

$$L_q = \lambda W_q.$$

So far we have assumed that we are given which class should get the highest priority, second highest, etc. This may be obvious in some settings such as a hospital emergency ward. However, in other settings such as a manufacturing system we may need to determine an optimal way of assigning priorities. To do that, consider there are K classes of customers and it costs the server C_j per unit time a customer of class j spends in the system (this can be thought of as the holding cost for class j customer). It turns out (we will show that in a problem next) if the objective is to minimize the total expected cost per unit time in the long run, then the optimal priority assignment is to give class i higher priority than class j if $C_i \mu_i > C_j \mu_j$ (for all i, j such that $i \neq j$). In other words, sort the classes in the decreasing order of the product $C_i \mu_i$ and assign first priority to the largest $C_i \mu_i$ and the last priority to the smallest $C_i \mu_i$ over all K classes. This is known as the $C\mu$ rule. Also note that if all the C_i values were equal, then this policy reduces to "serve the customer with the shortest expected processing time first." We derive the optimality of the $C\mu$ rule next.

Problem 46

Consider an $M/G/1$ queue with K classes with notations described earlier in this section and service discipline being nonpreemptive priority. Further, it costs the server C_j per unit time a customer of class j spends in the system and the objective is to minimize the total expected cost per unit time in the long run. Show that the optimal priority assignment is to give class i higher priority than class j if $C_i \mu_i > C_j \mu_j$ (provided $i \neq j$).

Solution

Let T_C be the average cost incurred per unit time if the priorities are $1, 2, \ldots, K$ from highest to lowest for the system under consideration. Since a cost C_n is incurred during the sojourn of a class n customer, the total cost incurred per class n customer on average in steady state is $C_n W_n$ (the reason for not using i or j but n is that i and j are reserved for something else). Also, class n customers arrive at rate λ_n resulting in an average cost per unit time due to class n customers being $\lambda_n C_n W_n$. Thus, we have

$$T_C = \sum_{n=1}^{K} \lambda_n C_n W_n.$$

We use an exchange argument for this problem. In essence, we would like to evaluate the total cost if instead of ordering the priorities from class-1 to K, we swap the priorities of two classes. For that we select a particular $i \in \{1, \ldots, K-1\}$. Also, we pick $j = i+1$ so that we only swap neighboring classes to make our analysis simple. It is crucial to realize that we are indeed selecting only one i and one j. For example, if $K = 5$, then i could be 2, and j would therefore be 3. Now, denote T_{Ce} as the average cost incurred per unit time by exchanging priorities for i and j. For example, say $K = 5$, $i = 2$, and $j = 3$. Then, T_C is the average cost per unit time when the priority order is 1-2-3-4-5 and T_{Ce} is the average cost per unit time when the priority order is 1-3-2-4-5. Now we proceed with a generic i and $j = i+1$ for the remainder of the analysis. We compute $T_c - T_{Ce}$ with the understanding that if it is positive, then we must switch i and j, otherwise we should stick with the original choice. Note that for some $n \in \{1, \ldots, K\}$

$$W_n = W_{nq} + E[S_n] = \frac{\frac{1}{2} \sum_{k=1}^{K} \lambda_k E[S_k^2]}{(1 - \alpha_n)(1 - \alpha_{n-1})} + \frac{1}{\mu_n}.$$

We first make a simplifying step

$$\frac{\lambda_n}{(1 - \alpha_n)(1 - \alpha_{n-1})} = \frac{\mu_n}{1 - \alpha_n} - \frac{\mu_n}{1 - \alpha_{n-1}},$$

using the notation of $\alpha_n = \rho_1 + \cdots + \rho_n$ with $\rho_k = \lambda_k / \mu_k$ for all $1 \leq k \leq n$. This gives us

$$\lambda_n C_n W_n = \frac{1}{2} C_n \sum_{k=1}^{K} \lambda_k E[S_k^2] \left(\frac{\mu_n}{1 - \alpha_n} - \frac{\mu_n}{1 - \alpha_{n-1}} \right) + \frac{\lambda_n C_n}{\mu_n}.$$

Next, note that while computing $T_c - T_{Ce}$, all the terms except the ith and jth terms in $\sum_n \lambda_n C_n W_n$ would be identical and cancel out. Using these results, the definitions of α_k and ρ_k, as well as some algebra, we can derive the following (presented without displaying the canceled terms $\sum_n \lambda_n C_n / \mu_n$):

$$
\frac{T_C - T_{Ce}}{\frac{1}{2}\sum_{n=1}^{K}\lambda_n E[S_n^2]} = \sum_{r=1}^{K}\left(\frac{C_r\mu_r}{1-\alpha_r} - \frac{C_r\mu_r}{1-\alpha_{r-1}}\right) - \sum_{r=1}^{i-1}\frac{C_r\mu_r}{1-\alpha_r} - \frac{C_j\mu_j}{1-\alpha_{i-1}-\rho_j}
$$

$$
- \frac{C_i\mu_i}{1-\alpha_{i-1}-\rho_j-\rho_i} - \sum_{r=i+2}^{K}\frac{C_r\mu_r}{1-\alpha_r} + \sum_{r=1}^{i-1}\frac{C_r\mu_r}{1-\alpha_{r-1}}
$$

$$
+ \frac{C_j\mu_j}{1-\alpha_{i-1}} + \frac{C_i\mu_i}{1-\alpha_{i-1}-\rho_j} + \sum_{r=i+2}^{K}\frac{C_r\mu_r}{1-\alpha_{r-1}}
$$

$$
= \frac{C_i\mu_i}{1-\alpha_i} + \frac{C_j\mu_j}{1-\alpha_i-\rho_j} - \frac{C_i\mu_i}{1-\alpha_{i-1}} - \frac{C_j\mu_j}{1-\alpha_i} - \frac{C_j\mu_j}{1-\alpha_{i-1}-\rho_j}
$$

$$
- \frac{C_i\mu_i}{1-\alpha_{i-1}-\rho_j-\rho_i} + \frac{C_j\mu_j}{1-\alpha_{i-1}} + \frac{C_i\mu_i}{1-\alpha_{i-1}-\rho_j}
$$

$$
= (C_i\mu_i - C_j\mu_j)\left[\frac{1}{1-\alpha_i} - \frac{1}{1-\alpha_{i-1}} + \frac{1}{1-\alpha_{i-1}-\rho_j}\right.
$$

$$
\left. - \frac{1}{1-\alpha_i-\rho_j}\right]
$$

$$
= \frac{\rho_i\rho_j(\alpha_j - 2)(C_i\mu_i - C_j\mu_j)}{(1-\alpha_i)(1-\alpha_{i-1})(1-\alpha_{i-1}-\rho_j)(1-\alpha_i-\rho_j)}.
$$

Given that $\alpha_j - 2 < 0$, we can make the following conclusion:

$$
\text{if } C_i\mu_i - C_j\mu_j > 0, \text{ then } T_C - T_{Ce} < 0.
$$

Therefore, if $C_i\mu_i < C_j\mu_j$, then we should switch the priorities of i and j since $T_C - T_{Ce} > 0$. In this manner, if we compare $C_i\mu_i$ and $C_j\mu_j$ for all pairs of neighbors, the final priority rule would converge to one that is in the decreasing order of $C_n\mu_n$. In other words, the optimal priority assignment is to give class i higher priority than class j if $C_i\mu_i > C_j\mu_j$ (provided $i \neq j$). Therefore, one should sort the product $C_i\mu_i$ and call the highest $C_i\mu_i$ as class-1 and the lowest as class K. ∎

5.2.3 *M/G/1* with Preemptive Resume Priority

Here we consider a slight modification to the *M/G/1* nonpreemptive priority considered in the previous subsection. The modification is to allow preemption during service. During the service of a customer, if another customer of higher priority arrives, then the customer in service is preempted and service begins for this new higher priority customer. When the pre-empted customer returns to service, service resumes from where it was preempted. This is a work-conserving discipline (however, if the service has to start from the beginning which is called preemptive repeat, then it is not work conserving because the server wasted some time serving). As we described earlier, if the service times are exponential, due to memoryless property, preemptive resume and preemptive repeat are the same. However, there is another case called preemptive identical which requires that the service that was interrupted is repeated with an identical service time (in the preemptive repeat mechanism, the service time is sampled again from a distribution). We do not consider those here and only concentrate on preemptive resume priority.

All the other preliminary materials for the nonpreemptive case also hold here for the preemptive resume policy (namely, multiclass $M/G/1$ with class-1 being highest priority and class K lowest). Also, customers within a class will be served according to FCFS policy. But a server will serve a customer of a particular class only when there is no customer of higher priority in the system. Upon arrival, customer of class i can preempt a customer of class j in service if $j > i$. Also, the total service time is unaffected by the interruptions, if any. Assume that the system is stable, that is, $\rho < 1$. Note that there could be more than one customer with unfinished (but started) service. Therefore, the results of Section 5.1.4 cannot be applied here. However, the service discipline is still work conserving and we will take advantage of that in our analysis. Further, note that the sojourn time of customers of class i is unaffected by customers of class j if $j > i$.

With that thought we proceed with our analysis. We begin by considering class-1 customers. Clearly, as far as class-1 customers are concerned, they can be oblivious of the lower class customers. Therefore, class-1 customers effectively face a standard single class $M/G/1$ system with arrival rate λ_1 and service time distribution $G_1(\cdot)$. Class-1 customers get served upon arrival if there are no other class-1 customers in the system, and they will wait only for other class-1 customers for their service to begin. Thus, from Pollaczek–Khintchine formula in Equation 4.6, we get the sojourn time of class-1 customers as

$$W_1 = \frac{1}{\mu_1} + \frac{\lambda_1 E(S_1^2)}{2(1 - \rho_1)}.$$

Next we turn to class-2. Since preemptive resume is a work-conserving discipline, the workload in the system at any time due to classes 1 and 2 alone is identical to that of a FCFS queue with only classes 1 and 2. In the preemptive resume framework, the classes 1 and 2 as a single group is unaffected by the dynamics of the other classes. Therefore, we can simply consider the two classes as a set and compare with another queue that has the same two classes but FCFS discipline. The steady-state expected workload under FCFS (for the two-class system) is

$$\frac{(\lambda_1 E(S_1^2) + \lambda_2 E[S_2^2])}{2} + \sum_{i=1}^{2} \rho_i W_{iq} = \frac{(\lambda_1 E(S_1^2) + \lambda_2 E[S_2^2])}{2}$$

$$+ (\rho_1 + \rho_2)\frac{1}{2}\frac{\lambda_1 E(S_1^2) + \lambda_2 E[S_2^2]}{1 - \rho_1 - \rho_2}.$$

This expression is equal to the expected workload of class-1 customers plus that of class-2 customers under the preemptive resume discipline. Because of PASTA, entering class-2 customers in steady state see this workload due to the two classes. If W_2 is the steady-state average sojourn time of a class-2 customer in the preemptive resume discipline, then W_2 equals the preceding average workload plus the expected workload due to all the customers of class-1 that arrived during W_2 (which equals $\rho_1 W_2$), plus the service time of this customer $1/\mu_2$. Thus, we can derive (details are shown for W_i subsequently for any $i > 1$)

$$W_2 = \frac{1}{\mu_2(1 - \rho_1)} + \frac{\sum_{j=1}^{2} \lambda_j E(S_j^2)}{2(1 - \rho_1 - \rho_2)(1 - \rho_1)}.$$

In a similar manner, we can derive the expressions for class-3. However, it is as straightforward to just show this for a generic class i in the K-class $M/G/1$ queue with preemptive resume priority. Recall that if only the first i classes of customers are considered, then the processing of these customers as a group is unaffected by the lower priority customers. The crux of the analysis is in realizing that the workload of this system (with only the top i classes) at all times is the same as that of an $M/G/1$ queue with FCFS and top i classes due to the work-conserving nature. Therefore, using the results for work-conserving systems, the performance analysis of this system is done. Consider an $M/G/1$ queue with only the first i classes and FCFS service. Let $\overline{W}(i)$ be the average workload in the system when the system is observed at an arbitrary time in steady state. Then using Equation 5.1 and plugging in the relation for the waiting time in the queue from the FCFS analysis (one must be cautious to use i and not K for the number of classes), we get

$$\overline{W}(i) = \frac{\sum_{j=1}^{i} \lambda_j E(S_j^2)}{2(1 - \alpha_i)}$$

where $\alpha_i = \rho_1 + \rho_2 + \cdots + \rho_i$ for $i > 0$ and $\alpha_0 = 0$.

The only reason we needed FCFS discipline is to obtain $\overline{W}(i)$. Now we resort back to the preemptive resume service discipline with i classes. Since the preemptive resume policy is also work conserving, $\overline{W}(i)$ is the average workload in the system with the first i classes alone. In addition, due to PASTA, $\overline{W}(i)$ will also be the average workload in the preemptive resume $M/G/1$ queue as seen by an arriving class-i customer. This in turn is also equal to the average workload due to the first i classes in the K-class system. Now consider an $M/G/1$ queue with all K classes where a customer of class i is about to enter in steady state. Then the sojourn time in the system for this customer depends only on the customers of classes 1 to i in the system upon arrival. Therefore, W_i can be computed by solving

$$W_i = \overline{W}(i) + \frac{1}{\mu_i} + \alpha_{i-1} W_i$$

as the mean sojourn time is equal to the expected workload upon arrival from all customers of classes 1 to i plus the mean service time of this class i customer plus the average service time of all the customers of classes 1 to $i - 1$ that arrived during the sojourn time. Substituting the expression for $\overline{W}(i)$ and rearranging terms, we have

$$W_i = \frac{1}{\mu_i(1 - \alpha_{i-1})} + \frac{\sum_{j=1}^{i} \lambda_j E(S_j^2)}{2(1 - \alpha_i)(1 - \alpha_{i-1})}.$$

Now, using W_i, other mean performance measures for the preemptive resume service discipline can be obtained as follows:

$$W_{iq} = W_i - E[S_i],$$

$$L_i = \lambda_i W_i,$$

$$L_{iq} = L_i - \rho_i.$$

The results for the individual classes can be used to obtain aggregate performance measures as follows:

$$L = L_1 + L_2 + \cdots + L_K,$$

$$W = \frac{L}{\lambda},$$

$$W_q = W - \frac{1}{\mu},$$

$$L_q = \lambda W_q.$$

As described earlier, the reader is encouraged to consider results in the literature for other priority queues such as preemptive repeat and preemptive identical. Next we present a case study to illustrate some minor variations to the models we have seen in this section. After that, we will move on to other policies in subsequent sections.

5.2.4 Case Study: Emergency Ward Planning

Between January 2009 and June 2011, several hospitals all over the United States have started to report their emergency room wait times on billboards. That information can also be accessed on smartphones through apps as well as by going to the hospital websites. One such hospital that got onto that bandwagon is the University Town Hospital (UTH). University Town is a small college town and UTH serves residents of University Town as well as other small towns around it. Nonetheless, UTH is facing some fierce competition from other hospitals and clinics in the area that have recently established many urgent care centers (including high acuity ones that deal with emergency patients). To make matters worse, emergency rooms (ER) have been getting bad press such as: ER visits reached an all time high of over 123 million in 2008 (it was 117 million in 2007); in addition, a government report showed that on many occasions patients waited nearly a half hour instead of being seen immediately at ERs.

Clearly, UTH's emergency ward needed a makeover fearing the loss of clientele. Feeling the pressure, toward the end of year 2010, upper management of UTH met to develop a public relations strategy for the sustainability of the emergency ward of UTH. They did not want to waste any more time and made two major decisions. The first decision was to install a billboard before January 2011 on Main Street in downtown University Town that would electronically display the current wait time for ER patients. The second decision was to roll out a campaign called *30-Minute ER Commitment* that would assure that patients would be guaranteed to be seen by an ER doctor within 30 min. Focus groups revealed that this would clearly establish UTH's commitment to patient satisfaction. In addition, many of the focus-group participants felt that the high-tech billboard technology would give the impression that UTH was equipped with the latest medical devices as well. By mid-April, the upper management was delighted to hear that the average wait time was only 14 min while the number of patients served per day in fact increased since November 2009.

However, in June 2011 the upper management of UTH was informed of a few local blogs which reported that patients waited over half an hour at the emergency ward to see a doctor. This was a cause for concern. The upper management immediately called upon the hospital management engineer. Her name is Jenna and she was charged to look into the issue. She had a week to produce a report and make a presentation. Jenna had worked on her Master's thesis on emergency ward hospital-bed scheduling. But ever since she has been employed at UTH, she has worked on other topics such as nurse scheduling and pharmaceutical inventories. However, Jenna loved queueing theory and was excited about a project on wait times. Without much adieu, Jenna went about gathering all the information she could as well as collected the necessary data for her analysis. The first thing she found out was that the billboard wait times were updated every 15 min (through an automatic RSS feed) and the displayed value was the average wait time over the past 2 h.

Jenna wanted to know how they computed wait times and she was told that the wait time for a patient is the time from when the patient checks in until when the patient is called by a clinical professional. Jenna immediately realized that it did not include the time the clinical professional spends seeing the patient. So it represented the waiting time in the queue and not the sojourn time. Jenna found out that the entire time spent by patients in the ER could even be several hours if a complicated surgery needs to be performed. However, she was glad that she did not have to focus on those issues. But what was concerning for her was whether someone with a heart failure had to wait on average for 14 min to see a clinical professional. She was reassured that when patients arrive at the emergency ward, they are immediately seen by a triage nurse. The nurse would determine the severity of the patient's illness or injury to determine if they would have to be seen immediately by a clinical professional. Priority was given to patients with true emergencies (this does not include life-threatening cases, pregnancies, etc., where patients are not seen by a clinical professional but are directly admitted to the hospital).

Upon speaking with the triage nurse, Jenna found out that there are essentially two classes of patients. One class is the set of patients with true emergencies and the second class is the remaining set of patients. Within a class, patients were served according to FCFS; however, the patients with true emergencies were given preemptive priority over those that did not have a true emergency. The triage nurse also whispered to Jenna that she would much rather have three classes instead of two. It was hard to talk a lot to the triage nurse because she was always busy. But Jenna managed to also find out that there are always two doctors (i.e., clinical professionals) at the emergency ward, and like Jenna saw during her Master's thesis days, the most crowded times were early evenings. Jenna next stopped at the information technology office to get historical data of the patients. A quick analysis revealed that patients arrived according to a Poisson process and the time a doctor took to see a patient was exponentially distributed. Interestingly, the time a doctor spent to see a patient was indifferent for the two classes of patients.

Jenna looked at her textbook for her course on waiting line models. She distinctly remembers studying preemptive queues. However, when she saw the book, she did not see anything about two-server systems (note that since the ward has two doctors, that would be the case here). Further, the book only had results for mean wait times and not distributions, which is something she thought was needed for her analysis. Nonetheless, she decided to go ahead and read that chapter carefully so that she gets ideas to model the system and analyze it. Jenna also checked the simulation software packages she was familiar with and none had an in-built preemptive priority option (all of them only had nonpreemptive). At this time Jenna realized that her only option was to model the system from scratch. She wondered if the two-server system was even work conserving. But she did feel there was hope since the interarrival times and service times were exponentially distributed. Also, the service times were class independent. "How hard can that be to analyze," she thought to herself.

Jenna started to model the system. Based on her data she wrote down that class-1 patients (with true emergencies) arrived to the emergency ward according to $PP(\lambda_1)$ and class-2 patients arrived according to $PP(\lambda_2)$. The service time for either class is $\exp(\mu)$. There are two servers that use FCFS within a class and class-1 has preemptive resume priority over class-2. In the event that there are two class-2 patients being served when a class-1 arrives, Jenna assumed that with equal probability one of the class-2 patients was selected to be preempted. Jenna first started to model the system as a CTMC $\{(X_1(t), X_2(t)), t \geq 0\}$ where for $i = 1, 2$, $X_i(t)$ is the number of class-i patients in the system. Then Jenna realized that there must be an easier way to model the system. She recalled how the $M/G/1$ queue with preemptive priority was modeled in her textbook. An idea immediately dawned on her.

5.2.4.1 Service Received by Class-1 Patients

Jenna realized that the class-1 patients are easy to analyze and they are the most important ones to analyze too. The main reason is that the hospital is extremely concerned that these patients who have a true emergency should not wait for too long. Thus, to model class-1 patients, note that they are oblivious to the class-2 patients. So Jenna modeled the class-1 patients using an $M/M/2$ queue with $PP(\lambda_1)$ arrivals and $\exp(\mu)$ service times and two servers. From the $M/M/2$ results in her queueing theory text, she was able to write down W_{1q}, the average time spent by class-1 patients in the queue prior to being seen by a clinical professional, as

$$W_{1q} = \left(\frac{\rho_1^2}{1 - \rho_1^2} \right) \frac{1}{\mu}$$

where $\rho_1 = \lambda_1/(2\mu)$ is the traffic intensity brought by class-1 patients. Based on Jenna's data, the arrival rate was about two class-1 patients per hour ($\lambda_1 = 0.0333$ per minute) and the average service time was about 17.5 min ($1/\mu = 17.5$ min) resulting in $\rho_1 = 0.2917$. Plugging into this formula, Jenna got $W_{1q} = 1.6271$ min which was reasonably close to what her data revealed.

At first it appears like a 1.63-min wait on average to see a doctor for a patient with a true emergency does not sound too bad. In addition, it is still fairly lower than the 30-min ER wait commitment. However, based on her data it showed that some patients (although rare) did wait for quite a while but there weren't enough sample points to make a meaningful statistical analysis for those patients. So Jenna decided to forge ahead with a queueing model. She let Y_q be the time spent waiting to begin service for a class-1 patient that arrives in steady state. She computed the LST as

$$E\left[e^{-sY_q}\right] = p_0 + p_1 + \sum_{i=2}^{\infty} \left(\frac{2\mu}{2\mu + s}\right)^{i-1} p_i$$

where p_i is the steady-state probability that the $M/M/2$ queue has i in the system. From the $M/M/2$ analysis, $p_0 = (1 - \rho_1)/(1 + \rho_1)$ and for $i \geq 1$, $p_i = 2p_0\rho_1^i$. Then she obtained

$$E\left[e^{-sY_q}\right] = p_0 \left[\frac{1 + 2\rho_1 + 2\rho_1\lambda_1}{2\mu - \lambda_1 + s}\right]$$

which upon inverting yielded

$$P\{Y_q \leq y\} = 1 - \frac{2\rho_1^2}{1 + \rho_1} e^{-(1-\rho_1)2\mu y}.$$

Plugging in the numbers, Jenna calculated the probability of a class-1 patient waiting for less than 30 min as

$$P\{Y_q \leq 30\} = 0.9884$$

which says that more than 1% of the patients with a true emergency would wait over 30 min to see a clinical professional. Jenna thought to herself that this could mean quite a few patients a month that would wait over 30 min, and it was not surprising to her that many would be blogging about it.

5.2.4.2 Experience for Class-2 Patients

For the patients that arrive at the emergency ward that do not have a true emergency, Jenna felt that it is not crucial to guarantee a less than 30-min wait. She was confident that no one was going to blog that they had a very bad cold and had to wait for 40 min in the emergency ward to see a doctor. However, she realized that it would also not be a good idea to dissuade those patients. In fact, in her opinion the main purpose of the billboard is to ensure that these patients consider the emergency ward as an alternative to urgent care facilities. In other words, Jenna felt that it is important to have a low overall average wait time so that based on the billboard display, some nonemergency patients would be lured to the emergency ward as opposed to visiting an urgent care facility. So she proceeded to compute the average time spent by the stable patients (i.e., ones without a true emergency) waiting before they see a clinical professional.

To model that, Jenna let $X(t)$ be the total number of patients in the system at time t; these include those that do and do not have a true emergency. The "system" in the previous sentence includes patients that are waiting to see a clinical professional as well as those that are being seen by a clinical professional. Since the service times for both times of patients are identically distributed $\exp(\mu)$, $X(t)$ would be stochastically identical to the number of customers in an $M/M/2$ queue with FCFS service, $PP(\lambda_1 + \lambda_2)$ arrivals, and $\exp(\mu)$ service. For such an $M/M/2$ queue with FCFS service, the steady-state average number in the system is $(\lambda_1 + \lambda_2)/\mu + (\rho^2/(1 - \rho^2))(\lambda_1 + \lambda_2)/\mu$ where $\rho = (\lambda_1 + \lambda_2)/(2\mu)$. Thus, it is also equal to the expected value of the total number of patients in the system in steady state, $L_1 + L_2$, where L_i for $i = 1, 2$ is the mean number of class-i patients in the system in steady state. Therefore,

$$L_1 + L_2 = \frac{\lambda_1 + \lambda_2}{\mu} + \left(\frac{\rho^2}{1 - \rho^2}\right)\frac{\lambda_1 + \lambda_2}{\mu}.$$

Jenna realized that L_1 can be computed as

$$L_1 = \frac{\lambda_1}{\mu} + \left(\frac{\rho_1^2}{1 - \rho_1^2}\right)\frac{\lambda_1}{\mu},$$

where $\rho_1 = \lambda_1/2\mu$ using the W_{1q} result discussed earlier and the fact that $L_1 = \lambda_1/\mu + \lambda_1 W_{1q}$. Thus, she calculated L_2 as

$$L_2 = \left(\frac{2\rho}{1 - \rho^2}\right) - \left(\frac{2\rho_1}{1 - \rho_1^2}\right).$$

Using that she wrote down, W_{2q}, the average time spent by class-2 patients waiting as

$$W_{2q} = \frac{L_2 - \rho_2}{\lambda_2}$$

where $\rho_2 = \lambda_2/\mu$. Based on Jenna's data, the arrival rate was about 2.5 class-2 patients per hour ($\lambda_2 = 0.0417$ per minute) and the average service time of 17.5 min ($1/\mu = 17.5$ min) resulting in $\rho_2 = 0.3646$. Plugging into this formula for W_{2q}, Jenna computed $W_{2q} = 31.2759$ min. Although this would mean that UTH cannot guarantee an "average" wait time of less than 30 min for stable patients, across all patients the average wait time was a little over 18 min which sounded reasonable. Jenna felt it was important to clarify that W_{2q} included time spent while being preempted by a class-1 patient, and not just the time to see a clinical professional for the first time.

5.2.4.3 Three-Class Emergency Ward Operation

Although Jenna had done a thorough analysis of the current state of the emergency ward, she was reminded of her conversation with the UTH upper management wanting recommendations too. So Jenna started to contemplate about what could change. When she checked online, Jenna found out that the statistics for true emergencies was somewhat less than 50% of the patients that go to ER. She guessed things were similar at UTH as well. But Jenna remembered what the triage nurse told her. It dawned on her that not all true emergencies need to be treated right away, some could wait a little while others cannot. So she decided to break the true emergency cases into two groups, one is *critical* cases when immediate treatment is a must for survival, and the other is *serious* cases where early treatment would be beneficial. About one in six of the true emergencies was a critical case.

Thus, using the same results as in the true emergency cases, Jenna calculated that for the critical cases, the mean arrival rate is 0.0056 patients per minute, bringing in a load of 0.0486, resulting in a mean wait time of 0.0415 min, and the probability of being seen within 30 min is 99.98% which is much more reasonable. Jenna thought she should recommend dividing the patients into three classes: critical, serious, and stable cases. Under very rare circumstances, if critical cases have to wait an unusually long time, then Jenna suggested that they be admitted to the hospital and get immediate care. Jenna also realized that the patients with serious conditions would experience a higher probability of being seen after 30 min than when they were clubbed together with the critical cases (i.e., the true emergencies). However, this is a risk worth taking, she felt. Jenna also recommended that UTH clarify on their website what the billboard average wait times truly indicate.

Jenna finished her report and made slides for her presentation. The upper management of UTH was impressed with Jenna's findings and also agreed with her recommendations. Jenna was thrilled and she was also very excited that her tools in queueing theory came very handy. In fact, she realized that rare events such as patients waiting for over 30 min cannot be reliably estimated using data or simulations. She met the triage nurse subsequently and told the nurse that her idea was acknowledged in the presentation, and it appears like the hospital was going to adopt it.

5.3 Evaluating Policies for Classification Based on Location: Polling Models

In the previous section, we mainly considered classes based on types, and priorities that fitted naturally for such systems. In this section, we consider classification based on physical location. Although we model in a general form allowing different "types" here too, in practice these are usually of a single type that contend for the common server (resource). For example, a four-way stop sign, a material handling device for storage and retrieval, a token bus in a wide area network, etc., are all examples of the kind of systems we are going to consider, namely, polling systems. Sometimes this is also referred to as round-robin scheduling or cyclic queues. Since these systems are classified based on location, it is natural to consider a queue for each location. The server goes from one location to another performing service in a cyclic manner. One of the unique aspects of polling systems is the presence of a switch-over time to go from one queue to another. Therefore, the system is not work conserving. It is also worthwhile to point out that we continue to assume that the arrival and service times are nonanticipative, specifically the realized service time is known only upon service completion. There are many types of polling systems, and depending on the application one can consider a suitable model. We present a few models to illustrate a flavor of the analysis techniques.

We begin by presenting the setting, notation, and some results (mostly based on Sarkar and Zangwill [94]) that are common to all polling systems considered in this section. We once again consider K classes of customers; however, each class has its own queue (note that in the previous section on priorities also, each class could have had its own queue, but since the switch-over time was zero, it was not mathematically different than having one queue). See Figure 5.2 where class-i customers arrive into queue i according to $PP(\lambda_i)$ ($i = 1, 2, \ldots, K$). Each class-i customer requires service times that are IID with mean $E[S_i] = 1/\mu_i$, second moment $E[S_i^2]$, and CDF $G_i(\cdot)$ (for $i = 1, 2, \ldots, K$). Further, we use $\rho_i = \lambda_i/\mu_i$. There is a single server, and each queue has infinite waiting rooms. The server polls queue 1 and serves it, then switches to queue 2 and serves it. In this manner, the server continues till queue K and then cycles back to queue 1. For this reason, the policy is called cyclic or polling or round-robin. The server spends a random time D_i to go from queue $i - 1$ to i. Throughout this section, the understanding is that if i equals 1, then $i - 1$ is K. We use $E[D_i]$ and $E[D_i^2]$ to represent the mean and second moment of D_i, respectively.

One thing we have not described here is what discipline the server adopts to switch from one queue to another. As one might expect, this could be done differently. We consider three disciplines that have been well-studied in the literature: (1) *exhaustive discipline* where the server completely serves a

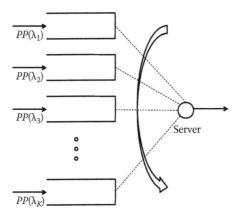

FIGURE 5.2
Schematic of a polling system with a single server.

queue and leaves it only when it becomes empty; (2) *gated discipline* where the server serves all the customers that arrived to that queue prior to the server arrival in that cycle; (3) *limited discipline* where a maximum fixed number can be served during each poll. Naturally, if the switch-over times are large (such as a setup time to manufacture a class of jobs), then one may favor the exhaustive discipline. Whereas if they are small, then it makes sense to consider a limited discipline (such as even a maximum of one customer per visit to the queue). The gated discipline falls in between. We assume that the server can see the contents of a queue only upon arrival. Hence, even if a queue is empty, the server would still spend the time to switch in and out of that queue. We also would like to point out that customers in a single queue (i.e., of a single class) are served according to FCFS.

In addition to this notation, we also require that not all mean switch-over times can be zero. That provides us with a result quite unique to polling systems which we describe next. Let $\rho = \rho_1 + \rho_2 + \cdots + \rho_K$ with $K \geq 2$. If the system is stable (we will describe the conditions later), then the long-run fraction of time the server is attending to customers is ρ. Likewise, $(1 - \rho)$ is the fraction of time spent switching in the long run. This is a relatively straightforward observation since the server is never at a queue idling when that queue is empty. However, to prove it rigorously one needs to consider times when the system regenerates and use results from regenerative processes to show that. Using that we can state the following result: if $E[C]$ is the expected time to complete a cycle (including service as well as switch-over times) in steady state, then

$$\frac{\sum_{i=1}^{K} E[D_i]}{E[C]} = 1 - \rho. \tag{5.3}$$

Although one is tempted to say that by taking the limit, all the switch-over times can become zero, but that one has to be careful because of the close tie with ρ. It turns out the zero switch-over time case is more complicated to handle than the one with at least one nonzero switch-over time. Next we analyze each of the types of queue emptying policies, that is, exhaustive, gated, and limited.

5.3.1 Exhaustive Polling

Consider Figure 5.2. If the server adopted an exhaustive polling scheme, then the server would poll a queue, serve all the customers back to back, and then proceed onto the next queue. As alluded to before, if the switch-over times are large, then it may be a good policy to adopt. Of course the downside is that the waiting time for a customer that narrowly missed a server would be too long. Such policies are indeed common in flexible manufacturing systems where there is a significant setup time to switch from making one product to another. With this motivation, we begin to analyze the exhaustive polling policy.

Let B_i^n be the random time the server spends at queue i in the nth cycle serving customers till the queue empties out in that cycle. A cycle time associated with queue i begins at the instant a server departs queue i (and proceeds to queue $i + 1$). Note that at this instant, queue i is empty which is the motivation to consider that time as the cycle time epoch. The cycle time ends the next time the server departs queue i. In other words, the cycle time associated with queue i in the nth cycle is represented as C_i^n which is the time between when the server departs queue i for the $(n-1)$th and nth time. Note that if we let the system be stationary at time 0 or we let $n \to \infty$, then $E[C_i^n]$ would just be $E[C]$, the average cycle time defined earlier. We can also write down C_i^n as

$$C_i^n = \sum_{j=1}^K D_j + \sum_{j=1}^i B_j^n + \sum_{j=i+1}^K B_j^{n-1}$$

since it is essentially the sum of the time the server spends in each queue plus the time switching. However, note that we have been careful to use $n - 1$ for queues greater than i and n for others so that the index $n - 1$ or n appropriately denotes the cycle number with respect to the server.

Using these definitions, we can immediately derive the following results. Given C_i^n, the expected number of arrivals of class-i customers during that cycle time is $\lambda_i C_i^n$. All these arrivals would be served before this cycle time is completed, and each one of them on average requires $1/\mu_i$ amount of service time. Therefore, the average amount of time spent by the server in queue i in the nth cycle, given C_i^n, is

$$E[B_i^n | C_i^n] = \frac{\lambda_i C_i^n}{\mu_i} = \rho_i C_i^n.$$

Taking expectations of this expression, we get

$$E[B_i^n] = \rho_i E[C_i^n].$$

By assuming the system is stationary at time 0 or by letting $n \to \infty$, we see that in steady state the server spends on average $\rho_i E[C]$ time serving customers in queue i in each cycle. In fact, using that we can derive Equation 5.3. It may be worthwhile for the reader to verify that. Further, that can also be used to deduce the necessary condition for stability which is

$$\rho < 1.$$

Now we are in a position to derive expressions for the performance measures of the system assuming it is stable.

Consider an arbitrary queue, say i (such that $1 \le i \le K$). As far as that queue is concerned, it is an $M/G/1$ queue with server vacations. The server goes on vacation if the system is empty after a service completion. If the system is empty upon the return of the server from vacation, the server goes on another vacation; otherwise the server begins service. In this system, the vacation time is essentially the time between when a cycle starts till the server returns to queue i, that is, $C_i^n - B_i^n$ in the nth server cycle. In steady state (letting $n \to \infty$) or assuming stationarity, we denote this as V_i, the vacation time corresponding to queue i. One of the exercise problems in Chapter 4 (*An M/G/1 queue with server vacations*) is to derive the steady-state number in the system L as

$$L_i = \rho_i + \frac{1}{2}\frac{\rho_i^2}{1 - \rho_i}(1 + \sigma_i^2 \mu_i^2) + \frac{m_i^{(2)}}{2m_i},$$

where m_i and $m_i^{(2)}$ are, respectively, the mean and second factorial moment of the number of arrivals during a vacation. Since the arrivals are according to a Poisson process with rate λ_i into queue i, we can write down $m_i = \lambda_i E[V_i]$ and $m_i^{(2)} = \lambda_i^2 E[V_i^2]$. Plugging that into L_i and writing in terms of W_{iq} as

$$W_{iq} = \frac{1}{2}\frac{\rho_i/\mu_i}{1 - \rho_i}(1 + \sigma_i^2 \mu_i^2) + \frac{E[V_i^2]}{2E[V_i]}.$$

We still need expressions for $E[V_i]$ and $E[V_i^2]$. It is relatively straightforward to obtain $E[V_i]$ using the fact that it is the limit as $n \to \infty$ $E[C_i^n - B_i^n]$ which can be computed as

$$E[C_i^n - B_i^n] = (1 - \rho_i)E[C_i^n].$$

As $n \to \infty$, we get

$$E[V_i] = (1 - \rho_i)E[C] = \left(\sum_{j=1}^{K} E[D_j] \right) \frac{1 - \rho_i}{1 - \rho}$$

where the last equality is from Equation 5.3. However, obtaining $E[V_i^2]$ is fairly involved. In the interest of space, we merely state the results from Takagi [101] without describing the details.

Let T_i^n be the station time for the server in queue i defined as

$$T_i^n = B_i^n + D_i.$$

In other words, this is the time between when the server leaves queue $i - 1$ and queue i. Define b_{ij} as the steady-state covariance of cycle times of queues i and j during consecutive visits, with the understanding that if $i = j$, it would be the variance. In other words, for all i and j

$$b_{ij} = \lim_{n \to \infty} \begin{cases} Cov(T_i^n, T_j^n) & \text{if } j > i, \\ Cov(T_i^{n-1}, T_j^n) & \text{if } j \le i. \end{cases}$$

Using the results in Takagi [101], we can obtain b_{ij} by solving the following sets of equations:

$$b_{ij} = \frac{\rho_i}{1 - \rho_i} \left(\sum_{k=i+1}^{K} b_{jk} + \sum_{k=1}^{j-1} b_{jk} + \sum_{k=j}^{i-1} b_{kj} \right), \quad \text{for } j < i$$

$$b_{ij} = \frac{\rho_i}{1 - \rho_i} \left(\sum_{k=i+1}^{j-1} b_{jk} + \sum_{k=j}^{K} b_{kj} + \sum_{k=1}^{i-1} b_{kj} \right), \quad \text{for } j > i$$

$$b_{ii} = \frac{Var(D_i)}{(1 - \rho_i)^2} + \frac{\rho_i}{1 - \rho_i} \left(\sum_{j=1}^{i-1} b_{ij} + \sum_{j=i+1}^{K} b_{ij} \right) + \frac{\lambda_i E[S_i^2]E[V_i]}{(1 - \rho_i)^3}.$$

These sets of equations can be solved using a standard matrix solver by writing down these equations in matrix form $[b_{ij}]$. Assuming that can be done, we can write down $Var(V_i)$ as

$$Var[V_i] = Var[D_i] + \frac{1 - \rho_i}{\rho_i} \left(\sum_{j=1}^{i-1} b_{ij} + \sum_{j=i+1}^{K} b_{ij} \right).$$

Using that we can obtain W_{iq}. Thereby we can also immediately write down $L_{iq} = \lambda_i W_{iq}$, $W_i = W_{iq} + 1/\mu_i$, and $L_i = \lambda_i W_i$. Of course we can also obtain the total number in the entire system in steady-state L as $L_1 + L_2 + \cdots + L_K$. Using that we could get metrics such as W, W_q, and L_q.

5.3.2 Gated Policy

In this section, we consider the gated policy. In a lot of ways, this is extremely similar to the exhaustive policy. However, there are some subtle differences in notation that must not be neglected. Once again, consider Figure 5.2. If the server adopted a gated polling scheme, then the server would poll a queue, serve only the customers that were in the system upon its arrival to the queue, and then proceed onto the next queue. As alluded to before, this policy balances the time wasted in switching against the waiting time for each class. Such policies are indeed common in multiaccess communication protocols such as token rings in local area networks (or other similar protocols in metropolitan and wide area networks). In a token ring multiaccess system, there are K nodes that are interested in transmitting information. The token ring polls each node (which has packets waiting in a queue). The token ring selects and transmits all the packets that were in the queue when it arrived, and then proceeds onto the next queue. With this motivation, we begin to analyze the gated polling policy.

Let B_i^n be the random time the server spends at queue i in the nth cycle serving customers that were in the queue when it arrived. A cycle time associated with queue i begins at the instant a server arrives at queue i (this is different from the cycle time definition we had for exhaustive service policy) and ends the next time this happens. Note that all customers that arrived in queue i during its cycle would be served one by one during the time the server spends in queue i. This is the reason we consider the time when a server arrives at queue i as the cycle time epoch. Thus, the cycle time associated with queue i in the nth cycle is represented as C_i^n, which is the time between when the server arrives to queue i for the $(n-1)$th and nth time. Note that if we let the system be stationary at time 0 or we let $n \to \infty$, then $E[C_i^n]$ would just be $E[C]$ the average cycle time defined earlier. We can also write down C_i^n as

$$C_i^n = \sum_{j=1}^{K} D_j + \sum_{j=1}^{i-1} B_j^n + \sum_{j=i}^{K} B_j^{n-1}$$

since it is essentially the sum of the time the server spends in each queue plus the time switching. However, note that we have been careful to use $n - 1$ for queues greater than or equal to i and n for others so that the index $n - 1$ or n appropriately denotes the cycle number with respect to the server.

Using these definitions, we can immediately derive the following results. Given C_i^n, the expected number of arrivals of class i customers during that cycle time is $\lambda_i C_i^n$. All these arrivals would be served during the server's sojourn in queue i, and each one of them on average requires $1/\mu_i$ amount of service time. Therefore, the average amount of time spent by the server in queue i in the nth cycle, given C_i^n, is

$$E[B_i^n | C_i^n] = \frac{\lambda_i C_i^n}{\mu_i} = \rho_i C_i^n.$$

Taking expectations of this expression, we get

$$E[B_i^n] = \rho_i E[C_i^n].$$

By assuming the system is stationary at time 0 or by letting $n \to \infty$, we see that in steady state the server spends on average $\rho_i E[C]$ time serving customers in queue i in each cycle. Further, the necessary condition for stability is

$$\rho < 1.$$

All the results derived so far are identical to those of exhaustive service policies. However, this is enabled only by a careful selection of how C_i^n is defined. It may hence be worthwhile to note the subtle differences in both policies. Now we are in a position to derive expressions for the performance measures of the system assuming it is stable.

Problem 47

Consider an arbitrary queue, say i (such that $1 \le i \le K$). Let $E[C_i]$ and $E[C_i^2]$, respectively, denote the steady-state mean and second moment of the cycle time C_i^n. Write down an expression for W_{iq} in terms of $E[C_i]$ and $E[C_i^2]$.

Solution

Let a customer arrive into queue i in steady state. From results of renewal theory, the remaining time for completion as well as the elapsed time since the start of the cycle in progress are both according to the equilibrium distribution of C_i. Therefore, the expected value of both the elapsed time since the start of the cycle in progress as well as the remaining time for the cycle in progress to end are equal to $E[C_i^2]/(2E[C_i])$. The customer in question would have to wait for the cycle in progress to end plus the service times of all the customers that arrived since the cycle in progress began. Therefore, the average waiting time in the queue for this customer is $E[C_i^2]/(2E[C_i]) + \lambda_i E[C_i^2]/(2E[C_i]\mu_i)$. The second term uses the fact that $\lambda_i E[C_i^2]/(2E[C_i])$ customers would have arrived on average since the cycle in progress began, and each of them requires on average $1/\mu_i$ service time. Thus, we have

$$W_{iq} = \frac{(1+\rho_i)E[C_i^2]}{2E[C_i]}.$$

There are certainly other ways to derive W_{iq}, one of which is given as an exercise problem. ∎

Of course, we still need expressions for $E[C_i]$ and $E[C_i^2]$. It is relatively straightforward to obtain $E[C_i]$ using the fact that it is indeed $E[C]$ we had computed earlier, the cycle time of the server in steady state. Therefore

$$E[C_i] = E[C] = \left(\sum_{j=1}^{K} E[D_j] \right) \frac{1}{1-\rho}$$

where the last equality is from Equation 5.3. However, obtaining $E[C_i^2]$ is fairly involved. In the interest of space, we merely state the results from Takagi [101] without describing the details.

Let T_i^n be the station time for the server in queue i defined (slightly different from the exhaustive polling case) as

$$T_i^n = B_i^n + D_{i+1}.$$

In other words, this is the time between when the server enters queue i and queue $i+1$. Define b_{ij} as the steady-state covariance of cycle times of queues i and j during consecutive visits, with the understanding that if $i = j$, it would be the variance. In other words, for all i and j

$$b_{ij} = \lim_{n \to \infty} \begin{cases} Cov(T_i^n, T_j^n) & \text{if } j > i, \\ Cov(T_i^{n-1}, T_j^n) & \text{if } j \leq i. \end{cases}$$

Using the results in Takagi [101], we can obtain b_{ij} by solving the following sets of equations:

$$b_{ij} = \rho_i \left(\sum_{k=i}^{K} b_{jk} + \sum_{k=1}^{j-1} b_{jk} + \sum_{k=j}^{i-1} b_{kj} \right), \quad \text{for } j < i$$

$$b_{ij} = \rho_i \left(\sum_{k=i}^{j-1} b_{jk} + \sum_{k=j}^{K} b_{kj} + \sum_{k=1}^{i-1} b_{kj} \right), \quad \text{for } j > i$$

$$b_{ii} = \text{Var}(D_{i+1}) + \rho_i \left(\sum_{j=1}^{i-1} b_{ij} + \sum_{j=i+1}^{K} b_{ij} \right) + \rho_i^2 \sum_{j=1}^{K} b_{ji} + \lambda_i E[S_i^2] E[C].$$

These sets of equations can be solved using a standard matrix solver by writing down these equations in matrix form $[b_{ij}]$. Assuming that can be done, we can write down $E(C_i^2)$ as

$$E[C_i^2] = \{E[C]\}^2 + \frac{1}{\rho_i} \left(\sum_{j=1}^{i-1} b_{ij} + \sum_{j=i+1}^{K} b_{ij} \right) + \sum_{j=1}^{K} b_{ji}.$$

Using that we can obtain W_{iq}. Thereby we can also immediately write down $L_{iq} = \lambda_i W_{iq}$, $W_i = W_{iq} + 1/\mu_i$, and $L_i = \lambda_i W_i$. Of course we can also obtain the total number in the entire system in steady-state L as $L_1 + L_2 + \cdots + L_K$. Using that we could get metrics such as W, W_q, and L_q.

5.3.3 Limited Service

In some sense, the exhaustive policy and gated policy considered in the previous sections were somewhat similar. However, the limited service polling we consider here is dramatically different in a lot of ways. Consider Figure 5.2 with an additional requirement that the server uses a limited-ℓ policy. The server would poll a queue, serve a maximum of ℓ customers, and then proceed onto the next queue. As explained earlier, this policy is appropriate when the switching times are small because of the large fraction of time spent in switching. Some systems with time division slotted scheduling can be modeled this way. Each node in a network is allowed to transmit at most ℓ packets when it receives its slot to transmit. As soon as the packets are transmitted, the scheduler moves to the next node without wasting any time idling. Unfortunately, the most general case of such a system is very difficult to analyze, and the waiting time expressions W_{iq} cannot be written just in terms of the arrival rate as well as the first two moments of the service times and switching times. It appears like we would need the whole distribution even for the small case of $K=2$. In that light, we will make several simplifying assumptions. First we let $\ell=1$, that is, at each poll if a queue is empty, the server immediately begins its journey to the next queue; otherwise the server serves one customer and then begins its journey to the next queue.

Let B_i^n be the random time the server spends at queue i in the nth cycle serving customers that were in the queue when it arrived. Of course B_i^n will either be equal to zero or equivalent of one class-i customer's service time. We drop the superscript n by either considering a stationary system or letting $n \to \infty$. Thus, B_i is the random variable corresponding to the time spent serving customers in queue i during a server visit in stationary or steady state. Recall that D_i is the random variable associated with the time to switch from queue $i-1$ to i. Let $E[D] = E[D_1]+E[D_2]+\cdots+E[D_K]$ so that $E[D]$ is the average time spent switching in each cycle. Since ρ_i is the long-run fraction of time the server spends in queue i, we have

$$\rho_i = \frac{E[B_i]}{E[D] + \sum_{j=1}^{K} E[B_j]}.$$

By summing over all i and rewriting, we get

$$\sum_{j=1}^{K} E[B_j] = E[D]\frac{\rho}{1-\rho}.$$

For all i, the average number of class-i arrivals in a cycle is $\lambda_i \left(E[D] + \sum_{j=1}^{K} E[B_j]\right)$, and that number should be less than 1 for stability because at most one class-i customer can be served in a cycle. Hence, the stability condition is $\lambda_i \left(E[D] + \sum_{j=1}^{K} E[B_j]\right) < 1$, and we can rewrite that as

$$\lambda_i E[D] < 1 - \rho$$

for all $i \in [1,2,\ldots,K]$. Further, the mean cycle time $E[C]$, as stated in Equation 5.3, can be verified from previous equation as

$$E[C] = E[D] + \sum_{j=1}^{K} E[B_j] = \frac{E[D]}{1-\rho}.$$

The next step is to obtain performance metrics such as W_{iq}. It turns out that unlike the exhaustive and gated cases where we could write down W_{iq} in terms of just the first two moments of the service times and switch-over times, here in the limited case we do not have the luxury. Except for $K = 2$, the exact analysis is quite intractable. However, we can still develop some relations between the various W_{iq} values known as *pseudo-conservation law*. Note that this is equivalent to the work conservation result for multiclass queues with at most one partially complete service. Note that although here too we have at most one partially complete service, because of the switch-overs the system is not work conserving. However, by suitably adjusting for the time spent switching, we can show that the amount of workload in the queue for the limited polling policy is the same as that of an equivalent $M/G/1$ queue with FCFS. The resulting pseudo-conservation law yields

$$\sum_{i=1}^{K} \rho_i \left(1 - \frac{\lambda_i R}{1 - \rho}\right) W_{iq} = \frac{\rho}{2(1 - \rho)} \sum_{i=1}^{K} \lambda_i E[S_i^2] + \frac{\rho}{2E[D]} \sum_{i=1}^{K} Var[D_i]$$

$$+ \frac{E[D]}{2(1 - \rho)} \left(\rho + \sum_{i=1}^{K} \rho_i^2\right).$$

Using this expression, the only case we can obtain W_{iq} is when the system is symmetric, that is, the parameters associated with each queue and switch-over time is identical to that of the others. Instead of using the subscript i, we use *sym* to indicate the symmetric case. Thus, the average time spent waiting in the queue before service in the symmetric case is

$$W_{sym,q} = \frac{K\lambda_{sym}E[S_{sym}^2] + E[D_{sym}](K + \rho) + Var[D_{sym}]K\lambda_{sym}}{2(1 - \rho - \lambda_{sym}KE[D_{sym}])} + \frac{Var[D_{sym}]}{2E[D_{sym}]}.$$

Thereby we can also immediately write down $L_{sym,q} = \lambda_{sym} W_{sym,q}$, $W_{sym} = W_{sym,q} + 1/\mu_{sym}$, and $L_{sym} = \lambda_{sym} W_{sym}$. Of course we can also obtain the total number in the entire system in steady-state L as KL_{sym}. Using that we could get metrics such as W, W_q, and L_q.

5.4 Evaluating Policies for Classification Based on Knowledge of Service Times

Consider an $M/G/1$ queue into which customers arrive according to a Poisson process with mean rate λ per unit time. Let S be a random variable denoting the service times with CDF $G(t) = P\{S \leq t\}$, mean $E[S]$, and second moment $E[S^2]$. However, as soon as a customer arrives into the system, it reveals its service time. In all the situations we have considered thus far, only when the service completes we know the realized service times (we called that nonanticipative). However, here the service times are declared upon arrival (which we call anticipative). In applications such as web servers, this is reasonable since we would know the file size of an arriving request, and hence its service time. Also in many flexible manufacturing systems, the processing times can be calculated as soon as the specifications are known from the request. Therefore, based on the knowledge of service times we consider each customer to belong to a different class indexed by the service time. For analytical tractability, we assume that the service time is a continuous random variable without any point masses. Thus, it results in a multiclass system with an uncountably infinite number of classes, where each class corresponds to service time.

Although the overall system is indeed a single class system, we treat it as a multiclass system by differentiating the classes based on their service time requirement. An arrival would be classified as class x if its service time requirement is x amount of time. Analogous to the mean waiting time before service for a discrete class-i customer defined as W_{iq} in the previous sections, here we define W_{xq}. The quantity W_{xq} is the time a customer of class x would wait in the queue on average, not including the service time (which is x). Likewise, W_x would indicate the corresponding sojourn time for this customer with x as the amount of service. Of course the quantities W_{xq} and W_x would depend on the scheduling policy. The overall sojourn time (W) as well as the overall time waiting in the queue (W_q) for the various policies can be computed as

$$W = \int_0^\infty W_x dG(x),$$

$$W_q = \int_0^\infty W_{xq} dG(x).$$

Next, we consider three scheduling policies and compute W_x or W_{xq} with the understanding that using these expressions we can obtain W and W_q. It is crucial to note that we are only considering policies that use the service time information (such as shortest processing time first without preemption, preemptive shortest job first, and shortest remaining processing time first). Also, all policies that do not use the service time information and have at most one job with partially completed service have the same W_{xq} for all x which is given by the Pollaczek–Khintchine formula (Equation 4.6). These include policies such as FCFS, LCFS (without preemption), random order of service, etc. Note that when we derived the Pollaczek–Khintchine formula in Chapter 4, the discrete time Markov chain (DTMC) corresponding to the number in the system as seen by a departing customer could not change whether we used FCFS, LCFS, or random order of service. Therefore, for policies such as FCFS, LCFS, and random order of service that do not use the service time information and have at most one job with partially completed service, we have (from Equation 4.6)

$$W_q = W_{xq} = W_x - x = \frac{\lambda}{2} \frac{E[S^2]}{1 - \lambda E[S]},$$

$$W = E[S] + W_q.$$

However, policies such as processor sharing and preemptive LCFS that do not use service time information but can have more than one partially completed service do not have these expressions for their performance measures (unless the service times are exponentially distributed). It turns out that preemptive LCFS and processor sharing do have the same W_x given by (but other policies that belong to the same category do not)

$$W_x = \frac{x}{1 - \lambda E[S]}.$$

With that we proceed to the three scheduling policies: shortest processing time first without preemption, preemptive shortest job first, and shortest remaining processing time first.

5.4.1 Shortest Processing Time First

Consider the setting described earlier where customers arrive according to a Poisson process, and upon arrival customers reveal their service time requirement by sampling from a distribution with CDF $G(\cdot)$, mean $E[S]$, and second moment $E[S^2]$. Assume that the system is stable, that is, $\lambda E[S] < 1$. In the shortest processing time first (SPTF) policy, we consider a scheduling discipline that gives a nonpreemptive priority to jobs with shorter processing time. In other words, the server always selects the customer with the shortest processing time. However, during the service of this customer, if there are others that arrive with shorter processing times, then (because we do not allow preemption) we consider them only after this job is completed. To implement this, the server stores jobs in the queue by sorting according to the service time. Therefore, this policy is the continuous analog of the nonpreemptive priority considered in Section 5.2.2. In fact we just use those results to derive W_{xq} here which we do next.

Consider the nonpreemptive priority discipline analyzed in Section 5.2.2. The number of classes K in that setting is uncountably infinite, and if the service time is in the interval $(x, x + dx)$, that customer belongs to class x. Note that if $x + dx < y$, then class x is given higher priority than y which is consistent with SPTF requirements. We would like to derive the expected time a customer with service time x waits to being serviced, W_{xq}. Recall from the expression in Section 5.2.2, we need expressions for λ_x, $E[S_x^2]$, α_x, and

α_{x-dx}. For an infinitesimal dx, note that $\lambda_x = \lambda \frac{dG(x)}{dx} dx$. This is because λ_x corresponds to the arrival rate of customers with service time in the interval $(x, x + dx)$ which equals λ times the probability that an arrival has service time in the interval $(x, x+dx)$, and that is exactly the PDF of service times at x multiplied by dx. Also, as $dx \to 0$, we have $E[S_x^2] \to x^2$ (since the service time in the interval $(x, x + dx)$ converges to a deterministic quantity x as $dx \to 0$). Finally, as $dx \to 0$, we need to compute α_x. By definition, if x were countable, then $\alpha_x = \sum_{z=0}^{x} \lambda_z E[S_z]$ which by letting $dx \to 0$ and using the result for λ_z, we get in the uncountable case $\alpha_x = \int_{z=0}^{x} \lambda z dG(z)$ realizing that $E[S_z] \to z$. Therefore, we have as $dx \to 0$, $\alpha_x \to \alpha_{x-dx} \to \int_0^x \lambda t dG(t)$.

Using the results for W_{iq} in Section 5.2.2 for class i corresponding to service time in $(x, x + dx)$, we get by letting $dx \to 0$

$$W_{xq} = \frac{\frac{1}{2} \int_{y=0}^{\infty} \lambda y^2 dG(y)}{\left(1 - \int_0^x \lambda t dG(t)\right)^2} = \frac{\frac{1}{2} \lambda E[S^2]}{(1 - \rho(x))^2},$$

where

$$\rho(x) = \int_0^x \lambda t dG(t).$$

Of course we can immediately compute W_x as $x + W_{xq}$, and thereby

$$W = \int_0^\infty W_x dG(x),$$

$$W_q = \int_0^\infty W_{xq} dG(x).$$

5.4.2 Preemptive Shortest Job First

The service-scheduling policy preemptive shortest job first (PSJF) considered here is a lot similar to what we just saw. The only difference is that preemption is allowed here. For the sake of completeness, we go ahead and describe it formally. Customers arrive according to a Poisson process, and upon arrival customers reveal their service time requirement by sampling from a distribution with CDF $G(\cdot)$, mean $E[S]$, and second moment $E[S^2]$. Assume that the system is stable, that is, $\lambda E[S] < 1$. In the PSJF policy, we consider a scheduling discipline that gives preemptive priority to jobs with shorter processing time. In other words, the server always selects the customer with the shortest processing time. However, during the service of this customer, if a customer with shorter processing time arrives, then that customer preempts the one in service. It is crucial to realize that the server only uses the initially declared service times for determining priorities but resumes from where the service was completed. To implement this, the server stores jobs in the queue by sorting according to the total service time and always serving customers on the top of the list. Therefore, this policy is the continuous analog of the preemptive resume priority considered in Section 5.2.3. In fact, we merely use those results to derive W_x here which we do next as a problem.

Problem 48

Derive an expression for the sojourn time for a request with service time x under PSJF.

Solution

Consider the preemptive resume priority discipline analyzed in Section 5.2.3. First, let the number of classes K in that setting go to infinite. To map from class i in that setting to class x here, if the service time is in the interval $(x, x + dx)$, that customer belongs to class x. Note that if $x + dx < y$, then class x is given higher preemptive priority than y which is consistent with PSJF requirements. We need to derive the expected time a customer with service time x spends in the system, W_x. Recall from the corresponding expression in Section 5.2.3, we need expressions for λ_x, μ_x, $E[S_x^2]$, α_x, and α_{x-dx}. For an infinitesimal dx, note that $\lambda_x = \lambda \frac{dG(x)}{dx} dx$. This is because λ_x corresponds to the arrival rate of customers with service time in the interval $(x, x+dx)$ which equals λ times the probability that an arrival has service time in the interval $(x, x+dx)$, and that is exactly the PDF of service times at x multiplied by dx. As $dx \to 0$, $\mu_x \to 1/x$ since the service time would just be x. Also, as $dx \to 0$, we have $E[S_x^2] \to x^2$ (since the service time in the interval $(x, x+dx)$ converges to a deterministic quantity x as $dx \to 0$). Finally, as $dx \to 0$, we need to compute α_x. By definition, $\alpha_x = \sum_{z=0}^{x} \lambda_z E[S_z]$ which by letting $dx \to 0$ and using the result for λ_z, we get $\alpha_x = \int_{z=0}^{x} \lambda z dG(z)$ realizing that $E[S_z] \to z$. Therefore, we have as $dx \to 0$, $\alpha_x \to \alpha_{x-dx} \to \int_0^x \lambda t dG(t)$.

Using the results for W_x in Section 5.2.3 for class i corresponding to service time being in $(x, x + dx)$, we get by letting $dx \to 0$

$$W_x = \frac{x}{1 - \int_{z=0}^{x} \lambda z dG(z)} + \frac{\int_{y=0}^{x} \lambda y^2 dG(y)}{2 \left(1 - \int_{z=0}^{x} \lambda z dG(z)\right)^2} = \frac{x}{1 - \rho(x)} + \frac{\frac{1}{2} \lambda \Psi(x)}{(1 - \rho(x))^2},$$

where

$$\Psi(x) = \int_0^x y^2 dG(y)$$

and

$$\rho(x) = \int_0^x \lambda t \, dG(t). \qquad \blacksquare$$

Of course we can immediately compute W_{xq} as $W_x - x$, and thereby

$$W = \int_0^\infty W_x \, dG(x),$$

$$W_q = \int_0^\infty W_{xq} \, dG(x).$$

5.4.3 Shortest Remaining Processing Time

The shortest remaining processing time (SRPT) scheme is very similar to the PSJF we just considered. The key difference is that instead of sorting customers according to the initial service times in PSJF, here they are sorted according to the remaining processing times. Hence the name SRPT. To clarify SRPT, customers arrive according to a Poisson process with rate λ. Upon arrival, customers reveal their service time requirement by sampling from a distribution with CDF $G(\cdot)$, mean $E[S]$, and second moment $E[S^2]$. Assume that the system is stable, that is, $\lambda E[S] < 1$. The SRPT policy is a scheduling discipline that gives a preemptive priority to jobs with shorter remaining processing time. In other words, the server always selects the customer with the shortest remaining processing time. However, during the service of this customer, if there are others that arrive with shorter processing times than what is remaining for this customer, then the customer in service gets preempted. To implement this, the server stores jobs in the queue by sorting according to the remaining service time. Unlike the previous two policies, the class of a customer in service keeps changing here in SRPT and, hence, cannot be analyzed like those. We adopt a different but related technique which has been directly adapted from Schrage and Miller [95].

Our objective is to compute W_x, the expected time from when an arrival with service time requirement of x occurs in steady state till that customer's service is complete, that is, mean sojourn time in steady state of a customer with service time x. To compute W_x, we divide the sojourn time into two intervals, one from the time of arrival till the customer starts getting served for the first time, and the second interval is from that time till the end of the sojourn. Hence, we write that as a sum

$$W_x = V_x + R_x,$$

where

V_x is the expected time for an arriving customer with service time x to begin processing by the server (note that until that time, the remaining processing time is equal to service time x)

R_x is the expected time from when this customer enters the server for the first time until service is completed (during this time, the server could get preempted by arriving customers with service time smaller than the remaining processing time for the customer in question)

We first obtain an expression for V_x and then for R_x. To obtain V_x, we consider a class x customer, that is, one with a service time requirement of x. The average arrival rate of customers with service times less than or equal to x is $\lambda G(x)$. In the long run, a fraction of time-equal to $\int_0^x \lambda u dG(u)$ there would be customers at the server with original service time less than or equal to x. This is essentially our definition of $\rho(x)$. Therefore, here too

$$\rho(x) = \int\limits_0^x \lambda u dG(u).$$

Further, the probability that an arriving customer will see the server with a customer whose remaining service time is less than x is $\beta(x)$, given by

$$\beta(x) = \rho(x) + \lambda x(1 - G(x))$$

where $\lambda x(1 - G(x))$ is the long-run fraction of time the server serves customers with remaining processing time less than x, although initially they had more than x service time to begin with. A *busy period of type x* is defined as the continuous stretch of time during which the server only processes customers with remaining processing time less than x. It is crucial to point out that if a class x customer arrives during a busy period of type x (note that this happens with probability $\beta(x)$), that customer waits till the busy period ends to begin service. Of course if this class x customer arrives at a time other than during a busy period of type x, the customer immediately gets served by the server. Therefore, by conditioning on whether or not an arriving class x customer encounters a busy period of type x and then unconditioning, we get

$$V_x = \beta(x) E[B_e(x)]$$

where $B_e(x)$ is the remaining time left in the busy period of type x.

Since $B_e(x)$ is the equilibrium random variable corresponding to $B(x)$, the length of the busy period of type x, from renewal theory we can write down $E[B_e(x)]$ as

$$E[B_e(x)] = \frac{E[B(x)^2]}{2E[B(x)]}.$$

Thus, all we need to compute are $E[B(x)^2]$ and $E[B(x)]$. For this we need another notation $\tau(x)$, the remaining service time of the job that initiated a type x busy period. Note that when a busy period of type x is initiated, there would be exactly one customer in the system with remaining processing time not greater than x, and this customer initiates the busy period. This can happen in two ways: (1) if a customer with service time t such that $t < x$ arrives when a busy period of type x is not in progress, then this will start a busy period of type x with $\tau(x) = t$ so that the probability that a given customer initiates a busy period of type x and the initiating customer has remaining processing time between t and $t + dt$ is $(1 - \beta(x))dG(t)$; (2) if a customer that has an original processing time greater than x initiates a type x busy period as soon as this customer's remaining time reaches x so that the probability that a given customer initiates a busy period of type x and the initiating customer has remaining processing time x is $(1 - G(x))$. Therefore, the probability that a customer initiates a busy period of type x is

$$\int_0^x (1 - \beta(x))dG(t) + (1 - G(x)) = 1 - G(x)\beta(x).$$

We also have

$$E[\tau(x)] = \frac{(1 - \beta(x)) \int_0^x t\, dG(t) + x(1 - G(x))}{1 - G(x)\beta(x)},$$

$$E[\tau(x)^2] = \frac{(1 - \beta(x)) \int_0^x t^2\, dG(t) + x^2(1 - G(x))}{1 - G(x)\beta(x)}.$$

The remaining customers that are served in a busy period of type x arrive after the busy period is initialized and have service times less than x. Let S_x be the service times of one such customer, then clearly we have

$$E[S_x] = \frac{\int_0^x t\, dG(t)}{G(x)},$$

$$E[S_x^2] = \frac{\int_0^x t^2\, dG(t)}{G(x)}.$$

It is crucial to note that the busy period distribution would be identical to the case of an $M/G/1$ with any work-conserving service-scheduling discipline for which customers that initiate the busy period has processing times according to $\tau(x)$, the customers that arrive during the busy period have processing times according to S_x, and arrivals according to $PP(\lambda G(x))$. Next we solve a problem where we select an appropriate scheduling discipline to compute the first two moments of the busy period.

Problem 49

Using the notation and description from previous text, show that

$$E[B(x)] = \frac{E[\tau(x)]}{1 - \rho(x)} = \frac{\beta(x)}{\lambda(1 - G(x)\beta(x))},$$

$$E\left[B(x)^2\right] = \frac{E\left[\tau(x)^2\right]}{\{1 - \rho(x)\}^2} + \lambda G(x)E[\tau(x)]\frac{E[S_x^2]}{(1 - \rho(x))^3}.$$

Solution

The busy period $B(x)$ can be computed by selecting an appropriate scheduling discipline. First serve the customer that initializes the busy period and this takes $\tau(x)$ time. During this time $\tau(x)$, say $N(\tau(x))$ new customers arrived with service times smaller than x according to a Poisson process with parameter $\lambda G(x)$. After $\tau(x)$ time, we serve the first of the $N(\tau(x))$ customers (if there is one) and all the customers that arrive during this service time that have service times smaller than x. Note that this time is identical to that of the busy period of an $M/G/1$ queue with $PP(\lambda G(x))$ arrivals and service times according to S_x. Once this "mini" busy period is complete we serve the second (if any) of the $N(\tau(x))$ for another mini busy period and then continue until all the $N(\tau(x))$ customers' mini busy periods are complete. Thus, we can write down

$$B(x) = \tau(x) + \sum_{i=1}^{N(\tau(x))} b_i(x), \tag{5.4}$$

where $b_i(x)$ is the ith mini busy period of an $M/G/1$ queue with $PP(\lambda G(x))$ arrivals and service times according to S_x. Note that $b_i(x)$ over all i are IID random variables. Based on one of the exercise problems of Chapter 4 on computing the moments of the busy period of an $M/G/1$ queue, we have

$$E[b_i(x)] = \frac{E[S_x]}{1 - \lambda G(x)E[S_x]},$$

$$E[b_i(x)^2] = \frac{E[S_x^2]}{(1 - \lambda G(x)E[S_x])^3}.$$

Also, $N(\tau(x))$ is a Poisson random variable such that $E[N(\tau(x))] = \lambda G(x)\tau(x)$ and $E[N(\tau(x))^2] = \lambda G(x)\tau(x) + \{E[N(\tau(x))]\}^2$. Using Equation 5.4 we can write down

$$E[B(x)|\tau(x)] = \tau(x) + E[N(\tau(x))]E[b_i(x)] = \frac{\tau(x)}{1 - \rho(x)},$$

$$E[B(x)^2|\tau(x)] = \tau(x)^2 + E[N(\tau(x))]E[b_i(x)^2] + 2\tau(x)E[N(\tau(x))]E[b_i(x)]$$

$$+ E[N(\tau(x))\{N(\tau(x)) - 1\}]\{E[b_i(x)]\}^2$$

$$= \tau(x)^2 + \lambda G(x)\tau(x)\frac{E[S_x^2]}{(1 - \lambda G(x)E[S_x])^3}$$

$$+ 2\tau(x)^2 \lambda G(x)\frac{E[S_x]}{1 - \lambda G(x)E[S_x]}$$

$$+ \{\lambda G(x)\tau(x)\}^2 \left\{\frac{E[S_x]}{1 - \lambda G(x)E[S_x]}\right\}^2$$

$$= \frac{\tau(x)^2}{\{1 - \rho(x)\}^2} + \lambda G(x)\tau(x)\frac{E[S_x^2]}{(1 - \rho(x))^3}.$$

Taking expectations (thereby unconditioning $\tau(x)$), we get

$$E[B(x)] = \frac{E[\tau(x)]}{1 - \rho(x)},$$

$$E[B(x)^2] = \frac{E[\tau(x)^2]}{\{1 - \rho(x)\}^2} + \lambda G(x)E[\tau(x)]\frac{E[S_x^2]}{(1 - \rho(x))^3}.$$

Also, using the expression for $E[\tau(x)]$ and the relationship

$$\beta(x) = \rho(x) + \lambda x(1 - G(x)),$$

we can rewrite $E[B(x)]$ as $\beta(x)/(\lambda(1 - G(x)\beta(x)))$. ∎

Plugging the expressions for $E[B(x)]$ and $E[B(x)^2]$ into $V_x = \beta(x)E[B(x)^2]/(2E[B(x)])$ and by appropriately substituting the relevant terms such as $E[\tau(x)]$ and $E[\tau(x)^2]$, we get

$$V_x = \frac{\lambda\left[\int_0^x t^2 dG(t) + x^2(1 - G(x))\right]}{2(1 - \rho(x))^2}.$$

Having computed V_x, next we compute R_x so that we can add them to get W_x. Recall that R_x is the average time for a customer that just started to get served (hence, remaining service time is x) to complete the service including possibly being interrupted by customers with service times smaller than the remaining service time for the customer in question. For this we discretize the service time x into infinitesimal intervals of length dt. Consider an arbitrary interval $t + dt$ to t. The expected time for the remaining service time to go from $t + dt$ to t is equal to dt plus the expected time for interruptions in service. The probability that a service will be interrupted from $t + dt$ to t is the probability that an arrival would occur in that time dt (which is equal to λdt), and that arrival would be one with service time smaller than t (which happens with probability $G(t)$). Thus, the probability of being interrupted is $\lambda dt G(t)$. The expected duration of the interruption (given there was one) can be computed as the busy period of an $M/G/1$ queue with only customers that have service times smaller than t from our original system. However, since our discipline is work conserving, the busy period would be identical to that of an equivalent FCFS queue which we know is equal to $E[S(t)]/(1 - \lambda(t)E[S(t)])$, where $E[S(t)]$ is the mean service time and $\lambda(t)$ the arrival rate, with the parameter t used to distinguish from other service times and arrival rates. We can show that $E[S(t)] = \int_0^t u dG(u)/G(t)$ since the CDF of the service time given it is smaller than t is $G(\cdot)/G(t)$. Likewise, $\lambda(t) = \lambda G(t)$, which is the rate at which customers with service times smaller than t arrive.

Hence, we can write down the expected time for the remaining service time to go from $t + dt$ to t as

$$dt + \lambda dt G(t) \frac{E[S(t)]}{1 - \lambda(t)E[S(t)]} = dt + \lambda dt G(t) \frac{\int_0^t u dG(u)/G(t)}{1 - \lambda G(t) \int_0^t u dG(u)/G(t)}$$

$$= dt + \frac{\rho(t) dt}{(1 - \rho(t))},$$

where $\rho(t) = \lambda \int_0^t u dG(u)$ as defined earlier. Before proceeding ahead to use this to obtain R_x, it is worthwhile to point out that in the previous expression we have ignored higher-order terms that would vanish as dt approaches 0. The key factor in that is that we have only considered a maximum of one interruption in the interval of length dt, and this is reasonable based on the definition of Poisson processes which states that the probability of two or more events in a Poisson process in time dt is of the order of $(dt)^2$ or higher. Thus, we can write down R_x by integrating the previous expression from 0 to x as

$$R_x = \int_0^x dt + \frac{\rho(t) dt}{(1 - \rho(t))} = \int_0^x \frac{dt}{(1 - \rho(t))}.$$

Therefore, we can immediately write down

$$W_x = V_x + R_x$$

$$= \frac{\lambda \left[\int_0^x t^2 dG(t) + x^2(1 - G(x)) \right]}{2(1 - \rho(x))^2} + \int_0^x \frac{dt}{(1 - \rho(t))}.$$

Of course we can then compute W_x as $x + W_{xq}$, and thereby

$$W = \int_0^\infty W_x dG(x).$$

5.5 Optimal Service-Scheduling Policies

So far, we have only considered obtaining performance measures for each class of a multiclass system for a *given* service-scheduling discipline. In this section, we turn the tables somewhat and attempt to go in the other direction, namely, obtaining the service-scheduling discipline (or policy) that would optimize a *given* metric, which could be a combination of performance measures. The objective of this section is purely for the sake of completeness and is not to illustrate methodologies to obtain the optimal scheduling policies. However, in the spirit of describing analysis, an outline of the methodology would be provided in each case.

5.5.1 Setting and Classification

In this section on determining the optimal service-scheduling policies for queueing systems, we first describe some settings and ground rules. We consider an infinite-sized $M/G/1$ queue with K classes. Class-i customers arrive into the queue according to $PP(\lambda_i)$ ($i = 1, 2, \ldots, K$). Each class-i customer requires service times that are IID with mean $E[S_i] = 1/\mu_i$, second moment $E[S_i^2]$, and CDF $G_i(\cdot)$ (for $i = 1, 2, \ldots, K$). We continue to use $\rho_i = \lambda_i/\mu_i$ and $\rho = \rho_1 + \cdots + \rho_K < 1$ for stability. We assume that the time required to switch serving one class to another is zero; hence, it would not matter whether or not we stored each class (or each customer for that matter) in a separate infinite capacity queue. We also only consider scheduling policies that are work conserving with the understanding that for most of the objectives we consider, a nonwork-conserving policy would perhaps fare only poorer. Also as we described earlier, the amount of workload in the system at any time would be unaffected by the scheduling discipline used. This is a feature that we would take advantage of in some instances. A natural question to ask is: What else is needed besides work conservation?

The scheduling policies that we will consider can broadly be classified into four categories depending on (1) whether there could be more that one customer in the system with partially completed service, and also (2) whether the service times are revealed upon arrival. Sometimes in the literature when there can be more than one customer with partially completed service under a certain policy, then that policy is also called preemptive for good reasons. But it is not just the strictly preemptive policies that are included; policies such as processor sharing (which are only partially preemptive) should also be considered in that group. Further, when the service times are revealed upon arrival, it is sometimes called anticipative (although anticipative could include a much broader class of policies, not just revealing service times upon arrival). Therefore, as examples of the four classes of policies we have:

1. *Maximum of one customer with partially completed service and service times not revealed upon arrival*: FCFS, nonpreemptive LCFS, random order of service, nonpreemptive priority, and round-robin. Among the classes of policies, this is minimal in terms of bookkeeping and information gathering.

2. *More than one customer with partially completed service and service times not revealed upon arrival*: Processor sharing, preemptive resume priority, preemptive LCFS, timed round-robin, and least-attained service (will explain this subsequently). Although we do not need to know the service times upon arrival, there is a lot of overhead in terms of bookkeeping for customers that have partially completed service.

3. *Maximum of one customer with partially completed service and service times revealed upon arrival*: SPTF. Although some sorting to determine the next customer to serve is needed and also some way to gage service time requirements upon arrival, these policies are still reasonably less intensive.

4. *More than one customer with partially completed service and service times revealed upon arrival*: PSJF and SRPT. Among the classes of policies, this is perhaps the most intensive in terms of bookkeeping and information gathering. However, one can gain a lot of insights from considering these policies as they significantly outperform other policies.

Having discussed the overheads in terms of bookkeeping and information gathering, next we briefly touch upon the notion of fairness across the various policies. What is considered *fair* depends on the application, the individual, and a score of other factors that are difficult to capture using physics. In human lines, if a server says "may I help the next person in line," clearly the one that arrived earliest among the ones waiting goes to the server. Thus, FCFS has been accepted as a fair policy especially when the workload each customer brings is unknown. Oftentimes in grocery stores where the workload each customer brings can be somewhat assessed, it is not uncommon that a person with a full cart lets someone behind them with one or two items to be served first (although this is never initiated by the cashier, presumably that is why there is a separate line for customers with fewer items in a grocery store). Thus, when the service times are known upon arrival, it appears like a fair thing to do is the sojourn times be proportional to the service time. For example, when there is a single class (i.e., $K = 1$), the mean sojourn time for a customer with service time x under processor sharing scheme is $x/(1 - \rho)$. Thus, the mean sojourn time is proportional to the service time, and it would not be terribly unreasonable to consider processor sharing as a "fair" policy. That is why many computer system CPUs adopt roughly a processor sharing regime where the CPU spends a small time (called time quantum) for each job and switches context to the next job. Interestingly, the preemptive LCFS policy also has mean sojourn time for a customer with service time x as $x/(1 - \rho)$. However, it is unclear if preemptive LCFS would be considered "fair" by the customers although it certainly is for the service provider. Therefore, while determining an optimal policy for a queueing system (which we will see next), it becomes crucial to consider the perspectives of both the customers and the service providers.

The optimal service-scheduling policy in a multiclass queueing system depends greatly on the choice of objective function. The issue of fairness becomes extraordinarily important if the customers can observe the queues, in which case FCFS or anything considered fair by the users must be adopted. However, for the rest of this chapter we assume that the customers cannot observe the queue; however, the service provider has real-time access to the queue (usually number in the system, sometimes service time requirements and amount of service complete). The service provider thus uses a scheduling policy that would optimize a performance measure that would strike a balance between the needs of the customers and the service provider. We consider one such objective function which is to minimize the mean number of customers in the system. In particular, let L_i^π represent the average number of class-i customers in the system in steady state under policy π (for $i = 1, \ldots, K$). Then, the objective function is to determine the optimal service scheduling among all work-conserving policies Π that minimizes the total number in the system. Hence, our objective function is

$$\min_{\pi \in \Pi} \sum_{i=1}^{K} L_i^\pi.$$

Using Little's law, the objective function is equivalent to minimizing the mean sojourn time (W^π) over all policies π, which is essentially the original objective function divided by a constant, that is, $\left\{ \sum_{i=1}^{K} L_i^\pi \right\} / (\lambda_1 + \cdots + \lambda_K)$. Toward the end of this section, we would mention other objectives that one could consider. Next we are ready to describe the optimal policies. We do that in two stages: first for the single class, that is, $K = 1$ and subsequently for a general K class system.

5.5.2 Optimal Scheduling Policies in Single Class Queues

Here we consider the single class case, that is, $K = 1$, to develop insights and intuition into the optimal policy. Since we only have one class, we drop the subscript and seek to find the policy π that minimizes L^π or W^π which are effectively equivalent. We have so far allowed π to be any work-conserving discipline. However, depending on restrictions imposed on the system (such as whether service times are known upon arrival, and whether more than one job with partially completed service is allowed), an appropriate optimal policy would need to be selected. We describe the optimal policies under various conditions next.

1. *Service times can be known upon arrival*: If it is possible to know the service time of each customer upon arrival, then the optimal policy that minimizes W is SRPT. For a discussion on performance analysis of SRPT, refer to Section 5.4.3. Here we give an idea of why SRPT is indeed optimal. Let $X(t)$ be the number of customers in the system at time t. We can use a sample path argument to show that SRPT yields the smallest $X(t)$ among all work-conserving policies (Π). Let $W(t)$ be the workload at time t, and it can be written as the sum of the remaining processing times of the $X(t)$ customers. Consider SRPT and any other work-conserving policy π so that the number of customers in the respective queues are $X(t)^{\text{SRPT}}$ and $X(t)^\pi$. We would like to show that $X(t)^{\text{SRPT}} \leq X(t)^\pi$. Pick any j such that $j \leq \min\{X(t)^{\text{SRPT}}, X(t)^\pi\}$. By the definition of SRPT, the sum of the j largest remaining processing times for policy π would be smaller than that of SRPT (since SRPT would have favored the remaining jobs). However, as described before, the sum of the remaining processing times of all jobs for either policy should be equal. This is possible only when there are more jobs in π than SRPT, that is, if $X(t)^{\text{SRPT}} \leq X(t)^\pi$. Therefore, clearly L would be smallest for SRPT (and thereby W). Note that this argument holds for any arrival and service process; hence, SRPT is optimal for a much wider class of systems, not just $M/G/1$ queues. If preemption is not allowed, then it is possible to show using a similar argument that SPTF is optimal.

2. *Service times unknown upon arrival; many jobs with partially complete service allowed*: Consider the case where it is not possible to know the service time of each customer upon arrival (i.e., nonanticipative); however, we allow more than one job to have partially complete

service. Then the optimal policy that minimizes W is what is known as Gittins index. To explain the Gittins index policy, we need some notation. Let a be the amount of attained service (i.e., amount of completed service) for an arbitrary customer in a queue. Then define $k(a, x)$ as

$$k(a, x) = \frac{\int_a^{a+x} dG(y)}{\int_a^{a+x}(1 - G(y))dy}. \tag{5.5}$$

This is indeed the hazard rate function when $x = 0$. We will use the relation subsequently. Define $\kappa(a)$ as

$$\kappa(a) = \max_{x \geq 0} k(a, x)$$

and $x^* = \arg\max_{x \geq 0} k(a, x)$, that is, the value of x that maximizes $k(a, x)$. Then, the Gittins index policy works as follows: at every instant of time, serve the customer with the largest $\kappa(a)$. To implement this, select the customer with the largest $\kappa(a)$ and serve this customer until whichever happens first among the following: (1) the customer is served for time x^*, (2) the customer's service is complete, or (3) a new customer arrives with a higher Gittins index. The proof that the Gittins index policy is fairly detailed is not presented here. However, the key idea is to target the application with the highest probability of completing service within a time x and at the same time have a low expected remaining service time. Although not exactly, the $k(a, x)$ measure roughly captures that. We saw earlier that if the service times are known upon arrival, then SRPT is the best, and when they are unknown, we do what best we can based on the information we have (which is what Gittins index policy does). Two special cases of Gittins index policy are discussed subsequently when the hazard rate functions are monotonic.

3. *Service times unknown upon arrival; only one job with partially complete service allowed*: Here we consider the case where it is not possible to know the service time of each customer upon arrival (i.e., nonanticipative), and it is not possible to have more than one partially complete service. Interestingly, in this restricted framework every policy would yield the same W. Therefore, all work-conserving policies in this restricted framework (such as FCFS, nonpreemptive LCFS, random order of service, etc.) are optimal. Thus, a policy such as FCFS which is generally fair would be ideal.

Before moving onto multiclass queues, we describe two special cases of Gittins index policies by framing it as a problem.

Problem 50

When the service times are unknown upon arrival and many jobs with partially complete service are allowed, then Gittins index policy is optimal. Show that if the service times are IFR, then the Gittins index policy reduces to FCFS-like policies that do not allow more than one job to be partially complete (although that is not a requirement). Also describe the special case optimal policy when the service times are DFR.

Solution

Refer to Aalto et al. [1] for a rigorous proof; we just provide an outline based on that paper here. From the definition of $k(a, x)$ given in Equation 5.5, we have

$$k(a, x) \int_0^x (1 - G(a + y)) dy = \int_0^x g(a + y) dy$$

where $g(y)$ is the PDF of the service times, that is, $dG(y)/dy$. By taking derivative with respect to x of this equation, we get

$$\frac{\partial k(a, x)}{\partial x} \int_0^x (1 - G(a + y)) dy + k(a, x)[1 - G(a + x)] = g(a + x).$$

We can rewrite this expression in terms of the hazard (or failure) rate function $h(y)$ defined as

$$h(y) = \frac{dG(y)/dy}{1 - G(y)}$$

which would yield

$$\frac{\partial k(a, x)}{\partial x} = \frac{[h(a + x) - k(a, x)]}{\int_0^x (1 - G(a + y)) dy} [1 - G(a + x)].$$

From this expression we can conclude that by keeping a a constant, $k(a, x)$ would be increasing (or decreasing, respectively) with respect to x if $h(a + x)$ is greater than (or less than, respectively) $k(a, x)$.

Recall that by service times being IFR (or DFR, respectively), we mean that $h(x)$ is increasing (or decreasing, respectively) with respect to x. Thus, if the service times are IFR (or DFR, respectively), $h(a+u) \leq$ (or $\geq$, respectively) $h(a + x)$ for all $u \leq x$. Since we can rewrite $k(a, x)$ in Equation 5.5 as

$$k(a, x) = \frac{\int_0^x h(a + u)[1 - G(a + u)] du}{\int_0^x (1 - G(a + u)) du}$$

clearly the RHS is $\leq$ (or $\geq$, respectively) $h(a+x)$ if the service times are IFR (or DFR, respectively). Therefore, if the service times are IFR, $k(a,x) \leq h(a+x)$, and if they are DFR, $k(a,x) \geq h(a+x)$. However, we showed earlier that $k(a,x)$ would increase (or decrease, respectively) with respect to x if $h(a+x)$ is greater than (or less than, respectively) $k(a,x)$. Hence, we can conclude that if the service times are IFR (or DFR, respectively), $k(a,x)$ would increase (or decrease, respectively) with respect to x. Thus, we can compute $\kappa(a) = \max_x\{k(a,x)\}$ when the service times are IFR and DFR. In particular, we can show the following:

- If service times are IFR

$$\kappa(a) = k(a,\infty) = \frac{1 - G(a)}{\int_0^\infty [1 - G(a + t)]dt}.$$

This can be used to show that $\kappa(a)$ increases with a (using a very similar derivative argument, left as an exercise for the reader). Since Gittins index policy always picks the job with the largest $\kappa(a)$, the optimal policy is to select the job with the largest attained service a. Consider a queue that is empty, since the discipline is work conserving, an arriving customer is immediately served. This customer is never interrupted because we always serve the customer that has the largest attained service. When this customer completes service if there are many jobs to choose from, any of them can be picked since they all have $a = 0$ and served. Again this customer is never interrupted until service is complete. Thus, any FCFS-like policy that does not allow more than one job to be partially complete is optimal. Note that all of these FCFS-like policies yield the same L or W.
- If service times are DFR

$$\kappa(a) = k(a,0) = h(a).$$

Since $h(a)$ is decreasing, clearly we have $\kappa(a)$ decreasing with respect to a. Since Gittins index policy always picks the job with the largest $\kappa(a)$, the optimal policy is to select the job with the smallest attained service a. This policy is known as least-attained service (LAS) or foreground-background (FB, also called forward-backward or even feedback). This policy is a little tricky and deserves some explanation. Consider a queue that is empty, since the discipline is work conserving, an arriving customer is immediately served. This customer is interrupted if a customer arrives because the arriving customer would have LAS. This new customer is served until either a time equal to how much the first customer was served elapses or another customer arrives. If it is the latter, then the new customer goes through what the previous new customer went through;

however, the former would result in going back and forth between the two customers that have equally attained service. Hence the name FB.   ■

5.5.3 Optimal Scheduling Policies in Multiclass Queues

Here we consider the more generic $M/G/1$ queue with K classes as described earlier in this section. Recall that our objective is to find the policy π that minimizes the total number in the system on average in steady state, that is, $\sum_{i=1}^{K} L_i^{\pi}$. We return to the only restriction on π being that it can be any work-conserving discipline. However, depending on restrictions imposed on the system (such as whether or not service times are known upon arrival, and whether or not more than one job with partially completed service is allowed), a different policy may be optimal. We describe the optimal policies under various conditions as follows:

1. *Service times can be known upon arrival*: If it is possible to know the service time of each customer upon arrival, then the optimal policy that minimizes $L = L_1 + \cdots + L_K$ is SRPT. This is identical to what we saw in the single class case, that is, $K = 1$ (the reader is encouraged to go over the SRPT explanation in the previous section for $K = 1$). However, that result should not be surprising which is the proof of why SRPT is optimal given it is so when $K = 1$. Consider another queueing system where customers arrive according to $PP(\lambda)$ such that $\lambda = \lambda_1 + \cdots + \lambda_K$, and when a customer arrives the service time is sampled from S_i (service time of a class-i customer in the original system) with probability λ_i/λ and revealed to the server. This new single class system with one aggregate class is identical to the original system with K classes. However, we know that SRPT is optimal for the new system if our objective is to minimize the number in the system (this is based on the single class results). Thus, SRPT would be optimal for the original K class system. For a discussion on performance analysis of SRPT, refer to Section 5.4.3. In the performance analysis, the parameters to use is the new queueing system with a single aggregated class. Before moving on to the next case, it is crucial to point out that if we considered other objectives (such as a weighted sum of queue lengths or sojourn times), then SRPT may not be optimal.

2. *Service times unknown upon arrival; many jobs with partially complete service allowed*: Consider the case where it is not possible to know the service time of each customer upon arrival (i.e., nonanticipative); however, we allow more than one job to have partially complete service. In the single class version ($K = 1$) in the previous section,

we saw that the Gittins index policy minimizes L in the case considered. Would that work here too for a general K? As it turns out, the functions used in the Gittins index (such as $k(a, x)$ and $\kappa(a)$) are class-dependent. Therefore, one has to be careful in writing down the Gittins index parameter for each customer in the system. However, if we are able to do that, then indeed the Gittins index policy would be optimal (see Theorems 1 and 2 in Aalto et al. [1] where what we refer to as Gittins index policy is what they call Gittins index quantum policy). The idea of the proof is similar to that when there is a single class and the reader is encouraged to refer to that. Now we explain the policy in the general K class case. Let a be the amount of attained service (i.e., amount of completed service) for an arbitrary customer in a queue. Then define $k_i(a, x)$ for a class-i customer in the system as

$$k_i(a, x) = \frac{\int_a^{a+x} dG_i(y)}{\int_a^{a+x} (1 - G_i(y)) dy}.$$

Define $\kappa_i(a)$ for a class-i customer as

$$\kappa_i(a) = \max_{x \geq 0} k_i(a, x)$$

and $x_i^* = \arg\max_{x \geq 0} k_i(a, x)$, that is, the value of x that maximizes $k_i(a, x)$. Then, the Gittins index policy works as follows: at every instant of time, serve the customer with the largest $\kappa_i(a)$ over all customers of all classes i. To implement this, from all the customers in the system (belonging to various classes) select the one with the largest $\kappa_i(a)$. Say this customer is of class j. Serve this class-j customer until whichever happens first among the following: (1) the customer is served for time x_j^*, (2) the customer's service is complete, or (3) a new customer arrives with a higher Gittins index.

Now we briefly discuss two special cases of Gittins index policy, that is, when the hazard rate functions are monotonic: (1) If the service time distributions of all K classes are DFR, then the Gittins index policy reduces to serving the job with the highest failure rate $h_i(a)$. Since the service times are DFR, within a class we always use LAS. However, across classes we need to compare the hazard rate (or failure rate) functions of the least attained service customer in each class and serve the one with the highest failure rate. An interesting case is when the failure rate functions $h_i(x)$ do not overlap, then we can order the classes according to hazard rate and use a preemptive priority policy (and LAS within a class) that assigns highest priority to the class with the highest hazard rate function. (2) If the service time distributions of all K classes are IFR, then the Gittins index policy reduces to serving the customer with the shortest expected remaining service time. Since the service times are IFR, within a class we always use most attained service (an example of which is FCFS, as we discussed earlier any policy with at most one customer with partially complete service would be optimal). However, across classes we need to compare the expected remaining service times of the most attained service customer in each class and serve the one with the smallest expected remaining service times. Note that we could have more than one customer with partially complete service across the classes (although within a class that is not possible).

3. *Service times unknown upon arrival; only one job with partially complete service allowed*: Here we consider the case where it is not possible to know the service time of each customer upon arrival (i.e., nonanticipative), and it is not possible to have more than one partially complete service. Since our objective is to find a policy π that minimizes $L^\pi = \sum_i L_i^\pi$, a natural choice is to give highest nonpreemptive priority to the class with shortest expected service times. Also within a class, the jobs can be served according to FCFS or any policy that is nonanticipative and allows at most one job with partially completed service. Before explaining why that policy is optimal, we can consider a slightly extended version. Say our objective is to minimize some weighted sum of the number in the system $\sum_{i=1}^{K} c_i L_i^\pi$ across all allowable policies π. In the previous case, we had all c_i values as one. The optimal policy is the nonpreemptive priority policy (see Section 5.2.2). Within a class, the server serves according to FCFS and across classes the server gives nonpreemptive priority to the class with the highest $c_i/E[S_i]$. This is precisely the $c - \mu$ rule we saw in Problem 46. An elegant proof that the nonpreemptive policy is optimal uses the concept *achievable region* approach nicely detailed in Green and Stidham [48]. Consider Problem 44 where we showed one of the strong conservation results that the sum

$$\sum_{i=1}^{K} \rho_i W_{iq}$$

is conserved over all work-conserving policies with only one job with partially completed service allowed. Our objective is to find a policy π that satisfies the previous strong conservation result and also minimizes $\sum_i c_i \lambda_i W_i^\pi$. We can easily write down the objective function in terms of W_{iq}; then the problem can be solved as a linear program with the optimal solution at one of the corner points of the polyhedron formed by $\sum_{i=1}^{K} \rho_i W_{iq} = K_c$, where K_c is a constant that can be computed for say FCFS. This polyhedron feasible region is the achievable region described earlier. The proof is to mainly show that the nonpreemptive policy is indeed one of the corner points and the one that minimizes the objective function.

There are innumerable other optimal scheduling problems considered in the literature. A closely related problem is the armed bandits problem which has received a lot of attention where researchers have developed index policies. An immediate extension to the previous problem is to consider more general cost functions, not just the weighted sum with linear weights. Another popular extension is to consider more complex costs structure (such as a combination of holding costs, penalty costs, rewards, etc.) which can typically be formulated as a stochastic dynamic program which yields switching-curve-like policy. Therefore, the server would look at (for example) the queue lengths of various classes and select the appropriate one. A natural extension that is much harder for performance analysis is multiserver systems (because they are usually not work conserving). However, the structure of the optimal policy can be derived for multiserver queues. Also, as a last word it is important to mention that the field of stochastic scheduling is mature with a lot of fascinating results and analysis techniques for which this section does not do justice in terms of coverage.

Reference Notes

Analysis of queues with multiple classes of customers can be approached from many angles as evident from the literature, namely, based on applications, based on objectives such as performance analysis versus optimization, and also theory versus practice. However, a common thread across the angles is the objective where the server needs to decide which class of customer to serve. For that reason, this is also frequently referred to as stochastic scheduling. The topic of stochastic scheduling has recently received a lot of attention after a surge of possibilities in computer network applications. Although the intention of this chapter was to provide a quick review of results from the last 50 years of work in single-station, single-server, multiclass queues, a large number of excellent pieces of work had to be left out. The main focus of this chapter has been to present analytical expressions for various performance measures under different service-scheduling policies. These are categorized based on how the customers are classified, that is, depending on type, location, or service times.

This chapter brings together some unique aspects of queues and the theoretical underpinnings for those can be found in several texts. For example, the fundamental notion of work conservation has been greatly influenced by some excellent texts on queueing such as Kleinrock [63], Heyman and Sobel [53], and Wolff [108]. Those books also address the first set of performance analysis in this chapter, namely, priority queues. For more details on work conservation for application to computer systems, refer to Gelenbe and Mitrani [45], and for a more theoretical treatment see Baccelli and Bremaud [7]. The next set of models for performance analysis namely for polling systems in following the pioneering work of Takagi [101]. Then, the third set has been motivated by some recent research on anticipative policies, especially when the service times are known upon arrival. An excellent resource for that is Harchol-Balter [51]. The last section of this chapter looks at finding the best scheduling policy that would minimize the number in the system. That section is based on a recent paper by Aalto et al. [1]. It nicely presents all the results on the optimality of various policies in single as well as multiclass $M/G/1$ queues.

Exercises

5.1 Consider a repair shop that undertakes repairs for K different types of parts. Parts of type i arrive to the repair shop according to a Poisson process with mean arrival rate λ_i parts per hour $(i = 1, \ldots, K)$. At a time only one part can be repaired in the shop with a given mean repair time τ_i hours and a given standard deviation of σ_i hours $(i = 1, \ldots, K)$. There is a single waiting room for all parts. This system can be modeled as a standard $M/G/1$ multiclass queue. Use $K = 5$ and the following numerical values:

i	λ_i	τ_i	σ_i
1	0.2	1.2	1.0
2	0.3	0.3	0.6
3	0.1	1.5	0.9
4	0.4	1.0	1.0
5	0.2	0.3	0.8

Compute for each type i ($i = 1, \ldots, 5$) the mean number of parts (L_i) and the average time in the system (W_i) under the following rules:

(a) FCFS

(b) Nonpreemptive priority (1 is highest and 5 is lowest priority)

(c) Preemptive resume priority (1 is highest and 5 is lowest priority)

5.2 Consider a single-server queue with two types of customers. Customers of type i arrive according to $PP(\lambda_i)$ for $i = 1, 2$. The two arrival processes are independent. The service times are $\exp(\mu)$ for both types. If the total number of customers (of both types) in the system is greater than or equal to K, customers of type 2 do not join the system, whereas customers of type 1 always join the system. State the condition of stability for this queueing system. Compute the expected waiting time for each type of customer in steady state, assuming they exist. Note that for type 2 customers, you are only required to obtain the mean waiting time for those customers that join the system.

5.3 A discrete time polling system consists of a single communication channel serving N buffers in cyclic order starting with buffer 1. At time $t = 0$, the channel polls buffer 1. If the buffer has any packets to transmit, the channel transmits one and then moves to buffer 2 at time $t = 1$. The same process repeats at each buffer until at time $t = N - 1$, the channel polls buffer N. Then at time $t = N$, the channel polls buffer 1 and the cycle repeats.

Now consider buffer 1. Let Y_t be the number of packets it receives during the interval $(t, t + 1]$. Assume that $\{Y_t, t \geq 0\}$ is a sequence of IID random variables with mean m and generating function $\psi(z)$. Let X_n be the number of packets available for transmission at buffer 1 when it is polled for the nth time. Model $\{X_n, n \geq 1\}$ as a DTMC. Show that $Nm < 1$ is the stability condition for buffer 1. Let $\pi_j = \lim_{n \to \infty} P\{X_n = j\}$. Compute the generating function $\Pi(z)$ corresponding to π_j, that is,

$$\Pi(z) = \sum_{i=0}^{\infty} \pi_i z^i.$$

Please note that you are only asked for $\Pi(z)$ and NOT the individual π_j values.

5.4 Consider an $M/M/1$ queue with nonpreemptive LCFS service discipline. Nonpreemptive means that a customer in service does not get replaced by a newly arriving customer. Show that the LST of the sojourn time in the system in terms of the LST of the busy period distribution $\tilde{B}(s)$ is

$$E[e^{-sY}] = (1 - \rho)\frac{\mu}{s + \mu} + \rho\tilde{B}(s)\frac{\mu}{s + \mu}$$

[Note: At time $t = 0$, let a customer arrive into an empty $M/M/1$ system with nonpreemptive LCFS service discipline. Assume that at time $t = T$, the system becomes empty for the first time. Then, T is a random variable known as the busy period. Then, $B(\cdot)$ is the CDF of the busy period. It is crucial to realize that the busy period does not depend on the service discipline as long as it is work conserving. So if we know $\tilde{B}(s)$ for FCFS discipline, then we can compute $E[e^{-sY}]$ for the nonpreemptive LCFS.]

5.5 Consider an $M/G/1$ queue with K classes. Using the expressions for W_{iq} for both FCFS and nonpreemptive priority service disciplines, show that $\sum_{i=1}^{K} \rho_i W_{iq}$ results in the same expression. In other words, this verifies that the amount of work in the waiting area is conserved and is equal to the expression previous. Further, for the special case of exponential service times for all K classes, show that the preemptive resume policy also yields the same expression for $\sum_{i=1}^{K} \rho_i W_{iq}$ by rewriting the expression for FCFS and nonpreemptive priority using exponential service times. Although the preemptive resume policy does not satisfy the condition that there can be at most only one customer with partially complete service, why does the result hold?

5.6 Answer the following multiple choice questions:

(i) Consider a stable $G/G/1$ queue with four classes (average arrival and service rates are λ_i and μ_i, respectively, for class i) using preemptive resume priority. What fraction of time in the long run is the server busy?

(a) $\frac{\lambda_1+\lambda_2+\lambda_3+\lambda_4}{\mu_1+\mu_2+\mu_3+\mu_4}$

(b) $\frac{\lambda_1}{\mu_1} + \frac{\lambda_2}{\mu_2} + \frac{\lambda_3}{\mu_3} + \frac{\lambda_4}{\mu_4}$

(c) $1 - \left(\frac{\lambda_1+\lambda_2+\lambda_3+\lambda_4}{\mu_1+\mu_2+\mu_3+\mu_4} \right)$

(d) $\frac{1}{\frac{\lambda_1}{\mu_1}+\frac{\lambda_2}{\mu_2}+\frac{\lambda_3}{\mu_3}+\frac{\lambda_4}{\mu_4}}$

(ii) Consider an $M/G/1$ queue with four classes (call them classes A, B, C, D) such that the average service time for them are 2, 3, 1, and 5 min, respectively. Also, the holding cost for retaining a customer of classes A, B, C, and D are, respectively, 3, 5, 2, and 5 dollars per item per minute. If we use a nonpreemptive priority, what should be the order of priority from highest to lowest?

(a) D, A, B, C

(b) C, A, B, D

(c) D, B, A, C

(d) C, B, A, D

(iii) Consider a stable $G/G/1$ queue with K classes. Which of the following quantities would be different for the FCFS discipline as compared to nonpreemptive priority?

(a) p_0 (probability of empty system)

(b) $\frac{W}{W_q}$

(c) Average time server is continuously busy (mean busy period)

(d) $L - L_q$

(iv) You are told that for a stable work-conserving single class $G/G/1$ queue in steady state, $W = 18$ min, server was busy 90% of the time, and average arrival rate is 30 per hour. Which of the following statements can you conclude from that?

(a) If the arrival process is Poisson, then service times are exponentially distributed

(b) $L = 540$

(c) Service discipline is FCFS

(d) $W_q = 81/5$ min

(v) Consider a stable $M/G/1$ queue with K classes. Which of the following quantities would be different for the FCFS discipline as compared to nonpreemptive priority?

(a) $\sum_{i=1}^{K} \lambda_i E(S_i^2)$

(b) $\sum_{i=1}^{K} L_i$

(c) $\sum_{i=1}^{K} \rho_i W_{iq}$

(d) $\sum_{i=1}^{K} \rho_i$

5.7 Consider a stable three-class $M/G/1$ queue with nonpreemptive priority (class-1 has highest priority and class-3 the lowest). The mean and standard deviation of the service times (in minutes) as well as the arrival rates (number per minute) for the three classes are given in the following table:

Class i	Arrival Rate	Mean Service Time	Std. Deviation Service Time
1	2	0.1	0.2
2	3	0.2	0.1
3	1	0.1	0.1

In steady state, what would be the expected remaining service time for the entity in service (if any) as seen by an arriving class-3 customer? Also compute the average time spent by a class-2 customer in the queue before beginning service (in steady state).

5.8 For the following TRUE or FALSE statements, give a brief reason why you picked the statements to be true or false.

(a) Consider a stable single class $M/G/1$ queue with nonpreemptive LCFS discipline. Is the following statement TRUE or FALSE? The average time in the system (W) would not change if the discipline is changed to FCFS.

(b) Let V be the variance of the workload at an arbitrary time in steady state for a multiclass $G/G/1$ queue. Is the following statement TRUE or FALSE? The quantity V would remain the same for any work-conserving service discipline (FCFS, preemptive resume, etc.).

(c) Consider a stable multiclass $M/M/1$ queue with different arrival and service rates for each class. Let L be the average number of entities in the system in steady state under nonpreemptive priority scheme. Is the following statement TRUE or FALSE? If we change the service discipline to preemptive resume priority but retain the priority order, then L would remain unchanged.

(d) Consider a single station queue with two classes. For $i = 1, 2$, arrivals for class i are according to $PP(\lambda_i)$ and service times for class i take $\exp(\mu_i)$ time. There is a single server that uses nonpreemptive priority across classes and FCFS within a class. Let $X_i(t)$ be the number of customers of class i in the system at time t. Is the following statement TRUE or FALSE? The stochastic process $\{(X_1(t), X_2(t)), t \geq 0\}$ is a CTMC.

5.9 Consider a stable $M/M/1$ queue with arrival rate λ and service rate μ. A modification is made to this system such that each customer upon entry must state if their service time is going to be greater than or smaller than a given constant quantity b (but the customer does not disclose the actual service time). If the service time is smaller than b, it is called class-1 or else class-2. For this system, if we give nonpreemptive priority to class-1, what would be the sojourn time for the two classes, W_1 and W_2? Also, is the aggregate sojourn time $1/(\mu - \lambda)$?

5.10 Let $\tilde{W}_i(s)$ be the LST of the waiting time before service in steady state for a class-i customer in a polling system using gated policy. Ferguson and Aminetzah [37] show that

$$\tilde{W}_i(s) = \frac{\tilde{C}_i(\lambda_i - \lambda_i \tilde{G}_i(s)) - \tilde{C}_i(s)}{E[C][s - \lambda_i + \lambda_i \tilde{G}_i(s)]}$$

where $\tilde{C}_i(\cdot)$ is the LST of the cycle time associated with queue i in steady state and all other variables are described in Section 5.3. Using the previous expression, show that

$$W_{iq} = \frac{(1 + \rho_i)E\left[C_i^2\right]}{2E[C]}.$$

5.11 For an $M/G/1$ queue with a single class, show that if service times are IFR, then the Gittins index policy parameter $\kappa(a)$ increases with a using the expression

$$\kappa(a) = \frac{1 - G(a)}{\int_0^\infty [1 - G(a + t)]dt}$$

where $G(\cdot)$ is the service time CDF.

5.12 Consider an $M/G/1$ queue with a single class operating LAS scheduling policy. Derive an expression for the sojourn time of a customer that arrives with a job requiring x amount of service.

6

Exact Results in Network of Queues: Product Form

So far we have only considered single-stage queues. However, in practice, there are several systems where customers go from one station (or stage) to other stations. For example, in a theme park, the various rides are the different stations and customers wait in lines at each station, get served (i.e., go on a ride), and randomly move to other stations. Several engineering systems such as production, computer-communication, and transportation systems can also be modeled as networks of multistage queues. Such networks, which have a queue at each node, are called *queueing networks*. Just like the single-station case, the analysis of queueing networks also has several nuances to consider, such as multiple classes, scheduling disciplines, and capacity constraints, to name a few. However, as one would expect, the most general case that captures all the nuances is intractable in terms of analytical modeling. To that end, we divide the analysis into two groups, one where exact results are derivable, which is the focus of this chapter, and the other where we rely on approximations (that may or may not be asymptotically exact), which we consider in the next chapter. In addition, for each group, we start with the most basic model and develop some of the generalized classifications.

We now describe the focus of this chapter. In this chapter, we mostly only consider single-class queueing networks except in the last section. In all the cases we discuss, the common thread is that we will develop a *product-form solution* for the joint distribution for the number of customers in each node of the network. Details of the product-form solution are in Section 6.2. Similar to the single-station case seen in the previous chapters, here too we start with Markovian networks. We first describe some results for acyclic open-queueing networks that leverage upon Poisson processes. The scenario in the first model is that there is at least one node into which customers arrive and at least one node from which customers depart the network. Such a queueing network is called an open-queueing network. By acyclic, we mean networks that do not have cycles. They are also called feed-forward networks. Subsequently, we delve into analyzing general networks with cycles that have Poisson arrivals and exponential service times. We will then consider closed-queueing networks that do not allow new customers to enter or the existing customers to leave (and thereby are necessarily cyclic). The simplest to analyze (possibly cyclic) open- and closed-queueing networks are called Jackson networks (due to Jackson [56]). After analyzing Jackson networks, we will consider other queueing networks that also exhibit the product form. As mentioned earlier, we begin with acyclic networks.

6.1 Acyclic Queueing Networks with Poisson Flows

In this section, we consider queueing networks where the underlying network structure is acyclic. By acyclic, we mean that there are no cycles in the directed network. Figure 6.1 depicts an example of an acyclic network. Sometimes acyclic networks are called feed-forward networks. In contrast, a cyclic network is one where there are one or more cycles in the directed network. Figure 6.2 depicts an example of a cyclic network where there are two cycles: 4-5-4 and 4-3-5-4. Cyclic networks are also called networks with feedback. In addition to requiring that there be no cycles in the directed network in this section, we also require that there be a single queue at each node. Therefore, what we essentially have is an acyclic queueing network. Since closed-queueing networks are cyclic in nature, all our acyclic networks belong to open-queueing networks. Thus the word "open" would be redundant if we call our system an acyclic open-queueing network. Consider an acyclic queueing network with N nodes. External arrivals into node i is according to a Poisson process with parameter λ_i for $1 \leq i \leq N$. Some λ_i values can be zero indicating no external arrival into that node.

If the queue in node i cannot accommodate a potential arrival, that customer is rejected from the network. However, if the customer does enter

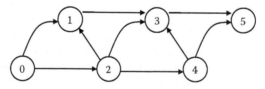

FIGURE 6.1
Example of an acyclic network.

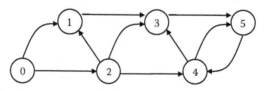

FIGURE 6.2
Example of a cyclic network.

TABLE 6.1

Conditions to be Satisfied at Node i

No. of Servers	Capacity	Service Time Distribution	Stability
s_i	∞	Exponential	Required
s_i	s_i	General	Not applicable
∞	∞	General	Not applicable

node i and gets served, then upon service completion, the customer joins node j with probability p_{ij} and leaves the network with probability r_i. The queue service in node i must satisfy one of the categories given in Table 6.1 (although we do not consider in this section, the results also hold if node i is a single-server queue with processor sharing discipline or LCFS with preemptive resume policy). If $N = 1$, the single node case, these correspond to $M/M/s$, $M/G/s/s$, and $M/G/\infty$ cases in the order provided in the table. With that in mind, our first step to analyze the acyclic queueing network is to characterize the output (or departure) process from the $M/M/s$, $M/G/s/s$, and $M/G/\infty$ queues, which would potentially act as input for a downstream node.

6.1.1 Departure Processes

The objective of this section is to characterize the output (or departure) process from the $M/M/s$, $M/G/s/s$, and $M/G/\infty$ queues in steady state. We begin with an $M/M/1$ queue and derive the interdeparture time distribution in steady state as a problem.

Problem 51

Consider an $M/M/1$ queueing system with $PP(\lambda)$ arrivals and $\exp(\mu)$ service time distribution. Assume that $\lambda < \mu$. Let U be a random variable that denotes the time between two arbitrarily selected successive departures from the system in steady state. Show by conditioning, whether or not the first of those departures has left the system empty, that U is an exponentially distributed random variable with mean $1/\lambda$.

Solution

Say a departure just occurred from the $M/M/1$ queue in steady state. Let X denote the time of the next arrival and Y denote the service time of the next customer. We would like to obtain the CDF of U, the time of the next departure. Define $F(x) = P(U \leq x)$. Also, let Z be a random variable such that $Z = 0$ if there are no customers in the system currently (notice that a departure just occurred), and $Z = 1$ otherwise. If $Z = 0$ then $U = X + Y$, otherwise

$U = Y$. Recall that $\pi_j = \pi_j^* = p_j$ for all j, that is, the probability there are j in the system as observed by a departing customer in steady state would be the same as that of an arriving customer as well as the steady-state probability that there are j in the system. We also know that $p_0 = 1 - \rho$ where $\rho = \lambda/\mu$. Therefore, we have the LST of $F(x)$ by conditioning on Z as

$$\tilde{F}(s) = E[e^{-sU}] = E[e^{-sU}|Z = 0]P(Z = 0) + E[e^{-sU}|Z = 1]P(Z = 1)$$

$$= E[e^{-s(X+Y)}](1 - \rho) + E[e^{-sY}]\rho$$

$$= \left(\frac{\lambda}{\lambda + s}\right)\left(\frac{\mu}{\mu + s}\right)\left(1 - \frac{\lambda}{\mu}\right) + \left(\frac{\mu}{\mu + s}\right)\left(\frac{\lambda}{\mu}\right) = \frac{\lambda}{\lambda + s}.$$

Inverting the LST, $F(x) = P(U \leq x) = 1 - e^{-\lambda t}$ or $U \sim \exp(\lambda)$. Thus, U is an exponentially distributed random variable with mean $1/\lambda$.   ■

Considering the above result, a natural question to ask is whether the departure process is a Poisson process with parameter λ. Notice in the above problem we have only shown that the marginal distribution for an arbitrary interdeparture time in steady state is $\exp(\lambda)$. It is more involved to show that the departure process from a stable $M/M/1$ queue in steady state is indeed a Poisson process. It is very nicely done in Burke [14] and the result is called *Burke's theorem*. We only present an outline of the proof by considering a reversible process that can easily be extended to $M/M/s$ queues. Recall the reversibility definition and properties from Chapter 2. Clearly, the queue length process $\{X(t), t \geq 0\}$ of an $M/M/s$ queue is a reversible process because it satisfies the condition for all $i \in S$ and $j \in S$,

$$p_i q_{ij} = p_j q_{ji}$$

which is a direct artifact of the balance equations resulting from the arc cuts for consecutive nodes i and j (otherwise $q_{ij} = 0$).

One of the implications of reversible processes is that if the system is observed backward in time, then one cannot tell the difference in the queue length process. Thus the departure epochs would correspond to the arrival epochs in the reversed process and vice versa. Therefore, the departure process would be stochastically identical to the arrival process, which is a Poisson process. In fact, for the $M/G/\infty$ and $M/G/s/s$ queues as well, the departures are according to a Poisson process. We had indicated in Chapter 4 that if we define an appropriate Markov process for the $M/G/s/s$ queue (note that when $s = \infty$ we get the $M/G/\infty$ queue, so the same result holds), then that process is reversible. Due to reversibility, the departures from the original system correspond to arrivals in the reversed system.

Therefore, the departure process from the $M/G/s/s$ queue is a Poisson process with rate $(1 - p_s)\lambda$ departures per unit time on average, where p_s is the probability that there are s in the system in steady state (i.e., zero when $s = \infty$), that is, the probability of a potential arrival is rejected.

It is indeed strange that for the stable $M/M/s$ queue and the $M/G/\infty$ queue, the output process is not affected by the service process. Of course, this is incredibly convenient in terms of analysis. Poisson processes have other extremely useful properties, such as superpositioning and splitting, that are conducive for analysis, which we will see next. Before that it is worthwhile to point out that in the $M/G/s/s$ case, through p_s the departure processes does depend on the mean service rate. However, the distribution of service times has no effect on the departure process in steady state. Having said that, except for a small example we will consider in the next section, until we reach Section 6.4.4 on loss networks, we will only consider infinite capacity stable queues and not consider any rejections.

6.1.2 Superpositioning and Splitting

We first describe two results from Poisson processes that we would use to analyze the performance of feed-forward networks. The first result is *superposition* of Poisson processes. When n Poisson processes $PP(\lambda_1)$, $PP(\lambda_2)$, ..., $PP(\lambda_n)$ are superimposed, then the aggregate counting process that keeps track of the number of events is a $PP(\lambda)$, where $\lambda = \lambda_1 + \lambda_2 + \cdots + \lambda_n$. In other words, when n Poisson processes are merged, the resultant process is Poisson with the rate equal to the sum of the rates of the merging processes. This result can be proved using the memoryless property of exponential random variables and minimum of exponentials (properties of exponential distribution are summarized in Section A.4). Denote the n merging Poisson processes as n streams such that the interevent times of stream i is according to $\exp(\lambda_i)$ for $1 \le i \le n$. Let X_i be the remaining time until the next event from stream i. Due to the memoryless property, $X_i \sim \exp(\lambda_i)$. The next event in the merged process occurs after $\min(X_1, X_2, \ldots, X_n)$ time and that is according to $\exp(\lambda)$. Hence, the interevent times in the merged process is according to $\exp(\lambda)$.

Next, we describe splitting of Poisson processes. Consider events occurring according to a Poisson process with parameter λ for some $\lambda > 0$. Independent of everything else, for $j = 1, \cdots, k$, an event is of type j with probability q_j so that $q_1 + \cdots + q_k = 1$. Such a splitting of each event into k types is called a Bernoulli splitting. In this case, type j events occur according to $PP(\lambda q_j)$. This can be proved using the sum of geometric number of IID exponentials property. From this description, events occur with interevent times according to $\exp(\lambda)$ and each event is of type j with probability q_j and not type j with probability $1 - q_j$. Thus the number of events till an even of type j occurs is according to a geometric distribution with mean $1/q_j$ events. Thus if Y_i is the ith interevent time of the original process and Z is the number

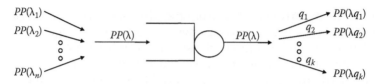

FIGURE 6.3
Merge, flow through a stable queue with s exponential servers and split.

of events until one of type j occurs (which is geometric with parameter q_j), then the time for that event is $Y_1 + \cdots + Y_Z$, which from the property "sum of geometric number of IID exponentials" is according to $\exp(\lambda q_j)$. Hence, type j events occur according to $PP(\lambda q_j)$. Note that Bernoulli splitting is critical and this would not work otherwise.

Now let us consider in a single framework, superposition, flow through a node (or queue) as well as splitting. Let n flows merge before entering a queue (i.e., n streams of customers arrive into the queue) so that each flow i is according to $PP(\lambda_i)$ for $1 \le i \le n$. The net customer-arrival process to the queue is $PP(\lambda)$, where $\lambda = \lambda_1 + \cdots + \lambda_n$. Now these merged customers encounter the queue with s identical servers and service times $\exp(\mu)$ such that $\lambda < s\mu$. The output from this queue is $PP(\lambda)$. If the exiting customers are split into k flows according to Bernoulli splitting with probabilities $q_1, \ldots, q_k$, then the resulting flow j (for $1 \le j \le k$) is $PP(\lambda q_j)$. This is illustrated in Figure 6.3. Of course, this would hold for all the cases given in Table 6.1 except in the last case one has to be careful to write down the rate of the departing Poisson process. Using these results, we can now analyze any acyclic queueing network with nodes satisfying conditions in Table 6.1 by going through merging, flow through a queue and splitting in a systematic manner. We illustrate this through an example problem next.

Problem 52

Consider the acyclic network in Figure 6.1. Say customers arrive externally at nodes 0, 2, and 4 according to Poisson processes with mean rates 13, 12, and 15 customers per hour, respectively. After service at nodes 3, 4, and 5, a certain fraction of customers exit the network. After service at all nodes i (such that $0 \le i \le 5$), customers choose with equal probabilities among the options available. For example, after service at node 4, with an equal probability of 1/3, customers choose nodes 3 or 5 or exit the network (while customers that complete service at nodes 0, 1, and 2 do not immediately exit the network). Further, node 0 has two servers but a capacity of 2 and generally distributed service times. Nodes 1, 3, and 5 are single-server nodes with exponentially distributed service times and infinite capacity. Node 2 is a two-server node with exponentially distributed service times and infinite capacity. Node 4 is an infinite-server node with generally distributed service times. Assume that the mean service times (in hours) at nodes 0, 1, 2, 3, 4, and 5 are 1/26,

1/15, 1/10, 1/30, 1/7, and 1/20, respectively. Compute the average number of customers in each node in steady state as well as the overall number in the network.

Solution

We consider node by node and derive L_j, the average number of customers in node j (for $0 \le j \le 5$).

Node 0: Arrivals to node 0 are according to $PP(13)$. Since there are two servers and capacity of 2, this node can be modeled as an $M/G/2/2$ queue. The probability that this node is full is p_2 and is given by

$$p_2 = \frac{\frac{1}{2} \left(\frac{13}{26} \right)^2}{1 + \frac{13}{26} + \frac{1}{2} \left(\frac{13}{26} \right)^2} = 1/13.$$

The output from node 0 is according to $PP(13(1 - p_2))$, which is $PP(12)$. Since exiting customers pick nodes 1 and 2 each with probability 0.5, departures from node 0 enter node 1 according to $PP(6)$ and node 2 according to $PP(6)$. Using the $M/G/s/s$ results, we get $L_0 = 12/26$.

Node 2: From node 0 customers enter node 2 according to $PP(6)$ and customers arrive externally according to $PP(12)$. Thus, total arrival into node 2 is according to $PP(18)$. Thus node 2 can be modeled as an $M/M/2$ queue. Using $M/M/s$ results, we have $L_2 = 180/19$. Since the service rate of each server is 10, the queue is stable and customers depart from this node according to $PP(18)$. Since a third of the departures go to node 1, a third to node 3, and a third to node 4, the process from node 2 to 1, 2 to 3, and 2 to 4 are all $PP(6)$.

Node 1: Arrivals to node 1 are a result of merging of two Poisson processes, $PP(6)$ from node 0 and $PP(6)$ from node 2. Thus, customers enter node 1 according to $PP(12)$. Since this is a single exponential server node with service rate 15, it can be modeled as an $M/M/1$ queue and $L_1 = 4$. Since this queue is stable, customers depart according to $PP(12)$, all of which go to node 3.

Node 4: From node 2 customers enter node 4 according to $PP(6)$ and customers arrive externally according to $PP(15)$. Thus total arrival into node 4 is according to $PP(21)$. Thus node 4 can be modeled as an $M/G/\infty$ queue. Using $M/G/\infty$ results, we have $L_4 = 3$. Customers depart from this node according to $PP(21)$. Since a third of the departures go to node 3, a third to node 5, and a third exit, the processes from node 4 to 3 and 4 to 5 are both $PP(7)$.

Node 3: Arrivals to node 3 are a result of merging of three Poisson processes, namely, $PP(12)$ from node 1, $PP(6)$ from node 2, and $PP(7)$ from node 4. Thus customers enter node 3 according to $PP(25)$. Since this is a single exponential server node with service rate 30, it can be modeled as an $M/M/1$ queue and $L_3 = 5$. Since this queue is stable, customers depart according to $PP(25)$, and with probability half of each departing customer goes to node 5 (or exit the network).

Node 5: Arrivals to node 5 are a result of merging of two Poisson processes, $PP(12.5)$ from node 3 and $PP(7)$ from node 4. Thus customers enter node 5 according to $PP(19.5)$. Since this is a single exponential server node with service rate 20, it can be modeled as an $M/M/1$ queue and $L_5 = 39$. Since this queue is stable, customers depart according to $PP(19.5)$, all of which exit the system.

Further, the overall average number in the network in steady state is $L = L_0 + L_1 + L_2 + L_3 + L_4 + L_5 = 60.935$. ∎

Interestingly, the above example does not fully illustrate the properties of Poisson-based acyclic queueing networks in all its glory. In fact, one of the most unique properties that these acyclic networks satisfy and that is not satisfied by the cyclic networks (that we will discuss in the next section) is given in the next remark.

Remark 11

Consider an acyclic network that has N nodes with external arrivals according to a Poisson process and conditions in Table 6.1 satisfied. Assume that the network is in steady state at time 0. Let $X_i(t)$ be the number of customers in node i at time t for $1 \le i \le N$ and $t \ge 0$. Then for any j and u such that $1 \le j \le N$, $j \ne i$, and $u \ge 0$, the two random quantities $X_i(t)$ and $X_j(u)$ are independent. ∎

This remark enables us to *decompose* the queueing network and study one node at a time (of course, the right order should be picked) without worrying about dependence between them. Further, if one were to derive the sojourn time distribution for a particular customer, then it would just be the sum of sojourn times across each node in its path, which are all independent. For example, a customer that enters node 2 in Problem 52, goes to node 3, then to 5, and exits the network would have a total sojourn time T equal to the sum of the times in nodes 2, 3, and 5, say T_2, T_3, and T_5, respectively. Then since T_2, T_3, and T_5 are independent, we can compute the LST of T as

$$E[e^{-sT}] = E[e^{-s(T_2+T_3+T_5)}] = E[e^{-sT_2}]E[e^{-sT_3}]E[e^{-sT_5}].$$

Clearly, the RHS can be computed from single-station analysis in the previous chapters.

It is worthwhile to point out that as described earlier in this section, besides the cases that can be captured in Table 6.1, all the results presented here are also satisfied for nodes with a single server, general IID service times with either processor sharing or LCFS preemptive resume discipline. Before concluding this section, we would like to add another remark. Since it is extremely convenient to analyze Poisson-based cyclic networks due to the decomposability, a natural question to ask is whether there are other queues where the departure processes can be characterized as Poisson, at least approximately. The following remark is from Whitt [104].

Remark 12

Consider a stable $G/G/m$ queue where many servers are busy. The arrivals are according to a renewal process and the servers are identical with service times according to a general distribution. Whitt [104] shows that the departures from such a queue is approximately a Poisson process. ∎

This remark is extremely useful especially while approximately analyzing some queueing networks. We consider approximate analysis in the next chapter. For the remainder of this chapter, we will consider networks that are not necessarily acyclic and derive the so-called product-form results. We start with Jackson networks in the next section. However, we wrap up this section with a case study.

6.1.3 Case Study: Automobile Service Station

Carfix is an automobile service station strategically located near three major interstate highways. It provides routine maintenance, services, and repairs such as tune-ups, brake work, exhaust issues, electrical system malfunctioning, oil and fluid changes, and tire work. Pretty much anything that needs to be fixed in a car, SUV, light truck, or van can be brought to Carfix and the technicians there are capable of taking care of it. The business at Carfix has been steady, and due to the proximity to major interstate highways, the clientele was not limited to local customers. With the economy going south (note that this case study was written in 2011), several dealerships in the area have started to offer lucrative service packages since new vehicle sales have plummeted. Thus, the demographics of Carfix's clients have shifted from being largely local a few years ago to now a vast majority of out-of-town customers.

One morning, Cleve, the owner of Carfix was playing a round of golf with his buddy Vineet who owned a couple of motels in town. As soon as Vineet asked Cleve how things were at Carfix, Cleve seized the moment and asked, "so Vineet, since all your clients are from out of town, do you really worry about customer service?" To that Vineet replied that several years ago, all he worried about was to make sure that the expectations of the parent company from which he franchised the motels were being met. However, in this day and age where customers can easily check reviews on their smart phones while they are traveling, it has become extremely important to provide excellent service individually (not just overall). Cleve thought to himself that a good number of his prospective clients are going to search on their smartphones and it would be of paramount importance to have good reviews.

Subsequently, Cleve brainstormed with Vineet opportunities for enhancing customer satisfaction. Based on that, three main ideas emerged: (1) perform a complementary multipoint inspection at the end of a service/repair for all customers (a vehicle inspector would need to be hired for that); (2) offer guarantees such as *if you do not get your vehicle back within some τ minutes, the service is free*; and (3) install an additional bay and hire an additional automotive technician. It was not clear to Cleve, what the benefits of these improvements would be. He knew for sure that idea (1) needs to be done because all the dealership service stations are providing that inspection, and to stay competitive, Carfix must also offer it. Cleve told Vineet that once he figures out the best option, he would be ready for another round of golf with Vineet to discuss what kind of discounts he could provide his customers to stay at one of Vineet's motels.

One of Cleve's nieces, Lauren, was a senior in Industrial Engineering who Cleve thought could help him analyze the various options. When Lauren heard about the problem, she was excited. She talked to her professor to find out if he would allow her and a couple of her friends to work on Cleve's problem for their capstone design course. The professor agreed, in fact he was elated because it would solve the mismatch he was encountering between the number of students and projects. When Lauren and her friends arrived at Carfix they found out that there was no historical data, so they spent a few hours collecting data. Interestingly, when the students were at Carfix, Cleve also had two inspector-candidates who were going to interview for the multipoint inspection position. For the interview, Cleve asked the two candidates to perform inspections on a few vehicles.

6.1.3.1 System Description and Model

On the basis of the data they collected as well as the discussions with the staff and technicians working at Carfix, Lauren and her friends abstracted the system in the following manner. Essentially, there are three stations in a tandem fashion. Vehicles arrived to the first station according to a Poisson process at an average rate of about 4.8 per hour. As soon as a vehicle arrives, a greeter would meet the driver of the vehicle, write down the problems or symptoms, and politely request the driver (and any passengers) to park the vehicle and wait in the lounge area. This whole process took 1–2 min and it appeared to Lauren's team that it was uniformly distributed. Also, Lauren found that although they only had one greeter, the drivers would never wait to be greeted. That is because if the greeter was busy when a new vehicle arrived, the cashier or Cleve himself would go and greet.

At the second stage, technicians would pick up the parked vehicle in the order they arrived and take them to their bay. There were four technicians, each with his or her own bay. The bays were equipped to handle all repairs and service operations. After completing the repair, the technicians would return the vehicles back to the parking area. The time between when a technician would pick up a vehicle from where it is parked till he or she would drop that vehicle off is exponentially distributed with mean about 36 min. Given the different types of services and repairs as well as the types of vehicles, it did not surprise Lauren and friends that a high-variability distribution such as exponential fitted well. Then in the third stage, the inspector would perform a multipoint inspection at the parking lot and send a report.

Among the two inspectors interviewed, inspector A had a mean of 6 min and a standard deviation of 4.24 min, whereas inspector B had a mean of 7 min and a standard deviation of 3.5 min. Although the number of sample points were small, Lauren and her friends felt comfortable using a gamma distribution for the inspection times. They also realized that it was not necessary to include the time at the cashier because what the customers really cared about was the time between when they dropped off the car and when they hear from Cleve or another staff member (on days Cleve is out golfing) that the repair or service is complete. So Lauren and friends decided on using a three-stage tandem system to represent Carfix's shop.

Since there was never a queue buildup, Lauren and friends modeled the first stage as an $M/G/\infty$ queue with $PP(\lambda)$ arrivals and $Unif(a, b)$ service times where $\lambda = 4.8$ per hour, $a = 1$ min, and $b = 2$ min. Clearly, the departures from the queue would be $PP(\lambda)$ and will act as arrivals to the second stage. They modeled the second stage as an $M/M/4$ queue with $PP(\lambda)$ arrivals and $\exp(\mu)$ service times, where $\mu = 5/3$ per hour. Since the departure from a stable $M/M/s$ queue is a Poisson process, they modeled the third stage as an $M/G/1$ queue with $PP(\lambda)$ arrivals, and mean and variance of service times depending on whether inspector A or B is used.

6.1.3.2 Analysis and Recommendation

With the preceding model, Lauren and friends decided to analyze the system stage by stage. At stage 1, since the system is an $M/G/\infty$ queue with $PP(\lambda)$ arrivals and $Unif(a, b)$ service times, clearly the sojourn time in stage 1 (they called it Y_1) would be uniformly distributed between 1 and 2 min. Also, the mean time spent by a vehicle in stage 1 is $W_1 = 1.5$ min and the LST of Y_1 is

$$E\left[e^{-sY_1}\right] = \frac{e^{-s}(1 - e^{-s})}{s}.$$

At this point, Lauren and friends were not sure if the LST was necessary but felt that it was good to keep it in case they were to compute the LST of the total sojourn time in the system. Then, using the fact that $\lambda = 4.8$ per hour, the average number of vehicles in stage 1 is $\lambda * 1.5/60 = 0.12$. Thus the steady-state probability that there are no more than two vehicles in stage 1 is $(1 + 0.12 + (0.12)^2/2!)e^{-0.12} = 0.9997$, which clearly justifies the use of the $M/G/\infty$ model. In fact, it shows how rarely Cleve would have to greet a customer (although it appears that the cashier would have to greet only one in 100 vehicles on average as the probability of zero or one vehicles at stage one is 0.9934, in reality it was a lot more often because the greeter was called upon to run odd jobs from time to time).

Lauren and friends next considered stage 2, which they modeled as an $M/M/4$ queue with $PP(\lambda)$ arrivals and $\exp(\mu)$ service. Plugging in $\lambda = 4.8$ per hour and $\mu = 5/3$ per hour, they got a traffic intensity $\rho = \lambda/(4\mu) = 0.72$ at stage 2. Clearly, this is stable albeit not a low-traffic intensity. It does appear like adding a new bay (thus, $M/M/5$ queue) would significantly reduce the traffic intensity to 0.576 resulting in better customer service. Starting with the present $M/M/4$ system, Lauren and friends calculated the mean sojourn time in stage 2 as $W_2 = 50.79$ min (then for the $M/M/5$ system they calculated that it would be 39.52 min). Also, with s servers, the sojourn time Y_2 had a CDF $F_2(y) = P\{Y_2 \leq y\}$ given by

$$F_2(y) = \sum_{j=0}^{s-1} p_j(1 - e^{-\mu y}) + p_0 \frac{(\lambda/\mu)^s}{s!} \left\{ \frac{s\mu}{(s-1)\mu - \lambda}(1 - e^{-\mu y}) \right.$$

$$\left. - \frac{s\mu^2}{(s\mu - \lambda)[(s-1)\mu - \lambda]}(1 - e^{-(s\mu - \lambda)y}) \right\}$$

for any $y \geq 0$, where

$$p_0 = \left[\sum_{n=0}^{s-1} \left\{ \frac{1}{n!} \left(\frac{\lambda}{\mu} \right)^n \right\} + \frac{(\lambda/\mu)^s}{s!} \frac{1}{1 - \lambda/(s\mu)} \right]^{-1}$$

and for $j = 1, \ldots, s$, $p_j = 1/j!(\lambda/\mu)^j p_0$. By plugging in the numbers for λ and μ, and using the current system with $s = 4$, Lauren and friends computed the probability $P\{Y_2 \leq y\}$ to be 0.6798 for $y = 1$ h and 0.8363 for $y = 1.5$ h. They also quickly computed that if Carfix were to consider $s = 5$ bays and workers, the probability that the sojourn time in stage 2 would be less than 1 and 1.5 h is 0.7815 and 0.9036, respectively.

Lauren and friends decided to forge ahead with stage 3, which they modeled as an $M/G/1$ queue. They realized that irrespective of whether Cleve decides to adopt 4 or 5 bays in stage 2, the output will still be $PP(\lambda)$, which would act as input to stage 3. Using the Pollaczek–Khintchine formula for $M/G/1$ queue, the mean sojourn in stage 3, W_3 is 10.1538 min for inspector A (mean inspection time 6 min and standard deviation 4.24 min), whereas it is 12.5682 min for inspector B (mean inspection time 7 min and standard deviation 3.5 min). Given that inspector A yielded a much better mean sojourn time, Lauren and friends recommended that inspector. Also, they fitted an Erlang distribution for the inspection time so that the LST of the CDF of the inspection time in minutes was $\tilde{G}(s) = 1/(1+3s)^2$. Thereby, the LST of the sojourn times Y_3 (in min) was calculated as

$$E[e^{-sY_3}] = \frac{(1-\rho_3)s\tilde{G}(s)}{s - \lambda(1-\tilde{G}(s))/60}$$

where the traffic intensity $\rho_3 = 0.1\lambda$ since the mean inspection time is 6 min, that is, 0.1 h. Inverting the LST, Lauren and friends computed the stage-3 sojourn time (in min) CDF as

$$P\{Y_3 \le y\} = 1.3724(1 - e^{-0.1252y}) - 0.3724(1 - e^{-0.4615y}).$$

Plugging in for $y = 20$ min, we get the probability of completing the inspection within 20 min (for inspector A) as 0.8879, which Lauren and friends felt was quite poor considering that the traffic intensity was only 0.48 and the mean inspection time was only 6 min.

Having completed the analysis for the three stages, Lauren and friends felt it was time to make recommendations. Recall the three main ideas that emerged out of the discussion between Cleve and Vineet. Lauren and friends responded to those suggestions as follows: (1) Since performing a multipoint inspection was almost necessary to stay competitive, it would be best to hire inspector A. However, it is important to note that it would result in an increase of the total time by over 10 min on average and about 11% of the customers could face over 20 min of wait. It was up to Cleve to determine if that was worth it. (2) Given that the mean sojourn time was a little over an hour (62.44 min) and the standard deviation was quite high, it did not make much sense to offer guarantees such as *if you do not get your vehicle back within 2 h, the service is free*. The rationale for that is if on average it would only take 1 h, the guarantee of 2 h might be misleading. Instead, the recommendation would be to provide an estimate based on the queues at various stations (Lauren offered to write a program to compute that estimate or use the CDFs to calculate the probabilities of completing the entire service within some time τ). (3) Installing the additional bay (i.e., the fifth one) and hiring an additional automotive technician would cut the mean sojourn times by over 11 min. However, that might become absolutely necessary if Carfix's demand increases. Ultimately, Lauren was able to perform the required what-if analysis so that Cleve could use a cost-benefit analysis to determine the best alternatives. Eventually, Cleve ended up implementing all the recommendations Lauren and friends made. As expected, it improved customer satisfaction as well as increased the demand. However, Cleve was well positioned for that higher demand.

6.2 Open Jackson Networks

An open Jackson network is a special type of open-queueing network where arrivals are according to Poisson processes and service times are exponentially distributed. Cycles are allowed in Jackson networks. In addition, open Jackson networks can be categorized by the following:

1. The open-queueing network consists of N service stations (or nodes).

2. There are s_i identical servers at node i (such that $1 \leq s_i \leq \infty$), for all i satisfying $1 \leq i \leq N$.

3. Service times of customers at node i are IID $\exp(\mu_i)$ random variables. They are independent of service times at other nodes.

4. There is infinite waiting room at each node, and stability condition (we will describe that later) is satisfied at every node.

5. Externally, customers arrive at node i according to a Poisson process with rate λ_i. All arrival processes are independent of each other and the service times. At least one λ_i must be nonzero.

6. When a customer completes service at node i, the customer departs the network with probability r_i or joins the queue at node j with probability p_{ij}. Note that $p_{ii} > 0$ is allowed and corresponds to rejoining queue i for another service. It is required that for all i, $r_i + \sum_{j=1}^{N} p_{ij} = 1$ as all customers after completing service at node i either depart the network or join another node (in particular, joining node j with probability p_{ij}). The routing of a customer does not depend on the state of the network.

7. Define $P = [p_{ij}]$ as the routing matrix of p_{ij} values. Assume that $I - P$ is invertible, where I is an $N \times N$ identity matrix. It turns out that the $I - P$ matrix is invertible if every customer would eventually exit the network after a finite sojourn.

To analyze the open Jackson network, one of the preliminary results is flow conservation and stability, which we describe next.

6.2.1 Flow Conservation and Stability

The results in this section can be applied to a much wider class of networks than the Jackson network. In particular, the requirement of exponential inter-arrival and service times is not necessary. The concepts of flow conservation and stability are closely linked to each other. In fact, for flow conservation we need stability, but condition of stability can be derived only after performing flow conservation. One of the most fundamental results of flow conservation is that if a queue is stable, then the long-run average arrival rate of customers is equal to the long-run departure rate of customers. In other words, every customer that arrives must depart, no customers can be created or lost. This is the essence of flow conservation at a queue (see Section 1.2). In a network where flows merge and split, we can again apply flow conservation at merge points and split points. Average flow rate before and after the merge or split points is equal.

With that in mind, the next objective is to obtain the average arrival (or for that matter departure) rates of customers into each node. Recall the notation and conditions for open Jackson networks. In steady state, the total average arrival rate into node j (external and internal put together) is denoted by a_j and is given by

$$a_j = \lambda_j + \sum_{i=1}^{N} a_i p_{ij} \quad \forall j = 1, 2, \ldots, N. \tag{6.1}$$

This result is due to the fact that the total arrival rate into node j equals the external arrival rate λ_j plus the sum of the departure rates from each node i (for $i = 1, \ldots, N$) times the fraction that are routed to j (i.e., p_{ij}). Therefore, let $a = (a_1, a_2, \ldots, a_N)$ be the resulting row vector that we need to obtain. We can rewrite the a_j Equation 6.1 in matrix form as $a = \lambda + aP$, where $\lambda = (\lambda_1, \lambda_2, \ldots, \lambda_N)$ is a row vector. Then a can be solved using

$$a = \lambda(I - P)^{-1}. \tag{6.2}$$

Note that for this result we require $(I - P)$ to be invertible. Unlike the acyclic networks where a_j values are easy to compute, when the networks are cyclic, one may have to rely on Equation 6.2.

Now that a_j values can be obtained for all j, we are in a position to state the stability condition. For any j (such that $1 \leq j \leq N$), node j is stable if

$$a_j < s_j \mu_j.$$

Further, the queueing network is stable if the stability condition is satisfied for every node of the network. Note that the fact we needed the queues to be stable to compute a_j does not cause any computational difficulties. Essentially we compute a_j, which does not require the knowledge of μ_i values.

Then we check if each queue is stable. If all queues are stable then we can conclude that the flow rates through node j are indeed a_j. For the open Jackson network described earlier, the objective is to derive the joint distribution as well as marginal distribution (and moments) of the number of customers in each queue. We do that next and describe a product form for the joint distribution and thereby derive the marginals.

6.2.2 Product-Form Solution

Consider an open Jackson network. Note that if the network is acyclic, then we have the acyclic queueing network results from Section 6.1. Thus the most interesting case is if the network has cycles. However, we present a unified method that covers both acyclic and cyclic open Jackson networks. To analyze Jackson networks, a natural question to ask is why not just use the Poisson property like we did for acyclic networks. As it turns out, the Poisson result does not hold when there is feedback. In particular for cyclic networks, if there is a node that is part of a cycle, then the output from that node is not according to a Poisson process. Disney and Kiessler [26] show the steady-state correlation between the queue lengths at two different nodes (that are part of a cycle) i and j at two different times u and t, that is, $X_i(t)$ and $X_j(u)$. In this reference, there is a fitting illustration on their correlation for different t and u values as well as an explanation for why the output process from such nodes that are parts of a cycle are not Poisson. Thus we need a method that does not use the Poisson property explicitly.

For that we begin by modeling the entire system as a stochastic process. For $i = 1, \ldots, N$, let $X_i(t)$ be the number of customers in node i of an open Jackson network at time t. Let $X(t)$ be a vector that captures a snapshot of the state of the network at time t and is given by $X(t) = [X_1(t), X_2(t), \ldots, X_N(t)]$. It is not hard to see that the N-dimensional stochastic process $\{X(t), t \geq 0\}$ is a CTMC since the interevent times are exponentially distributed. Let $p(x)$ be the steady-state probability that the CTMC is in state $x = (x_1, x_2, \ldots, x_N)$, that is,

$$p(x) = \lim_{t \to \infty} P\{X(t) = (x_1, x_2, \ldots, x_N)\}.$$

To obtain $p(x)$ we consider the balance equations (flow out equals flow in) for $x = (x_1, x_2, \ldots, x_N)$ just like we would do for any CTMC. To make our notation crisp, we write the balance equations in terms of e_i, which is a unit vector with one as the ith element and zeros everywhere else. For example, if $N = 4$, then $e_2 = (0, 1, 0, 0)$. Also for notational convenience we denote $p(x)$ as zero if any $x_j < 0$. Thus, the generic balance equation takes the form

$$p(x) \left[\sum_{i=1}^{N} \lambda_i + \sum_{i=1}^{N} \min(x_i, s_i) \mu_i \right]$$

$$= \sum_{i=1}^{N} p(x - e_i)\lambda_i + \sum_{i=1}^{N} p(x + e_i) r_i \min(x_i + 1, s_i) \mu_i$$

$$+ \sum_{j=1}^{N} \sum_{i=1}^{N} p(x + e_i - e_j) p_{ij} \min(x_i + 1, s_i) \mu_i.$$

To explain this briefly, note that the LHS includes all the transitions out of state x, which include any external arrivals or service completions. Likewise, the RHS includes all the transitions into state x, that is, external arrivals as well as service completions that lead to exiting the network or joining other nodes. If this is not clear, it may be worthwhile for the reader to try an example with a small number of nodes before proceeding further.

It is mathematically intractable to directly solve the balance equations to get $p(x)$ for all x except for special cases. However, since we know that there is a unique solution to the balance equations, if we find a solution then that is *the* solution. In that spirit, consider an acyclic open Jackson network for which from an earlier section we can compute $p(x)$. For the acyclic open Jackson network, node j (such that $1 \le j \le N$) would be an $M/M/s_j$ queue with $PP(a_j)$ arrivals, $\exp(\mu_j)$ service and s_j servers (if the stability condition at each node j is satisfied, that is, $a_j < s_j\mu_j$). Hence, it is possible to obtain the steady-state probability of having n customers in node j, which we denote as $\phi_j(n)$. Using the $M/M/s$ queue results in Chapter 2, we have

$$\phi_j(n) = \begin{cases} \frac{1}{n!} \left(\frac{a_j}{\mu_j} \right)^n \phi_j(0) & \text{if } 0 \le n \le s_j - 1 \\ \frac{1}{s_j! \, s_j^{n-s_j}} \left(\frac{a_j}{\mu_j} \right)^n \phi_j(0) & \text{if } n \ge s_j \end{cases} \tag{6.3}$$

where

$$\phi_j(0) = \left[\sum_{n=0}^{s_j-1} \left\{ \frac{1}{n!} (a_j/\mu_j)^n \right\} + \frac{(a_j/\mu_j)^{s_j}}{s_j!} \frac{1}{1 - a_j/(s_j\mu_j)} \right]^{-1}.$$

Since the number of customers in each node in steady state is independent of those in other nodes, we have for the acyclic open Jackson network

$$p(x) = \phi_1(x_1)\phi_2(x_2) \dots \phi_N(x_N).$$

This structure is what is called product form.

Since we have a solution for a special open Jackson network, as a first option we could try to see if such a $p(x)$ would be a solution to the generic open Jackson network (not just acyclic). If the $p(x)$ given here does not satisfy the balance equations, then we will try some other $p(x)$. However, if it does satisfy the balance equations, then we are done since there is a unique solution to the balance equations. Thus for a generic open Jackson network, we consider the following as a possible solution to the balance equations:

$$p(x) = \phi_1(x_1)\phi_2(x_2)\ldots\phi_N(x_N)$$

where $\phi_j(n)$ is given by Equation 6.3. We need to verify that this $p(x)$ satisfies the balance equation

$$p(x)\left[\sum_{i=1}^{N}\lambda_i + \sum_{i=1}^{N}\min(x_i,s_i)\mu_i\right]$$

$$= \sum_{i=1}^{N}p(x-e_i)\lambda_i + \sum_{i=1}^{N}p(x+e_i)r_i\min(x_i+1,s_i)\mu_i$$

$$+ \sum_{j=1}^{N}\sum_{i=1}^{N}p(x+e_i-e_j)p_{ij}\min(x_i+1,s_i)\mu_i.$$

For that, notice first of all if $p(x) = \phi_1(x_1)\phi_2(x_2)\ldots\phi_N(x_N)$, then

$$\frac{p(x)}{p(x\pm e_i)} = \frac{\phi_i(x_i)}{\phi_i(x_i\pm 1)},$$

$$\frac{p(x)}{p(x+e_i-e_j)} = \frac{\phi_i(x_i)\phi_j(x_j)}{\phi_i(x_i+1)\phi_j(x_j-1)},$$

for all i and j. In addition, from the definition of $\phi_i(n)$ in Equation 6.3, we can obtain the following:

$$a_i\phi_i(x_i-1) = \min(x_i,s_i)\mu_i\phi_i(x_i)$$

$$a_i\phi_i(x_i) = \min(x_i+1,s_i)\mu_i\phi_i(x_i+1)$$

with the additional condition that $\phi_i(n) = 0$, if $n < 0$.

Using these equations in the balance equation (by dividing it by $p(x)$), we get

$$\sum_{i=1}^{N} \lambda_i + \sum_{i=1}^{N} \min(x_i, s_i)\mu_i$$

$$= \sum_{i=1}^{N} \frac{\min(x_i, s_i)\mu_i\lambda_i}{a_i} + \sum_{i=1}^{N} a_i r_i + \sum_{j=1}^{N}\sum_{i=1}^{N} \frac{a_i}{a_j}\min(x_j, s_j)\mu_j p_{ij}$$

$$= \sum_{i=1}^{N} \frac{\min(x_i, s_i)\mu_i\lambda_i}{a_i} + \sum_{i=1}^{N} a_i r_i + \sum_{j=1}^{N} \frac{\min(x_j, s_j)\mu_j}{a_j} \sum_{i=1}^{N} a_i p_{ij}$$

$$= \sum_{i=1}^{N} \frac{\min(x_i, s_i)\mu_i\lambda_i}{a_i} + \sum_{i=1}^{N} a_i r_i + \sum_{j=1}^{N} \frac{\min(x_j, s_j)\mu_j}{a_j}(a_j - \lambda_j)$$

$$= \sum_{i=1}^{N} a_i r_i + \sum_{j=1}^{N} \min(x_j, s_j)\mu_j$$

where the third equation can be derived using $\sum_{i=1}^{N} a_i p_{ij} = a_j - \lambda_j$, which is directly from Equation 6.1. Since the other two terms cancel, $p(x) = \phi_1(x_1)\phi_2(x_2) \dots \phi_N(x_N)$ is the solution to the balance equation if

$$\sum_{i=1}^{N} \lambda_i = \sum_{i=1}^{N} a_i r_i.$$

This equation is true because (using the notation e as a column vector of ones) from Equation 6.2 we have

$$\lambda e = a(I - P)e,$$

$$\Rightarrow \sum_{i=1}^{N} \lambda_i = \sum_{i=1}^{N} a_i \left(1 - \sum_{j=1}^{N} p_{ij}\right),$$

$$\Rightarrow \sum_{i=1}^{N} \lambda_i = \sum_{i=1}^{N} a_i r_i.$$

Thus, $p(x) = \phi_1(x_1)\phi_2(x_2) \dots \phi_N(x_N)$ satisfies the balance equations for a generic open Jackson network. In other words, the steady-state joint probability distribution of having x_1 in node 1, x_2 in node 2,..., x_N in node N is equal to the $p(x)$, which is the product of the $\phi_j(x_j)$ values for all j. Hence, this result is known as *product form*. Now to get the marginal distribution that queue j has x_j customers in steady state for some j such that $1 \leq j \leq N$, all we have to do is sum over all x keeping x_j a constant. Thus the marginal probability that node j has x_j in steady state is given by $\phi_j(x_j)$. Notice that the joint probability is the product of the marginals. From a practical standpoint, this is extremely convenient because we can model each node j as though it is an $M/M/s_j$ queue (although in reality it may not be) with $PP(a_j)$ arrivals and $\exp(\mu_j)$ service times. Then we can obtain $\phi_j(x_j)$ and then get $p(x)$. Also, the steady-state expected number in node j, L_j, can be computed using the $M/M/s_j$ results as well. Then L can be computed by adding over all j from 1 to N. Similarly, it is possible to obtain performance measures at node j such average waiting time (W_j), time in queue not including service (W_{jq}), and number in queue not including service (L_{jq}), using the single-station $M/M/s$ queue analysis in Chapter 2. However, while computing something like sojourn time distribution (across a single node or the network), one has to be more careful. In the next remark we illustrate some of the issues that have been nicely described in Disney and Kiessler [26].

Remark 13

Consider an open Jackson network (with cycles) that are stationary (i.e., in steady state) at time 0. For this network the following results hold:

1. Although the marginal distribution of the steady-state number in node j behaves similar to that of an $M/M/s_j$ queue, the arrival process to node j may not be Poisson. However, the arrival point probabilities can still be computed using the moving units see time average (MUSTA) property in Serfozo [96]. Thus, it may be possible to derive the sojourn time distribution across the network.

2. At a given instant of time, the number of customers in each queue is independent of other queues. However, the number of customers in a particular queue at a given instant may not be independent of another queue at a different instant. Thus, while computing sojourn times across a network, since the time of entry into different nodes occur at different instances, the independence assumption may not be valid; however, the MUSTA property still holds. ∎

6.2.3 Examples

In this section, we present four examples to illustrate the approach to obtain performance measures in open Jackson networks, discuss design issues as well as point out some paradoxical situations. We present these examples as problems.

Problem 53

Bay-Gull Bagels is a bagel store in downtown College Taste-on. See Figure 6.4 for a schematic representation of the store as well as numerical values used for arrivals, service, as well as routing probabilities. Assume all arrival processes are Poisson and service times are exponential. Average arrival rates and average service times for each server are given in the figure. Assume there are infinite servers whenever a station says self-service. Also assume all queues have infinite capacity. Model the system as an open Jackson network and obtain the average number of customers in each of the five stations. Then state how many customers are in the system on an average.

Solution

The system can be modeled as an open Jackson network with $N = 5$ nodes or stations. Let the set of nodes be $\{B, S, D, C, E\}$ (as opposed to numbering them from 1 to 5) denoting the five stations: bagels, smoothies, drinks, cashier, and eat-in. Note that external arrival processes are Poisson and they occur at stations B, S, and D. The service times are exponentially distributed. Note the cycle D–C–E–D. Thus we have an open Jackson network with a cycle. There are three servers at node B, two servers at node S, ∞ servers at node

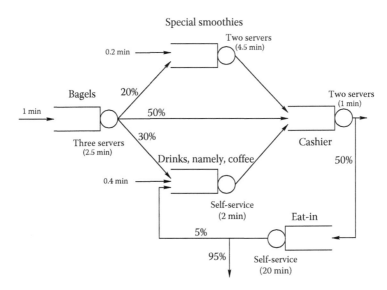

FIGURE 6.4
Schematic of Bay-Gull Bagels.

D, two servers at node *C*, and ∞ servers at node *E*. The external arrival rate vector $\lambda = [\lambda_B \, \lambda_S \, \lambda_D \, \lambda_C \, \lambda_E]$ is

$$\lambda = [1 \; 0.2 \; 0.4 \; 0 \; 0].$$

We assume that after getting served at the bagel queue, each customer chooses node *S*, *C*, and *D* with probabilities $p_{BS} = 0.2$, $p_{BC} = 0.5$, and $p_{BD} = 0.3$, respectively. Likewise, after getting served at *C*, customers go to node *E* or exit the system with equal probability, that is, $p_{CE} = r_C = 0.5$. Similarly, after service at node *E*, with probability $p_{ED} = 0.05$ enter the drinks node or exit the network with probability $r_E = 0.95$. Also, $p_{SC} = 1$ and $p_{DC} = 1$. Thus the routing probabilities in the order $\{B, S, D, C, E\}$ are

$$P = \begin{bmatrix} 0 & 0.2 & 0.3 & 0.5 & 0 \\ 0 & 0 & 0 & 1 & 0 \\ 0 & 0 & 0 & 1 & 0 \\ 0 & 0 & 0 & 0 & 0.5 \\ 0 & 0 & 0.05 & 0 & 0 \end{bmatrix}.$$

The effective arrival rate vector $a = (a_B \, a_S \, a_D \, a_C \, a_E)$ is given by

$$a = \lambda(I - P)^{-1} = [1 \; 0.4 \; 0.741 \; 1.641 \; 0.8205],$$

where *I* is the 5×5 identity matrix. Notice that stability condition is satisfied in each of the five queues.

Having modeled the system as an open Jackson network, next we obtain the average number of customers in each of the five stations. Recall the number of customers in any node in steady state is identical to that of an *M/M/s* queue with the same arrival rate, service rate, and number of servers. However, the actual arrivals may not be Poisson. With that we obtain L_i, the average number of customers in steady state for all *i* by analyzing each node as an *M/M/s* queue (using Chapter 2 results), as follows:

1. *Bagel queue*: Arrival rate is $a_B = 1$, service times are exp(1/2.5), and there are three servers. Using the results for *M/M/3*, we get $L_B = 6.01124$, the average number of customers in the Bagel station.

2. *Smoothies queue*: Arrival rate is $a_S = 0.4$, service times are exp(1/4.5), and there are two servers. Using the results for *M/M/2* we get $L_S = 9.47368$, the average number of customers in the Smoothies station.

3. *Drinks queue*: Arrival rate is $a_D = 0.741$, service times are exp(1/2), and there are ∞ servers. Using the results for *M/M/∞* we get $L_D = 1.482$, the average number of customers in the Drinks station.

4. *Cashier queue*: Arrival rate is $a_C = 1.641$, service times are exp(1), and there are two servers. Using the results for $M/M/2$ we get $L_C = 5.02173$, the average number of customers in the Cashier station.

5. *Eat-in queue*: Arrival rate is $a_E = 0.8205$, service times are exp(1/20), and there are ∞ servers. Using the results for $M/M/\infty$ we get $L_E = 16.41$, the average number of customers in the Eat-in station.

Thus the average number of customers inside Bay-Gull Bagels is $L_B + L_S + L_D + L_C + L_E = 38.4$ customers. ∎

Say in the previous problem, we were to find out the probability there are 3 customers in node B, 2 in S, 1 in D, 4 in C, and 20 in E. Then that is equal to the product $\phi_B(3)\phi_S(2)\phi_D(1)\phi_C(4)\phi_E(20)$ and can be computed using Equation 6.3. Note that $\phi_j(n)$ computation is indeed equal to the probability there are n in a steady-state $M/M/s_j$ queue with $PP(a_j)$ arrivals and exp(μ_j) service. Having described that, we next present an example that discusses some design issues by comparing various ways of setting up systems with multiple stages and servers.

Problem 54

Consider a system into which customers arrive according to a Poisson process with parameter λ. Each customer needs N stages of service and each stage takes exp(μ) amount of time. There are N servers in the system and N buffers for customers to wait. Assume that the buffers have infinite capacity and $\lambda < \mu$. There are two design alternatives to consider:

1. *Serial system*: Each set of buffer and server is placed serially in a tandem fashion as described in Figure 6.5. Each node corresponds to a different stage of service. Customers arrive at the first node according to $PP(\lambda)$. There is a single server that takes exp(μ) time to serve after which the customer goes to the second node. At the second node there is a single server that takes exp(μ) time. This continues until the Nth node and then the customer exits.

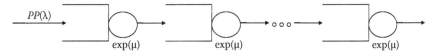

FIGURE 6.5
System of N buffers and servers in series.

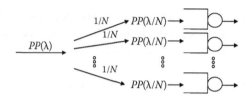

FIGURE 6.6
System of N buffers and servers in parallel.

2. *Parallel system*: Each set of buffer and server is placed in a parallel fashion as described in Figure 6.6. Each node corresponds to a single server that performs all N stages of service. Customers arrive according to $PP(\lambda)$ and they are split with equal probability of $1/N$ to one of the N nodes. There is a single server at each of the parallel nodes that takes a sum of N IID times each equal to $\exp(\mu)$ to serve each customer by performing all N stages. After completing service at a node, the customer exits the system.

By comparing the mean sojourn time for an arbitrary customer in steady state in both systems, determine whether the serial or parallel system is better.

Solution

Note that although the figures appear to be somewhat different, the resources of the system and service needs of customers are identical. In other words, in both systems we have N single-server queues each with an infinite buffer. Also, in both systems the customers experience N stages of service, each taking an $\exp(\mu)$ time. We now analyze the system in the same order they were presented.

1. The serial system is an open Jackson network (in fact an acyclic network) with N single-server nodes back to back. Service time at each node is $\exp(\mu)$. Such a serial system is called a pipeline system in the computer science literature and tandem network in the manufacturing literature. Clearly, each queue is an $M/M/1$ queue with $PP(\lambda)$ arrivals and $\exp(\mu)$ service. The average time in each node is thus $1/(\mu - \lambda)$. Thus, the mean sojourn time in the system is

$$W_{series} = \frac{N}{\mu - \lambda}.$$

2. The parallel system is a set of N single-server $M/G/1$ queues in parallel. The arrival rate into each of the $M/G/1$ queues is λ/N. Service time at each node is the sum of N $\exp(\mu)$ random variables. Therefore, the service times are according to an Erlang (or gamma) distribution with mean N/μ and variance N/μ^2. Since all

queues are stochastically identical, the mean sojourn time on any of them would be equal to the others'. Using the Pollaczek–Khintchine formula (Equation 4.6), we get the mean sojourn time in the system as

$$W_{parallel} = \frac{N}{\mu} + \frac{\lambda(N+1)}{2\mu(\mu-\lambda)} = \frac{2N\mu - N\lambda + \lambda}{2\mu(\mu-\lambda)}.$$

Comparing W_{series} and $W_{parallel}$, if $N > 1$ then $W_{series} > W_{parallel}$; however, if $N = 1$ then $W_{series} = W_{parallel}$. Thus the parallel system is better. ∎

Before proceeding onto the next example, it is worthwhile to comment on some practical considerations regarding the conclusion of Problem 54. In many situations one is faced with this type of decision. For example, consider fast restaurants such as those making submarine sandwiches or burritos. The process of making a customized sandwich or burrito can be broken down into N approximately equally time-consuming tasks, if there are N servers. In many of these restaurants, the servers are placed in a serial fashion and the customer walks through explaining what they want. However, there are some restaurants that adopt the parallel service mechanism where a single server walks the customer through all stages of the sandwich- or burrito-making process. We illustrate these cases in exercise problems at the end of this chapter. A question to ask is why are some restaurants going for the serial option when the parallel one appears better (at least in the exponential service time case). In most of these restaurants, space and resources are constraints rendering it impractical to do a parallel operation without servers being *blocked*. However, when space and resources are not constraining, some restaurants adopt the parallel system, but typically instead of N queues in parallel, they use a single queue. For the Problem 54 it can be shown that instead of using N parallel $M/G/1$ queues, if one were to use a single $M/G/N$ queue with the service times being according to an Erlang (or gamma) distribution with mean N/μ and variance N/μ^2, then the mean sojourn time can be reduced further.

Therefore, if there are no constraints in terms of space or resources, then it appears like we should pool all our servers and make them perform all tasks of an operation instead of pipelining our servers and ask them to perform a single task. Then why do we see assembly lines? This brings us to the second practical issue in terms of designing a multiserver system. Although from a customer's standpoint there are N tasks each taking a random $\exp(\mu)$ amount of time, the server might actually take a lot longer than the sum of those N random times. That is due to several reasons. First of all, the servers have to be trained in all the tasks, which is unrealistic. So typically a server would take a lot longer to perform all the tasks especially in a complex manufacturing environment. Second, it would either be too time consuming from a material handling standpoint as well (or too expensive if the time consumption is reduced). Therefore, in practice, sometimes a combination of serial and parallel tasking is used. Usually servers are trained in two to four tasks that they perform together. This not only improves the system performance but also reduces monotonous conditions. Having described that, the next two examples are paradoxes that further help understand issues in queueing networks.

Problem 55

Braess' paradox: In a network, does adding extra capacity always improve the system in terms of performance? Although it appears intuitive, adding extra capacity to a network when the moving entities selfishly choose their routes can in some cases worsen the overall performance! Illustrate this using an example.

Solution

Consider a network with nodes A, B, C, and D. There are directed arcs from A to B, B to D, A to C, and C to D. Customers arrive into node A according to a Poisson process with mean rate 2λ. The customers need to reach node D and they have two paths, one through B and the other through C, as shown in Figure 6.7. Along the arc from A to B there is a single-server queue with exponentially distributed service times (and mean $1/\mu$). Likewise, there is an identical queue along the arc from C to D. In addition, it takes a deterministic time of 2 units to traverse arcs AC and BD. Assume that

$$\mu > \lambda + 1.$$

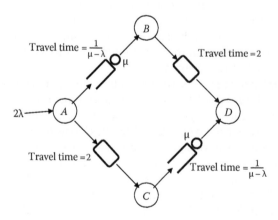

FIGURE 6.7
Travel times along arcs in equilibrium.

In equilibrium, each arriving customer to node A would select either of the two paths with equal probability. Thus the average travel time from A to D in equilibrium is

$$2 + \frac{1}{\mu - \lambda}.$$

Note that the constant time of 2 units can be modeled as an $M/G/\infty$ queue with deterministic service time of 2 units, then the output from that queue is still $PP(\lambda)$. Another way of seeing that would be that the departure process after spending 2 time units would be identical to the entering process, just shifted by 2 time units. Hence, it would have to be Poisson. Thus the time across either path is $2 + 1/(\mu - \lambda)$, where the second term is the sojourn time of an $M/M/1$ queue with $PP(\lambda)$ arrivals and $\exp(\mu)$ service.

Now a new path from B to C is constructed along which it would take a deterministic time of 1 unit to traverse. For the first customer that arrives into this new system described in Figure 6.8, this would be a shortcut because the new expected travel time would be $1 + 2/(\mu - \lambda)$, which is smaller than the old expected travel time given earlier under the assumption $\mu > \lambda + 1$. Soon, the customers would selfishly choose their routes so that in equilibrium, all three paths $A - B - D$, $A - C - D$, and $A - B - C - D$ have identical mean travel times. Actually the equilibrium splits would not be necessary to calculate, instead notice that each of the three routes would take 3 time units to traverse on average (this is the only way the three paths would have identical travel times). But the old travel time before the new capacity was added, $2 + 1/(\mu - \lambda)$, is actually less than 3 units under the assumption $\mu > \lambda + 1$. Thus adding extra capacity has actually worsened the average travel times! ∎

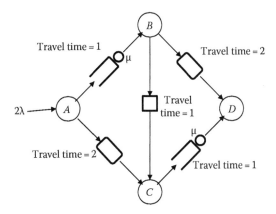

FIGURE 6.8
New travel times along arcs in equilibrium.

Problem 56

Can the computation of waiting times in a queueing system depend on the method? Consider a stable queue that gets customer arrivals externally according to a Poisson process with mean rate λ. There is a single server and infinite waiting room. The service times are exponentially distributed with mean $1/\mu$. At the end of service each customer exits the system with probability p and reenters the queue with probability $(1 - p)$. The system is depicted in Figure 6.9, for now ignore A, B, C, and D. We consider two models:

1. If the system is modeled as a birth and death process with birth rates λ and death rates $p\mu$, then $L = \lambda/(p\mu - \lambda)$ and $W = L/\lambda = 1/(p\mu - \lambda)$.
2. If the system is modeled as a Jackson network with 1 node and effective arrival rate λ/p and service rate μ, then $L = (\lambda/p)/(\mu - \lambda/p)$ and $W = L/(\lambda/p) = p/(p\mu - \lambda)$.

Clearly, the two methods give the same L but the W values are different! Explain.

Solution

Although this appears to be a paradox, that is really not the case. Let us revisit Figure 6.9 but now let us consider A, B, C, and D. The W from the first method (birth and death model) is measured between A and D, which is the total time spent by a customer in the system (going through one or more rounds of service). The W from the second method (Jackson network model) is measured between B and C, which is the time spent by a customer from the time he or she entered the queue until one round of service is completed. Note that the customer does a geometric number of such services (with mean $1/p$). Therefore, the total time spent on average would indeed be the same in either methods if we used the same points of reference, that is, A and D.   ∎

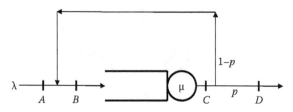

FIGURE 6.9
Points of reference.

The exercises at the end of the chapter describe several more examples of open Jackson networks. Next, we consider closed Jackson networks.

6.3 Closed Jackson Networks

Closed-queueing networks are networks where the number of customers at all times stays constant (say C). These C customers move from one node to another. There are no external arrivals to the system and no departures from the system. The network structure for closed-queueing networks is necessarily cyclic because customers cycle through the network visiting node by node and never exit the system. At every node there is a single queue with large enough capacity that no customers are rejected or dropped. There could be one or more servers at each node. Before describing further modeling details that are conducive for analysis, it may be worthwhile motivating the need to study such systems. Closed-queueing networks are popular in population studies, multiprogrammed computer systems, window flow control, Kanban, amusement parks, etc. The key feature is that the number of customers is kept a constant at least approximately. For example, in amusement parks, most customers usually arrive when the park opens and stay almost till the end of the day going from one ride to another. If one were to design, control, or schedule events in an amusement park, closed-queueing network analysis might be appropriate. Another example of such systems is if a new customer enters the network as soon as an existing customer leaves (a popular scheme in just-in-time manufacturing). Thus, although we do not have the same C customers in the system, the total is still a constant and can be analyzed suitably.

To analyze closed-queueing networks, we make some assumptions similar to those in the open-queueing networks. These assumptions would result in what is popularly known as closed Jackson network (mainly because the requirements are similar to those of the open Jackson network). However, some researchers also call these as Gordon–Newell networks because of a paper about 10 years after Jackson [56] by Gordon and Newell [47] that explicitly considers all the conditions mentioned here. The attributes of a closed Jackson network are as follows:

1. The network has N service stations and a total of C customers.

2. When a customer completes service at node i, the customer joins node j with probability p_{ij}. Note that customers do not leave the system from any node. We assume that the closed-queueing network is irreducible (i.e., there is a path from every node to every other node in the network).

3. The service rate at node i when there are n customers in that node is $\mu_i(n)$ with $\mu_i(0) = 0$ and $\mu_i(n) > 0$ for $1 \leq n \leq C$. The service times are exponentially distributed.

It may be worthwhile to elaborate on the last attribute. Note that we do not mention about the number of servers in node i. For example, if there are s_i servers in node i and service rate for each server is μ, then $\mu_i(n) = \min(n, s_i)\mu$ since the RHS is the aggregate rate at which service is completed. Now the representation is versatile enough to also allow state-dependent service. In other words, we could have one or more servers and the service rate depends on the number in the system. We will see subsequently that such a consideration can be made for the open Jackson network as well.

6.3.1 Product-Form Solution

Similar to the open Jackson network case, for the closed Jackson networks too we will develop a joint distribution for the steady-state number in each node. In fact, we will aim to show that here too the joint distribution has a *product form*. For that we begin by modeling the entire system as a stochastic process. For $i = 1, \ldots, N$, let $X_i(t)$ be the number of customers in node i of the closed Jackson network at time t. Let $X(t)$ be a vector that captures the state of the network at time t and is given by $X(t) = [X_1(t), X_2(t), \ldots, X_N(t)]$ with the condition that $X_1(t) + X_2(t) + \cdots + X_N(t) = C$ for all t. Using the definition of $X(t)$, the N-dimensional stochastic process $\{X(t), t \geq 0\}$ can be modeled as a CTMC since the interevent times are exponentially distributed. Let $p(x)$ be the steady-state probability that the CTMC is in state $x = (x_1, x_2, \ldots, x_N)$, that is,

$$p(x) = \lim_{t \to \infty} P\{X(t) = (x_1, x_2, \ldots, x_N)\}.$$

Of course, we require that $x_1 + x_2 + \cdots + x_N = C$, otherwise $p(x)$ would be zero. To obtain $p(x)$ we consider the balance equations for $x = (x_1, x_2, \ldots, x_N)$ just like we did for the open Jackson network. We write down the balance equations in terms of e_i, which is a unit vector with one as the ith element and zeros everywhere else. In addition, we denote $p(x)$ as zero if any $x_j < 0$. Thus the generic balance equation takes the form

$$p(x) \sum_{i=1}^{N} \mu_i(x_i) = \sum_{j=1}^{N} \sum_{i=1}^{N} p(x + e_i - e_j)p_{ij}\mu_i(x_i + 1).$$

In this balance equation, the LHS describes the total rate of all the transitions out of state x that includes all possible service completions. Likewise, the RHS includes all the transitions into state x, that is, service completions from various states that result in x.

Except for small C and N, solving the balance equations directly is difficult. However, like we saw in the open Jackson network case, here too we will guess a $p(x)$ solution and check if it satisfies the balance equations. If it does, then we are done since there is only one solution to the balance equations. As an initial guess for $p(x)$, we try the open-queueing network result itself. For that, recall from Equation 6.2 that $a(I - P) = \lambda$; however, λ is a vector of zeros since there are no external arrivals. Hence, we define a as the solution to $a(I - P) = 0$, in other words a solution to

$$a = aP.$$

This equation has a solution if the closed-queueing network is irreducible (which is an assumption we made earlier). Interestingly, the P matrix is similar to that of a DTMC with N states and a is similar to its stationary probabilities. However, we do not necessarily need a to be a probability vector that sums to one. Technically it is the left eigen vector of P that corresponds to eigen value of 1. Let a_j be the jth element of a. Understandably, being an eigen vector, a is not unique. However for all a, we have a_i/a_j a constant for every pair of i and j, and for that reason the a values are also called visit ratios. In that light, following $p(x)$ in the open Jackson network case, we define $\phi_j(n)$ very similar to that in Equation 6.3 as follows: $\phi_j(0) = 1$ and

$$\phi_j(n) = \prod_{k=1}^{n} \left(\frac{a_j}{\mu_j(k)} \right) \quad \text{for } n \geq 1. \tag{6.4}$$

Therefore, as an initial guess we try

$$p(x) = G(C)\phi_1(x_1)\phi_2(x_2)\ldots\phi_N(x_N) = G(C)\prod_{i=1}^{N} \phi_i(x_i),$$

where the normalizing constant $G(C)$ is chosen such that

$$\sum_{x:x_1+x_2+\cdots+x_N=C} p(x) = 1.$$

Next, we need to verify if the $p(x)$ here satisfies the balance equation

$$p(x)\sum_{i=1}^{N} \mu_i(x_i) = \sum_{j=1}^{N}\sum_{i=1}^{N} p(x + e_i - e_j)p_{ij}\mu_i(x_i + 1).$$

For that, first of all if $p(x) = G(C)\phi_1(x_1)\phi_2(x_2)\ldots\phi_N(x_N)$, then

$$\frac{p(x)}{p(x + e_i - e_j)} = \frac{\phi_i(x_i)\phi_j(x_j)}{\phi_i(x_i + 1)\phi_j(x_j - 1)},$$

for all i and j. In addition, from the definition of $\phi_i(n)$ in Equation 6.4, we can obtain the following:

$$a_i\phi_i(x_i - 1) = \mu_i(x_i)\phi_i(x_i)$$

$$a_i\phi_i(x_i) = \mu_i(x_i + 1)\phi_i(x_i + 1)$$

with the additional condition that $\phi_i(n) = 0$ if $n < 0$.

Using these equations in the balance equation (by dividing it by $p(x)$), we get

$$\sum_{i=1}^{N}\mu_i(x_i) = \sum_{j=1}^{N}\sum_{i=1}^{N}\frac{p(x + e_i - e_j)}{p(x)}\mu_i(x_i + 1)p_{ij}$$

$$= \sum_{j=1}^{N}\sum_{i=1}^{N}\frac{\phi_i(x_i + 1)\phi_j(x_j - 1)}{\phi_i(x_i)\phi_j(x_j)}\mu_i(x_i + 1)p_{ij}$$

$$= \sum_{j=1}^{N}\sum_{i=1}^{N}\frac{a_i}{\mu_i(x_i + 1)}\frac{\mu_j(x_j)}{a_j}\mu_i(x_i + 1)p_{ij}$$

$$= \sum_{j=1}^{N}\sum_{i=1}^{N}a_i\frac{\mu_j(x_j)}{a_j}p_{ij}$$

$$= \sum_{j=1}^{N}\frac{\mu_j(x_j)}{a_j}\sum_{i=1}^{N}a_ip_{ij}$$

$$= \sum_{j=1}^{N}\frac{\mu_j(x_j)}{a_j}a_j = \sum_{j=1}^{N}\mu_j(x_j)$$

where the penultimate equation can be derived using $\sum_{i=1}^{N}a_ip_{ij} = a_j$, which is directly from $a = aP$. Thus, $p(x) = G(C)\phi_1(x_1)\phi_2(x_2)\ldots\phi_N(x_N)$ satisfies the balance equations for a closed Jackson network.

In other words, the steady-state joint probability distribution of having x_1 in node 1, x_2 in node 2, $\ldots$, x_N in node N is equal to $p(x)$, which is the product of the $\phi_j(x_j)$ values for all j times a normalizing constant $G(C)$.

Hence, this result is also a *product form*. Note that for this result, similar to the other product-form cases that we will consider subsequently, the difficulty arises in computing the normalizing constant $G(C)$. In general, it is not computationally trivial. However, once $G(C)$ is obtained one can compute the marginal distribution that queue j has x_j customers in steady state for some j such that $1 \leq j \leq N$, all we have to do is sum over all x keeping x_j a constant. We proceed by first explaining a simple example.

Problem 57

Consider a closed Jackson network with three nodes and five customers, that is, $N=3$ and $C=5$. The network structure is depicted in Figure 6.10. Essentially all five customers behave in the following fashion: upon completing service at node 1, a customer rejoins node 1 with probability 0.5, or joins node 2 with probability 0.1, or joins node 3 with probability 0.4; upon completing service in node 2 or 3, a customer always joins node 1. Node 1 has a single server that serves at rate i if there are i customers at the node. Node 2 has two servers each with service rate 1. Node 3 has one server with service rate 2. Determine the joint as well as marginal probability distribution of the number of customers at each node in steady state. Also compute the average number in each node as well as the network in steady state.

Solution

Although it is not explicitly stated, the service times are exponentially distributed (since it is a closed Jackson network). For such a system, to compute the joint distribution of the steady-state number in each node we first solve for vector a in

$$a = aP$$

where P is the routing probability matrix that can be obtained from the problem statement as

$$P = \begin{bmatrix} 0.5 & 0.1 & 0.4 \\ 1 & 0 & 0 \\ 1 & 0 & 0 \end{bmatrix}.$$

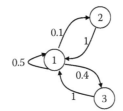

FIGURE 6.10
Closed Jackson network with $C=5$ customers.

Note that the network is irreducible. Therefore, solving for a we get

$$a = [a_1 \ a_2 \ a_3] = \begin{bmatrix} \dfrac{10}{15} & \dfrac{1}{15} & \dfrac{4}{15} \end{bmatrix}.$$

To obtain $\phi_j(n)$ described in Equation 6.4, we need expressions for $\mu_j(n)$. From the preceding problem description, we have

$$[\mu_1(1) \ \mu_1(2) \ \mu_1(3) \ \mu_1(4) \ \mu_1(5)] = [1 \ 2 \ 3 \ 4 \ 5]$$

since the single server at node 1 serves at rate n when there are n customers. Thus, $\mu_1(n) = n$ for all n, which is the service rate when there are n customers in node 1. Likewise, since node 2 has two servers each serving at rate 1, if there is only 1 customer in node 2, the service rate is 1; however, if there are 2 or more customers in that node, the net service rate at the node (which is the rate at which customers exit that node) is 2. Hence we have

$$[\mu_2(1) \ \mu_2(2) \ \mu_2(3) \ \mu_2(4) \ \mu_2(5)] = [1 \ 2 \ 2 \ 2 \ 2].$$

Since there is a single server serving at rate 2 in node 3, the service rate vector when there are n customers in node 3 is given by

$$[\mu_3(1) \ \mu_3(2) \ \mu_3(3) \ \mu_3(4) \ \mu_3(5)] = [2 \ 2 \ 2 \ 2 \ 2].$$

Using the preceding values of a_j and $\mu_j(n)$ for $j = 1, 2, 3$ and $n = 1, 2, 3, 4, 5$, we can compute $\phi_j(n)$ from Equation 6.4 as

$$\phi_j(n) = \prod_{k=1}^{n} \left(\frac{a_j}{\mu_j(k)} \right).$$

Also, for $j = 1, 2, 3$ we have $\phi_j(0) = 1$. Thereby for all combinations of x_1, x_2, and x_3 such that $x_1 + x_2 + x_3 = 5$ we can compute

$$q(x_1, x_2, x_3) = \phi_1(x_1) \phi_2(x_2) \phi_3(x_3).$$

Thus we can obtain the normalizing constant $G(C)$ using

$$\frac{1}{G(C)} = \sum_{x_1, x_2, x_3 : x_1 + x_2 + x_3 = 5} q(x_1, x_2, x_3).$$

Using that we have for all combinations of x_1, x_2, and x_3 such that $x_1 + x_2 + x_3 = 5$, the joint probability distribution

$$p(x_1, x_2, x_3) = G(C) q(x_1, x_2, x_3) = G(C) \phi_1(x_1) \phi_2(x_2) \phi_3(x_3).$$

TABLE 6.2

Example Joint Probability Distribution

x_1	x_2	x_3	$p(x_1, x_2, x_3)$	x_1	x_2	x_3	$p(x_1, x_2, x_3)$	x_1	x_2	x_3	$p(x_1, x_2, x_3)$
0	0	5	0.0077	0	1	4	0.0039	0	2	3	0.0010
0	3	2	0.0002	0	4	1	0.0001	0	5	0	0.00002
1	0	4	0.0386	1	1	3	0.0193	1	2	2	0.0048
1	3	1	0.0012	1	4	0	0.0003	2	0	3	0.0964
2	1	2	0.0482	2	2	1	0.0121	2	3	0	0.0030
3	0	2	0.1607	3	1	1	0.0803	3	2	0	0.0201
4	0	1	0.2009	4	1	0	0.1004	5	0	0	0.2009

It may be worthwhile to write a computer program to verify the following sample numerical results: $q(2,1,2) = 0.00026337$ and $q(5,0,0) = 0.0011$. Due to space restrictions other q values are not reported. However, note that $G(C) = 183.038$ and the joint probability $p(x_1, x_2, x_3)$ is described in Table 6.2.

Having obtained the joint probability distribution, next we obtain the marginal probabilities. Let $p_i(n)$ be marginal probability that node i has n customers (for $i = 1, 2, 3$ and $n = 0, 1, 2, 3, 4, 5$). Clearly,

$$p_1(x_1) = \sum_{x_2=0}^{5} \sum_{x_3=0}^{5} p(x_1, x_2, x_3),$$

$$p_2(x_2) = \sum_{x_1=0}^{5} \sum_{x_3=0}^{5} p(x_1, x_2, x_3),$$

$$p_3(x_3) = \sum_{x_1=0}^{5} \sum_{x_2=0}^{5} p(x_1, x_2, x_3).$$

For the preceding numerical values, the marginal probability vectors p_i (for $i = 1, 2, 3$) are

$$p_1 = [0.0129\ 0.0642\ 0.1597\ 0.2611\ 0.3013\ 0.2009]$$

$$p_2 = [0.7051\ 0.2521\ 0.0379\ 0.0045\ 0.0004\ 0.000015]$$

$$p_3 = [0.3247\ 0.2945\ 0.2140\ 0.1167\ 0.0424\ 0.0077]$$

where $p_i = [p_i(0)\ p_i(1)\ p_i(2)\ p_i(3)\ p_i(4)\ p_i(5)]$. Let L_i be the average number of customers in node i in steady state for $i = 1, 2, 3$. We can compute L_i as $\sum_n n p_i(n)$. Hence, we have $L_1 = 3.3764$, $L_2 = 0.3429$, and $L_3 = 1.2807$. The total number in the system in steady state is C, which is 5. It can also be obtained as $L_1 + L_2 + L_3$.   ∎

Next, we describe the probability that an arriving customer to node i would see j others in that queue. Subsequently, we will consider the single-server case where we can avoid explicitly computing $G(C)$. Then we will wrap up closed Jackson networks with an example in analysis of computer systems.

6.3.2 Arrivals See Time Averages (ASTA)

In Chapter 4 we discussed the relationship between the arrival point probabilities (i.e., the distribution of the system state for an arrival in steady state) and the corresponding long-run time-averaged fractions. In particular, we claimed for the $M/G/1$ queue using PASTA that $\pi_j = p_j$ for all j, where π_j is the probability that an arrival in steady state would see j others in the system versus p_j is the long-run fraction of time there are j customers in the system. However, for the $G/M/1$ where the arrivals are not necessarily Poisson, we showed that π_j is not equal to p_j but they are related ($p_j \propto \pi_{j-1}$ for all $j \geq 1$). In this section, we seek to obtain a relationship between the arrival point probabilities and the steady-state probabilities for the *closed Jackson network*. Similar to the $G/M/1$ queue, in the closed Jackson network as well, the arrival processes to nodes are not Poisson. So it is natural to not expect the arrival point probabilities and the steady-state probabilities to be equal (although the title of this section—ASTA—might mislead one to believe them to be equal). But would they be related? Let us find out.

Consider a closed Jackson network (with all the usual notation) in steady state. Let $\pi_j(x)$ be the probability that an arriving customer into node j in steady state will see x_k customers in node k for all k so that x is a vector of those x_k values. Note that $x_1 + x_2 + \cdots + x_k = C - 1$ since we are not counting the arriving customer into node j in the state description. Consider an infinitesimal time h units long and e_i an N-dimensional unit vector, that is, a one on the ith element and zero everywhere else. We can show that

$$\pi_j(x) = \lim_{h \to 0} \frac{\sum_{i=1}^N p_{ij}[\mu_i(x_i + 1)h + o(h)]p^C(x + e_i)}{\sum_y \sum_{i=1}^N p_{ij}[\mu_i(y_i + 1)h + o(h)]p^C(y + e_i)}$$

where
  $y = [y_1, y_2, \ldots, y_N]$ such that $y_1 + \cdots + y_N = C - 1$
  $o(h)$ is a set of terms of the order h such that $o(h)/h \to 0$ as $h \to 0$
  $p^C(x + e_i)$ is the usual $p(x + e_i)$ with C used for clarity to denote the total number of customers

We first explain the $\pi_j(x)$ equation.

The denominator of the $\pi_j(x)$ equation essentially is the probability that in an infinitesimal time interval of length h in steady state, an arrival occurs into queue j, although initially it may appear that the summation in the denominator does not capture all the states of the system because there are C customers and y is summed over $C - 1$ customers being in various queues. However, note that by summing over all possible states i where the extra customer could lie we have indeed taken care of all possibilities. In addition, we not only need a service completion in node i but that should also result in the customer moving to node j (with probability p_{ij}). Further, if there are n customers in node i, then with probability $\mu_i(n)h + o(h)$ a departure occurs in a time interval of length h. In other words, the terms inside the summations of the denominator account for the probability that the system state is $y + e_i$, and a service is completed for a customer in node i, and that customer joins node j. This is computed by conditioning first on the state, then on the completion. The numerator in a very similar fashion is the probability that an arrival occurs into queue j during an infinitesimal duration of length h when the state is $x + e_i$. Thus the ratio of the numerator to denominator in $\pi_j(x)$ would result in the probability that the system state is x (not counting the arrival) given an arrival occurrence. Of course, this arrival could in theory have come from any one of the N nodes, and hence we would have to consider all states $x + e_i$ over all i from which a departure would have occurred and then that customer joins node j.

Taking the limit and simplifying the $\pi_j(x)$ equation we get

$$\pi_j(x) = \frac{\sum_{i=1}^{N} p_{ij}\mu_i(x_i + 1)p^C(x + e_i)}{\sum_y \sum_{i=1}^{N} p_{ij}\mu_i(y_i + 1)p^C(y + e_i)}$$

$$= \frac{\sum_{i=1}^{N} p_{ij}G(C)\phi_1(x_1)\phi_2(x_2)\dots\phi_i(x_i + 1)\dots\phi_N(x_N)\mu_i(x_i + 1)}{\sum_y \sum_{i=1}^{N} p_{ij}\mu_i(y_i + 1)G(C)\phi_1(y_1)\phi_2(y_2)\dots\phi_i(y_i + 1)\dots\phi_N(y_N)}$$

$$= \frac{\sum_{i=1}^{N} p_{ij}\phi_1(x_1)\phi_2(x_2)\dots\phi_i(x_i)\dots\phi_N(x_N)a_i}{\sum_y \sum_{i=1}^{N} p_{ij}a_i\phi_1(y_1)\phi_2(y_2)\dots\phi_i(y_i)\dots\phi_N(y_N)}$$

since $\phi_i(n + 1) = \phi_i(n)a_i/\mu_i(n + 1)$, and canceling $G(C)$

$$= \frac{\phi_1(x_1)\phi_2(x_2)\dots\phi_N(x_N)\sum_{i=1}^{N} p_{ij}a_i}{\sum_y \phi_1(y_1)\phi_2(y_2)\dots\phi_N(y_N)\sum_{i=1}^{N} p_{ij}a_i}$$

by collecting the product form outside the summation

$$= \frac{\phi_1(x_1)\phi_2(x_2)\dots\phi_N(x_N)a_j}{\sum_y \phi_1(y_1)\phi_2(y_2)\dots\phi_N(y_N)a_j}$$

$$\sum_{i=1}^{N} p_{ij}a_i = a_j \text{ from } aP = a$$

$$= \frac{\phi_1(x_1)\phi_2(x_2)\dots\phi_N(x_N)}{\sum_y \phi_1(y_1)\phi_2(y_2)\dots\phi_N(y_N)}$$

canceling a_j from numerator and denominator

$$= G(C-1)\phi_1(x_1)\phi_2(x_2)\dots\phi_N(x_N)$$

since $\sum_y \phi_1(y_1)\phi_2(y_2)\dots\phi_N(y_N) = 1/G(C-1)$. Since $G(C-1)\phi_1(x_1)\phi_2(x_2)$ $\dots \phi_N(x_N) = p^{C-1}(x)$, that is, if the closed Jackson network had $C-1$ instead of C customers, then this is the probability that the state of this system in steady state is x. Therefore,

$$\pi_j(x) = p^{C-1}(x).$$

The preceding result is called ASTA because the RHS of the equation is the time-averaged probabilities and the LHS is as seen by arriving customers. In fact, to be more precise, if one were to obtain the distribution of the system state by averaging across those seen by arriving customers (note that arriving customers do not include themselves in the system state), then this is identical to a time-averaged distribution of the system state when there is one less customer. Furthermore, if one were to insert a "dummy" customer in the system to obtain the system state every time this customer enters a node, then it is possible to get the system state distribution without this dummy customer. Sometimes, one is not necessarily interested in the entire vector of states but just that of the entering node. This is the essence of the next remark, sometimes also known as *arrival theorem*.

Remark 14

In a closed Jackson network with C customers, for any n, the probability that there are n customers in node i, as seen at the time of arrival of an arbitrary customer to that node, is equal to the probability that there are n customers at this node with one less job in the network (i.e., $C-1$). ∎

This remark can be immediately derived using $\pi_j(x) = p^{C-1}(x)$ by summing over all x_j such that $j \neq i$. For example, if we were to modify Problem 57 so that there are $C=6$ customers, then the probability that an arriving customer will see two customers in node 3 is 0.214, which can be obtained by considering a network of $C=5$ customers (done in that Problem 57); computing p_3, which is the probability distribution of the number in node 3; and then using the term corresponding to two customers in the system.

Next we present another example of a closed Jackson network, which in the queueing literature is popularly known as finite population queue.

Problem 58

Consider a single-server queue where it takes $\exp(\mu)$ amount of time to serve a customer. Unlike most of the systems in the previous chapters, here we assume that there is a finite population of C customers. Each customer after completion of service returns to the queue after spending $\exp(\lambda)$ time outside. First model the system as a birth and death process, and obtain the steady-state probabilities. Then compute the arrival point probabilities. Subsequently, model the system as a closed Jackson network and compare the corresponding results.

Solution

The finite population single-server queue model is depicted in Figure 6.11. The top of the figure is the queue under consideration and the box in the bottom denotes the idle time before customers return to the queue. There are several applications of such a system. For example, in the client–server model with C clients that submit requests to a server. Once the server sends a response, after a think time the clients send another request and so on. The request service time for the server is $\exp(\mu)$ and different requests contend for the server. The think times for the clients are according to $\exp(\lambda)$. Another example is a bank that has C customers in total. Each customer visits the bank (with a single teller, although the model and analysis can easily be extended to multiple tellers), waits for service, gets served for $\exp(\mu)$ time, and revisits the bank after spending $\exp(\lambda)$ time outside.

To model this system as a birth and death process, let $X(t)$ denote the number of customers in the queue (including any at the server) at time t. When $X(t) = n$, there are $C - n$ customers outside the queue and hence the arrival time would be when the first of those $C - n$ customers enters the queue. Since each customer spends $\exp(\lambda)$ time outside, the arrival rate when $X(t) = n$ is $(C - n)\lambda$. Likewise, if there is a customer at the queue, the service rate is μ. Therefore, we can show the $\{X(t), t \geq 0\}$ process is a birth and death process with birth parameters $\lambda_n = (C - n)\lambda$ for $0 \leq n \leq C - 1$ and

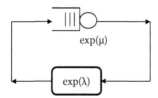

FIGURE 6.11
Finite population queue with C customers.

death parameters $\mu_n = \mu$ for $1 \leq n \leq C$. All other birth and death parameters are zero.

Next, we obtain the steady-state probabilities p_j, which is the probability that there are j in the queue (including any at the server) in steady state. For that we consider the birth and death process and write down the following balance equations using arc cuts: $C\lambda p_0 = \mu p_1$, $(C-1)\lambda p_1 = \mu p_2$, and $(C-2)\lambda p_2 = \mu p_3, \ldots, \lambda p_{C-1} = \mu p_C$. Rewriting the equations in terms of p_0 we get for all $i = 1, \ldots, C$

$$p_i = C(C-1)\ldots(C-i+1)\left(\frac{\lambda}{\mu}\right)^i p_0 = \binom{C}{i}i!\left(\frac{\lambda}{\mu}\right)^i p_0.$$

Solving for p_0 and substituting, we get for all i such that $0 \leq i \leq C$,

$$p_i = \frac{\binom{C}{i}i!\left(\frac{\lambda}{\mu}\right)^i}{\sum_{j=0}^{C}\binom{C}{j}j!\left(\frac{\lambda}{\mu}\right)^j}. \tag{6.5}$$

To compute the arrival point probabilities, we write the p_i as p_i^C to denote that the population size is C. This is in the same line as what we did for the closed-queueing network. Let π_j^* denote the probability that an arriving customer into the queue find j others in there. We can write down π_j^* in terms of E_i and A_h, which, respectively, denote the event that there are i customers in queue i (for all $i \in \{0, 1, \ldots, C\}$) at an arbitrary time in steady state and the event that an arrival occurs into the queue during an infinitesimal time of length h. By definition we have

$$\pi_j^* = \lim_{h \to 0} P\{E_j | A_h\}.$$

To obtain this conditional probability, we use Bayes' rule and then compute the conditional probabilities in the following manner:

$$\pi_j^* = \lim_{h \to 0} \frac{P\{A_h | E_j\}P(E_j)}{\sum_{i=0}^{C} P\{A_h | E_i\}P(E_i)}$$

$$= \lim_{h \to 0} \frac{\{(C-j)\lambda h + o(h)\}p_j^C}{\sum_{i=0}^{C}\{(C-i)\lambda h + o(h)\}p_i^C}$$

$$= \frac{(C-j)\lambda p_j^C}{\sum_{i=0}^{C}(C-i)\lambda p_i^C}.$$

To further simplify this expression, note from Equation 6.5 that

$$\frac{p_j^C}{p_j^{C-1}} = \left(\frac{C}{C-j}\right)\frac{p_0^C}{p_0^{C-1}}.$$

Therefore, we can write down the last expression for π_j^* as

$$\pi_j^* = \frac{(C-j)}{\sum_{i=0}^{C}(C-i)p_i^C}p_j^{C-1}\left(\frac{C}{C-j}\right)\frac{p_0^C}{p_0^{C-1}}$$

$$= \frac{Cp_0^C}{p_0^{C-1}\sum_{i=0}^{C}(C-i)p_i^C}p_j^{C-1}$$

$$= kp_j^{C-1}$$

where k is a constant that does not depend on j. But k must be 1 because $\sum_j \pi_j^* = \sum_j kp_j^{C-1} = 1$. Thus, we have

$$\pi_j^* = p_j^{C-1}, \tag{6.6}$$

that is, the probability that there are j in the queue as seen by an arriving customer is the same as the probability that there are j in the queue for a similar system with one less customer.

Next, we model the system in the problem as a closed Jackson network with $N=2$ nodes and C customers. We denote the single-server queue as node 1 and outside of the queue as node 2. The service rate at node 1 when there are n (such that $n>0$) customers in it is μ, hence $\mu_1(n)=\mu$. The service rate at node 2 when there are n customers in it is $n\lambda$ (which is essentially the rate at which a departure occurs when there are n in node 2). Thus, $\mu_2(n)=n\lambda$. The routing probability matrix P can be obtained from the problem statement as

$$P = \begin{bmatrix} 0 & 1 \\ 1 & 0 \end{bmatrix}.$$

The network is irreducible and a solution to $a=aP$ is $a=[1 \quad 1]$. Note that the elements of a do not need to sum to one, and in this case it is especially convenient to keep them both equal to 1.

Using the values of a_j and $\mu_j(n)$, for $j=1,2$ and $n=1,\ldots,C$ we can compute $\phi_j(n)$ from Equation 6.4 as

$$\phi_1(n) = \prod_{k=1}^{n} \left(\frac{a_j}{\mu_j(k)} \right) = \frac{1}{\mu^n}$$

and

$$\phi_2(n) = \prod_{k=1}^{n} \left(\frac{a_j}{\mu_j(k)} \right) = \frac{1}{n!\lambda^n}.$$

Also for $j=1,2$ we have $\phi_j(0)=1$. Further, since we only have two nodes, if one node has x_1, then necessarily the other node must have $x_2 = C - x_1$. Therefore, the joint probability distribution

$$p(x_1, C - x_1) = G(C)\phi_1(x_1)\phi_2(C - x_1) = G(C)\frac{1}{(C - x_1)!\lambda^{C-x_1}\mu^{x_1}}$$

$$= G(C)\frac{1}{C!\lambda^C}\binom{C}{x_1}x_1!\left(\frac{\lambda}{\mu}\right)^{x_1}.$$

Since the sum over all x_1 (from 0 to C) is one, we have

$$G(C) = \frac{C!\lambda^C}{\sum_{j=0}^{C}\binom{C}{j}j!\left(\frac{\lambda}{\mu}\right)^{j}}.$$

Hence, we have

$$p(x_1, C - x_1) = \frac{\binom{C}{x_1}x_1!\left(\frac{\lambda}{\mu}\right)^{x_1}}{\sum_{j=0}^{C}\binom{C}{j}j!\left(\frac{\lambda}{\mu}\right)^{j}}$$

which is identical to Equation 6.5. Thus we have arrived at the steady-state distribution of the number in the queue by modeling and analyzing in two different ways. Also note that the probability that an arriving customer to node 1 sees x_1 in node 1 and the remaining in node due to ASTA 2 is

$$\pi_1(x_1, C - x_1 - 1) = p^{C-1}(x_1, C - x_1 - 1)$$

where $p^{C-1}(x_1, C - x_1 - 1)$ is the steady-state probability that there are x_1 in queue 1 when there are a total of $C-1$ customers in the system. Note that this result is identical to Equation 6.6 that was derived by analyzing the system using a birth and death process.   ∎

6.3.3 Single-Server Closed Jackson Networks

Although for the two closed Jackson network examples thus far the computing $G(C)$ was not particularly difficult, as the number of nodes and customers
increase, this would become a very tedious task. There are a few algorithms
to facilitate that computation, one of which is provided in the exercises at
the end of this chapter. Nonetheless, there is one case where it is possible to circumvent computing $G(C)$. That case is a closed Jackson network
where there is a single server at all the nodes. Assume that for all i, there
is a single server at node i with service rate μ_i operating FCFS discipline at
every node. Then the mean performance measures can be computed without going through the computation of the normalizing constant $G(C)$. For
that we first need some notation (for anything not defined here the reader
is encouraged to refer to the closed Jackson network notation). Define the
following steady-state measures for all $i = 1, \ldots, N$ and $k = 0, \ldots, C$:

- $W_i(k)$: Average sojourn time in node i when there are k customers
  (as opposed to C) in the closed Jackson network
- $L_i(k)$: Average number in node i when there are k customers (as
  opposed to C) in the closed Jackson network
- $\lambda(k)$: Measure of average flow (sometimes also referred to as
  throughput) in the closed Jackson network when there are k customers (as opposed to C) in the network

We do not have an expression for any of the preceding measures and the
objective is to obtain them iteratively. However, before describing the iterative algorithm, we first explain the relationship between those parameters.

On the basis of the arrival theorem described in Remark 14, in a network
with k customers (such that $1 \le k \le C$), the expected number of customers that
an arrival to node i (for any $i \in \{1, \ldots, N\}$) would see is $L_i(k - 1)$. Note that
$L_i(k - 1)$ is the steady-state expected number of customers in node i when
there are $k - 1$ customers in the system. Thereby, the net mean sojourn time
experienced by that arriving customer in steady state is the average time to
serve all those in the system upon arrival plus that of the customer. Since the
average service time is $1/\mu_i$, we have

$$W_i(k) = \frac{1}{\mu_i}[1 + L_i(k - 1)].$$

Let a be the solution to $a = aP$ as usual with the only exception that the a_j
values sum to one here. Thus, the a_j values describe the fraction of visits
that are made into node j. The aggregate sojourn time weighted across the
network using the fraction of visits is given by $\sum_{i=1}^{N} a_i W_i(k)$ when there are
k customers in the network. One can think of an aggregate sojourn time as

the sojourn time for a customer about to enter a node. Hence, by conditioning on the node of entry as i (which happens with probability a_i) where the mean sojourn time is $W_i(k)$, we can get the result $\sum_{i=1}^{N} a_i W_i(k)$. Thereby we derive the average flow in the network using Little's law across the entire network as

$$\lambda(k) = \frac{k}{\sum_{i=1}^{N} a_i W_i(k)}$$

when there are k customers in the network. Essentially, $\lambda(k)$ is the average rate at which service completion occurs in the entire network, taken as a whole. Thereby applying Little's law across each node i we get

$$L_i(k) = \lambda(k) W_i(k) a_i$$

when there are k customers in the network.

Using these results we can develop an algorithm to determine $L_i(C)$, $W_i(C)$, and $\lambda(C)$ defined earlier. The inputs to the algorithm are N, C, P, and μ_i for all $i \in \{1,\ldots,N\}$. For the algorithm, initialize $L_i(0) = 0$ for $1 \le i \le N$ and obtain the a_i values for all $i \in \{1,\ldots,N\}$. Then for $k=1$ to C, iteratively compute for each i (such that $1 \le i \le N$):

$$W_i(k) = \frac{1}{\mu_i}[1 + L_i(k-1)],$$

$$\lambda(k) = \frac{k}{\sum_{i=1}^{N} a_i W_i(k)},$$

$$L_i(k) = \lambda(k) W_i(k) a_i.$$

We present an example in website server architecture to illustrate the preceding algorithm via a problem that is suitably adapted from Menasce and Almeida [81].

Websites, especially dealing with e-business, typically consist of a three-tier architecture described in Figure 6.12. Users access such websites through

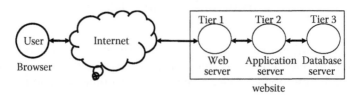

FIGURE 6.12
Three-tier architecture for e-business websites.

their browsers by connecting to the first tier, namely, the web server. The web server provides web pages and forms for users to enter requests or information. When the users enter the information and send back to the web server, it passes the information onto the application server (which is the second tier). The application server processes the information and communicates with the database server (in the third tier). The database server then searches its database and responds to the application server, which in turn passes it onto to the web server, which transmits to the user. For example, consider running a website for a used car dealership (with URL www.usedcar.com). When a user types www.usedcar.com on their browser, the request goes to the first tier for which the web server responds with the relevant web page. Say the user fills out a set of makes and models as well as desirable years of manufacture. When the user submits this form expecting to see the set of used cars available that meets his or her criteria, the web server passes the set of criteria to the application server (second tier). The application server processes this set of criteria to check if the form is filled with all the required fields and then submits to the database server (third tier). The database server queries the database of all cars available in the dealership that meet the criteria and then responds with the appropriate set. Having described some background for websites, we now describe a problem.

Problem 59

The bottleneck in many three-tier architecture websites is the database server that does not scale up to handle a large number of users simultaneously. Let us say that the database server can handle at most C connections simultaneously. During peak periods, one typically sees the database server handling its full capacity of C connections at every instant of time. In practice, every time one of the C connections is complete, instantaneously a new connection is added to the database server thus maintaining C connections throughout the peak period. Hence, we can model the database server system as a closed queueing network with C customers. The database server system consists of a processor and four disks as shown in Figure 6.13. All five nodes are single-server queues with exponential service times. Each customer after being processed at the processor goes to any of the four disks with equal probability. The average service time (in milliseconds) at the processor is 6 and at the four disks are 17, 28, 13, and 23, respectively. For $C = 16$ use the preceding algorithm to determine the expected number at each of the five nodes as well as the throughput of the database–server system. What happens to those metrics when C is 25, 50, 75, and 100?

Solution

Note that we have a Jackson network with $N = 5$ nodes and $C = 16$ customers (we will later consider other C values). The P matrix corresponding to the processor node and the four disks is

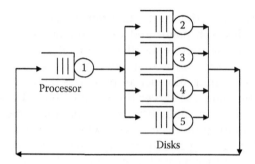

FIGURE 6.13
Closed queueing network inside database server.

$$P = \begin{bmatrix} 0 & 0.25 & 0.25 & 0.25 & 0.25 \\ 1 & 0 & 0 & 0 & 0 \\ 1 & 0 & 0 & 0 & 0 \\ 1 & 0 & 0 & 0 & 0 \\ 1 & 0 & 0 & 0 & 0 \end{bmatrix}.$$

We can obtain $a = [a_1 \ a_2 \ a_3 \ a_4 \ a_5]$ by solving for $a = aP$ and $a_1 + a_2 + a_3 + a_4 + a_5 = 1$ to get

$$a = [1/2 \ 1/8 \ 1/8 \ 1/8 \ 1/8].$$

We are also given in the problem statement that

$$\mu = [\mu_1 \ \mu_2 \ \mu_3 \ \mu_4 \ \mu_5] = [1/6 \ 1/17 \ 1/28 \ 1/13 \ 1/23].$$

By initializing $L_i(0) = 0$ for $i = 1, 2, 3, 4, 5$, we can go through the algorithm of iteratively computing for $k = 1$ to C (and all i such that $1 \le i \le 5$) using the following steps: $W_i(k) = 1/\mu_i[1 + L_i(k-1)]$, $\lambda(k) = k / \sum_{i=1}^{N} a_i W_i(k)$, and $L_i(k) = \lambda(k) W_i(k) a_i$. These computations are tabulated in Table 6.3. From the last row of that table note that with $C = 16$, the expected number of connections at nodes 1, 2, 3, 4, and 5 are 3.6023, 1.327, 7.208, 0.7815, and 3.0812, respectively. Also, the throughput of the database–server system is 0.2719 transactions per millisecond. If we were to increase C to 25, 50, 75, and 100, the corresponding values would be as described in Table 6.4. Note in the table that the throughput λ has practically leveled off and increasing C is mostly contributing only to longer queues in node 3 (i.e., disk 2), which is the bottleneck in this system. Also notice how all the other nodes have not only scaled very well but the contribution to the overall number in the system is becoming negligible. This will be the basis of bottleneck-based approximations that we will consider in the next chapter. ■

TABLE 6.3

The Single-Server Closed Jackson Network Iterations

k	$L_1(k)$	$L_2(k)$	$L_3(k)$	$L_4(k)$	$L_5(k)$	$\lambda(k)$
1	0.2286	0.1619	0.2667	0.1238	0.2190	0.0762
2	0.4631	0.3102	0.5570	0.2294	0.4403	0.1256
3	0.7018	0.4452	0.8714	0.3195	0.6621	0.1599
4	0.9433	0.5674	1.2102	0.3962	0.8829	0.1848
5	1.1860	0.6776	1.5737	0.4615	1.1013	0.2034
6	1.4284	0.7765	1.9620	0.5173	1.3158	0.2178
7	1.6692	0.8650	2.3754	0.5649	1.5255	0.2291
8	1.9073	0.9439	2.8138	0.6057	1.7294	0.2382
9	2.1414	1.0142	3.2772	0.6406	1.9266	0.2455
10	2.3706	1.0766	3.7657	0.6706	2.1165	0.2515
11	2.5940	1.1321	4.2790	0.6964	2.2985	0.2565
12	2.8109	1.1812	4.8169	0.7187	2.4723	0.2607
13	3.0207	1.2246	5.3792	0.7379	2.6376	0.2642
14	3.2228	1.2631	5.9654	0.7546	2.7942	0.2672
15	3.4167	1.2970	6.5752	0.7690	2.9421	0.2697
16	3.6023	1.3270	7.2080	0.7815	3.0812	0.2719

TABLE 6.4

When C Is Increased to 25, 50, 75, and 100

C	L_1	L_2	L_3	L_4	L_5	λ
25	4.8794	1.4786	13.8299	0.8418	3.9704	0.2818
50	5.9284	1.5435	37.0910	0.8660	4.5711	0.2856
75	5.9972	1.5454	61.9915	0.8667	4.5992	0.2857
100	5.9999	1.5455	86.9880	0.8667	4.6000	0.2857

6.4 Other Product-Form Networks

Besides the Jackson network, there are other product-form queueing networks, most of which are generalizations of the Jackson network. In some sense these networks relax few of the restrictive assumptions and still give rise to product-form steady-state distributions. Some of the generalizations include state-dependent arrivals and service, multiclass customers, deterministic routing, general service times, state-dependent routing, networks with losses, and networks with negative customers. In this section, we consider only a few of those generalizations, one at a time, to get an appreciation of each of their effects. However, in the literature one typically finds a combination of the preceding generalizations such as those in Baskett– Chandy–Muntz–Palacios (BCMP) networks, Kelly networks, Whittle networks, loss networks, and networks with signals, to name a few. The reader is encouraged to consider recent texts such by Serfozo [96], Chao et al. [18], or Chen and Yao [19] for a more rigorous and thorough treatment of this topic through a concept called *quasi-reversibility*. In addition, Bolch et al. [12] provides algorithms for performance analysis of product-form networks. With that introduction, we now describe four product-form networks.

6.4.1 Open Jackson Networks with State-Dependent Arrivals and Service

In the literature, state-dependent arrivals and/or services are also included under open Jackson networks, much like what we did for closed Jackson networks. There are some technicalities that we felt are better addressed if considered separately and hence we are including them here. Other than the arrivals and the service times, all conditions described in Section 6.2 hold (especially important is that there are N nodes in the network and the routing matrix is P). We first describe the notion of state-dependent service times, which is identical to what we saw for closed Jackson networks. Let the service rate at node i when there are n customers at that node be $\mu_i(n)$ with $\mu_i(0) = 0$. Also assume that the service rate does not depend on the states of the remaining nodes (toward the end of this section we will talk about Whittle networks where service rates could depend on the state at all nodes). Next we describe state-dependent arrivals.

Let $\lambda(n)$ be the total external arrival rate to the network as a whole when there are n customers in the entire network. A special case of $\lambda(n)$ is when it is λ for $n \leq C$ and 0 for $n > C$. This special case amounts to an open Jackson network with at most C customers (some texts such as Serfozo [96] make a distinction between this special case and the usual open Jackson network). Now back to the general $\lambda(n)$. When a customer arrives, then with probability u_i this incoming customer joins node i, independently of other customers for $i = 1, \ldots, N$. Therefore, external arrivals to node i are at rate $u_i\lambda(n)$. Recall that the service rate at node i when there are n_i customers at that node is given by $\mu_i(n_i)$ with $\mu_i(0) = 0$.

It is not hard to see that we can model the entire system as a CTMC. For $i = 1, \ldots, N$, let $X_i(t)$ be the number of customers in node i at time t. Let $X(t)$ be a vector that captures a snapshot of the state of the network at time t and is given by $X(t) = [X_1(t), X_2(t), \ldots, X_N(t)]$. Let $p(x)$ be the steady-state probability that the CTMC $\{X(t), t \geq 0\}$ is in state $x = (x_1, x_2, \ldots, x_N)$, that is,

$$p(x) = \lim_{t \to \infty} P\{X(t) = (x_1, x_2, \ldots, x_N)\}.$$

Although we would not derive the results (since it is extremely similar to those in the previous sections), we will next just describe $p(x)$ as a product form.

For that we first obtain the visit ratios that we call b_i for node i (for $i = 1, \ldots, N$). Clearly, b_j is the unique solution to

$$b_j = u_j + \sum_{i=1}^{N} b_i p_{ij}.$$

It can be crisply computed in vector form as

$$b = u[I - P]^{-1}$$

where
 u is a row vector of u_j values
 b a row vector of b_j values
 I the identity matrix
 P the routing matrix

Next, define the following for all $i \in \{1, \ldots, N\}$: $\phi_i(0) = 1$ and

$$\phi_i(n) = \prod_{j=1}^{n} \left(\frac{b_i}{\mu_i(j)} \right) \qquad \text{for } n \geq 1.$$

Define $\hat{x} = \sum_{i=1}^{N} x_i$. Using this notation it is possible to show that the steady-state probability $p(x)$ is given by

$$p(x) = c \prod_{i=1}^{N} \phi_i(x_i) \prod_{j=0}^{\hat{x}} \lambda(j),$$

where the normalizing constant c is

$$c = \left\{ \sum_{x} \prod_{i=1}^{N} \phi_i(x_i) \prod_{j=0}^{\hat{x}} \lambda(j) \right\}^{-1}.$$

Using the preceding joint distribution, it is possible to obtain certain performance measures. However, one of the difficulties is obtaining the normalizing constant c. This was a concern for the closed Jackson network as well. Once we obtain c, the marginal distribution at each node can be obtained.

That can be used to get the distribution of the number of customers in the system as well as the mean (and higher moments). Then using Little's law, the mean sojourn times (across a node and the network itself) can also be obtained.

An immediate extension to this model is to allow service rates to depend on the number of customers in each node of the network. Therefore, the service rate at node i when there are $x_1,\ldots,x_N$ customers, respectively, at nodes $1,\ldots,N$ instead of being $\mu_i(x_i)$, is now $\mu_i(x)$. A network with that extension is known as a *Whittle network* for which Serfozo [96] describes conditions for a product-form solution. In the next few sections, we will describe other networks where the steady-state probabilities can be represented as product form.

6.4.2 Open Jackson–Like Networks with Deterministic Routing

In the description for open Jackson networks in Section 6.2, at the end of service at a node (say i) each customer joins node j with probability p_{ij}. Further, the past route a customer has followed up to node i is of no use in predicting the next node. In this section, we modify that requirement. In particular, each customer entering the network has a fixed route to follow, which is deterministic and revealed upon arrival to the network. Note that a route is a collection of nodes representing the path to be followed by a customer on the route. The paths followed by various routes could be overlapping. Say there are R routes in a network, then each arriving customer belongs to one of those R routes. In fact, in many situations we say that the customers belong to R different classes depending on their prespecified routes. There are several examples: repair shops where there are R types of repairs that require items to visit a given set of workstations in a particular order, multiple types of requests in a computer system each requiring a set of resources in a particular order, and tasks to be performed in a multiagent software system with precedence constraints.

As a generic description, we restate the conditions described in Section 6.2 to reflect this situation as follows:

1. The system is an open-queueing network consisting of N service stations (or nodes).
2. There are R deterministic routes in the network and each customer arrives, follows one of the routes, and exits the network.
3. External arrivals are according to a Poisson process. In particular, customers following route r (for $r=1,\ldots,R$) arrive externally according to $PP(\lambda_r)$ into the first node of that route. These customers are sometimes said to belong to class r.
4. There are s_i identical servers at node i (such that $1 \le s_i \le \infty$), for all i satisfying $1 \le i \le N$.

5. Service time requirements of customers at node i are IID $\exp(\mu_i)$ random variables. They are independent of service times at other nodes and independent of routes.

6. The service discipline is one of the following (irrespective of the class or route of the customer): FCFS, random order of service with new customer selected at every arrival and service completion, or processor sharing (with $s_i = 1$ at node i).

7. There is infinite waiting room at each node and stability condition is satisfied at every node i, that is, the sum of the average arrival rate from all routes sharing node i must be less than $s_i \mu_i$.

8. When a customer of class r (i.e., in route r) completes service at node i, the customer joins node j if that is the next node in route r or exits the network if i is the last node in route r. The routes can have cycles (i.e., multiple visits to a node).

This is a special case of what is known in the literature as *Kelly networks*. To analyze such a system, define M as an $R \times N$ route–node incidence matrix such that an element, say M_{rj}, denotes the number of times node j would be visited in route r (in acyclic routes this would just be zero or one). For $i = 1, \ldots, N$, let $X_{ri}(t)$ be the number of customers belonging to route r in node i at time t. Let $X(t)$ be a vector that captures a snapshot of the state of the network at time t and is given by $X(t) = [X_{11}(t), X_{21}(t), \ldots, X_{ri}(t), \ldots, X_{RN}(t)]$. Let $p(x)$ be the steady-state probability of being in state $x = (x_{11}, x_{21}, \ldots, x_{RN})$, that is,

$$p(x) = \lim_{t \to \infty} P\{X(t) = (x_{11}, x_{21}, \ldots, x_{RN})\}.$$

Of course, in practice we would not have to consider such a large dimensionality for $X(t)$ considering that x_{rj} should be zero if $M_{rj} = 0$, that is, if route r does not include node j. Hence, let us define set $\mathcal{E}$ as the set of all possible x satisfying the criterion that $x_{rj} = 0$ if $M_{rj} = 0$ for all r and j. Therefore, $p(x) > 0$ if $x \in \mathcal{E}$ and $p(x) = 0$ otherwise.

Note that at every node, if the service discipline is either processor sharing (with $s_i = 1$) or random order of service with new customer selected at every arrival and service completion, the stochastic process $\{X(t), t \geq 0\}$ is a CTMC. However, if the discipline is FCFS, $\{X(t), t \geq 0\}$ is not a CTMC since the transition rates would be different if the history is known as opposed to when it is not known as the state information does not include the customer class under service. For the FCFS, case we would have to keep track of the type of customer in each position of every queue in the network that would form a CTMC. But any permutation within a queue would result in the same probability. Thus, adding across all permutations we can obtain $p(x)$ identical to that of the random order of service. Having said that, next we describe a product-form solution for $p(x)$.

To obtain a product-form expression for $p(x)$, we require that for all r and i (such that $1 \leq r \leq R$ and $1 \leq i \leq N$), the rate at which a class r customer completes service at node i when the state of the network is x such that $x \in \mathcal{E}$ is

$$\frac{x_{ri}}{y_i} \mu_i \min(y_i, s_i)$$

where y_i is the total number of customers in node i, that is,

$$y_i = \sum_{r=1}^{R} x_{ri}$$

(when y_i is zero, the service completion rate is zero as well). All three service disciplines mentioned earlier satisfy that condition, and others that do can also be included in the list of service disciplined allowed. With that understanding, we will next just describe $p(x)$ as a product form without going into details of the derivation.

Define the following for all $i \in \{1, \ldots, N\}$: $\phi_i(0, 0, \ldots, 0) = \beta_i$ and

$$\phi_i(n_1, n_2, \ldots, n_R) = \beta_i \bar{n}! \left(\prod_{j=1}^{\bar{n}} \frac{1}{\mu_i \min(j, s_i)} \right) \left(\prod_{r=1}^{R} \frac{\lambda_r^{n_r}}{n_r!} \right) \quad \text{for } \bar{n} \geq 1$$

where

$$\bar{n} = \sum_r n_r \quad \text{and} \quad \beta_j^{-1} = 1 + \sum_{k=1}^{\infty} \prod_{n=1}^{k} \frac{\left(\sum_{r=1}^{R} \lambda_r M_{rj} \right)}{\mu_j \min(n, s_i)}.$$

Using this notation it is possible to show that the steady-state probability $p(x)$ is given by

$$p(x) = \begin{cases} \prod_{i=1}^{N} \phi_i(x_{1i}, x_{2i}, \ldots, x_{Ri}) & \text{if } x \in \mathcal{E} \\ 0 & \text{otherwise.} \end{cases}$$

One can consider several extensions to this model. In fact, Kelly networks, on which the preceding analysis is based, are a lot more general. The reader is referred to the end of the next section on multiclass networks for a description of some of the possible extensions as they are common to both sections. After all what we saw here is just a special type of multiclass network. Before forging ahead, we present a small example to illustrate these results.

Problem 60

Consider the four-node network in Figure 6.14. There are three routes described using three types of arrows. Route-1 uses path 1-3-4-2, route-2 uses 4-2-1, and route-3 uses 2-3. Route-1 customers arrive according to a Poisson process with mean rate 4 per hour. Likewise, route-2 and route-3 customers arrive according to Poisson processes with mean rates of 2 and 3 per hour, respectively. Nodes 1, 2, and 3 have a single server that serves according to an exponential distribution with rates 8, 10, and 9 per hour, respectively. Node 4 has two servers each serving at rate 4 per hour. The service discipline is FCFS in nodes 1, 2, and 4 but processor sharing in node 3. What is the probability that there is one route-1 customer in each of the four nodes, two route-2 customers in node 4, two in node 2 and one in node 1, and three route-3 customers in node 2 and four in node 3?

Solution

The problem illustrates an example of a queueing network with fixed routes. Using the notation of this section, $N = 4$ nodes and $R = 3$ routes. Also, $(\lambda_1, \lambda_2, \lambda_3) = (4, 2, 3)$, $(\mu_1, \mu_2, \mu_3, \mu_4) = (8, 10, 9, 4)$, and $(s_1, s_2, s_3, s_4) = (1, 1, 1, 2)$. In the problem, the vector of x_{rj} values for route r and node j describes x given by $x = (x_{11}, x_{21}, x_{31}, x_{12}, x_{22}, x_{32}, x_{13}, x_{23}, x_{33}, x_{14}, x_{24}, x_{34})$. Using the numerical values in the problem, we have $x = (1, 1, 0, 1, 2, 3, 1, 0, 4, 1, 2, 0)$ for which we need to compute $p(x)$. Using the results in this section we have

$$p(x) = \phi_1(1, 1, 0)\phi_2(1, 2, 3)\phi_3(1, 0, 4)\phi_4(1, 2, 0).$$

For $i = 1, 2, 3, 4$, we can obtain $\phi_i(\cdot, \cdot, \cdot)$ as

$$\phi_1(1, 1, 0) = \beta_1(2!)\left(\prod_{j=1}^{2}\frac{1}{\mu_1}\right)(\lambda_1\lambda_2) = \beta_1\frac{1}{4}$$

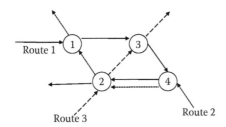

FIGURE 6.14
Queueing network with fixed routes.

$$\phi_2(1,2,3) = \beta_2(6!) \left(\prod_{j=1}^{6} \frac{1}{\mu_2} \right) \left(\frac{\lambda_1 \lambda_2^2 \lambda_3^3}{12} \right) = \beta_2 \frac{81}{3125}$$

$$\phi_3(1,0,4) = \beta_3(5!) \left(\prod_{j=1}^{5} \frac{1}{\mu_3} \right) \left(\frac{\lambda_1 \lambda_3^4}{24} \right) = \beta_3 \frac{20}{729}$$

$$\phi_4(1,2,0) = \beta_4(3!) \left(\prod_{j=1}^{3} \frac{1}{\min(j,2)\mu_4} \right) \left(\frac{\lambda_1 \lambda_2^2}{6} \right) = \beta_4 \frac{1}{32}.$$

Thus, the only thing left is to obtain the β_j values for $j = 1, 2, 3, 4$. Although it is possible to directly use the formula, it is easier if we realize that β_j is the probability that node j is empty in steady state. Since nodes 1, 2, and 3 are effectively $M/M/1$ queues with arrival rates 6, 9, and 7 as well as service rates 8, 10, and 9, respectively, we have $\beta_1 = 1/4$, $\beta_2 = 1/10$, and $\beta_3 = 2/9$. Likewise, node 4 is effectively an $M/M/2$ queue with arrival rate 6 and service rate 4 for each server. Hence, we have $\beta_4 = 1/7$. Thus, the probability that there is one route-1 customer in each of the four nodes, two route-2 customers in node 4, two in node 2 and one in node 1, and three route-3 customers in node 2 and four in node 3 is $\beta_1\beta_2\beta_3\beta_4(1/180000) = 1/22680000$. ∎

6.4.3 Multiclass Networks

Around the same time that Kelly [59] introduced networks with deterministic routing, a closely related multiclass network called BCMP (Baskett, Chandy, Muntz, and Palacios) network [8] was developed, named after the four authors of that manuscript. We present a special case of BCMP networks here, and toward the end of this section give an idea of what extensions are possible. Similar to the previous section, here too we state the conditions akin to Section 6.2 to reflect the situation under consideration:

1. The system is an open-queueing network consisting of N service stations (or nodes).
2. There are K classes of customers in the network and each class of customer moves randomly in the network.
3. External arrivals are according to a Poisson process. Class k customers arrive into the system from the outside into node i at rate λ_{ki}.
4. At each node i there is a single server (for all i such that $1 \leq i \leq N$).
5. Service time requirements of class k customers at node i are IID $\exp(\mu_{ki})$ random variables.

6. The service discipline is one of the following: FCFS (in which case we require μ_{ki} to be independent of k (i.e., all K class have the same service rate that we call μ_i), processor sharing, or LCFS with preemptive resume.

7. When a class k customer completes service at node i, the customer departs the network with probability r_{ki} or joins the queue at node j as a class ℓ customer with probability $p_{ki,\ell j}$. The routing of a customer does not depend on the state of the network.

8. There is infinite waiting room at each node and stability condition is satisfied at every node i.

As earlier, here we obtain the visit ratios $a_{\ell j}$ for class ℓ customers into node j (i.e., the effective arrival rate of class ℓ customers into node j). We solve the following set of simultaneous equations:

$$a_{\ell j} = \lambda_{\ell j} + \sum_{i=1}^{N} \sum_{k=1}^{K} a_{ki} p_{ki,\ell j}$$

for all ℓ (such that $1 \le \ell \le K$) and j (such that $1 \le j \le N$). For $i = 1, \ldots, N$ and $k = 1, \ldots, K$, let $X_{ki}(t)$ be the number of customers belonging to class k in node i at time t. Let $X(t)$ be a vector that captures a snapshot of the state of the network at time t and is given by $X(t) = [X_{11}(t), X_{21}(t), \ldots, X_{ki}(t), \ldots, X_{KN}(t)]$. Let $p(x)$ be the steady-state probability of being in state $x = (x_{11}, x_{21}, \ldots, x_{KN})$, that is,

$$p(x) = \lim_{t \to \infty} P\{X(t) = (x_{11}, x_{21}, \ldots, x_{KN})\}.$$

Note that the stochastic process $\{X(t), t \ge 0\}$ is not a CTMC if the discipline is FCFS (although for the other disciplines mentioned earlier, it would be a CTMC) since the transition rates would be different if the history is known as opposed to when it is not known as the state information does not include the customer class under service. For the FCFS case we would have to keep track of the class of customer in each position of every queue in the network that would form a CTMC. But any permutation within a queue would result in the same probability. Thus, adding across all permutations we can obtain $p(x)$. With that understanding in place, next we describe $p(x)$, which would be a product form, without going into details of the derivation.

Let Λ_i be the total arrival rate into node i aggregated over all classes, that is,

$$\Lambda_i = \sum_{k=1}^{K} a_{ki}.$$

Likewise, let Γ_i be the aggregate service rate at node i, that is,

$$\frac{1}{\Gamma_i} = \sum_{k=1}^{K} \frac{a_{ki}}{\mu_{ki}\Lambda_i}.$$

Note that if the discipline is FCFS, $\Gamma_i = \mu_{ki}$ for all k since μ_{ki} does not change with k. Next, define the following for all $i \in \{1,\dots,N\}$:

$$\phi_i(n_1, n_2, \dots, n_K) = \left(\frac{1-\Lambda_i}{\Gamma_i}\right)\left(\left(\sum_k n_k\right)!\right)\prod_{k=1}^{K}\left(\frac{a_{ki}^{n_k}}{n_k!\mu_{ki}^{n_k}}\right).$$

Using this notation it is possible to show that the steady-state probability $p(x)$ is given by

$$p(x) = \prod_{i=1}^{N} \phi_i(x_{1i}, x_{2i}, \dots, x_{Ki}).$$

To illustrate this result, we consider a numerical example adapted from Bolch et al. [12].

Problem 61

Consider an open-queueing network with $K=2$ classes and $N=3$ nodes. Node 1 uses processor sharing, while nodes 2 and 3 use LCFS preemptive resume policy. The service rates μ_{ki} for class k customers in node i are $\mu_{11}=8$, $\mu_{21}=24$, $\mu_{12}=12$, $\mu_{22}=32$, $\mu_{13}=16$, $\mu_{23}=36$. Arrivals for both classes occur externally into node 1 at rate 1 per unit time. The routing probabilities are $p_{12,11}=0.6$, $p_{22,21}=0.7$, $p_{11,12}=0.4$, $p_{12,13}=0.4$, $p_{21,22}=0.3$, $p_{22,23}=0.3$, $p_{11,13}=0.3$, $p_{13,11}=0.5$, $p_{21,23}=0.6$, $p_{23,21}=0.4$, $p_{13,12}=0.5$, $p_{23,22}=0.6$, $r_{11}=0.3$, and $r_{21}=0.1$. Note that class switching is not allowed in the network. Compute the probability that there are one class-1 and two class-2 customers in node 1, one class-1 and one class-2 customers in node 2, and zero class-1 and one class-2 customer in node 3.

Solution

Note that since the external arrival rate is 1 into node 1 for both classes, we have $\lambda_{11}=1$, $\lambda_{21}=1$, $\lambda_{12}=0$, $\lambda_{22}=0$, $\lambda_{13}=0$, and $\lambda_{23}=0$. Using those and solving for the simultaneous equations for the visit ratios, we get $a_{11}=3.3333$, $a_{21}=10$, $a_{12}=2.2917$, $a_{22}=8.0488$, $a_{13}=1.9167$, and $a_{23}=8.4146$. In fact, since no class switching is allowed, we can solve the simultaneous equations one class at a time. Therefore, we can compute the net arrival rate into node i aggregating over all the classes Λ_i as $\Lambda_1=13.3333$, $\Lambda_2=10.3404$, and $\Lambda_3=10.3313$. Likewise, we can obtain $\Gamma_1=16$, $\Gamma_2=23.3684$, and $\Gamma_3=29.2231$. Using the formulae for $\phi_i(n_1, n_2)$, we get $\phi_1(1,2) = 0.0362$, $\phi_2(1,1) = 0.0536$, and $\phi_3(0,1) = 0.1511$. Note that $\phi_i(n_1, n_2)$ actually gives us the probability that in node-1 there are n_1 class-1 customers and n_2 class-2 customers. Of course, the answer to the question given in the problem is the joint probability, which is the product form $p(1,2,1,1,0,1) = \phi_1(1,2)\phi_2(1,1)\phi_3(0,1) = 0.00029271$. ■

One can consider several extensions to the preceding model that would still give us product-form solutions. As a matter of fact, BCMP networks, when they were first introduced, considered a few more generalizations. For example, if a node uses FCFS discipline, any number of servers were allowed; infinite server queues with general service times is an option (not just FCFS, processor sharing and LCFS with preemptive resume); general service times were allowed for processor sharing and LCFS with preemptive resume nodes; also closed-queueing networks are analyzable. Subsequently, several research studies further generalized the BCMP networks to include state-dependent (local and network-wide) arrivals and service, state-dependent routing, networks with negative customers, networks with blocking, open networks with a limited total capacity, etc. Refer to some recent books on queueing networks such as by Serfozo [96], Chao et al. [18], and Chen and Yao [19] for results under those general cases. In fact, both multiclass queueing networks with fixed routing (Kelly networks) and BCMP networks are combined into a single framework (by using routing probabilities of 0 or 1 for Kelly networks). It is also worthwhile to consider algorithms for product-form networks (especially in the extensions, we would require the use of normalizing constants that are harder to obtain). That said, we conclude the product-form network analysis by describing loss networks next.

6.4.4 Loss Networks

Telephone networks and circuit-switched networks are typically modeled as loss networks where there is virtually no buffering (or waiting) at nodes. Calls or customers or requests are either accepted or rejected. When they are accepted, they traverse the network with zero delay but use up network resources while a call is in progress. This is akin to generalizing the $M/G/s/s$ queue (recall that the $M/G/s/s$ queue results in a reversible process) to a network of resources. In fact, queueing theory arguably started when A.K. Erlang studied this problem. But, we have so far mostly considered only delay networks and not loss networks. There are good reasons for including and for excluding *loss networks* from a chapter on queueing networks. By stating upfront that loss networks have very little similarity with everything seen thus far, we hope that the reader would pay careful attention to the description to follow.

Consider a network with N nodes and J arcs (or links). Let arc j (for $j = 1, \ldots, J$) have a capacity of C_j. To explain further, we use the telephone network example. Say each accepted telephone call takes 1 unit of arc capacity (this is about 60 kbps). Then on arc j you can have at most C_j calls simultaneously. Let $\mathcal{R}$ be the set of routes in the telephone network such that a route r is described by a set of arcs that are traversed. In this manner, we only focus on the arcs and not on the nodes. For all $r \in \mathcal{R}$, telephone calls requesting route r arrive according to a Poisson process with mean rate λ_r. A call requesting route r is blocked and lost if there is no capacity available on any of the links in the route. If the call is accepted, it uses up 1 unit of capacity in each of the arcs in the route. The holding time for accepted calls of class r is generally distributed with mean $1/\mu_r$.

Define $X_r(t)$ as the number of calls in progress on route r at time t, for all $r \in \mathcal{R}$. Let $R = |\mathcal{R}|$, the number of routes in the network. Let the R-dimensional vector $X(t)$ be $X(t) = (X_1(t), \ldots, X_r(t), \ldots, X_R(t))$. Then the steady-state distribution of the stochastic process $\{X(t), t \geq 0\}$ can be computed as a product form (note that the process is reversible). Let $p(x)$ be the steady-state probability of being in state $x = (x_1, x_2, \ldots, x_R)$, that is,

$$p(x) = \lim_{t \to \infty} P\{X(t) = (x_1, x_2, \ldots, x_R)\}.$$

Let us define set $\mathcal{E}$ as the set of all possible x satisfying the criterion that the total number of calls in each link is less than or equal to the capacity. Therefore, $p(x) > 0$ if $x \in \mathcal{E}$ and $p(x) = 0$, otherwise. We can write down $p(x)$ as a product form given by

$$p(x) = G(C_1, \ldots, C_J) \prod_{r=1}^{R} \frac{1}{x_r!} \left(\frac{\lambda_r}{\mu_r}\right)^{x_r}, \quad \forall x \in \mathcal{E}.$$

The normalizing constant $G(C_1, \ldots, C_J)$ can be obtained by

$$\sum_{x \in \mathcal{E}} p(x) = 1.$$

Next, we present an example to illustrate loss networks.

Problem 62

Consider the six-node network in Figure 6.15. There are four routes in the network. Route-1 is through nodes A–C–D–E, route-2 is through nodes A–C–D–F, route-3 is through nodes B–C–D–E, and route-4 is through nodes B–C–D–F. The capacities of the five arcs are described below the arcs. Each call on a route uses one unit of the capacity of all the arcs on the route. Calls on routes 1, 2, 3, and 4 arrive according to a Poisson process with respective rates 10, 16, 15, 20 per hour. The average holding time for all calls is 3 min.

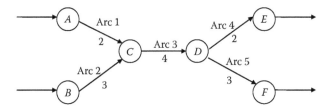

FIGURE 6.15
Loss network.

What is the probability that in steady state there is one call on each of the four routes?

Solution

For $r = 1, 2, 3, 4$, let $X_r(t)$ be the number of calls on route r. Let $x = (x_1, x_2, x_3, x_4)$, and we would like to compute $p(x)$. First we describe $\mathcal{E}$, the set of all feasible x values. Then

$$\mathcal{E} = \{(x_1, x_2, x_3, x_4) : x_1 + x_2 \le 2, x_1 + x_2 + x_3 + x_4 \le 4, x_3 + x_4 \le 3,$$

$$x_1 + x_3 \le 2, x_2 + x_4 \le 3\}.$$

There are a total of 37 elements in set $\mathcal{E}$. For any $x \in \mathcal{E}$, we have

$$p(x) = G(C_1, \ldots, C_5) \frac{1}{x_1!} \left(\frac{10}{20}\right)^{x_1} \frac{1}{x_2!} \left(\frac{16}{20}\right)^{x_2} \frac{1}{x_3!} \left(\frac{15}{20}\right)^{x_3} \frac{1}{x_4!} \left(\frac{20}{20}\right)^{x_4}.$$

We can compute $G(C_1, \ldots, C_5) = 1/14.4475$ and hence $p(1, 1, 1, 1) = 0.0208$, which is what we require. ∎

Reference Notes

We began this chapter with acyclic networks as well as open and closed Jackson networks. Most of the results here were adapted from Kulkarni [67]. In fact, many standard texts on queues would also typically contain a few chapters on these topics. The main emphasis of this chapter is product-form solutions. There is a strong connection between product-form queueing networks, the notion of reversibility, as well as insensitivity to the service time distribution. Note that all our product-form results use only the mean arrival time and mean service time at every node but not the entire distribution. Of course, the link in itself is quasi-reversibility that results in partial balance equations. We have not gone into any of those details in this chapter but the reader is encouraged to refer to other resources, in particular this is well explained in Kelly [59].

Further, except for a few portions of this chapter, we only considered models with exponential distributions (unless the result is insensitive to the distribution). However, it is critical to point out that everything is not perfect when we have exponential distributions. In fact, if we are in a finite capacity queue (with or without blocking), we do not have a nice structure. For finite capacity networks, the joint stationary distribution is not product form. Hence, an exact analysis of these networks is limited to very small networks. Refer to some recent books on queueing networks such as Serfozo [96], Chao et al. [18], and Chen and Yao [19] for the most recent advances on the topic of product-form solutions.

Exercises

6.1 Consider a queueing network of single-server queues shown in the Figure 6.16. Note that, external arrival is Poisson and service times are exponential. Derive the stability condition and compute (1) the expected number of customers in the network in steady state and (2) the fraction of time the network is completely empty in steady state.

6.2 Consider a seven-node single-server Jackson network where nodes 2 and 4 get input from the outside (at rate 5 per minute each on an average). Nodes 1 and 2 have service rates of 85, nodes 3 and 4 have service rates of 120, node 5 has a rate of 70, and nodes 6 and 7 have rates of 20 (all in units per minute). The routing matrix is given by

$$P = [p_{ij}] = \begin{pmatrix} \frac{1}{3} & \frac{1}{4} & 0 & \frac{1}{4} & 0 & \frac{1}{6} & 0 \\ \frac{1}{3} & \frac{1}{4} & 0 & \frac{1}{3} & 0 & 0 & 0 \\ 0 & 0 & \frac{1}{3} & \frac{1}{3} & \frac{1}{3} & 0 & 0 \\ \frac{1}{3} & 0 & \frac{1}{3} & 0 & \frac{1}{3} & 0 & 0 \\ 0 & 0 & 0 & \frac{4}{5} & 0 & 0 & \frac{1}{6} \\ \frac{1}{6} & 0 & \frac{1}{6} & \frac{1}{6} & \frac{1}{6} & \frac{1}{6} & 0 \\ 0 & \frac{1}{6} & \frac{1}{6} & \frac{1}{6} & \frac{1}{6} & 0 & \frac{1}{6} \end{pmatrix}.$$

FIGURE 6.16
Single-server queueing network.

Find the average number in the network and the mean delay at each node.

6.3 Consider a closed-queueing network of single-server stations. Let $\rho_i = a_i/\mu_i$. Show that the limiting joint distribution is given by

$$p(x) = \frac{1}{\gamma(C)} \prod_{i=1}^{N} \rho_i^{x_i} \quad \text{when } \sum x_i = C$$

where C is the number of customers in the system (note that the relationship between our usual normalizing constant $G(\cdot)$ and $\gamma(\cdot)$ used here is $\gamma(C) = 1/G(C)$). Also, show that the generating function $\tilde{G}(z)$ of $\gamma(C)$ is given by

$$\tilde{G}(z) = \sum_{C=0}^{\infty} \gamma(C)z^C = \prod_{i=1}^{N} \frac{1}{1 - \rho_i z}.$$

Now define $B_j(z)$ and $b_j(n)$ as follows:

$$B_j(z) = \prod_{i=1}^{j} \frac{1}{1 - \rho_i z} = \sum_{n=0}^{\infty} b_j(n)z^n \quad j = 1, 2, \ldots, N.$$

Thus, $B_N(z) = \tilde{G}(z)$. Show that

$$B_j(z) = \rho_j z B_j(z) + B_{j-1}(z) \quad j = 2, \ldots, N$$

with $B_1(z) = 1/(1 - \rho_1 z)$. Also, show that

$$b_1(n) = \rho_1^n \quad \text{with } b_j(0) = 1$$

and

$$b_j(n) = b_{j-1}(n) + \rho_j b_j(n-1).$$

Then one can use this recursion to compute $\gamma(C) = b_N(C)$. Thus, $\gamma(C)$ can be computed in $O(NC)$ time.

6.4 For a closed network of single-server nodes with C customers, show that

(a) $\lim\limits_{t\to\infty} P\{X_i(t) \geq j\} = \rho_i^j \dfrac{G(C)}{G(C-j)}.$

(b) $L_i = \lim\limits_{t\to\infty} E[X_i(t)] = \sum\limits_{j=1}^{C} \rho_i^j \dfrac{G(C)}{G(C-j)}.$

Hint: You may want to first derive the result that for a discrete random variable Z taking values 0, 1, 2, 3, ..., that is, $E[Z] = \sum\limits_{j=1}^{\infty} P(Z \geq j).$

6.5 A simple communication network consists of two nodes labeled A and B connected by two one-way communication links: line AB from A to B and line BA from B to A. There are 150 users at each node. The ith user ($1 \leq i \leq 150$) at node A (or B respectively) is denoted by A_i (or B_i respectively). User A_i has an interactive session set up with user B_i and it operates as follows. User A_i sends a message to user B_i. All the messages generated at node A wait in a buffer at node A for transmission on line AB in an FCFS fashion to appropriate users at node B. When user B_i receives the message, he or she spends a random amount of time, called think time, to generate a response to it. All the messages generated at node B wait in a buffer for transmission (in an FCFS fashion) to appropriate users at node A on line BA. When user A_i receives a response from user B_i, he or she goes into think mode and generates a response to it after a random amount of think time. This process of messages going back and forth continues forever. Suppose all think times are exponentially distributed with mean 2 min and that all the message transmission times are exponentially distributed with mean 0.5 s. Model this as a queueing network. What is the expected number of messages in the buffers at nodes A and B in steady state. This will require some careful computing, so if you like you can leave the result using summations.

6.6 The system in a restaurant called Takes-Makes can be modeled as a tandem queueing network of three stations (order placing, fixings, and cashier). Assume that customers arrive to the system according to a Poisson process with mean arrival of 1 per minute. The entering customers go straight to join a line to place their order. After placing their order, the customers join a line at the fixings station at the end of which they go to pay at the cashier. There is a single server at the order placing station, two servers at the fixings station and one server at the cashier station. The average service times at the order placing, fixings, and cashier stations are, respectively, 54, 90, and 48 s. Assume that service times at all stations are exponential. Compute the average number of customers in each station and the total in the system of three stations.

6.7 Consider a modification to the previous problem where the service times are now 48, 96, and 48 s, at the order placing, fixings, and cashier stations, respectively. How does the average time in the entire system of three stations in this modified scenario compare against that of the original scenario? Note that in the modified scenario, there would be more people waiting at the fixing station, which could be undesirable.

6.8 Consider another modification to Takes-Makes. What if each server took the order, did the fixings, and acted as the cashier? Assume that the system can be modeled as a single $M/M/4$ queue with average service time of 192 s ($54+90+48=192$). What is the average time in the system for each customer? Comment on the following: (a) compare against time in the system for scenarios in the previous two cases; (b) if the assumptions in the previous cases are true, would the service time distribution be exponential here; and (c) what are the physical implications and limitations (space, resources, etc.)?

6.9 A queueing network with six single-server stations is depicted in Figure 6.17. Externally arrivals occur according to a Poisson process with average rate 24 per hour. The service time at each station is exponentially distributed with mean (in minutes) 2, 3, 2.5, 2.5, 2, and 2, at stations A, B, C, D, E, and F, respectively. A percentage on any arc (i, j) denotes the probability that a customer after completing service in node i joins the queue in node j. Compute the average number of customers in the network. What would be the waiting time experienced by a customer that enters node A, goes to node C, and exits the system through node E?

6.10 A repair shop (Figure 6.18) can be modeled as a queueing network with four stations. Externally, products arrive at stations A and C according to Poisson processes with mean 10 products per day and 20 products per day, respectively. Products are repaired at one or

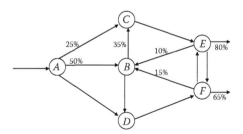

FIGURE 6.17
Schematic of a six-station network.

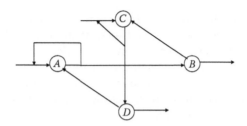

FIGURE 6.18
Schematic of a repair shop.

more stations among A, B, C, and D. The number of servers at stations A, B, C, and D are, respectively, 2, 3, 3, and 2. Average service times at stations A, B, C, and D are 0.08, 0.15, 0.075, and 0.08 days, respectively. Upon service completion, both at nodes A and C, with probability 0.2, the products immediately reenter those nodes. Also, 75% of the products after service in stations B and D exit the system. Compute the in-process inventory of the number of products in the system.

6.11 Answer the following TRUE or FALSE questions:

(a) Consider a closed Jackson network with N nodes and C customers. Is the following statement TRUE or FALSE? The interarrival times at any node can be modeled as IID exponential random variables.

(b) Consider a closed-queueing network with two identical stations and C customers in total. The service times both stations are according to $\exp(\mu)$ distribution. Each customer at the end of service joins either stations with equal probability. Let p_i be the steady-state probability there are i customers in one of the queues (and $C - i$ in the other). Is the following TRUE or FALSE?

$$p_i = \binom{C}{i} \frac{1}{2^C} \quad \text{for } i = 1, 2, \ldots, C$$

(c) Consider a closed Jackson network with two nodes and C customers. Let $X(t)$ be the number of customers in one of the nodes (the other node would have $C - X(t)$ customers). Is the following statement TRUE or FALSE? The CTMC $\{X(t), t \geq 0\}$ is reversible.

6.12 Answer the following multiple-choice questions:

(a) In a stable open Jackson network, which of the following is the average time spent by an entity in the network?

(i) $\sum_{i=1}^{N} L_i / \sum_{i=1}^{N} \lambda_i$

(ii) $\sum_{i=1}^{N} L_i / \lambda_i$

(iii) $\sum_{i=1}^{N} L_i / \sum_{i=1}^{N} a_i$

(iv) $\sum_{i=1}^{N} L_i / a_i$

(b) Consider an open Jackson network with two nodes. Arrivals occur externally according to $PP(\lambda)$ into each node. There is a single server that takes $\exp(\mu)$ service time and there is infinite waiting room at each node. At the end of service at a node, each customer chooses the other node or exits the system, both with probability $1/2$. What is the average number of customers in the entire network in steady state, assuming stability?

(i) $2\lambda/(\mu - \lambda)$

(ii) $2\lambda/(\mu - 3\lambda)$

(iii) $4\lambda/(\mu - 2\lambda)$

(iv) $4\lambda/(\mu - 4\lambda)$

6.13 Consider a series system of two single-server stations. Customers arrive at the first station according to a $PP(\lambda)$ and require $\exp(\mu_1)$ service time. Once service is completed, customers go to the second station where the service time is $\exp(\mu_2)$, and exit the system after service. Assume the following: both queues are of infinite capacity; $\lambda < \mu_1 < \mu_2$; no external arrivals into the second station; and both queues use FCFS and serve one customer at a time. Compute the LST of the CDF of the time spent in the entire series system.

6.14 Consider an open Jackson network with N nodes and a single-server queue in each node. We need to determine the optimal service rate μ_i at each node i for $i \in \{1, \ldots, N\}$ subject to the constraint $\mu_1 + \cdots + \mu_N \leq C$. The total available capacity is C, which is a given constant. Essentially we need to determine how to allocate the available capacity among the nodes. Use the objective of minimizing the total expected number in the system in steady state. Formulate a nonlinear program and solve it to obtain optimal μ_i values in terms of the net arrival rates a_j (for all j), which are also constants.

6.15 Consider a feed-forward tandem Jackson network of N nodes and the arrivals to the first node is $PP(\lambda)$. The service rate at node i is $\exp(\mu_i)$. We have a pool of S workers that need to be assigned one time to the N nodes. Formulate an optimization problem to determine s_i, the number of servers in node i (for $i \in \{1, \ldots, N\}$) so that $s_1 + \ldots s_N \leq S$, if the objective is to minimize the total expected number in the system in steady state. Describe an algorithm to derive the optimal allocation.

7

Approximations for General Queueing Networks

Jackson networks and their extensions that we saw in the previous chapter lent themselves very nicely for performance analysis. In particular, they resulted in a product-form solution that enabled us to decompose the queueing network so that individual nodes can be analyzed in isolation. However, a natural question to ask is: What if the conditions for Jackson networks are not satisfied? In this chapter, we are especially interested in analyzing general queueing networks where the interarrival times or service times or both can be according to general distributions. In particular, how do you analyze open queueing networks if each node cannot be modeled as an $M/M/s$, $M/G/c/c$, or $M/G/\infty$ queue? For example, the departure process from an $M/G/1$ FCFS queue is not even a renewal process, let alone a Poisson process. So if this set of customers departing from an $M/G/1$ queue join another queue, we would not know how to analyze that queue because we do not have results for queues where the arrivals are not according to a renewal process.

So how do we analyze general queueing networks? In the most general case, our only resort is to develop approximations. It is worthwhile to mention that one way is to use discrete-event simulations. There are several computer-simulation software packages that can be used to obtain queueing network performance measures numerically. At the time of writing this book, the commonly used packages are Arena and ProModel especially for manufacturing systems applications. However, although simulation methodology is arguably the most popular technique in the industry, it is not ideal for developing insights and intuition, performing quick what-if analysis, obtaining symbolic expressions that can be used for optimization and control, etc. For those reasons we will mainly consider analytical models that can suitably approximate general queueing networks. However, one of the objectives of the analytical approximations is that they must possess underlying theory, must be exact under special cases or asymptotic conditions and reasonably accurate under other conditions, and must be relatively easy to implement.

In that spirit we will consider, for example, approximations based on reflected Brownian motion (we will define and characterize this subsequently). One of the major benefits of reflected Brownian motion is that it can be modeled using just the mean and variance of the interarrival time as well as service time at each queue, hence it is easy to implement. Further, if the queueing network dynamics is indeed based on fluid and diffusion processes, then the marginal queue lengths are reflected Brownian motions. Further, the joint distribution of the queue lengths would be a *product form*. However, we will first describe the underlying theory that maps the general queueing network formulation into the reflected Brownian motion description. Then we will show the results are asymptotically exact and investigate the accuracy under other situations. We will begin with a simple FCFS network where there is only one class of customers and only one server per node of the network. Then we will extend to multiserver, multiclass, and other scheduling policies.

7.1 Single-Server and Single-Class General Queueing Networks

In this section, we consider a queueing network where all nodes have only a *single server* and there is only one class of customers. Customers arrive externally according to a renewal process and the service times at each node are according to general distributions. Notice that the description appears to be that of an open queueing network; however, if we let external arrival rates and external departure probabilities to be zero, then we indeed obtain a closed queueing network. Also, if the arrival processes are Poisson and service times exponential, we essentially have a Jackson network. For that reason, this network is sometimes referred to as the generalized Jackson network. Next we explicitly characterize these networks and set the notation as follows:

1. The queueing network consists of N service stations (or nodes).
2. There is one server at node i for all i satisfying $1 \leq i \leq N$.
3. Service times of customers at node i are IID random variables with mean $1/\mu_i$ and squared coefficient of variation $C_{S_i}^2$. They are independent of service times at other nodes.
4. There is infinite waiting room at each node and stability condition (we will describe that later) is satisfied at every node. Customers are served according to FCFS.
5. Externally, customers arrive at node i according to a renewal process with mean interarrival time $1/\lambda_i$ and squared coefficient of variation $C_{A_i}^2$. All arrival processes are independent of each other and the service times.
6. When a customer completes service at node i, the customer departs the network with probability r_i or joins the queue at node j with probability p_{ij}. Notice that $p_{ii} > 0$ is allowed and corresponds to

rejoining queue i for another service. It is required that for all i, $r_i + \sum_{j=1}^{N} p_{ij} = 1$ as all customers after completing service at node i either depart the network or join node j with probability p_{ij} for all $j \in [1, \ldots, N]$. The routing of a customer does not depend on the state of the network.

7. Define $P = [p_{ij}]$ as the routing matrix of p_{ij} values. Notice that if all λ_i and r_i values are zero, then we essentially have a closed queueing network in which case P is a stochastic matrix with all rows summing to one. Otherwise, we assume that $I - P$ is invertible (i.e., when we have an open queueing network), where I is the $N \times N$ identity matrix.

Our objective is to develop steady-state performance measures for such a queueing network.

7.1.1 G/G/1 Queue: Reflected Brownian Motion–Based Approximation

We begin by considering a special case of the queueing network described earlier. In particular, we let $N = 1$. Thus we essentially have a $G/G/1$ queue. Although we have developed bounds and approximations for the $G/G/1$ queue in Chapter 4, the key motivation here is to develop an approximation based on reflected Brownian motion so that it can be used to obtain a product-form solution in the more general case of N. We drop the subscript i from the previous notation since we have only one node. Hence arrivals occur according to a renewal process with interarrival times having a general distribution with mean $1/\lambda$ and squared coefficient of variation C_a^2. Likewise, the service times are IID random variables with mean $1/\mu$ and squared coefficient of variation C_s^2. Although at this time we assume that the interarrival times and service time distributions are known, our objective is to build an approximation for performance measures based only on λ, μ, C_a^2, and C_s^2 and not the entire distribution. We assume that the system is stable, that is, $\lambda < \mu$.

For the analysis, we first consider the arrival and service processes. Say $A(t)$ is the number of customers that arrived from time 0 to t, and $S(t)$ is the number of customers the server would process from time 0 to t if the server was busy during that whole time. Since the interarrival times and service times are IID, the processes $\{A(t), t \geq 0\}$ and $\{S(t), t \geq 0\}$ are renewal processes. We know from renewal theory that for large t, $A(t)$ is approximately normally distributed with mean λt and variance $\lambda C_a^2 t$. Likewise, for large t, $S(t)$ is normally distributed with mean μt and variance $\mu C_s^2 t$. Therefore, it is quite natural (as we will subsequently show) to analyze the system by approximating (for some large constant T) $\{A(t), t \geq T\}$ and $\{S(t), t \geq T\}$ as Gaussian processes, in particular, Brownian motions. This is a reasonable approximation, but a more rigorous approach to show that is to make a scaling argument appropriately as done in Chen and Yao [19] (see Chapter 8). However, before we proceed with the Brownian approximation, we first need to write down the relevant performance measures for the queue in terms of $A(t)$ and $S(t)$. We do that next.

We first describe some notation. Let $X(t)$ denote the number of customers in the $G/G/1$ queue at time t with $X(0) = x_0$, a given finite constant number of customers initially. To write down $X(t)$ in terms of $A(t)$ and $S(t)$, it is important to know how long the server was busy and idle during the time period 0 to t. For that, let $B(t)$ and $I(t)$, respectively, denote the total time the server has been busy and idle from time 0 to t. We emphasize that the server is work conserving, that means the server would be idle if and only if there are no customers in the system. Note that

$$B(t) + I(t) = t.$$

Thus we can write down an expression for $X(t)$ as

$$X(t) = x_0 + A(t) - S(B(t)), \tag{7.1}$$

since the total number in the system at time t equals all the customers that were present at time 0, plus all those that arrived in time 0 to t, minus those that departed in time 0 to t. Note that while writing the number of departures we need to be careful to use only the time the server was busy. Hence we get the preceding result.

Equation 7.1 is not conducive to obtain an expression for $X(t)$. Hence we rewrite as follows:

$$X(t) = U(t) + V(t)$$

where

$$U(t) = x_0 + (\lambda - \mu)t + (A(t) - \lambda t) - (S(B(t)) - \mu B(t)),$$
$$V(t) = \mu I(t).$$

Verify that the preceding result yields Equation 7.1 realizing that $B(t) + I(t) = t$. Note that we are ultimately interested in the steady-state distribution of $X(t)$; however, to do that we start by computing the expected value and variance of $U(t)$, for large t. Thus we have for large t

$$E[U(t)] = x_0 + (\lambda - \mu)t,$$
$$Var[U(t)] \approx \lambda C_a^2 t + \lambda C_s^2 t$$

since we have from renewal theory $E[A(t)] = \lambda t$, $E[S(B(t))|B(t)] = \mu B(t)$ for any $B(t)$. However, the variance result is a lot more subtle. Note that for a large t, the total busy period can be approximated as $B(t) \approx (\lambda/\mu)t$ since λ/μ is the fraction of time the server would be busy. Hence we write down $B(t) = (\lambda/\mu)t$ in the expression for $U(t)$ and then take the variance. However, since we know that $Var[A(t)] = \lambda C_a^2 t$ and $Var\left[S\left(\frac{\lambda}{\mu}t\right)\right] = \lambda C_s^2 t$, we get the preceding approximate result for $Var[U(t)]$. Note that $E[U(t)]$ is exact for any t, however, $Var[U(t)]$ is reasonable only for large t, that is, in the asymptotic case.

It is straightforward to see that for large t, if $A(t)$ and $S(t)$ are normally distributed random variables, then $U(t)$ is normally distributed with mean $x_0 + (\lambda - \mu)t$ and variance $\lambda(C_a^2 + C_s^2)t$. Therefore, if $\{A(t), t \geq T\}$ and $\{S(t), t \geq T\}$ can be approximated as Brownian motions for some large T, then from the description of $U(t)$, $\{U(t), t \geq T\}$ is also a Brownian motion with initial state x_0, drift $\lambda - \mu$, and variance $\lambda(C_a^2 + C_s^2)$.

Next we seek to answer the question: If $\{U(t), t \geq 0\}$ is a Brownian, then what about $\{X(t), t \geq 0\}$? To answer this we observe that the following relations ought to hold for all $t \geq 0$:

$$X(t) \geq 0, \tag{7.2}$$

$$\frac{dV(t)}{dt} \geq 0 \quad \text{with} \quad V(0) = 0 \tag{7.3}$$

and

$$X(t)\frac{dV(t)}{dt} = 0. \tag{7.4}$$

Before moving ahead, it may be worthwhile to explain the preceding relations. Condition (7.2) essentially holds because the number in the system, $X(t)$, cannot be negative. Likewise, since the amount of idle time $I(t)$ is a nondecreasing function of t, $V(t)$, which is $\mu I(t)$, is also a nondecreasing function of t, hence $dV(t)/dt \geq 0$. Of course $I(0)$ is zero based on the definition of $I(t)$, hence $V(0)$ is zero and we get condition (7.3). Equation 7.4 holds because whenever $X(t) > 0$, the idle time does not increase as the server is busy, hence $dV(t)/dt = 0$. However, if $X(t) = 0$, that is, the system is empty, the idle time increases, in other words $dV(t)/dt > 0$. Thus we have $X(t)dV(t)/dt = 0$.

We still have not addressed the issue of whether we know something about $U(t)$, what can we say about $X(t)$. In fact, if $U(t)$ is known there exists a unique $X(t)$ that satisfies conditions (7.2 through 7.4). We determine that by illustrating it as a problem.

Problem 63

Given $U(t)$, show that there exists a unique pair $X(t)$ and $V(t)$ such that $X(t) = U(t) + V(t)$, which satisfy conditions (7.2 through 7.4). Also show that the unique pair $X(t)$ and $V(t)$ can be written in terms of $U(t)$ as follows:

$$V(t) = \sup_{0 \le s \le t} \max\{-U(s), 0\}, \tag{7.5}$$

$$X(t) = U(t) + \sup_{0 \le s \le t} \max\{-U(s), 0\}. \tag{7.6}$$

Solution

We first show that $X(t)$ and $V(t)$ are a unique pair. For that, consider another pair $\hat{X}(t)$ and $\hat{V}(t)$ such that given a $U(t)$ for all t, $\hat{X}(t) = U(t) + \hat{V}(t)$ and the pair $\hat{X}(t)$ and $\hat{V}(t)$ satisfy conditions (7.2 through 7.4). Hence $\hat{X}(t) \ge 0$, $d\hat{V}(t)/dt \ge 0$ with $\hat{V}(0) = 0$, and $\hat{X}(t)d\hat{V}(t)/dt = 0$. If we show that the only way that can happen is if $X(t) = \hat{X}(t)$ (hence $V(t) = \hat{V}(t)$ because $U(t) = X(t) - V(t) = \hat{X}(t) - \hat{V}(t)$). For that, consider $1/2\{X(t) - \hat{X}(t)\}^2$ and write it as follows (the first equation is an artifact of integration and uses the fact that $X(0) = \hat{X}(0) = x_0$; the second equation is due to substituting $X(u) - \hat{X}(u)$ by $V(u) - \hat{V}(u)$ since $U(u) = X(u) - V(u) = \hat{X}(u) - \hat{V}(u)$ for all u; the last equation can be derived using condition (7.4), i.e., $X(u)dV(u) = 0$ and $\hat{X}(u)d\hat{V}(u) = 0$):

$$\frac{1}{2}\{X(t) - \hat{X}(t)\}^2 = \int_0^t \{X(u) - \hat{X}(u)\}d\{X(u) - \hat{X}(u)\},$$

$$= \int_0^t \{X(u) - \hat{X}(u)\}d\{V(u) - \hat{V}(u)\},$$

$$= -\int_0^t X(u)d\hat{V}(u) - \int_0^t \hat{X}(u)dV(u).$$

However, based on conditions (7.2) and (7.3), $X(u)$, $d\hat{V}(u)$, $\hat{X}(u)$, and $dV(u)$ are all ≥ 0. Thus we have

$$\frac{1}{2}\{X(t) - \hat{X}(t)\}^2 \le 0.$$

But the LHS is nonnegative. So the only way these result holds is if $X(t) = \hat{X}(t)$. Hence the pair $X(t)$ and $V(t)$ such that $X(t) = U(t) + V(t)$, which satisfy conditions (7.2 through 7.4), is unique.

Having shown that $X(t)$ and $V(t)$ is unique, we now proceed to show that $V(t)$ and $X(t)$ defined in Equations 7.5 and 7.6 satisfy $X(t) = U(t) + V(t)$ and the conditions (7.2 through 7.4). Subtracting Equation 7.5 from Equation 7.6, we get $X(t) = U(t) + V(t)$. Since $\max\{-U(s), 0\} \geq 0$ for all s, from the definition of $V(t)$, we have $V(t) \geq 0$. Thus if $U(t) \geq 0$, $X(t) = U(t) + V(t) \geq 0$. Now, if $U(t) < 0$, from the definition of the supremum we have $V(t) \geq -U(t)$ since $V(t) \geq -U(s)$ for all s such that $0 \leq s \leq t$ based on Equation 7.5. Since $V(t) \geq -U(t)$, $U(t) + V(t) \geq 0$, hence $X(t) \geq 0$. Thus condition (7.2) is verified. Next to show condition (7.3) is satisfied, we first show that since $U(0) = x_0$, which is nonnegative, we have $V(0) = \max\{-U(0), 0\} = 0$. Also, for any $dt \geq 0$ we have $V(t + dt) \geq V(t)$ since the supremum over time 0 to $t + dt$ must be greater than or equal to the supremum over any interval within 0 to $t + dt$, in particular, 0 to t. Thus we have

$$\frac{dV(t)}{dt} = \lim_{dt \to 0} \frac{V(t + dt) - V(t)}{dt} \geq 0$$

thereby satisfying condition (7.3). The only thing we are left to do is to show that $V(t)$ and $X(t)$ defined in Equations 7.5 and 7.6 satisfy Equation 7.4. Note that in Equation 7.5, if the supremum in the RHS occurs at $s = t$, then $dV(t)/dt > 0$, otherwise $dV(t)/dt = 0$. However, if the supremum indeed occurs at $s = t$, then $V(t) = -U(t)$, which means $V(t) + U(t) = 0$, implying $X(t) = U(t) + V(t) = 0$. Thus Equation 7.4 is satisfied since either $X(t) = 0$ or $dV(t)/dt = 0$. ∎

Based on the characteristics of this result, $X(t)$ is called the reflected process of $U(t)$ and $V(t)$ the regulator of $U(t)$. From the expression for $X(t)$ in Equation 7.6, we can conclude that if $\{U(t), t \geq 0\}$ is a Brownian motion with initial state x_0, drift θ, and variance σ^2, then $\{X(t), t \geq 0\}$ is a reflected Brownian motion (sometimes also called Brownian motion with reflecting barrier on the x-axis). To illustrate the Brownian motion and the reflected Brownian motion, we simulated a single sample path of $U(t)$ for 1000 time units sampled at discrete time points 1 time unit apart. Using numerical values for initial state $x_0 = 6$, drift -0.01, and variance 0.09, a sample path of $U(t)$ is depicted in Figure 7.1. Using the relation between $U(t)$ and $X(t)$ in Equation 7.6, we generated $X(t)$ values corresponding to $U(t)$. Although this is only a sample path, note from Figure 7.2, the reflected Brownian motion starts at x_0 and then keeps getting reflected at the origin and behaves like a Brownian motion at other points. Since the drift is negative, the reflected Brownian motion hits the origin (i.e., $X(t) = 0$) infinitely often. In the specific case of the $G/G/1$ queue, we showed that the $\{U(t), t \geq 0\}$ process especially for large t is a Brownian motion with drift $(\lambda - \mu)$ and variance $\lambda(C_a^2 + C_s^2)$. Then the $\{X(t), t \geq 0\}$ process is a corresponding reflected Brownian motion.

Having described an approximation for the number in the system process $\{X(t), t \geq 0\}$ as a reflected Brownian motion, we remark that it is rather awkward to approximate a discrete quantity $X(t)$ by a continuous process such

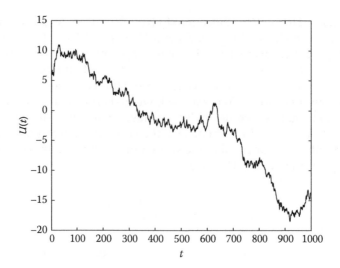

FIGURE 7.1
Simulation of Brownian motion $\{U(t), t \geq 0\}$.

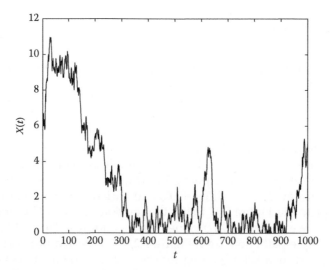

FIGURE 7.2
Generated reflected Brownian motion $\{X(t), t \geq 0\}$.

as a reflected Brownian motion. Bolch et al. [12] get around this by mapping the probability density function of the reflected Brownian motion to a probability mass function of the number in the system in steady state. That is certainly an excellent option. However, here we follow the literature on diffusion approximation or heavy-traffic approximations. In particular, we model the workload process (which is inherently continuous) as a reflected Brownian motion. For that, let $W(t)$ be the workload at time t, that is, the time it would take to complete service of every customer in the system at time t including any remaining service at the server. Chen and Yao [19] use the approximation

$$W(t) \approx \frac{X(t)}{\mu}$$

to relate the workload in the system to the number in the system. It is easy to see that if $\{X(t), t \geq 0\}$ is a reflected Brownian motion with initial state x_0, drift $\lambda - \mu$, and variance $\lambda(C_a^2 + C_s^2)$, then $\{W(t), t \geq 0\}$ is also a reflected Brownian motion with initial state x_0/μ, drift $(\lambda - \mu)/\mu$, and variance $\lambda(C_a^2 + C_s^2)/\mu^2$, when $W(t) = X(t)/\mu$. Next we state a problem to derive the steady-state distribution of $W(t)$ using the reflected Brownian motion. Subsequently, we will use the steady-state distribution to derive performance metrics for the $G/G/1$ queue.

Problem 64

Show that for any reflected Brownian motion $\{W(t), t \geq 0\}$ with drift θ (such that $\theta < 0$) and variance σ^2, the steady-state distribution is exponential with parameter $-2\theta/\sigma^2$.

Solution

Consider a reflected Brownian motion $\{W(t), t \geq 0\}$ with initial state w_0, drift θ, and variance σ^2. Let the cumulative distribution function $F(t, x; w_0)$ be defined as

$$F(t, x; w_0) = P\{W(t) \leq x | W(0) = w_0\}.$$

It is well known that $F(t, x; w_0)$ satisfies the following partial differential equation (which is also called forward Kolmogorov equation or Fokker–Planck equation):

$$\frac{\partial}{\partial t} F(t, x; w_0) = -\theta \frac{\partial}{\partial x} F(t, x; w_0) + \frac{\sigma^2}{2} \frac{\partial^2}{\partial x^2} F(t, x; w_0) \qquad (7.7)$$

with initial condition

$$F(0, x; w_0) = \begin{cases} 0 & \text{if } x < w_0 \\ 1 & \text{if } x \geq w_0 \end{cases}$$

and boundary conditions (based on the reflecting barrier on x-axis)

$$F(t, 0; w_0) = 0, \quad \text{whenever } w_0 > 0 \text{ and } t > 0.$$

Since we are only interested in the steady-state distribution of the reflected Brownian motion, we let $t \to \infty$. We denote $F(x)$ as the limiting distribution, that is,

$$F(x) = \lim_{t \to \infty} F(t, x; w_0).$$

Since there is no dependence on t, the partial derivative with respect to t is zero. Hence the differential Equation 7.7 reduces to

$$-\theta \frac{dF(x)}{dx} + \frac{\sigma^2}{2} \frac{d^2 F(x)}{dx^2} F(x) = 0$$

which can be solved by integrating once with respect to x and then using standard differential equation techniques to yield

$$F(x) = a e^{2\theta x/\sigma^2} - \frac{c}{\theta}$$

for some constants a and c that are to be determined. Using the boundary condition $F(0) = 0$ and the CDF property $F(\infty) = 1$ we get $c = -\theta$ and $a = -1$. Thus we have

$$F(x) = 1 - e^{2\theta x/\sigma^2}.$$

Therefore, any reflected Brownian motion with drift θ (such that $\theta < 0$) and variance σ^2 has a steady-state distribution that is exponential with parameter $-2\theta/\sigma^2$. ∎

For the $G/G/1$ queue described earlier, since we approximated the workload process $\{W(t), t \geq 0\}$ as a reflected Brownian motion with initial state $w_0 = x_0/\mu$, drift $\theta = (\lambda - \mu)/\mu$ and variance $\sigma^2 = \lambda(C_a^2 + C_s^2)/\mu^2$, we have the expected workload in steady state (using the preceding problem where we showed it is $\sigma^2/(2\theta)$) as:

$$\frac{\lambda(C_a^2 + C_s^2)}{2(1 - \rho)\mu^2}.$$

Therefore, if the $G/G/1$ queue is observed at an arbitrary time in steady state, the expected workload in the system would be approximately equal to the preceding quantity. Hence we can approximately equate that to the workload seen by an arriving customer in steady state. But the expected workload seen by an arriving customer in steady state is indeed W_q, the time in the system waiting for service to begin. In summary, we have an approximation for W_q of a $G/G/1$ queue as

$$W_q \approx \frac{\lambda(C_a^2 + C_s^2)}{2(1-\rho)\mu^2}. \tag{7.8}$$

We can immediately obtain L_q, W, and L using $L_q = \lambda W_q$, $W = W_q + 1/\mu$, and $L = \lambda W$. A good thing about the approximation (like the other $G/G/1$ approximations in Chapter 4) is that it only uses the mean and variance of the interarrival time and service time, but not the entire distribution. This makes it very convenient from a practical standpoint because one only needs to gather enough data for reasonably estimating the mean and variance. For that reason, this is also called a second-order approximation (a first-order approximation would be to use only the mean interarrival and service time, and use an $M/M/1$ expression) that uses the first two moments. However, a natural question to ask is: How good is the approximation (Equation 7.8)? For that, we first make the following remark and then subsequently address the issue further.

Remark 15

The expression for W_q in (Equation 7.8) is exact when the arrivals are Poisson. In other words, if we had an $M/G/1$ queue, then based on the preceding result

$$W_q = \frac{\lambda(1 + C_s^2)}{2(1-\rho)\mu^2}$$

since $C_a^2 = 1$. That gives us

$$L = \frac{\lambda}{\mu} + \frac{\lambda^2(1 + C_s^2)}{2(1-\rho)\mu^2}.$$

But that is exactly L as described in the Pollaczek–Khintchine formula (4.6). ∎

It has been reported extensively in the literature that the approximation for W_q given in Equation 7.8 is an excellent one. It is known to work remarkably well especially when the traffic intensity ($\rho = \lambda/\mu$) is neither too high nor too low (typically in the range $0.1 \leq \rho \leq 0.95$) and also the coefficient of variations are not too high (typically when $C_a^2 \leq 2$ and $C_s^2 \leq 2$). However, most of those results are based on using distributions such as Erlang, gamma, truncated normal, Weibull, hyperexponential, hypoexponential, and uniform for interarrival and service times. What if the interarrival times and service times were according to a Pareto distribution? Does the approximation (Equation 7.8) work well under the condition $0.1 \leq \rho \leq 0.95$, $C_a^2 \leq 2$, and $C_s^2 \leq 2$ where other distributions have shown promise? This motivates us to consider the next problem.

Problem 65

Simulate a $G/G/1$ queue with mean arrival rate $\lambda = 1$, with both interarrival times and service times both according to Pareto distributions. Generate 100 replications and in each run use 1 million customer departures to obtain the time in the system for various values of ρ, C_a^2, and C_s^2. Compare the simulation results against the approximation for W that can be derived from Equation 7.8.

Solution

For this problem, we are given λ, ρ, C_a^2, and C_s^2. Using Equation 7.8, we can derive an *analytical* expression (in terms of those four quantities) for the expected time in the system in steady state as

$$W = \frac{\rho}{\lambda} + \frac{\rho^2 (C_a^2 + C_s^2)}{2(1 - \rho)\lambda}.$$

Next, we can simulate a $G/G/1$ queue using the algorithm in Chapter 4 (Problem 37) when the interarrival times and the service times are according to Pareto distribution. For that, we first need the parameters of Pareto distributions. Let k_a and β_a be the parameters of the Pareto interarrival distribution, so that the CDF of interarrival times for $x \geq k_a$ is $1 - (k_a/x)^{\beta_a}$. Using the mean and variance of Pareto distributions, we can derive k_a and β_a as

$$\beta_a = 1 + \sqrt{1 + \frac{1}{C_a^2}},$$

$$k_a = \frac{\beta_a - 1}{\lambda \beta_a}.$$

Next, we can easily obtain the inverse distribution for the CDF as $F^{-1}(u) = k_a(1 - u)^{-1/\beta_a}$. In a similar manner, one can obtain k_s, β_s, and the inverse of the service time CDF by changing all the arrival subscripts from a to s and λ to $\mu = \lambda/\rho$.

We perform 100 replications of the simulation algorithm for each set of λ, ρ, C_a^2, and C_s^2 values. In each replication, we run the simulation till 1 million customers are served to obtain the average time in the system over the 1 million customers. Using the 100 sample averages we obtain a confidence

TABLE 7.1

Comparing Simulation Confidence Interval against Analytical
Approximation

ρ	C_a^2	C_s^2	Confidence Interval for W (via Simulations)	Analytical Approx. for W
0.9	1.00	2.00	(8.0194, 12.2170)	13.0500
0.6	1.00	2.00	(1.2347, 1.5313)	1.9500
0.9	0.49	2.00	(1.1135, 22.0137)	10.9845
0.9	2.00	2.00	(9.1947, 11.0142)	17.1000
0.6	2.00	2.00	(0, 7.7464)	2.4000
0.6	0.49	2.00	(1.0702, 1.7480)	1.7205
0.6	0.49	0.49	(0.7926, 0.8577)	1.0410
0.9	0.49	0.49	(3.8001, 3.9382)	4.8690
0.9	4.00	0.49	(5.8341, 5.9503)	19.0845
0.6	4.00	0.49	(0.8837, 0.9007)	2.6205

interval (three standard deviations on each side of the grand average across
the 100 sample averages). We tabulate the results in Table 7.1. Notice that
$\lambda = 1$ in all cases.   ∎

From the results of the previous problem it appears that in most cases
the analytically predicted W does not even fall within the confidence interval, leave alone being close to the grand average. However, it is not clear
from the preceding text whether that is because of the simulation with Pareto
distribution or the accuracy of the approximation. Looking at how wide
the confidence intervals are, it gives an indication that it might be an inac-
curacy in simulation. It is worthwhile further investigating this issue by
considering an $M/G/1$ queue where the service times are according to Pareto
distribution where we know the exact steady-state mean sojourn time. For a
similar situation where we run 100 replications of 1 million customers in
each replication for $\rho = 0.9$, $\lambda = 1$, $C_s^2 = 2$ with Pareto distribution for service times, we get a confidence interval of (10.2088, 14.2815). Although the
exact result using the Pollaczek–Khintchine formula in Equation 4.6 yields
$W = 13.05$, which is within the confidence interval, the grand average (of
12.2452) using the simulation runs is still significantly away considering it
is averaged over as many as 100 million customers. Another thing to notice
is that the grand average is smaller than the expected analytical value. One
reason for this is that the extremely rare event of seeing a humongous service time has not been realized in the simulations but has been accounted
for in the analytical models. There are some research papers that have
been addressing similar problems of determining how many simulation
runs would be needed to predict performance under Pareto distributions
with reasonable accuracy. Having said that, we will move along with the

understanding that the approximation (Equation 7.8) does work reasonably well for most problems especially when the load, variability, and distributions are reasonable.

7.1.2 Superpositioning, Splitting, and Flow through a Queue

Having developed an approximation based on reflected Brownian motion for a $G/G/1$ queue that uses only the first two moments of the interarrival times as well as service times, we are in a position to extend it to a network. For that, we go just a little beyond a single queue and consider the flows coming into a node as well as those that go out of a node. First consider a node as described in Figure 7.3. Note that the net arrival process to this queue is the superposition of flows from other nodes as well as from outside the network. The arrivals get processed according to FCFS in a single-server queue and then depart the queue. Upon departure, each customer joins different nodes (or exits the network) with certain probabilities. We assume that this is according to Bernoulli splitting, that is, each customer's choice of flows is independent of other customers as well as anything else in the system. The key question we aim to address is if we are given the first two moments of interarrival times for each of the superpositioning flows as well as those of the service times, can we obtain the first two moments of the interevent times of each of the split flows. Since every node in the network has such a structure (as described in Figure 7.3), one can immediately obtain the first two moments of the interarrival times at all the nodes of the network. Then by modeling each node as a $G/G/1$ queue we can obtain relevant performance measures.

7.1.2.1 Superposition

With that we begin with the first step, namely superposition of flows. Consider m flows with known characteristics that are superimposed into a single flow that act as arrivals to a queue. For $i = 1, \ldots, m$, let θ_i and C_i^2 be the average arrival rate of customers as well as the squared coefficient of variation of interarrival times on flow i. Likewise, let θ and C^2 be the effective arrival rate as well as the effective squared coefficient of variation of interarrival times obtained as a result of superposition of the m flows. Given

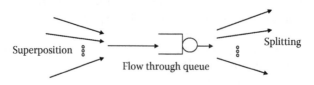

FIGURE 7.3
Modeling a node of the network.

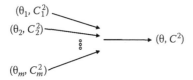

FIGURE 7.4
Superposition of flows.

$\theta_1, \ldots, \theta_m, C_1^2, \ldots, C_m^2$, the objective is to find θ and C^2. This is described in Figure 7.4. At this time, we assume that each of the m flows is an independent renewal process. In particular, for $i = 1, \ldots, m$, $\{N_i(t), t \geq 0\}$ is a renewal process with mean interrenewal time $1/\theta_i$ and squared coefficient of variation of interrenewal time C_i^2. Thus the counting process $\{N(t), t \geq 0\}$ that is a superposition of the m flows is such that

$$N(t) = N_1(t) + \cdots + N_m(t)$$

for all t. For large t we know from renewal theory that $N_i(t)$ for all $i \in \{1, \ldots, m\}$ is normally distributed with mean $\theta_i t$ and variance $\theta C_i^2 t$. Also, since $N(t)$ is the sum of m independent normally distributed quantities (for large t), $N(t)$ is also normally distributed. Taking the expectation and variance of $N(t)$ we get

$$E[N(t)] = E[N_1(t)] + \cdots + E[N_m(t)] = \left(\sum_{i=1}^{m} \theta_i \right) t,$$

$$Var[N(t)] = Var[N_1(t)] + \cdots + Var[N_m(t)] = \left(\sum_{i=1}^{m} \theta_i C_i^2 \right) t.$$

Working backward, if there is a renewal process $\{N(t), t \geq 0\}$ such that for large t, $N(t)$ is normally distributed with mean $E[N(t)]$ and variance $Var[N(t)]$, then the interrenewal times would have a mean $1/\theta$ where

$$\theta = \theta_1 + \cdots + \theta_m.$$

Further, the squared coefficient of variation of the interrenewal times is C^2 given by

$$C^2 = \sum_{i=1}^{m} \frac{\theta_i}{\theta} C_i^2.$$

Notice how this is derived by going backward, that is, originally we started with a renewal process with a mean and squared coefficient of variation of interrenewal times, and then derived the distribution of the counting process $N(t)$ for large t; but here we reverse that. It is crucial to notice that the actual interevent times of the aggregated superpositioned process is not IID, and hence the process is not truly a renewal process. However, we use the results as an approximation pretending the superimposed process to be a renewal process.

7.1.2.2 Flow through a Queue

Consider a $G/G/1$ queue where the interarrival times are according to a renewal process with mean $1/\theta_a$ and squared coefficient of variation C_a^2. Also, the service times are according to a general distribution with mean $1/\theta_s$ and squared coefficient of variation C_s^2. Thus given θ_a, C_a^2, θ_s, and C_s^2, our objective is to obtain the mean $(1/\theta_d)$ and squared coefficient of variation (C_d^2) of the interdeparture time in steady state assuming stability. These parameters are described pictorially in Figure 7.5. Of course, since the flow is conserved, average arrival rate must be equal to average departure rate. Hence we have

$$\theta_d = \theta_a.$$

Note that we have derived an expression for C_d^2 in Chapter 4 using mean value analysis (MVA) in Equation 4.16. We now rewrite that expression in terms of θ_a, θ_s, $\rho = \theta_a/\theta_s$, C_a^2, and C_s^2 as

$$C_d^2 = C_a^2 + 2\rho^2 C_s^2 + 2\rho(1-\rho) - 2\theta_a W(1-\rho).$$

Using the approximation in Equation 7.8 for W, we can rewrite it in terms of θ_a, θ_s, $\rho = \theta_a/\theta_s$, C_a^2, and C_s^2 as

$$W \approx \frac{1}{\theta_s} + \frac{\rho^2(C_a^2 + C_s^2)}{2\theta_a(1-\rho)}.$$

Hence we can write down an approximate expression for C_d^2 as

$$C_d^2 \approx (1-\rho^2)C_a^2 + \rho^2 C_s^2.$$

$$(\theta_a, C_a^2) \longrightarrow \boxed{}\!\!-\!\!O\!\!\longrightarrow (\theta_d, C_d^2)$$
$$(\theta_s, C_s^2)$$

FIGURE 7.5
Flow through a queue.

Before proceeding, it is worthwhile to take a moment to realize that based on the aforementioned expression, as the traffic intensity (or utilization) ρ goes close to one, the variability of the departures is similar to that of the service times. However, if ρ is low, the variability of the departures resemble those of the arrivals. Now, this is a fairly intuitive result since high utilization implies there is a line of customers waiting more often than not (hence the departures are close to the service variability), likewise, if the utilization is low, then most often than not customers are not waiting and the departures look like the arrivals.

7.1.2.3 Bernoulli Splitting

Consider the renewal process $\{N(t), t \geq 0\}$ with interrenewal times with mean $1/\theta$ and squared coefficient of variation C^2 as illustrated in the LHS of Figure 7.6. This renewal process is split according to a Bernoulli splitting process such that with probability p_i each event in the renewal process is classified as belonging to class i for $i = 1, \ldots, n$. Our objective is to derive for all i, the mean θ_i and coefficient of variation C_i^2 of interevent times of the split process of class i. For this, consider a particular split i. From the main renewal process $\{N(t), t \geq 0\}$ with probability p_i each event is marked as belonging to class i and with probability $1 - p_i$ it is marked as belonging to some other class (or split flow). Let $X_1, X_2, \ldots X_N$, denote the interevent times of the original renewal process $\{N(t), t \geq 0\}$. Likewise, let IID random variables $Y_1, Y_2, \ldots Y_N$, denote the interevent times of the split process along flow i. Note that Y_j is the sum of a geometrically distributed number of X_k values (since it is the number of events until a success is obtained) for all j and appropriately chosen k. In particular, $Y_1 = X_1 + X_2 + \cdots + X_N$ where N is a geometric random variable with probability of success of p_i. We are interested in the mean θ_i as well as squared coefficient of variation C_i^2 of the interevent times of the split process. Of course $E[Y_1] = 1/\theta_i$ and $Var[Y_1] = C_i^2/\theta_i^2$. Therefore, once we get $E[Y_1]$ and $Var[Y_1]$, we can immediately get θ_i and C_i^2 that we need.

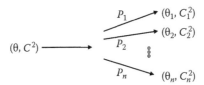

FIGURE 7.6
Bernoulli splitting of a flow.

Since $Y_1 = X_1 + X_2 + \cdots + X_N$, we can immediately compute $E[Y_1]$ and $Var[Y_1]$ by conditioning on N as follows:

$$E[Y_1] = E[E[Y_1|N]] = E\left[\frac{N}{\theta}\right] = \frac{1}{p_i\theta}$$

$$Var[Y_1] = E[Var[Y_1|N]] + Var[E[Y_1|N]]$$

$$= E\left[\frac{NC^2}{\theta^2}\right] + Var\left[\frac{N}{\theta}\right] = \frac{C^2}{(p_i\theta^2)} + \frac{(1-p_i)}{(p_i^2\theta^2)}.$$

Thus we have for all i such that $1 \le i \le n$

$$\theta_i = p_i\theta,$$
$$C_i^2 = C^2 p_i + 1 - p_i.$$

7.1.3 Decomposition Algorithm for Open Queueing Networks

Using what we have seen thus far, we are now in a position to put everything together in the context of a network of single-server nodes that we considered earlier in this section. Recall that we have a single-class network of N queues and in each queue or node there is a single server. Customers arrive externally according to a renewal process and the service times at each node are according to general distributions. It is worthwhile to recollect the notations that were described earlier. In particular, for $i = 1, \ldots, N$, service times of customers at node i are IID random variables with mean $1/\mu_i$ and squared coefficient of variation $C_{S_i}^2$. Likewise, externally customers arrive at node i according to a renewal process with mean interarrival time $1/\lambda_i$ and squared coefficient of variation $C_{A_i}^2$. When a customer completes service at node i, the customer departs the network with probability r_i or joins the queue at node j with probability p_{ij}. The crux of the analysis lies in the fact that if the arrival processes were truly Brownian motions and so were the service processes, then the number in each queue would be a reflected Brownian motion if the network is of feed-forward type. We do not prove this but refer the reader to Chen and Yao [19]. In fact, that would result in a product-form solution for the joint number in each node of the system. Therefore, we could use the approximate results for $G/G/1$ queues based on reflected Brownian motion to analyze this system. For that, we consider each node separately and approximate the net arrivals into each node as a renewal process using its mean and squared coefficient of variation. For that reason, such an analysis is called decomposition technique where we decompose the network into individual nodes and analyze them.

The main idea is that we substitute the random process $\{W_i(t), t \ge 0\}$ that denotes the workload in node i, by a reflected Brownian motion. Then the joint process $W_i(t)$ over all i yields a multidimensional reflected Brownian motion with a product-form steady-state distribution. However, we decompose the system and study the marginal distributions of the number of customers in each node by suitably approximating each node i as a $G/G/1$ queue. Of course we already know the service characteristics of each $G/G/1$ node since the service time i has mean $1/\mu_i$ and SCOV $C_{S_i}^2$. Also, the average aggregate arrival rate into node i, a_i can be computed relatively easily (using an identical result as that in Jackson networks). In particular, we obtain a_j for all $1 \leq j \leq N$ by solving

$$a_j = \sum_{i=1}^{N} p_{ij} a_i + \lambda_j. \tag{7.9}$$

Then for $1 \leq i, j \leq N$, if $P = [p_{ij}]$ an $N \times N$ matrix, then $[a_1 \ a_2 \ \ldots \ a_N] = [\lambda_1 \ \lambda_2 \ \ldots \ \lambda_N][I - P]^{-1}$. We assume that at every node stability is satisfied, that is, for all i such that $1 \leq i \leq N$,

$$a_i < \mu_i.$$

We also define the traffic intensity at node i as

$$\rho_i = \frac{a_i}{\mu_i}.$$

Once the a_i's are obtained, the only parameter left to compute to use in the $G/G/1$ result is the squared coefficient of variation of the interarrival times into node i that we denote as $C_{a,i}^2$. We use an approximation that the net arrivals into node i (for all i) is according to a renewal process and obtain an approximate expression for $C_{a,i}^2$ using the results for superposition, flow through a queue, and splitting that we saw in the previous section. But we require a feed-forward network for that so that we can perform superposition, flow, and splitting as we go forward in the network. However, as an approximation, we consider any generic network and show (see following problem) that for all j such that $1 \leq j \leq N$,

$$C_{a,j}^2 = \frac{\lambda_j}{a_j} C_{A_j}^2 + \sum_{i=1}^{N} \frac{a_i p_{ij}}{a_j} \{1 - p_{ij} + p_{ij}[(1 - \rho_i^2) C_{a,i}^2 + \rho_i^2 C_{S_i}^2]\}. \tag{7.10}$$

Therefore, there are N such equations for the N unknowns $C_{a,1}^2, \ldots, C_{a,N}^2$, which can be solved either by writing as a matrix form or by iterating starting with an arbitrary initial $C_{a,j}^2$ for each j. Once they are solved, we can use Equation 7.8 to obtain an approximate expression for the steady-state average number of customers in node j as

$$L_j \approx \rho_j + \frac{\rho_j^2 \left(C_{a,j}^2 + C_{S_j}^2\right)}{2(1 - \rho_j)}. \tag{7.11}$$

Note that this result is *exact* for the single-server Jackson network. Before progressing further, we take a moment to derive the expression for $C_{a,j}^2$ described in Equation 7.10 as a problem.

Problem 66

Using the terminology, notation, and expressions derived until Equation 7.10, show that for a feed-forward network

$$C_{a,j}^2 = \frac{\lambda_j}{a_j} C_{A_j}^2 + \sum_{i=1}^{N} \frac{a_i p_{ij}}{a_j} \{1 - p_{ij} + p_{ij}[(1 - \rho_i^2)C_{a,i}^2 + \rho_i^2 C_{S_i}^2]\}.$$

Solution

Consider the results for superposition of flows, flow through a queue, as well as Bernoulli splitting, all described in Section 7.1.2. Based on that we know that if node i has renewal interarrival times with mean $1/a_i$ and squared coefficient of variation $C_{a,i}^2$ (and service times with mean $1/\mu_i$ and squared coefficient of variation $C_{S_i}^2$), then the interdeparture times have mean $1/a_i$ and squared coefficient of variation $(1 - \rho_i^2)C_{a,i}^2 + \rho_i^2 C_{S_i}^2$. Since the probability that a departing customer from node i will join node j is p_{ij}, the interarrival times of customers from node i to node j has mean $a_i p_{ij}$ and squared coefficient of variation $1 - p_{ij} + p_{ij}[(1 - \rho_i^2)C_{a,i}^2 + \rho_i^2 C_{S_i}^2]$. Since the aggregate arrivals to node j is from all such nodes i as well as external arrivals, the effective interarrival times into node j has a squared coefficient of variation (defined as $C_{a,j}^2$) given by

$$C_{a,j}^2 = \frac{\lambda_j}{a_j} C_{A_j}^2 + \sum_{i=1}^{N} \frac{a_i p_{ij}}{a_j} \{1 - p_{ij} + p_{ij}[(1 - \rho_i^2)C_{a,i}^2 + \rho_i^2 C_{S_i}^2]\}$$

using the superposition results.   ∎

Next we describe an algorithm to obtain the performance measures of the general single-server open queueing network. Say we are given a stable single-server open queueing network (in the sense that every customer that arrives into the network exits it in a finite time) with N nodes and routing probabilities p_{ij}. The algorithm to obtain L_j for all $j \in \{1, 2, \ldots, N\}$ given mean and squared coefficient of variation of IID service times (i.e., $1/\mu_i$ and $C^2_{S_i}$) as well as mean and squared coefficient of variation of IID external interarrival times into node i (i.e., $1/\lambda_i$ and $C^2_{A_i}$) for all $i = 1, 2, \ldots, N$ is as follows:

1. Obtain aggregate customer arrival rate a_j into node j by solving Equation 7.9 for all j.
2. Using a_j, obtain the traffic intensity ρ_j at node j as $\rho_j = a_j/\mu_j$. Verify that all queues are stable, that is, $\rho_i < 1$ for all i.
3. Solve Equation 7.10 for all j and derive $C^2_{a,j}$ as a solution to the N simultaneous equations.
4. Using the derived values of ρ_j and $C^2_{a,j}$, plug them into Equation 7.11 and obtain L_j approximately for all j.

Before concluding this section, we present a numerical example to illustrate the algorithm for a specific single-server queueing network.

Problem 67

For a single-server open queueing network described in Figure 7.7, customers arrive externally only into node 0 and exit the system only from node 5. Assume that the interarrival times for customers coming externally has a mean 1 min and standard deviation 2 min. The mean and standard deviation of service times (in minutes) at each node is described in Table 7.2. Likewise, the routing probabilities p_{ij} from node i to j is provided

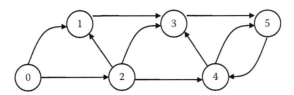

FIGURE 7.7
Single-server open queueing network.

TABLE 7.2

Mean and Standard Deviation of Service Times (min)

Node i	0	1	2	3	4	5
Mean $1/\mu_i$	0.8	1.25	1.875	1	1	0.5
Standard deviation $\sqrt{C^2_{S_i}/\mu_i^2}$	1	1	1	1	1	1

TABLE 7.3

Routing Probabilities from Node on Left to Node on Top

	0	1	2	3	4	5
0	0	0.5	0.5	0	0	0
1	0	0	0	1	0	0
2	0	0.2	0	0.3	0.5	0
3	0	0	0	0	0	1
4	0	0	0	0.2	0	0.8
5	0	0	0	0	0.4	0

in Table 7.3. Using this information compute the steady-state average number of customers at each node of the network as well as the mean sojourn time spent by customers in the network.

Solution

Since this is an open queueing network of single-server queues, to solve the problem we use the algorithm described earlier (with the understanding that the network is not a feed-forward network). Note the slight change from the original description where nodes moved from 1 to N; however, here it is from 0 to $N - 1$, where $N = 6$. From the problem description we can directly obtain P from Table 7.3 for nodes ordered $\{0, 1, 2, 3, 4, 5\}$ as

$$
P = \begin{bmatrix}
0 & 0.5 & 0.5 & 0 & 0 & 0 \\
0 & 0 & 0 & 1 & 0 & 0 \\
0 & 0.2 & 0 & 0.3 & 0.5 & 0 \\
0 & 0 & 0 & 0 & 0 & 1 \\
0 & 0 & 0 & 0.2 & 0 & 0.8 \\
0 & 0 & 0 & 0 & 0.4 & 0
\end{bmatrix}.
$$

Also, the external arrival rate is 1 per minute for node 0 and zero for all other nodes. Based on Equation 7.9 we have $a = [a_0 \ a_1 \ a_2 \ a_3 \ a_4 \ a_5] = [\lambda_0 \ \lambda_1 \ \lambda_2 \ \lambda_3 \ \lambda_4 \ \lambda_5][I - P]^{-1}$ with $\lambda_0 = 1$, and with $\lambda_i = 0$ for $i > 0$, we have $a = [1.0000 \ 0.6000 \ 0.5000 \ 0.9333 \ 0.9167 \ 1.6667]$ effective arrivals per minute into the various nodes. Using the service rates μ_i for every node i described in Table 7.2 we can obtain the traffic intensities for various nodes as $[\rho_0 \ \rho_1 \ \rho_2 \ \rho_3 \ \rho_4 \ \rho_5] = [0.8000 \ 0.7500 \ 0.9375 \ 0.9333 \ 0.9167 \ 0.8333]$. Clearly, all the nodes are stable; however, note how some nodes have fairly high traffic intensities.

Next we obtain the squared coefficient of variations for the effective interarrival times. For that, from the problem description we have the coefficient of variation for external arrivals as

$$
[C^2_{A_0} \ C^2_{A_1} \ C^2_{A_2} \ C^2_{A_3} \ C^2_{A_4} \ C^2_{A_5}] = [4 \ 0 \ 0 \ 0 \ 0 \ 0]
$$

and those of the service times we can easily compute from Table 7.2 as

$$[C_{S_0}^2 \ C_{S_1}^2 \ C_{S_2}^2 \ C_{S_3}^2 \ C_{S_4}^2 \ C_{S_5}^2] = [1.5625 \ 0.6400 \ 0.2844 \ 1.0000 \ 1.0000 \ 4.0000].$$

Thereby we use Equation 7.10 for all j to derive $C_{a,j}^2$ using the following steps: First obtain the row vector $\psi = [\psi_j]$ for all $i, j, k \in \{0, 1, 2, 3, 4, 5\}$ as

$$\psi_j = \lambda_j C_{A_j}^2 + [aP]_j - [a[p_{ik}^2]]_j + [[a_i \rho_i^2 C_{S_i}^2][p_{ik}^2]]_j$$

and then calculate

$$C_{a,j}^2 = \frac{[\psi[I - \mathrm{diag}(1 - \rho_i^2)[p_{ik}^2]]^{-1}]_j}{a_j}$$

where $\mathrm{diag}(b_k)$ is a diagonal matrix using elements of some vector (b_k). Using the preceding computation we can derive

$$\begin{bmatrix} \psi_j \\ a_j \end{bmatrix} = [4.0000 \quad 0.9750 \quad 1.0000 \quad 0.5461 \quad 1.4149 \quad 0.8716]$$

and

$$[C_{a,j}^2] = [4.0000 \quad 1.5819 \quad 1.7200 \quad 1.0107 \quad 1.5349 \quad 1.0308].$$

Now, using the values of ρ_j, $C_{a,j}^2$, and $C_{S_j}^2$ for $j = 0, 1, 2, 3, 4, 5$ in Equation 7.11 we can obtain the row vector of the mean number in each node in steady state as $L_0 = 9.7$, $L_1 = 3.2497$, $L_2 = 15.0312$, $L_3 = 14.0701$, $L_4 = 13.6969$, and $L_5 = 11.3143$. Also, the mean sojourn time spent by customers in the network in steady state is $\sum_j L_j / \sum_j \lambda_j = 67.0623$ min. ■

7.1.4 Approximate Algorithms for Closed Queueing Networks

Having seen a decomposition algorithm to obtain performance measures for single-server *open* queueing networks, a natural thing to wonder is what about closed queueing networks? As it turns out we can develop approximations for closed queueing networks as well. In fact, we will show two algorithms where we will decompose the network into individual nodes and solve each node using the $G/G/1$ results described earlier. Further, the analysis would rely heavily on the notation used for the open queueing networks case. Essentially, all the notation remains the same; however, for the sake of completion we describe them once again here. We consider a single-class closed network of N queues and in each queue or node there is a single server. There are C customers in total in the network, with no external arrivals or departures. For $i = 1, \ldots, N$, service times of customers at node i are IID random variables with mean $1/\mu_i$ and squared coefficient of variation $C_{S_i}^2$. When a customer completes service at node i, the customer joins the queue at node j with probability p_{ij} so that the routing matrix $P = [p_{ij}]$ has all rows summing to one. We present two algorithms based on Bolch et al. [12], one for large C called *bottleneck approximation* and the other for small C that uses MVA. There are other algorithms such as maximum entropy method (see Bolch et al. [12]) that are not described here.

7.1.4.1 Bottleneck Approximation for Large C

The main difficulty in extending the open queueing network decomposition algorithm to closed queueing network is that it is not easy to obtain the mean $(1/a_j)$ and squared coefficient of variation $(C_{a,j}^2)$ of the effective interarrival times for the queue in each node j. Of course once they are obtained, we just use those and the service time mean $(1/\mu_j)$ and squared coefficient of variation $(C_{S_j}^2)$ to determine L_j for every node j. In fact, the only quantity we need is a_j for every j, since once we have that we just need to make a minor adjustment to Equation 7.10. To obtain a_j the key idea in the bottleneck approximation is to first identify the bottleneck node, and once that happens we determine the traffic intensities of all other nodes relative to the bottleneck node.

To identify the bottleneck node, we obtain visit ratios v_j for all $1 \le j \le N$ by solving

$$v_j = \sum_{i=1}^{N} p_{ij} v_i \tag{7.12}$$

and normalizing using $v_1 + v_2 + \cdots + v_N = 1$. This is in essence solving for $[v_1 \; v_2 \; \ldots \; v_N]$ in $[v_1 \; v_2 \; \ldots \; v_N] = [v_1 \; v_2 \; \ldots \; v_N]P$ and $v_1 + v_2 + \cdots + v_N = 1$, where for $1 \le i, j \le N$, $P = [p_{ij}]$ is the $N \times N$ routing matrix. Once v_j is obtained for all j, the *bottleneck* node is the one with the largest v_j/μ_j, that is, node b is the bottleneck if

$$b = \arg \max_{j \in \{1,2,\ldots,N\}} \left\{ \frac{v_j}{\mu_j} \right\}.$$

Define λ as $\lambda = \mu_b/v_b$. We obtain the traffic intensities ρ_j (for all j) using the following approximation:

$$\rho_j = \frac{\lambda v_j}{\mu_j}.$$

Of course that would result in $\rho_b = 1$ for the bottleneck node and hence we have to be careful as we will see subsequently.

We can also immediately obtain $a_i = \lambda v_i$ for all i. Then, the only parameter left to compute to use in the $G/G/1$ result is the squared coefficient of variation of the interarrival times into node i that we denote as $C_{a,i}^2$ for all i. Since the external arrival rate is zero, this would just be a straightforward adjustment of Equation 7.10 for all j such that $1 \le j \le N$ as:

$$C_{a,j}^2 = \sum_{i=1}^{N} \frac{a_i p_{ij}}{a_j} \{1 - p_{ij} + p_{ij}[(1 - \rho_i^2)C_{a,i}^2 + \rho_i^2 C_{S_i}^2]\}. \tag{7.13}$$

Here too there are N such equations for the N unknowns $C_{a,1}^2, \ldots, C_{a,N}^2$, which can be solved either by writing as a matrix form or by iterating starting with an arbitrary initial $C_{a,j}^2$ for each j. Once they are solved, we can use Equation 7.8 to obtain an approximate expression for the steady-state average number of customers in node j for all $j \ne b$ as

$$L_j \approx \rho_j + \frac{\rho_j^2(C_{a,j}^2 + C_{S_j}^2)}{2(1 - \rho_j)}. \tag{7.14}$$

Note that if we used this equation for node b, the denominator would go to infinity. However, since the total number in the entire network is C, we can easily obtain L_b using

$$L_b = C - \sum_{j \ne b} L_j.$$

Next, we describe an algorithm to obtain the performance measures of the general single-server closed queueing network with a large number of customers C. Say we are given a single-server closed queueing network with N nodes and routing probabilities p_{ij}. The algorithm to obtain L_j for all $j \in \{1, 2, \ldots, N\}$ given mean and squared coefficient of variation of IID service times (i.e., $1/\mu_i$ and $C_{S_i}^2$) for all $i = 1, 2, \ldots, N$ is as follows:

1. Obtain visit ratios v_j into node j by solving Equation 7.12 for all j.
2. Identify the bottleneck node b as the node with the largest v_i/μ_i value among all $i \in [1, N]$.
3. Let $\lambda = \mu_b/v_b$ and obtain aggregate customer arrival rate a_j into node j as $a_j = \lambda v_j$ for all j.
4. Using a_j, obtain the traffic intensity ρ_j at node j as $\rho_j = a_j/\mu_j$.

5. Solve Equation 7.13 for all j and derive $C_{a,j}^2$ using the N simultaneous equations.

6. Using the derived values of ρ_j and $C_{a,j}^2$ for all $j \neq b$, plug them into Equation 7.14 and obtain L_j approximately.

7. Finally, $L_b = C - \sum_{i \neq b} L_i$.

7.1.4.2 MVA Approximation for Small C

Recall the MVA for product-form single-server closed queueing networks that we derived in Section 6.3.3. One can essentially go through that exact same algorithm with a modification in just one expression. However, for the sake of completeness we provide the entire analysis here. We first describe some notation (anything not defined here is provided prior to the bottle-neck approximation). Define the following steady-state measures for all $i = 1, \ldots, N$ and $k = 0, \ldots, C$:

- $W_i(k)$: Average sojourn time in node i when there are k customers (as opposed to C) in the closed queueing network
- $L_i(k)$: Average number in node i when there are k customers (as opposed to C) in the closed queueing network
- $\lambda(k)$: Measure of average flow (sometimes also referred to as throughput) in the closed queueing network when there are k customers (as opposed to C) in the network

We do not have an expression for any of the preceding metrics, and the objective is to obtain them iteratively. However, before describing the iterative algorithm, we first explain the relationship between those parameters.

As a first approximation, we assume that the arrival theorem described in Remark 14 holds here too. Thus in a network with k customers (such that $1 \leq k \leq C$) the expected number of customers that an arrival to node i (for any $i \in \{1, \ldots, N\}$) would see is $L_i(k-1)$. Note that $L_i(k-1)$ is the steady state expected number of customers in node i when there are $k - 1$ customers in the system. Further, the net mean sojourn time experienced by that arriving customer in steady state is the average time to serve all those in the system upon arrival plus that of the customer. Note that the average service time is $1/\mu_i$ for all customers waiting and $(1 + C_{S_i}^2)/(2\mu_i)$ for the customer in service (using the remaining time for an event in steady state for a renewal process). Thus we have

$$W_i(k) = \frac{1}{\mu_i} \left[\frac{1 + C_{S_i}^2}{2} + L_i(k-1) \right].$$

The preceding expression is a gross approximation because it assumes there is always a customer at the server at an arrival epoch, which is not true. However, since the server utilization is not known, we are unable to characterize the remaining service times more accurately and live with the preceding approximation hoping it would be conservative.

Let $v = [v_j]$ be a row vector, which is the solution to $v = vP$ and the v_j values sum to one. It is identical to the visit ratios in the *bottleneck* approximation given earlier. The aggregate sojourn time weighted across the network using the visit ratios is given by $\sum_{i=1}^{N} v_i W_i(k)$ when there are k customers in the network. Thereby we derive the average flow in the network using Little's law across the entire network as

$$\lambda(k) = \frac{k}{\sum_{i=1}^{N} v_i W_i(k)}$$

when there are k customers in the network. Thereby applying Little's law across each node i we get

$$L_i(k) = \lambda(k) W_i(k) v_i$$

when there are k customers in the network.

Using the preceding results, we can develop an algorithm to determine $L_i(C)$, $W_i(C)$, and $\lambda(C)$ defined earlier. The inputs to the algorithm are N, C, P, μ_i, and $C_{S_i}^2$ for all $i \in \{1, \ldots, N\}$. For the algorithm, initialize $L_i(0) = 0$ for $1 \le i \le N$ and obtain the v_i values for all $i \in \{1, \ldots, N\}$. Then for $k = 1$ to C, iteratively compute for each i (such that $1 \le i \le N$):

$$W_i(k) = \frac{1}{\mu_i} \left[\frac{1 + C_{S_i}^2}{2} + L_i(k-1) \right],$$

$$\lambda(k) = \frac{k}{\sum_{i=1}^{N} v_i W_i(k)},$$

$$L_i(k) = \lambda(k) W_i(k) v_i.$$

Next we present an example to illustrate the algorithms numerically for both small C and large C as well as compare them.

Problem 68

Consider a manufacturing system with five machines numbered 1, 2, 3, 4, and 5. The machines are in three stages as depicted in Figure 7.8. Four types of products are produced in the system, 32% are processed on machines 1, 2, and 4; 8% on machines 1, 2, and 5; 30% on machines 1, 3, and 4; and

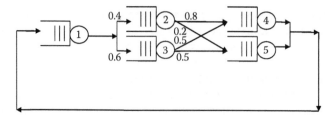

FIGURE 7.8
Single-server closed queueing network.

the remaining 30% on machines 1, 3, and 5. The manufacturing system is operated using a Kanban-like policy with a total of 30 products constantly in the system at all times. Whenever a product completes its three stages of production (i.e., on machines 1, 2 or 3, and 4 or 5), it leaves the system and immediately a new part enters the queue in machine 1. Also, because of the nature of the products and processes, the system uses Bernoulli splitting of routes. Thus when a product is processed on machine 1, it goes to either machine 2 or 3 with probabilities 0.4 or 0.6, respectively. Likewise, when a product is processed on machine 2, with probability 0.8 it is routed to machine 4 and with probability 0.2 to machine 5. Also, when a product is processed on machine 3, with probability 0.5 it is routed to machine 4 and with probability 0.5 to machine 5. Assume that all machines process in an FCFS fashion and there is enough waiting room at every machine to accommodate as many waiting products as necessary. Therefore, the system can be modeled as a single-server closed queueing network as described in Figure 7.8 with $N = 5$ and $C = 30$. Using the mean and standard deviation of the service times at each of the five machines in Table 7.4, compute the steady-state average number of products at each machine in steady state using the bottleneck approximation (assuming C is large) and MVA (assuming C is small).

Solution

We first set up the problem as a closed queueing network of single-server queues and then use both the bottleneck approximation (assuming C is large)

TABLE 7.4

Mean and Standard Deviation of Service Times (Min)

Machine i	1	2	3	4	5
Mean $1/\mu_i$	3	4	5	6	2
Standard deviation $\sqrt{C_{S_i}^2/\mu_i^2}$	6	2	1	3	2

and the MVA (assuming C is small). From the problem description we can directly obtain P ordered $\{1, 2, 3, 4, 5\}$ as

$$P = \begin{bmatrix} 0 & 0.4 & 0.6 & 0 & 0 \\ 0 & 0 & 0 & 0.8 & 0.2 \\ 0 & 0 & 0 & 0.5 & 0.5 \\ 1 & 0 & 0 & 0 & 0 \\ 1 & 0 & 0 & 0 & 0 \end{bmatrix}.$$

Based on Equation 7.12, we have $v = [v_1 \quad v_2 \quad v_3 \quad v_4 \quad v_5] = [0.3333 \; 0.1333 \; 0.2000 \; 0.2067 \; 0.1267]$ as the visit ratios into the various nodes.

7.1.4.2.1 Bottleneck Approximation

Using the service rates μ_i for every node i described in Table 7.4, we have $[v_i/\mu_i] = [1.0000 \; 0.5333 \; 1.0000 \; 1.2400 \; 0.2533]$. The highest v_i/μ_i is for machine 4, hence that is the bottleneck, that is, $b = 4$. Using $\lambda = \mu_4/v_4 = 0.8065$, we can get the approximate entering rate into each node j, a_j as λv_j, hence $a = [0.2688 \; 0.1075 \; 0.1613 \; 0.1667 \; 0.1022]$. Thus we have the relative traffic intensities for various nodes as

$$[\, \rho_1 \quad \rho_2 \quad \rho_3 \quad \rho_4 \quad \rho_5 \,] = [\, 0.8065 \quad 0.4301 \quad 0.8065 \quad 1.0000 \quad 0.2043 \,].$$

Next we obtain the squared coefficient of variations for the effective interarrival times. For that, we use the squared coefficient of variation of the service times, which we can easily compute from Table 7.4 as

$$[\, C_{S_1}^2 \quad C_{S_2}^2 \quad C_{S_3}^2 \quad C_{S_4}^2 \quad C_{S_5}^2 \,] = [\, 4.0000 \quad 0.2500 \quad 0.0400 \quad 0.2500 \quad 1.0000 \,].$$

Thereby we use Equation 7.13 for all j to derive $C_{a,j}^2$ using the following steps: First obtain the row vector $\psi = [\psi_j]$ for all $i, j, k \in \{1, 2, 3, 4, 5\}$ as

$$\psi_j = [aP]_j - [a[p_{ik}^2]]_j + [[a_i \rho_i^2 C_{S_i}^2][p_{ik}^2]]_j$$

and then calculate

$$C_{a,j}^2 = \frac{[\psi[I - \mathrm{diag}(1 - \rho_i^2)[p_{ik}^2]]^{-1}]_j}{a_j}$$

where $\mathrm{diag}(b_k)$ is a diagonal matrix using elements of some vector (b_k). Using the preceding computation, we can derive

$$\left[\frac{\psi_j}{a_j}\right] = [\, 0.1709 \quad 1.6406 \quad 1.9609 \quad 0.3706 \quad 0.5754 \,]$$

and

$$[C^2_{a,j}] = [\,0.5056 \quad 1.7113 \quad 2.0669 \quad 1.1213 \quad 0.9194\,].$$

Now, using the values of ρ_j, $C^2_{a,j}$, and $C^2_{S_j}$ for $j = 1,2,3,5$ in Equation 7.14, we can obtain the row vector of the mean number in each node in steady state as $L_1 = 8.3764$, $L_2 = 0.7484$, $L_3 = 4.3464$, and $L_5 = 0.2546$. We can obtain $L_4 = C - L_1 - L_2 - L_3 - L_5 = 16.2742$.

7.1.4.2.2 MVA Approximation

For the algorithm we initialized $L_i(0) = 0$ for $1 \le i \le 5$. We used the v_i values for all $i \in \{1, \ldots, 5\}$. Then for $k = 1$ to 30, iteratively compute for each i (such that $1 \le i \le 5$):

$$W_i(k) = \frac{1}{\mu_i}\left[\frac{1 + C^2_{S_i}}{2} + L_i(k-1)\right],$$

$$\lambda(k) = \frac{k}{\sum_{i=1}^{5} v_i W_i(k)},$$

$$L_i(k) = \lambda(k) W_i(k) v_i.$$

The approximation can be used to get L_i as $L_i(30)$. Writing a computer program we get $L_1 = 11.0655$, $L_2 = 0.4863$, $L_3 = 2.3016$, $L_4 = 15.8842$, and $L_5 = 0.2625$ with $\lambda(30) = 0.8207$, which is fairly close to the λ obtained in the bottleneck approximation. Note that the L_i values are relatively close. Upon running simulations with service times according to an appropriate gamma distribution, we get $L_1 = 5.5803$ (± 0.0250), $L_2 = 0.7811$ (± 0.0010), $L_3 = 3.6819$ (± 0.0119), $L_4 = 19.7360$ (± 0.0333), and $L_5 = 0.2205$ (± 0.0002) with the numbers in the brackets denoting width of 95% confidence intervals based on 100 replications. ∎

7.2 Multiclass and Multiserver Open Queueing Networks with FCFS

In Section 7.1, we considered only queueing networks with a single class of customers and every node in those networks had only a single server. In this section, we generalize the results in Section 7.1 to queueing networks with multiple classes of customers and at every node of these networks there could be more than one server. However, similar to Section 7.1, we consider only FCFS service discipline. In this section, the FCFS requirement is for two reasons. The first reason is to be able to derive expressions for the second moment of the interdeparture time from a queue (which is why Section 7.1 also used FCFS). The second reason is it would enable us to analyze each queue as an *aggregated* single-class queue. As described in Section 5.2.1 for multiclass $M/G/1$ queues, here too we can aggregate customers of all classes in a node in a similar fashion. Further, as we saw in Section 7.1.4, the approximations for general closed queueing networks were either rather naive or available only for networks with single-server queues. Therefore, we restrict our attention to only open queueing networks. With these introductory remarks, we proceed to analyze multiclass and multiserver open queueing networks with FCFS discipline.

7.2.1 Preliminaries: Network Description

As described earlier, we consider an open queueing network where each node can have one or more servers and there are many classes of customers in the networks. The classes of customers are differentiated due to the varying service time, arrival process, as well as routing requirement. Each class of customers arrive externally according to a renewal process and their service times at each node are according to general distributions. Further, although we assume routing is according to Bernoulli splitting, the routing probabilities could vary depending on the class of the customer. It is important to note that the notation is slightly different from before due to the large number of indices. Next we explicitly characterize these networks and set the notation for the "input" to the model as follows:

1. There are N service stations (or nodes) in the open queueing network. The outside world is denoted by node 0 and the other nodes are $1, 2, \ldots, N$. It is critical to point out that node 0 is used purely for notational convenience and we are not going to model it as a "node" for the purposes of analysis.

2. There are m_i servers at node i (such that $1 \le m_i \le \infty$), for all i satisfying $1 \le i \le N$.

3. The network has multiple classes of traffic and class switching is not allowed. Let R be the total number of classes in the entire network and each class has its unique external arrival process, service times at each node, as well as routing probabilities. They are explained next.

4. Externally, customers of class r (such that $r \in \{1, \ldots, R\}$) arrive at node i according to a renewal process such that the interarrival time has a mean $1/\lambda_{0i,r}$ and a squared coefficient of variation (SCOV) of $C_{0i,r}^2$. All arrival processes are independent of each other, the service times and the class.

5. Service times of class r customers at node i are IID with mean $1/\mu_{i,r}$ and SCOV $C^2_{S_{i,r}}$. They are independent of service times at other nodes.

6. The service discipline at all nodes is FCFS and independent of classes.

7. There is infinite waiting room at each node and stability condition (we will describe that later) is satisfied at every node.

8. When a customer of class r completes service at node i, the customer departs the network with probability $p_{i0,r}$ or joins the queue at node j (such that $j \in \{1, \ldots, N\}$) with probability $p_{ij,r}$. Note that $p_{ii,r} > 0$ is allowed and corresponds to rejoining queue i for another service. It is required that for all i and every r

$$\sum_{j=0}^{N} p_{ij,r} = 1$$

as all customers after completing service at node i either depart the network or join another node (say, j with probability $p_{ij,r}$). The routing of a customer does not depend on the state of the network. Also, every customer in the network, irrespective of the customer's class will eventually exit the network.

The preceding notations are summarized in Table 7.5 for easy reference. Our objective is to develop steady-state performance measures for such a

TABLE 7.5

Parameters Needed as Input for Multiserver and Multiclass Open Queueing Network Analysis

N	Total number of nodes
R	Total number of classes
i	Node index with $i = 0$ corresponding to external world, otherwise $i \in \{1, \ldots, N\}$
j	Node index with $j = 0$ corresponding to external world, otherwise $j \in \{1, \ldots, N\}$
r	Class index with $r \in \{1, \ldots, R\}$
$p_{ij,r}$	Fraction of traffic of class r that exits node i and join node j
m_i	Number of servers at node i for $i \geq 1$
$\mu_{i,r}$	Mean service rate of class r customers at node i
$C^2_{S_{i,r}}$	SCOV of service time of class r customers at node i
$\lambda_{0i,r}$	Mean external arrival rate of class r customers at node i
$C^2_{0i,r}$	SCOV of external interarrival time of class r customers at node i

queueing network. For example, we seek to obtain approximations for, say, the average number of customers of each class at every node in steady state.

7.2.2 Extending $G/G/1$ Results to Multiserver, Multiclass Networks

As we described earlier, our approach for the analysis is to aggregate the multiple classes into a single class and then analyze each queue in a decomposed manner as a single-class queue. Thus wherever appropriate, we borrow results from previous sections and chapters. We subsequently provide an algorithm in Section 7.2.3 that takes all the input metrics defined in Table 7.5 to obtain the mean number of customers of each class in each node of the network in steady state. If the reader would just like the algorithm, the remainder can be skipped and they could jump directly to Section 7.2.3.

To describe the analysis, we work our way backward from the performance measures all the way to the inputs given in Table 7.5. Note that the workload in queue i as seen by an arriving customer would not depend on the class of the customer. Also, the expected value of this workload in steady state would be the expected time this customer waits before beginning service. Hence we call this expected workload as W_{iq} in node i denoting the total time on average an arriving customer to node i would have to wait for service to begin considering FCFS service discipline (assuming system is in steady state and is stable). To compute W_{iq} we can model the queue in node i as a single aggregated class $G/G/m_i$ queue. In particular, if for the aggregated class, the interarrival time has mean $1/\lambda_i$ and SCOV $C_{A_i}^2$, and the service times are IID with mean $1/\mu_i$ and SCOV $C_{S_i}^2$, then using the $G/G/m_i$ approximation in Section 4.3 we have

$$W_{iq} \approx \frac{\alpha_{m_i}}{\mu_i} \left(\frac{1}{1 - \rho_i} \right) \left(\frac{C_{A_i}^2 + C_{S_i}^2}{2m_i} \right), \tag{7.15}$$

where $\rho_i = \lambda_i/(m_i\mu_i)$ and

$$\alpha_{m_i} = \begin{cases} \dfrac{\rho_i^{m_i} + \rho_i}{2} & \text{if } \rho_i > 0.7, \\[2mm] \rho_i^{\frac{m_i+1}{2}} & \text{if } \rho_i < 0.7. \end{cases}$$

Before proceeding it may be worthwhile to test the approximation for W_{iq} against simulations. We do that through a problem next.

Problem 69

Consider a single $G/G/m$ queue with interarrival and service times according to gamma distributions. Compare against simulations the expression for W_q given by

$$W_q \approx \frac{\alpha_m}{\mu} \left(\frac{1}{1 - \rho} \right) \left(\frac{C_A^2 + C_S^2}{2m} \right),$$

where $\rho = \lambda/(m\mu)$ and

$$\alpha_m = \begin{cases} \frac{\rho^m + \rho}{2} & \text{if } \rho > 0.7, \\ \rho_i^{\frac{m+1}{2}} & \text{if } \rho < 0.7. \end{cases}$$

Experiment with various cases $m = 2$ and $m = 9$, $\rho = 0.6$ and $\rho = 0.9$, SCOV of 0.25 and 4. For convenience let $\mu = 1$ for all experiments.

Solution

We consider all 16 cases of varying m, ρ, C_A^2, and C_S^2. For the simulations we perform 1 million service completions after an initial warm up period. The results are summarized in Table 7.6 where the simulations are a 95%

TABLE 7.6

Comparison of Simulation's 95% Confidence Interval against Analytical Approximation for W_q

Experiment	m	ρ	C_A^2	C_S^2	Approximation	Simulation
1	2	0.9	4	0.25	9.0844	9.4628 ± 0.0772
2	2	0.9	4	4	17.1000	16.5310 ± 0.1689
3	2	0.9	0.25	0.25	1.0688	0.9737 ± 0.0036
4	2	0.9	0.25	4	9.0844	7.9961 ± 0.0721
5	2	0.6	4	0.25	1.2345	1.4520 ± 0.0030
6	2	0.6	4	4	2.3238	2.4691 ± 0.0088
7	2	0.6	0.25	0.25	0.1452	0.0792 ± 0.0001
8	2	0.6	0.25	4	1.2345	0.9102 ± 0.0037
9	9	0.9	4	0.25	1.5199	1.9227 ± 0.0170
10	9	0.9	4	4	2.8609	3.0780 ± 0.0417
11	9	0.9	0.25	0.25	0.1788	0.1604 ± 0.0007
12	9	0.9	0.25	4	1.5199	1.2983 ± 0.0155
13	9	0.6	4	0.25	0.0459	0.1798 ± 0.0006
14	9	0.6	4	4	0.0864	0.19878 ± 0.0012
15	9	0.6	0.25	0.25	0.0054	0.0019 ± 0.00001
16	9	0.6	0.25	4	0.0459	0.0324 ± 0.0003

confidence interval on W_q over 50 replications and the approximation is the formula for W_q in the problem description. ∎

The approximation appears to be reasonable at least for the purposes of initial design and planning. In fact, in the literature it has been reported that the approximation for W_{iq} in Equation 7.15 is extremely effective in practice. Thus we will go ahead and use Equation 7.15 to derive our performance metrics in multiserver and multiclass queueing networks. However, note that none of the variables in the expression for W_{iq} described in Equation 7.15 are in terms of the input metrics defined in Table 7.5. Therefore, what is needed is a procedure to obtain those variables. For that, we consider not only superposition, splitting, and flow through a queue, but also the effect of multiple servers as well as multiple classes, which we describe first.

7.2.2.1 Flow through Multiple Servers

Besides the expression for W_{iq} given earlier, the effect of multiple servers is mainly in obtaining the SCOV of the departures from node i. Assuming that node i can be modeled as an aggregated single-class $G/G/m_i$ queue with interarrival time having mean $1/\lambda_i$ and SCOV $C_{A_i}^2$, and the service times IID with mean $1/\mu_i$ and SCOV $C_{S_i}^2$, then from Whitt [103] we can obtain the SCOV of the interdeparture times $C_{D_i}^2$. In particular, that expression is approximated as

$$C_{D_i}^2 = 1 + \frac{\rho_i^2 \left(C_{S_i}^2 - 1\right)}{\sqrt{m_i}} + \left(1 - \rho_i^2\right)\left(C_{A_i}^2 - 1\right)$$

where $\rho_i = \lambda_i/(m_i\mu_i)$. Note that the result is identical to that of a $G/G/1$ queue given in Section 7.1 if we let $m_i = 1$. Also, since the departures from $M/G/\infty$ and $M/M/m_i$ queues are Poisson, we can verify from above by using $C_{A_i}^2 = 1$ as well as letting $m_i \to \infty$ and $C_{S_i}^2 = 1$, respectively, for the two queues, we can show that $C_{D_i}^2$ is one in both cases.

7.2.2.2 Flow across Multiple Classes

Besides the parameters in Table 7.5, say the following are given (actually these are not known and would be solved recursively in the algorithm): $\lambda_{i,r}$, the mean arrival rate of class r customers to node i, and $C_{A_{i,r}}^2$, the SCOV of interarrival times of class r customers into node i. Using these we can obtain several metrics of interest. In particular, to obtain the traffic intensity $\rho_{i,r}$ of node i due to customers of class r can be computed as

$$\rho_{i,r} = \frac{\lambda_{i,r}}{m_i \mu_{i,r}}$$

since it is nothing but the ratio of class i arrival rate to the net service rate. Thus we can aggregate over all classes and obtain the effective traffic intensity into node i, ρ_i, as

$$\rho_i = \sum_{r=1}^{R} \rho_{i,r}.$$

It is worthwhile to point out that the condition for stability for node i is given by $\rho_i < 1$. We can also immediately obtain the aggregate mean arrival rate into note i, λ_i, as the sum of the arrival rate over all classes. Hence

$$\lambda_i = \sum_{r=1}^{R} \lambda_{i,r}.$$

Also, we can obtain the SCOV of the aggregate arrivals into node i (by aggregating over all classes)

$$C_{A_i}^2 = \frac{1}{\lambda_i} \sum_{r=1}^{R} C_{A_{i,r}}^2 \lambda_{i,r}.$$

This result can be derived directly from the SCOV of a flow as a result of superpositioning described in Section 7.1.2.

Having obtained all the expressions for the input to queue i, next we obtain the aggregate service parameters and the split output from node i. In particular, $\overline{\mu_i}$, the aggregate mean service rate of node i, can be obtained from its definition using

$$\overline{\mu_i} = \frac{1}{\sum_{r=1}^{R} \dfrac{\lambda_{i,r}}{\lambda_i} \dfrac{1}{m_i \mu_{i,r}}} = \frac{\lambda_i}{\rho_i}.$$

This result and the next one on the aggregate SCOV of service times across all classes at a node can be derived using that in Section 5.2.1 for $M/G/1$ queue with FCFS service discipline. Thus the aggregate SCOV of service time of node i, $C^2_{S_i}$, is given by

$$C^2_{S_i} = -1 + \sum_{r=1}^{R} \frac{\lambda_{i,r}}{\lambda_i} \left(\frac{\overline{\mu_i}}{m_i \mu_{i,r}} \right)^2 (C^2_{S_{i,r}} + 1).$$

Finally, the average flow rate for departures from node i into node j using splitting of flows can be computed. As defined earlier, $\lambda_{ij,r}$ is the mean departure rate of class r customers from node i that end up in node j. Since $p_{ij,r}$ is the fraction of traffic of class r that depart from node i join node j, we have

$$\lambda_{ij,r} = \lambda_{i,r} p_{ij,r}.$$

Next we are in a position to describe the SCOV due to superposition and splitting.

7.2.2.3 Superposition and Splitting of Flows

Besides the parameters in Table 7.5, say the following are given (actually these are not known and would be solved recursively in the algorithm): $\lambda_{ji,r}$, which is the mean arrival rate of class r customers from node j to node i, and $C^2_{ji,r}$, the SCOV of time between two consecutive class r customers that go from node j to node i. Using these we can obtain several metrics of interest. In particular, to obtain the effective class r arrival rate into node i, $\lambda_{i,r}$ can be computed by summing over all the flows from all nodes as

$$\lambda_{i,r} = \lambda_{0i,r} + \sum_{j=1}^{N} \lambda_{j,r} p_{ji,r}.$$

In fact, we would have to solve N such equations to obtain $\lambda_{i,r}$ for all i. In the single-class case recall that we solved that by inverting $I - P$ and multiplying that by external arrivals. Although something similar can be done here, care must be taken to ensure that the set of nodes that class r traffic traverses only must be considered (otherwise the generic $I - P$ is not invertible). Further, using the superposition of flows result in Section 7.1.2, we can derive $C^2_{A_{i,r}}$, the SCOV of class r interarrival times into node i as

$$C^2_{A_{i,r}} = \frac{1}{\lambda_{i,r}} \sum_{j=0}^{N} C^2_{ji,r} \lambda_{j,r} p_{ji,r}.$$

To obtain the SCOV of aggregate interarrival times into node i, $C^2_{A_i}$, we once again use the superposition result in Section 7.1.2 to get

$$C^2_{A_i} = \frac{1}{\lambda_i} \sum_{r=1}^{R} C^2_{A_{i,r}} \lambda_{i,r}.$$

We had seen earlier that we can obtain the effective SCOV of the departures from node i using $C^2_{S_i}$, the aggregate SCOV of service time of node i (also derived earlier), as

$$C^2_{D_i} = 1 + \frac{\rho_i^2 (C^2_{S_i} - 1)}{\sqrt{m_i}} + (1 - \rho_i^2)(C^2_{A_i} - 1).$$

Thus we can get $C^2_{ij,r}$, which is the SCOV of time between two consecutive class r customers going from node i to node j in steady state as

$$C^2_{ij,r} = 1 + p_{ij,r}(C^2_{D_i} - 1).$$

This result is directly from the splitting of flows departing a queue described in Section 7.1.2. With that we have defined all the parameters necessary for the algorithm to obtain steady-state performance measures for all classes of customers at all nodes.

7.2.3 QNA Algorithm

Here, we describe the queueing network analyzer (QNA) algorithm (based on Whitt [103]) to obtain the performance measures of general multiserver and multiclass open queueing networks. Say we are given all the parameters in Table 7.5. The algorithm to obtain $L_{j,r}$, the expected number of class r customers in node j in steady state for all $j \in \{1, 2, \ldots, N\}$ and $r \in \{1, 2, \ldots, R\}$, is described next. The parameters used in the algorithm for convenience are all shown in Table 7.7.

The decomposition algorithm essentially breaks down the network into individual nodes and analyzes each node as an independent $G/G/s$ queue with multiple classes (note that this is only FCFS and hence the multiple class aggregation is straightforward, similar to that for $M/G/1$ queues in Section 5.2.1). For the $G/G/s$ analysis, we require the mean arrival and service rates as well as the SCOV of the interarrival times and service times. The bulk of the algorithm in fact is to obtain them. There are three situations where this becomes tricky: when multiple streams are merged (superposition), when traffic flows through a node (flow), and when a single stream is divided into multiple streams (splitting). For convenience, we assume that just before entering a queue, the superposition takes place, which results in one stream.

TABLE 7.7

Parameters Obtained as Part of the QNA Algorithm

$\overline{\mu}_i$	Aggregate mean service rate of node i
$C^2_{S_i}$	Aggregate SCOV of service time of node i
$\lambda_{ij,r}$	Mean arrival rate of class r customers from node i to node j
$\lambda_{i,r}$	Mean class r arrival rate to node i (or mean departure rate from node i)
λ_i	Mean aggregate arrival rate to node i
$\rho_{i,r}$	Traffic intensity of node i due to customers of class r
ρ_i	Traffic intensity of node i across all classes
$C^2_{ij,r}$	SCOV of time between two consecutive class r customers going from node i to node j
$C^2_{A_{i,r}}$	SCOV of class r interarrival times into node i
$C^2_{A_i}$	Aggregate SCOV of interarrival times into node i
$C^2_{D_i}$	Aggregate SCOV of inter-departure times from node i
$L_{i,r}$	Expected number of class r customers in node i in steady state
W_{iq}	Expected time waiting before service in node i in steady state

Likewise, we assume that upon service completion, there is only one stream that gets split into multiple streams. The following are the three basic steps in the algorithm.

Step 1: Calculate the mean arrival rates, utilizations, and aggregate service rate parameters using the following:

$$\lambda_{i,r} = \lambda_{0i,r} + \sum_{j=1}^{N} \lambda_{j,r} p_{ji,r}$$

$$\lambda_{ij,r} = \lambda_{i,r} p_{ij,r}$$

$$\lambda_i = \sum_{r=1}^{R} \lambda_{i,r}$$

$$\rho_{i,r} = \frac{\lambda_{i,r}}{m_i \mu_{i,r}}$$

$$\rho_i = \sum_{r=1}^{R} \rho_{i,r} \quad \text{(condition for stability } \rho_i < 1 \forall i)$$

$$\overline{\mu}_i = \frac{1}{\sum_{r=1}^{R} \frac{\lambda_{i,r}}{\lambda_i} \frac{1}{m_i \mu_{i,r}}} = \frac{\lambda_i}{\rho_i}$$

$$C^2_{S_i} = -1 + \sum_{r=1}^{R} \frac{\lambda_{i,r}}{\lambda_i} \left(\frac{\overline{\mu}_i}{m_i \mu_{i,r}} \right)^2 (C^2_{S_{i,r}} + 1).$$

Step 2: Iteratively calculate the coefficient of variation of interarrival times at each node. Initialize all $C_{ij,r}^2 = 1$ for the iteration. Then until convergence performs (1), (2), and (3) cyclically.

(1) Superposition:

$$C_{A_{i,r}}^2 = \frac{1}{\lambda_{i,r}} \sum_{j=0}^{N} C_{ji,r}^2 \lambda_{j,r} p_{ji,r}$$

$$C_{A_i}^2 = \frac{1}{\lambda_i} \sum_{r=1}^{R} C_{A_{i,r}}^2 \lambda_{i,r}$$

(2) Flow:

$$C_{D_i}^2 = 1 + \frac{\rho_i^2 (C_{S_i}^2 - 1)}{\sqrt{m_i}} + (1 - \rho_i^2)(C_{A_i}^2 - 1)$$

(3) Splitting:

$$C_{ij,r}^2 = 1 + \frac{\lambda_{i,r}}{\lambda_i} p_{ij,r} (C_{D_i}^2 - 1).$$

Note that the splitting formula is exact if the departure process is a renewal process. However, the superposition and flow formulae are approximations. Several researchers have provided other expressions for the flow and superposition. As mentioned earlier, the preceding is QNA, described in Whitt [103].

Step 3: Obtain performance measures such as mean queue length and mean waiting times using standard $G/G/m$ queues. Treat each queue as an independent approximation. Choose α_{m_i} such that

$$\alpha_{m_i} = \begin{cases} \frac{\rho_i^{m_i} + \rho_i}{2} & \text{if } \rho_i > 0.7 \\ \\ \rho_i^{\frac{m_i+1}{2}} & \text{if } \rho_i < 0.7. \end{cases}$$

Then the mean waiting time for class r customers in the queue (not including service) of node i is approximately

$$W_{iq} \approx \frac{\alpha_{m_i}}{\mu_i} \left(\frac{1}{1 - \rho_i} \right) \left(\frac{C_{A_i}^2 + C_{S_i}^2}{2} \right).$$

Thus we can obtain the steady-state average number of class r customers in node i as

$$L_{i,r} = \frac{\lambda_{i,r}}{\mu_{i,r}} + \lambda_{i,r} W_{iq}.$$

Problem 70

Consider an e-commerce system where there are three stages of servers. In the first stage there is a single queue with four web servers; in the second stage there are four application servers, two of which are on the same node and share a queue; an in the third stage there are three database servers (two on one node sharing a queue and one on another node). The e-commerce system caters to two classes of customers but serve them in an FCFS manner (both across classes and within a class). This e-commerce system at the server end can be modeled as an $N=6$ node and $R=2$ class open queueing network with multiple servers. This multiserver and multiclass open queueing network is described in Figure 7.9. There are two classes of customers and both classes arrive externally only into node 1. Class-1 customers exit the system only from node 5 and class-2 customers only from node 6. Assume that the interarrival times for customers coming externally have a mean 1/3 units of time and standard deviation 2/3 time units for class-1 and a mean 1/6 units of time and standard deviation 1/4 units of time for class-2. The mean and standard deviation of service times (in the same time units as arrivals) at each node for each class are described in Table 7.10. Likewise, the routing probabilities $p_{ij,r}$ from node i to j is provided in Table 7.8 for $r=1$ (i.e., class-1) and Table 7.9 for $r=2$ (i.e., class-2). Using this information compute the steady-state average number of each class of customers at each node.

Solution

Based on the problem description we first cross-check to see that all input metrics described in Table 7.5 are given. Clearly, we have $N=6$ and $R=2$.

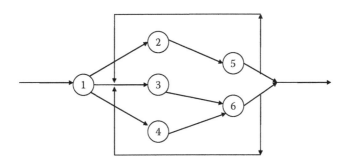

FIGURE 7.9
Multiserver and multiclass open queueing network.

TABLE 7.8

Class-1 Routing Probabilities $[p_{ij,1}]$ from Node on Left to Node on Top

	1	2	3	4	5	6
1	0	0.8	0.2	0	0	0
2	0	0	0	0	1	0
3	0	0	0	0	1	0
4	0	0	0	0	0	0
5	0	0	0.6	0	0	0
6	0	0	0	0	0	0

TABLE 7.9

Class-2 Routing Probabilities $[p_{ij,2}]$ from Node on Left to Node on Top

	1	2	3	4	5	6
1	0	0	0.1	0.9	0	0
2	0	0	0	0	0	0
3	0	0	0	0	0	1
4	0	0	0	0	0	1
5	0	0	0	0	0	0
6	0	0	0.75	0	0	0

TABLE 7.10

Number of Servers and Mean and Standard Deviation of Service Times for Each Class

Node i	1	2	3	4	5	6
No. of servers m_i	4	1	2	1	1	2
Class-1 service rate $\mu_{i,1}$	2	2.5	6	N/A	8	N/A
SCOV class-1 service $C_{S_{i,1}}^2$	2	0.64	1.44	N/A	0.81	N/A
Class-2 service rate $\mu_{i,2}$	4	N/A	20	6	N/A	15
SCOV class-2 service $C_{S_{i,2}}^2$	4	N/A	2	0.49	N/A	0.64

The routing probabilities (for all i and j) $p_{ij,1}$ and $p_{ij,2}$ are provided in Tables 7.8 and 7.9, respectively. Also, Table 7.10 lists m_i, $\mu_{i,1}$, $\mu_{i,2}$, $C_{S_{i,1}}^2$, and $C_{S_{i,2}}^2$ for $i = 1, 2, 3, 4, 5, 6$. Finally, $\lambda_{01,1} = 3$, $C_{01,1}^2 = 4$, $\lambda_{01,2} = 6$, and $C_{01,2}^2 = 2.25$, with all other $\lambda_{0i,r} = 0$ and $C_{0i,r}^2 = 0$. Now, we go through the three steps of the algorithm.

For step 1 note that class-1 customers go through nodes 1, 2, 3, and 5, whereas class-2 customers use nodes 1, 3, 4, and 5. Thus to solve

$$\lambda_{i,r} = \lambda_{0i,r} + \sum_{j=1}^{N} \lambda_{j,r} p_{ji,r}$$

for all $i \in [1,\ldots,6]$ and $r = 1,2$, we can simply consider the subset of nodes that class r customers traverse. Then using the approach followed in Jackson networks we can get the $\lambda_{i,r}$ values. However, in this example since there is only one loop in the network, one can obtain $\lambda_{i,r}$ in a rather straightforward fashion. In particular, we get

$$[\,\lambda_{1,1} \quad \lambda_{2,1} \quad \lambda_{3,1} \quad \lambda_{4,1} \quad \lambda_{5,1} \quad \lambda_{6,1}\,] = [\,3 \quad 2.4 \quad 5.1 \quad 0 \quad 7.5 \quad 0\,]$$

and

$$[\,\lambda_{1,2} \quad \lambda_{2,2} \quad \lambda_{3,2} \quad \lambda_{4,2} \quad \lambda_{5,2} \quad \lambda_{6,2}\,] = [\,6 \quad 0 \quad 18.6 \quad 5.4 \quad 0 \quad 24\,].$$

Using these results we can immediately obtain for all $i \in \{1,\ldots,6\}$, $j \in \{1,\ldots,6\}$, and $r = 1,2$

$$\lambda_{ij,r} = \lambda_{i,r} p_{ij,r}$$

using the preceding $\lambda_{i,r}$ values. Thus the aggregate arrival rate into node i across all classes, λ_i, can be obtained by summing over $\lambda_{i,r}$ for $r = 1,2$. Hence we have

$$[\,\lambda_1 \quad \lambda_2 \quad \lambda_3 \quad \lambda_4 \quad \lambda_5 \quad \lambda_6\,] = [\,9 \quad 2.4 \quad 23.7 \quad 5.4 \quad 7.5 \quad 24\,].$$

For all $i \in [1,\ldots,6]$ and $r=1,2$, we can write down $\rho_{i,r} = \lambda_{i,r}/m_i\mu_{i,r}$ and thereby obtain $\rho_i = \sum_{r=1}^{2} \rho_{i,r}$ as

$$[\,\rho_1 \quad \rho_2 \quad \rho_3 \quad \rho_4 \quad \rho_5 \quad \rho_6\,] = [\,0.75 \quad 0.96 \quad 0.89 \quad 0.9 \quad 0.9375 \quad 0.8\,].$$

Clearly, since $\rho_i < 1$ for all $i \in \{1,2,3,4,5,6\}$, all queues are stable. Further, the last computation in step 1 of the algorithm is to obtain the aggregate service rate and SCOV of service times at node i, which can be computed using

$$\overline{\mu_i} = \frac{\lambda_i}{\rho_i}$$

$$C_{S_i}^2 = -1 + \sum_{r=1}^{R} \frac{\lambda_{i,r}}{\lambda_i} \left(\frac{\overline{\mu_i}}{m_i \mu_{i,r}} \right)^2 (C_{S_{i,r}}^2 + 1).$$

For that, we get the following numerical values

$$[\overline{\mu_1} \quad \overline{\mu_2} \quad \overline{\mu_3} \quad \overline{\mu_4} \quad \overline{\mu_5} \quad \overline{\mu_6}] = [12 \quad 2.5 \quad 26.6292 \quad 6 \quad 8 \quad 30]$$

and

$$[C_{S_1}^2 \quad C_{S_2}^2 \quad C_{S_3}^2 \quad C_{S_4}^2 \quad C_{S_5}^2 \quad C_{S_6}^2] = [3.125 \quad 0.64 \quad 2.6291 \quad 0.49 \quad 0.81 \quad 0.64].$$

For step 2 of the algorithm we initialize all $C_{ij,r}^2 = 1$. Then, for all $i \in [1,\ldots,6]$ and $r = 1, 2$, we obtain

$$C_{A_{i,r}}^2 = \frac{1}{\lambda_{i,r}} \sum_{j=0}^{N} C_{ji,r}^2 \lambda_{j,r} p_{ji,r}$$

$$C_{A_i}^2 = \frac{1}{\lambda_i} \sum_{r=1}^{R} C_{A_{i,r}}^2 \lambda_{i,r}$$

$$C_{D_i}^2 = 1 + \frac{\rho_i^2 (C_{S_i}^2 - 1)}{\sqrt{m_i}} + (1 - \rho_i^2)(C_{A_i}^2 - 1).$$

Then we iteratively perform the preceding set of computations coupled with the following

$$C_{ij,r}^2 = 1 + p_{ij,r}(C_{D_i}^2 - 1)$$

till $C_{ij,r}^2$ converges for all $i, j \in [1,\ldots,6]$ and $r = 1, 2$. Upon convergence, we can obtain the aggregate SCOV of arrivals into node i, $C_{A_i}^2$ as

$$[C_{A_1}^2 \quad C_{A_2}^2 \quad C_{A_3}^2 \quad C_{A_4}^2 \quad C_{A_5}^2 \quad C_{A_6}^2] = [2.8333 \ 2.1198 \ 1.0449 \ 2.2598 \ 1.5487 \ 1.6753].$$

Finally, in step 3 of the algorithm, we obtain $\alpha_{m_i} = (\rho_i^{m_i} + \rho_i)/2$ since $\rho_i > 0.7$ for all i. Then using the approximation for W_{iq}, namely

$$W_{iq} \approx \frac{\alpha_{m_i}}{\mu_i} \left(\frac{1}{1 - \rho_i} \right) \left(\frac{C_{A_i}^2 + C_{S_i}^2}{2} \right)$$

for all $i \in \{1,\ldots,6\}$, we obtain

$$[W_{1q} \ W_{2q} \ W_{3q} \ W_{4q} \ W_{5q} \ W_{6q}] = [0.5295 \ 9.6637 \ 0.5203 \ 1.7474 \ 1.7312 \ 0.1282].$$

Upon running 50 simulation replications and using appropriate gamma distributions, we obtain a 95% confidence interval of corresponding W_{iq} values as

$$[0.6092 \pm 0.0037 \quad 21.8970 \pm 1.0294 \quad 0.4167 \pm 0.0033 \quad 2.2510 \pm 0.0182$$
$$2.1448 \pm 0.0615 \quad 0.1753 \pm 0.0006].$$

We can thereby obtain the steady-state average number of class r customers in node i using

$$L_{i,r} = \frac{\lambda_{i,r}}{\mu_{i,r}} + \lambda_{i,r} W_{iq}$$

as

$$[\,L_{1,1} \; L_{2,1} \; L_{3,1} \; L_{4,1} \; L_{5,1} \; L_{6,1}\,] = [\,3.0885 \; 24.1528 \; 3.5037 \; 0 \; 13.9212 \; 0\,]$$

and

$$[\,L_{1,2} \; L_{2,2} \; L_{3,2} \; L_{4,2} \; L_{5,2} \; L_{6,2}\,] = [\,4.6770 \; 0 \; 10.6083 \; 10.3359 \; 0 \; 4.6778\,].$$ ∎

Before moving ahead with other policies for serving multiclass traffic in queueing networks, we present a case study. This case study is based on the article by Chrukuri et al. [20] and illustrates an application of FCFS multiclass network approximation where many of the conditions required for the analysis presented earlier are violated. In particular, the system is (a) a multiclass queueing network with class switching, (b) a polling system with limited service discipline, and (c) there are finite-capacity queues with blocking. However, since these are not at the bottleneck node, the results are not terribly affected if we were to continue to use QNA. Further, the presentation of this case study is fairly different from the previous case studies in some ways.

7.2.4 Case Study: Network Interface Card in Cluster Computing

A network interface card (NIC) is a computer circuit board or card that is installed in a computer so that the computer can be connected to a network. Although personal computers and workstations on an Ethernet typically contain a NIC specifically designed for it, NICs are also used to interconnect clusters of computers or workstations so that the cluster can be used for high performance or massively parallel computations. This case study is regarding such an NIC used in cluster computing that adopts the virtual interface architecture (VIA). It contains a CPU called LANai (which is primarily responsible for coordinating data transfer between the computer where the NIC resides and the network, i.e., other nodes of the cluster); a host direct memory access (HDMA), which is used to transfer the data between the host memory and card buffer (SRAM); a Net Send DMA (NSDMA)

NIC

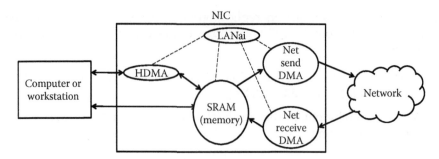

engine to transfer data from SRAM onto the network; and a Net Receive
DMA (NRDMA) engine to transfer data on to the SRAM from the network.
An example of such a VIA NIC is depicted in Figure 7.10 and its functioning
is described next.

The LANai goes through the following operations cyclically: polling the
doorbell queue to know if there is data that needs to be transferred, polling
the descriptor queue on SRAM that associates a doorbell with its data, and
polling the data queue. In addition, it programs NSDMA and NRDMA to
send and receive the data to and from the network, respectively. LANai polls
the doorbell queue and makes them available for HDMA to obtain the corresponding descriptors. Polled doorbells wait in a queue at HDMA to get
serviced on an FCFS basis. They are processed by the HDMA and the corresponding descriptors are stored in the descriptor queue on the SRAM. The
descriptors in this queue are polled by LANai and it makes them available
for HDMA to obtain the corresponding data. In the case of a send descriptor, LANai initiates the transfer of data from the host memory on to the data
queue on SRAM using HDMA. In the case of a receive descriptor, LANai
initiates the transfer of data from the network queue at NRDMA to the data
queue on SRAM using the NRDMA. LANai polls the data queue and if the
polled data is of type "send," it checks whether NSDMA is busy. If not, it
initiates the transfer of send data from SRAM data queue to NSDMA. If the
polled data is of type "receive," it initiates the transfer of data from SRAM
data queue to host memory using HDMA.

In summary, the operation of a VIA NIC can be modeled as a multiclass queueing network. An experiment was performed where only the send
messages were considered (without any data received from the network) to
measure interarrival times and service times. Based on this, the *send* process
is depicted in Figure 7.11. There are three stations in the queueing network
corresponding to the LANai, HDMA, and NSDMA. At the LANai there are
three queues. Entities arrive externally into one of the queues according to
$PP(\lambda)$, where λ is in units of per microsecond. It takes 22 µs to serve those

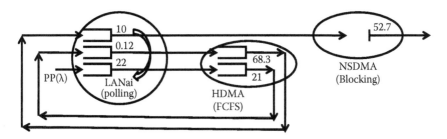

FIGURE 7.11
Multiclass queueing network model of an NIC with send.

entities. Then the entities go into the HDMA. Although there are two queues presented in the figure for illustration, there is really only one queue and entities are served according to FCFS. When an entity arrives for the first time it takes 21 μs to serve it and the entities go back to the LANai station where they are served in just 0.12 μs and they return to the HDMA for a second time, this time to be served in 68.3 μs. The entities return to the LANai. The LANai would spend 10 μs serving the entity if the NSDMA is idle, otherwise the LANai would continue polling its other queues. Note that the LANai knows if the NSDMA is idle or not because it also polls it to check if it is idle but the time is negligible. Once the entity reaches the idle NSDMA, it takes 52.7 μs to process and it exits the system. Note that all service times are deterministic.

In summary, the model of the system is that of a reentrant line. The first station LANai uses a limited polling policy where it polls each queue, serves at most one entity, and moves to the next queue. Also, entities in one of the queues (with 10 μs service time) can begin service only if the NSDMA station is idle. Thus the LANai would serve zero entities in that queue if the NSDMA is busy. Then the second station is HDMA, which uses a pure FCFS strategy. And the third station, NSDMA has no buffer, so it would get an entity only if it is idle, in other words, it blocks an entity in the corresponding LANai queue. Therefore, the system is a multiclass queueing network with reentrant lines (or class switching and deterministic routing). It uses a polling system with limited service discipline as opposed to FCFS at one of the nodes. There is a finite-capacity node that blocks one of the queues. Thus several of the conditions we saw in this section are violated. However, a quick glance would reveal that the bottleneck station is the HDMA. In particular, the utilizations of the NSDMA and HDMA are 52.7λ and 89.3λ, respectively. The utilization of the LANai is trickier to compute because the LANai could be idling due to being blocked by the NSDMA. Nonetheless, the fraction of time the LANai would have one or more entities would be only a little over 32.12λ.

Thus undoubtedly, the HDMA station would be the bottleneck. To analyze the system considering the significant difference in utilizations at the stations, we approximate using a three-station multiclass network where the policy is FCFS at all nodes, class-based routing, and all buffers (including the NSDMA) being of infinite capacity. Note that the bottleneck station, that is, HDMA, is relatively unaffected. Then we use the multiclass queueing network analysis presented in this section and obtain performance measures such as the number of entities in each station in steady state. We compared our results against simulations, which were done for the model depicted in Figure 7.11. We let $\rho_{HDMA} = 89.3\lambda$ be the traffic intensity at the bottleneck station, that is station 2 with HDMA. Then for $\rho_{HDMA} = 0.8032$ and $\rho_{HDMA} = 0.9825$, the probability that there were one or more entities in the LANai queues was 0.2935 and 0.364, respectively, which are both fairly close to 32.12λ as we had conjectured. In particular, they were 0.2891 and 0.3533, respectively. Despite the approximations, the analytical model was fairly close to the simulation results, especially for lower traffic intensities. For example, for $\rho_{HDMA} = 0.8032$ and $\rho_{HDMA} = 0.9825$, the total numbers in the HDMA queue were 1.5285 and 24.1981, respectively and in the LANai queues they were 0.0642 and 0.0980, respectively using the analytical models. The corresponding HDMA queue values in the simulations were 1.8653 and 30.499, respectively, and those of the LANai were 0.0854 and 0.1378, respectively. Although these are not terribly accurate, they are reasonably close. It is crucial to realize that the analytical model enables a user to quickly perform what-if analysis. In fact, for $\rho_{HDMA} = 0.8032$ and $\rho_{HDMA} = 0.9825$, the simulations took almost 21 and 38 min, whereas they were less than a second for the analytical model.

7.3 Multiclass and Single-Server Open Queueing Networks with Priorities

In Section 7.2, we considered multiclass open queueing networks with FCFS service discipline. A major change in this section is that we describe approximations when the service discipline is based on priorities across the various classes. In particular, we consider preemptive resume priority discipline described in Section 5.2.3 for the single node case. In the preemptive resume priority discipline the server would never process a low-priority job when there are high-priority jobs waiting. Thus if a high-priority job arrives when a low-priority job is being processed, the low-priority job gets preempted by the high-priority job. When there are no more high-priority jobs left, then the low-priority job's processing resumes. Further, within a class the jobs are processed according to FCFS rule. The first two sections that follow consider a global priority rule where the priority order of classes are maintained throughout the network. The last section allows local priorities where each node decides on a priority order among the different classes it processes.

It is worthwhile to point out that this entire section is devoted to *single-server* open queueing networks only. Wherever appropriate we will describe generalizations.

7.3.1 Global Priorities: Exponential Case

As described earlier, we consider an open queueing network where each node can have only one server. There are many classes of customers in the network. The classes of customers are differentiated due to the relative importance of the customers to the system as a whole. Hence we assume that there is a network-wide (i.e., global) priority order. We also allow classes to have varying service time, arrival locations, as well as routing requirements. Each class of customers arrive externally according to a Poisson process and their service times at each node are according to exponential distributions. Further, although we assume routing is according to Bernoulli splitting, the routing probabilities could vary depending on the class of the customer. Although the notation is very similar to the previous section, for the sake of completeness, we explicitly state them as follows:

1. There are N service stations (or nodes) in the open queueing network indexed $1, 2, \ldots, N$.

2. There is one server at node i, for all i satisfying $1 \leq i \leq N$.

3. The network has multiple classes of traffic and class switching is not allowed. Let R be the total number of classes in the entire network. There is a global priority order across the entire network with class-1 having highest priority and class R having lowest priority. Each class has its unique external arrival process, service times at each node, as well as routing probabilities. They are explained next.

4. Externally, customers of class r (such that $r \in \{1, \ldots, R\}$) arrive at node i according to a Poisson process such that the interarrival time has a mean $1/\lambda_{i,r}$. All arrival processes are independent of each other, the service times, and the class.

5. Service times of class r customers at node i are IID exponential random variables with mean $1/\mu_{i,r}$. They are independent of service times at other nodes.

6. The service discipline at all nodes is a static and global preemptive resume priority (with class a having higher priority than class b if $a < b$). Within a class the service discipline is FCFS.

7. There is infinite waiting room at each node and stability condition is satisfied at every node.

8. When a customer of class r completes service at node i, the customer joins the queue at node j (such that $j \in \{1, \ldots, N\}$) with probability $p_{ij,r}$. We require that $p_{ii,r} = 0$, although we eventually remark that it

is possible to easily extend to allowing customers to rejoin a queue immediately for another service. We assume that every customer in the network, irrespective of the customer's class will eventually exit the network. It is required that for all i and every r

$$\sum_{j=1}^{N} p_{ij,r} \le 1$$

as all customers after completing service at node i either depart the network or join another node (j). The routing of a customer does not depend on the state of the network.

Our objective is to develop steady-state performance measures for such a queueing network. For example, we seek to obtain approximations for, say, the average number of customers of each class at every node in the network in steady state. To analyze this queueing network, we use the MVA approach and make an approximation for what arriving customers see (akin to PASTA) to derive $L_{i,r}$, which as before is the expected number of class r customers in node i in steady state for all $i \in \{1, 2, \ldots, N\}$ and $r \in \{1, 2, \ldots, R\}$. In particular, the approximation we make is that an arriving customer to node i (irrespective of class) will see on average $L_{i,1}$ customers of class-1, $L_{i,2}$ customers of class-2, ..., $L_{i,R}$ customers of class R at that node.

Using that approximation, next we describe the algorithm (which is somewhat similar to PRIOMVA in Bolch et al. [12]) to obtain the performance measures of general single-server and multiclass open queueing networks with preemptive resume priorities. Say we are given N, R, $\lambda_{i,r}$, $p_{ij,r}$, and $\mu_{i,r}$ for all i, $j \in [1, \ldots, N]$ and $r \in [1, \ldots, R]$. The algorithm to obtain $L_{i,r}$, the expected number of class r customers in node i in steady state for all $i \in \{1, 2, \ldots, N\}$ and $r \in \{1, 2, \ldots, R\}$, is described next in terms of the metrics given earlier. For that, we define $a_{i,r}$ as the effective mean entering rate of class r customers into node i. However, at this moment we do not know $a_{i,r}$ or $L_{i,r}$, which are the two steps of the algorithm.

Step 1: Calculate the mean effective entering rates and utilizations using the following for all $i \in [1, \ldots, N]$ and $r \in [1, \ldots, R]$:

$$a_{i,r} = \lambda_{i,r} + \sum_{j=1}^{N} a_{j,r} p_{ji,r}$$

$$\rho_{i,r} = \frac{a_{i,r}}{\mu_{i,r}}.$$

To solve the first set of equations, if we select the subset of nodes that class r traffic visits, then we can create a traffic matrix $\hat{P}_r$ for that subset of nodes.

Then we can obtain the entering rate vector for that subset of nodes $(\hat{a}_r)$ in terms of the external arrival rate vector at those nodes $(\hat{\lambda}_r)$ as $\hat{a}_r = \hat{\lambda}_r (I - \hat{P}_r)^{-1}$, where I is the corresponding identity matrix. Thereby we can obtain $a_{i,r}$ for all $i \in [1, \ldots, N]$ and $r \in [1, \ldots, R]$. Also, the condition for stability is that

$$\sum_{r=1}^{R} \rho_{i,r} < 1$$

for every i.

Step 2: Sequentially compute for each node i (from $i = 1$ to $i = N$) and each r (from $r = 1$ to $r = R$)

$$L_{i,r} = \rho_{i,r} + a_{i,r} \sum_{k=1}^{r} \frac{L_{i,k}}{\mu_{i,k}} + L_{i,r} \sum_{k=0}^{r-1} \rho_{i,k}, \qquad (7.16)$$

where $\rho_{i,0} = 0$ for all i. Notice that it is important to solve for the r values sequentially because for the case $r = 1$ it is possible to derive $L_{i,1}$ using Equation 7.16, then for $r = 2$ to derive $L_{i,2}$ one needs $L_{i,1}$ in Equation 7.16, and so on.

Before illustrating the preceding algorithm using an example, we first explain the derivation of Equation 7.16 and also make a few remarks. Define $W_{i,r}$ as the sojourn time for a class r customer during a single visit to node i. Of course, due to Little's law we have for all $i \in [1, \ldots, N]$ and $r \in [1, \ldots, R]$

$$L_{i,r} = a_{i,r} W_{i,r}.$$

To write down $W_{i,r}$, consider a class r customer entering node i at an arbitrary time in steady state. The expected sojourn time for this customer equals the sum of the following three components: (1) the average time the customer waits for all customers of classes 1 to r that were in the system upon arrival to complete service (based on our approximation, the expected number of class k customers this arrival would see is $L_{i,k}$; hence the average time to serve all those class k customers is $L_{i,k}/\mu_{i,k}$), (2) during the sojourn time $W_{i,r}$ all arrivals of class k such that $k < r$ would have to be served (i.e., the $a_{i,k} W_{i,r}$ class k customers that arrive and each takes an average of $1/\mu_{i,k}$ of service time), and (3) the average time the customer spends being served (which is $1/\mu_{i,r}$). Hence we have

$$W_{i,r} = \sum_{k=1}^{r} \frac{L_{i,k}}{\mu_{i,k}} + \sum_{k=0}^{r-1} \left(W_{i,r} \frac{a_{i,k}}{\mu_{i,k}} \right) + \frac{1}{\mu_{i,r}}.$$

Using Little's law $L_{i,r} = a_{i,r} W_{i,r}$ we can rewrite this expression in terms of $L_{i,k}$ and obtain Equation 7.16. Having explained the algorithm, next we present a couple of remarks and then illustrate it using an example.

Remark 16

The algorithm is exact for the special cases of $N = 1$ with any R and $R = 1$ with any N. That is because for $N = 1$ and any R it reduces to a single station $M/G/1$ multiclass queues considered in Section 5.2.3. Also, for $R = 1$ and any N we get a Jackson network. It may be a worthwhile exercise to check the results for the preceding two special cases. Further, notice that under those special cases due to PASTA, arriving customers do see time-averaged number in the system. ∎

Remark 17

Although we required that $p_{ii,r}$ be zero for every i and r, the algorithm can certainly be used as an approximation even when $p_{ii,r} > 0$. Also, the algorithm can be seamlessly extended to non-preemptive priorities and closed queueing networks by suitably approximating what arrivals see. The results can be found in Bolch et al. [12]. ∎

Problem 71

Consider a small Internet service provider that can be modeled as a network with $N = 4$ nodes. There are $R = 3$ classes of traffic. Class-1 traffic essentially is control traffic that monitors the network states, and it is given highest priority with an external arrival rate of $\lambda_{i,1} = 0.5$ at every node i. Class-2 traffic arrives at node 1 at rate 3, then goes to server at node 2, and exits the network through node 4. Likewise, class-3 traffic arrives into node 1 at rate 2, then gets served at node 3, and exits the network after being served in node 4. Assume that all nodes have a single server, infinite waiting room, and the priority order is class-1 (highest) to 2 (medium) to 3 (lowest) at all nodes. The policy for priority is preemptive resume. Assume that external arrivals are according to Poisson processes and service times are exponentially distributed. The service rates (in the same time units as arrivals) at each node for each class are described in Table 7.11. Likewise, the routing probabilities $p_{ij,r}$ from node i to j are provided in Table 7.12 for $r = 1$ (i.e., class-1), Table

TABLE 7.11

Mean Service Rates for Each Class at Every Node

Node i	1	2	3	4
Class-1 service rate $\mu_{i,1}$	10	8	8	10
Class-2 service rate $\mu_{i,2}$	8	5	N/A	6
Class-3 service rate $\mu_{i,3}$	7	N/A	4	8

TABLE 7.12

Class-1 Routing Probabilities $[p_{ij,1}]$ from Node on Left to Node on Top

	1	2	3	4
1	0	0.2	0.2	0.2
2	0.25	0	0.25	0.25
3	0.25	0.25	0	0.25
4	0.2	0.2	0.2	0

TABLE 7.13

Class-2 Routing Probabilities $[p_{ij,2}]$ from Node on Left to Node on Top

	1	2	3	4
1	0	1	0	0
2	0	0	0	1
3	0	0	0	0
4	0	0	0	0

TABLE 7.14

Class-3 Routing Probabilities $[p_{ij,3}]$ from Node on Left to Node on Top

	1	2	3	4
1	0	0	1	0
2	0	0	0	0
3	0	0	0	1
4	0	0	0	0

7.13 for $r = 2$ (i.e., class-2), and Table 7.14 for $r = 3$ (i.e., class-3). Using this information compute the steady-state average number of each class of customers at each node.

Solution

To solve the problem we go through the two steps of the algorithm. For step 1, note that class-1 customers go through nodes 1, 2, 3, and 4; class-2 customers use nodes 1, 2, and 4; whereas class-3 customers use nodes 1, 3, and 4. Thus we solve

$$a_{i,r} = \lambda_{i,r} + \sum_{j=1}^{N} a_{j,r} p_{ji,r}$$

for all $i \in [1, \ldots, 4]$ and $r = 1, 2, 3$, to get

$$[a_{1,1} \quad a_{2,1} \quad a_{3,1} \quad a_{4,1}] = [1.5625 \quad 1.5 \quad 1.5 \quad 1.5625],$$

$$[a_{1,2} \quad a_{2,2} \quad a_{3,2} \quad a_{4,2}] = [3 \quad 3 \quad 0 \quad 3]$$

and

$$[a_{1,3} \quad a_{2,3} \quad a_{3,3} \quad a_{4,3}] = [2 \quad 2 \quad 0 \quad 2].$$

For all $i \in [1, \ldots, 4]$ and $r = 1, 2, 3$, we can write down $\rho_{i,r} = a_{i,r}/\mu_{i,r}$. Hence we get

$$[\rho_{1,1} \quad \rho_{2,1} \quad \rho_{3,1} \quad \rho_{4,1}] = [0.1563 \quad 0.1875 \quad 0.1875 \quad 0.1563],$$

$$[\rho_{1,2} \quad \rho_{2,2} \quad \rho_{3,2} \quad \rho_{4,2}] = [0.375 \quad 0.6 \quad 0 \quad 0.5]$$

and

$$[\rho_{1,3} \quad \rho_{2,3} \quad \rho_{3,3} \quad \rho_{4,3}] = [0.2857 \quad 0 \quad 0.5 \quad 0.25].$$

Notice that the condition for stability

$$\sum_{r=1}^{R} \rho_{i,r} < 1$$

is satisfied for every i.

Finally in step 2 of the algorithm, we obtain sequentially $L_{i,r}$ as

$$[L_{1,1} \quad L_{2,1} \quad L_{3,1} \quad L_{4,1}] = [0.1852 \quad 0.2308 \quad 0.2308 \quad 0.1852],$$

$$[L_{1,2} \quad L_{2,2} \quad L_{3,2} \quad L_{4,2}] = [0.9185 \quad 3.2308 \quad 0 \quad 1.6162]$$

and

$$[L_{1,3} \quad L_{2,3} \quad L_{3,3} \quad L_{4,3}] = [3.0179 \quad 0 \quad 1.7846 \quad 8.8081].$$

Upon running simulations with 50 replications, it was found that the results matched exactly for class-1, that is, $[L_{1,1} \quad L_{2,1} \quad L_{3,1} \quad L_{4,1}]$ since for class-1 the system is a standard open Jackson network. For classes 2 and 3, since the results are approximations, the 95% confidence interval for the *simulations* yielded

$$[\, L_{1,2} \;\; L_{2,2} \;\; L_{3,2} \;\; L_{4,2} \,] = [\, 0.9423 \pm 0.0022 \;\; 3.4073 \pm 0.1613 \;\; 0 \;\; 1.7050 \pm 0.0041 \,]$$

and

$$[\, L_{1,3} \;\; L_{2,3} \;\; L_{3,3} \;\; L_{4,3} \,] = [\, 3.1390 \pm 0.0178 \;\; 0 \;\; 2.3621 \pm 0.0108 \;\; 9.4006 \pm 0.0894 \,].$$

∎

7.3.2 Global Priorities: General Case

Here we briefly consider the case where interarrival and service times are according to general distributions. We essentially extend the approximations made in Section 7.3.1 in terms of what an arriving customer sees to the general case. Please refer to Section 7.3.1 as all the notation and assumptions remain the same. The only additional notation is that we define $\sigma_{i,r}^2$ as the variance of service times for class r customers in node i. All the parameters that are *given* as inputs to the model in Section 7.3.1 are inputs here as well, besides $\sigma_{i,r}^2$. We do not use the variance of interarrival times in our analysis and hence do not have a notation for that. For such a multi-priority and single-server open queueing network, our objective is to develop steady-state performance measures, especially the average number of customers of each class at every node in the network in steady state. The algorithm to obtain $L_{i,r}$, the expected number of class r customers in node i in steady state for all $i \in \{1, 2, \ldots, N\}$ and $r \in \{1, 2, \ldots, R\}$, is described next.

Step 1: Calculate the mean effective entering rates $a_{i,r}$ and utilizations $\rho_{i,r}$ using the following for all $i \in [1, \ldots, N]$ and $r \in [1, \ldots, R]$:

$$a_{i,r} = \lambda_{i,r} + \sum_{j=1}^{N} a_{j,r} p_{ji,r}$$

$$\rho_{i,r} = \frac{a_{i,r}}{\mu_{i,r}}.$$

To solve the first set of equations, if we select the subset of nodes that class r traffic visits, then we can create a traffic matrix $\hat{P}_r$ for that subset of nodes. Then we can obtain the entering rate vector for that subset of nodes ($\hat{a}_r$) in terms of the external arrival rate vector at those nodes ($\hat{\lambda}_r$) as $\hat{a}_r = \hat{\lambda}_r (I - \hat{P}_r)^{-1}$, where I is the corresponding identity matrix. Thereby we can obtain

$a_{i,r}$ for all $i \in [1, \dots, N]$ and $r \in [1, \dots, R]$. Also, the condition for stability is that

$$\sum_{r=1}^{R} \rho_{i,r} < 1$$

for every i. Note that this step is identical to step 1 in Section 7.3.1.

Step 2: Sequentially compute for each node i (from $i = 1$ to $i = N$) and each r (from $r = 1$ to $r = R$)

$$L_{i,r} = \rho_{i,r} + a_{i,r} \sum_{k=1}^{r} \frac{a_{i,k}}{2} \left(\sigma_{i,k}^2 + \frac{1}{\mu_{i,k}^2} \right) + a_{i,r} \sum_{k=1}^{r} \frac{L_{i,k} - \rho_{i,k}}{\mu_{i,k}} + L_{i,r} \sum_{k=0}^{r-1} \rho_{i,k} \quad (7.17)$$

where $\rho_{i,0} = 0$ for all i. Note that it is important to solve for the r values sequentially because for the case $r = 1$ it is possible to derive $L_{i,1}$ using Equation 7.17, then for $r = 2$ to derive $L_{i,2}$ one needs $L_{i,1}$ in Equation 7.17, and so on.

Before illustrating the preceding algorithm using an example, we first explain the derivation of Equation 7.17. Recall from Section 7.3.1 that $W_{i,r}$ is the sojourn time for a class r customer during a single visit to node i. Due to Little's law we have for all $i \in [1, \dots, N]$ and $r \in [1, \dots, R]$

$$L_{i,r} = a_{i,r} W_{i,r}.$$

To write down $W_{i,r}$, consider a class r customer entering node i at an arbitrary time in steady state. The expected sojourn time for this customer equals the sum of the following four components: (1) the average remaining service time (if there is one of class r or lesser), which can be computed using the fact that there is a probability $a_{i,k}/\mu_{i,k}$ that a class k customer is in service and its expected remaining service time is $\mu_{i,k}(\sigma_{i,k}^2 + 1/\mu_{i,k}^2)/2$, (2) the average time the customer waits for all customers of classes 1 to r that were in the queue (without any service started) upon arrival to complete service (based on our approximation, the expected number of class k customers this arrival would see is $L_{i,k} - \rho_{i,k}$; hence the average time to serve all those class k customers is $L_{i,k}/\mu_{i,k}$), (3) during the sojourn time $W_{i,r}$ all arrivals of class k such that $k < r$ would have to be served (i.e., the $a_{i,k}W_{i,r}$ class k customers that arrive and each takes an average of $1/\mu_{i,k}$ of service time), and (4) the average time the customer spends being served (which is $1/\mu_{i,r}$). Hence we have

$$W_{i,r} = \sum_{k=1}^{r} \frac{a_{i,k}}{2} \left(\sigma_{i,k}^2 + \frac{1}{\mu_{i,k}^2} \right) + \sum_{k=1}^{r} \frac{L_{i,k} - \rho_{i,k}}{\mu_{i,k}} + \sum_{k=0}^{r-1} \left(W_{i,r} \frac{a_{i,k}}{\mu_{i,k}} \right) + \frac{1}{\mu_{i,r}}.$$

Using Little's law $L_{i,r} = a_{i,r} W_{i,r}$ we can rewrite this expression in terms of $L_{i,k}$ and obtain Equation 7.17.

Problem 72

Consider an emergency ward of a hospital with four stations: reception where all arriving patients check in; triage where a nurse takes health-related measurements; lab area where blood work, x-rays, etc. are done; and an operating room. At each of the four stations, only one patient can be served at any time. Thus the system can be modeled as a single-server queueing network with $N=4$ nodes. There are $R=2$ classes of patients: class-1 corresponds to critical cases (and hence given preemptive priority) and class-2 corresponds to stable cases (hence lower priority). Both classes of patients arrive according to a Poisson process straight to node 1, that is, the reception. Note that node 2 is triage, node 3 is lab, and node 4 is the operating room. External arrival rate for class-1 and 2 patients are 0.001 and 0.08, respectively. The service rates (in the same time units as arrivals) and standard deviation of service time at each node for each class are described in Table 7.15. Likewise, the routing probabilities $p_{ij,r}$ from node i to j are provided in Table 7.16 for $r = 1$ (i.e., class-1) and Table 7.17 for $r = 2$ (i.e., class-2). Using this information compute the steady-state average number of each class of customers in the system as a whole.

TABLE 7.15

Mean Service Rates and Standard Deviation of Service Times for Each Class at Every Node

Node i	1	2	3	4
Class-1 service rate $\mu_{i,1}$	1	0.2	0.05	0.01
Class-1 service time std. dev. $\sigma_{i,1}$	1	2.5	10	20
Class-2 service rate $\mu_{i,2}$	0.5	0.1	0.025	0.02
Class-2 service time std. dev. $\sigma_{i,2}$	1	8	10	40

TABLE 7.16

Class-1 Routing Probabilities $[p_{ij,1}]$ from Node on Left to Node on Top

	1	2	3	4
1	0	1	0	0
2	0	0	0.4	0.3
3	0	0	0	0.5
4	0	0	0	0

TABLE 7.17

Class-2 Routing Probabilities $[p_{ij,2}]$ from Node on Left to Node on Top

	1	2	3	4
1	0	1	0	0
2	0	0	0.2	0.1
3	0	0	0	0.1
4	0	0	0	0

Solution

To solve the problem we go through the two steps of the algorithm. For step 1, note that both classes of customers have the possibility of going through all four nodes. Thus we solve

$$a_{i,r} = \lambda_{i,r} + \sum_{j=1}^{N} a_{j,r} p_{ji,r}$$

for all $i \in [1, \ldots, 4]$ and $r = 1, 2$, to get

$$[\, a_{1,1} \quad a_{2,1} \quad a_{3,1} \quad a_{4,1} \,] = [\, 0.001 \quad 0.001 \quad 0.0004 \quad 0.0005 \,]$$

and

$$[\, a_{1,2} \quad a_{2,2} \quad a_{3,2} \quad a_{4,2} \,] = [\, 0.08 \quad 0.08 \quad 0.016 \quad 0.0096 \,].$$

For all $i \in [1, \ldots, 4]$ and $r = 1, 2, 3$, we can write down $\rho_{i,r} = \lambda_{i,r}/\mu_{i,r}$. Using that we can write down

$$[\, \rho_{1,1} + \rho_{1,2} \quad \rho_{2,1} + \rho_{2,2} \quad \rho_{3,1} + \rho_{3,2} \quad \rho_{4,1} + \rho_{4,2} \,] = [\, 0.161 \quad 0.805 \quad 0.648 \quad 0.53 \,].$$

Notice that the condition for stability is satisfied for every i.

Finally, in step 2 of the algorithm, we obtain sequentially $L_{i,r}$ as

$$[\, L_{1,1} \quad L_{2,1} \quad L_{3,1} \quad L_{4,1} \,] = [\, 0.001 \quad 0.005 \quad 0.008 \quad 0.0514 \,]$$

and

$$[\, L_{1,2} \quad L_{2,2} \quad L_{3,2} \quad L_{4,2} \,] = [\, 0.1794 \quad 3.5182 \quad 1.2773 \quad 0.9889 \,].$$

Upon running simulations over 50 replications the results matched remarkably well that it is not worth reporting the numerical values of the 95% confidence intervals. Thus by summing over all i of $L_{i,r}$ for every r, we can state that in steady state there would be on average 0.0654 class-1 (critical condition) patients and 5.9638 class-2 patients in total in the system. ∎

7.3.3 Local Priorities

In this section, we consider the possibility of each node in a network to choose the priority order for serving customers. In other words, instead of a global priority, we have local priorities in each node. This topic has been studied extensively in the recent literature (albeit not under the title *local priorities*). Chen and Yao [19] provide an excellent treatment that includes a far more generalized version than what we present here. Further, Chen and Yao [19] also provide a very different method for approximately analyzing the performance, which is based on semi-martingale reflected Brownian motion (SRBM). Although we do not describe the SRBM approximation here, we leverage upon the state-space collapse argument they use as one of their approximations to study such systems.

We consider an open queueing network where each node can have only one server. There are many classes of customers in the network. The classes of customers are differentiated due to the routes taken. However, the nodes choose the priority order among the various classes. To keep the exposition simple, we assume that the nodes use a shortest-expected-processing-time-first policy. Within a class, the nodes serve according to FCFS. We also allow classes to have different service time at each node, arrival locations, as well as routing requirements. Except for the local priorities, almost everything else is similar to the global priority case. Hence refer to Section 7.3.1 for a complete list of notation (such as N, R, $p_{ij,r}$, λ_i, $\mu_{i,r}$, preemptive resume priority, and exponential interarrival, as well as service times). The only extra generalization we make here is that customers in a class can visit a particular node a deterministic number of times and each visit could have a different processing rate. Thus we denote $\mu_{i,r,k}$ as the service rate for all those class r customers in node i when they traverse it during the kth time (of course this is only if these customers go through that node more than once). We subsequently, illustrate this via a problem.

An interesting aspect that one encounters with the preceding description is that nodes may be unstable even if the traffic intensity is less than one at those nodes. Thus the traffic intensity condition is only necessary but not sufficient. We will address that issue of stability in Chapter 8 and here we just assume that the network is stable and all we are interested is in deriving performance measures. To analyze such a queueing network we use two approaches: one is the MVA approach we saw in Section 7.3.1 and the other is to use state-space collapse. Thus it is worthwhile to read through Section 7.3.1 before proceeding further.

7.3.3.1 MVA-Based Algorithm

Essentially, this is exactly the same as the algorithm in Section 7.3.1 with the only exception being the priority order in each node. Also because each class can visit a node more than once with a different service time during each visit, we encounter a few extra notations. Say we are given N, R, $\lambda_{i,r}$, $p_{ij,r}$, and $\mu_{i,r,k}$ for all $i, j \in [1, \ldots, N]$, $k \in [1, \ldots]$, and $r \in [1, \ldots, R]$. The algorithm to obtain $L_{i,r,k}$, the expected number of class r customers in node i during their kth visit to node i in steady state for all $i \in \{1, 2, \ldots, N\}$, $k \geq 1$ and $r \in \{1, 2, \ldots, R\}$, is described next in terms of the metrics given earlier. For that, we define $a_{i,r,k}$ as the effective mean entering rate of class r customers into node i for the kth time. However, at this moment we do not know $a_{i,r,k}$ or $L_{i,r,k}$, which are the two steps of the algorithm.

Step 1: Calculate the mean effective entering rates $a_{i,r,k}$. The technique is more or less similar to that in Section 7.3.1 except one has to be more careful with the possibility of multiple visits to a node. Then the utilization $\rho_{i,r,k}$ for all $i \in [1, \ldots, N]$, $r \in [1, \ldots, R]$ and appropriate k values is

$$\rho_{i,r,k} = \frac{a_{i,r,k}}{\mu_{i,r,k}}.$$

Also, the necessary condition for stability

$$\sum_{r=1}^{R} \sum_{k} \rho_{i,r,k} < 1$$

must be met for every i. It is critical to note that this condition may not be sufficient for stability.

Step 2: At node i let $q_i(r, k)$ be the priority given to class r traffic at node i when it enters it for the kth time. Thus if $q_i(r, k) < q_i(s, n)$, then class r traffic entering node i for the kth time is given higher priority than class s traffic that enters it for the nth time. Sequentially compute for each node i (from $i = 1$ to $i = N$) each r and appropriate k (in the exact priority order $q_i(r, k)$)

$$L_{i,r,k} = \rho_{i,r,k} + a_{i,r,k} \sum_{j=1}^{q_i(r,k)} \frac{L_{i,q_i^{-1}(j)}}{\mu_{i,q_i^{-1}(j)}} + L_{i,r,k} \sum_{j=0}^{q_i(r,k)-1} \rho_{i,q_i^{-1}(j)} \tag{7.18}$$

where $\rho_{i,0} = 0$ for all i and $q_i^{-1}(j)$ is the inverse function of $q_i(\cdot, \cdot)$ such that $q_i^{-1}(j) = (s, n)$ if $q_i(s, n) = j$. Note that it is important to solve for the (r, k) values in a sequence corresponding to the priority order $q_i(r, k)$ at each node i.

The preceding might be clearer once we present an example. Before that we describe an alternate technique.

7.3.3.2 State-Space-Collapse-Based Algorithm

In multiclass queueing networks with priorities, by performing a heavy-traffic scaling (see Chen and Yao [19]), one can see that only the lowest priority class has a nonempty queue. In fact, while running simulations with nodes with reasonably high traffic intensities, one sees that the queue lengths (not including any at server) of the lowest priority classes are significantly higher than other classes. Refer to the examples for multiclass priority queues/networks throughout this book for evidence of this phenomenon. Of course this would not be accurate if the traffic intensity due to the lowest priority class is negligible. Assuming that is not the case, the key approximation in this algorithm is to assume that at any time instant in steady state, the workload in the system equals to the workload of the lowest priority.

However, notice that the workload in the system at any time at a given node for a given arrival process to that node would not depend on the discipline. Of course in our case, the arrival processes do depend on the discipline, so this workload conservation across the network would only be a heuristic approximation. By making that approximation we can study the network as though it is a multiclass FCFS network and obtain the steady-state workload. Then we can say that this workload is equal to that of the lowest priority class and all other classes at the node have zero workload. Of course we can always use QNA in Section 7.2.3 to derive the workload in each node in steady state for the equivalent FCFS queueing network. The only concern is that QNA does not allow the general routing considered here (especially the multiple visits to a node and each visit having a different service time). We can get around that quite easily as we will show in an example to follow. Next we describe the algorithm.

Step 1: Modify the multiclass queueing network so that the assumptions of QNA in Section 7.2.3 are satisfied. Then assume that the service is FCFS at all nodes. Go through QNA till almost the last step where we compute

$$W_{iq} \approx \frac{\alpha_{m_i}}{\overline{\mu_i}} \left(\frac{1}{1 - \rho_i} \right) \left(\frac{C_{A_i}^2 + C_{S_i}^2}{2} \right).$$

This indeed is the steady-state expected workload in the system at node i. The terms in the RHS are presented just to recognize where to stop in the QNA. For the state-space collapse, we assume that any arriving customer would see a workload of W_{iq}, all of which is due to the lowest priority customers at node i.

Step 2: At node i let $a_{i,r}$ and $\rho_{i,r}$ be the entering rate and traffic intensity due to a class r customer, respectively, at that node. Since we do not allow more than one entry (with different service times) of each class into a node, we do not need the k subscript we used for the MVA-based algorithm. However, the total number of classes may have increased. We can obtain $a_{i,r}$ and $\rho_{i,r}$ using the QNA in step 1. Now, we switch back to the preemptive resume priority policy. Let $\rho_{i,\hat{r}}$ be the sum of traffic intensities of all classes strictly higher priority than r at node i. We consider the following two cases:

- If r is not the lowest priority in node i, then $L_{i,r} \approx \rho_{i,r}/(1 - \rho_{i,\hat{r}})$ due to the state-space collapse assumption.
- If r is the lowest priority in node i, then $L_{i,r} \approx \rho_{i,r} + a_{i,r}W_{iq}/(1 - \rho_{i,\hat{r}})$ since due to state-space collapse, we assume that all the workload belongs to this lowest class.

We next describe an example problem to illustrate these algorithms.

Problem 73

Consider a manufacturing system with three single-server workstations A, B, and C, as described in Figure 7.12. There are three classes of traffic. Class-1 jobs arrive externally into node A according to $PP(\lambda_{A,1})$ with $\lambda_{A,1} = 5$ jobs per day. They get served in node A, then at node B, and then they come back to node A for another round of service before exiting the network. Class-2 jobs arrive externally into node B according to $PP(\lambda_{B,2})$ with $\lambda_{B,2} = 4$ jobs per day. After service in node B, with probability 0.75 a class-2 job joins node C for service and then exits the network, whereas with probability 0.25 some

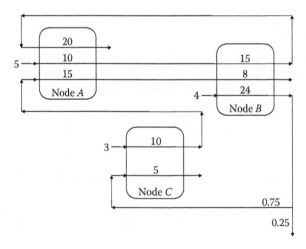

FIGURE 7.12
Single-server queueing network with local priorities.

class-2 jobs exit the network after service in node B. Class-3 jobs arrive externally into node C according to $PP(\lambda_{C,3})$ with $\lambda_{C,3} = 3$ jobs per day. They get served in node C, then at node A, and then node B before exiting the network. The service times to process a job at every node is exponentially distributed. The service rates are described in Figure 7.12 in units of number of jobs per day. In particular, the server in node A serves class-1 jobs during their first visit at rate $\mu_{A,1,1} = 10$ and second visit at rate $\mu_{A,1,2} = 20$, whereas it serves class-3 jobs at rate $\mu_{A,3,1} = 15$. Likewise, from the figure, at node B, we have $\mu_{B,1,1} = 15$, $\mu_{B,2,1} = 24$, and $\mu_{B,3,1} = 8$, and at node C, we have $\mu_{C,2,1} = 5$ and $\mu_{C,3,1} = 10$. The server at each node uses a preemptive resume priority scheme with priority order determined using the shortest-expected-processing-time-first rule. Thus each server gives highest priority to the highest $\mu_{\cdot,\cdot,\cdot}$. For such a system compute the steady-state expected number of each class of customer at every node.

Solution

We will stick to the notation used throughout this section, although it is worthwhile to point out that it may be easier to map the eight three-tuples of (node, class, visit number) to eight single-dimension quantities as done in most texts and articles. Note that the priority order (highest to lowest) is $(A, 1, 2)$, $(A, 3, 1)$, and $(A, 1, 1)$ in node A; $(B, 2, 1)$, $(B, 1, 1)$, and $(B, 3, 1)$ in node B; and $(C, 3, 1)$ and $(C, 2, 1)$ in node C. Since the flows are relatively simple in this example we can quickly compute the effective entering rates as $a_{A,1,2} = a_{A,1,1} = a_{B,1,1} = 5$, $a_{A,3,1} = a_{B,3,1} = a_{C,3,1} = 3$, $a_{B,2,1} = 4$, and $a_{C,2,1} = 3$ (due to the Bernoulli splitting only 75% of class-2 reach node C). Since the utilization (or relative traffic intensities) $\rho_{i,r,k}$ can be computed as $a_{i,r,k}/\mu_{i,r,k}$, we have $\rho_{A,1,2} = 0.25$, $\rho_{A,3,1} = 0.2$, $\rho_{A,1,1} = 0.5$, $\rho_{B,2,1} = 1/6$, $\rho_{B,1,1} = 1/3$, $\rho_{B,3,1} = 0.375$, $\rho_{C,3,1} = 0.3$, and $\rho_{C,2,1} = 0.6$ (they are presented so that the traffic intensities at the same node are together and within each node they are presented from the highest to lowest priority). Note that the necessary condition for stability is satisfied since the effective traffic intensity at nodes A, B, and C are $\rho_A = 0.95$, $\rho_B = 0.875$, and $\rho_C = 0.9$, respectively. Now, we proceed using the two different algorithms.

MVA-based algorithm: Note that we have already completed step 1 of the algorithm. In step 2, we just explain the $q_i(r, k)$ and $q_i^{-1}(j)$ but do not use them explicitly. For example, for node A, $q_A(1, 2) = 1$ being the highest priority at node A. Likewise, $q_A(3, 1) = 2$ and $q_A(1, 1) = 3$. Also, for node B, $q_B^{-1}(1) = (2, 1)$, $q_B^{-1}(2) = (1, 1)$, and $q_B^{-1}(3) = (3, 1)$. Thus using Equation 7.18, we get

$$L_{A,1,2} = \frac{\rho_{A,1,2}}{1 - \rho_{A,1,2}} = 0.3333,$$

$$L_{A,3,1} = \frac{\rho_{A,3,1} + a_{A,3,1}L_{A,1,2}/\mu_{A,1,2}}{1 - \rho_{A,1,2} - \rho_{A,3,1}} = 0.4545,$$

$$L_{A,1,1} = \frac{\rho_{A,1,1} + a_{A,1,1}(L_{A,1,2}/\mu_{A,1,2} + L_{A,3,1}/\mu_{A,3,1})}{1 - \rho_{A,1,2} - \rho_{A,3,1} - \rho_{A,1,1}} = 14.6970,$$

$$L_{B,2,1} = \frac{\rho_{B,2,1}}{1 - \rho_{B,2,1}} = 0.2,$$

$$L_{B,1,1} = \frac{\rho_{B,1,1} + a_{B,1,1}L_{B,2,1}/\mu_{B,2,1}}{1 - \rho_{B,2,1} - \rho_{B,1,1}} = 0.75,$$

$$L_{B,3,1} = \frac{\rho_{B,3,1} + a_{B,3,1}(L_{B,2,1}/\mu_{B,2,1} + L_{B,1,1}/\mu_{B,1,1})}{1 - \rho_{B,2,1} - \rho_{B,1,1} - \rho_{B,3,1}} = 4.4,$$

$$L_{C,3,1} = \frac{\rho_{C,3,1}}{1 - \rho_{C,3,1}} = 0.4286,$$

$$L_{C,2,1} = \frac{\rho_{C,2,1} + a_{C,2,1}L_{C,3,1}/\mu_{C,3,1}}{1 - \rho_{C,3,1} - \rho_{C,2,1}} = 7.2857.$$

State-space-collapse-based algorithm: We begin by considering an equivalent single-server queueing network described in Figure 7.12, however, with each node operating using an FCFS discipline. Using the $a_{i,r,k}$ values, we know the arrival rates at every node. To compute the SCOV as for the arrivals as well as for the departures we systematically use QNA. Of course we know that the arriving SCOV $C_{A,1,1}^2 = C_{B,2,1}^2 = C_{C,3,1}^2 = 1$ since the arrivals of class-1, 2, and 3 into nodes A, B, and C are according to a Poisson process. For the other five arriving SCOVs, namely $C_{A,1,2}^2$, $C_{B,1,1}^2$, $C_{C,2,1}^2$, $C_{A,3,1}^2$, and $C_{B,3,1}^2$, we iteratively compute them by computing the departure SCOV carefully as described in QNA. Once we know the effective SCOV of arrivals and service times (which can also be computed using QNA), we can compute the steady-state average workload at the three nodes using the QNA FCFS result as $W_{Aq} = 1.3111$, $W_{Bq} = 0.5015$, and $W_{Cq} = 1.2382$.

Reverting to the preemptive resume priority case, we can derive the expected number of each type of job in the system in steady state using the state-space-collapse algorithm as

$$L_{A,1,2} = \rho_{A,1,2} = 0.25,$$

$$L_{A,3,1} = \frac{\rho_{A,3,1}}{1 - \rho_{A,1,2}} = 0.2667,$$

$$L_{A,1,1} = \frac{\rho_{A,1,1} + a_{A,1,1}W_{Aq}}{1 - \rho_{A,1,2} - \rho_{A,3,1}} = 12.828,$$

$$L_{B,2,1} = \rho_{B,2,1} = 0.1667,$$

$$L_{B,1,1} = \frac{\rho_{B,1,1}}{1 - \rho_{B,2,1}} = 0.4,$$

$$L_{B,3,1} = \frac{\rho_{B,3,1} + a_{B,3,1}W_{Bq}}{1 - \rho_{B,2,1} - \rho_{B,1,1}} = 3.7589,$$

$$L_{C,3,1} = \rho_{C,3,1} = 0.3,$$

$$L_{C,2,1} = \frac{\rho_{C,2,1} + a_{C,2,1} W_{Cq}}{1 - \rho_{C,3,1}} = 6.1635.$$

Upon running 50 replications of simulations, the 95% confidence interval for the corresponding values are as follows:

$$L_{A,1,2} = 0.3308 \pm 0.0002,$$

$$L_{A,3,1} = 0.3801 \pm 0.0005,$$

$$L_{A,1,1} = 13.328 \pm 0.2076,$$

$$L_{B,2,1} = 0.2 \pm 0.0002,$$

$$L_{B,1,1} = 0.8916 \pm 0.0012,$$

$$L_{B,3,1} = 3.8873 \pm 0.0180,$$

$$L_{C,3,1} = 0.4284 \pm 0.0006,$$

$$L_{C,2,1} = 7.2532 \pm 0.0606.$$ ∎

Reference Notes

One of the key foundations of this chapter is approximations based on modeling nodes in a queueing network using a reflected Brownian motion. We began this chapter by giving a flavor for how the Brownian motion argument is made in a single node $G/G/1$ setting and derived an approximation for the number in the system. The analysis relies heavily on the excellent exposition in Chen and Yao [19] as well as Whitt [105]. A terrific resource for key elements of Brownian motion (that have been left out in this chapter) is Harrison [52]. Subsequently, we extend the single node to an open network of single-server nodes assuming each queue behaves as if it were a reflected Brownian motion. Then we provide approximations based on Bolch et al. [12] for closed queueing networks. The second portion of this chapter on multiclass and multiserver general open queueing networks is entirely from Whitt [103]. Finally, the topic of queueing network with priorities is mainly based out of Chen and Yao [19]. Although it is critical to point out that there are several research studies mainly focused on aspects of stability, they will be dealt with in the next chapter. The semi-martingale reflected Brownian motion offers the ability to derive approximations for the performance measures including the state-space collapse.

Exercises

7.1 Consider a stable $G/G/1$ queue with arrival rate 1 per hour, traffic intensity 0.8, $C_a^2 = 1.21$, and $C_s^2 = 1.69$. Obtain W_q using the reflected Brownian motion approximation in Equation 7.8 and also using the approximations in Chapter 4. Compare the approximations against simulations. For the simulations use either gamma or hyperexponential distribution.

7.2 Consider the queueing network of single-server queues shown in Figure 7.13 under a special case of $N = 3$ nodes, $p = 0.6$, and $\lambda = 1$ per minute. Note that external arrival is Poisson. Service times at node i are according to gamma distribution with mean $1/(i + 2)$ minutes and SCOV $i - 0.5$. Compute an approximate expression for the expected number of customers in each node of the network in steady state.

7.3 Consider the queueing network of single-server queues shown in Figure 7.13. Consider a special case of $N = 3$ nodes, $p = 1$, $\lambda = 0$ (hence a closed queueing network), and $C = 50$ customers. Service times at node i are according to gamma distribution with mean $1/(i + 2)$ minutes and SCOV $i - 0.5$. Compute an approximate expression for the expected number of customers in each node of the network in steady state using both the bottleneck approximation (for large C) and the MVA approximation (for small C).

7.4 A queueing network with six single-server stations is depicted in Figure 7.14. Externally, arrivals occur into node A according to a renewal process with average rate 24 per hour and SCOV 2. The service time at each station is generally distributed with mean (in minutes) 2, 3, 2.5, 2.5, 2, and 2, and SCOV of 1.5, 2, 0.25, 0.36, and 1, 1.44, respectively, at stations A, B, C, D, E, and F. A percentage on any arc (i, j) denotes the probability that a customer after completing service in node i joins the queue in node j. Compute the average number of customers in each node of the network in steady state.

7.5 Consider a seven-node single-server queueing network where nodes 2 and 4 get arrivals from the outside (at rate 5 per minute each on an average). Nodes 1 and 2 have service rates of 85,

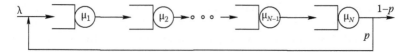

FIGURE 7.13
Single-server queueing network.

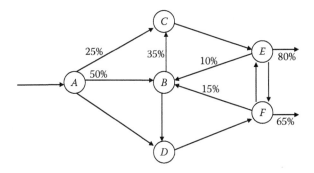

FIGURE 7.14
Schematic of six-station network.

nodes 3 and 4 have service rates of 120, node 5 has a rate of 70, and nodes 6 and 7 have rates of 20 (all in units per minute). Assume that all external interarrival times have SCOV of 0.64 and all service times have SCOV of 1.44. The routing matrix is given by

$$
P = [p_{ij}] = \begin{pmatrix}
\frac{1}{3} & \frac{1}{4} & 0 & \frac{1}{4} & 0 & \frac{1}{6} & 0 \\
\frac{1}{3} & \frac{1}{4} & 0 & \frac{1}{3} & 0 & 0 & 0 \\
0 & 0 & \frac{1}{3} & \frac{1}{3} & \frac{1}{3} & 0 & 0 \\
\frac{1}{3} & 0 & \frac{1}{3} & 0 & \frac{1}{3} & 0 & 0 \\
0 & 0 & 0 & \frac{4}{5} & 0 & 0 & \frac{1}{6} \\
\frac{1}{6} & 0 & \frac{1}{6} & \frac{1}{6} & \frac{1}{6} & \frac{1}{6} & 0 \\
0 & \frac{1}{6} & \frac{1}{6} & \frac{1}{6} & \frac{1}{6} & 0 & \frac{1}{6}
\end{pmatrix}.
$$

Using an approximation, find the average number of customers in each node of the network in steady state.

7.6 Recall the Bay-Gull Bagels problem depicted in Figure 6.4. Ignore the routing probability values shown in the figure. Use abbreviations B, S, D, C, and E to denote the five stations. There are three classes of customers. Class-1 customers arrive at rate 0.2 per minute into node S; they go through nodes S and C, then immediately exit the network (without eating in). Class-2 customers arrive at node B at rate 1 and then undergo routing exactly according to all the probabilities described in the figure. Class-3 customers arrive at node D externally at rate 0.4 per minute; they go through nodes D and C, then immediately exit the network (without eating in). Use QNA to obtain the average number of customers of each class at each of the five nodes of the network in steady state. Assume that the service times are not class dependent with SCOV of 0.25 at all nodes, and external arrivals are according to Poisson process for all nodes.

7.7 A production system consists of six work centers and two classes of parts that get produced. Denote the six work centers as nodes A, B, C, D, E, and F each of which contains a single server. Machine A receives requests for class-1 parts at an average rate of 4 per hour. Likewise, machine B receives requests for class-2 parts at an average rate of 6 per hour. Once machines A and B process their parts one by one, they drop them in a buffer for painting station C to pick them. The painting is done according to FCFS and then a robot in node D picks up the parts one by one and then delivers them to inspection stations E and F. Class-1 parts are inspected at E and class-2 are inspected at F. All parts after inspection exit the system. Compute the long-run expected number of parts of each class in each node in steady state. For this, assume that the SCOV of the request arrivals of class-1 and class-2 are 1.5 and 2.4, respectively. Also, the mean service time (in hours) and SCOV are described in Table 7.18.

7.8 Consider a sensor network with six single-server nodes. Node 1 is the gateway node to which requests arrive externally and responses are sent out to the external world. There are two classes of requests. Assume that each externally arriving request is class-1 with probability 0.2 and class-2 with probability 0.8. The mean and standard deviation of service times (in seconds) at each node for each class are described in Table 7.19. Likewise, the routing probabilities $p_{ij,r}$ from node i to j is provided in Table 7.20 for $r = 1$ (i.e., class-1) and Table 7.21 for $r = 2$ (i.e., class-2). Assume that class-1 has preemptive resume priority over class-2 in the entire network. First of all, determine the maximum external arrival rate that can be sustained in the network so that no node has a total traffic intensity larger than 0.95 (both classes combined). Assume that the interarrival times for requests coming externally has an SCOV of 1.44 for both classes. Using this maximum arrival

TABLE 7.18

Mean and Standard Deviation of Service Times for Each Class

Node	A	B	C	D	E	F
Class-1 mean service time	0.2	N/A	0.1	0.05	0.22	N/A
SCOV class-1 service time	0.25	N/A	0.64	2	0.75	N/A
Class-2 mean service time	N/A	0.15	0.05	0.1	N/A	0.14
SCOV class-2 service time	N/A	0.36	0.49	2.25	N/A	0.81

TABLE 7.19

Mean and Standard Deviation of Service Times for Each Class

Node i	1	2	3	4	5	6
Class-1 service time	2	2	3	1	2	4
SCOV class-1 service	1	1.44	1.69	1	0.81	0.49
Class-2 service time	3	1	2	3	4	2
SCOV class-2 service	0.25	1	0.81	0.49	2	0.64

TABLE 7.20

Class-1 Routing Probabilities $[p_{ij,1}]$ from Node on Left to Node on Top

	1	2	3	4	5	6
1	0	0.1	0.1	0.1	0.1	0.1
2	0.2	0	0.2	0.2	0.2	0.2
3	0.3	0.4	0	0.1	0.1	0.1
4	0.1	0.1	0.1	0	0.2	0.5
5	0.2	0.1	0.5	0.1	0	0.1
6	0.3	0.3	0.2	0.1	0.1	0

TABLE 7.21

Class-2 Routing Probabilities $[p_{ij,2}]$ from Node on Left to Node on Top

	1	2	3	4	5	6
1	0	0.5	0	0	0	0
2	0	0	1	0	0	0
3	0	0	0	1	0	0
4	0	0	0	0	1	0
5	0	0	0	0	0	1
6	1	0	0	0	0	0

rate, compute the steady-state average number of each class of customers at each node.

7.9 Consider the previous problem. If the service discipline is FCFS (instead of preemptive priority for class-1), solve that problem using QNA.

7.10 Consider the Rybko–Stolyar–Kumar–Seidman network in Figure 7.15. There are two classes of traffic. Class-1 jobs arrive externally into node A according to $PP(\lambda_A)$ with $\lambda_A = 7$ jobs per hour. They get served in node A, then at node B, and then they

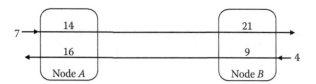

FIGURE 7.15
Rybko–Stolyar–Kumar–Seidman-type network.

exit the network. Class-2 jobs arrive externally into node B according to $PP(\lambda_B)$ with $\lambda_B = 4$ jobs per hour. After service in node
B, class-2 jobs get served at node A and then exit the network.
The service rates are described in Figure 7.15 in units of number of jobs per hour. In particular, the server in node A serves
class-1 jobs at rate $\mu_{A,1} = 14$, whereas it serves class-2 jobs at
rate $\mu_{A,2} = 16$. Likewise, from the figure, at node B, we have
$\mu_{B,1} = 21$ and $\mu_{B,2} = 9$. There is a single server at each node.
The servers use a preemptive resume priority scheme with priority order determined using shortest expected processing time first
rule. Thus each server gives highest priority to the highest $\mu_{.,.}$ in
that node. For such a system compute the steady-state expected
number of each class of customer at every node. Use both MVA
and state-space collapse technique. **Note:** The necessary condi-
tions for stability of this network are that $\lambda_A/\mu_{A,1} + \lambda_B/\mu_{A,2} < 1$
and $\lambda_B/\mu_{B,1} + \lambda_B/\mu_{B,2} < 1$. However, the sufficient condition
(assuming class-2 has high priority in node A and class-1 has high
priority in node B) is that $\lambda_A/\mu_{B,1} + \lambda_B/\mu_{A,2} < 1$.

8

Fluid Models for Stability, Approximations, and Analysis of Time-Varying Queues

In this chapter and in the next two chapters, we will consider the notion of fluid models or fluid queues. However, there is very little commonality between what is called fluid queues here and what we will call fluid queues in the next two chapters. In fact they have evolved in the literature rather independently, although one could fathom putting them together in a unified framework. We will leave them in separate chapters in this book with the understanding that in this chapter we are interested in the *fluid limit* of a *discrete queueing network* whereas in the next two chapters we will directly consider queueing networks with fluid entities (as opposed to discrete entities) flowing through them. Another key distinction is that the resulting fluid network in this chapter is deterministic lending itself straightforward ways to determine stability of queueing networks, develop performance measures approximately, as well as study transient and time-varying queues. In the next section, we will study *stochastic* fluid networks.

Deterministic fluid models have been applied to many other systems besides queueing networks. In pure mathematics the deterministic fluid models are called *hydrodynamic limits* and in physics they fall under *mean field theory*. The key idea is to study systems using only mean values by scaling metrics appropriately. We begin by considering a single queue with a single server to explain the deterministic fluid model concept and flush out details such as *functional strong law of large numbers* which is a concept central to the theory developed. Subsequently, we will use fluid limits in a network setting to analyze stability, obtain performance metrics approximately, and finally to study nonstationary queues under transient conditions.

8.1 Deterministic Fluid Queues: An Introduction

The objective of this section is to provide an introduction to deterministic fluid queues which is also called fluid limits of discrete queues. For that we first consider the simplest system, namely a single queue with one server. That would enable us to quickly compare against results we have already seen in this book and thereby develop an intuition for the material. Subsequently, in future sections, we will consider refinements to this simple system which cannot typically be analyzed using anything we have seen thus far in this book. For example, stability of queueing networks with local priorities, queues with abandonments and retrials, and queues with time-varying parameters are a few such examples that we will consider in the latter sections of this chapter. With that motivation in mind for what is to come in the future, we describe the simple system of a single server queue.

8.1.1 Single Queue with a Single Server

Consider a queue with an infinite waiting room where entities arrive one by one. There is a single server that serves customers one by one using an FCFS discipline. This is *not* necessarily a $G/G/1$ queue as we do not make assumptions such as IID interarrival times and IID service times. The most ideal example of such a system is a single-server node in the middle of a queueing network where arrivals are not necessarily according to a renewal process and in fact service times also do not have to be IID. Having said that, it is worthwhile to point out that for the sake of simplicity of explaining the results, we will frequently consider the example of a $G/G/1$ queue. In fact, some of the proofs for the most general results are much harder to show although they only require some mild assumptions such as the interarrival times and service times need to have finite mean and variance. In summary, the $G/G/1$ queue setting is provably *sufficient* to derive the results in this section but it is not *necessary*.

It is crucial to point out that the system is a discrete and stochastic queue with a single server. We are yet to take the fluid limit of this system. To do that, we first restate the notation used in Section 7.1.1 with the caveat that we do not require a renewal arrival process and IID service times as done in that section. Let $A(t)$ be the number of entities that arrive from time 0 to t. Also let $S(t)$ be the number of entities the server would process from time 0 to t if the server was busy during that whole time. Further, denote the average arrival rate as λ and the average service rate as μ. In other words,

$$\lambda = \lim_{t \to \infty} \frac{A(t)}{t} \quad \text{and}$$

$$\mu = \lim_{t \to \infty} \frac{S(t)}{t}.$$

With that description we are now ready to take the fluid limits of the discrete system described here. Define $\overline{A}_n(t)$ as

$$\overline{A}_n(t) = \frac{A(nt)}{n}$$

for any $n > 0$ and $t \geq 0$. We now study what happens to $\overline{A}_n(t)$ as $n \to \infty$. In other words, we scale the stochastic process $\{A(t), t \geq 0\}$ by a factor n and take the limit as $n \to \infty$. We first illustrate the fluid scaling by means of an example followed by the theoretical underpinnings in the next section. To demonstrate the power of the fluid limits, for our illustrations we consider an arrival process with a high coefficient of variation under the belief that for smaller coefficient of variations the result would only be more powerful. Further we only consider a fairly small t because for larger t values the results may not be all that surprising. In addition, we consider the second node of a tandem network where the arrivals are not truly IID.

Problem 74

Consider a $G/G/1$ queue with interarrival times as well as service times according to Pareto distributions. The coefficient of variation for interarrival times is 5 and for the service time it is equal to 2. The average arrival rate is 1 per unit time and the average service rate is 1.25 per unit time. The departures from this queue act as arrivals to a downstream queue. Let $A(t)$ be the number of entities that arrive at the downstream node during $(0, t]$. For $t = 0$ to 10 time units, graph three sample paths of $\overline{A}_n(t) = A(nt)/n$ versus t for $n = 1, 10, 100,$ and 1000.

Solution

It is crucial to note that the $A(t)$ process is the arrivals to the *downstream* node which is the same as the departures from the $G/G/1$ node described in the question. Also the average arrival rate is $\lambda = 1$. By writing a simulation using the algorithm in Problem 37 in Chapter 4, we can obtain sample paths of the output process from the $G/G/1$ queue, in particular the number of departures during any interval of time. Using this for various values of $n = 1$, $10, 100,$ and 1000, we can plot three sample paths of $\overline{A}_n(t) = A(nt)/n$ versus t as shown in Figure 8.1(a)–(d).

From the figure, note that in (a) where $n = 1$, the sample paths are piecewise constant graphs for the number of arrivals until that time. We expect to get about 10 arrivals during $t = 10$ time units. Also notice in (a) that the sample paths are quite varying. Now, when $n = 10$ as seen in (b), the sample paths are still piecewise constant but they are closer than in case (a). We see this trend more prominent in case (c) where the sample paths have closed in and the piecewise constant graph has started to look more like a constant graph. Finally in (d) when $n = 1000$ which for this example is sufficiently large, the sample paths merge with one another thereby the entire stochastic process in the limit goes to a deterministic limit.

Notice that the numerical example is for small t, however, the convergence would only be faster if we used larger t values. Thus we can conclude that $\overline{A}_n(t)$ converges to λt as n grows to infinite. In addition, if we were to choose a smaller coefficient of variation we would see that the convergence is much faster as a matter of fact. ∎

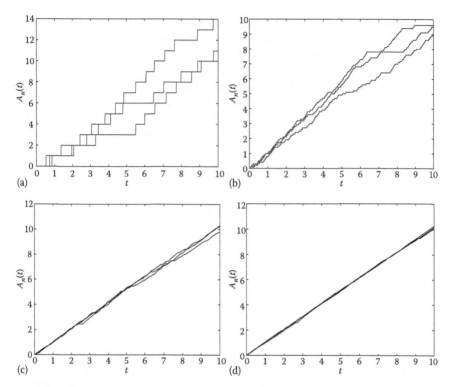

FIGURE 8.1

Sample paths of scaled arrival process $\bar{A}_n(t)$. (a) $n = 1$, (b) $n = 10$, (c) $n = 100$, and (d) $n = 1000$.

The preceding result would not be particularly surprising if $\{A(t), t \geq 0\}$ was a renewal process since for a renewal process with average interrenewal time $1/\lambda$ for a large t, $A(t)$ would be approximately normally distributed with mean λt and variance $\lambda C_a^2 t$, where C_a^2 is the squared coefficient of variation of interarrival times. For that reason in the previous problem the convergence was tested for an arrival process that is not a renewal process. Further, it is worthwhile noting that if one were to scale the service process $\{S(t), t \geq 0\}$, similar graphs can be obtained and the process $\bar{S}_n(t) = S(nt)/n$ would converge to μt as n grows to infinity. In fact the fluid limit can be taken for any stochastic process, not just counting processes. The theory underlying this is presented in the next section, which extends the well-known strong law of large numbers (see Section A.5 and Resnick [90]) to functionals.

8.1.2 Functional Strong Law of Large Numbers and the Fluid Limit

One version of the well-known strong law of large numbers (SLLN) is that if $X_1, X_2, \ldots$ are IID random variables with mean m, then $S_n = \frac{X_1 + X_2 + \cdots + X_n}{n}$

converges almost surely to m. In other words, $S_n \to m$ as $n \to \infty$ almost surely. For the sake of completeness, the other version of SLLN (although we will not use in this section) is that the fraction $\frac{I(X_1 \leq x) + I(X_2 \leq x) + \cdots + I(X_n \leq x)}{n}$ converges to $P(X_i \leq x)$ almost surely as $n \to \infty$ for some arbitrary i, where $I(\mathcal{A})$ is an indicator function that is one when $\mathcal{A}$ is true and zero otherwise. Back to the version we are interested, that is, $S_n \to m$ as $n \to \infty$. That is what is known as point-wise convergence of the random variable S_n.

However, in the previous section we saw that if $A(t)$ and $S(t)$ are the number of arrivals and number of service completions (if the server was busy during that whole time) in time $(0, t]$, respectively, then $\overline{A}_n(t) = A(nt)/n \to \lambda t$ and $\overline{S}_n(t) = A(nt)/n \to \mu t$. Basically the entire sequence of stochastic processes $\{\overline{A}_n(t), t \geq 0\}$ converges to the deterministic process λt. Likewise, the entire sequence of stochastic processes $\{\overline{S}_n(t), t \geq 0\}$ converges to the deterministic process μt. Such convergence of sequences of stochastic processes are called functional strong law of large numbers (FSLLN). Notice that in SLLN just a sequence of random variables converged to a deterministic quantity, but in FSLLN entire stochastic processes converge. However, Whitt [105] states that SLLN and FSLLN are actually equivalent (especially in the renewal context where it is not hard to see).

With that said, we have derived a deterministic system where arrivals occur at rate λ and service occurs at rate μ. In other words, the deterministic fluid model or fluid scaling process of the arrival process corresponds to a fluid that flows at a constant rate λ, thus in time t exactly λt amount of fluid flows into the system. Likewise the service capacity is μ per unit time. But that is a little more subtle since there may not be enough fluid to drain out at rate μ per unit time. In summary, we started with a discrete stochastic queue and created a sequence of stochastic processes that converge to a deterministic fluid queue. The deterministic fluid model or fluid scaled queue is a queue into which fluid arrives at rate λ per unit time continuously and the service rate is a deterministic μ per unit time. A picture converting the discrete stochastic queue to a fluid queue is depicted in Figure 8.2. The picture is depicted as though $\lambda < \mu$ and there was some fluid at $t = 0$ (we will show later when we scale that this fluid level must be zero) and the snapshot was taken at some small t. This leads us to the next question about how the contents of the fluid queue looks like for all t and for cases of both $\lambda < \mu$ and $\lambda \geq \mu$.

Actually the contents of the fluid queue at any time is rather easy to guess. What needs to be shown is that the stochastic process tracking the number of entities in the discrete system over time converges to that established by the fluid queue. For that, we let $X(t)$ be the number of discrete entities in the original stochastic queue. Let $\overline{X}_n(t) = X(nt)/n$ be the scaled process that we are interested in taking the limit as $n \to \infty$. For that we let $X(0) = x_0$ such that x_0 is a given finite constant number of entities initially (refer to Chen and Yao [19] for the more general case of x_0 being infinite and scaled appropriately).

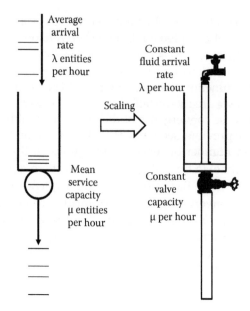

FIGURE 8.2
Snapshot of a discrete stochastic queue and a scaled deterministic fluid queue.

Thus we have $\overline{X}_n(0) = x_0/n$ which in the limit goes to zero. All the results in this section use $0 \le x_0 < \infty$.

To derive the scaled process we use a reflection mapping argument where the steps are identical to that in Section 7.1.1 with similar notation as well. Let $B(t)$ and $I(t)$ denote the total time the server has been busy and idle, respectively, from time 0 to t. The corresponding fluid limits by definition are $\overline{B}_n(t) = B(nt)/n$ and $\overline{I}_n(t) = I(nt)/n$. Of course $B(t) + I(t) = t$ and $\overline{B}_n(t) + \overline{I}_n(t) = t$. We can apply scaling to Equation 7.1 and obtain

$$\overline{X}_n(t) = \frac{x_0}{n} + \overline{A}_n(t) - \overline{S}_n\big(\overline{B}_n(t)\big).$$

This equation can be rewritten as

$$\overline{X}_n(t) = \overline{U}_n(t) + \overline{V}_n(t)$$

where

$$\overline{U}_n(t) = \frac{x_0}{n} + (\lambda - \mu)t + \big(\overline{A}_n(t) - \lambda t\big) - \big(\overline{S}_n\big(\overline{B}_n(t)\big) - \mu\overline{B}_n(t)\big),$$

$$\overline{V}_n(t) = \mu\overline{I}_n(t).$$

$$(8.1)$$

Refer to Section 7.1.1 for a detailed explanation.

Then, by scaling the expressions for $U(t)$ and $V(t)$ defined in Section 7.1.1, we know that given $\overline{U}_n(t)$, there exists a *unique* pair $\overline{X}_n(t)$ and $\overline{V}_n(t)$ such that $\overline{X}_n(t) = \overline{U}_n(t) + \overline{V}_n(t)$, which satisfies the following three conditions (obtained by rewriting conditions (7.2), (7.3), and (7.4) by scaling for any $t \geq 0$):

$$\overline{X}_n(t) \geq 0,$$

$$\frac{d\overline{V}_n(t)}{dt} \geq 0 \quad \text{with } \overline{V}_n(0) = 0 \text{ and}$$

$$\overline{X}_n(t) \frac{d\overline{V}_n(t)}{dt} = 0.$$

We also showed in Section 7.1.1 that the unique pair $\overline{X}_n(t)$ and $\overline{V}_n(t)$ can be written in terms of $\overline{U}_n(t)$ as

$$\overline{V}_n(t) = \sup_{0 \leq s \leq t} \max\{-\overline{U}_n(s), 0\}, \tag{8.2}$$

$$\overline{X}_n(t) = \overline{U}_n(t) + \sup_{0 \leq s \leq t} \max\{-\overline{U}_n(s), 0\}. \tag{8.3}$$

Since $\overline{B}_n(t) + \overline{I}_n(t) = t$, as $n \to \infty$, $\overline{B}_n(t)$ is bounded. Thus as $n \to \infty$, $(\overline{S}_n(\overline{B}_n(t)) - \mu\overline{B}_n(t)) \to 0$. By taking the limit $n \to \infty$, Equation 8.1 results in $\overline{U}_n(t) \to (\lambda - \mu)t$. Substituting that in Equation 8.2, we get $\overline{V}_n(t)$ converges to $\max\{-(\lambda - \mu)t, 0\}$ as $n \to \infty$. Using that in Equation 8.3, we get $\overline{X}_n(t) \to (\lambda - \mu)t + \max\{-(\lambda - \mu)t, 0\}$ as $n \to \infty$. Since for any real scalar a, we know that $a + \max\{-a, 0\} = \max\{a, 0\}$, we get as $n \to \infty$,

$$\overline{X}_n(t) \to \max\{(\lambda - \mu)t, 0\}. \tag{8.4}$$

Notice that we do not make any assumptions about whether λ is less than or greater than μ. But we can see that the scaled fluid queue length converges to $\max\{(\lambda - \mu)t, 0\}$, which is zero if $\mu > \lambda$ and $(\lambda - \mu)t$ if $\lambda \geq \mu$. Now consider the fluid queue with deterministic arrivals at rate λ and capacity μ. We assume that the initial fluid level zero as in the scaled process the discrete stochastic queue with x_0 entities initially converges to zero initial level due to scaling x_0/n. The amount of fluid in this queue at time t would be zero if $\mu > \lambda$ and $(\lambda - \mu)t$ is $\lambda \geq \mu$. Thus we have shown that not only the arrival process and service process but also the number in the system in the discrete stochastic case converges to that of the deterministic fluid case. Chen and Yao [19] show the workload process also converges accordingly. However, it is crucial to realize that all we have shown is that by scaling the discrete stochastic queue we get its limit as a deterministic fluid queue.

But how do we use this fluid limit? We will see that in the remainder of this chapter especially in the context of stability of networks, approximations, and analyzing time-varying systems.

8.2 Fluid Models for Stability Analysis of Queueing Networks

The main objective of this section is to use fluid models to determine the stability of queueing networks. For example, consider the single-server single-class infinite-waiting-area queue described in the previous section. We know that the system is stable if $\lambda < \mu$ where λ and μ are the average arrival rate and average service rate, respectively. Also as $t \to \infty$, the contents in the deterministic fluid model (by scaling the original discrete stochastic system, as shown in the previous section) shoot off to infinity if $\lambda > \mu$. So if the fluid system is stable, then the discrete stochastic model is stable. We ask the question: can we extend this idea to a network of queues? We will see later in this section that the discrete stochastic network is stable if the deterministic fluid network is stable.

That certainly sounds appealing from a mathematical standpoint. But in Chapter 7, we never worried about it. We just checked if the resulting mean arrival rate in every node of a network is smaller than the average service rate, and if all nodes satisfy that then the network is stable. So why bother with this cumbersome task of taking the fluid limit to check for stability? This is an extremely reasonable question to ask and until the recent past no one questioned that approach. However, a flurry of research indicated that the condition of all nodes having arrival rates smaller than service rates is certainly *necessary* but not sufficient. The fluid model will help us indicate if the system would be unstable, despite the necessary conditions being satisfied. That is the motivation to study them. But before we present the fluid model, we will see in the next section examples of networks where the necessary condition for stability is indeed not sufficient.

8.2.1 Special Multiclass Queueing Networks with Virtual Stations

In this section we consider three queueing networks and run simulations to see if they are stable. In all cases the naive condition that the arrival rates into every node must be less than the service capacity at that node will be satisfied. However, we will show that the simulations indicate the networks are indeed unstable thus motivating the need to find any "additional" conditions for stability. We will also present that by means of a concept called "virtual station," which would be responsible for the need of additional stability conditions. We begin by stating an example from Dai [25].

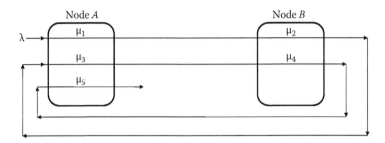

FIGURE 8.3
Reentrant line example.

Consider a multiclass network depicted in Figure 8.3 (Dai [25]). Jobs enter node A at average rate λ, after service (mean service time of $1/\mu_1$) they go to node B for a service that takes $1/\mu_2$ time on average. Then they reenter node A for another round of service (with mean $1/\mu_3$) and then go to node B again for a service that takes an average time of $1/\mu_4$. They finally visit node A for a service that takes an average $1/\mu_5$ time before exiting the system. Such a system is called a queueing network with reentrant lines and is typical in semiconductor manufacturing facilities. Although there is only a single flow, for ease of explanation we call the first visit to node A as class-1, first visit to node B as class-2, second visit to node A as class-3, second visit to node B as class-4, and final visit to node A as class-5. Notice that the subscripts of the service rates match the respective classes.

There is a single server at node A that uses a priority order class-5 (highest) then class-3 and then class-1 (lowest priority). Likewise, there is a single server at node B as well that gives higher priority to class-2 than class-4. However, at all nodes jobs within a class are served FCFS. Also, the priorities are *preemptive resume* (see Section 5.2.3 for a definition) type. We are not specifying the probability distributions for the interarrival times or service times for two reasons: (i) stability can be determined by just knowing their means; (ii) we do not want to give the impression that the interarrival times or the service times are IID. Having said that, when we run simulations we do need to specify distributions and for that reason we would do so in the examples.

The stability conditions in terms of traffic intensities at nodes A and B are

$$\frac{\lambda}{\mu_1} + \frac{\lambda}{\mu_3} + \frac{\lambda}{\mu_5} < 1 \quad \text{and} \quad \frac{\lambda}{\mu_2} + \frac{\lambda}{\mu_4} < 1 \tag{8.5}$$

respectively. The question is whether these conditions are sufficient to ensure that the system is stable. To answer that we consider a numerical problem next where these conditions are satisfied.

Problem 75

Consider the reentrant line in Figure 8.3. Let the arrivals be according to a Poisson process with mean $\lambda = 1$. Also all service times are exponentially distributed with $\mu_1 = 10$, $\mu_2 = 2$, $\mu_3 = 8$, $\mu_4 = 2.5$, and $\mu_5 = 1.5$. Verify that conditions in (8.5) are satisfied. Then simulate the system for about 2000 time units with an initially empty system to obtain the number of jobs in each of the two nodes A and B over time. Also state if either servers is underutilized for the duration of the simulation.

Solution

We can immediately verify that conditions in (8.5) are satisfied because

$$\frac{\lambda}{\mu_1} + \frac{\lambda}{\mu_3} + \frac{\lambda}{\mu_5} = 0.89167 < 1 \quad \text{and} \quad \frac{\lambda}{\mu_2} + \frac{\lambda}{\mu_4} = 0.9 < 1.$$

We start with an empty system and simulate arrivals and service according to the description given. By keeping track of the number of jobs in each of the nodes, we plot Figure 8.4(a) and (b). In Figure 8.4a, notice how the number of jobs in node A rises and then falls, then rises higher and then crashes to zero with the high queue length periods increasing in size. A similar trend can also be observed in Figure 8.4b. However, a curious finding is the fact that when there are jobs in one node, the other is more or less empty. In other words, notice that the number in node A and that in node B are negatively correlated. Further, if we add the number of jobs in nodes A and B, we can plot the total number of jobs in the entire system and that is given in Figure 8.5. Although it is true from Figure 8.4a and b that the number in each queue hits zero often but if one were to add the number of jobs in node A to that in node B, the total number of jobs shows an increasing trend (Figure 8.5). We can hence conclude that the system is indeed unstable because the total number in the entire system is showing a rising trend over time. Also, in terms of the utilization during the course of simulation we find the following. Although the utilization of node A is close to the traffic intensity of 0.89167, the utilization of node B is only 0.7836 which is significantly lower than the traffic intensity of 0.9 that we expect to see. ∎

Next we investigate the reason behind why we see the strange behavior of the system becoming unstable and the nodes not reaching their expected utilization in Problem 75. Let $X_i(t)$ be the number of class-i jobs in the system at time t for $i = 1, 2, 3, 4, 5$. In particular, consider jobs of class-2 and class-5. Could node A be working on a class-5 job at the same time when node B is working on a class-2 job? Say that is possible and there is one job of class-2 and one job of class-5 in the system. Since class-2 and class-5 get preemptive priorities at their respective nodes, both must be in service. However, the service times could not have started simultaneously. So let us say that

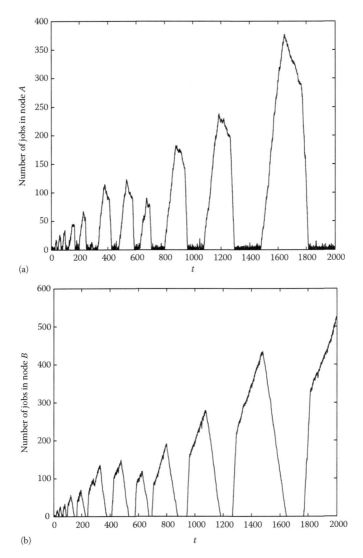

FIGURE 8.4
Sample path of number of jobs over time in each node of the reentrant line example. (a) Number of jobs in A versus t. (b) Number of jobs in node B versus t.

the class-2 job was in the system when the class-5 job entered and began service (the argument would not change if we went the other way around). But that is impossible because the class-5 job would have been a class-4 job that completed service (but the server would be processing class-2 and hence a class-4 cannot have completed). In other words, before becoming a class-2 and a class-5 job, they were class-1 and class-4 jobs, respectively, and a

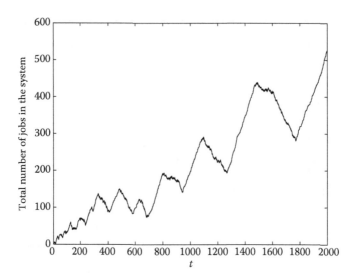

FIGURE 8.5
Total number in the entire system in the reentrant line example.

class-1 job cannot be completed when there is a class-5 job in the system and a class-4 job cannot be completed when there is a class-2 job in the system.

Hence we make the crucial observation that

$$X_2(t)X_5(t) = 0$$

for all t especially if we started with an empty system at $t = 0$. That means that the system can never process a class-2 and a class-5 job simultaneously. Therefore, if the load brought by class-2 and class-5 jobs is too high, then the system will not be stable. In Problem 75, notice that

$$\frac{\lambda}{\mu_2} + \frac{\lambda}{\mu_5} = 1.167 > 1.$$

The system cannot spend a fraction λ/μ_2 time serving class-2 and another λ/μ_5 fraction of time serving class-5 since both cannot be served simultaneously. Thus a crucial condition for stability is

$$\frac{\lambda}{\mu_2} + \frac{\lambda}{\mu_5} < 1. \tag{8.6}$$

This condition can also be interpreted as though there is a "virtual station" into which class-2 and class-5 flow that needs to be stable.

Remark 18

The conditions for the network represented in Figure 8.3 (with priority policy described earlier) to be stable are

$$\frac{\lambda}{\mu_2} + \frac{\lambda}{\mu_5} < 1,$$

$$\frac{\lambda}{\mu_1} + \frac{\lambda}{\mu_3} + \frac{\lambda}{\mu_5} < 1 \quad \text{and}$$

$$\frac{\lambda}{\mu_2} + \frac{\lambda}{\mu_4} < 1.$$

■

At this juncture a natural question to ask is if reentrant lines or the priority policy is needed to observe such virtual stations and have additional conditions for stability. In the next two examples we will relax one of those two conditions. First we present the example that is popularly known as Kumar–Seidman–Rybko–Stolyar network in the literature. Sometimes it is also referred to as Rybko–Stolyar–Kumar–Seidman network and is depicted in Figure 8.6. Around the same time Kumar and Seidman as well as Rybko and Stolyar wrote articles considering the network in Figure 8.6. The only difference is that Kumar and Seidman considered a deterministic system whereas Rybko and Stolyar a stochastic system.

Since we are only interested in the average rates, the deterministic and the stochastic versions are identical from a stability standpoint. Class-1 jobs enter node A externally at an average rate of λ_1 per unit time. They get served at node A for an average time of $1/\mu_1$ and then go to node B where they are called class-2. Class-2 jobs get served for an average time of $1/\mu_2$ and exit the system. Class-3 jobs arrive externally at an average rate of λ_3 per unit time into node B. They require an average processing time of $1/\mu_3$ and upon completion they go to node A and get served for an average time of $1/\mu_4$ (as class-4) before exiting the network. There is a single server at node A and a

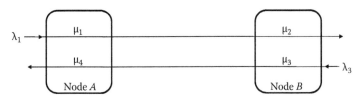

FIGURE 8.6
Rybko–Stolyar–Kumar–Seidman network.

single server at node B. Notice that this is not a reentrant line. However, like the previous example, here too we consider a preemptive resume priority scheme. Class-2 and class-4 jobs are given higher priority at their respective nodes. This is natural because by giving priority to them, we could purge jobs out of the system (with the hope that it would reduce the number of jobs in the system).

The stability conditions in terms of traffic intensities at nodes A and B are

$$\frac{\lambda_1}{\mu_1} + \frac{\lambda_3}{\mu_4} < 1 \quad \text{and} \quad \frac{\lambda_1}{\mu_2} + \frac{\lambda_3}{\mu_3} < 1 \tag{8.7}$$

respectively. We once again are interested to determine whether these conditions are sufficient to ensure that the system is stable. For that, we describe a numerical problem in which these conditions are satisfied. It is crucial to reiterate that contrary to what is presented in the example, it is not necessary for the arrivals and/or service to be stochastic; also it is not necessary for the interarrival time or the service times to be IID.

Problem 76

Consider the network in Figure 8.6. Let the arrivals be according to a Poisson process with mean $\lambda_1 = \lambda_3 = 1$. Also all service times are exponentially distributed with $\mu_1 = 5$, $\mu_2 = 10/7$, $\mu_3 = 4$, and $\mu_4 = 4/3$. Verify that conditions in (8.7) are satisfied. Then simulate the system for about 4000 time units with an initially empty state to obtain the number of jobs in each of the two nodes A and B over time.

Solution

It is relatively straightforward to verify that conditions in (8.7) are satisfied since

$$\frac{\lambda_1}{\mu_1} + \frac{\lambda_3}{\mu_4} = 0.95 < 1 \quad \text{and} \quad \frac{\lambda_1}{\mu_2} + \frac{\lambda_3}{\mu_3} = 0.95 < 1.$$

Starting with an empty system, we simulate arrivals and service according to distributions described. We plot graphs of the number of jobs in nodes A and B in Figure 8.7(a) and (b), respectively. Notice that in both nodes the number of jobs rises sharply and falls to zero in cycles and each cycle grows bigger and bigger. Also when one node has a large number of jobs, the other is small (or even zero) and vice versa. Also if we monitor the total number of jobs in the entire network, as done in Figure 8.8, we see that the system is unstable because the total number in the entire system is constantly rising. ∎

The reason for instability is very similar to that we saw in Problem 75, although here too the traffic intensity conditions (8.7) are satisfied. Let $X_2(t)$

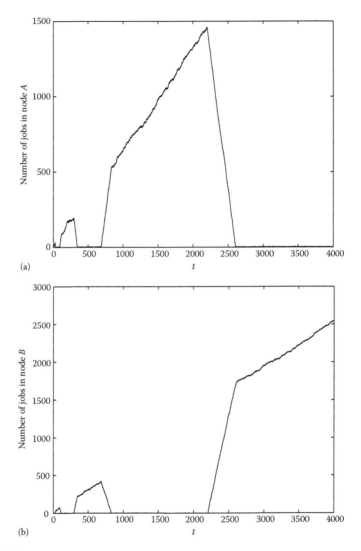

(a)

(b)

FIGURE 8.7
Number in each node of Kumar–Seidman–Rybko–Stolyar network example. (a) Number of jobs in node A versus t. (b) Number of jobs in node B versus t.

and $X_4(t)$ be the number of class-2 and class-4 jobs, respectively, in the system at time t. It is impossible for node A to be working on a class-4 job at the same time when node B is working on a class-2 job. This is because with respect to the class-2 and class-4 jobs, if they were both being served simultaneously, the previous event would have been start of a class-2 job or start of a class-4 job. For that it would be necessary for a class-1 or a class-3 job to have been completed respectively. But that is impossible because the respective

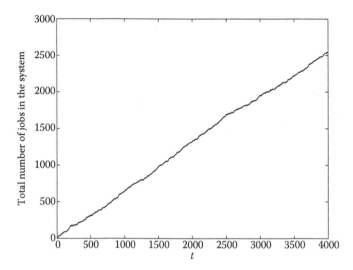

FIGURE 8.8
Number of jobs in the entire Kumar–Seidman–Rybko–Stolyar network.

nodes would be working on the higher priority jobs. In other words, before becoming a class-2 job, a job would have been a class-1 job that would have just completed. But for a class-1 job to be complete there could be no class-4 jobs in the system. So there would be a class-2 job in the system only if there is no class-4 job in the system. Likewise, we can see using an identical argument that there would be a class-4 job in the system only if there are no class-2 jobs in the system.

Hence we conclude that

$$X_2(t)X_4(t) = 0$$

for all t if we started with an empty system at $t = 0$. This means that the system *as a whole* cannot process a class-2 and a class-4 job simultaneously. Therefore, if the load brought by class-2 and class-4 jobs is too high, then the system will not be stable. In Problem 76, notice that

$$\frac{\lambda_1}{\mu_2} + \frac{\lambda_3}{\mu_4} = 1.45 > 1.$$

The system cannot spend a fraction λ_1/μ_2 time serving class-2 and another λ_3/μ_4 fraction of time serving class-4 since both cannot be served simultaneously. Thus a crucial condition for stability is

$$\frac{\lambda_1}{\mu_2} + \frac{\lambda_3}{\mu_4} < 1. \tag{8.8}$$

This condition is as though there exists a *virtual station* into which class-2 and class-4 flow and that station also needs to have a traffic intensity of less than 1.

Remark 19

The conditions for the network represented in Figure 8.6 (with priority policy described earlier) to be stable are

$$\frac{\lambda_1}{\mu_2} + \frac{\lambda_3}{\mu_4} < 1,$$

$$\frac{\lambda_1}{\mu_1} + \frac{\lambda_3}{\mu_4} < 1 \quad \text{and}$$

$$\frac{\lambda_1}{\mu_2} + \frac{\lambda_3}{\mu_3} < 1. \qquad \blacksquare$$

To illustrate that the network is stable if the conditions in Remark 19 are satisfied, we consider a set of numerical values different from those in Problem 76. Although $\lambda_1 = \lambda_3 = 1$, we have $\mu_1 = 2$, $\mu_2 = 2.5$, $\mu_3 = 20/11$, and $\mu_4 = 20/9$. Notice that the conditions in Remark 19 are satisfied. In particular, similar to the numerical values in Problem 76, here too

$$\frac{\lambda_1}{\mu_1} + \frac{\lambda_3}{\mu_4} = 0.95 \quad \text{and} \quad \frac{\lambda_1}{\mu_2} + \frac{\lambda_3}{\mu_3} = 0.95.$$

However,

$$\frac{\lambda_1}{\mu_2} + \frac{\lambda_3}{\mu_4} = 0.85.$$

For this set of numerical values we simulate the system and obtain the total number of customers in this stable system over time in Figure 8.9. By contrasting with that of the unstable network in Figure 8.8, notice how the number in the system does not blow up and keeps hitting zero from time to time. Thus clearly the standard traffic intensity conditions are only necessary but not sufficient. Having made a case for that we present one final example of an unstable network.

We present as a last example an FCFS network with reentrant lines. This network is depicted in Figure 8.10 and is identical to the example considered in Chen and Yao [19]. Chen and Yao [19] describe this network as a Bramson network since it is a simplification of the network considered by Bramson [13]. Like the previous examples here too we only describe the network in terms of the average rates. Also, the deterministic and the stochastic versions

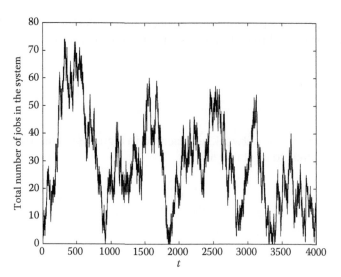

FIGURE 8.9
Number of jobs in a stable Kumar–Seidman–Rybko–Stolyar network.

are identical from a stability standpoint. Class-1 jobs enter node A externally at an average rate of λ per unit time. They get served at node A for an average time of $1/\mu_1$ and then go to node B where there are called class-2. Class-2 jobs get served for an average time of $1/\mu_2$ and go for another round of service at node B as class-3 jobs. Class-3 jobs take an average $1/\mu_3$ time for service and convert to class-4 jobs at the end of service. Class-4 jobs are also served at node B at an average rate of μ_4 per unit time. Upon service completion, class-4 jobs convert to class-5 and get served at node A before exiting the system. Average class-5 service time is $1/\mu_5$. There is a single server at node A and a single server at node B and each server at their respective nodes use FCFS discipline. Notice that this is indeed a reentrant line. However, the main difference is this is FCFS (and not priority scheme as we saw in the two previous examples).

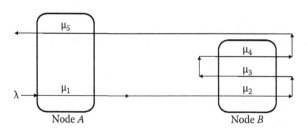

FIGURE 8.10
Network with FCFS at all stations.

The stability conditions in terms of traffic intensities at nodes A and B are

$$\frac{\lambda}{\mu_1} + \frac{\lambda}{\mu_5} < 1 \quad \text{and} \quad \frac{\lambda}{\mu_2} + \frac{\lambda}{\mu_3} + \frac{\lambda}{\mu_4} < 1 \tag{8.9}$$

respectively. We once again are interested in determining whether these conditions are sufficient to ensure that the system is stable. For that, we describe a numerical problem in which these conditions are satisfied. The numerical values used are identical to that in Chen and Yao [19]. It is crucial to reiterate that contrary to what is presented in the example, it is not necessary for the arrivals and/or service to be stochastic; also it is not necessary for the interarrival time or the service times to be IID, leave alone exponential distribution.

Problem 77

Consider the network in Figure 8.10. Let the arrivals be according to a Poisson process with mean rate $\lambda = 1$. Also all service times are exponentially distributed with $1/\mu_1 = 0.02$, $1/\mu_2 = 0.8$, $1/\mu_3 = 0.05$, $1/\mu_4 = 0.05$, and $1/\mu_5 = 0.88$. Verify that conditions in (8.9) are satisfied. Then simulate the system for about 50,000 time units with an initially empty state to obtain the number of jobs in each of the two nodes A and B over time.

Solution

It is relatively straightforward to verify that conditions in (8.9) are satisfied since

$$\frac{\lambda}{\mu_1} + \frac{\lambda}{\mu_5} = 0.9 < 1 \quad \text{and} \quad \frac{\lambda}{\mu_2} + \frac{\lambda}{\mu_3} + \frac{\lambda}{\mu_4} = 0.9 < 1.$$

Starting with an empty system, we simulate arrivals and services according to distributions described in the problem. Since we have simulated for a much larger time compared to the previous two examples, Figure 8.11(a) and (b) clearly indicate, the cyclic and increasing pattern in terms of the number of jobs in nodes A and B, respectively. Also Figure 8.12 shows the total number of customers in the system and from that we can certainly conclude that the system is unstable. That is because the total number in the entire system has a rising trend. Although it is true that the number in each queue hits zero from time to time, but if one were to add the number of jobs in node A to that in node B, the total number of jobs continuously increases. However, a curious finding is the fact that when one node has a lot of jobs, the other one is relatively empty. ∎

The reason for instability is very similar to that we saw in the previous two problems, that is, Problems 75 and 76. However, the virtual station

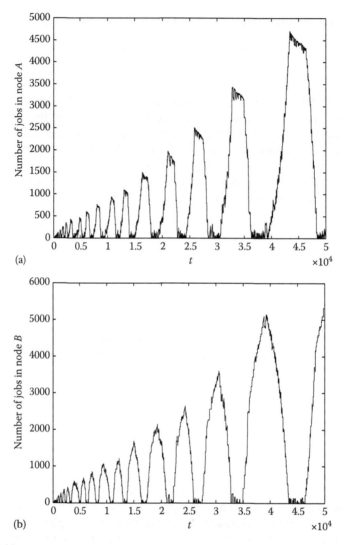

FIGURE 8.11
Sample path for number in each node of FCFS network example. (a) Number of jobs in node A
versus t. (b) Number of jobs in node B versus t.

condition is a lot more subtle and hard to explain. However, here too the
traffic intensity conditions (8.9) are satisfied. But the servers end up idling
for longer than they can afford and keep catching up. As that happens the
queue piles up and this causes a cascading effect. Nonetheless it is not easy
to write down the explicit sufficient conditions for stability. As one would
expect, for larger networks it would indeed be more complicated to test for
stability using virtual stations. Hence we use fluid models to analyze the
stability, which is the focus of the remainder of this section.

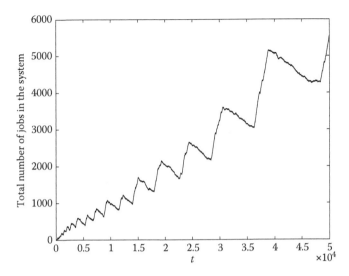

FIGURE 8.12
Total number in the entire system in the FCFS network example.

8.2.2 Stable Fluid Network Implies Stable Discrete Network

Recall that in Section 8.1, we considered a single queueing station with a single server and entities flow through it exactly once. For that discrete and stochastic system, we showed that if we scaled it appropriately (called fluid scaling), then it would result in a deterministic fluid queue. More importantly, the number in the discrete and stochastic system, would in the limit converge to that of the fluid queue. In other words, if we started with a discrete system with average arrival rate λ and average service capacity μ, then we can construct a fluid queue into which fluid flows deterministically at rate λ and gets emptied out from an orifice with capacity μ per unit time. The number in the system in the original stochastic discrete queue will converge to that of the deterministic fluid queue. As we saw earlier, one of the benefits of that is that if the deterministic fluid queue is stable then stochastic discrete queue will also be stable.

Here we extend that notion to an entire network stating that if a deterministic fluid network is stable, then the stochastic discrete network (that was scaled to obtain the fluid network) is also stable. For that, we first describe how to convert a discrete queueing network into a fluid network; then we state that if we can show the fluid network is stable, we are done; finally we briefly describe how to show our fluid network is stable. Our treatment here is rather preliminary and is only meant to get a flavor for the concept. In fact to begin with, our notion of what is *stable* is rather vague. Technically what we mean by stable for the stochastic discrete network is that the network is "positive Harris recurrent." Also, what we mean by stable for the fluid network we will address only subsequently (and contrast it against the notion of "weakly stable"). At this time we just say "stable" to not get distracted by technical details. There are some excellent texts and monographs that interested readers are encouraged to consider for a fully rigorous treatment of this subject. They include Dai [25], Meyn [82], Chen and Yao [19], and Bramson [13], to name a few.

We first describe the network setting. It is crucial to realize that the notation is somewhat different from those in the previous chapters. The setting as well as converting from a discrete network to a fluid network has been adapted from Meyn [82]. Consider a network with many single-server nodes or stations. Henceforth we will use the terms node and station interchangeably. There could be one or more queues or buffers at each station (we use the terms buffers and queues interchangeably). The key difference in the notation in this section is that the flow, routing, and service are with respect to the buffers and *not* the nodes unlike previous chapters. However, as always, the flow in this network is discrete and stochastic in terms of arrivals and service. But the routing from buffer to buffer is deterministic. Next we explicitly characterize these networks and describe the inputs for our analysis as follows:

1. The network consists of N service stations (or nodes).

2. There is one server at node i for all i such that $1 \leq i \leq N$.

3. There are ℓ buffers in the entire network and at least one in each node. Clearly, $\ell \geq N$. Throughout the network, buffers are numbered 1, 2, ..., ℓ. The one-to-one relationship between nodes and buffers are described in matrix C which is a node-buffer incidence matrix. Thus if node i has buffer j, then $C_{ij} = 1$ for $i \in [1, \ldots, N]$ and $j \in [1, \ldots, \ell]$. The matrix C is $N \times \ell$, so that the rows correspond to the nodes and columns correspond to buffers.

4. For all $j \in [1, \ldots, \ell]$, service times of customers (or jobs) at buffer j have a mean $1/\mu_j$ if they are processed in isolation. In other words, if the server is processing a job from buffer j, then the service rate is μ_j if that is the only job the server is processing.

5. We assume that the service discipline or policy used in the network can be specified. The only requirements are that the policy be: (a) nonidling or work conserving, that is, if any of the buffers in a node is nonempty, the server would not be idle; (b) head-of-the-line service at every buffer, that is, at most one job in a buffer can have partially completed service. Condition (a) necessarily requires that every server processes jobs at its full capacity, even if it is processor sharing. Condition (b) does not preclude having processor sharing across buffers.

6. There is infinite waiting room at each buffer.

7. Externally, customers or jobs arrive at buffer j at an average rate of λ_j per unit time. All external arrivals are independent of each other, the service times and the network state.

8. For all $i \in [1,\ldots,\ell]$ and $j \in [1,\ldots,\ell]$, when a customer completes service at buffer i, the customer either goes to another buffer j (such that $j \neq i$) or departs the network. The deterministic routing matrix R is defined as $R_{ij} = 1$ if after completing service customers at buffer i join buffer j. We assume that $I - R$ is invertible, where I is the $\ell \times \ell$ identity matrix.

Our objective is to determine whether such a queueing network is stable.

To address that objective, our first step is to provide some examples to clarify this network setting since it is somewhat different from before. Then we will show how to convert such a discrete and stochastic network into a fluid and deterministic one. Finally, we will state that if the fluid network is stable, then so will the discrete one. With that said, we first present an example problem.

Problem 78

Consider a network with single servers in each node that has buffers as depicted in Figure 8.13. There are three products that flow in the network. The buffers have infinite size. One product has a deterministic route of buffers 1, 2, and 3 before exiting the network; another goes through buffers 4, 5, and then 6; and the last one enters buffer-7, gets served, and exits after being served at buffer-8. The external arrival rates and the service rates are provided. Say the servers at each node use a preemptive priority policy giving highest priority to the shortest expected processing time among all types of jobs waiting at its node. Assume that $\mu_i < \mu_j$ if $i < j$. Can this system be modeled using the network setting described earlier?

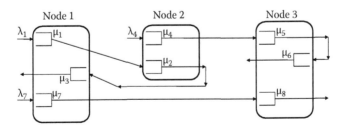

FIGURE 8.13
Network with nodes with multiple buffers and deterministic routes.

Solution

The system can be modeled using the network setting described as follows. The network has $N = 3$ service stations (or nodes) called node 1, node 2, and node 3. There is one server at each of the nodes 1, 2, and 3. There are $\ell = 8$ buffers in the entire network and at least one in each node. Notice that $\ell > N$. Using the buffer numbers we can see that $C_{11} = C_{13} = C_{17} = 1$ since node 1 has buffers 1, 3, and 7. Likewise, we have $C_{22} = C_{24} = 1$ and $C_{35} = C_{36} = C_{38} = 1$ for the same reason. Thus we have the C matrix as

$$
C = \begin{bmatrix}
1 & 0 & 1 & 0 & 0 & 0 & 1 & 0 \\
0 & 1 & 0 & 1 & 0 & 0 & 0 & 0 \\
0 & 0 & 0 & 0 & 1 & 1 & 0 & 1
\end{bmatrix}.
$$

The service rates at the buffers are specified in Figure 8.13. Also, the service discipline is described in the problem statement (although we would not use that here, we just verify that it is nonidling). There is infinite waiting room at each buffer.

External arrival rate of customers into buffers 1, 2, 3, 4, 5, 6, 7, and 8 are λ_1, 0, 0, λ_4, 0, 0, λ_7, and 0, respectively. The routing matrix from buffer to buffer is given by

$$
R = \begin{bmatrix}
0 & 1 & 0 & 0 & 0 & 0 & 0 & 0 \\
0 & 0 & 1 & 0 & 0 & 0 & 0 & 0 \\
0 & 0 & 0 & 0 & 0 & 0 & 0 & 0 \\
0 & 0 & 0 & 0 & 1 & 0 & 0 & 0 \\
0 & 0 & 0 & 0 & 0 & 1 & 0 & 0 \\
0 & 0 & 0 & 0 & 0 & 0 & 0 & 0 \\
0 & 0 & 0 & 0 & 0 & 0 & 0 & 1 \\
0 & 0 & 0 & 0 & 0 & 0 & 0 & 0
\end{bmatrix}
$$

which can be computed as $R_{ij} = 1$ whenever there is an arc from buffer i to buffer j in Figure 8.13 and $R_{ij} = 0$ otherwise. Note that $I - R$ is invertible. ∎

In a similar fashion, it is relatively straightforward to model the networks in Figures 8.3, 8.6, and 8.10 using this setting to obtain C and R for them (see Problem 79). However, in all the examples there is a notion of individual flows (such as the three flows in Figure 8.13) in the network. That is actually not necessary for the network setting. One could envision a feed-forward subnetwork where several flows merge to form a super-flow. In other words, the kind of networks that are allowed are fairly generic. Also, in some sense the networks we considered in earlier chapters could be envisioned as special cases. In fact, some researchers call each buffer as a separate class (like we did in the previous section) and hence such a network would be a multiclass network with deterministic routing and reentrant lines. Another way to consider this network is that there is a resource constraint that forces each server to work on multiple buffers. For example, each queue could be corresponding to a buffer of a machine and each node an operator. So each operator is responsible for a set of machines and the operator switches between jobs on all the machines he or she is assigned to work on. Thus the whole node can be thought of as either a machine or a resource.

With that motivation, the next question to ask is how do we convert such a discrete and stochastic network into a fluid and deterministic network by scaling it appropriately? It turns out that the procedure is relatively straightforward where we decompose the network into individual *buffers* and replace the discrete arrivals by fluids and valves for emptying buffers. Thus for the fluid model, all we need to state is the fluid entering rate and emptying rate for every buffer at all times. That would specify our fluid model. We explain next how a deterministic and fluid model of a stochastic discrete network looks like. For that we first consider a small example to explain and subsequently generalize it to any network. Recall the Rybko–Stolyar–Kumar–Seidman network in Figure 8.6. Priority is given to buffer-4 at node A and buffer-2 at node B. The fluid model of that network would be constructed in the following manner. Fluid would arrive at constant rates λ_1 and λ_3 continuously into buffers 1 and 3, respectively. If buffer-2 is nonempty, then node B would drain it at rate μ_2. Notice that if buffer-2 is empty, that does not mean it is not getting any inputs, it is just that the input rate is smaller than μ_2. So if buffer-2 is empty, then whatever capacity buffer-2 is not using will be used to drain buffer-3. Likewise, at node A, if buffer-4 is nonempty then all of the node's capacity will be used to drain buffer-4. However, if buffer-4 is empty, then the node will offer just the necessary amount of capacity to buffer-4 to ensure it continues to be empty, and the remaining capacity to drain out buffer-1.

We formalize that mathematically. Let $\zeta_j(t)$ be the processing capacity allocated to buffer j for $j = 1, 2, 3, 4$ for the network in Figure 8.6 (we will subsequently define $\zeta_j(t)$ more precisely for a generic network). For example, if at time t node A is draining a nonempty buffer-4, then $\zeta_4(t) = 1$ and $\zeta_1(t) = 0$. However, at time t if buffer-4 is empty but it gets arrivals at rate a_4 and buffer-1 is nonempty, then $\zeta_4(t) = a_4/\mu_4$ and $\zeta_1(t) = 1 - a_4/\mu_4$ (we need the condition $a_4/\mu_4 < 1$ for buffer-4 to be empty). Finally if both buffers 1 and 4 are empty at time t and arrival rates into them are λ_1 and a_4, respectively, then $\zeta_4(t) = a_4/\mu_4$ and $\zeta_1(t) = \lambda_1/\mu_1$ (we need the condition $\lambda_1/\mu_1 + a_4/\mu_4 < 1$ for both buffers 1 and 4 to be empty). In a similar fashion, one could consider node B and describe $\zeta_2(t)$ and $\zeta_3(t)$. With that one could decompose the network into individual buffers and write down the arrival as well as emptying rates for each buffer at time t, as described in Table 8.1.

Therefore, notice that it is relatively straightforward to convert a discrete stochastic network into a fluid deterministic one. Now we formalize

TABLE 8.1

Arrival and Drainage Rates in Fluid-Scaled Network

Buffer	Arrival Rate at Time t	Draining Rate at Time t
1	λ_1	$\mu_1 \zeta_1(t)$
2	$\mu_1 \zeta_1(t)$	$\mu_2 \zeta_2(t)$
3	λ_3	$\mu_3 \zeta_3(t)$
4	$\mu_3 \zeta_3(t)$	$\mu_4 \zeta_4(t)$

that for a generic discrete stochastic network with N nodes, ℓ buffers, node-buffer incidence matrix C, and buffer-to-buffer routing matrix R. Also, for all $j \in \{1, \ldots, \ell\}$, λ_j is the external average arrival rate into buffer j and $1/\mu_j$ is the average service time for a job in buffer j. This can be converted into a fluid deterministic network and decomposed into individual buffers so that all we need to specify is the input and drainage rate of each buffer. To explain the conversion process, we use some extra notation to keep the presentation less cumbersome. Let J_i be the set of buffers in node i, that is, $J_i = \{j : C_{ij} = 1\}$ for all $i \in [1, \ldots, N]$. In Figure 8.13, for example, $J_1 = \{1, 3, 7\}$, $J_2 = \{2, 4\}$, and $J_3 = \{5, 6, 8\}$. Likewise, let $s(j)$ be the node where buffer j resides. Again, in the example in Figure 8.13, $s(3) = 1$ since buffer-3 is in node 1, and $s(8) = 3$ since buffer-8 is in node 3. This gives us a mapping between buffers and nodes.

For a given buffer j such that $j \in \{1, \ldots, \ell\}$, let $z_j(t)$ be the cumulative time allocated by node $s(j)$ to process buffer j in time $(0, t]$. For all $t \geq 0$, let $\zeta_j(t)$ be the right derivative of $z_j(t)$ and is written as

$$\zeta_j(t) = \frac{d^+}{dt} z_j(t).$$

Since $z_j(t)$ although continuous is not differentiable everywhere, as a convention we use its right derivative (especially in nondifferentiable points) to define $\zeta_j(t)$. Like before, $\zeta_j(t)$ is indeed the processing capacity allocated to buffer j by node $s(j)$. With that said we can model buffer j as a fluid queue with arrival rate $\lambda_j + \sum_{i=1}^{\ell} \mu_i \zeta_i(t) R_{ij}$ at time t and drainage rate $\mu_j \zeta_j(t)$ at time t. The capacity allocations $\zeta_j(t)$ for buffer j is closely linked to the scheduling policy used as well as the contents of all the buffers. However they should satisfy some generic conditions such as

$$\sum_{j \in J_i} \zeta_j(t) \leq 1$$

for all $i \in \{1, \ldots, N\}$. This inequality would be an equality if at least one of the buffers in the set J_i of node i is nonempty.

That said, we can conclude that it is possible to convert a discrete stochastic queueing network into a fluid deterministic queueing network. It is also relatively straightforward to see that the scaled arrival process and service process converge to the exact same fluid limits described in the fluid model. However, it is a little more involved to show that by scaling the queue length, busy period, remaining workload process, etc., one would obtain the corresponding quantities in the fluid network. The reader is referred to Dai [25] as well as Chen and Yao [19] for a detailed description, proof and mapping of all the processes from the discrete to the fluid model by scaling. The technique follows a similar argument to that made for the single buffer case in Section 8.1. Now, since the discrete model's queue length with scaling converges to that of the fluid model, we can make some arguments relating to their respective stability. In particular, it is possible to show that if the fluid queue is stable, then so is the discrete stochastic one. For that we first characterize the different forms of stability for both the discrete and the fluid cases. Then we describe what to predict for the stability of the discrete queue given our assessment of the fluid queue.

For that we first define $X_j(t)$ as the number of jobs in buffer j at time t (including any jobs being served) in the discrete stochastic network such that $1 \leq j \leq \ell$. We also define $x_j(t)$ as the amount of fluid in buffer j at time t in the fluid deterministic network (obtained using the conversion described earlier) such that $1 \leq j \leq \ell$. For all $j \in [1, \ell]$, Dai [25] as well as Chen and Yao [19] show that as $n \to \infty$,

$$\frac{X_j(nt)}{n} \to x_j(t).$$

This is what we meant in the previous paragraph that not only do the arrival and service process converge to their fluid limits but so do the number in each buffer. Notice that $x_j(t)$ is a deterministic quantity. In some articles $x_j(t)$ is also written as $\overline{X}_j(t)$ to specifically denote the fluid limit. Next we define various "degrees" of stability for the discrete as well as the fluid network in terms of $X_j(t)$ and $x_j(t)$ for all $j \in [1, \ell]$.

For the discrete stochastic network, we define two "degrees" of stability as follows:

- *Stable*: A discrete stochastic network is called stable if $\sum_j X_j(t) < \infty$ for all t, especially as $t \to \infty$. For that one typically shows that the stochastic process $\{X(t), t \to \infty\}$ is *positive Harris recurrent*, where $X(t) = (X_1(t), \ldots, X_\ell(t))$.
- *Rate stable*: A discrete stochastic network is called rate-stable if for every buffer j, the steady-state departure rate equals the steady-state "effective" arrival rate obtained by solving the flow balance. To mathematically state that, let $D_j(t)$ be the number of jobs that depart buffer j in time $(0, t]$. Also let $a = [a_1 \ \ldots \ a_\ell]$ be a row vector

of effective arrival rates that can be obtained as $a = \bar{\lambda}(I - R)^{-1}$ where $\bar{\lambda} = [\lambda_1 \quad \lambda_2 \quad \ldots \quad \lambda_\ell]$, I the identity matrix and R the routing matrix. If for every j, we have $D_j(t)/t \to a_j$ almost surely as $t \to \infty$, then the network is rate stable. Note that $\sum_j X_j(t)$ could be ∞ as $t \to \infty$.

Now, for the fluid deterministic network, we define two corresponding "degrees" of stability as follows:

- *Stable*: A fluid deterministic network is called stable if there exists a finite time δ so that $\sum_j x_j(t) = 0$ for all $t > \delta$, given any finite initial fluid level $x_1(0), \ldots, x_\ell(0)$. That means that if the fluid queue started with any initial level of fluid, it would eventually all drain out.
- *Weakly stable*: A fluid deterministic network is called weakly stable if $\sum_j x_j(t) = 0$ for all $t > 0$, given that the network is empty initially, that is, $\sum_j x_j(0) = 0$. That means that if the fluid queue started out empty, it would remain empty throughout.

The reason we presented the degrees of stability for the discrete and fluid networks in a "corresponding" fashion is that as the title of this section states, the discrete network is stable if the fluid network is stable in a corresponding manner. We formalize this in the next remark.

Remark 20

If the fluid deterministic network is weakly stable, then the discrete stochastic network is rate stable. Also, if the fluid deterministic network is stable, then the discrete stochastic network is positive Harris recurrent (hence stable). ∎

To explain this remark as well as stability notions, let us consider the simplest example of a single buffer on a single node, as done in Section 8.1. The deterministic fluid model has an inflow rate λ and an orifice capacity μ. If $\lambda < \mu$, no matter how much fluid there was in the system initially, as long as it was finite, the buffer would empty in a finite time. Therefore, the fluid model is stable if $\lambda < \mu$. Remark 20 states that if the fluid model is stable then the original discrete queue is stable. This can be easily verified because we know that the discrete stochastic system is stable (or positive Harris recurrent) if $\lambda < \mu$. Now if $\lambda = \mu$, the fluid queue would remain at the initial level at all times. Thus if there is a nonzero initial fluid level, then the time to empty is infinite. But if the initial fluid level is zero, then it would remain zero throughout. Hence when $\lambda = \mu$, the queue is only weakly stable but not stable. Thus when $\lambda = \mu$ the discrete queue is only rate stable. Of course if $\lambda > \mu$ the fluid queue is unstable and so is the discrete queue.

Although in the preceding simple example it was not necessary to invoke the fluid model, in a larger network it may be. The understanding is that it is easier to determine if the fluid deterministic network is stable, weakly stable, or unstable than the corresponding notions for the discrete stochastic network. However, once we know the stability of the fluid deterministic network, we can immediately state that for the discrete stochastic network. We will see in the next section how to assess the stability of a fluid deterministic network.

8.2.3 Is the Fluid Model of a Given Queueing Network Stable?

We saw in the previous section how to scale and convert a discrete stochastic queueing network into a fluid deterministic one. We also saw there that if the fluid deterministic network was stable (or weakly stable), then the corresponding discrete stochastic network is stable (or rate stable). What remains is to check if the fluid deterministic network is stable, weakly stable, or unstable, and we would be done. As described in Dai [25], that is not particularly easy although it is usually better than checking if the discrete stochastic network is stable. In particular, the fluid network stability would almost have to be done on a case-by-case basis. There are two steps to follow. In the first step, start with an empty fluid queue and see if any queue builds up. If no queue builds up then the fluid network is weakly stable, otherwise it is unstable (in which case we are done). If the fluid network is weakly stable, then as a second step try an arbitrary initial fluid level. The usual candidate is $x_1(0) = 1$ and $x_j(0) = 0$ for all $j \neq 1$. If there does not exist a finite emptying time δ after which the system would continue to be empty, then the system is only weakly stable. However, if one can show there exists a finite emptying time, then consider a more generic initial state and if there exists an emptying time that is finite, then we have a stable system. We illustrate this using networks we have seen earlier in the next remark.

Remark 21

The corresponding fluid networks in Problems 75, 76, and 77 are all weakly stable since if we started with an empty system they would remain empty. But they are not stable since if we have a finite nonzero amount of fluid in the buffer initially, then the time to empty becomes infinite. As evident from the simulations, the discrete stochastic networks are all not positive Harris recurrent but they are rate stable. ∎

The preceding technique shows how to check if a fluid network is stable given a particular numerical setting. What if the arrival rates are smaller or if the processing rates are faster? It is cumbersome to check for each case the network's stability. However, if we knew the necessary and sufficient conditions for a fluid network to be stable, that would be helpful. We describe them next. To obtain the necessary conditions, let $a = [a_1 \ \ldots \ a_\ell]$ be a row vector of effective arrival rates that can be obtained as $a = \bar{\lambda}(I - R)^{-1}$ where $\bar{\lambda} = [\lambda_1 \ \lambda_2 \ \ldots \ \lambda_\ell]$, I the identity matrix, and R the routing matrix. Then for every buffer j, we define $\rho_j = a_j/\mu_j$ and the row vector $\rho = [\rho_1 \ \ldots \ \rho_\ell]$. If ρ^T is the transpose of ρ and $\hat{e}$ is an $N \times 1$ column of ones, then the necessary conditions for the fluid model to be stable is

$$C\rho^T < \hat{e}.$$

Usually, if these necessary conditions are satisfied, then the fluid model is at least weakly stable with allocation rates at buffer j $\zeta_j(t) = \rho_j$ for all t. But those conditions may not be sufficient to ensure stability (beyond weak stability). We would address the sufficient conditions later but first explain the necessary conditions with an example problem.

Problem 79

Consider the networks in Figures 8.3, 8.6, and 8.10. For all three networks, derive the necessary conditions for stability which would result in the fluid models to be at least weakly stable, if not stable?

Solution

For each of the Figures 8.3, 8.6, and 8.10 using their respective R and C, as well as λ_j and μ_j values for each buffer j, we derive the conditions for the fluid models to be weakly stable in the following manner.

- For the network in Figure 8.3, we have $N = 2$, $\ell = 5$, $\lambda_1 = \lambda$, and $\lambda_2 = \lambda_3 = \lambda_4 = \lambda_5 = 0$. Thus $\bar{\lambda} = [\lambda \ 0 \ 0 \ 0 \ 0]$. Also, the routing matrix is

$$R = \begin{bmatrix} 0 & 1 & 0 & 0 & 0 \\ 0 & 0 & 1 & 0 & 0 \\ 0 & 0 & 0 & 1 & 0 \\ 0 & 0 & 0 & 0 & 1 \\ 0 & 0 & 0 & 0 & 0 \end{bmatrix}$$

and the node-buffer incidence matrix is

$$C = \begin{bmatrix} 1 & 0 & 1 & 0 & 1 \\ 0 & 1 & 0 & 1 & 0 \end{bmatrix}.$$

The effective arrival rate vector is $a = \bar{\lambda}(I - R)^{-1} = [\lambda\ \lambda\ \lambda\ \lambda\ \lambda]$. Thus we can obtain $\rho = \left[\frac{\lambda}{\mu_1}\ \frac{\lambda}{\mu_2}\ \frac{\lambda}{\mu_3}\ \frac{\lambda}{\mu_4}\ \frac{\lambda}{\mu_5}\right]$. The conditions for the fluid model to be at least weakly stable are $C\rho^T < \hat{e}$, which results in

$$\frac{\lambda}{\mu_1} + \frac{\lambda}{\mu_3} + \frac{\lambda}{\mu_5} < 1 \quad \text{and} \quad \frac{\lambda}{\mu_2} + \frac{\lambda}{\mu_4} < 1.$$

These are indeed identical to the conditions in (8.5). For any nonidling or work-conserving policy, if the fluid model started out with an empty system, then the allocation rates at buffer j (for $j = 1, 2, 3, 4, 5$) would be $\zeta_j(t) = \lambda/\mu_j$ at all times t. That would result in the buffers being empty at all times, hence the system would be weakly stable. However, it is worthwhile pointing out that if the fluid model's initial state was nonempty in at least one buffer, then the necessary conditions are not sufficient to ensure stability.

- For the network in Figure 8.6, we have $N = 2$, $\ell = 4$, and $\bar{\lambda} = [\lambda_1\ 0\ \lambda_3\ 0]$. Also, the routing matrix is

$$R = \begin{bmatrix} 0 & 1 & 0 & 0 \\ 0 & 0 & 0 & 0 \\ 0 & 0 & 0 & 1 \\ 0 & 0 & 0 & 0 \end{bmatrix}$$

and the node-buffer incidence matrix is

$$C = \begin{bmatrix} 1 & 0 & 0 & 1 \\ 0 & 1 & 1 & 0 \end{bmatrix}.$$

The effective arrival rate vector is $a = \bar{\lambda}(I - R)^{-1} = [\lambda_1\ \lambda_1\ \lambda_3\ \lambda_3]$. Thus we can obtain $\rho = \left[\frac{\lambda_1}{\mu_1}\ \frac{\lambda_1}{\mu_2}\ \frac{\lambda_3}{\mu_3}\ \frac{\lambda_3}{\mu_4}\right]$. The conditions for the fluid model to be at least weakly stable are $C\rho^T < \hat{e}$ which results in

$$\frac{\lambda_1}{\mu_1} + \frac{\lambda_3}{\mu_4} < 1 \quad \text{and} \quad \frac{\lambda_1}{\mu_2} + \frac{\lambda_3}{\mu_3} < 1.$$

These are indeed identical to the conditions in (8.7). For any nonidling or work-conserving policy, if the fluid model started out with an empty system, then the allocation rates at buffer j (for $j = 1, 2, 3, 4$) would be $\zeta_j(t) = a_j/\mu_j$ at all times t. That would result in the buffers being empty at all times, hence the system would be weakly stable. However, it is worthwhile pointing out that if buffers 2 and 4 were given priorities at the respective nodes, then we saw in Problem 76 that the preceding conditions were not sufficient to ensure stability. One can show that if we started with an initial fluid level of 1 in buffer-1 and zero in all other buffers, then the fluid system would never empty if $\lambda_1/\mu_2 + \lambda_3/\mu_4 > 1$. In fact we will show subsequently that the sufficient condition to ensure stability is $\lambda_1/\mu_2 + \lambda_3/\mu_4 < 1$.

- For the network in Figure 8.10, we have $N=2$, $\ell=5$, $\lambda_1 = \lambda$, and $\lambda_2 = \lambda_3 = \lambda_4 = \lambda_5 = 0$. Thus $\bar{\lambda} = [\lambda\ 0\ 0\ 0\ 0]$. Also, the routing matrix is

$$R = \begin{bmatrix} 0 & 1 & 0 & 0 & 0 \\ 0 & 0 & 1 & 0 & 0 \\ 0 & 0 & 0 & 1 & 0 \\ 0 & 0 & 0 & 0 & 1 \\ 0 & 0 & 0 & 0 & 0 \end{bmatrix}$$

and the node-buffer incidence matrix is

$$C = \begin{bmatrix} 1 & 0 & 0 & 0 & 1 \\ 0 & 1 & 1 & 1 & 0 \end{bmatrix}.$$

The effective arrival rate vector is $a = \bar{\lambda}(I - R)^{-1} = [\lambda\ \lambda\ \lambda\ \lambda\ \lambda]$. Thus we can obtain $\rho = \left[\frac{\lambda}{\mu_1}\ \frac{\lambda}{\mu_2}\ \frac{\lambda}{\mu_3}\ \frac{\lambda}{\mu_4}\ \frac{\lambda}{\mu_5} \right]$. The conditions for the fluid model to be at least weakly stable are $C\rho^T < \hat{e}$ which results in

$$\frac{\lambda}{\mu_1} + \frac{\lambda}{\mu_5} < 1 \quad \text{and} \quad \frac{\lambda}{\mu_2} + \frac{\lambda}{\mu_3} + \frac{\lambda}{\mu_4} < 1.$$

These are indeed identical to the conditions in (8.9). For any nonidling or work-conserving policy (such as FCFS), if the fluid model started out with an empty system, then the allocation rates at buffer j (for $j = 1, 2, 3, 4, 5$) would be $\zeta_j(t) = \lambda/\mu_j$ at all times t. That would result in the buffers being empty at all times, hence the system would be weakly stable. ∎

Having described the necessary conditions for stability, our next goal is to obtain the sufficient conditions. Unfortunately, unlike the necessary conditions, the sufficient conditions cannot be stated in a generic fashion and would have to be addressed on a case-by-case basis. However, knowledge of the dynamics of the network would certainly aid in the process of obtaining some conditions and all we need to do is to check if those conditions are sufficient. As an example, recall the virtual station conditions described in Section 8.2.1. Are those virtual station conditions sufficient to ensure stability or would more conditions be needed? To answer that question, we consider a specific example, namely the Kumar–Seidman–Rybko–Stolyar network in

Figure 8.6. Assume that the necessary conditions

$$\frac{\lambda_1}{\mu_1} + \frac{\lambda_3}{\mu_4} < 1 \quad \text{and} \quad \frac{\lambda_1}{\mu_2} + \frac{\lambda_3}{\mu_3} < 1$$

are satisfied. We would like to check if in addition the virtual station condition $\lambda_1/\mu_2 + \lambda_3/\mu_4 < 1$ is sufficient to ensure stability.

To check that we begin with an initial buffer level of $x_1(0)$, $x_2(0)$, $x_3(0)$, and $x_4(0)$. Using that if we can show there exists a finite time δ so that $x_1(t) = x_2(t) = x_3(t) = x_4(t) = 0$ for all $t > \delta$ under the condition $\lambda_1/\mu_2 + \lambda_3/\mu_4 < 1$, then we are done. We assume that

$$\mu_1 > \mu_2 \quad \text{and} \quad \mu_3 > \mu_4,$$

otherwise the condition $\lambda_1/\mu_2 + \lambda_3/\mu_4 < 1$ would always be satisfied if the necessary conditions are satisfied. That is because if $\mu_1 \leq \mu_2$ then $\lambda_1/\mu_2 + \lambda_3/\mu_4 \leq \lambda_1/\mu_1 + \lambda_3/\mu_4 < 1$ (similarly when $\mu_3 \leq \mu_4$). Hence we make that assumption to avoid the trivial solution.

Without loss of generality we assume that all four buffers are nonempty initially with the understanding that other cases can be handled in a similar fashion. Node 2 would drain buffer-2 at rate μ_2 and node 1 would drain buffer-4 at rate μ_4 since buffers 2 and 4 have priority. This would continue until one of buffers 2 or 4 becomes empty. Say that is buffer-4 (the argument would not be different if it was buffer-2). Now that buffer-4 is empty and is not receiving any inputs from buffer-3 to process, buffer-1 can now be drained at rate μ_1 and at the same time buffer-2 is being drained at μ_2. Notice that buffers 1 and 3 have been getting input fluids at rates λ_1 and λ_3, respectively, since time $t = 0$. Also currently, buffer-2 is getting input at rate μ_1. Since $\mu_2 < \mu_1$, contents in buffer-2 would only grow while that in buffer-1 would shrink until buffer-1 becomes empty. Now we have buffers 1 and 4 empty and other two nonempty.

However, since buffer-1 gets input at rate λ_1, its departure is also λ_1. Since buffer-2 now has a smaller input rate than output, it will drain out all its fluid and become empty. Thus buffers 1, 2, and 4 are now empty. Now buffer-3 would start draining at rate μ_3. Since $\mu_3 > \mu_4$, buffer-4 would now start building up. Because of that buffer-1 would stop draining and it would also start accumulating. But buffer-2 would continue to remain empty. Thus the next event is buffer-3 would empty out. At this time buffers 1 and 4 would be nonempty. But buffer-4 would now receive input only at rate λ_3 from buffer-3, which would result in buffer-4 draining out but buffer-1 would continue building up. This would continue till buffer-4 becomes empty at which time the only nonempty buffer would be buffer-1. At this time buffer-1 would start draining at rate μ_1 into buffer-2 which in turn would drain at a slower rate μ_2. Thus buffer-1 would drain out, buffer-4 would remain empty, while buffers 2 and 3 would accumulate. This would continue till

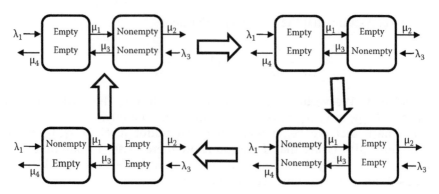

FIGURE 8.14
Cycling through buffer conditions in fluid model of Rybko–Stolyar–Kumar–Seidman network.

buffer-1 becomes empty. Thus buffers 2 and 3 are nonempty while buffers 1 and 4 are empty. This is the same situation as the beginning of this paragraph. In essence this process would cycle through until all buffers empty, as depicted in Figure 8.14.

Notice that irrespective of the initial finite amount of fluid in the four buffers, the system would reach one of the four conditions in Figure 8.14. Then it would cycle through them. A natural question to ask is: would the cycle continue indefinitely or would it eventually lead to an empty system and stay that way? Since this is a deterministic system, if we could show that if in every cycle the total amount of fluid strictly reduces, then the system would eventually converge to an empty one. Therefore, all we need to show is if we started in one of the four conditions in Figure 8.14, then the next time we reach it there would be lesser fluid in the system. Say we start in the state where buffer-3 is nonempty (with a units of fluid) and buffers 1, 2, and 4 are empty. If we show that the next time we reach that same situation, the amount of fluid in buffer-3 would be strictly less than a, then the condition that enables that would be sufficient for the fluid model to be stable. Using that argument we present the next problem which can be used to show that the condition $\lambda_1/\mu_2 + \lambda_3/\mu_4 < 1$ is sufficient for the fluid model of the network in Figure 8.6 to be stable.

Problem 80

Consider the fluid model of the network in Figure 8.6. Assume that the necessary conditions $\lambda_1/\mu_1 + \lambda_3/\mu_4 < 1$ and $\lambda_1/\mu_2 + \lambda_3/\mu_3 < 1$ are satisfied. Also assume that $\mu_1 > \mu_2$ and $\mu_3 > \mu_4$. Let the initial fluid levels be $x_1(0) = x_2(0) = x_4(0) = 0$ and $x_3(0) = a$ for some $a > 0$. Further, let T be the first passage time defined as the next time that buffers 1, 2, and 4 are empty, and buffer-3 is nonempty, that is,

$$T = \min\{t > 0 : x_1(t) = x_2(t) = x_4(t) = 0, x_3(t) \geq 0\}.$$

Show that the condition

$$\lambda_1/\mu_2 + \lambda_3/\mu_4 < 1$$

is sufficient to ensure that $x_3(T) < a$.

Solution

Notice that we begin with the north-east corner of Figure 8.14 with $x_1(0) = x_2(0) = x_4(0) = 0$ and $x_3(0) = a$. Buffer-3 would start emptying out while buffers 1 and 2 would start filling up. At time $t_1 = a/(\mu_3 - \lambda_3)$ we would reach the south-east corner of Figure 8.14 with $x_1(t_1) = \lambda_1 t_1$, $x_2(t_1) = x_3(t_1) = 0$, and $x_4(t_1) = (\mu_3 - \mu_4)t_1$. After time t_1 buffer-4 would start emptying out while buffer-1 would continue to build up. Then at time $t_2 = t_1 + (\mu_3 - \mu_4)t_1/(\mu_4 - \lambda_3)$ buffer-4 would become empty and we would reach the south-west corner of Figure 8.14. At time t_2 the system state would be $x_1(t_2) = \lambda_1 t_2$ and $x_2(t_2) = x_3(t_2) = x_4(t_2) = 0$. Immediately after time t_2, buffer-1 would start emptying while buffers 2 and 3 would fill up. This would continue until $t_3 = t_2 + \lambda_1 t_2/(\mu_1 - \lambda_1)$. At time t_3, the system state would be in the north-west corner of Figure 8.14 with $x_1(t_3) = 0$, $x_2(t_3) = (t_3 - t_2)(\mu_1 - \mu_2)$, $x_3(t_3) = \lambda_3(t_3 - t_2)$, and $x_4(t_1) = 0$. Soon after time t_3, buffer-2 would start emptying while buffer-3 would continue to grow till we reach the north-east corner of Figure 8.14 when the first passage time occurs. As given in the problem statement, that time is T which can be computed as $T = t_3 + (t_3 - t_2)(\mu_1 - \mu_2)/(\mu_2 - \lambda_1)$. At time T we have $x_3(T) = \lambda_3(T - t_2)$ and $x_1(T) = x_2(T) = x_4(T) = 0$.

We need to show that if $\lambda_1/\mu_2 + \lambda_3/\mu_4 < 1$, then $x_3(T) < a$. In other words, we need to find the condition that ensures $\lambda_3(T - t_2) < a$. For that, let us write down $\lambda_3(T - t_2) < a$ in terms of the problem parameters working our way through the definitions of other terms defined earlier (substituting for T, t_3, t_2, and finally t_1) as follows:

$$\lambda_3(T - t_2) < a,$$

$$\Rightarrow \quad \lambda_3(t_3 - t_2)(\mu_1 - \lambda_1)/(\mu_2 - \lambda_1) < a,$$

$$\Rightarrow \quad \lambda_3\lambda_1 t_2/(\mu_2 - \lambda_1) < a,$$

$$\Rightarrow \quad \frac{\lambda_3\lambda_1 t_1(\mu_3 - \lambda_3)}{(\mu_2 - \lambda_1)(\mu_4 - \lambda_3)} < a,$$

$$\Rightarrow \quad \frac{\lambda_3\lambda_1 a}{(\mu_2 - \lambda_1)(\mu_4 - \lambda_3)} < a.$$

If we cancel out a which is positive on both sides and rewrite the expression we get

$$\lambda_1/\mu_2 + \lambda_3/\mu_4 < 1$$

which indeed is the sufficient condition for stability.   ∎

From this we could also show that

$$T = \alpha a$$

where

$$\alpha = \frac{\mu_2}{(\mu_2 - \lambda_1)(\mu_4 - \lambda_3)}$$

for any $a > 0$. So if we started with a amount if fluid in buffer-3 and all other buffers empty, then after time $T = \alpha a$ we would have βa amount of fluid in buffer-3 and all other buffers empty, where

$$\beta = \frac{\lambda_3\lambda_1}{(\mu_2 - \lambda_1)(\mu_4 - \lambda_3)}.$$

Now if we started with βa, then after time $\alpha\beta a$ we would have $\beta^2 a$ amount of fluid in buffer-3 and all other buffers empty. In this manner if we were to continue, then the total time to empty the system (that started with a amount of fluid in buffer-3 and all other buffers empty) is

$$\alpha a + \beta\alpha a + \beta^2\alpha a + \beta^3\alpha a + \cdots = \frac{\alpha a}{1 - \beta} = \frac{\mu_2 a}{\mu_2\mu_4 - \lambda_1\mu_4 - \lambda_3\mu_2}.$$

This shows that if the condition $\lambda_1/\mu_2 + \lambda_3/\mu_4 < 1$ is satisfied, then the amount of fluid in the system converges to zero in a finite time. Thus we can see that if we started with some arbitrary amount of fluid $x_1(0)$, $x_2(0)$, $x_3(0)$, and $x_4(0)$ in buffers 1, 2, 3, and 4, respectively, then there exists a finite time δ after which the system would remain empty. Therefore, under that condition the fluid network is stable. That guarantees that the corresponding stochastic discrete network originally depicted in Figure 8.6 would also be stable. In a similar manner one could derive the sufficient conditions for stability of other deterministic fluid networks and thereby the corresponding stochastic discrete network on a case-by-case basis.

Having said that, it is important to point out that there are other ways to derive the conditions for stability. In particular, *Lyapunov functions* provide an excellent way to check if fluid networks are stable. Although we do not go into details of Lyapunov functions in this book, it is worthwhile describing them for the sake of completeness. Lyapunov functions have been used extensively to study the stability of deterministic dynamical systems.

They can hence be immediately used to study the stability of deterministic fluid networks. Consider a fluid queue with $x_j(t)$ being the amount of fluid in buffer j at time t for all $j \in \{1, \ldots, \ell\}$. Let x_t be the ℓ-dimensional vector of fluid levels at time t, that is, $x_t = [x_1(t) \ \ldots \ x_\ell(t)]$. A Lyapunov function $\mathcal{L}(x_t)$ is such that $\mathcal{L} : R_+^\ell \to R_+$ and satisfies the condition $\mathcal{L}(x_t) = 0$ only when $x_j(t) = 0$ for all j. Some examples of Lyapunov functions are: $\mathcal{L}(x_t) = \sum_j a_j x_j(t)$ such that $a_j > 0$ for all j; $\mathcal{L}(x_t) = \max_j x_j(t)$; $\mathcal{L}(x_t) = \sum_j x_j^2(t)$; and combinations such as $\mathcal{L}(x_t) = \max_{b_1, \ldots, b_\ell} \sum_j b_j x_j(t)$ such that $b_j > 0$ for all j. Lyapunov functions $\mathcal{L}(x_t)$ usually increase with increase in $x_j(t)$ for any j.

Although there are many possible Lyapunov functions to choose from, all we need is one appropriate one. In particular, if we can find a Lyapunov function $\mathcal{L}(x_t)$ such that $\frac{d\mathcal{L}(x_t)}{dt} < 0$ for all values of $\mathcal{L}(x_t)$ greater than some finite constant value, then the fluid model is *stable*. Again, the choice of Lyapunov functions can be made on a case-by-case basis. Popular Lyapunov function choices are quadratic and piecewise constant functions. In fact even the *time for a fluid system to empty* can be used as a Lyapunov function. For many deterministic fluid networks, using Lyapunov functions is indeed the preferred way to show stability; however, sometimes it is just difficult to find the right function. As it turns out, even for the stochastic discrete queueing networks one can use the *Foster–Lyapunov* criterion that works in a similar fashion. One could use the Foster–Lyapunov criterion to show stability of both discrete time (see Meyn [82]) and continuous time networks. With that said, in the next section we move to a different application of fluid models.

8.3 Diffusion Approximations for Performance Analysis

In this section, we consider queueing systems for which exact closed-form algebraic expressions for performance measures are either analytically intractable or too cumbersome to use. Our objective here is to develop reasonable approximations for such systems that are simple to use for design (such as in an optimization framework) and control. The $G/G/1$ queue is one such example for which we have in previous chapters developed approximations for some performance measures. Notice that those approximations can be used to develop an intuition for how parameters (such as the arrival rate) affect performance measures (such as the mean sojourn time). They can also be used in optimization contexts where, for example, a constraint could be that the mean sojourn time must be less than a stipulated value.

However, unlike the $G/G/1$ queue, for multiserver queues it is harder to use one of the previous methods (such as MVA). Although we have presented approximations for the $G/G/s$ queues in previous chapters, it turns out that they are all based on *diffusion approximations*, which is the main technique we consider in this section. What is interesting is that in many situations it is more appealing to use the $G/G/s$ approximation than the exact result for even an $M/M/s$ queue! The reason is that the exact $M/M/s$ queue is not easy to use. For example, if one were to design (or control) the number of servers, the mean sojourn time formula is a complicated expression in terms of s that one would rather use the simpler $G/G/s$ approximation. To add to the mix if we were to also consider abandonments, retrials, and server breakdowns, diffusion approximations may be the only alternative even for Markovian systems.

With that motivation in the next few sections we present a brief introduction to diffusion approximations without delving into great detail with respect to all the technical aspects. There is a rich literature with some excellent books and articles on this topic. The objective of this section is to merely provide a framework, perhaps some intuition and also fundamental background for the readers to access the vast literature on diffusion approximation (which is also sometimes referred to as heavy-traffic approximations especially in queues). For technical details on weak convergence, which is the foundation of diffusion approximations, readers are referred to Glynn [46] and Whitt [105]. We merely present the scaling procedure which results in what is called *diffusion limit*. Similar to the fluid limit we presented in Section 8.1, next we present the diffusion limit which is based on Chen and Yao [19]. Subsequently, we will describe diffusion approximations for multiserver queues.

8.3.1 Diffusion Limit and Functional Central Limit Theorem

In this section, we use simulations to develop an intuition for a diffusion process and then describe diffusion limit of a stochastic process. The approach is somewhat similar to that of the fluid limit we considered in Section 8.1. Here too we begin by considering $A(t)$, the number of entities that arrived into a system from time 0 to t. The average arrival rate is λ which is

$$\lambda = \lim_{t \to \infty} \frac{A(t)}{t}.$$

Recall that to obtain the fluid limit of the discrete arrival process $\{A(t), t \geq 0\}$, we defined $\overline{A}_n(t)$ as

$$\overline{A}_n(t) = \frac{A(nt)}{n}$$

for any $n > 0$ and $t \geq 0$. We showed that $\overline{A}_n(t) \to \lambda t$ as $n \to \infty$ which we called the fluid limit.

In a similar fashion, here we define $\hat{A}_n(t)$ as

$$\hat{A}_n(t) = \sqrt{n}\left[\overline{A}_n(t) - \lambda t\right] = \frac{A(nt) - n\lambda t}{\sqrt{n}}$$

for any $n > 0$ and $t \geq 0$. We would like to study $\hat{A}_n(t)$ as $n \to \infty$ which we will call the diffusion scaling (because the resulting process is a diffusion process). We first illustrate the diffusion scaling using the same example as in Section 8.1. Recall that to illustrate the strength of the results we consider (i) an arrival process with an extremely high coefficient of variation, (ii) a fairly small t, and (iii) analyze arrivals to the second node of a tandem network (hence arrivals are not IID).

Problem 81

Consider a $G/G/1$ queue with interarrival times as well as service times according to Pareto distributions. The coefficient of variation for interarrival times is 5 and for the service time it is equal to 2. The average arrival rate is 1 per unit time and the average service rate is 1.25 per unit time. The departures from this queue act as arrivals to a downstream queue. Let $A(t)$ be the number of entities that arrive at the downstream node during $(0, t]$. For $t = 0$ to 10 time units, graph three sample paths of $\hat{A}_n(t) = \frac{A(nt) - n\lambda t}{\sqrt{n}}$ versus t for $n = 1, 10, 100$, and 1000.

Solution

It is crucial to note that the $A(t)$ process is the arrivals to the *downstream* node which is the same as the departures from the $G/G/1$ node described in the question. Also the average arrival rate is $\lambda = 1$. By writing a simulation using the algorithm in Problem 37 in Chapter 4, we can obtain sample paths of the output process from the $G/G/1$ queue, in particular the number of departures during any interval of time. Using this for various values of $n = 1, 10, 100$, and 1000, we can plot three sample paths of $\hat{A}_n(t) = \frac{A(nt) - n\lambda t}{\sqrt{n}}$ versus t, as shown in Figure 8.15(a)–(d).

From the figure, note that in (a) where $n = 1$, the sample paths are similar to the workload process with jumps and constant declining sample paths, except the values go below zero. When $n = 10$ as seen in (b), the sample paths are still similar but they are closer than in case (a) because we have about 100 arrivals as opposed to 10 in case (a). We see this trend more prominent in case (c) where the sample paths have closed in and the jumps are not so prominent. Finally in (d) when $n = 1000$ which for this example is sufficiently large, the sample paths essentially look like Brownian motions. There are a couple of things to notice. Unlike the fluid limits, the diffusion limit does not go to a deterministic value but it appears to be a normal random variable (and the whole process converges to a Brownian motion). Also, the range

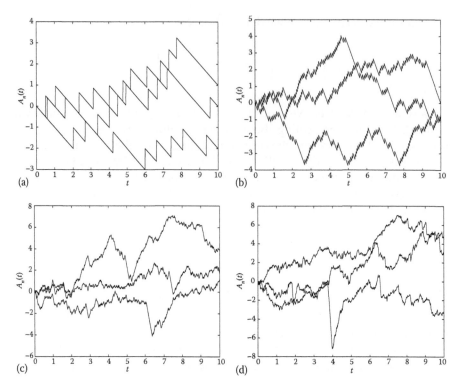

FIGURE 8.15
Sample paths of scaled arrival process $\hat{A}_n(t)$. (a) $n = 1$. (b) $n = 10$. (c) $n = 100$. (d) $n = 1000$.

of values for all four cases are more or less the same. In other words, the variability does not appear to depend on n. ∎

   While this result appears to be reasonable for a general arrival process, for the description of the key results we only consider renewal processes (refer to Whitt [105] for a rigorous description in the more general case of arrival processes, not necessarily renewal). Now, let $\{A(t), t \geq 0\}$ be a renewal process with average interrenewal time $1/\lambda$ and squared coefficient of variation of interarrival time C_a^2. We know that for a large n and any t, $A(nt)$ would be approximately normally distributed with mean λnt and variance $\lambda C_a^2 nt$. Thus for a large n, $\hat{A}_n(t) = \frac{A(nt) - n\lambda t}{\sqrt{n}}$ would be normally distributed with mean 0 and variance $\lambda C_a^2 t$. Also, based on the previous example, we can conjecture that the stochastic process $\{\hat{A}_n(t), t \geq 0\}$ converges to a Brownian motion with drift 0 and variance term λC_a^2 as $n \rightarrow \infty$. The theory that supports this conjecture is an extension of the well-known central limit theorem to functionals.

This functional central limit theorem (FCLT) is called Donsker's theorem which we present next.

To describe FCLT, we suppose that Z_1, Z_2, Z_3, ..., is a sequence of IID random variables with finite mean m and finite variance σ^2. Let S_n be the partial sum described as

$$S_n = Z_1 + Z_2 + \cdots + Z_n$$

for any $n \geq 1$. Central limit theorem essentially states that as $n \to \infty$, $\frac{S_n - mn}{\sigma\sqrt{n}}$ converges to a standard normal random variable. In practice, for large n one approximates S_n as a normal random variable with mean nm and variance $n\sigma^2$. Donsker's theorem essentially generalizes this to functionals thus resulting in the FCLT. Define $Y_n(t)$ as

$$Y_n(t) = \frac{S_{\lfloor nt \rfloor} - m\lfloor nt \rfloor}{\sigma\sqrt{n}} = \frac{1}{\sigma\sqrt{n}} \sum_{i=1}^{\lfloor nt \rfloor} [Z_i - m]$$

for any $t \geq 0$ where $\lfloor z \rfloor$ is the floor function denoting the greatest integer less than or equal to z. Donsker's FCLT states that as $n \to \infty$, the entire stochastic process $\{Y_n(t), t \geq 0\}$ converges to the standard Brownian motion (also knows as the Weiner process). In other words if the stochastic process $\{B(t), t \geq 0\}$ is a standard Brownian motion, that is, a Brownian motion with drift 0 and variance term 1, then as $n \to \infty$, $\{Y_n(t), t \geq 0\}$ converges in distribution to $\{B(t), t \geq 0\}$.

Notice that we have shown FCLT for only the partial sum S_n. Next we consider the counting process or renewal process $\{N(t), t \geq 0\}$ where $N(t) = \max\{k \geq 0 : S_k \leq t\}$, that is, the number of renewals in the time interval $(0, t]$. Define $R_n(t)$ as

$$R_n(t) = \frac{N(nt) - nt/m}{(\sigma/m)\sqrt{n/m}}$$

for any $t \geq 0$. Chen and Yao [19] show that by applying Donsker's theorem and random change theorem as $n \to \infty$, the stochastic process $\{R_n(t), t \geq 0\}$ also converges to the standard Brownian motion. To develop an intuition it may be worthwhile to show that for large t, $N(t)$ is a normal random variable with mean t/m and variance $\sigma^2 t/m^3$ (see Exercises at the end of the chapter). In summary, if $\{B(t), t \geq 0\}$ is a Brownian motion with drift 0 and variance term 1, then as $n \to \infty$, $\{R_n(t), t \geq 0\}$ converges in distribution to $\{B(t), t \geq 0\}$. Now, we put this in perspective with respect to the arrival process $\{A(t), t \geq 0\}$ which is a renewal process with average interrenewal time $1/\lambda$ and squared coefficient of variation of interarrival time C_a^2. Using the preceding result we can verify our conjecture that $\{\hat{A}_n(t), t \ge 0\}$ defined earlier converges to a Brownian motion with drift 0 and variance term λC_a^2 as $n \to \infty$.

It is not difficult to see that similar to the arrival process, the service time process when scaled in a similar fashion also converges to a Brownian motion. Thus the next natural step is to use the results in a $G/G/1$ setting where the average arrival rate is λ and the SCOV of the interarrival times is C_a^2, and the service rate is μ with service time SCOV C_s^2. The analysis would be identical to that in Section 7.1.1. There we showed the results using the normal approximation which would follow in a very similar fashion, albeit more rigorous, if we modeled the underlying stochastic processes as Brownian motions. For sake of completeness we simply restate those results here. As the traffic intensity ρ (recall that $\rho = \lambda/\mu$) approaches 1, the workload in the system converges to a reflected Brownian motion with drift $(\lambda - \mu)/\mu$ and variance term $\lambda \left(C_a^2 + C_s^2 \right)/\mu^2$. Thus the steady-state distribution of the workload is exponential with parameter γ per unit time, where $\gamma = \frac{2(1-\rho)\mu^2}{\lambda \left(C_a^2 + C_s^2 \right)}$. Since an arriving customer in steady state would wait for a time equal to the workload for service to begin, the waiting time before service is also according to $\exp(\gamma)$ when $\rho \approx 1$.

Although we did not explicitly state in Section 7.1.1, this is an extremely useful result. For example we could answer questions such as what is the probability that the service for an arriving customer would begin within the next t_0 time (answer: $1 - e^{-\gamma t_0}$). This is also extremely useful in designing the system. For example if the quality-of-service metric is that not more than 5% of the customers must wait longer than 5 time units (e.g., minutes), then we can write that constraint as $e^{-\gamma 5} \le 0.05$. Thus it is possible to obtain approximate expressions for the distribution of waiting times and sojourn times using the diffusion approximation when the traffic intensity is close to one (for that reason these approximations are also referred to as *heavy-traffic approximations*). That said, in the next two sections we will explore the use of diffusion approximations in multiserver queue settings. However, the approach, scaling, and analysis are significantly different from what was considered for the $G/G/1$ case.

8.3.2 Diffusion Approximation for Multiserver Queues

As the title suggests, here we consider diffusion approximations specifically for multiserver queues and the methodology is somewhat different from that for single-server queues. We first explain the diffusion approximation in a rather crude fashion and then describe the multiserver setting in more detail. The main goal in a diffusion approximation or diffusion scaling is to consider a stochastic process $\{Z(t), t \ge 0\}$ whose transient or steady-state distribution we are interested in (usually this is the number in the system process $\{X(t), t \ge 0\}$ but we will keep it more generic here). The process is scaled by a factor "n" across time and $\sqrt{n}$ across "space" so that we define $\hat{Z}_n(t)$ as

$$\hat{Z}_n(t) = \frac{Z(nt) - \overline{Z}(nt)}{\sqrt{n}}$$

for any $n > 0$ and $t \geq 0$. The term $\overline{Z}(nt)$ is the deterministic fluid model (potentially different from the fluid scaling we saw earlier in this chapter) of the stochastic process $\{Z(t), t \geq 0\}$. Usually, $\overline{Z}(nt) = E[Z(nt)]$ or a heuristic approximation for it. However, if that is not possible, then the usual fluid scaling (via a completely different scale) can be applied, that is, $\overline{Z}(nt) = \overline{Z}_\ell(nt)$ where the RHS is the usual fluid limit where we let the fluid scale $\ell \to \infty$.

Assuming that the deterministic fluid model ($\overline{Z}(nt)$) can be computed, the main objective here is to study $\hat{Z}_n(t)$. In particular, by applying the scaling "n," the analysis is to show that as $n \to \infty$, the stochastic process $\{\hat{Z}_n(t), t \geq 0\}$ converges to a diffusion process (that is the reason this method is called diffusion approximation or diffusion scaling). A diffusion process is a continuous-time stochastic process with almost surely continuous sample paths and satisfies the Markov property. Examples of diffusion processes are Brownian motion, Ornstein–Uhlenbeck process, Brownian bridge process, branching process, etc. It is beyond the scope of this book to show the convergence of the stochastic process $\{\hat{Z}_n(t), t \geq 0\}$ as $n \to \infty$ to a diffusion process $\{\hat{Z}_\infty(t), t \geq 0\}$. However, we do provide an intuition and interested readers are referred to Whitt [105] for technical details. The key idea of diffusion approximation is to start by using the properties of $\{\hat{Z}_\infty(t), t \geq 0\}$, such as the distribution of $\hat{Z}_\infty(\infty)$. Then for large n, $\hat{Z}_n(\infty)$ is approximately equal in distribution to $\hat{Z}_\infty(\infty)$. Thereby, we can approximately obtain a distribution for $Z(\infty)$ using

$$Z(\infty) = \overline{Z}(\infty) + \sqrt{n}\hat{Z}_n(\infty)$$

such that $\hat{Z}_n(\infty) \approx \hat{Z}_\infty(\infty)$.

Having said that, the issue that is unclear is what n is or how it should be picked. In other words, under what scale of time (and/or space), the diffusion approximation is appropriate. Notice that we pick a large n for the approximation, but does it have any physical significance? Turns out for the multiserver queue with arrival rate λ, service rate μ, and s servers, there are two possibilities. One could choose $n = 1/(1 - \rho)^2$, where $\rho = \lambda/s\mu$ or $n = s$. In other words, the diffusion approximation works well when the traffic is heavy or there are a large number of servers (or even both). That is essentially what we would consider for the rest of this section. For the sake of illustration and development of intuition, we only consider an $M/M/s$ queue and interested readers are referred to the literature, especially Whitt [105], for the $G/G/s$ case. In $M/M/s$ queues, Markov property leads to diffusion processes, however, in the $G/G/s$ case although the marginal distribution at any time in steady state converges to Gaussian, the process itself may not be a diffusion (since Markov property would not be satisfied). Nonetheless there is merit in considering the $M/M/s$ case. Thus for the remainder of this section we only consider $M/M/s$ queues, that is, Poisson arrivals (at rate λ) and exponential service times (with mean $1/\mu$ at every server).

For such an $M/M/s$ queue, let $X(t)$ be the number of customers in the system at time t. We are interested in applying diffusion scaling to the stochastic process $\{X(t), t \geq 0\}$. Further, define $\hat{X}_n(t)$ as

$$\hat{X}_n(t) = \frac{X(nt) - \overline{X}(nt)}{\sqrt{n}}$$

for any $n > 0$ and $t \geq 0$. As a heuristic approximation for the deterministic fluid model $\overline{X}(nt)$, we consider the steady-state number in the system L for the $M/M/s$ queue which from Section 2.1 is

$$L = \frac{\lambda}{\mu} + \frac{p_0(\lambda/\mu)^s \lambda}{s! s\mu[1 - \lambda/(s\mu)]^2}$$

where

$$p_0 = \left[\sum_{n=0}^{s-1} \left\{ \frac{1}{n!}(\lambda/\mu)^n \right\} + \frac{(\lambda/\mu)^s}{s!} \frac{1}{1 - \lambda/(s\mu)} \right]^{-1}.$$

Thus we use the heuristic approximation $\overline{X}(nt) = L$. Since $L = E[X(\infty)]$, the approximation is exact for large (nt). That said, we now study the diffusion-scaled process $\{\hat{X}_n(t), t \geq 0\}$ where

$$\hat{X}_n(t) = \frac{X(nt) - L}{\sqrt{n}}$$

by increasing n (in all our numerical experiments we let $X(0) = \lfloor L \rfloor$). Our objective here is to show that as n increases, $\{\hat{X}_n(t), t \geq 0\}$ converges to a diffusion process. For that we consider three different sets of experiments corresponding to the three different scalings for n.

8.3.2.1 Fix s, Increase λ

In this scaling, we consider a sequence of $M/M/s$ queues where μ and s are held a constant and only λ is increased so that ρ approaches 1. We use the

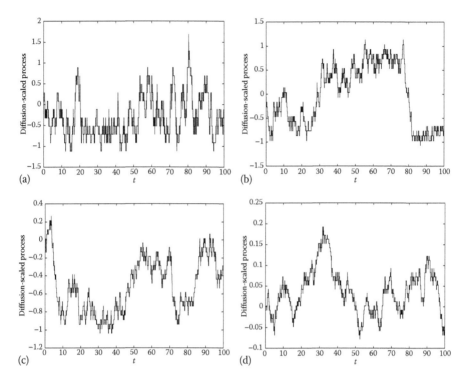

FIGURE 8.16
Sample paths of scaled process $\hat{X}_n(t/n)$ vs. t for $\mu = 1$ and $s = 4$. (a) $\lambda = 3.2$, $\rho = 0.8$, $n = 25$. (b) $\lambda = 3.6$, $\rho = 0.9$, $n = 100$. (c) $\lambda = 3.8$, $\rho = 0.95$, $n = 400$. (d) $\lambda = 3.96$, $\rho = 0.99$, $n = 10,000$.

scale $n = 1/(1 - \rho)^2$ so that n increases as ρ increases. We plot

$$\hat{X}_n(t/n) = \frac{X(t) - L}{\sqrt{n}}$$

versus t for various increasing values of λ in Figure 8.16. Notice that the diffusion-scaled process is a little different and not scaled across time (we use t/n as opposed to t). From Figure 8.16 it is clear the $\{\hat{X}_n(t/n), t \geq 0\}$ process converges to a diffusion process as n is scaled. This would be more powerful if we were to have scaled time as well, that is, plotted $\hat{X}_n(t)$ instead of $\hat{X}_n(t/n)$. One could use this scaling when the system has high traffic intensity but not a large number of servers.

8.3.2.2 Fix ρ, Increase λ and s

In this scaling, we consider a sequence of $M/M/s$ queues where μ and ρ are held a constant but λ and s are increased so that s approaches ∞. We use the

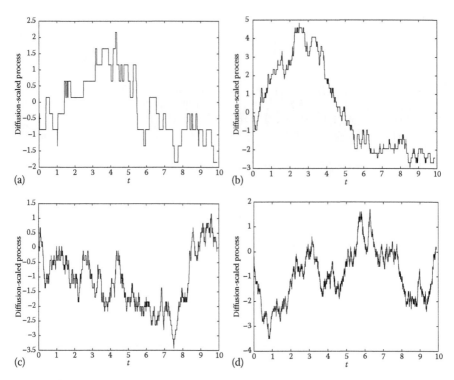

FIGURE 8.17
Sample paths of scaled process $\hat{X}_n(t/n)$ vs. t for $\mu=1$ and $\rho=0.9$. (a) $\lambda=3.6$, $s=4$, $n=4$. (b) $\lambda=14.4$, $s=16$, $n=16$. (c) $\lambda=36$, $s=40$, $n=40$. (d) $\lambda=90$, $s=100$, $n=100$.

scale $n=s$ so that n increases as s increases. We plot

$$\hat{X}_n(t/n) = \frac{X(t) - L}{\sqrt{n}}$$

versus t for various increasing values of λ and s in Figure 8.17. Notice that the diffusion-scaled process is a little different and not scaled across time (we use t/n as opposed to t). From Figure 8.17 it is clear the $\{\hat{X}_n(t/n), t \geq 0\}$ process converges to a diffusion process as n is scaled. This would be more powerful if we were to have scaled time as well, that is, plotted $\hat{X}_n(t)$ instead of $\hat{X}_n(t/n)$. One could use this scaling when the system has a large number of servers but not a very high traffic intensity.

8.3.2.3 Fix β, increase λ and s

In this scaling, we consider a sequence of $M/M/s$ queues where only μ is held a constant but λ and s are increased so that ρ approaches 1. In particular we consider the Halfin–Whitt regime (due to Halfin and Whitt [50]) in which $\rho \to 1$ but β is held a constant where

$$\beta = (1 - \rho)\sqrt{s}.$$

We use the scale $n = s$ (the choice of $n = 1/(1 - \rho)^2$ would have also worked) so that n increases as s increases. We plot

$$\hat{X}_n(t/n) = \frac{X(t) - L}{\sqrt{n}}$$

versus t for various increasing values of λ and s in Figure 8.18. Notice that the diffusion-scaled process is a little different and not scaled across time (we use t/n as opposed to t). From Figure 8.18 it is clear the $\{\hat{X}_n(t/n), t \geq 0\}$ process converges to a diffusion process as n is scaled. This would be more powerful if we were to have scaled time as well, that is, plotted $\hat{X}_n(t)$ instead of $\hat{X}_n(t/n)$. One could use this scaling when the system has both high traffic

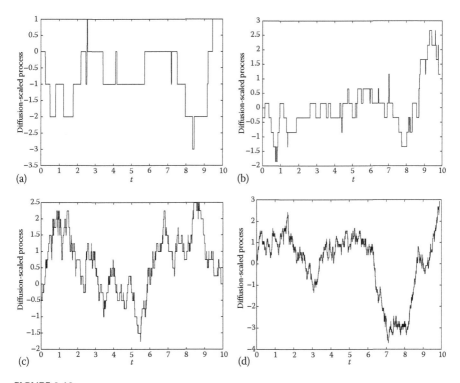

FIGURE 8.18
Sample paths of scaled process $\hat{X}_n(t/n)$ vs. t for $\mu = 1$ and $\beta = 0.2$. (a) $\lambda = 0.8$, $\rho = 0.8$, $n = s = 1$. (b) $\lambda = 3.6$, $\rho = 0.9$, $n = s = 4$. (c) $\lambda = 15.2$, $\rho = 0.95$, $n = s = 16$. (d) $\lambda = 62.4$, $\rho = 0.975$, $n = s = 64$.

intensity and a large number of servers (which is typical in inbound call centers).

In summary, based on the experimental evidence, it is not hard to see that the stochastic process $\{\hat{X}_n(t), t \geq 0\}$ converges to a diffusion process as n increases. Also, as $n \to \infty$ and $t \to \infty$, $E[\hat{X}_n(t)]$ converges to zero since $E[X(nt)]$ would be L. However, it is more tricky to obtain a distribution for $\hat{X}_n(t)$ as $t \to \infty$ and $n \to \infty$. Halfin and Whitt [50] show that if we define

$$\hat{X}_n(t) = \frac{X(nt) - s}{\sqrt{n}}$$

and use the scale $n = s$, then

$$\lim_{n \to \infty} P\{\hat{X}_n(\infty) \geq 0\} = \theta$$

where θ is a constant satisfying $0 < \theta < 1$. We provide an intuition for that first and subsequently describe θ. For $\hat{X}_n(t) \geq 0$, the process converges to a reflected Brownian motion with negative drift (since for $X(nt) \geq s$, the CTMC is a birth and death process with constant parameters λ and $s\mu$ which is a random walk that converges to a Brownian motion upon scaling). However, for $\hat{X}_n(t) < 0$, the process converges to an Ornstein–Uhlenbeck process (since for $X(nt) < s$, the CTMC is a birth and death process with parameters λ and $X(nt)\mu$ which is a random walk that converges to an Ornstein–Uhlenbeck process upon scaling). Halfin and Whitt [50] state that this argument also holds for $G/M/s$ queues as well.

Therefore, from this realization it is possible to obtain the probability that $\hat{X}_n(t) \geq 0$ as $t \to \infty$ and $n \to \infty$. Thus under the scaling $n = s$,

$$P\{X(\infty) \geq s\} \to \theta$$

and Halfin and Whitt [50] show that

$$\theta = \left[1 + \sqrt{2\pi}\beta\phi(\beta)\exp(\beta^2/2)\right]^{-1}$$

where $\phi(x)$ is the probability a standard normal random variable is less than x. Thus with probability θ, a customer arriving to the system in steady state will experience any delay. It is crucial to notice that with the other two scalings we considered earlier as well, it is possible to obtain θ. In those cases, θ would just be 0 or 1. In particular, if we fix s and increase λ then θ approaches 1 (that is, an arriving request with probability 1 will be delayed for service to begin). Whereas if we fix ρ but increase λ and s, then θ approaches zero (that is, an arriving customer with probability 1 will find a free server).

Whether or not an arriving customer would have to wait for service to begin leads us to a related topic of abandonments where customers abandon the queue if their service does not begin in a reasonable time. This is the focus of the next section.

8.3.3 Efficiency-Driven Regime for Multiserver Queues with Abandonments

In this section, we consider queues with customer *abandonments*. There are two main types of abandonments: balking and reneging. When a customer arrives to a queueing system but decides to abandon it without joining (usually because the queue is long), then we say that the customer is *balking*. Balking typically occurs only in queues where customers can actually observe the queues (such as in restaurants, post offices, banks, etc.). Thus it is not very crucial to address balking in systems like inbound call centers, computer systems, manufacturing, etc. However, when a customer enters the system and waits for a while but decides to abandon it because service has not begun, then that customer is *reneging* from the queue. In this section when we say abandonments, we mean reneging as the analysis is motivated by inbound call centers.

Consider a single-stage queueing system with s identical servers. Customers arrive according to a Poisson process with mean arrival rate λ per second. The service times are exponentially distributed with mean $1/\mu$ seconds. The traffic intensity is $\rho = \lambda/s\mu$. So far the system looks like an $M/M/s$ queue which is also known as the Erlang-C model (note: the $M/M/s/k$ queue is known as Erlang-A model and $M/M/s/s$, the Erlang-B model). However, we extend the model to allow for customer abandonments. Each customer has a patience time so that if the time in the queue before service exceeds the patience time then the customer would abandon. We assume that the patience times are IID random variables that are exponentially distributed with mean patience time $1/\alpha$ seconds. This system is represented as an $M/M/s/\infty + M$ queue. The number in the system at time t, $X(t)$, can be used to model $\{X(t), t \geq 0\}$ as a birth and death process with birth parameters λ and death parameters $\min(s, i)\mu + (i - s)^+\alpha$ when $X(t) = i$. However, the analysis does not result in simple expressions that can be used in design and control. Hence we consider diffusion approximations.

One can derive diffusion approximations of such systems in three regimes: (i) $\rho < 1$ is called quality-driven regime because it strives for quality of service (typically $\lambda \to \infty$ and $s \to \infty$ and $\lambda/(s\mu) \to \rho < 1$); (ii) $\rho > 1$ is called efficiency-driven (ED) regime because it strives for efficient use of servers (typically $\lambda \to \infty$ and $s \to \infty$ and $\lambda/(s\mu) \to \rho > 1$); (iii) $\rho \approx 1$ is called quality- and efficiency-driven (QED) regime as it strives for both (especially as $\lambda \to \infty$, $s \to \infty$ such that $\sqrt{s}(1 - \rho) \to \beta$, also known as Halfin–Whitt limiting regime). Notice that because of the abandonments (assume $\alpha > 0$) all regimes would result in stable queues (in fact the β in QED could also be negative). Here we only consider the analysis of the ED regime by closely following the results in Whitt [106]. In fact the references in Whitt [106] point to articles that consider the other two regimes.

To obtain the diffusion limits for the ED regime of the $M/M/s+M$ model, we begin with the Erlang-A $M/M/s/K+M$ model. The waiting space $K-s$ will be chosen to be large enough and scaled in a manner that would approach infinity. Thus we consider a sequence of $M/M/s/K+M$ queues indexed by s, the number of servers which we would use to scale. In particular, let λ_s and K_s be the scaled arrival rate and system capacity. However the service rate μ and abandonment rate α are not scaled. Also, the traffic intensity ρ is not scaled and remains fixed for the entire sequence of queues with $\rho > 1$ (since the regime is ED). Let

$$q = \frac{\mu(\rho - 1)}{\alpha}. \tag{8.10}$$

We perform the following scaling

$$\lambda_s = \rho s \mu \tag{8.11}$$

$$K_s = s(\eta + 1) \tag{8.12}$$

for some $\eta > q$. Equation 8.12 is to ensure that asymptotically no arriving customers are rejected due to a full system (see Whitt [106]). With this scaling we proceed with the diffusion scaling.

Let $X_s(t)$ be the number of customers in the system at time t when there are s servers. Define the diffusion term

$$\hat{X}_s(t) = \frac{X_s(t) - \overline{X}_s(t)}{\sqrt{s}}$$

where $\overline{X}_s(t)$ is a deterministic model of $X_s(t)$. We use a heuristic approximation for the deterministic quantity $\overline{X}_s(t)$ which is where the $X_s(t)$ tends to linger around in the ED regime. In particular, we select $\overline{X}_s(t)$ as an "equilibrium" point where the system growth rate equals the shrinkage rate. Thus we have $\overline{X}_s(t)$ as the solution to

$$\lambda_s = s\mu + [\overline{X}_s(t) - s]\alpha$$

by making the realization that $\overline{X}_s(t)$ must be greater than s (as $\rho > 1$ results in $\lambda_s > \min\{i, s\}\mu$ for any $i \geq 0$). Thus we have

$$\overline{X}_s(t) = \frac{\lambda_s - s\mu}{\alpha} + s = (1 + q)s \tag{8.13}$$

where the last equality is by substitution for λ_s in Equation 8.11 and using Equation 8.10 for q. Thus we represent the diffusion term as

$$\hat{X}_s(t) = \frac{X_s(t) - s(1+q)}{\sqrt{s}} \tag{8.14}$$

for all $t \geq 0$.

Whitt [106] shows that the stochastic process $\{\hat{X}_s(t), t \geq 0\}$ as $s \to \infty$ converges to an Ornstein–Uhlenbeck diffusion process. In state x, the infinitesimal mean or state-dependent drift of the Ornstein–Uhlenbeck process is $-\alpha x$ and infinitesimal variance $2\mu\rho$. Further, the steady-state distribution of $\hat{X}_s(\infty)$ converges to a normal distribution with mean 0 and variance $\rho\mu/\alpha$. Next we explain that briefly. We showed earlier in this section via simulations how processes like $\{\hat{X}_s(t), t \geq 0\}$ converge to diffusion processes (hence the term diffusion limit) as $s \to \infty$. Thus that is not a surprising result. Further, it is possible to show a weak convergence of the birth and death process $\{X_s(t), t \geq 0\}$ to an Ornstein–Uhlenbeck process by appropriately scaling (akin to how the constant birth and death parameter converges to a Brownian motion). That is because beyond state $s(1+q)$ since the death rate exceeds the birth rate, the process gets pulled back to $s(1+q)$. Likewise, below state $s(1+q)$ where the birth rate exceeds the death rate, the process gets pushed up to $s(1+q)$. This results in a convergence to the Ornstein–Uhlenbeck process centered around $s(1+q)$. Also, the steady-state distribution of an Ornstein–Uhlenbeck diffusion process centered at zero (with mean drift rate $-m$ and infinitesimal variance v) is zero-mean normal with variance equal to $v/(2m)$. Notice that the drift rate of $-m$ implies that the drift in state x is $-mx$. That said, the only things remaining to be shown are that the drift rate for our process is $m = \alpha$ and infinitesimal variance equal to $2\mu\rho$. This is the focus of the next problem.

Problem 82

Show that the Ornstein–Uhlenbeck diffusion process that results from scaling $\{\hat{X}_s(t), t \geq 0\}$ as $s \to \infty$ has a drift of $m(x) = -\alpha x$ and infinitesimal variance $v(x) = 2\mu\rho$ for any feasible state x.

Solution

Notice that since $X_s(t)$ takes on any integer value $k \geq 0$, $\hat{X}_s(t)$ would correspondingly take on discrete values $[k - s(1+q)]/\sqrt{s}$ for $k = 0, 1, 2, \ldots$ but we are ultimately interested in any real-valued state x. For that we first consider an arbitrary real value x and a sequence of x_s values for $\hat{X}_s(t)$ for each s so that $x_s \to x$ as $s \to \infty$. Assuming that s is sufficiently large, we can consider the following choice for x_s so that the preceding condition is met:

$$x_s = \frac{\lfloor s(1+q) + x\sqrt{s} \rfloor - s(1+q)}{\sqrt{s}}.$$

For any s, the infinitesimal mean (corresponding to the drift) $m_s(x_s)$ can be computed as follows (with the first equation being the definition):

$$m_s(x_s) = \lim_{h \to 0} E[(\hat{X}_s(t+h) - \hat{X}_s(t))/h | \hat{X}_s(t) = x_s],$$

$$= \lim_{h \to 0} E[(X_s(t+h) - X_s(t))/(h\sqrt{s}) | X_s(t) = \sqrt{s}x_s + s(1+q)],$$

$$= \lim_{h \to 0} \frac{\lambda_s h - \mu s h - \alpha h (\sqrt{s}x_s + sq) + o(h)}{h\sqrt{s}},$$

for any $x_s \geq -q\sqrt{s}$ where $o(h)$ is a collection of terms of order less than h such that $o(h)/h \to 0$ as $h \to 0$.

By taking the limit $h \to 0$ and substituting for λ_s using Equation 8.11 we get

$$m_s(x_s) = \mu\rho\sqrt{s} - \mu\sqrt{s} - \alpha x_s - \alpha q\sqrt{s} = -\alpha x_s$$

by substituting for q in Equation 8.10. Now we let $s \to \infty$ resulting in $m_s(x_s) \to m(x)$ such that

$$m(x) = -\alpha x.$$

It is worthwhile pointing out that this expression was derived assuming that $x_s \geq -q\sqrt{s}$. But what if $x_s < -q\sqrt{s}$? It turns out that for sufficiently large s it is not even feasible to reach states x_s that are smaller than $-q\sqrt{s}$. Even if one were to reach such a state, the calculation would result in $m(x) = \infty$ which would imply an instantaneous drift to a higher x. However, that is the reason the problem is worded as $m(x) = \alpha x$ for any *feasible* state x. Whitt [106] shows that $P\{X_s(\infty) \leq s\} \to 0$ as $s \to \infty$ using a fluid model which implies that in steady state there is not chance for x_s to be less than zero (leave alone less than $-q\sqrt{s}$).

Next we consider the infinitesimal variance. For any s, the infinitesimal variance $v_s(x_s)$ can be computed as follows (with the first equation being the definition):

$$v_s(x_s) = \lim_{h \to 0} E[(\hat{X}_s(t+h) - \hat{X}_s(t))^2/h | \hat{X}_s(t) = x_s],$$

$$= \lim_{h \to 0} E[(X_s(t+h) - X_s(t))^2/(hs) | X_s(t) = \sqrt{s}x_s + s(1+q)],$$

$$= \lim_{h \to 0} \frac{\lambda_s h + \mu s h + \alpha h (\sqrt{s}x_s + sq) + o(h)}{hs},$$

for any $x_s \geq -q\sqrt{s}$ where $o(h)$ is a collection of terms of order less than h such that $o(h)/h \to 0$ as $h \to 0$ but different from the $o(h)$ defined in $m_s(x_s)$.

By taking the limit $h \to 0$ and substituting for λ_s using Equation 8.11 we get

$$v_s(x_s) = \mu\rho + \mu + \alpha x_s/\sqrt{s} + \alpha q = 2\rho\mu + \alpha x_s/\sqrt{s}$$

where the second equality uses expressions for q from Equation 8.10. Now we let $s \to \infty$ resulting in $v_s(x_s) \to v(x)$ such that

$$v(x) = 2\mu\rho. \qquad \blacksquare$$

Thus the steady-state distribution of the diffusion process, that is, $\hat{X}_s(\infty)$ as $s \to \infty$ converges to a normal distribution with mean 0 and variance $\rho\mu/\alpha$. This involves a rigorous argument taking stochastic process limits appropriately (see Whitt [106] and Whitt [105] for further details). Therefore, as an approximation we can use for fairly large s values that $\hat{X}_s(\infty)$ is approximately normally distributed with mean 0 and variance $\rho\mu/\alpha$. Hence using Equation 8.14 we can state that $X_s(\infty)$ is approximately normally distributed with mean $s(1+q)$ and variance $s\rho\mu/\alpha$. Assuming a reasonably significant abandonment rate α so that $\frac{\mu}{\alpha} << s$ we can see that $X_s(\infty)$ would be greater than s with a very high probability (approximately 1). Hence we can write down $L_q \approx L - s$ using our usual definition of L and L_q being the steady-state number of customers in the system and in the queue waiting for service to begin, respectively. Thus we have $L_q \approx sq$ since $L \approx s(1+q)$. Now define P_{ab} as the probability that an arriving customer in steady state would abandon without service. Using Little's law for abandoning customers, we have

$$L_q = \lambda_s P_{ab} \frac{1}{\alpha}.$$

Using the fact that $L_q \approx sq = s\mu(\rho-1)/\alpha$ and $\lambda_s = s\rho\mu$, we can compute

$$P_{ab} \approx \frac{\rho-1}{\rho}.$$

To illustrate the results we use an example from Whitt [106]. Consider an inbound call center with $s = 100$ servers each with service rate $\mu = 1$ customer per minute. The arrival rate is $\lambda = 110$ customers per minute and the average abandonment time is $1/\alpha = 10$ min for each customer. For such a system $\rho = 1.1$ which gives rise to $L_q \approx sq = s\mu(\rho-1)/\alpha = 100$ and $P_{ab} \approx \frac{\rho-1}{\rho} = 10/11$. These approximations match extremely closely with the exact results. Using the fact that $X_s(\infty)$ approximately equals s plus the number of customers waiting to begin service, the variance of $X_s(\infty)$ must be equal to the variance of the number of customers waiting to begin service. Using that we can obtain for the preceding numerical example that the standard deviation of the number of customers waiting to begin service is approximately $\sqrt{s\rho\mu/\alpha} = 33.1662$, which is remarkably close to the exact result of 33.1 customers. The key point to make is that these approximations are conducive to use in design and control (as opposed to the exact results). This is typical for most diffusion approximations where the results are surprisingly simple although the process to obtain them are fairly intensive. That said, in the next section we leverage upon both fluid and diffusion approximations for transient analysis.

8.4 Fluid Models for Queues with Time-Varying Parameters

In this section, we focus on transient analysis of queueing systems where the parameters such as arrival rate, service rate, and number of servers are time varying. We only consider systems that are modulated by Poisson processes. However, the Poisson processes can be deterministically or stochastically nonhomogeneous over time. As an example, consider the $M_t/M/s_t$ queue that is useful to model inbound call centers where the expected arrival rate is time varying (say λ_t at time t) and the number of servers or agents is also time varying (say s_t at time t). The mean service rate μ remains a constant. Out ultimate goal is to derive the mean and variance of $X(t)$ which is the number of customers in the system at time t. For that we first describe the notation for a generic nonhomogeneous Poisson process and then consider the two nonhomogeneous Poisson processes that modulate the dynamics of the $M_t/M/s_t$ example system.

In terms of notation, let $\Lambda(t)$ be the expected number of events in a generic nonhomogeneous Poisson process in time $(0, t]$. Then the nonhomogeneous Poisson process is written as $N(\Lambda(t))$ which is the *random* number of events from time 0 to t. For the $M_t/M/s_t$ queue described here, the arrival process is the classical nonhomogeneous Poisson process $N_a(\cdot)$ for which we use the subscript "a" to denote "arrival." We can write down the nonhomogeneous arrival process as

$$N_a\left(\int_0^t \lambda_u du\right).$$

The other set of events that are responsible for the dynamics of an $M_t/M/s_t$ queue are the departures. We let $N_d(\cdot)$ to denote the departure process, which is also a Poisson process that is not only time-homogeneous but also state dependent. Thus the nonhomogeneous departure process is

$$N_d \left(\int_0^t \min\{X(u), s_u\} \mu du \right).$$

Thus, for this $M_t/M/s_t$ queue, we can write down $X(t)$ in terms of the initial state of the system $X(0)$, as well as the two nonhomogeneous Poisson processes $N_a(\cdot)$ and $N_d(\cdot)$ as

$$X(t) = X(0) + N_a \left(\int_0^t \lambda_u du \right) - N_d \left(\int_0^t \min\{X(u), s_u\} \mu du \right).$$

To this say we add two additional situations: (i) customers renege after $\exp(\beta)$ time if their service does not start; (ii) there is a new stream of customers that arrive according to a homogeneous Poisson process with parameter α but could balk upon arrival resulting in a queue joining probability $q_{X(t)}$ if the customer arrives at time t and sees $X(t)$ others in the system. Clearly the reneging occurs according to a nonhomogeneous Poisson process, let us call it $N_r(\cdot)$. Likewise, let $N_b(\cdot)$ denote the nonhomogeneous Poisson process corresponding to the second stream of customers that potentially balk. For this modified system we can write down $X(t)$ as

$$X(t) = X(0) + N_a \left(\int_0^t \lambda_u du \right) - N_d \left(\int_0^t \min\{X(u), s_u\} \mu du \right)$$

$$- N_r \left(\beta \int_0^t \max\{X(u) - s_u, 0\} du \right) + N_b \left(\alpha \int_0^t q_{X(u)} du \right).$$

With that example and extension we next consider a generic single-station queueing system with time-varying parameters and state-dependent transitions modulated by k Poisson processes (where k is an arbitrary number which is equal to 2 and 4, respectively, in the previous example and its extension). Let $X(t)$ be the number of customers in this queueing system at time t. We would like to write down a generic expression for $X(t)$ so that we can describe a theory for that. In that light, assume that we can write down an integral equation for $X(t)$ as

$$X(t) = X(0) + \sum_{i=1}^{k} l_i Y_i \left(\int_0^t f_i(s, X(s)) ds \right), \tag{8.15}$$

where $Y_i(\cdot)$ is an independent nonhomogeneous Poisson process, $l_i = \pm 1$, and $f_i(\cdot, \cdot)$ is a continuous function for all $i \in \{1, 2, \ldots, k\}$. It is a worthwhile exercise to explicitly write down l_i, $Y_i(\cdot)$, and $f_i(\cdot, \cdot)$ for the previous example and its extension before forging ahead. After having done that we ask the question: Using fluid and diffusion approximations can we write down expressions for $E[X(t)]$ and $Var[X(t)]$? The answer is *yes* and that will be the focus of the rest of this section.

8.4.1 Uniform Acceleration

To obtain $E[X(t)]$ for $X(t)$ defined in Equation 8.15, we use the concept of fluid scaling. The approach presented here was developed initially in Kurtz [71] and further fine-tuned by Mandelbaum et al. [77]. For the approach to work, we make some assumptions that are fairly mild for most queueing systems. We first assume that the number of modulating Poisson processes k is finite. Then we assume that for any i, t, and x, $|f_i(t, x)| \leq C_i(1 + x)$ for some $C_i < \infty$ and $T < \infty$ such that $t \leq T$. Next we define the function $F(x, t)$ as

$$F(t, x) = \sum_{i=1}^{k} l_i f_i(t, x). \tag{8.16}$$

We assume that there exists a constant $M < \infty$ such that $|F(t, x) - F(t, y)| \leq M|x - y|$ for all $t \leq T$ and $T < \infty$. It is crucial to point out that we will not explicitly state the assumptions in the results to follow, but they are all necessary to be satisfied.

Now we are ready to obtain the fluid limits for the stochastic process $\{X(t), t \geq 0\}$. For this we perform a fluid scaling that results in a deterministic process. Recall $X(t)$ defined in Equation 8.15. Consider a sequence of stochastic processes $\{X_n(t), t \geq 0\}$ indexed by n so that $X_n(t)$ is obtained by scaling $X(t)$ as follows:

$$X_n(t) = \frac{nX_n(0) + \sum_{i=1}^{k} l_i Y_i \left(\int_0^t n f_i(s, X_n(s)) ds \right)}{n} \tag{8.17}$$

with $nX_n(0) = X(0)$ so that the initial state is also scaled. The scaled process $\{X_n(t), t \geq 0\}$ is obtained essentially by taking n times faster rates of events. Such a scaling is also called *uniform acceleration* in the literature (see Massey and Whitt [79]). Like in all the fluid models we have seen in this chapter, here too as $n \to \infty$, the scaled process $\{X_n(t), t \geq 0\}$ converges to a deterministic process almost surely.

The result once again is an artifact of functional strong law of large numbers (FSLLN), which leads to what is called the *strong approximation*. In particular, as described in Kurtz [71],

$$\lim_{n \to \infty} X_n(t) = \bar{X}(t)$$

almost surely, where $\bar{X}(t)$ is a deterministic quantity which results in a deterministic process $\{\bar{X}(t), t \geq 0\}$. If we obtain $\bar{X}(t)$, our fluid model is complete. As a heuristic we consider $\bar{X}(t) = \lim_{n \to \infty} E[X_n(t)]$ which is reasonable when the assumptions described earlier are satisfied. Thus taking expectations on both sides of Equation 8.17 we get

$$E[X_n(t)] = \frac{nX_n(0) + \sum_{i=1}^{k} l_i E\left[Y_i\left(\int_0^t nf_i(s, X_n(s))ds\right)\right]}{n}$$

$$= \frac{nX_n(0) + \sum_{i=1}^{k} l_i E\left[\int_0^t nf_i(s, X_n(s))ds\right]}{n} \tag{8.18}$$

$$= X_n(0) + \sum_{i=1}^{k} l_i E\left[\int_0^t f_i(s, X_n(s))ds\right] \tag{8.19}$$

where Equation 8.18 is due to the Poisson process property recalling that the expected value of a nonhomogeneous Poisson process $N(\Lambda(t))$ is $\Lambda(t)$. If we know the distribution of $X_n(s)$ then we can write down Equation 8.19, but we consider a nonparametric approach. For that we use the Lipschitz property of the function $f_i(\cdot, \cdot)$ due to which

$$|E[f_i(s, X_n(s))] - f_i(s, E[X_n(s)])| \leq ME[|X_n(s) - E[X_n(s)]|].$$

If we let $n \to \infty$ in this expression, then the RHS goes to zero. Thus we have

$$\lim_{n \to \infty} E[f_i(s, X_n(s))] = \lim_{n \to \infty} f_i(s, E[X_n(s)]).$$

Using this result in Equation 8.19 and letting $n \to \infty$, we can rewrite Equation 8.19 as

$$\lim_{n \to \infty} E[X_n(t)] = X_n(0) + \sum_{i=1}^{k} l_i \int_0^t \lim_{n \to \infty} f_i(s, E[X_n(s)])ds.$$

Since we consider $\bar{X}(t) = \lim_{n \to \infty} E[X_n(t)]$, using the previous equation we can write down $\bar{X}(t)$ as the solution to the equation

$$\bar{X}(t) = \bar{X}(0) + \sum_{i=1}^{k} l_i \int_{0}^{t} f_i(s, \bar{X}(s)) ds. \tag{8.20}$$

Note that using the previous expression it is possible to solve numerically for $\bar{X}(t)$ for any t. Thus for large n one can approximate $X_n(t)$ as the deterministic quantity $\bar{X}(t)$. But what is the connection to $E[X(t)]$ that we alluded to earlier in this section? As it turns out, that would have to be done on a case-by-case basis. We illustrate that process using an example next.

Problem 83

Consider an $M_t/M/s_t$ system that models an inbound call center. The constant mean service rate for this call center is $\mu = 4$ customers per hour. Table 8.2 describes the expected arrival rate λ_t per hour and number of servers (s_t) by discretizing into eight hourly intervals. Develop a fluid scaling or uniform acceleration for this system by numerically describing the deterministic process $\{\bar{X}(t), 0 \leq t \leq 8\}$. Compare against a simulated sample path of the number in the system process $\{X(t), 0 \leq t \leq 8\}$. Assume that $X(0) = 80$, that is, at time zero there are already 80 customers in the system. Also, obtain an approximation for $E[X(t)]$ and compare against simulations by creating 100 replications and averaging them.

Solution

The problem description is that of a call center where the arrival rate and number of servers (that is, representatives or call handlers) are time varying. However, within an hour we assume that they are held a constant (the

TABLE 8.2

Hourly Arrival Rate and
Staffing at a Call Center

t	λ_t	s_t
$(0, 1]$	400	110
$(1, 2]$	440	120
$(2, 3]$	500	130
$(3, 4]$	720	170
$(4, 5]$	800	220
$(5, 6]$	720	200
$(6, 7]$	600	140
$(7, 8]$	400	120

analytical model does not need this assumption, it is there just for the insights). It may be worthwhile going through Table 8.2. In essence during the first hour 400 customers are expected and the number of servers is 110. Likewise, during the seventh hour 600 customers are expected and the number of servers is 140. Notice that in time intervals $(3,4]$ and $(6,7]$ there is an overload situation, that is, the arrival rate is larger than the service capacity, $\lambda_t > s_t \mu$. However, since this is a transient analysis, that is not much of an issue but worth watching out for.

That said we now consider the $M_t/M/s_t$ system with $\mu = 4$ and λ_t and s_t from Table 8.2. Let $X(t)$ be the number of customers in the system at time t with $X(0) = 80$. We can rewrite Equation 8.15 as

$$X(t) = X(0) + Y_1\left(\int_0^t \lambda_u du\right) - Y_2\left(\int_0^t \mu \min\{s_u, X(u)\}du\right), \qquad (8.21)$$

where $Y_1(\cdot)$ and $Y_2(\cdot)$ are the nonhomogeneous Poisson arrival and departure processes, respectively. For some large n and all $t \in [0, 8]$, let

$$a_t = \frac{\lambda_t}{n} \quad \text{and} \quad r_t = \frac{s_t}{n}.$$

We will pretend r_t is an integer for the fluid scaling but that will not be necessary for the limiting deterministic process that will be defined subsequently. Define $X_n(t)$ as

$$X_n(t) = \frac{nX_n(0) + Y_1\left(\int_0^t na_u du\right) - Y_2\left(\int_0^t \mu n \min\{r_u, X_n(u)\}du\right)}{n},$$

where $X_n(0) = X(0)/n$. Notice that this equation is identical in form to that of the scaled process in Equation 8.17. More crucially notice that this equation is also identical to Equation 8.21 if we let

$$X(t) = nX_n(t)$$

for all t.

As $n \to \infty$, $X_n(t)$ converges to $\bar{X}(t)$ which is given by the solution to

$$\bar{X}(t) = \bar{X}(0) + \left(\int_0^t a_u du\right) - \left(\int_0^t \mu \min\{r_u, \bar{X}(u)\}du\right), \qquad (8.22)$$

as described in Equation 8.20. Before proceeding ahead, it is important to realize that the theory is developed in an opposite direction. In particular, in theory we begin with r_t and a_t known and then scale the system by a huge factor n and say that the scaled process $\{X_n(t), t \geq 0\}$ converges to its fluid limit $\{\bar{X}(t), t \geq 0\}$. However, in this problem we begin with λ_t and s_t, select an n and then figure r_t and a_t. Thus the approximation would work well when λ_t is significantly larger than μ and s_t significantly larger than 1. In fact, the choice of n is actually irrelevant.

We arbitrarily select $n = 50$ (any other choice would not change the results). Then we solve for $\bar{X}(t)$ by performing a numerical integration for Equation 8.22 via first principles in calculus. Using this fluid scaling we plot an approximation for $X(t)$ using $X(t) \approx n\bar{X}(t)$ in Figure 8.19. To actually compare against a sample path of $X(t)$, we also plot a simulated sample path of $X(t)$ in that same figure (see the jagged line). The smooth line in that figure corresponds to $n\bar{X}(t)$. The crucial thing to realize is that figure would not change if a different n was selected. Notice how remarkably closely the simulated graph follows the fluid limit giving us confidence it is performing well.

However, the next thing to check is whether $E[X(t)]$ is close to $\bar{X}(t)$. For that consider Figure 8.20. The smooth line is $n\bar{X}(t)$ which is identical to that in Figure 8.19. By performing 100 simulations $E[X(t)]$ can be estimated for every t. The estimated $E[X(t)]$ is the jagged line in Figure 8.20. The two graphs are incredibly close suggesting that the approximation is fairly reasonable. The three ways this fit could improve even further: (1) if λ_t and s_t were much higher; (2) if we used a parametric approach instead of Lipschitz to resolve Equation 8.19; and (3) if we used well over 100

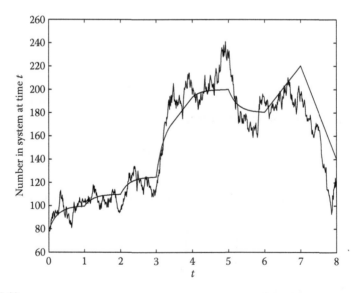

FIGURE 8.19

Number in the system after fluid scaling (smooth line) vs. single simulation sample path (jagged line).

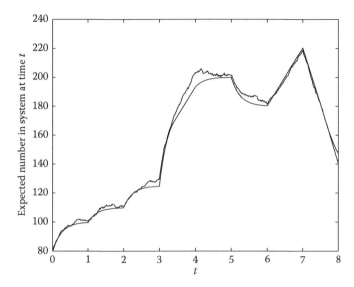

FIGURE 8.20
Mean number in the system using 100 replications of simulation (jagged line) vs. fluid approximation (smooth line).

replications. Nonetheless it is remarkable how good the approximation is for $E[X(t)]$ which gives a lot of credibility for this transient analysis. ■

Similar to the previous example it is possible to develop fluid models in other situations as well, especially including balking and reneging is fairly straightforward. Making the service time time-varying is a relatively straightforward extension. In summary, it is possible to develop a reasonable approximation for $E[X(t)]$ by performing a fluid scaling, if that is appropriate for the system (that is, in the example above we saw we need λ_t to be significantly larger than μ and s_t be significantly larger than 1). The fluid scaling uses the fact that $X_n(t)$ approaches $\bar{X}(t)$ as $n \to \infty$. The next natural question to ask (considering our approach in the previous section) is what happens to the difference of $X_n(t) - \bar{X}(t)$, by scaling it by $\sqrt{n}$? Does it lead to a diffusion process? And can we use that to approximate $Var[X(t)]$? Those are precisely what are dealt with in the next section.

8.4.2 Diffusion Approximation

In this section, we consider a diffusion approximation for $X_n(t)$ described in the previous section in Equation 8.17. That would enable us to obtain an approximate distribution for $X_n(t)$ for any t (hence transient analysis) especially when n is large. We will subsequently, via an example, connect that to

$X(t)$ (defined in Equation 8.15) like we did in the previous section. Now, for the distribution of $X_n(t)$, we consider the "usual" diffusion scaling. Define the scaled process $\{\hat{X}_n(t), t \geq 0\}$ where $\hat{X}_n(t)$ is given by

$$\hat{X}_n(t) = \sqrt{n}(X_n(t) - \bar{X}(t)).$$

Besides the assumptions made in the previous section, we also require that F satisfies

$$\left| \frac{d}{dx} F(t,x) \right| \leq M,$$

for some finite M and $0 \leq t \leq T$. Kurtz [71] shows that under those conditions

$$\lim_{n \to \infty} \hat{X}_n(t) = \hat{X}(t)$$

where $\{\hat{X}(t), t \geq 0\}$ is a diffusion process. In addition, Kurtz [71] shows that $\hat{X}(t)$ is the solution to

$$\hat{X}(t) = \sum_{i=1}^{k} l_i \int_0^t \sqrt{f_i(s, \bar{X}(s))} dW_i(s) + \int_0^t F'(s, \bar{X}(s)) \hat{X}(s) ds, \qquad (8.23)$$

$W_i(\cdot)$'s are independent standard Brownian motions, and $F'(t,x) = dF(t,x)/dx$. It is crucial to note that this result requires that $F(t,x)$ is differentiable everywhere with respect to x but that is often not satisfied in many queueing models. There are few ways to get around this which is the key fine-tuning by Mandelbaum et al. [77] that we alluded to earlier. However, here we take the approach in Mandelbaum et al. [78] which states that as long as the deterministic process $\bar{X}(t)$ does not linger around the nondifferentiable point or points, Equation 8.23 would be good to use.

With that understanding we now describe the diffusion approximation for $X_n(t)$. For that we use the result in Ethier and Kurtz [33] which states that if $\hat{X}(0)$ is a constant or a Gaussian random variable, then $\{\hat{X}(t), t \geq 0\}$ is a Gaussian process. Since a Gaussian process is characterized by its mean and variance, we only truly require the mean and variance of $\hat{X}(t)$ which can be obtained from Equation 8.23. We will subsequently use that but for now note that we have a diffusion approximation. In particular for a large n,

$$X_n(t) \approx \bar{X}(t) + \frac{\hat{X}(t)}{\sqrt{n}}. \qquad (8.24)$$

Since $\hat{X}(t)$ is Gaussian when $\hat{X}(0)$ is a constant (which is a reasonable assumption), we have $X_n(t)$ to approximately be a Gaussian random variable when n is large. Thus we only require $E[X_n(t)]$ and $Var[X_n(t)]$ to characterize $X_n(t)$.

Assuming that $\bar{X}(0) = X_n(0) = nX(0)$ where $X(0)$ is a deterministic known constant quantity, we can see that $\hat{X}(0) = 0$. Further, by taking the expected value and variance of approximate Equation 8.24, we get

$$E[X_n(t)] \approx \bar{X}(t) + \frac{E[\hat{X}(t)]}{\sqrt{n}}, \tag{8.25}$$

$$Var[X_n(t)] \approx \frac{Var[\hat{X}(t)]}{n}. \tag{8.26}$$

However, we had shown earlier that $E[X_n(t)] = \bar{X}(t)$ as $n \to \infty$. Using that or by showing $E[\hat{X}(t)] = 0$ for all t since $\hat{X}(0) = 0$ using Equation 8.23, we can say that $E[X_n(t)] \approx \bar{X}(t)$. Now, for $Var[\hat{X}(t)]$ we use the result in Arnold [6] for linear stochastic differential equations. In particular by taking the derivative with respect to t of Equation 8.23 and using the result in Arnold [6], we get $Var[\hat{X}(t)]$ as the solution to the differential equation:

$$\frac{dVar[\hat{X}(t)]}{dt} = \sum_{i=1}^{k} f_i(t, \bar{X}(t)) + 2F'(t, \bar{X}(t))Var[\hat{X}(t)], \tag{8.27}$$

with initial condition $Var[\hat{X}(0)] = 0$. Once we solve for this ordinary differential equation, we can obtain $Var[\hat{X}(t)]$ which we can use in Equation 8.26 to get $Var[X_n(t)]$ and subsequently $Var[X(t)]$. We illustrate this by means of an example, next.

Problem 84

Consider the $M_t/M/s_t$ system that models an inbound call center described in Problem 83. This is a continuation of that problem and it is critical to go over that before proceeding ahead. Using the results of Problem 83, develop a diffusion model for that system. Then, obtain an approximation for $Var[X(t)]$ and compare against simulations by creating 100 replications and obtaining sample variances.

Solution

Several of the details are based on the solution to Problem 83. Recall $X(t)$, the number in the system at time t is described in Equation 8.21 as

$$X(t) = X(0) + Y_1\left(\int_0^t \lambda_u du\right) - Y_2\left(\int_0^t \mu \min\{s_u, X(u)\} du\right),$$

where $Y_1(\cdot)$ and $Y_2(\cdot)$ are the nonhomogeneous Poisson arrival and departure processes, respectively. For some large n and all $t \in [0, 8]$, let

$$a_t = \frac{\lambda_t}{n} \quad \text{and} \quad r_t = \frac{s_t}{n}.$$

Recall that $X_n(t)$ is

$$X_n(t) = \frac{nX_n(0) + Y_1\left(\int_0^t na_u du\right) - Y_2\left(\int_0^t \mu n \min\{r_u, X_n(u)\}du\right)}{n},$$

where $X_n(0) = X(0)/n$. In fact,

$$X(t) = nX_n(t)$$

for all t, hence

$$Var[X(t)] = n^2 Var[X_n(t)].$$

As $n \to \infty$, $X_n(t)$ converges to $\bar{X}(t)$ which is given by the solution to

$$\bar{X}(t) = \bar{X}(0) + \left(\int_0^t a_u du\right) - \left(\int_0^t \mu \min\{r_u, \bar{X}(u)\}du\right).$$

Likewise, as $n \to \infty$, $\hat{X}_n(t)$ defined as

$$\hat{X}_n(t) = \sqrt{n}(X_n(t) - \bar{X}(t))$$

converges to $\hat{X}(t)$ such that $\{\hat{X}(t), t \geq 0\}$ is a diffusion process. All we need is $Var[\hat{X}(t)]$. To obtain $Var[\hat{X}(t)]$, we rewrite Equation 8.27 by appropriately substituting for terms to obtain $Var[\hat{X}(t)]$ as the solution to

$$\frac{dVar[\hat{X}(t)]}{dt} = a_t + \mu \min\{r_t, \bar{X}(t)\} - 2I(r_t \geq \bar{X}(t))\mu Var[\hat{X}(t)],$$

where the indicator function $I(A)$ is one if A is true and zero if A is false. This ordinary differential equation can be solved by numerically integrating it to get $Var[\hat{X}(t)]$ for all t in $0 \leq t \leq 8$.

From Equation 8.26 we have $Var[X_n(t)] \approx Var[\hat{X}(t)]/n$ and from an earlier equation we know $Var[X(t)] = n^2 Var[X_n(t)]$. Hence we get the approximation

$$Var[X(t)] \approx nVar[\hat{X}(t)].$$

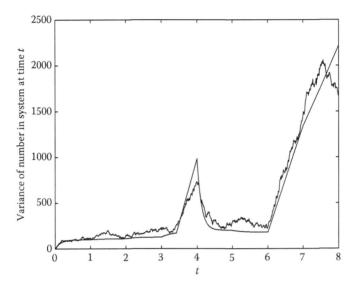

FIGURE 8.21
Variance of number in system using 100 replications of simulation (jagged line) vs. diffusion approximation (smooth line).

Using this diffusion scaling we plot an approximation for $Var[X(t)]$ in Figure 8.21 which is the smooth line. By performing 100 simulations $Var[X(t)]$ can be estimated for every t. The estimated $Var[X(t)]$ is the jagged line in Figure 8.21. The two graphs are reasonably close suggesting that the approximation is fairly reasonable. ∎

As evident from the example problem, fluid and diffusion approximations are useful for transient analysis of systems with high arrival rates and a large number of servers. Although we restricted ourselves to $M_t/M/s_t$ queues, they can seamlessly be extended to queues with balking and reneging, as briefly explained earlier. Further, Mandelbaum et al. [77] describe how to extend to a wider variety of situations such as Jackson networks with possibly state-dependent routing, queues with retrials, and priority queues, all of which include the possibility of abandonments. Some of these will be described in the exercises.

Reference Notes

In the last two decades, one of the most actively researched topics in the analysis-of-queues area is arguably the concept of fluid scaling. Fluid limits results are based on some phenomenally technical underpinnings that use stochastic process limits. This chapter does not do any justice in that regard and interested readers are encouraged to refer to Whitt [105]. Our objective of this chapter was to present some background material that would familiarize a reader to this approach to analyze queues using fluid models. The author is extremely thankful to several colleagues that posted handouts on the world wide web that were tremendously useful while preparing this material. In particular, Balaji Prabhakar's handouts on fluid models, Lyapunov functions, and Foster–Lyapunov criterion; Varun Gupta's gentle introduction to fluid and diffusion approximations; Gideon Weiss' treatment of stability of fluid networks; and John Hasenbein's collection of topics on fluid queues were all immensely useful in developing this manuscript.

As the title suggests, this chapter is divided into three key pieces: stability, fluid-diffusion approximations, and time-varying queues. Those topics have somewhat independently evolved in the literature and usually do not appear in the same chapter of any book. Thus it is worthwhile describing them individually from a reference notes standpoint. The topic of stability of multiclass queueing networks is a fascinating one, years ago most researchers assumed that the first-order traffic intensity conditions were sufficient for stability. However, in case of multiclass queueing networks with reentrant lines using deterministic routing and/or local priorities among classes, more conditions are necessary for stability due to the virtual station condition. This is articulated nicely in several books and monographs with numerous examples. In particular, this chapter benefited greatly from: Chen and Yao [19] with the clear exposition of fluid limits and all the excellent multiclass network examples; Dai [25] for describing the fluid networks and conditions for stability; Meyn [82] for the explanation of how to go about showing a fluid network is stable; and Bramson [13] for the technical details and plenty of examples.

Moving on to the next section, it was a somewhat familiar topic for this book, that is, fluid and diffusion approximations. In fact, we used those approximations in Chapter 4 without proof. However, only in Chapter 7 we showed by approximations based on reflected Brownian motion how one could get good approximations for general queues. However, those methods were similar to the traditional (Kobayashi [64]) diffusion approximation. On the other hand, this section provides a diffusion approximation by scaling, in fact first by performing a fluid scaling and subsequently a diffusion one. There are numerous books and monographs that go into great details regarding fluid and diffusion scaling. This chapter benefited greatly from Whitt [105] as well as Chen and Yao [19], especially in terms of constructing scaled processes. However, for an overview of the mathematical details and example of diffusion processes, Glynn [46] is an excellent resource. This chapter also benefited from several articles such as Halfin and Whitt [50] as well as Whitt [106].

The last section of this chapter was on uniform acceleration and strong approximations. This topic was the focus of the author's student Young

Myoung Ko's doctoral dissertation. Most of the materials in that section are from Young's thesis, in particular from preliminary results of his papers. The pioneering work on strong approximations was done by Kurtz [71]. The concept also appears in the book by Ethier and Kurtz [33]. Subsequently, Mandelbaum et al. [77] described some of the difficulties in using the strong approximation results available in the literature due to issues regarding differentiability while obtaining the diffusion limits. The limits, both fluid and diffusion, are based on the topic of uniform acceleration that can be found in Massey and Whitt [79]. Numerical studies that circumvent the differentiability requirement can be found in Mandelbaum et al. [78]. Young Myong Ko has found ways to significantly improve both fluid and diffusion approximations so that they are extremely accurate (article forthcoming).

Exercises

8.1 Consider an extension to the Rybko–Stolyar–Kumar–Seidman network in Figure 8.6 with a node C in between A and B. The first flow gets served in nodes A, C, and then B whereas the second flow has a reverse order. Priority is given to the second flow in node C, hence in terms of priority it is identical to that of A. Obtain the condition for stability and verify using simulations. This network is taken from Bramson [13] which has the figure and the stability condition.

8.2 Solve Problem 76 assuming that: (i) interarrival times and service times are deterministic constants; (ii) interarrival times are the same as in Problem 76 but service times are according to a gamma distribution with coefficient of variation 2.

8.3 Show that the virtual station condition described in Section 8.2.1 along with the necessary conditions are sufficient to ensure stability for the network in Figure 8.3. Follow a similar argument as the one for the Rybko–Stolyar–Kumar–Seidman network outlined in Section 8.2.3.

8.4 Let $Z_1, Z_2, Z_3, \ldots$, be a sequence of IID random variables with finite mean m and finite variance σ^2. Define S_n as the partial sum

$$S_n = Z_1 + Z_2 + \cdots + Z_n$$

for any $n \geq 1$. Next, consider the renewal process $\{N(t), t \geq 0\}$ where $N(t) = \max\{k \geq 0 \ : \ S_k \leq t\}$. Show that for a very large t, $N(t)$ is a normal random variable with mean t/m and variance $\sigma^2 t/m^3$.

8.5 Redraw sample paths of scaled process $\hat{X}_n(t)$ versus t (not $\hat{X}_n(t/n)$ versus t) for the same range of t in Figures 8.16 through 8.18.

8.6 Consider an $M/M/s$ queue with arrival rate $\lambda = 10$ per minute and number of servers $s = 5$. For $\mu = 2.5$, 2, and 1.8 (all per minute) plot $\hat{X}_n(t)$ versus t for $t \in [0,1]$ minutes using $n = 2, 20, 200$, and 2000, where $\hat{X}_n(t)$ is defined in Section 8.4.2 in terms of $X_n(t)$ and $\bar{X}(t)$. Use multiple sample paths to illustrate the diffusion approximation. Also use $X(0) = 0$.

8.7 Consider a finite population queueing system with $s = 50$ servers each capable of serving at rate $\mu = 5$ customers per hour. Assume service times are exponentially distributed. Also upon service completion each customer spends an exponential time with mean $1\,h$ before returning to the queueing system. Assume that there are 400 customers in total but at time 0, the queue is empty. Obtain an approximation for the mean and variance of the number of customers in the system during the first hour. Perform 100 simulations and evaluate the approximations.

8.8 Consider the following extension to Problem 83. In addition to all the details in the problem description, say that customers renege from the queue (that is, abandon before service starts) after an $\exp(\beta)$ if their service does not begin. Use $1/\beta = 5\,min$. Use fluid and diffusion approximations and numerically obtain $E[X(t)]$ and $Var[X(t)]$. Compare the results by performing 100 simulations.

8.9 Solve the previous problem under the following additional condition (note that reneging still occurs): some customers access the call center from their web browser using Internet telephony. These customers arrive according to a homogeneous Poisson process at rate $\alpha = 10$ per hour. Only these customers also have real time access to their position in the wait line (e.g., position-1 implies next in line for service). However, because of that if there are i total customers waiting for service to begin, then with probability $(0.9)^i$ an arriving customer joins the system (otherwise the customer would balk).

8.10 Consider an $M_t/M/s_t/s_t$ queue, that is, there is no waiting room. If an arriving customer finds all servers busy, then the customer retries after $\exp(\theta)$ time. At this time we say that the customer is in an *orbit*. Assume that λ_t alternates between 100 and 120 each hour for a four hour period. Also $\mu = 1$ and s_t during the four hour-long slots are: 90, 125, 125, and 150, respectively. Compute approximately the mean and variance of the number of customers in the queue as well as in the orbit. Care is to be taken to derive expressions when $X(t)$ is a 2D vector.

9

Stochastic Fluid-Flow Queues: Characteristics and Exact Analysis

In the previous chapter, we saw deterministic fluid queues where the flow rates were mostly constant and toward the end we saw a case where the rates varied deterministically over time. In this and the next chapter, we focus on stochastic fluid queues where flow rates are piecewise constant and vary stochastically over time. We consider only the flow rates from a countable set. On a completely different note, in some sense the diffusion limits we saw in previous chapters can be thought of as a case of flow rates from an uncountable set that are continuously varying (as opposed to being piecewise constant). Thus from a big-picture standpoint, metaphorically the models in this chapter fall somewhere between the deterministically time-varying fluid queues and diffusion queues. However, here we will not be presenting any formal scaling of any discrete queueing system to result in these fluid queues. We focus purely on performance analysis of these queues to obtain workload distributions.

For the performance analysis we start by describing a queueing system where the entities are fluids. For example, a sink in a kitchen or bathroom can be used to explain the nuances. Say there is a fictitious tap or faucet that has a countable number of settings (as opposed to a continuous set which is usually found in practice). At each discrete *setting*, water flows into the sink at a particular rate. Typically the sojourn time in each setting is random and the setting changes stochastically over time. This results in a piecewise constant flow rate that changes randomly over time. The sink itself is the queue or buffer that holds fluid (in this case water), which flows into it. The drain is analogous to a server that empties the fluid off the sink (however, unlike a real bathtub or sink, here we assume the drainage rates are not affected by the weight of the fluid). For our performance analysis, we assume that we know the stochastic process that governs the input to the sink as well as the drainage. Using that our aim is to obtain the probability distribution of the amount of fluid in the sink.

Naturally, these models can be used in hydrology such as analyzing dams, reservoirs, and water bodies, as well as in process industries such as chemicals and petrochemicals. However, a majority of the results presented here have been motivated by applications in computer, communication, and information systems. Interestingly these are truly discrete systems, but there are so many discrete entities that flow in an extremely small amount of time that it is conducive to model them as fluids. In some sense what we are really doing is not analyzing the system at the granularity of a discrete entity (usually packet or cell) but at a higher granularity where we are concerned when the flow rate changes. This also makes it rather convenient since at the packet level one sees long-range dependence (Leland et al. [74]) in interarrival times as packets belonging to the same file, for example, are going to arrive back to back. With that motivation we first describe some introductory remarks next including details of various applications, followed by some performance analysis.

9.1 Introduction

The objective of this section is to provide some introductory remarks regarding stochastic fluid-flow queues. We begin by contrasting fluid-flow queues against discrete queues to understand their fundamental differences as well as underlying assumptions. Once we put things in perspective, we describe some more applications and elaborate on others described previously. Then we go over some preliminary material such as inputs that go into the performance analysis as well as a description of the condition for stability. We conclude the section with a characterization of the stochastic flow-rate processes that govern the flow of fluids.

9.1.1 Discrete versus Fluid Queues

Here we seek to put things in perspective by comparing and contrasting fluid-flow queues against queues with discrete entities. It may be worthwhile revisiting this aspect after becoming more comfortable with fluid queues such as at the end of this chapter. However, in some sense the two are hard to compare because they are typically considered at different levels of abstraction, use different aspects of variability/randomness, and the state information used in modeling are significantly different. Thus while making the comparison we will consider a specific system with the disclaimer that fluid queues have a much wider set of applications as we will see in the next section.

Consider a web server that responds to requests for files. Typically the files are broken down into small packets and sent from the server to a gateway which transmits the packets through the Internet to the users making the requests. Say we are interested in modeling the queueing system at the gateway. The gateway processes information at a rate of c kb/s and sends it into the Internet. We define the *workload* at the gateway as the amount of information (in kb) to be processed and forwarded. Our earlier definition of workload was in units of time which can easily be done by dividing the current definition of workload by c.

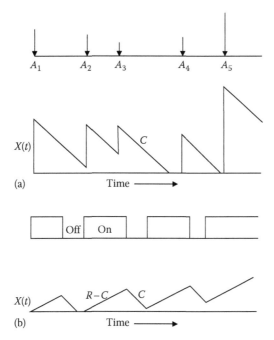

FIGURE 9.1
Comparing workloads in (a) discrete and (b) fluid queues.

The key difference between discrete queues and fluid queues is that in the discrete case the entire file arrives instantaneously, thus creating a jump or discontinuity in the workload process. Whereas in the fluid case, the file arrives gradually over time at a certain flow rate. Therefore, the workload process would not have jumps in fluid queues, that is, they are continuous. For both the discrete and the continuous cases, let $X(t)$ be the amount of fluid (i.e., workload) in the buffer at time t. Figure 9.1 illustrates this difference between the $X(t)$ process versus t in the discrete and fluid queues. In particular it is crucial to point out that although in practice the entire file does not actually arrive instantaneously but in the discrete model, Figure 9.1a, we essentially assume the arrival time of the file $(A_1, A_2, \ldots)$ as when the entire file completely arrives at the gateway.

In many systems, this is a necessity as the entire entity is necessary for processing to begin. Notice that with every arrival, the workload jumps up by an amount equal to the file size. The workload is depleted at rate c. We have seen such a workload process in the discrete queues, the only differences here are that (i) the notation for $X(t)$ is not what we used for the discrete case; and (ii) the workload depletion rate is c and not 1 like is usually done. However, what enables fluid model analysis is the fact that the gateway does not have to wait for the entire file to arrive. As soon as the first packet of the file arrives, it sends it off without waiting for the whole file to arrive. Alternatively one could think of the discrete queue as a "bulk" arrival of a batch of packets whereas in the fluid queue this batch arrives slowly over time. Since the batch arrives back to back, modeling at the "packet" level is tricky. In a similar fashion, although not represented in the workload process, we consider a discrete entity's departure from the system when all of its service is completed which is not the case for fluids. In fact thus the concept of sojourn times at the granularity of a whole file is not so easy in fluid queues.

Now we describe the workload process for the fluid queue. In particular, refer to Figure 9.1b. We assume that information flows into the system as an on–off fluid. We will explain on–off fluid subsequently, however, for the purposes of this discussion it would suffice to think of an "on" time as when there is one or more files back to back that gets processed by the server at rate c. When there is information, it flows in at rate R. However, when there is no information flow, we call that period as "off." From the figure, notice that the workload gradually increases at rate $R - c$ when the source is on. It is because fluid enters at rate R and is removed at rate c, resulting in an effective growth rate of $R - c$. Also, when the fluid entry is off, the workload reduces at rate c (provided there is workload to be processed, otherwise it would be zero). Notice that in Figure 9.1 there is no relationship between the discrete case's arrival times and file sizes against the fluid case's on and off times.

In summary, fluid-flow models are applicable in systems where the workload arrives gradually over time (as opposed to instantaneously) and we are interested in aggregate performance such as workload distribution. In most cases the entities are themselves either fluids or can be approximated as fluids. Under these situations the analysis based on fluid models would be extremely accurate and in comparison the discrete models would be way off. In particular, when the discrete entities arrive in a back-to-back fashion, it is extremely conducive to model them as fluids as opposed to discrete point masses with constant interarrival times. We will next see some applications where fluid models would be appropriate for their analysis.

9.1.2 Applications

Here we present some scenarios where stochastic fluid-flow models can be used for performance analysis. The idea is to give a flavor for the kind of systems that can be modeled using fluid queues. We begin by presenting some examples from computer and communication systems where information flow is modeled as fluids. Then we present an example from supply chain, followed by one in hydrology and finally a transportation setting. As described earlier, most of the fluid model results have been motivated by applications in computer-communication networks. Thus it is worthwhile to describe a few examples at different granularities of space, size, and time. However, what is common in all the cases is that entities arrives in a bursty fashion, that is, things happen in bursts at different rates as opposed to one by one in a fairly uniform fashion.

For example, consider a CPU on a computer that processes tasks from a software agent that is part of a multi-agent system. An agent is a complex software that can autonomously send, receive, and process information on behalf of a user. The agent sends tasks to its CPUs to aid in decision-making. In Aggarwal et al. [3] we consider software agents that are hosted on computers that perform military logistic operations. We show (see the first figure in Aggarwal et al. [3]) that the tasks are indeed generated in bursts by the agents and submitted to the CPU for processing. The burtsy nature is due to the fact that the agent receives a trigger externally and submits a set of jobs to the CPU to determine its course of action. The CPU does not wait for all the tasks of the agent to arrive to begin processing but processes them as they arrive. It is important to notice that these tasks are fairly atomic in nature that can be processed independently rather quickly with roughly similar processing times. This makes it ideal to analyze as fluid queues.

Fluid models have been successfully used in modeling packet-level traffic in both wired (also called wire-line) and wireless networks. Irrespective of whether it is end systems such as computers and servers, or inside the network such as routers, switches, and relays, or both such as multi-hop wireless nodes and sensors, information flow in the form of packets can be nicely modeled as fluids. In all these systems information stochastically flows into buffers in the form of tiny packets. These packets are processed and the stored packets are forwarded to a downstream node in the network. This process called store-and-forward of packets results in an extremely efficient network. Contrast this to the case where entire files are transferred hop by hop as a whole (as opposed to packetizing them); a significant amount of time would be wasted just waiting for entire files to arrive. In the store-and-forward case, some packets of a file would already be at the destination while other packets still at the origin, even if the origin and destination are in two extremes of the world.

At a much coarser granularity, consider users that access a server farm for web or other application processing. The users enter a session and within a session they send requests (usually through browsers for web applications) and receive responses. The users alternate between periods of activity and quiet times during a session. Also the servers can process requests independent of other requests. Thus within a session requests arrive in a bursty fashion much like a bunch of packets that are part of a file in the previous example. These requests are stored in a buffer and processed one by one by the server. One can model each user as a source that toggles between bursting requests and idling in a stochastic fashion. That can nicely be analyzed using fluid queues. In summary, there are several computer and communication systems that can be modeled using fluid queues. They key elements are: bursty traffic, the ability to process smaller elements of the traffic, and finally (although not emphasized earlier) the smaller elements must have similar processing requirements. Those would result in fluid queues being appropriate tools for analysis. Such features are prevalent in other systems as well, some of which we present in the following.

Consider a single bay of a cross-dock where trucks arrive and the contents (in boxes or palettes) are removed from the truck. After that the truck leaves and another truck pulls into the bay and the process continues. We focus on the flow of the contents (boxes or palettes) from the truck at the bay into the warehouse. Whenever there is a truck at the bay, contents flow out of the truck at a certain rate (equal to how fast the truck can be emptied). However when there is no truck, there is no flow of contents. Thus the input can be modeled as an *on–off source* (we will see that subsequently). The crucial point is that one does not wait until all the contents are emptied to start processing them. Instead as and when contents are removed from the truck, they are placed in a buffer and processed one by one. Since the processing starts well before all the contents are emptied from the truck, it would be better to use a fluid model as opposed to a discrete batch arrival $G^{[X]}/G/1$ queue to model the system.

Next we present two applications from civil engineering. The first is from hydrology. Fluid models are a natural fit to model water flow in dams and reservoirs. The reservoir is the buffer and water is drawn from it on a daily basis in a controlled fashion. The reservoir receives water input from various sources. The input is stochastic and at any time the input rate can be approximated as a discrete quantity. This lends itself nicely to model using fluid queues. Another civil engineering example is in transportation, in particular roadways. Although in these systems the queues or buffers are rather abstract, one routinely measures flow rates and capacities of roadways. Cars and other vehicles are typically modeled as fluids since they move in a back-to-back fashion during congestion with periods of large gaps due to signal effects. These are usually used in ramp metering strategies in highways, signal light controls, as well as designing capacities and tolls.

In summary, there are several applications of fluid queues, some of which have been presented in this section. In fact there are applications that are not queues but these analysis methods can be effectively used. For example, in reliability the fluid input to the system can be thought of as continuous stressors and the output capacity is related to continuous maintenance. Thus the fluid level in the buffer can be thought of as the condition of the system that goes up and down due to maintenance and stressors. Similar to reliability of physical systems these models are also extremely useful in insurance risk analysis as well as warranty reserves. Another set of applications that comes to mind is to model processing times, travel times, project completion times, etc. Akin to modeling lifetimes in reliability, these times can also be modeled using results from fluid queues. In particular, some of the transient analysis and first passage time analysis are useful in these applications. With that motivation, in the next section we formally define a fluid queue that we will subsequently analyze in the following sections.

9.1.3 Preliminaries for Performance Analysis

Here we present a preliminary setup for performance analysis of a single fluid queue. Consider a buffer that is capable of holding B amount of fluid workload. We call such a buffer as one of size B if B is finite, otherwise we call it an infinite-sized buffer. Fluid flows into the buffer for a random time at a constant rate sampled from a countable set. For example, fluid flows at rate $r(1)$ units per second (such as kbps) for a random amount of time t_1, then flows at rate $r(2)$ units per second for a random amount of time t_2, and so on. This behavior can be captured as a discrete-state stochastic process that jumps from one state to another whenever the traffic flow rate changes. This can be formalized as a stochastic process $\{Z(t), t \geq 0\}$ that is in state $Z(t)$ at time t. Fluid flows into the buffer at rate $r(Z(t))$ at time t.

We typically call the stochastic process $\{Z(t), t \geq 0\}$ that drives the traffic generation as an environment process and the origin of fluid we call a source. Fluid is removed from the buffer by a channel that has a fixed capacity c (Figure 9.2). Let $X(t)$ be the amount of fluid in the buffer (i.e., buffer content) at time t. The dynamics of the buffer content process $\{X(t), t \geq 0\}$ if $B = \infty$, is described by

$$\frac{dX(t)}{dt} = \begin{cases} r(Z(t)) - c & \text{if } X(t) > 0 \\ \{r(Z(t)) - c\}^+ & \text{if } X(t) = 0. \end{cases}$$

where $\{x\}^+ = \max(x, 0)$. The dynamics when $B < \infty$ is

$$\frac{dX(t)}{dt} = \begin{cases} \{r(Z(t)) - c\}^+ & \text{if } X(t) = 0 \\ r(Z(t)) - c & \text{if } X(t) > 0 \\ \{r(Z(t)) - c\}^- & \text{if } X(t) = B \end{cases}$$

where $\{x\}^- = \min(x, 0)$.

Next we describe the stability condition when $B = \infty$. We assume that $Z(\infty)$ is the state of the environment in steady state. If the expected arrival rate in steady state is lesser than the drainage capacity c, then we expect the fluid queue to be stable. This intuitive result has been shown in Kulkarni

FIGURE 9.2
Buffer with environment process $Z(t)$ and output capacity c. (From Gautam, N., Quality of service metrics, in *Frontiers in Distributed Sensor Networks*, S.S. Iyengar and R.R. Brooks, Eds., Chapman & Hall/CRC Press, Boca Raton, FL, 2004, pp. 613–628. With permission.)

and Rolski [66]. In other words, the buffer content process $\{X(t), t \geq 0\}$ (when $B = \infty$) is stable if the mean traffic arrival rate in steady state is less than c, that is,

$$E\{r(Z(\infty))\} < c. \tag{9.1}$$

If this stability condition is satisfied, then $X(t)$ has a limiting distribution.

For both $B = \infty$ and $B < \infty$, we are interested in the limiting distribution of $X(t)$, that is,

$$\lim_{t \to \infty} P\{X(t) \leq x\} = P\{X \leq x\}$$

such that $X(t) \to X$ as $t \to \infty$. This is the main performance metric that we will consider. Recall that $X(t)$ is the workload in the buffer (with appropriate units such as kb) at time t. To obtain $P\{X \leq x\}$ we require the characteristics of the stochastic process $\{Z(t), t \geq 0\}$ as well as the piecewise-constant arrival rates in each state $r(Z(t))$, the buffer size B and channel capacity c. In the next section we present some of the environment processes that can be effectively characterized and analyzed.

9.1.4 Environment Process Characterization

Almost any discrete-valued stochastic process $\{Z(t), t \geq 0\}$ can be used for fluid queues. However, the following is a collection of processes that are conducive for analysis. The first is a continuous-time Markov chain (CTMC), followed by an alternating renewal process, and finally a semi-Markov process (SMP).

9.1.4.1 CTMC Environmental Processes

The environmental process that is conducive for exact analysis to obtain $P(X \leq x)$ is the CTMC. In the next section we describe the CTMC environment process in great detail. However, for sake of completeness and to contrast with other processes, we briefly describe it here. Let $\{Z(t), t \geq 0\}$ be an irreducible, finite state CTMC with ℓ states and generator matrix Q. When the CTMC is in state i, traffic flows at rate $r(i)$. Let p_i be the stationary distribution that the CTMC is in state i such that $[p_1 \; p_2 \; \ldots \; p_\ell]Q = [0 \; 0 \ldots \; 0]$ and $\sum_{i=1}^{\ell} p_i = 1$. If $B = \infty$, the stability condition is $\sum_{i=1}^{\ell} p_i r(i) < c$.

9.1.4.2 Alternating Renewal Environmental Processes

The environment process for several of the applications we described earlier can be modeled as an alternating renewal process, in particular an on–off process. The environmental process $\{Z(t), t \geq 0\}$ alternates between on and off states. The on times form a renewal process and so do the off times. However, the on and off times in a cycle do not have to be independent. A source that generates such an alternating renewal environment process is also known as an on–off source. In this and the next chapter we will encounter on–off sources quite frequently. When the source is on, fluid flows into the buffer at rate r and no fluid flows when the source is off.

The on times (or also called "up" times) are distributed according to a general CDF $U(\cdot)$. The mean on time τ_U can be calculated as

$$\tau_U = \int_0^\infty t \, dU(t).$$

Likewise, the off times (or "down" times) are according to a general distribution with CDF $D(\cdot)$. The mean off time τ_D can be calculated in a similar manner as

$$\tau_D = \int_0^\infty t \, dD(t).$$

For the rest of this book we assume that the CDFs $U(\cdot)$ and $D(\cdot)$ are such that we can either compute their LSTs directly or they can be suitably approximated as phase-type distributions whose LSTs can be computed.

When the buffer size $B = \infty$, the system would be stable if

$$\frac{r\tau_U}{\tau_U + \tau_D} < c.$$

9.1.4.3 SMP Environmental Processes

The most "general" type of discrete-state environment process we will consider is the SMP. Notice that the CTMC, alternating renewal process, and even the DTMC are special cases of the SMP. Thus by developing methods that can be used to derive performance metrics when the environment is an SMP, we can cover the other stochastic processes as well. For that reason we next describe the SMP environment process in some detail. Consider an SMP $\{Z(t), t \geq 0\}$ on state space $\{1, 2, ..., \ell\}$. Fluid is generated at rate $r(i)$ at time t when the SMP is in state $Z(t) = i$. Let S_n denote the time of the nth jump epoch in the SMP with $S_0 = 0$. Define Z_n as the state of the SMP immediately after the nth jump, that is,

$$Z_n = Z(S_n+).$$

Let

$$G_{ij}(x) = P\{S_1 \le x; Z_1 = j | Z_0 = i\}.$$

The kernel of the SMP is

$$G(x) = [G_{ij}(x)]_{i,j=1,\dots,\ell}.$$

Note that $\{Z_n, n \ge 0\}$ is a DTMC, which is embedded in the SMP. Assume that this DTMC is irreducible and recurrent with transition probability matrix

$$P = G(\infty).$$

Let

$$G_i(x) = P\{S_1 \le x | Z_0 = i\} = \sum_{j=1}^{\ell} G_{ij}(x)$$

and the expected time the SMP spends in state i be

$$\tau_i = E(S_1 | Z_0 = i).$$

Let

$$\pi_i = \lim_{n \to \infty} P\{Z_n = i\}$$

be the stationary distribution of the DTMC $\{Z_n, n \ge 0\}$. It is given by the unique nonnegative solution to

$$[\pi_1\ \pi_2\ \dots\ \pi_\ell] = [\pi_1\ \pi_2\ \dots\ \pi_\ell]P \quad \text{and} \quad \sum_{i=1}^{\ell} \pi_i = 1.$$

The stationary distribution of the SMP is thus given by

$$p_i = \lim_{t \to \infty} P\{Z(t) = i\} = \frac{\pi_i \tau_i}{\sum_{m=1}^{\ell} \pi_m \tau_m}.$$

If $B = \infty$, the stability condition is

$$\sum_{i=1}^{\ell} p_i r(i) < c.$$

With that description we are now ready for performance analysis of fluid queues.

9.2 Single Buffer with Markov Modulated Fluid Source

Our objective in this section is to derive performance measures for the single buffer stochastic fluid-flow system as depicted in Figure 9.2. We assume that the input to the buffer is governed by a CTMC (this is also referred to in the literature as a Markov modulated fluid source or MMFS). This is the only case for which we obtain algebraic expressions for the steady-state buffer content distribution. We rely on bounds and approximations for other input (i.e., environment) processes which would be the focus of the next chapter. But, if we can model other environment processes using appropriate phase-type distributions, then (as we will see at the end of this section) we can use the CTMC models described here for analyzing those queues. Thus in some sense the CTMC case is general enough. With that motivation we first describe the set of notation to be used. Then we show Kolmogorov differential equations for our measures of interest. Thereafter we will illustrate their solutions and describe some examples at the end of the section.

9.2.1 Terminology and Notation

Here we describe the notation and terminology for the single buffer fluid queue with MMFS. Although some of these have been defined earlier and some will be defined again as appropriate, the main goal is to have a collection of notation as well as terminology in one location for ease of future reference. We begin with the notion of a *buffer* which is essentially a waiting area or queue. In this and the next chapter we assume that the entities flowing in and out of the buffer are fluids. Let B be the size of the buffer, that is, the maximum amount of fluid that can be held in the buffer. Depending on the application, the units for B would be appropriately defined, such as liters and bytes. In most of our analysis we will consider only infinite-size buffers, that is, $B = \infty$, however, whenever B is finite, it would be specifically mentioned.

Fluid enters this buffer stochastically from a "source" which we say is *modulated* or *governed* or *driven* by a process called the environment process which is essentially a stochastic process. Here we denote the input to the buffer of size B as driven by a random discrete-state stochastic process $\{Z(t), t \geq 0\}$. When the environment is in state $Z(t)$, fluid arrives to the buffer at rate $r(Z(t))$. The units of $r(Z(t))$ would depend on the applications and would take the form liters per second, bytes per second, etc. We assume that the environment process $\{Z(t), t \geq 0\}$ is an ergodic CTMC with state space

$S = \{1, 2, \ldots, \ell\}$. The number of states ℓ is finite, that is, $\ell < \infty$. The infinitesimal generator matrix for the CTMC $\{Z(t), t \geq 0\}$ is $Q = [q_{ij}]$, which is an $\ell \times \ell$ matrix. Let p_i be the steady-state probability that the environment is in state i, that is,

$$p_i = \lim_{t \to 0} P\{Z(t) = i\}$$

for all $i \in S$. Since the CTMC is ergodic, we can compute the steady-state probability row vector $\bar{p} = [p_1 \; p_2 \; \ldots \; p_\ell]$ by solving for

$$\bar{p}Q = [0 \; 0 \; \ldots \; 0] \quad \text{and} \quad \sum_{i=1}^{\ell} p_i = 1.$$

Having described the buffer and its input, next we consider the output. The output capacity of the buffer is c. This means that whenever there is fluid in the buffer it gets removed at rate c. However, if the buffer is empty and the input rate is smaller than c, then the output rate would be same as the input rate. For that reason it is called output capacity as the actual output rate could be smaller than c. The units of c are the same as that of $r(Z(t))$. Thus both would be in terms of liters per second or bytes per second, etc. The term *output capacity* is also sometimes referred to as channel capacity or processor capacity. Unless stated otherwise, we assume that c remains a constant. Before proceeding ahead it maybe worthwhile familiarizing with the input, buffer, and output using Figure 9.2.

Next we describe the buffer contents and its dynamics. Let $X(t)$ be the amount of fluid in the buffer at time t. We first assume that $B = \infty$ (and later relax that assumption). Whenever $X(t) > 0$ and $Z(t) = i$, $X(t)$ either increases at rate $r(i) - c$ or decreases at rate $c - r(i)$ depending on whether $r(i)$ is greater or lesser than c, respectively. To capture that we define the drift $d(i)$ when the CTMC $\{Z(t), t \geq 0\}$ is in state i (i.e., $Z(t) = i$) as

$$d(i) = r(i) - c$$

for all $i \in S$. Therefore, we have

$$\frac{dX(t)}{dt} = d(i)$$

if $X(t) > 0$ and $Z(t) = i$ for any $t \geq 0$. Next, when $X(t) = 0$ it stays at 0 as long as the drift is non-positive, that is,

$$\frac{dX(t)}{dt} = 0$$

if $X(t) = 0$, $Z(t) = i$, and $r(i) \leq c$ for any $t \geq 0$. However, as soon as the drift becomes positive, the buffer contents would start increasing from 0.

Given the dynamics of $X(t)$, a natural question to ask is if $X(t)$ would drift off to infinity (considering $B = \infty$). As it turns out, based on Equation 9.1, we can write down the stability condition as

$$\sum_{i=1}^{\ell} r(i)p_i < c. \tag{9.2}$$

In other words, the LHS of this expression is the steady-state average input rate by conditioning on the state of the environment and unconditioning. We need the average input rate to be smaller than the output capacity. Another way of stating the stability condition is

$$\sum_{i=1}^{\ell} d(i)p_i < 0$$

since the average drift must be negative.

Next, we present an example to illustrate various notations described previously (they are also summarized in Table 9.1). For the examples as well as the analysis to follow, we describe some matrices. We define the drift matrix D as

$$D = diag[d(i)]$$

TABLE 9.1

List of Notations

B	Buffer size (default is $B = \infty$)
$Z(t)$	State of environment (CTMC) modulating buffer input at time t
S	State space of CTMC $\{Z(t), t \geq 0\}$
ℓ	Number of states in S, i.e., $\ell = |S|$ and is finite
q_{ij}	Transition rate from state i to j in CTMC $\{Z(t), t \geq 0\}$
Q	Infinitesimal generator matrix, i.e., $Q = [q_{ij}]$
p_i	Stationary probability for the ergodic CTMC $\{Z(t), t \geq 0\}$
$r(i)$	Fluid arrival rate when $Z(t) = i$
R	Rate matrix, i.e., $R = diag[r(i)]$
c	Output capacity of the buffer
$d(i)$	Drift in state $Z(t) = i$, i.e., $d(i) = r(i) - c$
D	Drift matrix, i.e., $D = diag[d(i)] = R - cI$
$X(t)$	Amount of fluid in the buffer at time t

implying that it is a diagonal matrix with diagonal entries corresponding to the drift. We can also write down $D = R - cI$ where R is the rate matrix given by

$$R = diag[r(i)]$$

and I is the $\ell \times \ell$ identity matrix. As an example, consider a buffer with $B = \infty$, $c = 12$ kbps, and environment CTMC $\{Z(t), t \geq 0\}$ with $\ell = 4$ states, $S = \{1, 2, 3, 4\}$, and

$$Q = \begin{bmatrix} -10 & 2 & 3 & 5 \\ 0 & -4 & 1 & 3 \\ 1 & 1 & -3 & 1 \\ 1 & 2 & 3 & -6 \end{bmatrix}.$$

It is not hard to see that the CTMC is ergodic with $\bar{p} = [0.0668 \ 0.2647 \ 0.4118 \ 0.2567]$. The fluid arrival rates in states 1, 2, 3, and 4 are 20, 15, 10, and 5 kbps, respectively. In other words, $r(1) = 20$, $r(2) = 15$, $r(3) = 10$, and $r(4) = 5$ with

$$R = \begin{bmatrix} 20 & 0 & 0 & 0 \\ 0 & 15 & 0 & 0 \\ 0 & 0 & 10 & 0 \\ 0 & 0 & 0 & 5 \end{bmatrix}.$$

Thus we have the drift matrix as

$$D = \begin{bmatrix} 8 & 0 & 0 & 0 \\ 0 & 3 & 0 & 0 \\ 0 & 0 & -2 & 0 \\ 0 & 0 & 0 & -7 \end{bmatrix}.$$

Notice that the system is stable since $\sum_{i=1}^{4} r(i)p_i = 10.7086$, which is less than $c = 12$. For this numerical example, a sample path of $Z(t)$ and $X(t)$ is depicted in Figure 9.3(a) with $X(0) = 0$ and $Z(0) = 1$. Notice that when the drift is positive, fluid increases in the buffer and when the drift is negative, fluid is nonincreasing. It is also important to pay attention to the slope (although not drawn to scale) as they correspond to the drift.

Now we consider the case when the buffer size B is finite, that is, $B < \infty$. Most of what we have described thus far holds. We just state the differences here. In particular, when the buffer is full, that is, $X(t) = B$, if the drift is positive, then fluid enters the buffer at rate c and a fraction of fluid is dropped at rate $r(Z(t)) - c$. This would result in the $X(t)$ process staying at B until the drift becomes negative. However, when $X(t) < B$, the dynamics are identical

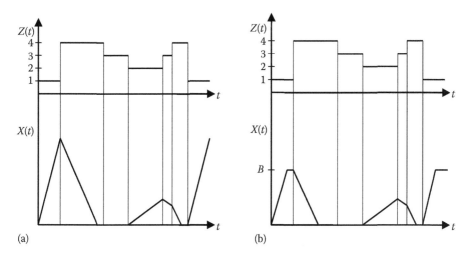

FIGURE 9.3
(a) Sample path of environment process and buffer contents when $B = \infty$ and (b) Sample paths when $B < \infty$.

to the infinite size case. The only other thing is that the system is always stable when $B < \infty$. Hence the stability condition described earlier is irrelevant. To illustrate the $B < \infty$ case, we draw a sample path of $Z(t)$ and $X(t)$ for the same example considered earlier, that is, $c = 12$ kbps and $\{Z(t), t \geq 0\}$ has $\ell = 4$ states, $S = \{1, 2, 3, 4\}$, and

$$
Q = \begin{bmatrix}
-10 & 2 & 3 & 5 \\
0 & -4 & 1 & 3 \\
1 & 1 & -3 & 1 \\
1 & 2 & 3 & -6
\end{bmatrix}
$$

with $r(1) = 20$, $r(2) = 15$, $r(3) = 10$, and $r(4) = 5$. The sample path is illustrated in Figure 9.3(b). Notice that for the sake of comparison, the $Z(t)$ sample paths are identical in Figures 9.3(a) and (b). However, when the $X(t)$ process reaches B in Figure 9.3(b), it stays flat till the system switches to a negative drift state.

The next step is to analyze the process $\{X(t), t \geq 0\}$. In other words, we would like to capture the dynamics of $X(t)$ and characterize it by deriving a probability distribution for $X(t)$ at least as $t \to \infty$. For that it is important to notice that unlike most of the random variables we have seen thus far, $X(t)$ is a mixture of discrete and continuous parts. Notice from Figures 9.3(a) and (b) that $X(t)$ has a mass (i.e., takes on a discrete value) at 0. Also if $B < \infty$, then $X(t)$ has a mass at B as well. Everywhere else $X(t)$ takes on continuous values. This we will see that $X(t)$ will have a mixture of discrete and continuous parts with a mass at 0 and possibly at B (if $B < \infty$). With that in mind we proceed with analyzing the dynamics of $X(t)$ next.

9.2.2 Buffer Content Analysis

The main objective here is to obtain a distribution for $X(t)$ and since $X(t)$ describes the amount of fluid in the buffer at time t, this is called *buffer content analysis*. All the notation and terminology used here are described in Section 9.2.1. Unless specified explicitly, for most of the analysis, B could either be finite or be infinite. For the buffer content analysis, we define for all $j \in S$ the joint distribution

$$F_j(t, x) = P\{X(t) \leq x, Z(t) = j\}. \tag{9.3}$$

We seek to obtain an analytical expression for $F_j(t, x)$ for all $j \in S$. For that we consider some $j \in S$ and derive the following expressions:

$$F_j(t + h, x) = P\{X(t + h) \leq x, Z(t + h) = j\} \tag{9.4}$$

$$= \sum_{i \in S} P\{Z(t + h) = j | X(t + h) \leq x, Z(t) = i\} P\{X(t + h) \leq x, Z(t) = i\} \tag{9.5}$$

$$= \sum_{i \in S} P\{Z(t + h) = j | Z(t) = i\} P\{X(t + h) \leq x | Z(t) = i\} P\{Z(t) = i\} \tag{9.6}$$

$$= (1 + q_{jj}h) P\{X(t + h) \leq x | Z(t) = j\} P\{Z(t) = j\}$$

$$+ \sum_{i \in S, i \neq j} q_{ij} h P\{X(t + h) \leq x | Z(t) = i\} P\{Z(t) = i\} + o(h). \tag{9.7}$$

Before proceeding ahead, it is worthwhile explaining the steps in the preceding derivation. Equation 9.4 is by directly replacing t by $t+h$ in the definition of $F_j(t, x)$ in Equation 9.3. By conditioning on $Z(t)$ as well as bringing $X(t+h)$ in the conditional part we get Equation 9.5. The step could have been broken down into two steps to arrive at the same result. It may be fruitful to verify that the conditional probabilities lead to the LHS. Next, since $Z(t)$ process is a CTMC, we can remove $X(t + h)$ in the conditional argument and also write the second term in product form to get Equation 9.6. Then, Equation 9.7 is obtained by writing down the CTMC transition probability from state i to j in time h as $q_{ij}h + o(h)$ if $i \neq j$ and $1 + q_{ij}h + o(h)$ if $i=j$, where $o(h)$ is a collection of terms of higher order than h such that $o(h)/h \to 0$ as $h \to 0$.

However, before taking the limit, we first consider an infinitesimally small $h > 0$ (although this is not rigorous, it makes explanation much easier). The amount of fluid that would have arrived into the buffer in time h when the source is in state i is $r(i)h$. Likewise the amount of fluid that would have departed the buffer in time h is ch. Then we have $P\{X(t+h) \leq x | Z(t) = i\} = P\{X(t) \leq x - (r(i) - c)h | Z(t) = i\} + \hat{o}(h)/h$ where $\hat{o}(h)$ is another collection of terms of higher order than h. However, for the sake of not being too messy, we will drop the $\hat{o}(h)$ term as it would vanish anyway when we take the limit $h \to 0$. Then we can write down Equation 9.7 as

$$F_j(t+h,x) = (1 + q_{jj}h)P\{X(t) \leq x - (r(j) - c)h | Z(t) = j\}P\{Z(t) = j\}$$

$$+ \sum_{i \in S, i \neq j} q_{ij}hP\{X(t) \leq x - (r(i) - c)h | Z(t) = i\}P\{Z(t) = i\} + o(h).$$

We can immediately rewrite this equation by converting the conditional probabilities to joint probabilities as

$$F_j(t+h,x) = (1 + q_{jj}h)P\{X(t) \leq x - (r(j) - c)h, Z(t) = j\}$$

$$+ \sum_{i \in S, i \neq j} q_{ij}hP\{X(t) \leq x - (r(i) - c)h, Z(t) = i\} + o(h).$$

Using the definition of $F_j(t,x)$ in Equation 9.3 we can then write down this equation as

$$F_j(t+h,x) = (1 + q_{jj}h)F_j(t, x - (r(j) - c)h)$$

$$+ \sum_{i \in S, i \neq j} q_{ij}hF_i(t, x - (r(i) - c)h) + o(h).$$

We can rearrange the equation by subtracting $F_j(t,x)$ on both sides and dividing by h to get

$$\frac{F_j(t+h,x) - F_j(t,x)}{h} = \frac{F_j(t, x - (r(j) - c)h) - F_j(t,x)}{h}$$

$$+ q_{jj}F_j(t, x - (r(j) - c)h)$$

$$+ \sum_{i \in S, i \neq j} q_{ij}F_i(t, x - (r(i) - c)h) + \frac{o(h)}{h}.$$

We rewrite the expression as

$$\frac{F_j(t+h,x) - F_j(t,x)}{h} + (r(j) - c)\frac{F_j(t, x - (r(j) - c)h) - F_j(t,x)}{-(r(j) - c)h}$$

$$= \sum_{i \in S} q_{ij} F_i(t, x - (r(i) - c)h) + \frac{o(h)}{h}.$$

Now we take the limit as $h \to 0$ and the above equation results in the following partial differential equation:

$$\frac{\partial F_j(t,x)}{\partial t} + (r(j) - c)\frac{\partial F_j(t,x)}{\partial x} = \sum_{i \in S} q_{ij} F_i(t,x). \tag{9.8}$$

To make our representation compact, we define the row vector $F(t,x)$ as

$$F(t,x) = [F_1(t,x)\ \ F_2(t,x)\ \ \ldots\ \ F_\ell(t,x)]. \tag{9.9}$$

Then the vector $F(t,x)$ satisfies the following partial differential equation

$$\frac{\partial F(t,x)}{\partial t} + \frac{\partial F(t,x)}{\partial x}D = F(t,x)Q, \tag{9.10}$$

where D is the drift matrix. Verify that the jth vector element of the equation is identical to that of Equation 9.8. Having described the partial differential equation, the next step is to write down the initial and boundary conditions. We assume that $X(0) = x_0$ and $Z(0) = z_0$ for some given finite and allowable x_0 and z_0. Hence the initial conditions for all $j \in S$ are

$$F_j(0,x) = \begin{cases} 1 & \text{if } j = z_0 \text{ and } x \geq x_0 \\ 0 & \text{otherwise.} \end{cases}$$

Also, the boundary conditions are

$$F_j(t,0) = 0 \quad \text{if } d(j) > 0$$

and (if $B < \infty$)

$$F_j(t,B) = P\{Z(t) = j\} \quad \text{if } d(j) < 0.$$

The first boundary condition states if j is a state of the environment such that the drift is positive (i.e., fluid arrival rate is greater than emptying rate), then the probability the buffer would be empty when the environment is in state j is zero. Thus if $d(j) > 0$, then $P\{X(t) = 0, Z(t) = j\} = 0$ which essentially is the same as $F_j(t,0)=0$ since $X(t)$ is nonnegative. Notice that if the drift is negative at time t (i.e., $Z(t)=j$ and $d(j)<0$), there is a nonzero probability of having $X(t)=0$. In other words, $X(t)$ has a mass at zero and when $X(t)=0$ the drift is negative.

The second boundary condition is a little more involved. However, it applies only when the buffer size is finite, that is, $B<\infty$. Just like how $X(t)$ has a mass at zero, it would also have a mass at B, that is, $P\{X(t)=B\}$ would be non-zero if the drift at time t is positive, that is, $d(Z(t))>0$. Thus as x approaches B from below, $F_j(t,x)$ governed by the partial differential equation would not include the mass at zero and would truly be representing $P\{X(t)<B, Z(t)=j\}$. However, if the drift is negative, there would be no mass at B for the $X(t)$ process. Thus if $d(j)<0$, $P\{X(t)<B, Z(t)=j\}$ would be equal to $P\{X(t)\le B, Z(t)=j\}$ which would just be $P\{Z(t)=j\}$ since $X(t)$ is bounded by B. When $d(j)>0$ we will have $F_j(t,B)+P\{X(t)=B, Z(t)=j\}=P\{Z(t)=j\}$. For that reason we do not have the second boundary condition for all j but only for $d(j)<0$. We will revisit this case subsequently through an example. But it is worthwhile pointing out that this indeed was not an issue when $X(t)=0$ but only when $X(t)=B$ because of the way the CDF is defined as a right-continuous function.

That said, now we have a partial differential equation for the unknown vector $F(t,x)$ with initial and boundary conditions. The next step is to solve it and obtain $F(t,x)$. There are two approaches. One is to use a numerical approach which is effective when numerical values are available for all the parameters. There are software packages that can solve such partial differential equations. The second approach is to analytically solve for $F(t,x)$ that we explain briefly. Let $\tilde{F}_j(w,x)$ be the LST of $F_j(t,x)$ defined as

$$\tilde{F}_j(w,x) = \int_0^\infty e^{-wt}\frac{\partial F(t,x)}{\partial t}dt.$$

Also the row vector of LSTs being the LST of the individual elements, hence

$$\tilde{F}(w,x) = [\tilde{F}_1(w,x)\ \ \tilde{F}_2(w,x)\ \ \ldots\ \ \tilde{F}_\ell(w,x)].$$

Thus taking the LST of both sides of Equation 9.10 we get

$$w\tilde{F}(w,x) - wF(0,x) + \frac{d\tilde{F}(w,x)}{dx}D = \tilde{F}(w,x)Q,$$

which is an ordinary differential equation in x for $\tilde{F}(w,x)$ once we use the initial condition for $F(0,x)$. However, except for certain special cases, in general solving this ordinary differential equation and inverting the LST to get

$F(t, x)$ as a closed-form algebraic expression is nontrivial. Hence we next consider the steady-state distribution of $X(t)$, that is, as $t \to \infty$ for which we can indeed obtain a closed-form algebraic expression.

9.2.3 Steady-State Results and Performance Evaluation

The objective of this section is to obtain a closed-form expression for the distribution for $X(t)$ as $t \to \infty$. For this steady-state result we use the notation and terminology described in Sections 9.2.1 and 9.2.2. Unless specified explicitly, for most of the analysis, B could either be finite or be infinite. For the steady-state analysis we define for all $j \in S$ the joint distribution

$$F_j(x) = \lim_{t \to \infty} F_j(t, x) = \lim_{t \to \infty} P\{X(t) \le x, Z(t) = j\}. \tag{9.11}$$

We would like to obtain a closed-form analytical expression for $F_j(x)$ for all $j \in S$. For that we require the system to be stable if $B = \infty$, and the condition of stability is

$$\sum_{i \in S} p_i d(i) < 0.$$

Assuming that is satisfied, since in steady state the system would be stationary, we have

$$\lim_{t \to \infty} \frac{\partial F_j(t, x)}{\partial t} = 0$$

for all $j \in S$. Hence we can write down an ordinary differential equation by taking the limit $t \to \infty$ of Equation 9.10 as

$$\frac{dF(x)}{dx} D = F(x)Q, \tag{9.12}$$

where

$$F(x) = [F_1(x) \quad F_2(x) \quad \dots \quad F_\ell(x)]. \tag{9.13}$$

Since this is steady-state analysis, initial conditions would not matter. The boundary conditions reduce to

$$F_j(0) = 0 \quad \text{if } d(j) > 0$$

and (if $B < \infty$)

$$F_j(B) = p_j \quad \text{if } d(j) < 0.$$

Note that matrix ordinary differential equations of the form $\frac{dG(x)}{dx} = G(x)A$ to obtain row vector $G(x)$ in terms of square matrix A (of appropriate size) can be solved as $G(x) = \overline{K}\exp(Ax)$ where $\exp(Ax) = I + Ax + A^2x^2/2! + A^3x^3/3! + \cdots$ and $\overline{K}$ is a constant row vector that can be obtained by using the boundary conditions. To obtain $\exp(Ax)$ we can write down $A = VEV^{-1}$, where E is a diagonal matrix with eigenvalues of A and the columns of V correspond to the right eigenvectors of A. Then we get $\exp(Ax) = Vdiag[e^{E_{ii}x}]V^{-1}$, where E_{ii} is the ith eigenvalue of A, that is, the ith diagonal element of E, and $diag[e^{E_{ii}x}]$ is a diagonal matrix with the ith entry corresponding to $e^{E_{ii}x}$. Motivated by that, for the matrix differential equation (9.12)

$$\frac{dF(x)}{dx}D = F(x)Q,$$

we try as solution

$$F(x) = e^{\lambda x}\phi,$$

where ϕ is a $1 \times \ell$ row vector. The solution $(F(x) = e^{\lambda x}\phi)$ works if and only if

$$\phi(\lambda D - Q) = [0 \quad 0 \quad \dots \quad 0]$$

since $\frac{dF(x)}{dx}D$ would be $\phi\lambda De^{\lambda x}$ and $F(x)Q$ would be $\phi Qe^{\lambda x}$. Essentially ϕ is a left eigenvector and λ an eigenvalue both of which need to be determined. For that we first solve the characteristic equation:

$$det(\lambda D - Q) = 0, \tag{9.14}$$

where $det(A)$ is the determinant of square matrix A. Upon solving the equation, we would get the eigenvalues λ. Then using $\phi(\lambda D - Q) = [0 \quad 0 \quad \dots \quad 0]$ for each solution λ, we can obtain the corresponding left eigenvectors ϕ.

Before forging ahead, we describe some properties and notation. We partition the state space S into three sets, S_+, S_0, and S_-, that denote the states where the drift is positive, zero, and negative, respectively. Also, ℓ_+, ℓ_0, and ℓ_- are the number of states with positive, zero, and negative drift, respectively, such that $\ell_+ + \ell_0 + \ell_- = \ell$. Thus we have

$$S_+ = \{i \in S : d(i) > 0\},$$

$$S_0 = \{i \in S : d(i) = 0\},$$

$$S_- = \{i \in S : d(i) < 0\},$$

$$\ell_+ = |S_+|,$$

$$\ell_0 = |S_0|,$$

$$\ell_- = |S_-|.$$

Using that we can write down some properties. Firstly notice that Equation 9.14 would have $\ell_+ + \ell_-$ solutions $\{\lambda_i, i = 1, 2, \ldots, \ell_+ + \ell_-\}$. The $\ell_+ + \ell_-$ solutions could include multiplicities. The crucial property is that when $\sum_{i \in S} p_i d(i) < 0$, exactly ℓ_+ of the λ_i values have negative real parts, one is zero, and, $\ell_- - 1$ have positive real parts.

For sake of convenience we number the λ_i's as

$$Re(\lambda_1) \leq Re(\lambda_2) \leq \cdots \leq Re(\lambda_{\ell_+}) < Re(\lambda_{\ell_+ + 1})$$

$$= 0 < Re(\lambda_{\ell_+ + 2}) \leq \cdots \leq Re(\lambda_{\ell_+ + \ell_-}), \qquad (9.15)$$

where $Re(\omega)$ is the real part of a complex number ω. Using this specific order we are now ready to state the general solution to the differential equation (9.12) as

$$F(x) = \sum_{i=1}^{\ell_+ + \ell_-} a_i e^{\lambda_i x} \phi_i, \qquad (9.16)$$

where a_i values are some constants that need to be obtained (recall that we know how to compute λ_i and ϕ_i for all i).

To compute a_i values, we explicitly consider two cases depending on whether the size of the buffer is infinite or finite. Hence we have the following:

- If $B = \infty$ with $\sum_{i \in S} p_i d(i) < 0$, then a_i values are given by the solution to

$$a_j = 0 \quad \text{if } Re(\lambda_j) > 0 \qquad (9.17)$$

$$a_{\ell_+ + 1} = 1/(\phi_{\ell_+ + 1} \bar{1}) \qquad (9.18)$$

$$\sum_{i=1}^{\ell_+ + 1} a_i \phi_i(j) = 0 \quad \text{if } j \in S_+, \qquad (9.19)$$

where $\bar{1}$ is a column vector of 1s and $\phi_i(j)$ is the jth element of vector ϕ_i. We now explain the preceding set of equations. Consider Equation 9.16. Recall that each element of $F(x)$, that is, the LHS, is a joint probability, hence must necessarily be between zero and one. If we let $x \to \infty$, the only way the elements of $F(x)$ would be a probability measure is if $a_j = 0$ for all λ_j where $Re(\lambda_j) > 0$. This results in Equation 9.17. Also notice that if we let $x \to \infty$ in Equation 9.16, all we would have is $F(\infty) = a_{\ell_+ + 1} \phi_{\ell_+ + 1}$ since $\lambda_{\ell_+ + 1} = 0$ and the terms with $Re(\lambda_j) < 0$ would vanish as $x \to \infty$. However, $F(\infty)\bar{1} = 1$ since $F(\infty)\bar{1}$ is equal to $\lim_{t \to \infty} P\{X(t) \le \infty\}$ which must be one if the system is stable. Hence we get Equation 9.18 since $a_{\ell_+ + 1}\phi_{\ell_+ + 1}\bar{1} = 1$. Finally, Equation 9.19 is directly an artifact of the boundary condition $F_j(0) = 0$ if $d(j) > 0$ for all $j \in S$ which is equivalent to $\sum_{i=1}^{\ell_+ + 1} a_i \phi_i(j) = 0$ if $j \in S_+$ since only the first $\ell_+ + 1$ of the a_i values are nonzero and the elements in S_+ are all those with positive drift, that is, $d(j) > 0$.

- If $B < \infty$, then a_i values are given by the solution to

$$\sum_{i=1}^{\ell_- + \ell_+} a_i \phi_i(j) = 0 \quad \text{if } j \in S_+ \tag{9.20}$$

$$\sum_{i=1}^{\ell_- + \ell_+} a_i \phi_i(j) e^{\lambda_i B} = p_j \quad \text{if } j \in S_-, \tag{9.21}$$

where $\phi_i(j)$ is the jth element of vector ϕ_i. Equation 9.20 is due to the boundary condition $F_j(0) = 0$ if $d(j) > 0$ for all $j \in S$ which is equivalent to $\sum_{i=1}^{\ell_- + \ell_+} a_i \phi_i(j) = 0$ if $j \in S_+$ since the elements in S_+ are all those with positive drift, that is, $d(j) > 0$. Likewise, Equation 9.21 is due to the boundary condition $F_j(B) = p_j$ if $d(j) < 0$ for all $j \in S$ which is equivalent to $\sum_{i=1}^{\ell_- + \ell_+} a_i \phi_i(j) e^{\lambda_i B} = p_j$ if $j \in S_-$ since the elements in S_- are all those with negative drift, that is, $d(j) < 0$.

9.2.4 Examples

In this section, we present a few examples to illustrate the approach to obtain buffer content distribution and describe relevant insights. For that we require characteristics of the environment process, namely, the generator matrix Q and rate matrix R as well as buffer characteristics such as size B and output capacity c. Then using Q, R, B, and c we can obtain the joint distribution $F_j(x)$ as well as the marginal limiting distribution of $X(t)$. The notation, terminology, and methodology used here are described in Sections 9.2.1, 9.2.2, and 9.2.3. All but the last example are steady-state analyses, and they are all presented in a problem-solution format.

Problem 85

Consider the example described in Section 9.2.1 where there is an infinite-sized buffer with output capacity $c = 12$ kbps and the input is driven by an environment CTMC $\{Z(t), t \geq 0\}$ with $\ell = 4$ states, $S = \{1, 2, 3, 4\}$, and

$$
Q = \begin{bmatrix}
-10 & 2 & 3 & 5 \\
0 & -4 & 1 & 3 \\
1 & 1 & -3 & 1 \\
1 & 2 & 3 & -6
\end{bmatrix}
$$

and fluid arrival rates in states 1, 2, 3, and 4 are 20, 15, 10, and 5 kbps, respectively. Obtain the joint distribution vector $F(x)$ as well as a graph of the CDF $\lim_{t \to \infty} P\{X(t) \leq x\}$ versus x.

Solution

For this problem we have Q described earlier,

$$
R = \begin{bmatrix}
20 & 0 & 0 & 0 \\
0 & 15 & 0 & 0 \\
0 & 0 & 10 & 0 \\
0 & 0 & 0 & 5
\end{bmatrix},
$$

$B = \infty$, and $c = 12$. The drift matrix is

$$
D = \begin{bmatrix}
8 & 0 & 0 & 0 \\
0 & 3 & 0 & 0 \\
0 & 0 & -2 & 0 \\
0 & 0 & 0 & -7
\end{bmatrix}.
$$

The steady-state probability vector for the environment process is $[p_1 \quad p_2 \quad p_3 \quad p_4] = [0.0668 \quad 0.2647 \quad 0.4118 \quad 0.2567]$. The system is stable since $\sum_{i=1}^{4} D_{ii} p_i < 0$.

Notice that $S_+ = \{1, 2\}$ since states 1 and 2 have positive drift. Likewise $S_- = \{3, 4\}$ since states 3 and 4 have negative drift. Also since there are no zero-drift states, S_0 is a null set. Also $\ell_+ = 2$ and $\ell_- = 2$. Thus by solving Equation 9.14 we would get two λ values with negative real parts, one λ value would be zero and one with positive real part. We check that first. We solve for λ in the characteristic equation

$$
det(\lambda D - Q) = 0,
$$

to obtain

$$[\lambda_1 \ \lambda_2 \ \lambda_3 \ \lambda_4] = [-1.3733 \ -0.5994 \ 0 \ 1.7465].$$

Notice that the λ_i values are ordered according to Equation 9.15. Then using $\phi(\lambda D - Q) = [0 \quad 0 \quad 0 \quad 0]$ for each solution λ, we can obtain the corresponding left eigenvectors $\phi_1 = [-0.2297 \ 0.9600 \ 0.1087 \ 0.1179]$, $\phi_2 = [0.1746 \ 0.7328 \ 0.5533 \ 0.3555]$, $\phi_3 = [0.1201 \ 0.4754 \ 0.7396 \quad 0.4610]$, and $\phi_4 = [-0.0317 \ -0.0660 \ -0.9741 \ 0.2138]$.

Thereby using Equation 9.16 we can write down $F(x)$ as

$$F(x) = a_1 e^{\lambda_1 x} \phi_1 + a_2 e^{\lambda_2 x} \phi_2 + a_3 e^{\lambda_3 x} \phi_3 + a_4 e^{\lambda_4 x} \phi_4.$$

All we need to compute are a_1, a_2, a_3, and a_4. For that we use Equations 9.17 through 9.19. In particular, from Equation 9.17, we have $a_4 = 0$ since $Re(\lambda_4) > 0$; based on Equation 9.18, $a_3 = 1/(\phi_3 \bar{1}) = 0.5568$; from Equation 9.19 we get $a_1 \phi_1(1) + a_2 \phi_2(1) + a_3 \phi_3(1) = -0.2297 a_1 + 0.1746 a_2 + 0.0668 = 0$ and $a_1 \phi_1(2) + a_2 \phi_2(2) + a_3 \phi_3(2) = 0.96 a_1 + 0.7328 a_2 + 0.2647 = 0$ which can be solved to obtain $a_1 = 0.0082$ and $a_2 = -0.3720$. Hence we have

$$F(x) = 0.0082[-0.2297 \ 0.9600 \ 0.1087 \ 0.1179] e^{-1.3733x}$$

$$- 0.3720[0.1746 \ 0.7328 \ 0.5533 \ 0.3555] e^{-0.5994x}$$

$$+ 0.5568[0.1201 \ 0.4754 \ 0.7396 \ 0.4610]$$

$$= [0.0668 - 0.0019 e^{-1.3733x} - 0.0650 e^{-0.5994x}$$

$$0.2647 + 0.0079 e^{-1.3733x} - 0.2726 e^{-0.5994x}$$

$$0.4118 + 0.0009 e^{-1.3733x} - 0.2058 e^{-0.5994x}$$

$$0.2567 + 0.0010 e^{-1.3733x} - 0.1323 e^{-0.5994x}].$$

Notice that the first component of this expression corresponds to $[0.0668 \quad 0.2647 \quad 0.4118 \quad 0.2567]$ which would be $F(\infty)$ and it is equal to $[p_1 \ p_2 \ p_3 \ p_4]$. Although theoretically this is not surprising, the calculations lead to this because when we solve $[p_1 \quad p_2 \quad p_3 \quad p_4]Q = [0 \ 0 \ 0 \ 0]$ we are essentially obtaining the left eigenvector which is $[p_1 \ p_2 \ p_3 \ p_4]$ and the eigenvalue of 0.

Next, we have the CDF

$$\lim_{t \to \infty} P\{X(t) \le x\} = P\{X \le x\} = F(x)\bar{1} = 1 + 0.0079 e^{-1.3733x} - 0.6757 e^{-0.5994x}$$

for all $x \ge 0$ by letting $X(t) \to X$ as $t \to \infty$. A graph of the CDF is depicted in Figure 9.4. Notice that there is a mass at $x = 0$ which can also be obtained

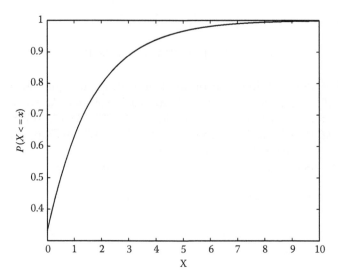

FIGURE 9.4
Graph of $P\{X \leq x\}$ vs. x for the infinite buffer case (Problem 85).

by letting $x=0$ in $1 + 0.0079e^{-1.3733x} - 0.6757e^{-0.5994x}$ that would result in $P\{X=0\}=0.3322$. ∎

To present the similarities and differences between the cases when the buffer size is infinite versus finite, in the next problem we consider the exact same numerical values as the previous problem, except for the size of the buffer. That is described next.

Problem 86

Consider Problem 85 with the only exception that the buffer size is finite with $B=2$. Obtain the joint distribution vector $F(x)$ for $0 \leq x < B$ as well as the distribution for $X(t)$ as $t \to \infty$.

Solution

Recall that the analysis in Section 9.2.3 does not make any assumptions about B until obtaining the constants a_i. Thus from the solution to Problem 85 we have (for $0 \leq x < B$)

$$F(x) = a_1 e^{\lambda_1 x}\phi_1 + a_2 e^{\lambda_2 x}\phi_2 + a_3 e^{\lambda_3 x}\phi_3 + a_4 e^{\lambda_4 x}\phi_4,$$

where

$$[\lambda_1 \ \lambda_2 \ \lambda_3 \ \lambda_4] = [-1.3733 \ -0.5994 \ 0 \ 1.7465],$$
$$\phi_1 = [-0.2297 \ 0.9600 \ 0.1087 \ 0.1179],$$

$$\Phi_2 = [0.1746 \; 0.7328 \; 0.5533 \; 0.3555],$$

$$\Phi_3 = [0.1201 \; 0.4754 \; 0.7396 \; 0.4610],$$

and

$$\Phi_4 = [-0.0317 \; -0.0660 \; -0.9741 \; 0.2138].$$

All we need to compute are $a_1, a_2, a_3,$ and a_4. For that we use Equations 9.20 and 9.21.

From Equation 9.20 we get $a_1\phi_1(1) + a_2\phi_2(1) + a_3\phi_3(1) + a_4\phi_4(1) = -0.2297a_1 + 0.1746a_2 + 0.1201a_3 - 0.0317a_4 = 0$ and $a_1\phi_1(2) + a_2\phi_2(2) + a_3\phi_3$ $(2) + a_4\phi_4(2) = 0.96a_1 + 0.7328a_2 + 0.4754a_3 - 0.0660a_4 = 0$. Likewise, from Equation 9.21 we get $a_1\phi_1(3)e^{\lambda_1 B} + a_2\phi_2(3)e^{\lambda_2 B} + a_3\phi_3(3)e^{\lambda_3 B} + a_4\phi_4(3)e^{\lambda_4 B} = 0.0070a_1 + 0.1669a_2 + 0.7396a_3 - 32.0307a_4 = p_3$ and $a_1\phi_1(4)e^{\lambda_1 B} + a_2\phi_2(4)e^{\lambda_2 B} + a_3\phi_3(4)e^{\lambda_3 B} + a_4\phi_4(4)e^{\lambda_4 B} = 0.0076a_1 + 0.1072a_2 + 0.4610a_3 + 7.0293a_4 = p_4$. Using the fact that $p_3 = 0.4118$ and $p_4 = 0.2567$, these four equations can be solved to obtain $a_1 = 0.0097$, $a_2 = -0.4397$, $a_3 = 0.6581$, and $a_4 = 0.000050924$.

Next using the notation $X(t) \to X$ as $t \to \infty$, we have the distribution of X given by

$$P\{X \le x\} = F(x)\bar{1} = 0.0093e^{-1.3733x} - 0.7986e^{-0.5994x}$$

$$+ 1.1820 - 0.000043698e^{1.7465x} \quad \text{if } 0 < x < B,$$

$$P\{X = x\} = F(0)\bar{1} = 0.3926 \quad \text{if } x = 0,$$

$$P\{X = x\} = 1 - F(B)\bar{1} = 0.0597 \quad \text{if } x = B.$$

A graph of $P(X \le x)$ versus x is depicted in Figure 9.5. Notice the mass at $x = 0$. Also at $x = B$ the CDF is not 1 and the gap explains the mass of X at B. ∎

Having described two examples in a numerical fashion, our next example will be a symbolic calculation. It is based on one of the simplest, yet widely used models which is the exponential on–off environment process that is also known as *CTMC on–off source*.

Problem 87

(CTMC on–off source) Consider a source that inputs fluid into an infinite-size buffer. The source toggles between on and off states. The on times are according to $\exp(\alpha)$ and off times according to $\exp(\beta)$. Traffic is generated at rate r when the source is in the on-state and no traffic is generated when the source is in the off-state. Assume that $r > c$, where c is the usual output

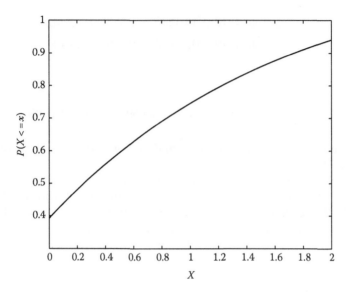

FIGURE 9.5
Graph of $P\{X \leq x\}$ vs. x for the finite buffer case of Problem 86.

capacity. Find the condition of stability. Assuming that the stability condition is satisfied what is the steady-state distribution of the buffer contents in terms of r, c, α, and β?

Solution

The environment process $\{Z(t), t \geq 0\}$ is a CTMC with $\ell = 2$ states and $S = \{1, 2\}$ with 1 representing off and 2 representing the on-state. Therefore,

$$Q = \begin{bmatrix} -\beta & \beta \\ \alpha & -\alpha \end{bmatrix} \text{ and } R = \begin{bmatrix} 0 & 0 \\ 0 & r \end{bmatrix}.$$

With $r > c$, the drift matrix is

$$D = \begin{bmatrix} -c & 0 \\ 0 & r-c \end{bmatrix}.$$

The steady-state probabilities for the environment process are $p_1 = \frac{\alpha}{\alpha+\beta}$ and $p_2 = \frac{\beta}{\alpha+\beta}$. The system is stable if $0p_1 + rp_2 < c$. Thus the stability condition (notice that $B = \infty$) is

$$r\beta/(\alpha + \beta) < c.$$

State 1 has negative drift and state 2 has positive drift. Hence by solving for Equation 9.14 we would get one λ value with negative real part and one

λ value would be zero. To confirm that we solve for λ in the characteristic equation

$$det(\lambda D - Q) = 0,$$

which yields

$$(-\lambda c + \beta)(\lambda r - \lambda c + \alpha) - \alpha\beta = 0.$$

By rewriting this equation in terms of the unknown λ we get

$$c(r - c)\lambda^2 - (\beta(r - c) - \alpha c)\lambda = 0.$$

Therefore, we have the two solutions to λ as

$$\lambda_1 = \beta/c - \alpha/(r - c),$$

$$\lambda_2 = 0.$$

From the stability condition $r\beta/(\alpha + \beta) < c$ we have $r\beta - c(\alpha + \beta) < 0$ and dividing that by the positive quantity $c(r - c)$, we get $\beta/c - \alpha/(r - c) < 0$. Hence $\lambda_1 < 0$. Thus we have verified that one λ value has negative real part and the other one is zero.

Next, using $\phi(\lambda D - Q) = [0\ 0]$ for each λ, we can obtain the corresponding left eigenvectors as $\phi_1 = [(r - c)/c\ \ 1]$ and $\phi_2 = [\alpha/(\alpha + \beta)\ \ \beta/(\alpha + \beta)]$. Thereby using Equation 9.16 we can write down $F(x)$ as

$$F(x) = a_1 e^{\lambda_1 x}\phi_1 + a_2 e^{\lambda_2 x}\phi_2.$$

All we need to compute are a_1 and a_2. For that we use Equations 9.18 and 9.19. From Equation 9.18, $a_2 = 1/(\phi_2 \bar{1}) = 1$. Also, from Equation 9.19 we get $a_1\phi_1(2) + a_2\phi_2(2) = 0$ since that corresponds to state with a positive drift and that results in $a_1 = -\frac{\beta}{\alpha + \beta}$. Hence we have

$$F(x) = [F_1(x)\ \ F_2(x)] = \left[\frac{\alpha c - (r - c)\beta e^{\lambda_1 x}}{c(\alpha + \beta)} \quad \frac{\beta}{\alpha + \beta}(1 - e^{\lambda_1 x})\right].$$

Therefore, the limiting distribution of the buffer content process is

$$\lim_{t \to \infty} P\{X(t) \leq x\} = F_1(x) + F_2(x) = 1 - \frac{\beta r}{c(\alpha + \beta)}e^{\lambda_1 x}, \tag{9.22}$$

where

$$\lambda_1 = \beta/c - \alpha/(r - c). \tag{9.23}$$

From Equation 9.22 we can quickly write down

$$\lim_{t \to \infty} P\{X(t) > x\} = \frac{\beta r}{c(\alpha + \beta)} e^{\lambda_1 x}. \tag{9.24}$$

In particular notice that

$$\lim_{t \to \infty} P\{X(t) > 0\} = \frac{\beta r}{c(\alpha + \beta)}$$

which makes sense since in a cycle of one busy period and one idle period on average a quantity proportional to $r\beta/(\alpha + \beta)$ fluid arrives but that fluid is depleted during a busy period at rate c. Thus the ratio of the mean busy period to the mean cycle of one busy period to an idle period must equal $r\beta/[c(\alpha + \beta)]$ which is also the equation for the fraction of time there is a non-zero amount of fluid in the buffer. ∎

On–off environment processes are a popular way to characterize information flow because when there is information to transmit, it flows at the speed of the cable and when there is no information to transmit, nothing flows. So information flow can be nicely modeled as an on–off process. However, the on and off times may not be exponentially distributed. But since any distribution can be approximated as a phase-type distribution, we could always use the analysis presented here, thus this approach is rather sound and powerful. We illustrate that in the next problem.

Problem 88

Consider an on–off source that generates fluid into a buffer of infinite size with output capacity 8 units per second. The on times are IID random variables with CDF $U(t) = 1 - 0.6e^{-3t} - 0.4e^{-2t}$ and the off times are IID Erlang random variables with mean 0.5 and variance 1/12 in appropriate time units compatible with the on times. When the source is on, fluid is generated at rate 16 units per second and no fluid is generated when the source is off. Compute the probability that there would be more than 10 units of fluid in the buffer in steady state.

Solution

Notice that the on times correspond to a two-phase hyperexponential distribution. So the on time would be exp(3) with probability 0.6 and it would be exp(2) with probability 0.4, which can be deduced from $U(t)$. The off times correspond to the sum of three IID exp(6) random variables. Thus we can write down the environment process $\{Z(t), t \geq 0\}$ as an $\ell = 5$ state CTMC with states 1 and 2 corresponding to on and states 3, 4, and 5 corresponding to off. Thus the Q matrix corresponding to $S = \{1, 2, 3, 4, 5\}$ is

$$Q = \begin{bmatrix} -3 & 0 & 3 & 0 & 0 \\ 0 & -2 & 2 & 0 & 0 \\ 0 & 0 & -6 & 6 & 0 \\ 0 & 0 & 0 & -6 & 6 \\ 3.6 & 2.4 & 0 & 0 & -6 \end{bmatrix}.$$

The rate matrix is

$$R = \begin{bmatrix} 16 & 0 & 0 & 0 & 0 \\ 0 & 16 & 0 & 0 & 0 \\ 0 & 0 & 0 & 0 & 0 \\ 0 & 0 & 0 & 0 & 0 \\ 0 & 0 & 0 & 0 & 0 \end{bmatrix}.$$

Using that and the fact that $c = 8$, we have the drift matrix

$$D = \begin{bmatrix} 8 & 0 & 0 & 0 & 0 \\ 0 & 8 & 0 & 0 & 0 \\ 0 & 0 & -8 & 0 & 0 \\ 0 & 0 & 0 & -8 & 0 \\ 0 & 0 & 0 & 0 & -8 \end{bmatrix}.$$

Since $B = \infty$, we need to first check if the buffer is stable. For that we obtain $[p_1 \quad p_2 \quad p_3 \quad p_4 \quad p_5] = [0.2222 \quad 0.2222 \quad 0.1852 \quad 0.1852 \quad 0.1852]$. We have $\sum_{j=1}^{5} D_{ii} p_i = -0.8889 < 0$, hence the system is stable.

The rest of the analysis proceeds very similar to Problem 85 with the only exception being the final expression to compute, which here is the probability that there would be more than 10 units of fluid in the buffer in steady state. Nevertheless for the sake of completion we go through the entire process. Notice that $S_+ = \{1, 2\}$ since states 1 and 2 have positive drift. Likewise $S_- = \{3, 4, 5\}$ since states 3, 4, and 5 have negative drift. Also since there are no zero-drift states, S_0 is a null set. Also $\ell_+ = 2$ and $\ell_- = 3$. Thus by solving Equation 9.14 we would get two λ with negative real parts, one λ value would be zero and two with positive real parts. We solve for λ in the characteristic equation

$$det(\lambda D - Q) = 0,$$

to obtain

$$[\lambda_1 \quad \lambda_2 \quad \lambda_3 \quad \lambda_4 \quad \lambda_5] = [-0.3227 \quad -0.0836 \quad 0 \quad 1.0156 - 0.3765i \quad 1.0156 + 0.3765i]$$

with λ_i values ordered according to Equation 9.15. Then using $\phi(\lambda D - Q) = [0 \quad 0 \quad \dots \quad 0]$ for each solution λ, we can obtain the corresponding left eigenvectors

$$\phi_1 = [0.8678 \quad -0.4163 \quad 0.2064 \quad 0.1443 \quad 0.1009],$$

$$\phi_2 = [0.5038 \quad 0.5881 \quad 0.4030 \quad 0.3626 \quad 0.3263],$$

$$\phi_3 = [0.4949 \quad 0.4949 \quad 0.4124 \quad 0.4124 \quad 0.4124],$$

$$\phi_4 = [-0.2334 + 0.0632i \quad -0.1686 + 0.0501i \quad 0.0980 - 0.2752i \quad 0.2741 + 0.3886i \quad -0.7740],$$

and

$$\phi_5 = [-0.2334 - 0.0632i \quad -0.1686 - 0.0501i \quad 0.0980 + 0.2752i \quad 0.2741 - 0.3886i \quad -0.7740].$$

Thereby using Equation 9.16 we can write down $F(x)$ as

$$F(x) = a_1 e^{\lambda_1 x} \phi_1 + a_2 e^{\lambda_2 x} \phi_2 + a_3 e^{\lambda_3 x} \phi_3 + a_4 e^{\lambda_4 x} \phi_4 + a_5 e^{\lambda_5 x} \phi_5.$$

We still need to compute a_1, a_2, a_3, a_4, and a_5. For that we use Equations 9.17 through 9.19. In particular, from Equation 9.17, we have $a_4 = 0$ and $a_5 = 0$ since $Re(\lambda_4) > 0$ and $Re(\lambda_5) > 0$. Based on Equation 9.18, $a_3 = 1/(\phi_3 \bar{1}) = 0.4491$. From Equation 9.19 we get $a_1 \phi_1(1) + a_2 \phi_2(1) + a_3 \phi_3(1) = 0.8678 a_1 + 0.5038 a_2 + 0.2222 = 0$ and $a_1 \phi_1(2) + a_2 \phi_2(2) + a_3 \phi_3(2) = -0.4163 a_1 + 0.5881 a_2 + 0.2222 = 0$ which can be solved to obtain $a_1 = -0.0260$ and $a_2 = -0.3963$. Hence we have all the elements of $F(x)$.

Also, we have the CDF

$$\lim_{t \to \infty} P\{X(t) \leq x\} = F(x)\bar{1} = a_1 e^{\lambda_1 x} \phi_1 \bar{1} + a_2 e^{\lambda_2 x} \phi_2 \bar{1} + a_3 e^{\lambda_3 x} \phi_3 \bar{1}$$

$$= 1 - 0.0235 e^{-0.3227x} - 0.8654 e^{-0.0836x}$$

for all $x \geq 0$. Thus the probability that there would be more than 10 units of fluid in the buffer in steady state is $P(X > 10) = 0.0235 e^{-3.227} + 0.8654 e^{-0.836} = 0.3761$. ∎

For this problem we have for any $x \geq 0$, $P(X > x) = 0.0235 e^{-0.3227x} + 0.8654 e^{-0.0836x}$. A plot of $P(X > x)$ is depicted in Figure 9.6 for $0 \leq x \leq 30$. Notice how slowly the probability reduces with x. Also note that the first term, namely, $0.0235 e^{-0.3227x}$ hardly contributes to $P(X > x)$. In the next chapter we will leverage upon this to obtain tail probabilities $P(X > x)$ for large x. However, for the rest of this chapter we will be concerned with the

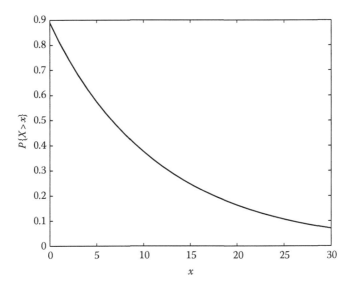

FIGURE 9.6
Graph of $P\{X > x\}$ vs. x for the infinite buffer case of Problem 88.

entire distribution. We conclude our examples with a problem on transient analysis adapted from Kharoufeh and Gautam [62].

Problem 89

The speed of a particular vehicle on a highway is modulated by a five-state CTMC $\{Z(t), t \geq 0\}$ with $S = \{1, 2, 3, 4, 5\}$. When the CTMC is in state i, the speed of the vehicle is $V_i = 75/i$ miles per hour for $i \in S$. The infinitesimal generator matrix of the CTMC is

$$
Q = \begin{bmatrix}
-919.75 & 206.91 & 264.85 & 238.67 & 209.32 \\
223.01 & -971.71 & 301.98 & 232.73 & 213.98 \\
343.04 & 277.78 & -1283.57 & 392.72 & 270.03 \\
353.91 & 232.27 & 213.69 & -1059.47 & 259.59 \\
370.92 & 200.89 & 216.80 & 225.60 & -1014.21
\end{bmatrix}
$$

in units of h^{-1}. Assume that the CTMC is in state 1 at time 0. Obtain a method to compute the CDF of the amount of time it would take for the vehicle to travel 1 mile. Also numerically compute for sample values of t the probability that the vehicle would travel one mile before time t. Verify the results using simulations.

Solution

Let $T(x)$ be the random time required for the vehicle to travel a distance x miles. We are interested in $P\{T(x) \leq t\}$ for $x = 1$; however, we provide an approach for a generic x. Now let $X(t)$ be the distance the vehicle traveled in time t. A crucial observation needs to be made which is that the events $\{T(x) \geq t\}$ and $\{X(t) \leq x\}$ are equivalent. In other words, the event that a vehicle travels in time t a distance less than or equal to x is the same as saying that the time taken to reach distance x is greater than or equal to t. Therefore, $P\{T(x) \geq t\} = P\{X(t) \leq x\}$ and hence the CDF of $T(x)$ is $P\{T(x) \leq t\} = 1 - P\{X(t) \leq x\}$.

Next, we will show a procedure to compute $P\{X(t) \leq x\}$, and thereby obtain $P\{T(x) \geq t\}$. Notice that, $X(t)$ can be thought of as the amount of fluid in a buffer at time t with $X(0) = 0$ modulated by an environment process $\{Z(t), t \geq 0\}$ with $Z(0) = 1$. The buffer size is infinite ($B = \infty$) and the output capacity $c = 0$. Fluid flows in at rate $r(Z(t)) = V_{Z(t)} = 75/Z(t)$ at time t. Essentially the amount of fluid at time t corresponds to the distance traveled by the vehicle at time t.

Now define the following joint probability distribution,

$$F_j(t, x) = P\{X(t) \leq x, Z(t) = j\}, \quad \forall j \in S$$

which is identical to that of Equation 9.3. If we know $F_j(t, x)$, then we can immediately obtain the required $P\{T(x) \leq t\}$ using

$$P\{T(x) \leq t\} = 1 - P\{X(t) \leq x\} = 1 - \sum_{i \in S} F_i(t, x).$$

Using the representation in Equation 9.9 we define the row vector $F(t, x)$ as

$$F(t, x) = [F_1(t, x) \; F_2(t, x) \; \cdots \; F_\ell(t, x)].$$

Based on Equation 9.10 we know that the vector $F(t, x)$ satisfies the partial differential equation

$$\frac{\partial F(t, x)}{\partial t} + \frac{\partial F(t, x)}{\partial x} D = F(t, x)Q,$$

with initial condition for all $i \in S$

$$F_i(0, x) = A_i(x) = P\{Z(0) = i\}.$$

Note that D is the drift matrix (essentially diagonal matrix of $r(i) - c$ values with $r(i) = 75/i$ and $c = 0$).

To solve the partial differential equation, we require some additional notation. Let $F_i^*(s_2, x)$ be the Laplace transform (LT) of $F_i(t, x)$ with respect to t, that is,

$$F_i^*(s_2, x) = \int_0^\infty e^{-s_2 t} F_i(t, x) dt.$$

Also, let $\tilde{F}_i^*(s_2, s_1)$, $i \in S$ denote the Laplace–Stieltjes transform (LST) of $F_i^*(s_2, x)$ with respect to x, that is,

$$\tilde{F}_i^*(s_2, s_1) = \int_0^\infty e^{-s_1 x} dF_i^*(s_2, x).$$

Writing in matrix form, we have $\tilde{F}^*(s_2, s_1) = \left[\tilde{F}_i^*(s_2, s_1) \right]_{i \in S}$ as the 1×5 row vector of transforms. Likewise $F^*(s_2, x) = \left[F_i^*(s_2, x) \right]_{i \in S}$ is the row vector of Laplace transforms of $F_i(t, x)$ with respect to t.

To solve the partial differential equation, we take the LT and then the LST of the partial differential equation to get an expression in the transform space

$$\tilde{F}^*(s_2, s_1) = \tilde{A}(s_1)(s_1 D + s_2 I - Q)^{-1}$$

where $\tilde{A}(s_1)$ is a 1×5 row vector of the LSTs of the initial condition. There are several software packages that can be used to numerically invert this transform. As additional complication is the 2D nature. Readers are referred to Kharoufeh and Gautam [62] for a numerical inversion algorithm as well as a list of references for different inversion techniques. Using $x = 1$ and the initial condition $X(0) = 0$ and $Z(0) = 1$ giving rise to

$$\tilde{A}(s_1) = [\ 1 \quad 0 \quad 0 \quad 0 \quad 0\ ]$$

we can obtain numerically an expression for $F(t, x)$ for various values of t. Thus we can obtain $P\{T(x) \leq t\}$ for $x = 1$ as described in Table 9.2. An empirical CDF based on 100,000 observations of travel time was generated via Monte-Carlo simulation methods, is also presented in the last column of the table.   ∎

Having described an example of transient analysis, it provides a nice transition to the next topic which is the first passage time analysis in fluid queues. We are interested in the first time the fluid reaches a particular level from a given initial level.

TABLE 9.2

Travel Time CDF to Traverse $x = 1$
Mile for Sample t Values

t	$P\{T(x) \le t\}$	$P\{T(x) \le t\}$
(min)	Inversion	Simulation
1.25	0.0786	0.0777
1.47	0.3335	0.3352
1.70	0.6859	0.6865
1.92	0.9141	0.9136
2.14	0.9873	0.9872
2.37	0.9991	0.9991
2.59	1.0000	0.9999
2.81	1.0000	1.0000

Source: Kharoufeh, J.P. and Gautam, N.,
             Transp. Sci., 38(1), 97, 2004. With
             permission.

9.3 First Passage Times

Consider a single buffer fluid model as described in Figure 9.2. We continue
to use the notation and terminology in Section 9.2.1. The input to this buffer
is driven by an environment process $\{Z(t), t \ge 0\}$ which is an ℓ-state CTMC
with state space S and infinitesimal generator matrix Q. The output capacity
of the buffer is c and the drift matrix is D. We assume for convenience that
all diagonal elements of the drift matrix are nonzero (i.e., for all $j \in S$ we have
$r(j) \ne c$). The assumption is purely for the purposes of presentation but does
not pose any difficulties analytically (see exercises at the end of the chapter).
Let $X(t)$ be the amount of fluid in the buffer at time t. In this section we are
interested in answering the question: Given $X(0) = x$ and $Z(0) = i$, what is
the distribution of the random time for the buffer content to reach a level
of a or b for the first time? This is sometimes known as first passage time
or hitting time. Three cases are relevant (all other cases can be written in
terms of these three cases): $a \le x \le b$, $a = b \le x$, and $a = b \ge x$. We generally
assume that the buffer is infinite-sized, that is, $B = \infty$. However, only the
case $a = b \le x$ requires the system to be stable. In fact for the other cases B can
even be finite, as long as $B > b$. Although we have considered the three cases,
we will mainly describe the analysis for $a \le x \le b$ and present remarks for the
other two cases toward the end of the next section.

Define the first passage time T as

$$T = \inf\{t > 0 : X(t) = a \quad \text{or} \quad X(t) = b\}. \tag{9.25}$$

Thus T denotes the time it takes for the buffer content to reach a or b for the first time. Our objective is to obtain a distribution for T as a CDF (i.e., $P\{T \le t\}$) or its LST (i.e., $E[e^{-wT}]$). For that let

$$H_{ij}(x, t) = P\{T \le t, \ Z(T) = j \mid X(0) = x, \ Z(0) = i\}$$

for $i, j \in S$; $a \le x \le b$ and $t \ge 0$. Define the $\ell \times \ell$ matrix $H(x, t) = [H_{ij}(x, t)]$ so that $H_{ij}(x, t)$ corresponds to the element in row i and column j in $H(x, t)$. We follow a procedure very similar to that in Section 9.2.2 and obtain a partial differential equation in the next section.

9.3.1 Partial Differential Equations and Boundary Conditions

In this section, we derive a partial differential equation very similar to Equation 9.10 for $H(x, t)$ defined there (the $H(x, t)$ here is different and in fact a matrix). Consider the term $H_{ij}(x, \ t + h)$ where h is a small positive real number. We can write $H_{ij}(x, \ t + h)$ as

$$H_{ij}(x, \ t + h) = P\{T \le t + h, \ Z(T) = j \mid X(0) = x, \ Z(0) = i\}.$$

By conditioning and unconditioning on $Z(h) = k$, we can show that

$$H_{ij}(x, \ t + h) = \sum_{k \in S, k \neq i} H_{kj}(x + h(r(i) - c), \ t) \ q_{ik} h$$

$$+ \ H_{ij}(x + h(r(i) - c), \ t) \ (1 + q_{ii}h) \ + \ o(h),$$

where $o(h)$ represents higher order terms of h. Subtracting $H_{ij}(x, t)$ on both sides and rearranging terms, we get

$$\frac{H_{ij}(x, \ t + h) - H_{ij}(x, t)}{h}$$

$$= \frac{H_{ij}(x + (r(i) - c)h, \ t) - H_{ij}(x, t)}{h} + \sum_{k \in S} H_{kj}(x + h(r(i) - c), \ t) \ q_{ik} + o(h)/h.$$

Taking the limit as $h \to 0$, since $\frac{o(h)}{h} \to 0$, we have

$$\frac{\partial H_{ij}(x, \ t)}{\partial t} = (r(i) - c) \frac{\partial H_{ij}(x, \ t)}{\partial x} + \sum_{k \in S} H_{kj}(x, \ t) \ q_{ik}.$$

Writing that in a matrix form results in

$$\frac{\partial H(x, t)}{\partial t} - D\frac{\partial H(x, t)}{\partial x} = QH(x, t). \tag{9.26}$$

Next we describe the boundary conditions for all $i \in S$ and $j \in S$ (with $a \leq x \leq b$) as follows:

$$H_{jj}(b, t) = 1 \quad \text{if } r(j) > c, \tag{9.27}$$

$$H_{jj}(a, t) = 1 \quad \text{if } r(j) < c, \tag{9.28}$$

$$H_{ij}(b, t) = 0 \quad \text{if } r(i) > c \text{ and } i \neq j, \tag{9.29}$$

$$H_{ij}(a, t) = 0 \quad \text{if } r(i) < c \text{ and } i \neq j. \tag{9.30}$$

The first boundary condition (i.e., Equation 9.27) is so because if the initial buffer content is b and source is in state j (assuming $r(j) > c$), then essentially the first passage time has occurred. Therefore, the probability that the first passage time occurs before time t and the source is in state j when it occurred is 1. The second boundary condition (Equation 9.28) is based on the fact that if the initial buffer content is a and the source is in state j such that $r(j) < c$, then the first passage time is zero. Hence, the probability that the first passage time happens before time t and the source is in state j when it occurred is 1. The third boundary condition (i.e., Equation 9.29) is due to the fact that although the first passage time is zero, the probability that the source is state j when the first passage time occurs is zero (since at time $t = 0$ the source is state i with $r(i) > c$ and $i \neq j$). For exactly the same reason, the last boundary condition (Equation 9.30) is the way it is, that is, the first passage time is zero but it cannot occur when the source is state j, given that the source was in state i at time $t = 0$ with $r(i) < c$ and $i \neq j$.

Next we solve the partial differential equation (PDE), that is, Equation 9.26. First we take the LST across the PDE with respect to t. That reduces to the following ordinary differential equation (ODE):

$$D\frac{d\tilde{H}(x, w)}{dx} = (wI - Q)\tilde{H}(x, w) \tag{9.31}$$

where $\tilde{H}(x, w)$ is the LST of $H(x, t)$ with respect to t and that in turn equals the LST of each element of $H(x, t)$. Not only is the ODE easier to solve, but we can also immediately obtain the LST of the CDF of the first passage time T. We first solve the ODE. For that let $S_1(w), \ldots, S_\ell(w)$ be ℓ scalar solutions to the characteristic equation

$$det(DS(w) - wI + Q) = 0.$$

For each $S_j(w)$ we can find column vectors $\phi_j(w)$ that satisfy

$$S_j(w)D\phi_j(w) = (wI - Q)\phi_j(w).$$

Thus given w, the $S_j(w)$ values are ℓ eigenvalues and $\phi_1(w), \ldots, \phi_\ell(w)$ are the corresponding right eigenvectors. Using those we can write down the solution to Equation 9.31.

The solution to this ODE is given by

$$\tilde{H}_{\cdot j}(x, w) = a_{1,j}(w)e^{S_1(w)x}\phi_1(w) + a_{2,j}(w)e^{S_2(w)x}\phi_2(w)$$

$$+ \cdots + a_{\ell,j}(w)e^{S_\ell(w)x}\phi_\ell(w), \tag{9.32}$$

where $a_{i,j}(w)$ values are constants to be determined and $\tilde{H}_{\cdot j}(x, w)$ is a column vector such that

$$\tilde{H}_{\cdot j}(x, w) = \begin{bmatrix} \tilde{H}_{1j}(x, w) \\ \tilde{H}_{2j}(x, w) \\ \vdots \\ \tilde{H}_{\ell j}(x, w) \end{bmatrix}.$$

We can obtain the ℓ^2 $a_{i,j}(w)$ values using the ℓ^2 equations corresponding to the LST of the boundary condition equations (9.27) through (9.30) which for all $i \in S$ and $j \in S$ are

$$\tilde{H}_{jj}(b, w) = 1 \quad \text{if } r(j) > c,$$

$$\tilde{H}_{jj}(a, w) = 1 \quad \text{if } r(j) < c,$$

$$\tilde{H}_{ij}(b, w) = 0 \quad \text{if } r(i) > c \text{ and } i \neq j,$$

$$\tilde{H}_{ij}(a, w) = 0 \quad \text{if } r(i) < c \text{ and } i \neq j.$$

Thereby, using Equation 9.32 we can write down the LST of the distribution of T. In particular, given $X(0) = x_0$ and $Z(0) = z_0$, the LST of the first passage time distribution can be computed as

$$E[e^{-wT}] = \sum_{j=1}^{\ell} \tilde{H}_{z_0 j}(x_0, w). \tag{9.33}$$

Although in most instances this equation cannot be inverted to get the CDF of T, one can quickly get moments of T. Specifically for $r = 1, 2, 3, \ldots,$

$$E[T^r] = (-1)^r \frac{d^r E[e^{-wT}]}{dw^r}$$

at $w = 0$. We can also obtain the probability that the first passage time ends in state j. In particular

$$P\{Z(T) = j | X(0) = x_0, Z(0) = z_0\} = H_{z_0 j}(x_0, \infty) = \tilde{H}_{z_0 j}(x_0, 0)$$

since the CDF in the limit $t \to \infty$ is equivalent to its LST in the limit $w \to 0$. In the next section we present some examples to illustrate the approach. We now present two remarks for the cases when $a = b \leq x$ and $a = b \geq x$.

Remark 22

For the case $a = b \leq x$ we require that the stability condition is satisfied, that is, $\sum_{j \in S} d(j) p_j < 0$ where $d(j)$ is the drift when the environment process is in state j and p_j is the stationary probability that the environment is in state j. The analysis is exactly the same as the case $a \leq x \leq b$, especially the definition of T in Equation 9.25 (of course the "or" is redundant) and the PDE in Equation 9.26. The only exception is that the boundary conditions would now be

$$H_{jj}(a, t) = 1 \quad \text{if } r(j) < c$$

$$H_{ij}(a, t) = 0 \quad \text{if } r(i) < c \text{ and } i \neq j$$

since the first passage time would be zero if we started at a in a state with negative drift. However, we cannot solve for all $a_{i,j}(w)$ values in Equation 9.32 using the given boundary conditions as there are not enough equations as unknowns. For that we use some additional conditions including $a_{i,j}(w) = 0$ if $S_i(w) > 0$ since as $x \to \infty$ we require the condition for $H_{ij}(x, t)$ to be a joint probability distribution. In addition, it is worthwhile noting that since a first passage time can only end in a state with negative drift

$$H_{ij}(x, t) = 0 \quad \text{if } r(j) > c$$

for any x. ∎

Remark 23

For the case $a = b \geq x$ as well the analysis is exactly the same as the case $a \leq x \leq b$, especially the definition of T in Equation 9.25 (again the "or" being redundant) and the PDE in Equation 9.26. The only exception is that the boundary conditions would now be

$$H_{jj}(b, t) = 1 \quad \text{if } r(j) > c$$

$$H_{ij}(b, t) = 0 \quad \text{if } r(i) > c \text{ and } i \neq j$$

since the first passage time would be zero if we started at b in a state with positive drift. However, we cannot solve for all $a_{i,j}(w)$ values in Equation 9.32 using the given boundary conditions as there are not enough equations as unknowns. For that notice that if the fluid level reached zero in state i (for that $r(i)$ must be less than c) then it stays at zero till the environment process changes state to some state $k \neq i$. Thus the first passage time is equal to the stay time in state i plus the remaining time from k till the first passage time starting in k. By conditioning on k and unconditioning, we can write down in LST format for all $i \in S$ such that $r(i) < c$

$$\tilde{H}_{ij}(0, w) = \sum_{k \in S, k \neq i} \tilde{H}_{kj}(0, w) \left(\frac{q_{ik}}{-q_{ii}} \right) \frac{q_{ik}}{q_{ik} + w}.$$

In addition, it is worthwhile noting that since a first passage time can only end in a state with positive drift

$$H_{ij}(x, t) = 0 \quad \text{if } r(j) < c$$

for any x. ∎

In the next section, we present some examples to illustrate the procedures developed here, to describe some closed-form solutions for special cases as well as applying to some problem instances that are extensions to those presented here.

9.3.2 Examples

To explain some of the nuances described in the previous section on first passage times, we consider a few examples here. They are presented in a problem–solution format.

Problem 90

Consider a reservoir from which water is emptied out at a constant rate of 10 units per day. Water flows into the reservoir according to a CTMC

$\{Z(t), t \geq 0\}$ with $S = \{1, 2, 3, 4, 5\}$ and

$$
Q = \begin{bmatrix}
-1 & 0.4 & 0.3 & 0.2 & 0.1 \\
0.4 & -0.7 & 0.1 & 0.1 & 0.1 \\
0.5 & 0.4 & -1.1 & 0.1 & 0.1 \\
0.2 & 0.3 & 0.3 & -1 & 0.2 \\
0.3 & 0.3 & 0.3 & 0.3 & -1.2
\end{bmatrix}.
$$

When $Z(t) = i$, the inflow rate is $4i$. It is considered nominal if there is between 20 and 40 units of water in the reservoir. However if the amount of water is over 40 units or below 20 units it is considered excessive or concerning, respectively. At time $t = 0$, there is 30 units of water and $Z(0) = 2$. How many days from $t = 0$ do you expect the water level to become excessive or concerning? What is the probability that at the end of a nominal spell the water level would be excessive?

Solution

Let $X(t)$ be the amount of water in the reservoir at time t with $X(0) = 30$ units. Let T be the first passage time to water levels of either $a = 20$ units or $b = 40$ units, whichever happens first. Based on the definition $H_{ij}(x, t) = P\{T \leq t, Z(T) = j | X(0) = x, Z(0) = i\}$, we can write down its LST with respect to t as

$$
\tilde{H}_{ij}(x, w) = \int_0^\infty e^{-wt} \frac{\partial H_{ij}(x, t)}{\partial t} dt.
$$

To compute the expected number of days from $t = 0$ for the water level to become excessive or concerning, all we need is

$$
E[T | X(0) = 30, Z(0) = 2] = (-1) \frac{d}{dw} \sum_{j=1}^{5} \tilde{H}_{2j}(30, w)
$$

at $w = 0$. For that we can compute $\frac{d}{dw} \tilde{H}_{ij}(x, w)$ at $w = 0$ by taking a very small $h > 0$ and obtaining $\frac{\tilde{H}_{ij}(x, h) - \tilde{H}_{ij}(x, 0)}{h}$. Before explaining how to obtain that, consider the other question, that is, the probability that at the end of a nominal spell the water level would be excessive. In other words, we need

$$
P\{Z(T) \in \{3, 4, 5\} | X(0) = 30, Z(0) = 2\} = \sum_{j=3}^{5} H_{2j}(30, \infty) = \sum_{j=3}^{5} \tilde{H}_{2j}(30, 0).
$$

Thus if we know for all i and j the values of $\tilde{H}_{ij}(x,h)$ for some small h and $\tilde{H}_{ij}(x,0)$, we can immediately compute both $E[T|X(0)=30, Z(0)=2]$ and $P\{Z(T) \in \{3,4,5\}|X(0)=30, Z(0)=2\}$.

To compute $\tilde{H}_{ij}(x,h)$ and $\tilde{H}_{ij}(x,0)$, we can write down from Equation 9.32, for $j=1,2,3,4,5$,

$$
\begin{bmatrix}
\tilde{H}_{1j}(x,w) \\
\tilde{H}_{2j}(x,w) \\
\tilde{H}_{3j}(x,w) \\
\tilde{H}_{4j}(x,w) \\
\tilde{H}_{5j}(x,w)
\end{bmatrix}
= a_{1,j}(w)e^{S_1(w)x}\phi_1(w)
$$
$$
+ a_{2,j}(w)e^{S_2(w)x}\phi_2(w)
$$
$$
+ \cdots + a_{5,j}(w)e^{S_5(w)x}\phi_5(w), \tag{9.34}
$$

where $a_{i,j}(w)$, $S_j(w)$, and $\phi_j(w)$ values need to be determined especially for $w=0$ and $w=h$ for some small h.

We can obtain $S_1(w), \ldots, S_5(w)$ as the five scalar solutions to the characteristic equation

$$
det(DS(w) - wI + Q) = 0.
$$

In particular, we get $S_1(0) = -0.4020$, $S_2(0) = 0.5317$, $S_3(0) = 0$, $S_4(0) = 0.0146$, and $S_5(0) = 0.1758$. When $h = 0.000001$, we have $S_3(h) = -3.0042 \times 10^{-6}$. However, the values of $S_j(h)$ for $j = 1, 2, 4, 5$ are too close to the respective $S_j(0)$ values and hence they are not reported here.

Then, for each $S_j(w)$ we can find column vectors $\phi_j(w)$ that satisfy

$$
S_j(w)D\phi_j(w) = (wI - Q)\phi_j(w).
$$

Thus we get

$$
\phi_1(0) = \begin{bmatrix} 0.3067 \\ -0.9393 \\ -0.1234 \\ -0.0783 \\ -0.0479 \end{bmatrix}, \quad
\phi_2(0) = \begin{bmatrix} -0.0675 \\ -0.0590 \\ -0.9832 \\ 0.1425 \\ 0.0705 \end{bmatrix}, \quad
\phi_3(0) = \begin{bmatrix} -0.4472 \\ -0.4472 \\ -0.4472 \\ -0.4472 \\ -0.4472 \end{bmatrix},
$$

$$
\phi_4(0) = \begin{bmatrix} 0.4064 \\ 0.4159 \\ 0.4359 \\ 0.4774 \\ 0.4939 \end{bmatrix}, \quad \text{and} \quad
\phi_5(0) = \begin{bmatrix} -0.0801 \\ -0.0637 \\ -0.1184 \\ -0.8042 \\ 0.5733 \end{bmatrix}.
$$

Here, the values of $\phi_j(h)$ for $j = 1, 2, 3, 4, 5$ and small h are too close to the respective $\phi_j(0)$ values and hence they are not reported.

To obtain the $a_{i,j}(w)$ values for $i = 1, 2, 3, 4, 5$, $j = 1, 2, 3, 4, 5$, and for two sets namely $w = 0$ and $w = h$, we solve the following 25 boundary condition equations for each w:

$$\tilde{H}_{jj}(40, w) = 1 \quad \text{for } j = 3, 4, 5 ,$$

$$\tilde{H}_{jj}(20, w) = 1 \quad \text{for } j = 1, 2 ,$$

$$\tilde{H}_{ij}(40, w) = 0 \quad \text{for } i = 3, 4, 5 \text{ and all } j \neq i ,$$

$$\tilde{H}_{ij}(20, w) = 0 \quad \text{for } i = 1, 2 \text{ and all } j \neq i.$$

Thus we get for $w = 0$,

$$
\begin{bmatrix}
a_{1,1}(0) & a_{2,1}(0) & a_{3,1}(0) & a_{4,1}(0) & a_{5,1}(0) \\
a_{1,2}(0) & a_{2,2}(0) & a_{3,2}(0) & a_{4,2}(0) & a_{5,2}(0) \\
a_{1,3}(0) & a_{2,3}(0) & a_{3,3}(0) & a_{4,3}(0) & a_{5,3}(0) \\
a_{1,4}(0) & a_{2,4}(0) & a_{3,4}(0) & a_{4,4}(0) & a_{5,4}(0) \\
a_{1,5}(0) & a_{2,5}(0) & a_{3,5}(0) & a_{4,5}(0) & a_{5,5}(0)
\end{bmatrix}
$$

$$
=
\begin{bmatrix}
-2416.2 & 0.1129 \times 10^{-9} & -4.5941 & -2.3770 & 0.0541 \times 10^{-3} \\
2516.0 & 0.0369 \times 10^{-9} & -1.5004 & -0.7763 & 0.0177 \times 10^{-3} \\
-9.2 & -0.5470 \times 10^{-9} & 0.3563 & 0.2910 & -0.0490 \times 10^{-3} \\
-36.2 & 0.2120 \times 10^{-9} & 1.4369 & 1.1714 & -0.6466 \times 10^{-3} \\
-54.4 & 0.1851 \times 10^{-9} & 2.0652 & 1.6909 & 0.6238 \times 10^{-3}
\end{bmatrix} .
$$

Also, for $w = h = 0.000001$, we get

$$
\begin{bmatrix}
a_{1,1}(h) & a_{2,1}(h) & a_{3,1}(h) & a_{4,1}(h) & a_{5,1}(h) \\
a_{1,2}(h) & a_{2,2}(h) & a_{3,2}(h) & a_{4,2}(h) & a_{5,2}(h) \\
a_{1,3}(h) & a_{2,3}(h) & a_{3,3}(h) & a_{4,3}(h) & a_{5,3}(h) \\
a_{1,4}(h) & a_{2,4}(h) & a_{3,4}(h) & a_{4,4}(h) & a_{5,4}(h) \\
a_{1,5}(h) & a_{2,5}(h) & a_{3,5}(h) & a_{4,5}(h) & a_{5,5}(h)
\end{bmatrix}
$$

$$
=
\begin{bmatrix}
-2416.2 & 0.1129 \times 10^{-9} & -4.5929 & -2.3757 & 0.0541 \times 10^{-3} \\
2516.0 & 0.0369 \times 10^{-9} & -1.5000 & -0.7759 & 0.0177 \times 10^{-3} \\
-9.2 & -0.5470 \times 10^{-9} & 0.3562 & 0.2909 & -0.0490 \times 10^{-3} \\
-36.2 & 0.2120 \times 10^{-9} & 1.4364 & 1.1709 & -0.6466 \times 10^{-3} \\
-54.4 & 0.1851 \times 10^{-9} & 2.0645 & 1.6901 & 0.6238 \times 10^{-3}
\end{bmatrix} .
$$

Now using $a_{i,j}(w)$, $S_j(w)$, and $\phi_j(w)$ values for $i = 1, 2, 3, 4, 5$, $j = 1, 2, 3, 4, 5$ at $w = 0$ and $w = h$ in Equation 9.34 we can compute $\tilde{H}_{ij}(x, w)$. In particular for $x = 30$ (which is what we need here) we get

$$
\begin{bmatrix}
\tilde{H}_{1,1}(30,0) & \tilde{H}_{1,2}(30,0) & \tilde{H}_{1,3}(30,0) & \tilde{H}_{1,4}(30,0) & \tilde{H}_{1,5}(30,0) \\
\tilde{H}_{2,1}(30,0) & \tilde{H}_{2,2}(30,0) & \tilde{H}_{2,3}(30,0) & \tilde{H}_{2,4}(30,0) & \tilde{H}_{2,5}(30,0) \\
\tilde{H}_{3,1}(30,0) & \tilde{H}_{3,2}(30,0) & \tilde{H}_{3,3}(30,0) & \tilde{H}_{3,4}(30,0) & \tilde{H}_{3,5}(30,0) \\
\tilde{H}_{4,1}(30,0) & \tilde{H}_{4,2}(30,0) & \tilde{H}_{4,3}(30,0) & \tilde{H}_{4,4}(30,0) & \tilde{H}_{4,5}(30,0) \\
\tilde{H}_{5,1}(30,0) & \tilde{H}_{5,2}(30,0) & \tilde{H}_{5,3}(30,0) & \tilde{H}_{5,4}(30,0) & \tilde{H}_{5,5}(30,0)
\end{bmatrix}
$$

$$
=
\begin{bmatrix}
0.5618 & 0.1776 & 0.0249 & 0.1047 & 0.1309 \\
0.5096 & 0.1844 & 0.0289 & 0.1197 & 0.1574 \\
0.4463 & 0.1481 & 0.0427 & 0.1612 & 0.2017 \\
0.2879 & 0.0955 & 0.0628 & 0.3251 & 0.2287 \\
0.2420 & 0.0799 & 0.0574 & 0.1811 & 0.4396
\end{bmatrix}.
$$

The values of $\tilde{H}_{ij}(30, h)$ for some small h do not differ in the first four significant digits from the corresponding $\tilde{H}_{ij}(30, 0)$ values for $i = 1, 2, 3, 4, 5$ and $j = 1, 2, 3, 4, 5$.

Thus we have the expected number of days from $t = 0$ (with initial water level $X(0) = 30$) for the water level to become excessive or concerning as

$$
E[T|X(0) = 30, Z(0) = 2] = (-1)\frac{d}{dw}\sum_{j=1}^{5}\tilde{H}_{2j}(30, w)|_{w=0}
$$

$$
= -\lim_{h \to 0}\sum_{j=1}^{5}\frac{\tilde{H}_{2j}(30, h) - \tilde{H}_{2j}(30, 0)}{h} \approx 5.4165
$$

days by using $h = 0.000001$. Likewise, the probability that at the end of a nominal spell the water level would be excessive (given the initial condition) is

$$
P\{Z(T) \in \{3, 4, 5\}|X(0) = 30, Z(0) = 2\}
$$

$$
= \sum_{j=3}^{5} H_{2j}(30, \infty) = \sum_{j=3}^{5}\tilde{H}_{2j}(30, 0) = 0.306. \qquad \blacksquare
$$

Next we consider a problem that leverages off the previous problem but uses the conditions in Remark 22. The objective is to provide a contrast against the previous problem, however, under a similar setting.

Problem 91

Consider the setting in Problem 90 where the first passage time ends with the water level becoming excessive and the environment in one of the three positive drift states 3, 4, or 5 with probabilities 0.0945, 0.3911, or 0.5144, respectively (these are the probabilities that the first passage time would end in states 3, 4, or 5 given that it ended with water level becoming excessive). Compute how long the water level will stay excessive before becoming nominal?

Solution

We let $t=0$ as the time when the water level just crossed over from nominal to excessive. Using the same notation as in Problem 90 for $X(t)$ and $Z(t)$, we have $X(0) = 40$, $P\{Z(0) = 3\} = 0.0945$, $P\{Z(0) = 4\} = 0.3911$, and $P\{Z(0) = 5\} = 0.5144$. Let T be the time when the water level crosses back to becoming nominal, that is,

$$T = \inf\{t > 0 : X(t) = 40\}.$$

To compute $E[T]$, we follow the analysis in Remark 22. We first require the stability condition to be satisfied. The steady-state probabilities for the environment process $\{Z(t), t \geq 0\}$ are $p_1 = 0.2725$, $p_2 = 0.3438$, $p_3 = 0.1652$, $p_4 = 0.1315$, and $p_5 = 0.0870$. Using the drift in state j, $d(j) = 4j - 10$ for $j = 1, 2, 3, 4, 5$, we have $\sum_{j=1}^{5} d(j)p_j = -0.3328$ which is less than zero and hence the system is stable. Now, to compute $E[T]$ we use $H_{ij}(x,t) = P\{T \leq t, Z(T) = j | X(0) = x, Z(0) = i\}$ in particular its LST with respect to t, $\tilde{H}_{ij}(x, w)$.

To compute $E[T]$, the expected number of days from $t=0$ for the water level to return to nominal values, we use

$$E[T] = (-1)\frac{d}{dw} \sum_{i=3}^{5} \sum_{j=1}^{5} \tilde{H}_{ij}(40, w) P\{Z(0) = i\}$$

at $w = 0$. To compute $\frac{d}{dw}\tilde{H}_{ij}(x, w)$ at $w = 0$ here too we consider a very small $h > 0$ and obtain it approximately as $\frac{\tilde{H}_{ij}(x,h) - \tilde{H}_{ij}(x,0)}{h}$. Now, to evaluate $\tilde{H}_{ij}(x, h)$ and $\tilde{H}_{ij}(x, 0)$, we can write down from Equation 9.32, for $j = 1, 2, 3, 4, 5$,

$$\begin{bmatrix} \tilde{H}_{1j}(x, w) \\ \tilde{H}_{2j}(x, w) \\ \tilde{H}_{3j}(x, w) \\ \tilde{H}_{4j}(x, w) \\ \tilde{H}_{5j}(x, w) \end{bmatrix} = a_{1,j}(w)e^{S_1(w)x}\phi_1(w)$$

$$+ a_{2,j}(w)e^{S_2(w)x}\phi_2(w) + \cdots + a_{5,j}(w)e^{S_5(w)x}\phi_5(w), \quad (9.35)$$

where $a_{i,j}(w)$, $S_j(w)$, and $\phi_j(w)$ values need to be determined for $w=0$ and $w=h$ for some small h.

We can obtain $S_j(w)$ for $j=1,2,3,4,5$ as the scalar solutions to the characteristic equation

$$det(DS(w) - wI + Q) = 0.$$

But this is identical to that in Problem 90. Likewise $\phi_j(w)$ can be computed as the column vectors that satisfy

$$S_j(w)D\phi_j(w) = (wI - Q)\phi_j(w)$$

which is also identical to that in Problem 90. Thus refer to Problem 90 for $\phi_j(w)$ and $S_j(w)$ for $j=1,2,3,4,5$ at $w=0$ and $w=h$. What remains in Equation 9.35 are the $a_{i,j}(w)$ values for $w=0$ and $w=h$. For that refer back to the approach in Remark 22. First of all

$$\begin{bmatrix} \tilde{H}_{1j}(x,w) \\ \tilde{H}_{2j}(x,w) \\ \tilde{H}_{3j}(x,w) \\ \tilde{H}_{4j}(x,w) \\ \tilde{H}_{5j}(x,w) \end{bmatrix} = \begin{bmatrix} 0 \\ 0 \\ 0 \\ 0 \\ 0 \end{bmatrix}$$

for $j=3,4,5$ since the first passage time can never end in states $3,4$, or 5 as the drift is positive in those states (only when the drift is negative, it is possible to cross over into a particular buffer content level from above). Thus we need only $a_{i,j}(w)$ values for $i=1,2,3,4,5$ and $j=1,2$. But $a_{i,j}(w)=0$ for all i where $S_i(w) > 0$, otherwise as $x \to \infty$, the expression for $\tilde{H}_{ij}(x,w)$ in Equation 9.35 would blow up. Hence, we have $a_{2,j}(w)=0$, $a_{4,j}(w)=0$, and $a_{5,j}(w)=0$ for $j=1,2$. Thus all we are left with is to obtain $a_{1,1}(w)$, $a_{1,2}(w)$, $a_{3,1}(w)$, and $a_{3,2}(w)$. For that we have four boundary conditions, namely,

$$\tilde{H}_{11}(40,\ w) = 1,$$

$$\tilde{H}_{22}(40,\ w) = 1,$$

$$\tilde{H}_{12}(40,\ w) = 0, \quad \text{and}$$

$$\tilde{H}_{21}(40,\ w) = 0.$$

Thus we have

$$
\begin{bmatrix}
\tilde{H}_{11}(40,w) \\
\tilde{H}_{21}(40,w) \\
\tilde{H}_{31}(40,w) \\
\tilde{H}_{41}(40,w) \\
\tilde{H}_{51}(40,w)
\end{bmatrix}
= a_{1,1}(w)e^{40S_1(w)}\phi_1(w) + a_{3,1}(w)e^{40S_3(w)}\phi_3(w) =
\begin{bmatrix}
1 \\
0 \\
\cdot \\
\cdot \\
\cdot
\end{bmatrix}
$$

and

$$
\begin{bmatrix}
\tilde{H}_{12}(40,w) \\
\tilde{H}_{22}(40,w) \\
\tilde{H}_{32}(40,w) \\
\tilde{H}_{42}(40,w) \\
\tilde{H}_{52}(40,w)
\end{bmatrix}
= a_{1,2}(w)e^{40S_1(w)}\phi_1(w) + a_{3,2}(w)e^{40S_3(w)}\phi_3(w) =
\begin{bmatrix}
0 \\
1 \\
\cdot \\
\cdot \\
\cdot
\end{bmatrix}
$$

where the $\vdots$ in the column vector denotes unknown quantities. Once we know $a_{1,1}(w)$, $a_{1,2}(w)$, $a_{3,1}(w)$, and $a_{3,2}(w)$, the unknown quantities can be obtained. Solving for the four equations at $w=0$ and $w=h=0.000001$ we get $a_{1,1}(0)=7.7343 \times 10^6$, $a_{3,1}(0)=-1.6856$, $a_{1,2}(0)=-7.7343 \times 10^6$, and $a_{3,2}(0)=-0.5505$; also $a_{1,1}(h)=7.7344 \times 10^6$, $a_{3,1}(h)=-1.6858$, $a_{1,2}(h)=-7.7344 \times 10^6$, and $a_{3,2}(h)=-0.5505$.

Now using $a_{i,j}(w)$, $S_i(w)$, and $\phi_i(w)$ values for $i=1,3$, $j=1,2$ at $w=0$ and $w=h$ in Equation 9.35 we can compute $\tilde{H}_{ij}(x,w)$. In particular for $x=40$ (which is what we need here) we get

$$
\begin{bmatrix}
\tilde{H}_{11}(40,0) & \tilde{H}_{12}(40,0) \\
\tilde{H}_{21}(40,0) & \tilde{H}_{22}(40,0) \\
\tilde{H}_{31}(40,0) & \tilde{H}_{32}(40,0) \\
\tilde{H}_{41}(40,0) & \tilde{H}_{42}(40,0) \\
\tilde{H}_{51}(40,0) & \tilde{H}_{52}(40,0)
\end{bmatrix}
=
\begin{bmatrix}
1 & 0 \\
0 & 1 \\
0.6548 & 0.3452 \\
0.6910 & 0.3090 \\
0.7153 & 0.2847
\end{bmatrix}.
$$

The values of $\tilde{H}_{ij}(40,h)$ for some small h do not differ in the first four significant digits from the corresponding $\tilde{H}_{ij}(40,0)$ values for $i=1,2,3,4,5$ and $j=1,2$, hence they are not reported.

Thus we have the expected number of days from $t=0$ (with initial water level $X(0)=40$ as well as initial environmental conditions $P\{Z(0)=3\}=0.0945$, $P\{Z(0)=4\}=0.3911$, and $P\{Z(0)=5\}=0.5144$) for the water level to become nominal as

$$E[T] = (-1)\frac{d}{dw}\sum_{i=3}^{5}\sum_{j=1}^{2}\tilde{H}_{ij}(40,w)|_{w=0}P\{Z(0) = i\}$$

$$= -\lim_{h\to 0}\sum_{i=3}^{5}\sum_{j=1}^{2}\frac{\tilde{H}_{ij}(40,h) - \tilde{H}_{ij}(40,0)}{h}P\{Z(0) = i\}$$

which is approximately 32.7935 days by using $h=0.000001$.   ∎

Having considered the case in Remark 22, next we solve a problem that uses the conditions in Remark 23. It is worthwhile contrasting against the previous two problems considering they are under a similar setting.

Problem 92

Consider the setting in Problem 90 with the exception that at time $t=0$ we just enter the concerning level and $Z(0)=1$. What is the expected sojourn time for the water level to stay at the concerning level before moving to nominal?

Solution

At $t=0$ water level just crosses over from nominal to concerning. Using the same notation as in Problem 90 for $X(t)$ and $Z(t)$, we have $X(0)=20$ and $Z(0)=1$. Let T be the time when the water level crosses back to becoming nominal from concerning, that is,

$$T = \inf\{t > 0 : X(t) = 20\}.$$

To compute $E[T]$, we follow the analysis in Remark 23 and use $H_{ij}(x,t) = P\{T \le t, Z(T) = j | X(0) = x, Z(0) = i\}$ in particular its LST with respect to t, $\tilde{H}_{ij}(x,w)$. To obtain $E[T]$, the expected number of days from $t=0$ for the water level to return to nominal values, we use

$$E[T] = (-1)\frac{d}{dw}\sum_{j=1}^{5}\tilde{H}_{1j}(20,w)$$

at $w=0$. To compute $\frac{d}{dw}\tilde{H}_{ij}(x,w)$ at $w=0$, here too we consider a very small $h>0$ and obtain it approximately as $\frac{\tilde{H}_{ij}(x,h)-\tilde{H}_{ij}(x,0)}{h}$. Now, to evaluate $\tilde{H}_{ij}(x,h)$

and $\tilde{H}_{ij}(x,0)$, we can write down from Equation 9.32, for $j=1,2,3,4,5$,

$$
\begin{bmatrix}
\tilde{H}_{1j}(x,w) \\
\tilde{H}_{2j}(x,w) \\
\tilde{H}_{3j}(x,w) \\
\tilde{H}_{4j}(x,w) \\
\tilde{H}_{5j}(x,w)
\end{bmatrix}
= a_{1,j}(w)e^{S_1(w)x}\phi_1(w)
$$

$$
+ a_{2,j}(w)e^{S_2(w)x}\phi_2(w) + \cdots + a_{5,j}(w)e^{S_5(w)x}\phi_5(w), \quad (9.36)
$$

where $a_{i,j}(w)$, $S_j(w)$, and $\phi_j(w)$ values need to be determined for $w=0$ and $w=h$ for some small h.

We can obtain $S_j(w)$ for $j=1,2,3,4,5$ as the scalar solutions to the characteristic equation

$$
det(DS(w) - wI + Q) = 0.
$$

But this is identical to that in Problem 90. Likewise $\phi_j(w)$ can be computed as the column vectors that satisfy

$$
S_j(w)D\phi_j(w) = (wI - Q)\phi_j(w)
$$

which is also identical to that in Problem 90. Thus refer to Problem 90 for $\phi_j(w)$ and $S_j(w)$ for $j=1,2,3,4,5$ at $w=0$ and $w=h$. What remains in Equation 9.35 are the $a_{i,j}(w)$ values for $w=0$ and $w=h$. For that refer back to the approach in Remark 23. First of all

$$
\begin{bmatrix}
\tilde{H}_{1j}(x,w) \\
\tilde{H}_{2j}(x,w) \\
\tilde{H}_{3j}(x,w) \\
\tilde{H}_{4j}(x,w) \\
\tilde{H}_{5j}(x,w)
\end{bmatrix}
=
\begin{bmatrix}
0 \\
0 \\
0 \\
0 \\
0
\end{bmatrix}
$$

for $j=1,2$ since the first passage time can never end in states 1 or 2 as the drift is negative in those states (only when the drift is positive is it possible to cross over into a particular buffer content level from below). Thus we need only $a_{i,j}(w)$ values for $i=1,2,3,4,5$ and $j=3,4,5$.

Of those 15 unknown $a_{i,j}(w)$ values, 9 can be obtained through the following boundary conditions:

$$\tilde{H}_{33}(20, w) = 1, \tilde{H}_{34}(20, w) = 0, \tilde{H}_{35}(20, w) = 0,$$

$$\tilde{H}_{44}(20, w) = 1, \tilde{H}_{43}(20, w) = 0, \tilde{H}_{45}(20, w) = 0,$$

$$\tilde{H}_{55}(20, w) = 1, \tilde{H}_{53}(20, w) = 0, \quad \text{and} \quad \tilde{H}_{54}(20, w) = 0.$$

For the remaining six unknowns we use for $j = 3, 4, 5$

$$\tilde{H}_{1j}(0, w) = \sum_{k=2}^{5} \tilde{H}_{kj}(0, w) \left(\frac{q_{1k}}{-q_{11}} \right) \frac{q_{1k}}{q_{1k} + w}$$

$$\tilde{H}_{2j}(0, w) = \tilde{H}_{1j}(0, w) \left(\frac{q_{21}}{-q_{22}} \right) \frac{q_{21}}{q_{21} + w} + \sum_{k=3}^{5} \tilde{H}_{kj}(0, w) \left(\frac{q_{2k}}{-q_{22}} \right) \frac{q_{2k}}{q_{2k} + w},$$

where q_{ij} corresponds to the element in the ith row and jth column of Q. Solving the 15 equations we get for $w = 0$,

$$
\begin{bmatrix}
a_{1,3}(0) & a_{2,3}(0) & a_{3,3}(0) & a_{4,3}(0) & a_{5,3}(0) \\
a_{1,4}(0) & a_{2,4}(0) & a_{3,4}(0) & a_{4,4}(0) & a_{5,4}(0) \\
a_{1,5}(0) & a_{2,5}(0) & a_{3,5}(0) & a_{4,5}(0) & a_{5,5}(0)
\end{bmatrix}
$$

$$
=
\begin{bmatrix}
0.0093 \times 10^{-3} & -0.2211 \times 10^{-4} & -0.2109 & -0.0033 & -0.0014 \\
0.1326 \times 10^{-3} & 0.1117 \times 10^{-4} & -0.8945 & -0.0466 & -0.0208 \\
-0.1419 \times 10^{-3} & 0.1094 \times 10^{-4} & -1.1307 & 0.0500 & 0.0223
\end{bmatrix}.
$$

Also, for $w = h = 0.000001$, the values of $a_{i,j}(h)$ are the same as that when $w = 0$ to the first few significant digits. Hence we do not present that here.

Now using $a_{i,j}(w)$, $S_i(w)$, and $\phi_i(w)$ values for $i = 1, 2, 3, 4, 5$ and $j = 3, 4, 5$ at $w = 0$ and $w = h$ in Equation 9.36 we can compute $\tilde{H}_{ij}(x, w)$. In particular for $x = 20$ (which is what we need here) we get

$$
\begin{bmatrix}
\tilde{H}_{13}(20,0) & \tilde{H}_{14}(20,0) & \tilde{H}_{15}(20,0) \\
\tilde{H}_{23}(20,0) & \tilde{H}_{24}(20,0) & \tilde{H}_{25}(20,0) \\
\tilde{H}_{33}(20,0) & \tilde{H}_{34}(20,0) & \tilde{H}_{35}(20,0) \\
\tilde{H}_{43}(20,0) & \tilde{H}_{44}(20,0) & \tilde{H}_{45}(20,0) \\
\tilde{H}_{53}(20,0) & \tilde{H}_{54}(20,0) & \tilde{H}_{55}(20,0)
\end{bmatrix}
=
\begin{bmatrix}
0.1583 & 0.3995 & 0.4422 \\
0.1497 & 0.3913 & 0.4590 \\
1 & 0 & 0 \\
0 & 1 & 0 \\
0 & 0 & 1
\end{bmatrix}.
$$

The values of $\tilde{H}_{ij}(20, h)$ for some small h do not differ in the first four significant digits from the corresponding $\tilde{H}_{ij}(20, 0)$ values for $i = 1, 2, 3, 4, 5$ and $j = 3, 4, 5$, hence they are not reported.

Thus we have the expected number of days from $t = 0$ (with initial water level $X(0) = 20$ as well as initial environmental condition $Z(0) = 1$) for the water level to become nominal as

$$E[T] = (-1)\frac{d}{dw}\sum_{j=3}^{5}\tilde{H}_{1j}(20, w)|_{w=0} = -\lim_{h \to 0}\sum_{j=3}^{5}\frac{\tilde{H}_{1j}(20, h) - \tilde{H}_{1j}(20, 0)}{h}$$

which is approximately 27.602 days by using $h = 0.000001$.   ■

In fact in the previous problem we can also immediately write down the time spent in critical state as 25.1175 days if we were to start in state 2 (instead of 1 in the previous problem). Having seen a set of numerical problems, we next focus on some exponential on–off source cases where we can obtain closed-form algebraic expressions.

Problem 93

Consider an exponential on–off source that inputs fluid into a buffer. The on times are according to $\exp(\alpha)$ and off times according to $\exp(\beta)$. When the source is on, fluid enters the buffer at rate r and no fluid enters the buffer when the source is off. The output capacity is c. Assume that initially there is x amount of fluid in the buffer. Define the first passage time as the time it would take for the buffer contents to reach level x^* or 0, whichever happens first with $x^* \geq x \geq 0$. Let states 1 and 2 represent the source being off and on, respectively. For $i = 1, 2$, find the LST of the first passage time given that the environment is initially in state i. Also for $i = 1, 2$, find the probability that the first passage time occurs with x^* or 0 amount of fluid, given that initially the environment is in state i.

Solution

The setting is identical to that in Problem 87. Recall that the environment process $\{Z(t), t \geq 0\}$ is a CTMC with

$$Q = \begin{bmatrix} -\beta & \beta \\ \alpha & -\alpha \end{bmatrix} \quad \text{and} \quad D = \begin{bmatrix} -c & 0 \\ 0 & r - c \end{bmatrix}.$$

Next we use the definition of $H_{ij}(x, t)$ and its LST $\tilde{H}_{ij}(x, w)$ in Section 9.3 for $i = 1, 2$ and $j = 1, 2$. Then the solution to Equation 9.31 is given by

$$\begin{bmatrix} \tilde{H}_{11}(x,w) \\ \tilde{H}_{21}(x,w) \end{bmatrix} = a_{11}(w)e^{S_1(w)x}\phi_1(w) + a_{21}(w)e^{S_2(w)x}\phi_2(w),$$

$$\begin{bmatrix} \tilde{H}_{12}(x,w) \\ \tilde{H}_{22}(x,w) \end{bmatrix} = a_{12}(w)e^{S_1(w)x}\phi_1(w) + a_{22}(w)e^{S_2(w)x}\phi_2(w).$$

For $i=1,2$ and $j=1,2$ we now obtain closed-form expressions for $S_i(w)$, $\phi_i(w)$, and $a_{ij}(w)$ for any w. To obtain $S_1(w)$ and $S_2(w)$, we solve characteristic equation

$$det(DS(w) - wI + Q) = 0.$$

The two roots of the characteristic equation yield

$$S_1(w) = \frac{-\hat{b} - \sqrt{\hat{b}^2 + 4w(w + \alpha + \beta)c(r - c)}}{2c(r - c)},$$

$$S_2(w) = \frac{-\hat{b} + \sqrt{\hat{b}^2 + 4w(w + \alpha + \beta)c(r - c)}}{2c(r - c)},$$

where $\hat{b} = (r - 2c)w + (r - c)\beta - c\alpha$.

Then, for each $S_j(w)$ such that $j=1,2$, we can find column vectors $\phi_j(w)$ that satisfy

$$S_j(w)D\phi_j(w) = (wI - Q)\phi_j(w).$$

For $i = 1,2$ we have

$$\phi_i(w) = \begin{bmatrix} \frac{w+\alpha-S_i(w)(r-c)}{\alpha} \\ 1 \end{bmatrix} = \begin{bmatrix} \frac{\beta}{w+\beta+S_i(w)c} \\ 1 \end{bmatrix}.$$

For convenience we write down in terms of $\psi_i(w)$ for $i=1,2$ so that

$$\psi_i(w) = \frac{\beta}{w + \beta + S_i(w)c}$$

and thus

$$\phi_i(w) = \begin{bmatrix} \psi_i(w) \\ 1 \end{bmatrix}.$$

Finally, solving for $a_{11}(w)$, $a_{21}(w)$, $a_{12}(w)$, and $a_{22}(w)$ using the LST of the boundary conditions $\tilde{H}_{22}(x^*, w) = 1$, $\tilde{H}_{11}(0, w) = 1$, $\tilde{H}_{21}(x^*, w) = 0$, and $\tilde{H}_{12}(0, w) = 0$, resulting in

$$a_{11}(w) = e^{S_2(w)x^*}/\delta(w),$$

$$a_{12}(w) = -\psi_2(w)/\delta(w),$$

$$a_{21}(w) = -e^{S_1(w)x^*}/\delta(w),$$

$$a_{22}(w) = \psi_1(w)/\delta(w), \tag{9.37}$$

where $\delta(w) = e^{S_2(w)x^*}\psi_1(w) - e^{S_1(w)x^*}\psi_2(w)$.

Now that we have expressions for $\tilde{H}_{ij}(x, w)$ for $i = 1, 2$ and $j = 1, 2$, the LST of the first passage time given that the environment is initially in state 1 (i.e., off) can be computed as

$$\tilde{H}_{11}(x, w) + \tilde{H}_{12}(x, w) = (a_{11}(w) + a_{12}(w))e^{S_1(w)x}\psi_1(w)$$
$$+ (a_{21}(w) + a_{22}(w))e^{S_2(w)x}\psi_2(w).$$

Likewise, the LST of the first passage time given that the environment is initially in state 2 (i.e., on) can be computed as

$$\tilde{H}_{21}(x, w) + \tilde{H}_{22}(x, w) = (a_{11}(w) + a_{12}(w))e^{S_1(w)x} + (a_{21}(w) + a_{22}(w))e^{S_2(w)x}.$$

Also for $i = 1, 2$, the probability that the first passage time occurs with 0 amount of fluid, given that initially the environment is in state i is $\tilde{H}_{i1}(x, 0)$. Likewise for $i = 1, 2$, the probability that the first passage time occurs with x^* amount of fluid, given that initially the environment is in state i is $\tilde{H}_{i2}(x, 0)$. To compute $\tilde{H}_{ij}(x, 0)$ for $i = 1, 2$ and $j = 1, 2$ when we let $w = 0$, we need to be careful about the fact that $\sqrt{\hat{b}^2} = |\hat{b}|$. Notice that if $r\beta < c(\alpha + \beta)$, then $\hat{b} < 0$, otherwise $\hat{b} \geq 0$.

Assume that $r\beta < c(\alpha + \beta)$, which would be necessary if we require the queue to be stable (note that it is straightforward to write down the expressions to follow even for the case $r\beta \geq c(\alpha + \beta)$ but is not presented here). Continuing with the assumption that $r\beta < c(\alpha + \beta)$, we can see by letting $w = 0$ that

$$S_1(0) = 0 \quad \text{and} \quad S_2(0) = \frac{c\alpha - \beta(r - c)}{c(r - c)}.$$

Hence, we have $\psi_1(0) = 1$ and $\psi_2(0) = \frac{\beta(r-c)}{c\alpha}$. Finally, using $\delta(0) = e^{S_2(0)x^*}\psi_1(0) - e^{S_1(0)x^*}\psi_2(0)$,

$$a_{11}(0) = e^{S_2(0)x^*}/\delta(0) = \frac{e^{S_2(0)x^*}}{e^{S_2(0)x^*} - \frac{\beta(r-c)}{c\alpha}},$$

$$a_{12}(0) = -\psi_2(0)/\delta(0) = \frac{\frac{-\beta(r-c)}{c\alpha}}{e^{S_2(0)x^*} - \frac{\beta(r-c)}{c\alpha}},$$

$$a_{21}(0) = -e^{S_1(0)x^*}/\delta(0) = \frac{-1}{e^{S_2(0)x^*} - \frac{\beta(r-c)}{c\alpha}},$$

$$a_{22}(0) = \psi_1(0)/\delta(0) = \frac{1}{e^{S_2(0)x^*} - \frac{\beta(r-c)}{c\alpha}}.$$

Hence, we can compute for $i = 1, 2$ and $j = 1, 2$ the values of $\tilde{H}_{ij}(x, 0)$ as

$$\tilde{H}_{11}(x,0) = \left(e^{S_2(0)x^*} - e^{S_2(0)x}\frac{\beta(r-c)}{c\alpha}\right) \frac{1}{e^{S_2(0)x^*} - \frac{\beta(r-c)}{c\alpha}},$$

$$\tilde{H}_{21}(x,0) = \left(e^{S_2(0)x^*} - e^{S_2(0)x}\right) \frac{1}{e^{S_2(0)x^*} - \frac{\beta(r-c)}{c\alpha}},$$

$$\tilde{H}_{12}(x,0) = \left(e^{S_2(0)x} - 1\right)\frac{\beta(r-c)}{c\alpha} \frac{1}{e^{S_2(0)x^*} - \frac{\beta(r-c)}{c\alpha}},$$

$$\tilde{H}_{22}(x,0) = \left(e^{S_2(0)x} - \frac{\beta(r-c)}{c\alpha}\right) \frac{1}{e^{S_2(0)x^*} - \frac{\beta(r-c)}{c\alpha}}. \qquad \blacksquare$$

In the next example, we continue with the setting of the previous example, however, with the restriction of the first passage time occurring only when the amount of fluid in the buffer becomes empty.

Problem 94

Consider an exponential on–off source that inputs fluid into an infinite-sized buffer. The on times are according to $\exp(\alpha)$ and off times according to $\exp(\beta)$. When the source is on, fluid enters the buffer at rate r and no fluid enters the buffer when the source is off. The output capacity is c. Assume that the system is stable. Define the first passage time as the time it would take for the buffer contents to empty for the first time given that initially there is x amount of fluid in the buffer and the source is in state i, for $i = 1$ and 2 representing off and on, respectively. Then using that result derive the LST of the busy period distribution, that is, the consecutive period of time there is nonzero fluid in the buffer.

Solution

Notice that the setting is identical to that of Remark 22 where $a = b \leq x$ with $a = b = 0$. Also, the stability condition is that $r\beta < c(\alpha + \beta)$. Following the procedure in Remark 22, define T as the first time the amount of fluid in the buffer reaches 0 and thereby $H_{ij}(x, t) = P\{T \leq t, Z(T) = j | X(0) = x, Z(0) = i\}$. Using the results in Problem 93 we can write down $\tilde{H}_{ij}(x, w)$, the LST of $H_{ij}(x, t)$ as follows:

$$\left[\begin{array}{c} \tilde{H}_{11}(x, w) \\ \tilde{H}_{21}(x, w) \end{array} \right] = a_{11}(w)e^{S_1(w)x}\phi_1(w) + a_{21}(w)e^{S_2(w)x}\phi_2(w),$$

with $\tilde{H}_{12}(x, w) = \tilde{H}_{22}(x, w) = 0$. Solving the characteristic equation

$$det(DS(w) - wI + Q) = 0,$$

we can obtain $S_1(w)$ and $S_2(w)$ as

$$S_1(w) = \frac{-\hat{b} - \sqrt{\hat{b}^2 + 4w(w + \alpha + \beta)c(r - c)}}{2c(r - c)},$$

$$S_2(w) = \frac{-\hat{b} + \sqrt{\hat{b}^2 + 4w(w + \alpha + \beta)c(r - c)}}{2c(r - c)},$$

where $\hat{b} = (r - 2c)w + (r - c)\beta - c\alpha$. Also the column vectors $\phi_j(w)$ that satisfy

$$S_j(w)D\phi_j(w) = (wI - Q)\phi_j(w)$$

are for $i = 1, 2$ are

$$\phi_i(w) = \left[\begin{array}{c} \frac{w + \alpha - S_i(w)(r - c)}{\alpha} \\ 1 \end{array} \right] = \left[\begin{array}{c} \frac{\beta}{w + \beta + S_i(w)c} \\ 1 \end{array} \right].$$

To solve for $a_{11}(w)$ and $a_{21}(w)$, we use the LST of the boundary conditions $\tilde{H}_{11}(0, w) = 1$ and $\tilde{H}_{12}(0, w) = 0$. Of course, $\tilde{H}_{12}(0, w) = 0$ any way, hence that boundary condition is not useful. We use the additional condition that $a_{ij}(w) = 0$ if $S_i(w) > 0$ for $i = 1, 2$ and $j = 1, 2$. Since $S_1(w) \leq 0$ and $S_2(w) > 0$, we have $a_{21}(w) = 0$. Thus the only term that is nonzero is $a_{11}(w)$ which is given by

$$a_{11}(w) = \frac{w + \beta + S_1(w)c}{\beta}$$

since $\tilde{H}_{11}(0, w) = 1$. Thus we have

$$\begin{bmatrix} \tilde{H}_{11}(x, w) \\ \tilde{H}_{21}(x, w) \end{bmatrix} = e^{S_1(w)x} \begin{bmatrix} 1 \\ \frac{w + \beta + S_1(w)c}{\beta} \end{bmatrix}.$$

Now we compute the LST of the busy period distribution. Notice that a busy period begins with the environment in state 2 with zero fluid in the buffer (i.e., $x = 0$) and ends when the environment is in state 1. Hence the LST of the busy period distribution is $\tilde{H}_{21}(0, w) = \frac{w + \beta + S_1(w)c}{\beta}$. ∎

Next we consider the case where we start with x amount of fluid and find the distribution for the time it would take for the buffer contents to reach x^* with $x^* \geq x$.

Problem 95

Consider an on–off source with on times according to $\exp(\alpha)$ and off times according to $\exp(\beta)$. When the source is on, fluid enters the buffer at rate r and no fluid enters the buffer when the source is off. The output capacity is c. Define the first passage time as the time it would take for the buffer contents to reach a level x^* for the first time given that initially there is x amount of fluid in the buffer (such that $x \leq x^*$) and the source is in state i, for $i = 1$ and 2 representing off and on, respectively. Derive the LST of the first passage time.

Solution

This setting is identical to that of Remark 23 where $a = b \geq x$ with $a = b = x^*$. We define T as the first time the amount of fluid in the buffer reaches x^* and thereby $H_{ij}(x, t) = P\{T \leq t, Z(T) = j | X(0) = x, Z(0) = i\}$. Using the results in Problem 93 we can solve

$$\det(DS(w) - wI + Q) = 0,$$

to obtain $S_1(w)$ and $S_2(w)$ as

$$S_1(w) = \frac{-\hat{b} - \sqrt{\hat{b}^2 + 4w(w + \alpha + \beta)c(r - c)}}{2c(r - c)},$$

$$S_2(w) = \frac{-\hat{b} + \sqrt{\hat{b}^2 + 4w(w + \alpha + \beta)c(r - c)}}{2c(r - c)},$$

where $\hat{b} = (r - 2c)w + (r - c)\beta - c\alpha$. Also since the column vectors $\phi_j(w)$ must satisfy

$$S_j(w)D\phi_j(w) = (wI - Q)\phi_j(w)$$

they are for $i = 1, 2$ as follows:

$$\phi_i(w) = \begin{bmatrix} \frac{w + \alpha - S_i(w)(r-c)}{\alpha} \\ 1 \end{bmatrix} = \begin{bmatrix} \frac{\beta}{w + \beta + S_i(w)c} \\ 1 \end{bmatrix}.$$

Thus we can write down $\tilde{H}_{ij}(x, t)$, the LST of $H_{ij}(x, t)$ as $\tilde{H}_{11}(x, w) = \tilde{H}_{21}(x, w) = 0$, and

$$\begin{bmatrix} \tilde{H}_{12}(x, w) \\ \tilde{H}_{22}(x, w) \end{bmatrix} = a_{12}(w)e^{S_1(w)x}\phi_1(w) + a_{22}(w)e^{S_2(w)x}\phi_2(w).$$

To solve for $a_{i2}(w)$ for $i = 1, 2$, we use the LST of the boundary conditions $\tilde{H}_{22}(x^*, w) = 1$ and $\tilde{H}_{21}(x^*, w) = 0$. Of course, $\tilde{H}_{21}(x, w) = 0$ any way, hence that boundary condition is not useful. We use the additional condition that

$$\tilde{H}_{12}(0, w) = \tilde{H}_{22}(0, w)\frac{\beta}{\beta + w}.$$

These together yield the following:

$$a_{12}(w)e^{S_1(w)x^*} + a_{22}(w)e^{S_2(w)x^*} = 1 \text{ (since } \tilde{H}_{22}(x^*, w) = 1),$$

$$\frac{\beta}{\beta + w}(a_{12}(w) + a_{22}(w)) = a_{12}(w)\frac{\beta}{w + \beta + S_1(w)c}$$

$$+ a_{22}(w)\frac{\beta}{w + \beta + S_2(w)c}$$

$$\text{(since } \tilde{H}_{12}(0, w) = \tilde{H}_{22}(0, w)\tfrac{\beta}{\beta + w}).$$

Solving for $a_{12}(w)$ and $a_{22}(w)$, we get

$$a_{12}(w) = \left[e^{S_1(w)x^*} - e^{S_2(w)x^*} \left\{ \frac{S_1(w)}{S_2(w)} \left(\frac{w + \beta + S_2(w)c}{w + \beta + S_1(w)c} \right) \right\} \right]^{-1},$$

$$a_{22}(w) = \left[e^{S_2(w)x^*} - e^{S_1(w)x^*} \left\{ \frac{S_2(w)}{S_1(w)} \left(\frac{w + \beta + S_1(w)c}{w + \beta + S_2(w)c} \right) \right\} \right]^{-1}.$$

Thus the LST of the first passage time $\tilde{H}_{ij}(x,t)$ is given by $\tilde{H}_{11}(x,w) = \tilde{H}_{21}(x,w) = 0$, and

$$\left[\begin{array}{c} \tilde{H}_{12}(x,w) \\ \tilde{H}_{22}(x,w) \end{array} \right] = a_{12}(w)e^{S_1(w)x}\phi_1(w) + a_{22}(w)e^{S_2(w)x}\phi_2(w),$$

where $a_{12}(w)$, $a_{22}(w)$, $\phi_1(w)$, $\phi_2(w)$, $S_1(w)$, and $S_2(w)$ are described earlier. ∎

Next we present some examples that are extensions of standard single buffer fluid models that can be solved using the techniques derived in the previous examples. First we present the fluid equivalent to the M-policy (also called m-policy) that is popular in discrete queues. Essentially in discrete queues the M-policy is one where a server starts serving a queue as soon as there are M customers in the system; it serves it until the queue is empty. Then it stops serving until the number of customers become M, and the cycle continues. We now present the fluid version of that. Such mechanisms are common in computer systems where the processor switches between high- and low-priority jobs.

Problem 96

Consider an infinite-sized buffer with fluid input according to an on–off source that has on and off times exponentially distributed with parameters α and β, respectively. When the source is on fluid flows in at rate r and no fluid flows in when off. As soon as the buffer content reaches level a, fluid is removed from the buffer at rate c. When the buffer becomes empty, the output valve is shut. It remains shut until the buffer content reaches a. In other words the output also behaves like an alternating on–off sink. Assume that $\frac{r\beta}{\alpha + \beta} < c < r$. Obtain LSTs of the consecutive time the buffer is drained at rate c as well as the time the buffer takes to reach a. Also derive the expected on and off times for the output valve or sink.

Solution

Let T_1 be the time for the output channel to empty the contents in the buffer starting with a. Also let T_2 be the time for the buffer contents to rise from 0 to a during which there is no output. Mathematically we denote

$$T_1 = \inf\{t > 0 \ : \ X(t) = 0 | X(0) = a\}$$

and

$$T_2 = \inf\{t > 0 \ : \ X(t) = a | X(0) = 0\}.$$

First we obtain the distribution of T_1 and then that of T_2. Let $O_1(t)$ be the CDF of the random variable T_1 such that

$$O_1(t) = P\{T_1 \leq t\}.$$

Define $\tilde{O}_1(w)$ as the LST of $O_1(t)$ such that

$$\tilde{O}_1(w) = E[e^{-wT_1}].$$

Due to the definition of T_1, the source is "on" initially with a amount of fluid in the buffer, so that T_1 is the first passage time to reach zero amount of fluid in the buffer. For that we can directly substitute for expressions in Problem 94 to obtain the results. The LST $\tilde{O}_1(w)$ is

$$\tilde{O}_1(w) = \begin{cases} \frac{w+\beta+cs_0(w)}{\beta} e^{a\, s_0(w)} & \text{if } w \geq w^* \\ \infty & \text{otherwise,} \end{cases} \tag{9.38}$$

where

$$w^* = \left(2\sqrt{c\alpha\beta(r-c)} - r\beta - c\alpha - c\beta\right)/r, \tag{9.39}$$

$$s_0(w) = \frac{-b - \sqrt{b^2 + 4w(w+\alpha+\beta)c(r-c)}}{2c(r-c)} \tag{9.40}$$

and $b = (r - 2c)w + (r - c)\beta - c\alpha$. Note that the LST is defined for all $w \geq w^*$ where w^* is essentially the point where $s_0(w)$ becomes imaginary because the term inside the square-root is negative. However, the fact that $w^* < 0$ ensures that this would not be a problem for $w \geq 0$.

Now let $O_2(t)$ be the CDF of the random variable T_2 such that

$$O_2(t) = P\{T_2 \leq t\}.$$

Define $\tilde{O}_2(s)$ as the LST of $O_2(t)$ such that

$$\tilde{O}_2(s) = E[e^{-sT_2}].$$

During time T_2, the output from the buffer is zero. Therefore, the buffer's contents $X(t)$ is nondecreasing. Thus we have

$$O_2(t) = P\{T_2 \leq t\} = P\{X(t) > a | X(0) = 0\}. \tag{9.41}$$

For all $t \in [0, \infty)$, let $Z(t) = 1$ denote that the source is off and $Z(t) = 2$ denote that the source is on at time t. Define for $i = 1, 2$

$$H_i(x, t) = P\{X(t) \leq x, Z(t) = i\}.$$

Also define the vector $H(x, t) = [H_1(x, t) \quad H_2(x, t)]$. Then $H(x, t)$ satisfies the following partial differential equation

$$\frac{\partial H(x, t)}{\partial t} + \frac{\partial H(x, t)}{\partial x} R = H(x, t) Q \tag{9.42}$$

with initial conditions $H_1(x, 0) = 1$ and $H_2(x, 0) = 0$, where

$$R = \begin{bmatrix} 0 & 0 \\ 0 & r \end{bmatrix} \text{ and } Q = \begin{bmatrix} -\beta & \beta \\ \alpha & -\alpha \end{bmatrix}.$$

Now, taking the Laplace transform of Equation 9.42 with respect to t, we get

$$sH^*(x, s) - H(x, 0) + \frac{\partial H^*(x, s)}{\partial x} R = H^*(x, s) Q. \tag{9.43}$$

Due to the initial conditions following Equation 9.42, Equation 9.43 reduces to

$$sH^*(x, s) - [1 \quad 0] + \frac{\partial H^*(x, s)}{\partial x} R = H^*(x, s) Q.$$

Taking the LST of this equation with respect to x yields

$$s\tilde{H}^*(w, s) - [1 \quad 0] + w\tilde{H}^*(w, s)R - wH^*(0, s)R = \tilde{H}^*(w, s)Q.$$

Since $P\{X(t) \leq 0, Z(t) = 1\} = 0$, we have $H^*(0, s) = \begin{bmatrix} H_1^*(0, s) & 0 \end{bmatrix}$ and, therefore, $wH^*(0, s)R = [0 \quad 0]$. Hence this equation reduces to

$$\tilde{H}^*(w, s) = [1 \quad 0][sI + wR - Q]^{-1}.$$

Plugging in for R and Q, and taking the inverse of the matrix yields

$$\tilde{H}^*(w, s) = \frac{1}{wr(s + \beta) + \alpha s + \beta s + s^2}[s + wr + \alpha \quad \beta]. \tag{9.44}$$

However,

$$\tilde{O}_2(s) = 1 - sH_1^*(a, s) - sH_2^*(a, s).$$

Therefore, inverting the transform in Equation 9.44 with respect to w, and substituting in this equation, yields

$$\tilde{O}_2(s) = \frac{\beta}{\beta + s} e^{-a \frac{\alpha s + \beta s + s^2}{rs + r\beta}}. \tag{9.45}$$

Next we use the relations $E[T_1] = -\frac{d\tilde{O}_1(w)}{dw}$ at $w = 0$ and $E[T_2] = -\frac{d\tilde{O}_2(s)}{ds}$ at $s = 0$. From that, the mean of T_1 and T_2 are given by

$$E[T_1] = \frac{r + a(\alpha + \beta)}{c\alpha + c\beta - r\beta}, \tag{9.46}$$

$$E[T_2] = \frac{r + a(\alpha + \beta)}{r\beta}. \tag{9.47}$$

■

Notice that the ratio $E[T_1]/E[T_2]$ is independent of a. This indicates that no matter what a is, the ratio of time spent by the sink in on and off times remains the same. Also $\frac{E[T_1]}{E[T_1 + T_2]} = \frac{r\beta}{c(\alpha + \beta)}$. This is not surprising because in every on–off cycle of the sink an average of $E[T_1 + T_2] \frac{r\beta}{\alpha + \beta}$ fluid enters the buffer all of which exit the buffer during time whose mean is $E[T_1]$, hence the average amount of fluid exiting a buffer in one cycle is $E[T_1]c$. Hence we have $\frac{E[T_1]}{E[T_1 + T_2]} = \frac{r\beta}{c(\alpha + \beta)}$.

Problem 97

Consider an exponential on–off source with on times according to $\exp(\alpha)$ and off times according to $\exp(\beta)$. When the source is on, fluid enters at rate r into an infinite-sized buffer and no fluid flows into the buffer when the source is off. Fluid is drained from the buffer using two rates according to a threshold policy. When the amount of fluid in the buffer is less than x^* the output capacity is c_1, whereas if there is more than x^* fluid, it is removed at rate $c_1 + c_2$. For such a buffer, derive an expression for

$$\lim_{t \to \infty} P\{X(t) > \hat{B}\}$$

where $0 < x^* < \hat{B} < \infty$. Assume that $r > c_1 + c_2 > c_1 > r\beta/(\alpha + \beta)$.

Solution

The aim is to compute the limiting distribution (as $t \to \infty$) of $X(t)$, the amount of fluid in the buffer at time t. Note that the output capacity is c_1 when $X(t) < x^*$ and it is $c_1 + c_2$ when $X(t) \geq x^*$. The output capacity of the system can be modeled as an alternating renewal process that stays at c_1 for a random time T_1 and switches to $c_1 + c_2$ for a random time T_2 before moving back to c_1. Essentially T_1 is the first passage time for the buffer content to reach x^* (from below) given that initially the source is off and there is x^* amount of fluid. During the entire time T_1, the output capacity is c_1 and hence we can directly use the results from Problem 95. Likewise T_2 is the first passage time for the amount of fluid to reach x^* (from above) given that initially there is x^* amount of fluid and the source is on. Also, during the time T_2, the output capacity is $c_1 + c_2$, which enables us to use results from Problem 94.

To compute the limiting probability (as $t \to \infty$) that $X(t)$ is greater than $\hat{B}$ we condition on the region above or below x^* to obtain

$$P\{X(t) > \hat{B}\} = P\{X(t) > \hat{B}|X(t) > x^*\}P\{X(t) > x^*\}$$

$$+ P\{X(t) > \hat{B}|X(t) \leq x^*\}P\{X(t) \leq x^*\}$$

$$\Rightarrow \lim_{t \to \infty} P\{X(t) > \hat{B}\}$$

$$= \lim_{t \to \infty} P\{X(t) > \hat{B}|X(t) > x^*\}P\{X(t) > x^*\}. \qquad (9.48)$$

Since the output capacity is an alternating renewal process (alternating between c_1 and $c_1 + c_2$ for T_1 and T_2, respectively) that switches states every time the threshold x^* is crossed, we have

$$\lim_{t \to \infty} P\{X(t) > x^*\} = \frac{E[T_2]}{E[T_1] + E[T_2]}. \qquad (9.49)$$

We will subsequently obtain closed-form expressions for $E[T_1]$ and $E[T_2]$ using the results from Problems 95 and 94, respectively. Thus to complete the RHS of Equation 9.48, we now derive an expression for $P\{X(t) > \hat{B}|X(t) > x^*\}$ in the limit as $t \to \infty$.

For that, consider the same CTMC on–off source that inputs fluid into another buffer whose output capacity is a constant $c_1 + c_2$. Let the amount of fluid in that buffer at time t be $\hat{X}(t)$. Then we have

$$P\{X(t) > \hat{B}|X(t) > x^*\} = P\{\hat{X}(t) > \hat{B} - x^*|\hat{X}(t) > 0\}$$

since the $X(t)$ process being above x^* is stochastically equivalent to the $\hat{X}(t)$ process being above 0. We can immediately write down

$$P\{\hat{X}(t) > \hat{B} - x^* | \hat{X}(t) > 0\} = P\{\hat{X}(t) > \hat{B} - x^*, \hat{X}(t) > 0\}/P\{\hat{X}(t) > 0\}$$

$$= P\{\hat{X}(t) > \hat{B} - x^*\}/P\{\hat{X}(t) > 0\}.$$

However, we know from Problem 87 (CTMC on–off source) Equation 9.24 that

$$\lim_{t \to \infty} P\{\hat{X}(t) > x\} = \frac{\beta r}{(c_1 + c_2)(\alpha + \beta)} e^{\lambda_1 x}$$

where $\lambda_1 = \beta/(c_1 + c_2) - \alpha/(r - c_1 - c_2)$ from Equation 9.23. Hence we have

$$\lim_{t \to \infty} \frac{P\{\hat{X}(t) > \hat{B} - x^*\}}{P\{\hat{X}(t) > 0\}} = e^{\lambda_1(\hat{B} - x^*)}.$$

Thus we can write down

$$\lim_{t \to \infty} P\{X(t) > \hat{B} | X(t) > x^*\} = e^{\lambda_1(\hat{B} - x^*)}, \qquad (9.50)$$

where $\lambda_1 = \beta/(c_1 + c_2) - \alpha/(r - c_1 - c_2)$.

Thus the only things that remain are expressions for $E[T_1]$ and $E[T_2]$. We do that one by one. Recall that T_1 is the first passage time to reach fluid level x^* from below given that initially there is x^* amount of fluid and the source is off. Using the LST of the first passage time distribution in Problem 95, we can see that the LST $E[e^{-wT_1}]$ is equal to $\tilde{H}_{12}(x^*, w)$ defined in that problem. Hence, we can write down

$$E[e^{-wT_1}] = \frac{S_2(w)\beta e^{S_1(w)x^*} - S_1(w)\beta e^{S_2(w)x^*}}{S_2(w)(w + \beta + S_1(w)c_1)e^{S_1(w)x^*} - S_1(w)(w + \beta + S_2(w)c_1)e^{S_2(w)x^*}}$$

where

$$S_1(w) = \frac{-\hat{b} - \sqrt{\hat{b}^2 + 4w(w + \alpha + \beta)c_1(r - c_1)}}{2c_1(r - c_1)},$$

$$S_2(w) = \frac{-\hat{b} + \sqrt{\hat{b}^2 + 4w(w + \alpha + \beta)c_1(r - c_1)}}{2c_1(r - c_1)},$$

with $\hat{b} = (r - 2c_1)w + (r - c_1)\beta - c_1\alpha$. To compute $E[T_1]$ we can take the derivative of $E[e^{-wT_1}]$ with respect to w and let $w = 0$. Also, we need to use an analysis similar to that done in Problem 93, to get $S_1(0) = 0$ and $S_2(0) = \frac{c_1\alpha - \beta(r - c_1)}{c_1(r - c_1)}$. For that we require the assumption that $r\beta < c_1(\alpha + \beta)$

which is stated in the problem statement. Notice that if $c_2 = 0$ then $S_2(0) = -\lambda_1$. By systematically taking the derivative and letting $w = 0$, we have

$$E[T_1] = -\frac{d}{dw} E[e^{-wT_1}]|_{w=0}$$

$$= \frac{1}{\beta} + \frac{c_1}{\beta} \left(\frac{\alpha + \beta}{c_1(\alpha + \beta) - r\beta} \right) (e^{S_2(0)x^*} - 1), \qquad (9.51)$$

where $S_2(0) = \frac{c_1\alpha - \beta(r-c_1)}{c_1(r-c_1)} > 0$.

Now we derive $E[T_2]$. Recall that T_2 is the first passage time to reach fluid level x^* from above given that initially there is x^* amount of fluid and the source is on. During this time the output capacity is $c_1 + c_2$. Notice that this first passage time is the same as the busy period of a buffer with CTMC on–off source input and output capacity $c_1 + c_2$. Thus using the LST of the busy period distribution described toward the end of the solution to Problem 94, we can see that

$$E[e^{-wT_2}] = \frac{w + \beta + S_0(w)(c_1 + c_2)}{\beta}$$

where

$$S_0(w) = \frac{-\bar{b} - \sqrt{\bar{b}^2 + 4w(w + \alpha + \beta)(c_1 + c_2)(r - c_1 - c_2)}}{2(c_1 + c_2)(r - c_1 - c_2)},$$

with $\bar{b} = (r - 2(c_1 + c_2))w + (r - c_1 - c_2)\beta - (c_1 + c_2)\alpha$. To compute $E[T_2]$ we take the derivative of $E[e^{-wT_2}]$ to get

$$E[T_2] = -\frac{d}{dw} E[e^{-wT_2}]|_{w=0}$$

$$= -\frac{1}{\beta} + \frac{c_1 + c_2}{\beta} \left(\frac{\alpha + \beta}{(c_1 + c_2)(\alpha + \beta) - r\beta} \right). \qquad (9.52)$$

Thus by consolidating Equations 9.48 through 9.52, and rearranging the terms, we get

$$\lim_{t\to\infty} P\{X(t) > \hat{B}\}$$

$$= \frac{r\beta}{\alpha + \beta} \left[\frac{c_1(\alpha + \beta) - r\beta}{c_1[(c_1 + c_2)(\alpha + \beta) - r\beta]e^{S_2(0)x^*} - c_2r\beta} \right] e^{\lambda_1(\hat{B}-x^*)}$$

where $S_2(0) = \frac{c_1\alpha - \beta(r-c_1)}{c_1(r-c_1)}$ and $\lambda_1 = \beta/(c_1 + c_2) - \alpha/(r - c_1 - c_2)$. ∎

In the next example, we will consider a few aspects that will give a flavor for the analysis in the next chapter. In particular, we will consider: (i) multiple sources that superpose traffic into a buffer, (ii) a network situation where the departure from one node acts as input to another node, and (iii) a case of non-CTMC-based environment process.

Problem 98

Consider two infinite-sized buffers in tandem as shown in Figure 9.7. Input to the first buffer is from N independent and identical exponential on–off sources with on time parameter α, off time parameter β and rate r. The output from the first buffer is directly fed into the second buffer. The output capacities of the first and second buffers are c_1 and c_2, respectively. What is the stability condition? Assuming that is satisfied, characterize the environment process governing input to the second buffer.

Solution

Let $Z_1(t)$ be the number of sources that are in the "on" state at time t. Clearly $\{Z_1(t), t \geq 0\}$ is a CTMC with $N + 1$ states. When $Z_1(t) = i$ the input rate is ir. For notational convenience we assume that c_1 is not an integral multiple of r. Thus every state in the CTMC $\{Z_1(t), t \geq 0\}$ has strictly positive or strictly negative drifts. Let

$$\ell = \left\lceil \frac{c_1}{r} \right\rceil.$$

Thus whenever $Z_1(t) \in \{0, \ldots, \ell - 1\}$, the drift is negative, that is, the first buffer's contents would be nonincreasing. Likewise, whenever $Z_1(t) \in \{\ell, \ldots, N\}$, the drift is positive, that is, the first buffer's contents would be increasing. The first buffer is stable if the average fluid arrival rate in steady state is lesser than the service capacity, that is, $\frac{Nr\beta}{\alpha+\beta} < c_1$. If the first buffer is stable then the steady-state average departure rate from that buffer is also $Nr\beta/(\alpha + \beta)$. Thus the second buffer is also stable if $\frac{Nr\beta}{\alpha+\beta} < c_2$. Although not needed for this problem's analysis, unless $c_2 < c_1$, the second

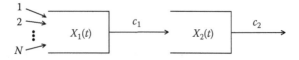

FIGURE 9.7
Tandem buffers with multiple identical sources. (From Gautam, N. et al., *Prob. Eng. Inform. Sci.*, 13, 429, 1999. With permission.)

buffer would always be empty. Thus, we assume that

$$\frac{Nr\beta}{\alpha + \beta} < c_2 < c_1.$$

Now we are ready to characterize the environment process governing input to the second buffer. Let $X_1(t)$ be the amount of fluid in the first buffer at time t. Notice that if $X_1(t) = 0$, then the output rate from the first buffer is equal to the input rate to the first buffer. This happens (with nonzero probability measure) only when the drift is negative. However, if $X_1(t) > 0$, the output rate is equal to c_1. Thus the output rate from the first buffer is from the set $\{0, r, 2r, \ldots, (\ell - 1)r, c_1\}$. Hence we can think of the input to the second buffer as though governed by a stochastic process with $\ell + 1$ discrete states $\{0, 1, \ldots, \ell\}$. Define

$$Z_2(t) = \begin{cases} Z_1(t) & \text{if } X_1(t) = 0 \\ \ell & \text{if } X_1(t) > 0, \end{cases}$$

where $Z_1(t)$ is the number of sources on at time t. The stochastic process $\{Z_2(t), t \geq 0\}$ is an SMP on state space $\{0, 1, \ldots, \ell\}$ with kernel

$$G(t) = [G_{ij}(t)]$$

which is to be derived. By definition, $G_{ij}(t)$ is the joint probability that given the current state is i the next state is j and the sojourn time in i is less than t.

For $i = 0, 1, \ldots, \ell - 1$ and $j = 0, 1, \ldots, \ell$, it is relatively straightforward to obtain $G_{ij}(t)$ as follows (since the sojourn times in state i are exponentially distributed):

$$G_{ij}(t) = \begin{cases} \frac{i\alpha}{i\alpha + (N-i)\beta} \left(1 - \exp\{-(i\alpha + (N-i)\beta)t\}\right) & \text{if } j = i - 1 \\ \frac{(N-i)\beta}{i\alpha + (N-i)\beta} \left(1 - \exp\{-(i\alpha + (N-i)\beta)t\}\right) & \text{if } j = i + 1 \\ 0 & \text{otherwise}. \end{cases}$$

To quickly explain this result, for $i = 0, 1, \ldots, \ell - 1$, when the SMP is in state i, there are i sources in the "on" state and $N - i$ in the "off" state. Thus the sojourn time in state i is until one of the sources changes states, hence it is exponentially distributed with parameter $i\alpha + (N - i)\beta$. Also the next state is $i + 1$ if one of the off sources switch to on which happens with probability $(N - i)\beta/(i\alpha + (N - i)\beta)$. Likewise the next state is $i - 1$ if one of the on sources switch to off which happens with probability $i\alpha/(i\alpha + (N - i)\beta)$. Notice that since the numerator would be zero when $i = 0$, we do not have to be concerned about the case $j = i - 1$ when $i = 0$.

The only tricky part in the kernel is to describe $G_{\ell j}(t)$. For that, we define a first passage time in the $\{X_1(t), t \geq 0\}$ process as

$$T = \min\{t > 0 : X_1(t) = 0\}.$$

Then for $j = 0, 1, \ldots, \ell - 1$, we have

$$G_{\ell j}(t) = P\{T \leq t, Z_1(T) = j | X_1(0) = 0, Z_1(0) = \ell\}.$$

We now derive an expression for $\tilde{G}_{\ell j}(s)$, the LST of $G_{\ell j}(t)$. Note that

$$G_{\ell \ell}(t) = 0.$$

Let

$$
Q = \begin{bmatrix}
-N\beta & N\beta & 0 & \cdots & 0 & 0 \\
\alpha & -\alpha - (N-1)\beta & (N-1)\beta & 0 & \cdots & 0 \\
\cdot & \cdot & \cdot & \cdot & \cdot & \cdot \\
\cdot & \cdot & \cdot & \cdot & \cdot & \cdot \\
\cdot & \cdot & \cdot & \cdot & \cdot & \cdot \\
0 & 0 & \cdots & (N-1)\alpha & -(N-1)\alpha - \beta & \beta \\
0 & 0 & \cdots & 0 & N\alpha & -N\alpha
\end{bmatrix}
$$

and

$$R = \mathrm{diag}(0, r, 2r, \ldots, Nr).$$

Let $s_k(w)$ and $\chi^k(w)$ be the kth eigenvalue and corresponding eigenvector respectively of $R^{-1}(wI - Q)$. As we described earlier, there are ℓ states with negative drift. Without loss of generality we let $s_0, s_1, \ldots, s_{\ell-1}$ be the ℓ negative eigenvalues and $\chi^0, \chi^1, \ldots, \chi^{\ell-1}$ be the corresponding eigenvectors written in that form suppressing they are functions of w for compact notation. Define

$$H_{ij}(x, t) = P\{T \leq t, Z_1(T) = j \mid X(0) = x, Z_1(0) = i\}$$

for $i, j = 0, 1, \ldots, N$. Then the LST of $H_{ij}(x, t)$ is

$$\tilde{H}_j(x, w) = \sum_{k=0}^{\ell-1} a_{kj} e^{s_k x} \chi^k.$$

Then the LST of the kernel element we require can be computed as

$$\tilde{G}_{\ell j}(w) = \tilde{H}_{\ell j}(0, w).$$

In other words,

$$\tilde{G}_{\ell 0}(w) = a_{00}\chi_\ell^0 + a_{10}\chi_\ell^1 + \cdots + a_{\ell-10}\chi_\ell^{\ell-1}$$

$$\tilde{G}_{\ell 1}(w) = a_{01}\chi_\ell^0 + a_{11}\chi_\ell^1 + \cdots + a_{\ell-11}\chi_\ell^{\ell-1}$$

and so on

$$\tilde{G}_{\ell\ell-1}(w) = a_{0\ell-1}\chi_\ell^0 + a_{1\ell-1}\chi_\ell^1 + \cdots + a_{\ell-1\ell-1}\chi_\ell^{\ell-1}.$$

Representing in matrix notations we have

$$\tilde{G}_{\ell j}(w) = [A \ \chi^*]_j,$$

where

$$\chi^* = [\chi_\ell^0 \ \chi_\ell^1 \ \cdots \ \chi_\ell^{\ell-1}]'.$$

To solve for the constants a_{ij} we use

$$\begin{pmatrix} a_{00} & a_{10} & \cdots & a_{\ell-10} \\ a_{01} & a_{11} & \cdots & a_{\ell-11} \\ \vdots & \vdots & \ddots & \vdots \\ a_{0\ell-1} & a_{1\ell-1} & \cdots & a_{\ell-1\ell-1} \end{pmatrix} \begin{pmatrix} \chi_0^0 & \chi_1^0 & \cdots & \chi_{\ell-1}^0 \\ \chi_0^1 & \chi_1^1 & \cdots & \chi_{\ell-1}^1 \\ \vdots & \vdots & \ddots & \vdots \\ \chi_0^{\ell-1} & \chi_1^{\ell-1} & \cdots & \chi_{\ell-1}^{\ell-1} \end{pmatrix} = I_{\ell \times \ell}.$$

In matrix notations

$$A \chi = I,$$

where

$$\chi = \begin{bmatrix} \overline{\chi}^0 & \overline{\chi}^1 & \cdots & \overline{\chi}^{\ell-1} \end{bmatrix}'$$

and $\overline{\chi}^k$ is an ℓ-dimensional vector obtained by truncating χ^k to its first ℓ elements. Therefore,

$$\tilde{G}_{\ell j}(w) = [\chi^{-1} \ \chi^*]_j.$$

Thus the input to the second buffer is governed by stochastic process $\{Z_2(t), t \geq 0\}$ which is an SMP on state space $\{0, 1, \ldots, \ell\}$ with kernel

$$G(t) = [G_{ij}(t)]$$

whose LST we describe above. When the environment is in state $Z_2(t)$ at time t, the fluid enters the second buffer at rate $\min(Z_2(t)r, c_1)$.

We can also compute the sojourn time τ_i in state i, for $i = 0, 1, \ldots, \ell$ as

$$\tau_i = \begin{cases} \frac{1}{i\alpha + (N-i)\beta} & \text{if } i = 0, 1, \ldots, \ell-1 \\ \sum_{j=1}^{\ell-1} \tilde{G}'_{\ell j}(0) & \text{if } i = \ell, \end{cases}$$

where $\tilde{G}'_{\ell j}(0)$ is the derivative of $\tilde{G}_{\ell j}(s)$ with respect to s evaluated at $s = 0$. Then for $i = 0, 1, \ldots, \ell$

$$p_i = \lim_{t \to \infty} P\{Z_2(t) = i\} = \frac{a_i \tau_i}{\sum_{k=0}^{\ell} a_k \tau_k}, \qquad (9.53)$$

where

$$a = a\,G(\infty) = a\,\tilde{G}(0). \qquad \blacksquare$$

Reference Notes

Stochastic fluid flow models or fluid queues have been around for almost three decades but have not received the attention that the deterministic fluid queues have received from researchers. In fact this may be the first textbook that includes two chapters on fluid queues. Pioneering work on fluid queues was done by Debasis Mitra and colleagues. In particular, the seminal article by Anick, Mitra, and Sondhi [5] is a must read for anyone interested in the area of fluid queues. At the end of the next chapter, there is a more extensive reference to the fluid queue literature. This chapter has been mainly an introductory one and we briefly describe the references relevant to its development.

The single buffer fluid model setting, notation, and characterization described in Sections 9.1.3, 9.1.4, and 9.2.1 are adapted from Kulkarni [69]. The main buffer content analysis for CTMCs in Sections 9.2.2 and 9.2.3 first appeared in Anick, Mitra, and Sondhi [5] and the version in this chapter has been mostly derived from Vidhyadhar Kulkarni's course notes. Section 9.3 on first passage time analysis with examples is from a collection of papers including Narayanan and Kulkarni [84], Aggarwal et al. [3], Gautam et al. [39], Kulkarni and Gautam [70], and Mahabhashyam et al. [76].

There are several extensions to the setting considered in this section. Some of these we will see in the next chapter. It is worthwhile mentioning about others that we will not see. In particular, Kulkarni and Rolski

[66] extend the fluid model analysis to continuous state space processes such as the Ornstein Uhlenbeck process. Krishnan et al. [65] consider fractional Brownian motion driving input to a buffer. Kella [58] uses Levy process inputs to analyze non-product form stochastic fluid networks. From a methodological standpoint other techniques are possible. For example, Ahn and Ramaswami [4] consider matrix-analytic methods for transient analysis, steady-state analysis, and first passage times of both finite- and infinite-sized buffers.

Exercises

9.1 Consider an infinite-sized buffer into which fluid entry is modulated by a six-state CTMC $\{Z(t), t \geq 0\}$ with $S = \{1, 2, 3, 4, 5, 6\}$ and

$$
Q = \begin{bmatrix}
-1 & 1 & 0 & 0 & 0 & 0 \\
1 & -2 & 1 & 0 & 0 & 0 \\
0 & 1 & -2 & 1 & 0 & 0 \\
0 & 0 & 1 & -2 & 1 & 0 \\
0 & 0 & 0 & 1 & -2 & 1 \\
0 & 0 & 0 & 0 & 1 & -1
\end{bmatrix}.
$$

The constant output capacity is $c = 18$ kbps and the fluid arrival rates in states 1, 2, 3, 4, 5, and 6 are 30, 25, 20, 15, 10, and 5 kbps, respectively. Obtain the joint probability that in steady state there is more than 20 kb of fluid in the buffer and the CTMC is in state 1. Also write down an expression for $\lim_{t \to \infty} P\{X(t) \leq x\}$.

9.2 Consider a finite-sized buffer whose input is modulated by a five-state CTMC $\{Z(t), t \geq 0\}$ with $S = \{1, 2, 3, 4, 5\}$ and

$$
Q = \begin{bmatrix}
-3 & 2 & 1 & 0 & 0 \\
2 & -5 & 2 & 1 & 0 \\
1 & 2 & -6 & 2 & 1 \\
0 & 0 & 2 & -4 & 2 \\
0 & 0 & 1 & 2 & -3
\end{bmatrix}.
$$

The output capacity for the buffer is $c = 15$ kbps and the fluid arrival rates in states 1, 2, 3, 4, and 5 are 20, 16, 12, 8, and 4 kbps, respectively. Consider three values of buffer size B in kb, namely, $B = 2$, $B = 4$, and $B = 6$. For the three cases obtain the probability there is more than 1 kb of fluid in the buffer in steady state. Also obtain the fraction of fluid that is lost because of a full buffer in all three cases.

9.3 Consider an infinite-sized buffer whose input is modulated by a five-state CTMC $\{Z(t), t \geq 0\}$ with $S = \{1, 2, 3, 4, 5\}$ and

$$Q = \begin{bmatrix} -20 & 8 & 6 & 4 & 2 \\ 4 & -15 & 4 & 4 & 3 \\ 1 & 4 & -10 & 3 & 2 \\ 3 & 3 & 3 & -12 & 3 \\ 1 & 3 & 5 & 7 & -16 \end{bmatrix}.$$

The output capacity for the buffer is $c = 4$ kbps and the fluid arrival rate is $(7 - i)$ kbps in state i for all $i \in S$. Derive an expression for $\lim_{t \to \infty} P\{X(t) \leq x\}$ for $x \geq 0$.

9.4 "Leaky Bucket" is a control mechanism for admitting data into a network. It consists of a data buffer and a token pool, as shown in Figure 9.8. Tokens in the form of fluid are generated continuously at a fixed rate γ into the token pool of size B_T. The new tokens are discarded if the token pool is full. External data traffic enters the infinite-sized data buffer in fluid form from a source modulated by an environmental process $\{Z(t), t \geq 0\}$ which is an ℓ-state CTMC. Data traffic is generated at rate $r(Z(t))$ at time t. If there are tokens in the token pool, the incoming fluid takes an equal amount of tokens and enters the network. If the token pool is empty then the fluid waits in the infinite-sized data buffer for tokens to arrive. Let $X(t)$ be the amount of fluid in the data buffer at time t and $Y(t)$ the amount of tokens in the token buffer at time t. Assume that at time $t = 0$ the token buffer and data buffer are both empty, that is, $X(0) = Y(0) = 0$. Draw a sample path of $Z(t)$, $X(t)$, $Y(t)$, and the output rate $R(t)$. What is the stability condition? Assuming stability, using the results in this chapter derive an expression

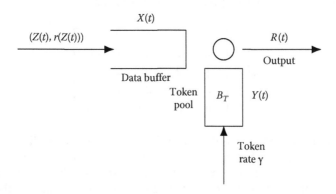

FIGURE 9.8
Single leaky bucket. (From Gautam, N., *Telecommun. Syst.*, 21(1), 35, 2002. With permission.)

for the distribution of the amount of fluid in the data buffer in steady state.

9.5 A sensor operates under three sensing environments: harsh, medium, and mild that we call states 1, 2, and 3, respectively. The environment toggles between the three states according to a CTMC with $S = \{1,2,3\}$ and in units of "per day,"

$$Q = \begin{bmatrix} -20 & 16 & 4 \\ 10 & -40 & 30 \\ 2 & 8 & -10 \end{bmatrix}.$$

There is a battery of 1 kJ charge that is used to run the sensor. The battery charge depletes continuously at rates 0.5, 0.2, and 0.1 kJ per day if operated under harsh, medium, and mild conditions, respectively. The battery is replaced instantaneously with a fully charged battery if the charge depletes to zero or the battery has not been replaced in 5 days, whichever happens first. What fraction of replacements are due to the charge depleting completely?

9.6 Reconsider one of the problems described earlier. There is a buffer whose input is modulated by a five-state CTMC $\{Z(t), t \geq 0\}$ with $S = \{1,2,3,4,5\}$ and

$$Q = \begin{bmatrix} -20 & 8 & 6 & 4 & 2 \\ 4 & -15 & 4 & 4 & 3 \\ 1 & 4 & -10 & 3 & 2 \\ 3 & 3 & 3 & -12 & 3 \\ 1 & 3 & 5 & 7 & -16 \end{bmatrix}.$$

The output capacity for the buffer is $c = 4$ kbps and the fluid arrival rate is $(7 - i)$ kbps in state i for all $i \in S$. Initially there is 0.1 kb of fluid and the environment is in state 3. Define the first passage time as the time to reach a buffer level of 0.2 or 0 kb, whichever happens first. Obtain the LST of the first passage time and derive the mean first passage time.

9.7 Consider a wireless sensor node that acts as a source that generates fluid at a constant rate r. This fluid flows into a buffer of infinite size. The buffer is emptied by a channel that toggles between capacity c and 0 for $\exp(\gamma)$ and $\exp(\delta)$ time, respectively. Assume that $c > r$ and that the system is stable. Derive an expression for the steady-state distribution of buffer contents. Also, characterize the output rate process from the queue (notice that the output rate is 0, r, or c).

9.8 An exponential on–off source inputs fluid into an infinite-sized buffer. The on and off time parameters are as usual α and β, respectively. There are two thresholds x_1 and x_2 such that $0 < x_1 < x_2 < \infty$. Let $X(t)$ be the amount of fluid in the buffer at time t. At time t, the output capacity is c_1 if $X(t) < x_1$, it is c_2 if $x_1 \leq X(t) < x_2$ and c_3 if $X(t) \geq x_2$. Assume that the system is stable and $c_1 < c_2 < c_3 < r$, where r is the usual arrival rate when the source is on. Derive an expression to determine the fraction of time the output capacity is c_i for $i = 1, 2, 3$. Use that to determine the probability that in steady state the amount of fluid in the buffer exceeds B, where $B > x_2$.

9.9 Starting with the LSTs, derive the expressions for $E[T_1]$ in Equations 9.46 and 9.51 as well as $E[T_2]$ in Equations 9.47 and 9.52 by taking the derivative of the respective LSTs and taking limits appropriately.

9.10 A node in a multi-hop wireless network has an infinite-sized buffer. The input to the buffer is modulated by a standard exponential on–off source with on and off parameters α and β, respectively. Also fluid arrival rate is r when the source is on. The channel capacity is also modulated by an exponentially distributed alternating renewal process. The channel capacity toggles between c_1 and c_2 for $\exp(\gamma)$ and $\exp(\delta)$ time, respectively. Assume that $r > c_2 > c_1$. What is the condition of stability? Assuming the system to be stable, obtain an expression for the steady-state distribution of buffer contents.

10

Stochastic Fluid-Flow Queues: Bounds and Tail Asymptotics

In Chapter 9, we saw single-stage stochastic fluid queues with continuous-time Markov chain (CTMC) environment processes governing the inputs, and the output channel capacities remaining a constant c. There we showed how to obtain steady-state workload distributions as well as perform first passage time analysis for those queues. What if we were to extend to general environment processes, considering varying output capacities, multiclass queues, multistage queues (i.e., networks), and merging fluid flows? The objective of this chapter is precisely to study the performance analysis under those extended conditions. However, it is intractable to obtain closed-form expressions, and we will show here how to derive bounds and approximations for those cases.

10.1 Introduction and Preliminaries

In this chapter, we begin by considering an infinite-sized buffer into which fluid enters at discrete rates governed by an environment process $\{Z(t), t \geq 0\}$. When the environment process is in state $Z(t)$ at time t, fluid enters the buffer at rate $r(Z(t))$. Fluid is removed from the buffer from an orifice with capacity c. To explain the notion of capacity, let $X(t)$ be the amount of fluid in the buffer at time t. If $X(t) > 0$, then fluid exits the buffer at rate c. However, if $X(t) = 0$, then fluid exits the buffer at rate $r(Z(t))$ ignoring the fact that there could be instantaneous times (hence measure zero) when $X(t) = 0$ and $r(Z(t)) > c$. This system is depicted in Figure 10.1. This is identical to the setting in Section 9.1.3. However, notice that at this time we do not make any assumption regarding the stochastic process $\{Z(t), t \geq 0\}$ other than the fact that $Z(t)$ is discrete.

Our objective is to first obtain bounds and approximations for $\lim_{t \to \infty} P\{X(t) > x\}$ under this setting. This would also enable us to compare against exact analysis when the environment process is a CTMC. After that, we will consider varying output capacities, multiclass queues, multistage queues, and multisource queues. For all those extensions, the bounds and approximations can be used suitably as we will see subsequently. However,

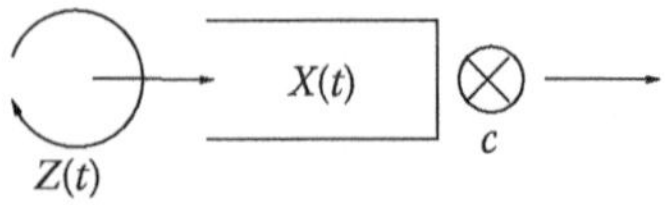

FIGURE 10.1
Single buffer with a single environment process and output capacity c (From Gautam, N., Quality of service metrics. In: *Frontiers in Distributed Sensor Networks*, S.S. Iyengar and R.R. Brooks (eds.), Chapman & Hall/CRC Press, Boca Raton, FL, pp. 613–628, 2004.)

to obtain bounds and approximations, we first need to find a way to summarize the input process to the queue. For that, we use the concept of asymptotic logarithmic moment generating function (ALMGF), which is also closely related to the notion of *effective bandwidths*. We consider that next.

10.1.1 Inflow Characteristics: Effective Bandwidths and ALMGF

To obtain bounds and approximations for buffer content distributions, we first need to summarize the characteristics of traffic flow into a buffer in an effective manner. For that we use ALMGF and effective bandwidths that we describe in the following text. It is crucial to point out here that throughout this section we will not consider the buffer at all. Our focus will just be on the input environment process. We will show in the next section how to use the ALMGF and effective bandwidths to obtain bounds and approximations. Interestingly, the notion and analysis can also be used under discrete queueing situations to obtain bounds and approximations for the workload distribution. However, we will mainly focus on the fluid case with occasional reference to discrete.

We describe the notation used and we deliberately restate some of the definitions so that they can all be in one place for easy reference. Let $A(t)$ be the total amount of (fluid or discrete) workload generated by a source (or environment process) over time $(0, t]$. The unit of $A(t)$ is in appropriate workload units (kB or liters or even time units especially for discrete). For the analysis to follow, we only consider a fluid model. Consider an ergodic stochastic process $\{Z(t), t \geq 0\}$ that models the flow of fluid workload. As defined earlier, $r(Z(t))$ is the rate at which workload is generated at time t.

Using $r(Z(t))$ we can write down the total amount of fluid generated as

$$A(t) = \int_0^t r(Z(u))du.$$

This uses the fact that $A(t)$ is the cumulative workload generated from time 0 to t, and $r(Z(u))$ is the instantaneous rate at which workload is generated. Thus, from the first principles of integration we have the expression for $A(t)$. Note that $\{A(t), t \geq 0\}$ is a stochastic process that is nondecreasing in t. Next we will summarize the $\{A(t), t \geq 0\}$ process so that it is conducive for performance analysis subsequently. We describe two notions, ALMGF and effective bandwidths, that are written in terms of $A(t)$.

The ALMGF of the workload process is defined as

$$h(v) = \lim_{t \to \infty} \frac{1}{t} \log E\{\exp(vA(t))\}$$

where $v \geq 0$ is a parameter much like the parameters in moment generating functions, LSTs, etc. We assume that $\{A(t), t \geq 0\}$ satisfies the Gärtner–Ellis conditions, that is, $h(v)$ as defined earlier exists, is finite for all real v, and is differentiable. Using the preceding expression for $h(v)$, it is possible to show that $h(v)$ is an increasing, convex function of v (see Figure 10.2) and for all $v \geq 0$,

$$r^{mean} \leq h'(v) \leq r^{peak}$$

where
$r^{mean} = E(r(Z(\infty)))$ is the mean traffic flow rate
$r^{peak} = \sup_z\{r(z)\}$ is the peak traffic flow rate
$h'(v)$ denotes the derivative of $h(v)$ with respect to v

The *effective bandwidth* of the workload process is defined as

$$eb(v) = \lim_{t \to \infty} \frac{1}{vt} \log E\{\exp(vA(t))\} = \frac{h(v)}{v}.$$

Thus, the ALMGF and effective bandwidths are related much similar to relationship between LST and the Laplace transform. There are benefits of both

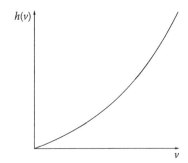

FIGURE 10.2
Graph of $h(v)$ versus v.

and thus we will continue using both. Using the definition of $eb(v)$, it can be shown that $eb(v)$ is an increasing function of v and

$$r^{mean} \leq eb(v) \leq r^{peak}.$$

Also,

$$\lim_{v \to 0} eb(v) = r^{mean} \quad \text{and} \quad \lim_{v \to \infty} eb(v) = r^{peak}.$$

These properties are depicted in Figure 10.3. Although we will see their implications subsequently, it is worthwhile explaining the properties related to r^{mean} and r^{peak}, which are the mean and peak input rates, respectively. Essentially, $eb(v)$ summarizes the workload generation rate process. Two obvious summaries are r^{mean} and r^{peak} that correspond to the average case and worst-case scenarios. The $eb(v)$ parameter captures those as well as everything in between. This would become more apparent when we consider the workload flowing into an infinite-sized buffer (with X denoting the steady-state buffer content level assuming it exists). If the output capacity of this buffer is $eb(v)$, then as $x \to \infty$, $P\{X > x\} \to e^{-vx}$. Naturally, the output capacity of the buffer must be greater than r^{mean} to ensure stability and it must be less than r^{peak} to have any fluid buildup. We expect the probability of the buffer content being greater than x to be higher when the output capacity is closer to r^{mean} than when it is closer to r^{peak}. That intuition can be verified.

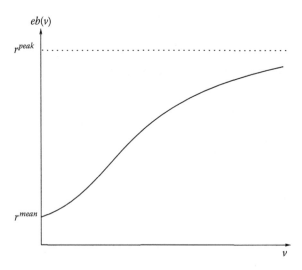

FIGURE 10.3
Graph of $eb(v)$ versus v.

Having defined ALMGF and the effective bandwidth (as well as briefly alluding to how we will use it for describing steady-state buffer content distribution), the next question is if we are given the environment process $\{Z(t), t \geq 0\}$ and the workload rates $r(Z(t))$ for all t, can we obtain expressions for $h(v)$ and $eb(v)$? That will be the focus of the following section.

10.1.2 Computing Effective Bandwidth and ALMGF

In this section, we show how to obtain the effective bandwidth and ALMGF starting with the environment process. In particular, we will show that for three types of environment processes: CTMCs, alternating renewal process, and semi-Markov processes (SMPs) (for a detailed description of these environment processes, see Section 9.1.4).

10.1.2.1 Effective Bandwidth of a CTMC Source

Elwalid and Mitra [30] and Kesidis et al. [61] use eigenvalue techniques to show how to compute the effective bandwidths of sources that are modulated by CTMCs. The description to follow is based on those articles. Let $\{Z(t), t \geq 0\}$ be an irreducible, finite state CTMC with generator matrix Q. When the CTMC is in state i, the source generates fluid at rate $r(i)$. Let R be a diagonal matrix with elements corresponding to the fluid rates at those states. In other words, $R = \text{diag}[r(i)]$. These descriptions are identical to the notations in Section 9.2.1. Let $e(M)$ denote the largest real eigenvalue of a square matrix M. Then,

$$h(v) = e(Q + vR). \tag{10.1}$$

Next, we show through Problem 99 we next show how to derive the previous expression.

Problem 99

Derive the expression for $h(v)$ in Equation 10.1 using the definition of $h(v)$ for a CTMC environment process $\{Z(t), t \geq 0\}$ with state space S, generator matrix $Q = [q_{ij}]$, and fluid rate matrix R.

Solution

From the definition of $h(v)$ given by

$$h(v) = \lim_{t \to \infty} \frac{1}{t} \log E\big[e^{vA(t)}\big],$$

we consider

$$g_i(t) = E\left[e^{vA(t)} \middle| Z(0) = i\right]$$

for some $i \in S$. We can immediately write down the following for some infinitesimally small positive h:

$$g_i(t+h) = E\left[e^{vA(t+h)} \middle| Z(0) = i\right]$$

$$= \sum_{j \in S} E\left[e^{vA(t+h)} \middle| Z(h) = j, Z(0) = i\right] P\{Z(h) = j \middle| Z(0) = i\}$$

$$= \sum_{j \in S} e^{vr(i)h} E\left[e^{vA(t)} \middle| Z(0) = j\right] q_{ij}h + e^{vr(i)h} E\left[e^{vA(t)} \middle| Z(0) = i\right] + o(h)$$

where $o(h)$ are terms of the order higher than h such that $o(h)/h \to 0$ as $h \to 0$. Before proceeding, we explain the last equation. First of all, $P\{Z(h)=j|Z(0)=i\} = q_{ij}h + o(h)$ if $i \neq j$ and $P\{Z(h)=i|Z(0)=i\} = 1 + q_{ii}h + o(h)$ using standard CTMC transient analysis results. Also, from time 0 to h when the CTMC is in state i, $r(i)h$ amount of fluid is generated. Thus, $A(t + h)$ is stochastically identical to $A(t) + r(i)h$ assuming that at time h the environment process toggles from i to j. Using that we can rewrite $g_i(t + h)$ in the previous equation as

$$g_i(t+h) = \sum_{j \in S} e^{vr(i)h} g_j(t) q_{ij}h + e^{vr(i)h} g_i(t) + o(h).$$

Subtracting $g_i(t)$ from both sides of the equation, dividing by h, and letting $h \to 0$, we get

$$\frac{dg_i(t)}{dt} = g_i(t)vr(i) + \sum_{j \in S} g_j(t)q_{ij}$$

using the fact that $e^{vr(i)h} = 1 + vr(i)h + o(h)$. We can write the preceding differential equation in vector form as

$$\frac{dg(t)}{dt} = [Rv + Q]g(t)$$

where $g(t)$ is a $|S| \times 1$ column vector of $g_i(t)$ values. The equation is similar to several differential equations derived in Chapter 9. From there we can see that the generic solution is of the form

$$g(t) = \sum_{j \in S} a_j e^{\lambda_j t} \psi_j$$

where

a_j values are scalar constants

λ_j and ψ_j are the jth eigenvalue and corresponding right eigenvector, respectively, of $(Rv + Q)$

Let $\theta = e(Rv+Q)$ be the largest real eigenvalue of $(Rv+Q)$. We can rewrite $g(t)$ as follows:

$$g(t) = e^{\theta t} \sum_{j \in S} a_j e^{(\lambda_j - \theta)t} \psi_j.$$

Notice that as $t \to \infty$, $e^{(\lambda_j - \theta)t} \to 0$ if $\lambda_j \neq \theta$ and $e^{(\lambda_j - \theta)t} \to 1$ if $\lambda_j = \theta$. By considering π_0 as the $1 \times |S|$ row-vector corresponding to the initial probability of the environment, that is, $\pi_0 = [P\{Z(0) = i\}]$, clearly we have from the definition of $h(v)$ that

$$h(v) = \lim_{t \to \infty} \frac{1}{t} \log[\pi_0 g(t)].$$

Using the expression for $g(t)$, we can immediately write down the following:

$$h(v) = \lim_{t \to \infty} \frac{1}{t} \log \left[e^{\theta t} \pi_0 \sum_{j \in S} a_j e^{(\lambda_j - \theta)t} \psi_j \right]$$

$$= \lim_{t \to \infty} \frac{1}{t} \left\{ \log \left[e^{\theta t} \right] + \log \left[\pi_0 \sum_{j \in S} a_j e^{(\lambda_j - \theta)t} \psi_j \right] \right\}$$

$$= \theta + \lim_{t \to \infty} \frac{1}{t} \log \left[\pi_0 \sum_{j \in S} a_j e^{(\lambda_j - \theta)t} \psi_j \right]$$

$$= \theta + 0.$$

For the last expression, we do need some j for which $\lambda_j = \theta$ so that the summation itself does not go to zero or infinite as $t \to \infty$. Since $\theta = e(Rv+Q)$, we have $h(v) = e(Rv + Q)$. ∎

Of course, $eb(v)$ can be immediately obtained as $eb(v) = h(v)/v$. We now illustrate $eb(v)$ for a CTMC on-off source.

Problem 100

Consider an on-off source with on-times according to $\exp(\alpha)$ and off-times according to $\exp(\beta)$. Traffic is generated at rate r when the source is in the on-state and no traffic is generated when the source is in the off-state. Obtain a closed-form algebraic expression for $eb(v)$ for such a source.

Solution

The environment process $\{Z(t), t \geq 0\}$ is a two-state CTMC, where the first state corresponds to the source being off and the second state corresponds to the source being on. Therefore,

$$R = \begin{bmatrix} 0 & 0 \\ 0 & r \end{bmatrix} \quad \text{and} \quad Q = \begin{bmatrix} -\beta & \beta \\ \alpha & -\alpha \end{bmatrix}.$$

Define matrix $M = Q + vR$ since we need to compute the largest eigenvalue of $Q + vR$. Thus,

$$M = \begin{bmatrix} -\beta & \beta \\ \alpha & rv - \alpha \end{bmatrix}.$$

The eigenvalues of M are given by the solution to the characteristic equation

$$|M - \lambda I| = 0,$$

which yields

$$(rv - \alpha - \lambda)(-\beta - \lambda) - \beta\alpha = 0.$$

It can be rewritten in the form of quadratic equation (in terms of λ) as

$$\lambda^2 + (\beta + \alpha - rv)\lambda - \beta rv = 0,$$

the solution to which is

$$\lambda = \frac{rv - \alpha - \beta \pm \sqrt{(rv - \alpha - \beta)^2 + 4\beta rv}}{2}.$$

Using the larger of the two eigenvalues of M, we obtain the effective bandwidth as

$$eb(v) = \frac{rv - \alpha - \beta + \sqrt{(rv - \alpha - \beta)^2 + 4\beta rv}}{2v}. \tag{10.2}$$

■

For the preceding CTMC source, we have the mean rate as

$$r^{mean} = \frac{r\beta}{\alpha + \beta}$$

and the peak rate as

$$r^{peak} = r.$$

By taking the limits in Equation 10.2 as $v \to 0$ and $v \to \infty$, we can see that $eb(0) = r^{mean}$ and $eb(\infty) = r^{peak}$. In fact, the result for $h'(v)$ can be obtained in exactly the same fashion. Next, we present a numerical example and graphically illustrate $eb(v)$ for various v values.

Problem 101

Water flows into the reservoir according to a CTMC $\{Z(t), t \geq 0\}$ with $S = \{1, 2, 3, 4, 5\}$ and

$$Q = \begin{bmatrix} -1 & 0.4 & 0.3 & 0.2 & 0.1 \\ 0.4 & -0.7 & 0.1 & 0.1 & 0.1 \\ 0.5 & 0.4 & -1.1 & 0.1 & 0.1 \\ 0.2 & 0.3 & 0.3 & -1 & 0.2 \\ 0.3 & 0.3 & 0.3 & 0.3 & -1.2 \end{bmatrix}.$$

When $Z(t) = i$, the inflow rate is $4i$. Graph $eb(v)$ versus v for the water flow.

Solution

For the CTMC source we have

$$R = \begin{bmatrix} 4 & 0 & 0 & 0 & 0 \\ 0 & 8 & 0 & 0 & 0 \\ 0 & 0 & 12 & 0 & 0 \\ 0 & 0 & 0 & 16 & 0 \\ 0 & 0 & 0 & 0 & 20 \end{bmatrix}.$$

Also $r^{peak} = 20$ and $r^{mean} = \sum_{i=1}^{5} 4ip_i = 9.6672$ since the solution to $[p_1 \ p_2 \ p_3 \ p_4 \ p_5]$ $Q = [0 \ 0 \ 0 \ 0 \ 0]$ and $p_1 + p_2 + p_3 + p_4 + p_5 = 1$ is $[p_1 \ p_2 \ p_3 \ p_4 \ p_5] = [0.2725 \ 0.3438 \ 0.1652 \ 0.1315 \ 0.0870]$. Using $eb(v) = e(Q/v + R)$, we plot $eb(v)$ versus v in Figure 10.4.   ■

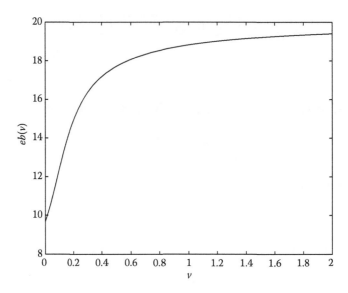

FIGURE 10.4
Graph of $eb(v)$ versus v for Problem 101.

Next, we present an approach to compute the effective bandwidth of an SMP source (after that we will show for an alternating on-off source by considering it as a special case of SMP).

10.1.2.2 Effective Bandwidth of a Semi-Markov Process (SMP) Source

Kulkarni [68] shows how to compute the effective bandwidths of sources that are modulated by Markov regenerative processes (MRGP). Since the SMP is a special type of MRGP, we follow the same analysis. Let $\{Z(t), t \geq 0\}$ be an ℓ-state SMP. Recall from the SMP environment characterization in Section 9.1.4 that S_1 is the time of the first jump epoch in the SMP. Also recall that Z_n is the state of the SMP after the nth jump. Define

$$\Lambda_{ij}(u, v) = E\{e^{-(u-r(i)v)S_1}I(Z_1 = j) \mid Z_0 = i\},$$

for $i, j = 1, 2, \ldots, \ell$ and $-\infty < u, v < \infty$, where $I(\cdot)$ is the indicator function. Notice that we can write down $\Lambda_{ij}(u, v)$ in terms of the kernel of the SMP $G(x) = [G_{ij}(x)]$, where

$$G_{ij}(x) = P\{S_1 \leq x; Z_1 = j \mid Z_0 = i\}.$$

In particular, if the LST of $G_{ij}(x)$ is $\tilde{G}_{ij}(w)$, that is,

$$\tilde{G}_{ij}(w) = \int_0^\infty e^{-wx} dG_{ij}(x),$$

then we have

$$\Lambda_{ij}(u,v) = \tilde{G}_{ij}(u - r(i)v). \tag{10.3}$$

Having characterized $\Lambda_{ij}(u,v)$, we write it in the matrix form as

$$\Lambda(u,v) = [\Lambda_{ij}(u,v)],$$

which is an $\ell \times \ell$ matrix. Using the same notation as in the CTMC case, we define $e(\Lambda(u,v))$ as the largest real-positive eigenvalue of $\Lambda(u,v)$.
   Define

$$e^*(v) = \sup_{\{u>0:e(\Lambda(u,v))<\infty\}} \{e(\Lambda(u,v))\}$$

and

$$u^*(v) = \inf\{u > 0 : e(\Lambda(u,v)) < \infty\}.$$

Then, for a given v,

(a) If $e^*(v) \geq 1$ (see Figure 10.5(a)), $h(v)$ is a unique solution to $e(\Lambda(h(v),v)) = 1$.
(b) If $e^*(v) < 1$ (see Figure 10.5(b)), $h(v) = u^*(v)$.

Thereby, the effective bandwidth can be obtained as $eb(v) = h(v)/v$.

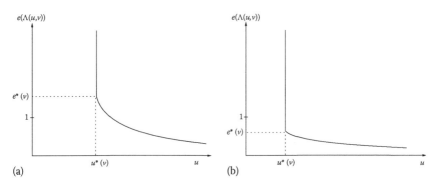

FIGURE 10.5
(a) $e(\Lambda(u,v))$ versus u and (b) $e(\Lambda(u,v))$ versus u.

We now explain computing the ALMGF and the effective bandwidth for SMP sources using an example. The example will be of the type where $e^*(v) \geq 1$ for all v due to the type of distributions used. We will see examples in the following sections for SMP sources, where we encounter $e^*(v) < 1$.

Problem 102

Fluid flows into a buffer according to a three-state SMP $\{Z(t), t \geq 0\}$ with state space $\{1, 2, 3\}$. The elements of the kernel of this SMP are given as follows: $G_{12}(t) = 1 - e^{-t} - te^{-t}$, $G_{21}(t) = 0.4(1 - e^{-0.5t}) + 0.3(1 - e^{-0.2t})$, $G_{23}(t) = 0.2(1 - e^{-0.5t}) + 0.1(1 - e^{-0.2t})$, $G_{32}(t) = 1 - 2e^{-t} + e^{-2t}$, and $G_{11}(t) = G_{13}(t) = G_{22}(t) = G_{31}(t) = G_{33}(t) = 0$. Also, the flow rates in the three states are $r(i) = i$ for $i = 1, 2, 3$. Graph $eb(v)$ versus v for $v \in [0, 3]$.

Solution

For the SMP source we have the LST of the kernel as

$$\tilde{G}(w) = \begin{bmatrix} 0 & \frac{1}{(w+1)^2} & 0 \\ \frac{0.2}{0.5+w} + \frac{0.06}{0.2+w} & 0 & \frac{0.1}{0.5+w} + \frac{0.02}{0.2+w} \\ 0 & \frac{2}{(2+w)(1+w)} & 0 \end{bmatrix}.$$

Now, using the elements of the LST of the kernel, we can easily write down for $i = 1, 2, 3$ and $j = 1, 2, 3$,

$$\Lambda_{ij}(u, v) = \tilde{G}_{ij}(u - r(i)v).$$

Notice that the LSTs are such that $\tilde{G}_{ij}(w)$ would shoot off to infinity only if their denominators (if any) become zero. However, the shooting off to infinity would not be sudden but gradual. Hence, for all v, $e^*(v) \geq 1$ (in fact $e^*(v) = \infty$ and $u^*(v) = \max\{r(1)v-1, r(2)v-0.2, r(3)v-1, 0\}$ because, for example, at $u = r(1)v - 1$, the denominator of $\tilde{G}_{12}(w)$ goes to zero, and so on for the other LSTs). Thus, to compute $eb(v)$, all we need is the unique solution to $e(\Lambda(veb(v), v)) = 1$.

Next, we explain how to numerically obtain the unique solution. Essentially, for a given v, $e(\Lambda(u, v))$ decreases with respect to u from $u^*(v)$ to infinity. Using that and the bounds on $eb(v)$, that is, $r^{mean} \leq eb(v) \leq r^{peak}$ for all v, we can perform a binary search for $eb(v)$ between $\max\{r^{mean}, u^*(v)/v\}$ and r^{peak} to find the unique solution to $e(\Lambda(h(v), v)) = 1$. Notice that $r^{peak} = r(3) = 3$ and $r^{mean} = \sum_{i=1}^{3} ip_i = 1.8119$, since the stationary distribution of the SMP can be computed as $[p_1 \ p_2 \ p_3] = 1/(\pi_1\tau_1 + \pi_2\tau_2 + \pi_3\tau_3)[\pi_1\tau_1 \ \pi_2\tau_2 \ \pi_3\tau_3]$ $= 1/(0.35 \times 2 + 0.5 \times 3.2 + 0.15 \times 1.5)[0.35 \times 2 \ 0.5 \times 3.2 \ 0.15 \times 1.5] = [0.2772 \ 0.6337 \ 0.0891]$.

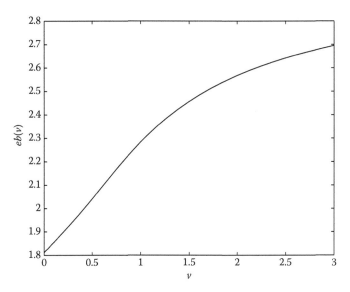

FIGURE 10.6
Graph of $eb(v)$ versus v for Problem 102.

Using $eb(v)$ as the solution to $e(\Lambda(h(v),v)) = 1$, we plot $eb(v)$ versus v for $v \in [0,3]$ in Figure 10.6. The ALMGF $h(v)$ can be obtained as $veb(v)$. ■

10.1.2.3 Effective Bandwidth of a General On/Off Source

Information flow in computer-communication networks are frequently modeled as general on/off flows because when measurements are made on a channel, either information is flowing at the channel speed of r kbps or nothing is flowing. With that motivation we consider the general on/off source, although we see that it is indeed a special case of an SMP. But for simplicity, we model it as an alternating renewal process. We will also see in a numerical example how this would be a special case of a CTMC source if we can model the general on/off process as a CTMC.

Consider a source that generates traffic according to an alternating renewal process. In particular, we consider an on-off source where the on and off times are independent. When the source is on, fluid is generated at rate r, and no fluid is generated when the source is off. Let U be a random variable that denotes an on-time and D be a random variable that denotes an off-time (the letters U and D denote up and down but the more appropriate notion is on and off in this context). We first directly use results from MRGPs, in particular regenerative processes. Say at time 0 the source just transitioned from off to on. Let S_1 be the time when such an event occurs once again. Then, S_1 is called a regeneration epoch and $S_1 = U + D$. Let F_1 be the total amount of fluid generated during S_1, and hence we have $F_1 = rU$. Define

$$\Lambda(u,v) = E[e^{-uS_1+vF_1}] = E[e^{-u(U+D)+vrU}] = E[e^{(-u+rv)U}]E[e^{-uD}].$$

Since $\Lambda(u,v)$ is a scalar, $e(\Lambda(u,v)) = \Lambda(u,v)$ in scalar notation.

Now, we could have derived the preceding using SMP, we would have unnecessarily have had to deal with a 2×2 matrix and take things from there. Nonetheless, it is worthwhile to check to see if the results match by going the SMP way. In fact, from here on we continue with the SMP approach. Following the technique in the SMP case earlier, define

$$e^*(v) = \sup_{\{u>0:\Lambda(u,v)<\infty\}} \{\Lambda(u,v)\}$$

and

$$u^*(v) = \inf\{u > 0 : \Lambda(u,v) < \infty\}.$$

For a given v,

(a) If $e^*(v) \geq 1$, $h(v)$ is a unique solution to $e(\Lambda(h(v),v)) = 1$.
(b) If $e^*(v) < 1$, $h(v) = u^*(v)$.

Thereby, the effective bandwidth can be obtained as $eb(v) = h(v)/v$. Next we present two examples to illustrate the approach to obtain ALMGF and effective bandwidths for general on/off sources. However, in both examples we will only see the case where $e^*(v) \geq 1$ (in a subsequent section we will consider the $e^*(v) < 1$ case).

Problem 103

Consider an on-off source that generates fluid so that the on-times are IID random variables with CDF $U(t) = 1 - 0.6e^{-3t} - 0.4e^{-2t}$ and the off-times are IID Erlang random variables with mean 0.5 and variance $1/12$ in appropriate time units compatible with the on-times. When the source is on, fluid is generated at the rate of 16 units per second; and no fluid is generated when the source is off. This is identical to the source described in Problem 88. Graph $h(v)$ versus v for $v \in [0,1]$.

Solution

For the on/off source, we have the LST of the on-times as

$$E[e^{-wU}] = \int_0^\infty e^{-wt} dU(t) = \frac{1.8}{3+w} + \frac{0.8}{2+w}.$$

Likewise, the LST of the off-times is

$$E[e^{-wD}] = \int_0^\infty e^{-wt} dD(t) = \left(\frac{6}{6+w}\right)^3$$

since the CDF of the off-times $D(t) = 1 - e^{-6t} - 6te^{-6t} - 18t^2e^{-6t}$ for all $t \geq 0$. Using these we can write down

$$\Lambda(u,v) = E[e^{(-u+rv)U}]E[e^{-uD}] = \left[\frac{1.8}{3 - rv + u} + \frac{0.8}{2 - rv + u}\right]\left(\frac{6}{6+u}\right)^3$$

with $r = 16$.

Notice that the LSTs are such that $\Lambda(u,v)$ would shoot off to infinity only if the denominator becomes zero. Also, the shooting off to infinity would not be abrupt but gradual. Hence, for all v, $e^*(v) \geq 1$ (in fact $e^*(v) = \infty$ and $u^*(v) = \max\{rv - 2, 0\}$ because for all $u > rv - 2$, the denominator of $\Lambda(u,v)$ is nonzero). Thus, to compute $h(v)$ all we need is the unique solution to $\Lambda(h(v), v) = 1$. To numerically obtain the unique solution, note that for a given v, $\Lambda(u,v)$ decreases with respect to u from $u^*(v)$ to infinity. Using that and the bounds on $eb(v)$, that is, $r^{mean} \leq eb(v) \leq r^{peak}$ for all v, we can perform a binary search for $h(v)$ between $\max\{vr^{mean}, u^*(v)\}$ and vr^{peak} to find the unique solution to $\Lambda(h(v), v) = 1$. Here we have $r^{peak} = r = 16$ and $r^{mean} = rE[U]/(E[U] + E[D]) = 7.1111$. Using $h(v)$ as the solution to $\Lambda(h(v), v) = 1$, we plot $h(v)$ versus v for $v \in [0, 1]$ in Figure 10.7. ∎

To contrast the methodologies, in the next exercise we consider the same problem as the previous one, but use the CTMC source approach. Although we expect the same results, the approach is quite different.

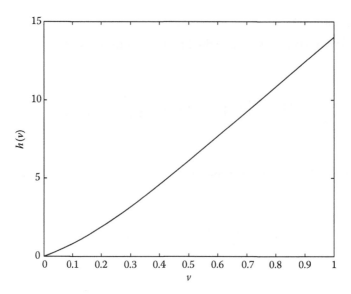

FIGURE 10.7
Graph of $eb(v)$ versus v for Problem 103.

Problem 104

Consider Problem 103 and obtain $h(v)$ for $v=0.5$ and $v=1$ using the CTMC source results.

Solution

We follow the analysis outlined in the solution to Problem 88. Notice that the on-times correspond to a two-phase hyperexponential distribution. Hence the on-time would be $\exp(3)$ with probability 0.6 and it would be $\exp(2)$ with probability 0.4, which can be deduced from $U(t)$. The off-times correspond to the sum of three IID $\exp(6)$ random variables. Thus, we can write down the environment process $\{Z(t), t \geq 0\}$ as an $\ell=5$ state CTMC with states 1 and 2 corresponding to on and states 3, 4, and 5 corresponding to off. Thus, the Q matrix corresponding to $S=\{1,2,3,4,5\}$ is

$$
Q = \begin{bmatrix}
-3 & 0 & 3 & 0 & 0 \\
0 & -2 & 2 & 0 & 0 \\
0 & 0 & -6 & 6 & 0 \\
0 & 0 & 0 & -6 & 6 \\
3.6 & 2.4 & 0 & 0 & -6
\end{bmatrix}.
$$

The rate matrix is

$$R = \begin{bmatrix} 16 & 0 & 0 & 0 & 0 \\ 0 & 16 & 0 & 0 & 0 \\ 0 & 0 & 0 & 0 & 0 \\ 0 & 0 & 0 & 0 & 0 \\ 0 & 0 & 0 & 0 & 0 \end{bmatrix}.$$

Since $h(v) = e(Q + vR)$, we obtain the eigenvalues of $(Q + vR)$. For $v = 0.5$, we get the eigenvalues as -9.3707, $-4.4931 + 3.4689i$, $-4.4931 - 3.4689i$, 5.2364, and 6.1205. When there are some eigenvalues that are complex, software packages when asked to compute the maximum might compute the maximum of the absolute value. However, in this case, that would return the wrong value. The correct approach is to determine the maximum for the *real* part. If one does that here, we get $h(0.5) = 6.1205$, which matches exactly with the approach used in Problem 103. There is a perfect match for $h(1)$ as well yielding a value 14.0226, again with two eigenvalues with imaginary parts. It is worthwhile to spend a few moments contrasting the two methods, that is, the one used in this problem and that of Problem 103. ∎

There are several other stochastic process sources for which one can obtain the effective bandwidths. As described earlier for an MRGP and regenerative process, we can use the results in Kulkarni [68]. See Krishnan et al. [65] for the calculation of effective bandwidths for traffic modeled by fractional Brownian motion. In fact, we can even obtain effective bandwidths for discrete sources. We do not present any of those here but the reader is encouraged to do a literature search to find out more about those cases. Next we consider some quick extensions.

10.1.3 Two Extensions: Traffic Superposition and Flow through a Queue

To extend our analysis to follow for fluid queues' workload distribution under cases of networks and multiple classes, we need to obtain the effective bandwidths under those conditions. The first case is to obtain the effective bandwidth in fluid queues with more than one source. The second one is the effective bandwidth of the output from a queue.

10.1.3.1 Superposition of Multiple Sources

Consider a single buffer that admits fluid from K independent sources (Figure 10.8). So far, we have only seen the special case $K = 1$. Here, each source k (for $k = 1, \ldots, K$) is driven by an independent environment process $\{Z^k(t), t \geq 0\}$. Note that $Z^k(t)$ can be thought of as the state of the kth input source $(k = 1, 2, \ldots, K)$ at time t. When source k is in state $Z^k(t)$, it generates

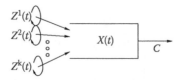

FIGURE 10.8
Single infinite-sized buffer with multiple input sources. (From Gautam, N. et al., *Prob. Eng. Inform. Sci.*, 13, 429, 1999. With permission.)

fluid at rate $r^k(Z^k(t))$ into the buffer. Let $eb_k(v)$ be the effective bandwidth of source k such that

$$eb_k(v) = \lim_{t \to \infty} \frac{1}{vt} \log E\{\exp(vA_k(t))\}$$

where

$$A_k(t) = \int\limits_0^t r^k(Z^k(u))du.$$

If the stochastic process $\{Z^k(t), t \geq 0\}$ is an SMP (or one of the processes that is tractable to get the effective bandwidths), then we can obtain $eb_k(v)$. Then the net effective bandwidth of the fluid arrival into the buffer due to all the K sources is, say, $eb(v)$. Since the net fluid input rate is just the sum of the input rates of the K superpositioned sources, we have $A(t) = A_1(t) + \cdots + A_K(t)$. By definition

$$eb(v) = \lim_{t \to \infty} \frac{1}{vt} \log E\{\exp(vA(t))\} = \lim_{t \to \infty} \frac{1}{vt} \log E[\exp(v\{A_1(t) + \cdots + A_K(t)\})].$$

Since the sources are independent, we have

$$eb(v) = \lim_{t \to \infty} \frac{1}{vt} \log E[\exp(vA_1(t))]E[\exp(vA_2(t))]\ldots E[\exp(vA_K(t))]$$

$$= \lim_{t \to \infty} \frac{1}{vt} \sum_{k=1}^{K} \log E[\exp(vA_k(t))].$$

Thus, we have

$$eb(v) = \sum_{k=1}^{K} eb_k(v).$$

In summary, the effective bandwidth of a superposition of K independent sources is equal to the sum of the effective bandwidths of each source.

10.1.3.2 Effective Bandwidth of the Output from a Queue

Here we consider fluid that flows through a queue and is removed at a maximum rate of c. The question we ask is what is the output effective bandwidth? The motivation is that in many applications, the output traffic from a buffer acts as input traffic for a downstream node in a network. Typically, it may not be possible to characterize some output processes as tractable stochastic processes and compute the effective bandwidths. Hence if we have a relationship between the input and output effective bandwidths, we do not need to in fact characterize the traffic process of the output.

To derive the ALMGF and effective bandwidth of the output from a buffer, we consider an infinite-sized buffer driven by a random environment process $\{Z(t), t \geq 0\}$. In principle, this could be one source or a superposition of multiple sources. But for ease of exposition we present as though it is one source similar to the one described in Figure 10.1. When the environment is in state $Z(t)$, fluid enters the infinite-sized buffer at rate $r(Z(t))$. Let $A(t)$ be the total amount of fluid generated by a source into the buffer over time $(0, t]$, and

$$A(t) = \int_0^t r(Z(u))du.$$

The *ALMGF* of the input traffic is

$$h_A(v) = \lim_{t \to \infty} \frac{1}{t} \log E\{\exp(vA(t))\}.$$

The *effective bandwidth* of the input traffic is

$$eb_A(v) = \lim_{t \to \infty} \frac{1}{vt} \log E\{\exp(vA(t))\}.$$

We assume that the buffer is stable, that is,

$$E[r(Z(\infty))] < c.$$

We ask the question: What is the ALMGF as well as the effective bandwidth of the output traffic from the buffer? For that let $D(t)$ be the total output from the buffer over $(0, t]$. By definition, the ALMGF of the output is

$$h_D(v) = \lim_{t \to \infty} \frac{1}{t} \log E\{\exp(vD(t))\}.$$

Recall that

$$r^{mean} \leq h'(v) \leq r^{peak}$$

for any ALMGF $h(v)$. For an infinite-sized stable buffer, since r^{mean} would be the same for the input and output due to no loss, we have

$$r^{mean} \leq h'_A(v) \leq r^{peak}$$

and

$$r^{mean} \leq h'_D(v) \leq c.$$

Figure 10.9 pictorially depicts this where there is a v^*, which is the value of v for which $h'_A(v) = c$. In other words, for $v > v^*$, $h_D(v)$ essentially follows the tangent at point v^*. We can write down the relationship between $h_A(v)$ and $h_D(v)$ as

$$h_D(v) = \begin{cases} h_A(v) & \text{if } 0 \leq v \leq v^* \\ h_A(v^*) - cv^* + cv & \text{if } v > v^*, \end{cases}$$

where v^* is obtained by solving for v in the equation,

$$\frac{d}{dv}[h_A(v)] = c.$$

The implications are similar for the effective bandwidths as well. Specifically, if the effective bandwidth of the output traffic from the buffer is

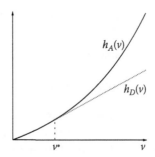

FIGURE 10.9
Relationship between ALMGF of the input and the output of a buffer. (From Kulkarni, V.G. and Gautam, N., *Queueing Syst. Theory Appl.*, 27, 79, 1997. With permission.)

$$eb_D(v) = \lim_{t \to \infty} \frac{1}{vt} \log E\{\exp(vD(t))\},$$

then to obtain the relationship between $eb_D(v)$ and $eb_A(v)$, we can write down the effective bandwidth $eb_D(v)$ of the output as

$$eb_D(v) = \begin{cases} eb_A(v) & \text{if } 0 \le v \le v^* \\ c - \frac{v^*}{v}\{c - eb_A(v^*)\} & \text{if } v > v^*. \end{cases}$$

Notice from the preceding expression that $eb_D(v) \le eb_A(v)$. That is mainly because $D(t) \le A(t)$ for all t if the queue was empty initially. Further, the peak rate for the input is r^{peak} whereas it is c for the output. Since $r^{peak} > c$, there would be values of v when $eb_D(v)$ will be lesser than $eb_A(v)$ (as for very large v, the effective bandwidths approach their respective peak values, i.e., r^{peak} for input and c for output). The main implication of $eb_D(v) \le eb_A(v)$ is that as fluid is passed from queue to queue, it will eventually get more and more smooth approaching closer to the mean rate.

For more details regarding effective bandwidths of output processes, refer to Chang and Thomas [16], Chang and Zajic [17], and de Veciana et al. [23]. In the following section, we will find out how to use effective bandwidths to obtain bounds and approximations for the steady-state buffer contents.

10.2 Performance Analysis of a Single Queue

Consider the single buffer fluid queue described in Figure 10.1. Recall that the buffer is of size infinity, and this buffer-size assumption will hold for the entire chapter. Fluid enters the buffer at discrete rates governed by a discrete-state environment process $\{Z(t), t \ge 0\}$. When the environment process is in state $Z(t)$ at time t, fluid enters the buffer at rate $r(Z(t))$. The output capacity of the buffer is a constant c. Let $X(t)$ be the amount of fluid in the buffer at time t. We assume that the system is stable, that is,

$$E[r(Z(\infty))] < c.$$

Then as $t \to \infty$, the buffer contents converge to a stationary distribution, in particular $X(t) \to X$. The objective of this section is to obtain the probability distribution of X. Specifically, we would like to obtain expressions for $P\{X > x\}$ in the form of bounds or approximations if exact ones are intractable.

A common feature for the bounds and approximations of $P\{X > x\}$ is that all of them use the effective bandwidth of the fluid input. In fact, the structure of all the approximations and bounds would also be somewhat similar. The key differences among the different bounds and approximations are the values of x for which the methods are valid and whether the result is conservative. By *conservative*, we mean that our expression for $P\{X > x\}$ is higher than the true $P\{X > x\}$. The reason we call that conservative is if one were to design a system based on our expression for $P\{X > x\}$, then what is actually observed in terms of performance would only be better. With that notion, we first summarize the various methods for bounds and approximations. Later we describe them in detail.

- *Exact computation:*
  Expressions for $P\{X > x\}$ of the type

  $$P\{X > x\} = \sum_i b_i e^{-\eta_i x}.$$

  Based on: Anick et al. [5], Elwalid and Mitra [28, 29], and Kulkarni [69].
  Described in: Section 9.2.3.
  Valid for: Any x and any CTMC environment process $\{Z(t), t \geq 0\}$ or environment processes that can easily be modeled as CTMCs.
  Drawback: Not easily extendable to other environment processes.
- *Effective bandwidth approximation:*
  Estimates of the tail probabilities of $P\{X > x\}$ using just the effective bandwidth calculations as

  $$P\{X > x\} \approx e^{-\eta x}.$$

  Based on: Elwalid and Mitra [30], Kesidis et al. [61], Krishnan et al. [65], and Kulkarni [68].
  Described in: Section 10.2.1.
  Valid for: Large x and a wide variety of stochastic processes.
  Drawback: Could be off by an order of magnitude for not-so-large x.
- *Chernoff dominant eigenvalue approximation:*
  Improvement to the effective bandwidth approximation for $P\{X > x\}$ as

  $$P\{X > x\} \approx L e^{-\eta x},$$

  where L is an adjustment using the loss probability in a bufferless system.

Based on: Elwalid et al. [31, 32].
Described in: Section 10.2.2.
Valid for: Medium to large x and many stationary stochastic processes.
Drawback: Not always conservative.

- *Bounds for buffer content distribution:*
Bounds on $P\{X > x\}$ of the form

$$C_* e^{-\eta x} \le P\{X > x\} \le C^* e^{-\eta x}.$$

Based on: Palmowski and Rolski [87, 88] and Gautam et al. [39].
Described in: Section 10.2.3.
Valid for: Any x and any SMP environment process $\{Z(t), t \ge 0\}$.
Drawback: Computationally harder than other methods.

Notice that the η described in the exponent of the last three methods are in fact equal. So essentially the methods eventually only differ in the constant that multiplies $e^{-\eta x}$. However, the approaches are somewhat different and their scopes are different too. We will see in the following sections. We already discussed the computation of $P\{X > x\}$ for CTMC environment processes in Chapter 9. The others are described in the following.

10.2.1 Effective Bandwidths for Tail Asymptotics

In Section 10.1, we saw how to compute the effective bandwidths of environment processes or sources. Here we will use effective bandwidths to characterize the tail of the buffer content distribution. For that consider the stochastic fluid flow queue in Figure 10.1. There is a discrete-state environment process $\{Z(t), t \ge 0\}$ that governs input to the infinite-sized buffer. When the environment is in state $Z(t)$, fluid enters the buffer at rate $r(Z(t))$. We assume that the Gärtner–Ellis conditions (see Section 10.1.1) are satisfied. The output channel capacity is a fixed constant c. Let $X(t)$ be the amount of fluid in the buffer at time t. We assume that the queue is stable, that is, $E[r(Z(\infty))] < c$. Under conditions of stability, $X(t) \to X$ as $t \to \infty$. We now describe how to compute the tail distribution of x, that is, $P\{X > x\}$ as $x \to \infty$. This is also called tail asymptotics (hence the title of this section).

Consider we can compute the effective bandwidth of the input to the buffer $eb(v)$. Using results from large deviations, for large values of x (specifically as $x \to \infty$),

$$P\{X > x\} \approx e^{-\eta x}, \tag{10.4}$$

where η is the solution to

$$eb(\eta) = c. \tag{10.5}$$

It is worthwhile recalling that $eb(v)$ is an increasing function of v that increases from r^{mean} to r^{peak} as v goes from 0 to ∞. Also owing to stability and nontriviality, we have $r^{mean} < c < r^{peak}$. So there would be a unique solution to Equation 10.5.

A reasonable algorithm to use is to pick an arbitrary initial v and see if $eb(v)$ is less than or greater than c. If it is less, we can keep increasing v until $eb(v) > c$. Then we can search for η between 0 and v. Since the effective bandwidth increases from 0 to a value greater than c as we go from 0 to v, we can find η by considering it as $v/2$. If $eb(v/2)$ is less than c, then we can search between $v/2$ and v, otherwise we can search between $v/2$ and v. We can proceed in this way by updating the upper and lower limits on η until it converges. Such an algorithm is called a binary search. Of course, one could develop more efficient algorithms. Next we present some examples to illustrate the approach of computing the tail distribution, provide some intuition for the results, and extend the results to different conditions.

Problem 105

Consider an infinite-sized buffer with output capacity c and input regulated by an on-off source with on-times according to $\exp(\alpha)$ and off-times according to $\exp(\beta)$. Traffic is generated at rate r, when the source is in the on-state, and no traffic is generated when the source is in the off-state. Assume that $r\beta/(\alpha+\beta) < c < r$. Using the effective bandwidth approximation, develop an expression for the tail distribution. Compare that with the exact expression for the buffer contents.

Solution

Recall from Equation 10.2 that the effective bandwidth of such a CTMC on-off source is

$$eb(v) = \frac{rv - \alpha - \beta + \sqrt{(rv - \alpha - \beta)^2 + 4\beta rv}}{2v}.$$

Solving for η in $eb(\eta) = c$, we get

$$\frac{\eta - \alpha - \beta + \sqrt{(\eta - \alpha - \beta)^2 + 4r\beta\eta}}{2\eta} = c$$

which results in

$$\eta = -\frac{\beta}{c} + \frac{\alpha}{(r-c)}.$$

Therefore, for large values of x (specifically as $x \to \infty$),

$$P\{X > x\} \approx e^{-\eta x},$$

where

$$\eta = -\frac{\beta}{c} + \frac{\alpha}{(r-c)}.$$

From the exact analysis in Equation 9.22 in Problem 87, we can see that

$$P\{X > x\} = \frac{\beta r}{c(\alpha + \beta)} e^{\lambda x},$$

where

$$\lambda = \frac{\beta}{c} - \frac{\alpha}{(r-c)}.$$

Notice that since $\lambda = -\eta$, the approximation is off by only a factor $r\beta/(c(\alpha + \beta))$, which would be reasonable as x becomes very large. ∎

Next we describe another CTMC environment process, where we again know the exact steady-state buffer content distribution. The objective is to see what transpires in a CTMC with more states.

Problem 106

Consider the system described in Problem 85, where there is an infinite-sized buffer with output capacity $c = 12$ kbps and the input is driven by a four-state CTMC with

$$Q = \begin{bmatrix} -10 & 2 & 3 & 5 \\ 0 & -4 & 1 & 3 \\ 1 & 1 & -3 & 1 \\ 1 & 2 & 3 & -6 \end{bmatrix}.$$

Also (in units of kbps),

$$R = \begin{bmatrix} 20 & 0 & 0 & 0 \\ 0 & 15 & 0 & 0 \\ 0 & 0 & 10 & 0 \\ 0 & 0 & 0 & 5 \end{bmatrix}.$$

Using the effective bandwidth approximation, derive an expression for $\lim_{t \to \infty} P\{X(t) > x\}$ and compare it against the exact value.

Solution

We have already verified in Problem 85 that the system is stable with $r^{mean} = 0.0668 \times 20 + 0.2647 \times 15 + 0.4118 \times 10 + 0.2567 \times 5 = 10.708$ kbps. Also, $r^{peak} = 20$ kbps. The effective bandwidth of a CTMC environment process with infinitesimal generator matrix Q and rate matrix R can be computed from Equation 10.1 as

$$eb(v) = e\left(\frac{Q}{v} + R\right),$$

where $e(Q/v + R)$ is the largest real eigenvalue of matrix $Q/v + R$. Using a computer program, we can solve for η in $eb(\eta) = c$ by performing a binary search arbitrarily starting with $v = 1$ (for which $eb(1) = 12.8591$). The search results in $\eta = 0.5994$ per kB. Therefore, for large values of x (specifically as $x \to \infty$),

$$\lim_{t \to \infty} P\{X(t) > x\} \approx e^{-0.5994x}.$$

From the exact analysis in Problem 85, we can write

$$\lim_{t \to \infty} P\{X(t) > x\} = 0.6757e^{-0.5994x} - 0.0079e^{-1.3733x}$$

for all $x \geq 0$. The second term in this example is practically negligible for almost any x value. Thus, the approximation is off about 0.6757, which would be reasonable when x is large and we would get the right order of magnitude. ∎

We now formalize what we saw in the previous two problems. Consider an infinite-sized stable queue with output capacity c and CTMC environment process with ℓ states of which m have drift lesser than or equal to zero (i.e., $\ell - m$ states with strictly positive drifts). Exact analysis (using the notation in Section 9.2.2) yields

$$P\{X > x\} = 1 - \sum_{j=1}^{\ell-m+1} F(x,j) = \sum_{i=1}^{\ell-m} k_i e^{\lambda_i x},$$

where each k_i is in terms of all the ϕ_j and a_j values. However, the effective bandwidth approximation for large values of x yields

$$P\{X > x\} \approx e^{-\eta x}$$

and it turns out that

$$\eta = - \max_{i:Re(\lambda_i)<0} \lambda_i.$$

In other words, η corresponds to the most dominant exponent of the exact analysis term. That makes sense because as x grows, the effect of the other terms would become negligible. The only issue is the factor k_i that multiplies the exponent. In the following section, we will see how to derive an approximation for k_i. However, for the rest of this section we continue with the effective bandwidth approximation with the understanding that we are mainly interested in only an approximate order of magnitude analysis. For example, in some applications a design requirement is for the tail probability to be of the caliber of 10^{-9}. For something as small as that, these analytical models may be the only feasible approaches. Next we will show how to use the analysis in the context of a well-known discrete queue, namely, the $M/M/1$ queue, and derive the steady-state workload tail distribution.

Problem 107

Consider a stable $M/M/1$ queue with arrival rate λ and service rate μ. Derive an expression for the steady-state workload distribution. Then use the effective bandwidth approximation to obtain the probability that the steady-state workload is greater than x for some very large x.

Solution

Let W_ℓ be a random variable denoting the steady-state workload in the system for an $M/M/1$ queue. By conditioning on the steady-state number in the system, we can write down the LST:

$$E[e^{-sW_\ell}] = \sum_{j=0}^{\infty} \left(1 - \frac{\lambda}{\mu}\right) \left(\frac{\lambda}{\mu}\right)^j \left(\frac{\mu}{\mu+s}\right)^j$$

$$= \left(1 - \frac{\lambda}{\mu}\right) \left(\frac{\mu+s}{\mu+s-\lambda}\right)$$

$$= \left(1 - \frac{\lambda}{\mu}\right) \left(1 + \frac{\lambda}{\mu+s-\lambda}\right).$$

The first equation is essentially by conditioning on the steady-state number in the system being j (which happens with probability $(1-\lambda/\mu)(\lambda/\mu)^j$), which results in a workload equal to the sum of j $\exp(\mu)$ random variables or an Erlang random variable with j phases. Inverting the LST we get the CDF of the steady-state workload as

$$P\{W_\ell \leq x\} = \left(1 - \frac{\lambda}{\mu}\right)\left(1 + \frac{\lambda}{\mu-\lambda}(1 - e^{-(\mu-\lambda)x})\right) = 1 - \frac{\lambda}{\mu}e^{-(\mu-\lambda)x}.$$

Thus, we have, for any x,

$$P\{W_\ell > x\} = \frac{\lambda}{\mu}e^{-(\mu-\lambda)x}.$$

Before proceeding, it is worthwhile to verify the result. In particular, we can verify that $P\{W_\ell = 0\} = 1 - \lambda/\mu$, which makes sense since the probability of zero workload is the same as the probability of having zero customers in the system. Also, $E[W_\ell] = \lambda/(\mu(\mu - \lambda))$, which also makes sense since $E[W_\ell]$ must be equal to W_q (as an arriving customer will see W_q workload on average) and it must also be equal to L/μ since on average there are L customers each with mean workload $1/\mu$. Having verified the results, we now consider the effective bandwidth approximation.

By definition, $A(t)$ is the total amount of workload that arrived into the buffer in time $(0, t]$. We consider $c = 1$. If S_i is the service time of customer i, then

$$A(t) = \sum_{i=1}^{N(t)} S_i,$$

where

$S_i \sim \exp(\mu)$

$N(t)$ is the number of events (i.e., arrivals) in time $(0, t]$ of a Poisson process

We can immediately write

$$E[e^{vA(t)}] = E[E[e^{vA(t)}|N(t)]] = E\left[\left(\frac{\mu}{\mu - v}\right)^{N(t)}\right]$$

since $E[e^{vS_i}] = \mu/(\mu - v)$. Also, by computing the generating function for a Poisson random variable $N(t)$, we can write

$$E[e^{vA(t)}] = e^{-(1-z)\lambda t},$$

where $z = \mu/(\mu - v)$. Thus, the effective bandwidth is

$$eb(v) = \lim_{t \to \infty} \frac{1}{vt} \log E[e^{vA(t)}] = \frac{(z-1)\lambda}{v} = \frac{\lambda}{\mu - v}.$$

Solving for η in $eb(\eta) = c$ with $c = 1$ we get

$$\eta = \mu - \lambda.$$

Thus, we have the tail distribution of the workload for very large x as

$$P\{W_\ell > x\} \approx e^{-\eta x},$$

where $\eta = \mu - \lambda$. Notice from the exact analysis that $P\{W_\ell > x\} = (\lambda/\mu)\, e^{-\eta x}$. Thus, similar to the fluid queue, here too the exponent term agrees perfectly, which would make the approximation excellent as x grows. ∎

Before proceeding, we quickly extend the effective bandwidth approximation to single buffers with "multiple" independent sources as described in Figure 10.8. Consider a single buffer that admits fluid from K independent sources such that each source k (for $k = 1, \ldots, K$) is driven by a random environment process $\{Z^k(t), t \geq 0\}$. When source k is in state $Z^k(t)$, it generates fluid at rate $r^k(Z^k(t))$ into the buffer. We still consider $X(t)$ to be the amount of fluid in the buffer at time t. The buffer has infinite capacity and is serviced by a channel of constant capacity c. The dynamics of the buffer-content process $\{X(t), t \geq 0\}$ is described by

$$\frac{dX(t)}{dt} = \begin{cases} \sum_{k=1}^{K} r^k(Z^k(t)) - c & \text{if } X(t) > 0 \\ \left\{\sum_{k=1}^{K} r^k(Z^k(t)) - c\right\}^+ & \text{if } X(t) = 0 \end{cases}$$

where $\{x\}^+ = \max(x, 0)$. The buffer-content process $\{X(t), t \geq 0\}$ is stable if

$$\sum_{k=1}^{K} E\{r^k(Z^k(\infty))\} < c.$$

Let $eb_k(v)$ be the effective bandwidths of source k such that

$$eb_k(v) = \lim_{t \to \infty} \frac{1}{vt} \log E\{\exp(vA_k(t))\},$$

where

$$A_k(t) = \int_0^t r^k(Z^k(u))du.$$

Assuming stability, let η be the unique solution to

$$\sum_{k=1}^{K} eb_k(\eta) = c.$$

The effective bandwidth approximation for large values of x yields

$$P\{X > x\} \approx e^{-\eta x}.$$

Next we present an example of using effective bandwidths in a feed-forward in-tree network that uses multiple sources as well as output effective bandwidths to illustrate our methodology. A network is called an in-tree network if the output from one node only goes to at most one other node, whereas the input to a node can come from several other nodes.

Problem 108

A wireless sensor system with seven nodes form a feed-forward in-tree network as depicted in Figure 10.10. Every node of the network has an infinite-sized buffer (denoted by $B_1, \ldots, B_7$) and information flows in and out of those buffers. Except for nodes 5 and 7 that only transmit sensed information, all other nodes sense as well as transmit information. We model sensed information to arrive into buffers $B_1, B_2, B_3, B_4,$ and B_6 as independent and identically distributed exponential on-off fluids with parameters α per second, β per second, and r kBps. For $i = 1, \ldots, 7$, the output capacity of buffer B_i is c_i kBps. The sensor network operators would prefer not to have more than b kBps of information stored in any buffer at any time in steady state. Using the effective bandwidth approximation, derive approximations for the probability of exceeding b kB of information in each of the seven buffers. Obtain

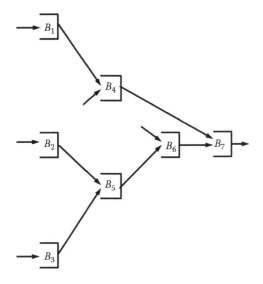

FIGURE 10.10
In-tree network.

numerical values for the approximate expressions using $\alpha = 5$, $\beta = 1$, $r = 6$, $c_1 = c_2 = c_3 = 2$, $c_4 = c_6 = 5$, $c_5 = 3$, $c_7 = 8$, and $b = 14$.

Solution

Notice that the sources are identical purely for ease of exposition and that is not necessary at all for the analysis. However, it is crucial that the sources are independent. We denote $eb_s(v)$ as the effective bandwidth of the source of sensed information. We have five such sources that input traffic into buffers B_1, B_2, B_3, B_4, and B_6. Since each of them is an exponential on-off source with parameters α, β, and r, the effective bandwidth of the sources from Equation 10.2 is

$$eb_s(v) = \frac{rv - \alpha - \beta + \sqrt{(rv - \alpha - \beta)^2 + 4\beta rv}}{2v}.$$

Also, the mean fluid-flow rate is

$$r_s^{mean} = \frac{r\beta}{(\alpha + \beta)}.$$

The condition, for stability for buffers $B_1, \ldots, B_7$ are $r_s^{mean} < c_1$, $r_s^{mean} < c_2$, $r_s^{mean} < c_3$, $2r_s^{mean} < c_4$, $2r_s^{mean} < c_5$, $3r_s^{mean} < c_6$, and $5r_s^{mean} < c_7$, respectively. Verify that for the numerical values $\alpha = 5$, $\beta = 1$, $r = 6$, $c_1 = c_2 = c_3 = 2$, $c_4 = c_6 = 5$, $c_5 = 3$, and $c_7 = 8$, the stability conditions are satisfied.

Now, let $eb_j^{in}(v)$ and $eb_j^{out}(v)$, respectively, be the net effective bandwidth flowing into and out of buffer B_j, for $j=1,\ldots,7$. The relationship between $eb_j^{in}(v)$ and $eb_j^{out}(v)$ for $j=1,\ldots,7$ is

$$eb_j^{out}(v) = \begin{cases} eb_j^{in}(v) & \text{if } 0 \le v \le v_j^* \\ c_j - \dfrac{v_j^*}{v}\left\{c_j - eb_j^{in}\left(v_j^*\right)\right\} & \text{if } v > v_j^*, \end{cases}$$

where v_j^* is obtained by solving for v in the equation

$$\frac{d}{dv}\left[v\, eb_j^{in}(v)\right] = c_j.$$

Also, from the network structure, we have the following relations:

$$eb_1^{in}(v) = eb_s(v),$$

$$eb_2^{in}(v) = eb_s(v),$$

$$eb_3^{in}(v) = eb_s(v),$$

$$eb_4^{in}(v) = eb_s(v) + eb_1^{out}(v),$$

$$eb_5^{in}(v) = eb_2^{out}(v) + eb_3^{out}(v),$$

$$eb_6^{in}(v) = eb_s(v) + eb_5^{out}(v),$$

$$eb_7^{in}(v) = eb_4^{out}(v) + eb_6^{out}(v).$$

Assuming stability, for $j=1,\ldots,7$, let X_j be the amount of fluid in buffer j in steady state. We need to obtain approximations for $P\{X_j > b\}$ for $j=1,\ldots,7$. On the basis of effective bandwidth approximation, we have $j=1,\ldots,7$

$$P\{X_j > b\} \approx e^{-\eta_j b},$$

where η_j is the unique solution to

$$eb_j^{in}(\eta_j) = c_j.$$

Thus, we solve for η_j for all $j=1,\ldots,7$ using numerical values $\alpha=5$, $\beta=1$, $r=6$, $c_1=c_2=c_3=2$, $c_4=c_6=5$, $c_5=3$, and $c_7=8$. We can solve for η in $eb_s(\eta)=4$ to obtain $\eta=\eta_1=\eta_2=\eta_3=0.75$ since $eb_1^{in}(v)=eb_2^{in}(v)=eb_3^{in}(v)=eb_s(v)$ and $c_1=c_2=c_3=4$.

Next, to compute $eb_j^{out}(v)$ for $j = 1, 2, 3$, we first need v_j^*. For $j = 1, 2, 3$, we can solve for v in

$$\frac{d}{dv}\left[v\, eb_j^{in}(v)\right] = c_j$$

to get

$$v_j^* = \frac{\beta}{r}\left(\sqrt{\frac{c_j\alpha}{\beta(r - c_j)}} - 1\right) + \frac{\alpha}{r}\left(1 - \sqrt{\frac{\beta(r - c_j)}{c_j\alpha}}\right) = 0.4031.$$

At buffer B_4 we have

$$eb_4^{in}(v) = eb_s(v) + eb_1^{out}(v) = \begin{cases} eb_s(v) + eb_1^{in}(v) & \text{if } 0 \le v \le v_1^* \\ eb_s(v) + c_1 - \frac{v_1^*}{v}\left\{c_1 - eb_1^{in}\left(v_1^*\right)\right\} & \text{if } v > v_1^*, \end{cases}$$

with $v_1^* = 0.4031$. Solving for $eb_4^{in}(\eta_4) = c_4$, we get $\eta_4 = 0.7301$.

Likewise, at buffer B_5 we have

$$eb_5^{in}(v) = eb_2^{out}(v) + eb_3^{out}(v)$$

$$= \begin{cases} eb_2^{in}(v) + eb_3^{in}(v) & \text{if } 0 \le v \le v_2^* = v_3^* \\ c_2 - \frac{v_2^*}{v}\left\{c_2 - eb_2^{in}\left(v_2^*\right)\right\} + c_3 - \frac{v_3^*}{v}\left\{c_3 - eb_3^{in}\left(v_3^*\right)\right\} & \text{if } v > v_2^* = v_3^*, \end{cases}$$

with $v_2^* = v_3^* = 0.4031$. Solving for $eb_5^{in}(\eta_5) = c_5$, we get $\eta_5 = 0.5738$.

At buffer B_6 we have

$$eb_6^{in}(v) = eb_s(v) + eb_5^{out}(v) = \begin{cases} eb_s(v) + eb_5^{in}(v) & \text{if } 0 \le v \le v_5^* \\ eb_s(v) + c_5 - \frac{v_5^*}{v}\left\{c_5 - eb_5^{in}\left(v_5^*\right)\right\} & \text{if } v > v_5^*, \end{cases}$$

with $v_5^* = 0.2363$. Solving for $eb_6^{in}(\eta_6) = c_6$, we get $\eta_6 = 0.8348$.

Finally, for the analysis of buffer B_7, we first obtain $v_4^* = 0.5407$ and $v_6^* = 0.4031$. Then we have

$$eb_7^{in}(v) = eb_4^{out}(v) + eb_6^{out}(v)$$

$$= \begin{cases} eb_4^{in}(v) + eb_6^{in}(v) & \text{if } 0 \le v \le v_6^*, \\ eb_4^{in}(v) + c_6 - \frac{v_6^*}{v}\left\{c_6 - eb_6^{in}\left(v_6^*\right)\right\} & \text{if } v_6^* < v \le v_4^*, \\ c_4 - \frac{v_4^*}{v}\left\{c_4 - eb_4^{in}\left(v_4^*\right)\right\} + c_6 - \frac{v_6^*}{v}\left\{c_6 - eb_6^{in}\left(v_6^*\right)\right\} & \text{if } v > v_4^*. \end{cases}$$

Solving for $eb_7^{in}(\eta_7) = c_7$, we get $\eta_7 = 0.6185$.

Thus, we have $P\{X_j > b\}$ for $j = 1, \dots, 7$ using the effective bandwidth approximation given by

$$P\{X_j > b\} \approx e^{-\eta_j b},$$

with $b = 14$. Using the values of η_j described earlier for $j = 1, \dots, 7$, we can write $P\{X_1 > b\} = P\{X_2 > b\} = P\{X_3 > b\} \approx 2.7536 \times 10^{-5}$, $P\{X_4 > b\} \approx 3.6383 \times 10^{-5}$, $P\{X_5 > b\} \approx 3.2451 \times 10^{-4}$, $P\{X_6 > b\} \approx 8.4007 \times 10^{-6}$, and $P\{X_7 > b\} \approx 1.7356 \times 10^{-4}$.   ∎

10.2.2 Chernoff Dominant Eigenvalue Approximation

Recall that in the previous section, we presented a reasonable approximation for the steady-state buffer content distribution of the form $P(X > x) \approx e^{-\eta x}$. However, in several examples we saw that although for large x the order of magnitude is reasonable, the approximation still had scope for fine-tuning. In particular, the examples using CTMC sources and $M/M/1$ queue where we can obtain exact results, we saw that a suitable multiplicative factor would boost the approximation significantly. This is the motivation for the results in this section called the Chernoff dominant eigenvalue (CDE) approximation, which is essentially a refinement of the effective bandwidth approximation.

Consider the single buffer fluid model with multiple input sources described in Figure 10.8. In essence, the CDE approximation yields

$$P\{X > x\} \approx L e^{-\eta x},$$

where L can be thought of as the fraction of the fluid that would be lost in steady state if there was no buffer and η is the solution to

$$\sum_{k=1}^{K} e b_k(\eta) = c.$$

Notice that η is the same exponent as in the effective bandwidth approximation. Thus, the only additional computation in the CDE approximation is L, which we describe next.

The input sources are characterized by a function $m_k(w)$, which is similar to the ALMGF ($h_k(v)$), and is defined as

$$m_k(w) = \lim_{t \to \infty} \log E\{\exp(w r^k(Z^k(t)))\}.$$

Let

$$s^* = \sup_{w \geq 0} \left\{ c\,w - \sum_{k=1}^{K} m_k(w) \right\},$$

and w^* be obtained by solving

$$\sum_{k=1}^{K} m'_k(w^*) = c,$$

where $m'_k(w)$ denotes the derivative of $m_k(w)$ with respect to w. Then the Chernoff estimate of L is

$$L \approx \frac{\exp(-s^*)}{w^* \sigma(w^*)\sqrt{2\pi}},$$

where

$$\sigma^2(w^*) = \sum_{k=1}^{K} m''_k(w^*),$$

with $m''_k(w)$ denoting the second derivative of $m_k(w)$ with respect to w. The main problem is in computing $m_k(w)$. If $\{Z^k(t), t \geq 0\}$ can be modeled as a stationary and ergodic process with state space S_k and stationary probability vector, p_k, then

$$m_k(w) = \log \left\{ \sum_{j \in S_k} p_k^j\, e^{w\, r^k(j)} \right\}.$$

Next, we present a few examples to study the effectiveness of the CDE approximation as well as illustrate the methods to compute L described earlier.

Problem 109

Consider a source modulated by an ℓ-state irreducible CTMC $\{Z(t), t \geq 0\}$ with infinitesimal generator

$$Q = [q_{ij}]$$

and stationary distribution vector $p = [p_\ell]$ satisfying

$$pQ = 0 \quad \text{and} \quad \sum_{l=1}^{\ell} p_l = 1.$$

When the CTMC is in state i, the source generates fluid at rate r_i. This source inputs traffic into an infinite capacity buffer with output channel capacity c. Assume that $\sum_{l=1}^{\ell} p_i r_i < c < \max_l r_l$. Using CDE approximation develop an expression for $P(X > x)$. Then, illustrate the approach for the numerical example of Problem 106.

Solution

Using the CDE approximation, we have

$$P(X > x) \approx L e^{-\eta x},$$

where η is the unique solution to

$$e\left(R + \frac{Q}{\eta}\right) = c,$$

with $e(M)$ denoting the largest eigenvalue of matrix M. To compute L, consider function $m(w)$ (which is essentially $m_k(w)$ but we only have $K = 1$) given by

$$m(w) = \log\left\{\sum_{j=1}^{\ell} p_j e^{w r_j}\right\}.$$

We can thus compute w^* as the solution to

$$\frac{\sum_{j=1}^{\ell} p_j r_j e^{w^* r_j}}{\sum_{j=1}^{\ell} p_j e^{w^* r_j}} = c.$$

Also,

$$s^* = cw^* - m(w^*)$$

and

$$\sigma^2(w^*) = \frac{\left(\sum_{j=1}^{\ell} p_j e^{w^* r_j}\right)\left(\sum_{j=1}^{\ell} p_j r_j^2 e^{w^* r_j}\right) - \left(\sum_{j=1}^{\ell} p_j r_j e^{w^* r_j}\right)^2}{\left(\sum_{j=1}^{\ell} p_j e^{w^* r_j}\right)^2}.$$

Using this, we can approximately estimate L as

$$L \approx \frac{\exp(-s^*)}{w^* \sigma(w^*)\sqrt{2\pi}}.$$

To illustrate the preceding bounds, we consider the system in Problem 106 with an infinite-sized buffer whose output capacity is $c = 12$ kbps and input driven by a four-state CTMC that has

$$Q = \begin{bmatrix} -10 & 2 & 3 & 5 \\ 0 & -4 & 1 & 3 \\ 1 & 1 & -3 & 1 \\ 1 & 2 & 3 & -6 \end{bmatrix}.$$

Also (in units of kbps),

$$R = \begin{bmatrix} 20 & 0 & 0 & 0 \\ 0 & 15 & 0 & 0 \\ 0 & 0 & 10 & 0 \\ 0 & 0 & 0 & 5 \end{bmatrix}.$$

Recall from Problem 106 that the steady-state probability vector is $p = [0.0668\ 0.2647\ 0.4118\ 0.2567]$ and η that solves $e(R + Q/\eta) = c$ is $\eta = 0.5994$ per kB. To express L, we numerically obtain $w^* = 0.0649$, $\sigma^2(w^*) = 20.3605$, $s^* = 0.0423$, thereby $L = 0.5208$. Thus, from the CDE approximation, we have

$$P(X > x) \approx 0.5208 e^{-0.5994x}. \qquad \blacksquare$$

Problem 110

Consider two sources that input traffic into an infinite-sized buffer with output capacity $c = 10$. The first source is identical to that in Problem 102 and the second source is identical to that in Problem 103. In other words, from source-1, fluid flows into the according to a three-state SMP $\{Z_1(t), t \geq 0\}$ with state space $\{1, 2, 3\}$. The buffer elements of the kernel of this SMP are $G_{12}(t) = 1 - e^{-t} - te^{-t}$, $G_{21}(t) = 0.4(1 - e^{-0.5t}) + 0.3(1 - e^{-0.2t})$, $G_{23}(t) = 0.2(1 - e^{-0.5t}) + 0.1(1 - e^{-0.2t})$, $G_{32}(t) = 1 - 2e^{-t} + e^{-2t}$, and $G_{11}(t) = G_{13}(t) = G_{22}(t) = G_{31}(t) = G_{33}(t) = 0$. Also, the flow rates in the three states are $r(i) = i$ for $i = 1, 2, 3$. Source-2 is an on-off source $\{Z_2(t), t \geq 0\}$ with state space $\{u, d\}$ that generates fluid so that the on-times are IID random variables with CDF $U(t) = 1 - 0.6e^{-3t} - 0.4e^{-2t}$ and the off-times are IID Erlang random variables with mean 0.5 and variance 1/12 in appropriate time units compatible with the on-times. When the source is on, fluid is generated at the rate of 16 units per second and no fluid is generated when the source is off. Using CDE approximation, develop an expression for the steady-state buffer content distribution.

Solution

Let $eb_1(v)$ and $eb_2(v)$ be the effective bandwidths of sources 1 and 2, respectively. We have devised methods to compute $eb_1(v)$ and $eb_2(v)$ in Problems 102 and 103, respectively. Then, using the CDE approximation we have the steady-state distribution of the buffer contents X as

$$P(X > x) \approx L e^{-\eta x},$$

where η is the unique solution to

$$eb_1(\eta) + eb_2(\eta) = c.$$

Solving for η we get

$$\eta = 0.0965,$$

with $eb_1(\eta) = 1.8535$ and $eb_2(\eta) = 8.1465$. To compute L, consider functions $m_k(w)$ (for $k = 1, 2$). We have

$$m_1(w) = \log \left\{ p_1^1 e^{w\, r(1)} + p_1^2 e^{w\, r(2)} + p_1^3 e^{w\, r(3)} \right\}$$

and

$$m_2(w) = \log \left\{ p_2^u e^{w\, r_u} + p_2^d e^{w\, r_d} \right\},$$

where $r(i) = i$ for $i = 1, 2, 3$, $\left[p_1^1 \; p_1^2 \; p_1^3 \right] = [0.2772 \; 0.6337 \; 0.0891]$ as described in Problem 102, $r_u = 16$, $r_d = 0$, and $[p_u \; p_d] = [4/9 \; 5/9]$ based on Problem 103. We can compute w^* as the solution to

$$\frac{p_1^1 r(1) e^{w^* r(1)} + p_1^2 r(2) e^{w^* r(2)} + p_1^3 r(3) e^{w^* r(3)}}{p_1^1 e^{w^* r(1)} + p_1^2 e^{w^* r(2)} + p_1^3 e^{w^* r(3)}} + \frac{p_2^u r_u e^{w^* r_u} + p_2^d r_d e^{w^* r_d}}{p_2^u e^{w^* r_u} + p_2^d e^{w^* r_d}} = c$$

using a binary search. Thus, we have $w^* = 0.0168$ for the numerical values stated in the problem. Also,

$$s^* = c w^* - m_1(w^*) - m_2(w^*) = 0.0091$$

and

$$\sigma^2(w^*) = \frac{\left(\sum_{i=1}^{3} p_1^i\, e^{w^*\, r(i)}\right)\left(\sum_{i=1}^{3} p_1^i r(i)^2\, e^{w^*\, r(i)}\right) - \left(\sum_{i=1}^{3} p_1^i r(i)\, e^{w^*\, r(i)}\right)^2}{\left(\sum_{i=1}^{3} p_1^i\, e^{w^*\, r(i)}\right)^2}$$

$$+ \frac{\left(p_2^u\, e^{w^*\, r_u} + p_2^d\, e^{w^*\, r_d}\right)\left(p_2^u r_u^2\, e^{w^*\, r_u} + p_2^d r_d^2\, e^{w^*\, r_d}\right) - \left(p_2^u r_u\, e^{w^*\, r_u} + p_2^d r_d\, e^{w^*\, r_d}\right)^2}{\left(p_2^u\, e^{w^*\, r_u} + p_2^d\, e^{w^*\, r_d}\right)^2}$$

$$= 64.2977.$$

Using this, we can approximately estimate L as

$$L \approx \frac{\exp(-s^*)}{w^*\sigma(w^*)\sqrt{2\pi}} = 1.1708.$$

Thus, from the CDE approximation, we have

$$P(X > x) \approx 1.1708 e^{-0.0965x},$$

especially for large values of x. ∎

Notice that the procedure for obtaining L does not depend on the stochastic process governing the sources once we know the steady-state distribution. This makes it convenient since a single approach can be used for any discrete stochastic process. However, as described earlier, the method is an approximation that is usually suitable for the tail distribution. Next we will consider an approach for bounds on the entire distribution, not just the tail.

10.2.3 Bounds for Buffer Content Distribution

The CDE approximation for the steady-state buffer content distribution that we presented in the previous section was of the form $P(X > x) \approx Le^{-\eta x}$. Although it was more accurate than the effective bandwidth approximation, the CDE approximation is not necessarily conservative and one cannot predict its accuracy a priori. To address that here, we describe bounds for $P(X > x)$ that works for all x, not just for large x. Similar to the case in CDE approximation, here too the setting is a single buffer fluid model with multiple input sources described in Figure 10.8. In particular, we consider an infinite capacity buffer into which fluid is generated from K sources according to environment processes $\{Z^k(t), t \geq 0\}$ for $k = 1, 2, \ldots, K$ that are independent SMPs. When the SMP $\{Z^k(t), t \geq 0\}$ for some $k \in \{1, 2, \ldots, K\}$ is in state i, fluid is generated into the buffer at rate $r_k(i)$. The SMP $\{Z^k(t), t \geq 0\}$ for all $k \in \{1, 2, \ldots, K\}$ has a state space $S_k = \{1, 2, \ldots, \ell_k\}$ and kernel $G^k(x) = \left[G_{ij}^k(x) \right]$.

Using that we can calculate the expected time spent in state i, which we call τ_i^k. Also assume that we can compute p^k, the stationary vector of the kth SMP $\{Z^k(t), t \geq 0\}$, where

$$p_i^k = \lim_{t \to \infty} P\{Z^k(t) = i\}.$$

Further, using the SMP source characteristics, say we can compute the effective bandwidth of the kth source, $eb_k(v)$. Then, as always, we let η be the solution to

$$\sum_{k=1}^{K} eb_k(\eta) = c.$$

Now we describe how to compute bounds for the steady-state buffer content distribution as follows:

$$C_* e^{-\eta x} \leq \lim_{t \to \infty} P\{X(t) > x\} \leq C^* e^{-\eta x} \quad \forall\, x \geq 0,$$

where
   $X(t)$ is the amount of fluid in the buffer at time t
   C_* and C^* are constants that we describe how to compute next

We only present an overview of the results; interested readers are referred to Gautam et al. [39]. The approach uses an argument based on an exponential change of measure.

Let $\tilde{G}_{ij}^k(s)$ be the Laplace Stieltjes transform (LST) of $G_{ij}^k(x)$. For a given $v > 0$, define

$$\chi_{ij}^k(v, u) = \tilde{G}_{ij}^k(-v(r_k(i) - u)),$$

$$\chi^k(v, u) = \left[\chi_{ij}^k(v, u) \right].$$

Denote $\Phi^k(\eta) = \chi^k(\eta, eb_k(\eta))$. Let h^k be the left eigenvector of $\Phi^k(\eta)$ corresponding to the eigenvalue 1, that is,

$$h^k = h^k \Phi^k(\eta). \tag{10.6}$$

Also define the following:

$$P^k(i,j) = [G^k(\infty)]_{ij},$$

$$H^k = \sum_{i=1}^{\ell_k} \frac{h_i^k}{\eta(r_k(i) - eb_k(\eta))} \left(\sum_{j=1}^{\ell_k} \left(\phi_{ij}^k(\eta) \right) - 1 \right), \tag{10.7}$$

$$\Psi_{min}^k(i,j) = \inf_x \left\{ \frac{h_i^k e^{-\eta(r_k(i) - eb_k(\eta))x} \int_x^\infty e^{\eta(r_k(i) - eb_k(\eta))y} dG_{ij}^k(y)}{(p_i^k/\tau_i^k) \int_x^\infty dG_{ij}^k(y)} \right\}, \tag{10.8}$$

$$\Psi_{max}^k(i,j) = \sup_x \left\{ \frac{h_i^k e^{-\eta(r_k(i) - eb_k(\eta))x} \int_x^\infty e^{\eta(r_k(i) - eb_k(\eta))y} dG_{ij}^k(y)}{(p_i^k/\tau_i^k) \int_x^\infty dG_{ij}^k(y)} \right\}. \tag{10.9}$$

Then the limiting distribution of the buffer content process is

$$C_* e^{-\eta x} \le \lim_{t \to \infty} P(X(t) > x) \le C^* e^{-\eta x}, \quad \forall x \ge 0,$$

where

$$C^* = \frac{\prod_{k=1}^K H^k}{\min_{\mathcal{A}} \prod_{k=1}^K \Psi_{min}^k(i_k, j_k)}, \quad C_* = \frac{\prod_{k=1}^K H^k}{\max_{\mathcal{A}} \prod_{k=1}^K \Psi_{max}^k(i_k, j_k)},$$

$$\mathcal{A} = \left\{ (i_1, j_1), (i_2, j_2), \ldots, (i_K, j_K) : \right.$$

$$\left. i_k, j_k \in S_k, \sum_{k=1}^K r_k(i_k) > c \text{ and } \forall k, \ P^k(i_k, j_k) > 0 \right\}.$$

In these expressions, the only unknown terms are Ψ_{max}^k and Ψ_{min}^k. Next, we describe how to compute them for some special cases. For that, *we drop k with the understanding that all the expressions pertain to k.*

First consider a nonnegative random variable Y with CDF $G_{ij}(x)/G_{ij}(\infty)$ and density

$$g_{ij}(x) = \frac{dG_{ij}(x)}{dx} \frac{1}{G_{ij}(\infty)}.$$

The failure rate function of Y is defined by

$$\lambda_{ij}(x) = \frac{g_{ij}(x)}{1 - G_{ij}(x)/[G_{ij}(\infty)]}.$$

Then Y is said to be an increasing failure rate (IFR) random variable if

$$\lambda_{ij}(x) \uparrow x$$

and Y is said to be a decreasing failure rate (DFR) random variable if

$$\lambda_{ij}(x) \downarrow x.$$

It is possible to obtain closed form algebraic expressions for $\Psi_{max}(i,j)$ and $\Psi_{min}(i,j)$, if random variable Y with distribution $G_{ij}(x)/G_{ij}(\infty)$ is an IFR or DFR random variable. The following result describes how to compute $\Psi_{max}(i,j)$ and $\Psi_{min}(i,j)$ in those cases. Let x^* and x_* be such that

$$x^* = \arg\sup_x \left\{ \frac{h_i \int_x^\infty e^{\eta(r_i-c)y} dG_{ij}(y)}{(p_i/\tau_i)e^{\eta(r_i-c)x} \int_x^\infty dG_{ij}(y)} \right\}$$

and

$$x_* = \arg\inf_x \left\{ \frac{h_i \int_x^\infty e^{\eta(r_i-c)y} dG_{ij}(y)}{(p_i/\tau_i)e^{\eta(r_i-c)x} \int_x^\infty dG_{ij}(y)} \right\}.$$

Then $\Psi_{max}(i,j)$ and $\Psi_{min}(i,j)$ occur at x values given by Table 10.1 with the understanding

$$\lambda_{ij}(\infty) = \lim_{x\to\infty} \lambda_{ij}(x).$$

TABLE 10.1

Computing $\Psi_{max}(i,j)$ and $\Psi_{min}(i,j)$

	IFR		DFR	
	$r_i > c$	$r_i \le c$	$r_i > c$	$r_i \le c$
x^*	0	∞	∞	0
$\Psi_{max}(i,j)$	$\dfrac{\tilde\Phi_{ij}(-\eta(r_i-c))\tau_i h_i}{p_{ij}p_i}$	$\dfrac{\tau_i h_i \lambda_{ij}(\infty)}{p_i(\lambda_{ij}(\infty)-\eta(r_i-c))}$	$\dfrac{\tau_i h_i \lambda_{ij}(\infty)}{p_i(\lambda_{ij}(\infty)-\eta(r_i-c))}$	$\dfrac{\tilde\Phi_{ij}(-\eta(r_i-c))\tau_i h_i}{p_{ij}p_i}$
x_*	∞	0	0	∞
$\Psi_{min}(i,j)$	$\dfrac{\tau_i h_i \lambda_{ij}(\infty)}{p_i(\lambda_{ij}(\infty)-\eta(r_i-c))}$	$\dfrac{\tilde\Phi_{ij}(-\eta(r_i-c))\tau_i h_i}{p_{ij}p_i}$	$\dfrac{\tilde\Phi_{ij}(-\eta(r_i-c))\tau_i h_i}{p_{ij}p_i}$	$\dfrac{\tau_i h_i \lambda_{ij}(\infty)}{p_i(\lambda_{ij}(\infty)-\eta(r_i-c))}$

To illustrate the bounds for the steady-state distribution, we consider a few examples. In all examples, we assume the buffer to be stable and $X(t) \to X$ as $t \to \infty$. We first show examples for a single source (i.e., $K = 1$) with a single buffer and then generalize it to many sources as well as simple networks. We begin with a CTMC source and compare it against the exact results.

Problem 111

Consider a source modulated by an ℓ-state irreducible CTMC $\{Z(t), t \geq 0\}$ with infinitesimal generator

$$Q = [q_{ij}]$$

and stationary distribution p satisfying

$$pQ = 0 \quad \text{and} \quad \sum_{l=1}^{\ell} p_l = 1.$$

When the CTMC is in state i, the source generates fluid at rate r_i. Let

$$R = \text{diag}[r_i].$$

This source inputs traffic into an infinite capacity buffer with output channel capacity c. Develop upper and lower bounds for $P(X > x)$. Then, for the numerical example of Problem 106, compare the bounds against exact results.

Solution

We obtain η by solving

$$eb(\eta) = c,$$

where $eb(\cdot)$ is the effective bandwidth of the CTMC source and it is given by

$$eb(v) = e\left(R + \frac{Q}{v}\right).$$

Next, note that $\{Z(t), t \geq 0\}$ is a special case of an SMP with kernel

$$G(x) = [G_{ij}(x)]$$

$$= \begin{cases} \left[\frac{q_{ij}}{q_i}(1 - e^{-q_i x})\right] & \text{if } i \neq j \\ 0 & \text{if } i = j \end{cases}$$

where $q_i = -q_{ii} = \sum_{j \neq i} q_{ij}$. The expected amount of time the CTMC spends in state i is

$$\tau_i = \frac{1}{q_i}.$$

From Equation 10.3, we have

$$\Lambda_{ij}(u, v) = \begin{cases} \frac{q_{ij}}{q_i - v r_i + u} & \text{if } i \neq j \\ 0 & \text{if } i = j \end{cases}$$

For a given v,

$$e^*(v) = \sup_{\{u > 0 : e(\Lambda(u,v)) < \infty\}} \{e(\Lambda(u, v))\}$$

approaches infinity. Therefore, $v^* = \infty$. Thus, we have

$$\phi_{ij}(\eta) = \begin{cases} \frac{q_{ij}}{q_i - \eta(r_i - c)} & \text{if } i \neq j \\ 0 & \text{if } i = j \end{cases}$$

since $eb(\eta) = c$ and the remaining terms can be obtained from $\tilde{G}_{ij}(c\eta - r_i\eta)$. Equation 10.6 without the superscript k reduces to

$$h_j = \sum_{i=1}^{\ell} h_i \phi_{ij}(\eta)$$

$$= \sum_{i \neq j} h_i \frac{q_{ij}}{q_i - \eta(r_i - c)}. \tag{10.10}$$

Using Equation 10.7, we have

$$H = \sum_{i=1}^{\ell} \frac{h_i}{\eta(r_i - c)} \left(\sum_{j=1}^{\ell} (\phi_{ij}(\eta)) - 1 \right)$$

$$= \sum_{i=1}^{\ell} \frac{h_i}{\eta(r_i - c)} \left(\frac{q_i}{q_i - \eta(r_i - c)} - 1 \right)$$

$$= \sum_{i=1}^{\ell} \frac{h_i}{q_i - \eta(r_i - c)}. \tag{10.11}$$

From Equation 10.9 without the superscript k, we have for $i \neq j$ and $q_{ij} > 0$

$$\Psi_{max}(i,j) = \sup_x \left\{ \frac{h_i e^{-\eta(r_i-c)x} \int_x^\infty e^{\eta(r_i-c)y} dG_{ij}(y)}{(p_i/\tau_i) \int_x^\infty dG_{ij}(y)} \right\}$$

$$= \sup_x \left\{ \frac{h_i e^{-\eta(r_i-c)x} \int_x^\infty e^{\eta(r_i-c)y} e^{-q_{iy}} dy}{p_i q_i \int_x^\infty e^{-q_{iy}} dy} \right\}$$

$$= \sup_x \left\{ \frac{1}{p_i} \frac{h_i}{q_i - \eta(r_i - c)} \right\}$$

$$= \inf_x \left\{ \frac{1}{p_i} \frac{h_i}{q_i - \eta(r_i - c)} \right\}$$

$$= \Psi_{min}(i,j).$$

The preceding result should come as no surprise since an exponential random variable has a constant hazard rate function.

Define

$$g = [g_i] = \left[\frac{h_i}{q_i - \eta(r_i - c)} \right].$$

Alternatively, we can compute g using the fact that g satisfies

$$g \left(R + \frac{Q}{\eta} \right) = cg,$$

since (using Equation 10.10)

$$\left[g \left(R + \frac{Q}{\eta} \right) \right]_j = \left(r_j - \frac{q_j}{\eta} \right) g_j + \sum_{i \neq j} \frac{g_i q_{ij}}{\eta}$$

$$= \left(r_j - \frac{q_j}{\eta} \right) \frac{h_j}{q_j - \eta(r_j - c)} + \sum_{i \neq j} \frac{q_{ij}}{\eta} \frac{h_i}{q_i - \eta(r_i - c)}$$

$$= \left(r_j - \frac{q_j}{\eta} \right) \frac{h_j}{q_j - \eta(r_j - c)} + \frac{h_j}{\eta}$$

$$= \frac{h_j}{q_j - \eta(r_j - c)} c$$

$$= c g_j.$$

At any rate, we can write down bounds for the steady-state buffer content distribution as

$$C_* e^{-\eta x} \leq P\{X > x\} \leq C^* e^{-\eta x},$$

where

$$C^* = \frac{H}{\min_{i:r_i>c,j:p_{ij}>0}\{\Psi_{min}(i,j)\}}$$

$$= \frac{\sum_{i=1}^{\ell} \frac{h_i}{q_i - \eta(r_i - c)}}{\min_{i:r_i>c}\left\{\frac{1}{p_i}\frac{h_i}{q_i - \eta(r_i - c)}\right\}}$$

$$= \frac{\sum_{i=1}^{\ell} g_i}{\min_{i:r_i>c} \frac{g_i}{p_i}}$$

and

$$C_* = \frac{H}{\max_{i:r_i>c,j:p_{ij}>0}\{\Psi_{max}(i,j)\}}$$

$$= \frac{\sum_{i=1}^{\ell} \frac{h_i}{q_i - \eta(r_i - c)}}{\max_{i:r_i>c}\left\{\frac{1}{p_i}\frac{h_i}{q_i - \eta(r_i - c)}\right\}}$$

$$= \frac{\sum_{i=1}^{\ell} g_i}{\max_{i:r_i>c} \frac{g_i}{p_i}}.$$

To illustrate the preceding bounds, we consider the system in Problem 106 with an infinite-sized buffer whose output capacity is $c = 12$ kbps and input driven by a four-state CTMC that has

$$Q = \begin{bmatrix} -10 & 2 & 3 & 5 \\ 0 & -4 & 1 & 3 \\ 1 & 1 & -3 & 1 \\ 1 & 2 & 3 & -6 \end{bmatrix}.$$

Also (in units of kbps),

$$R = \begin{bmatrix} 20 & 0 & 0 & 0 \\ 0 & 15 & 0 & 0 \\ 0 & 0 & 10 & 0 \\ 0 & 0 & 0 & 5 \end{bmatrix}.$$

Recall from Problem 106 that the steady-state probability vector is $p = [0.0668\ 0.2647\ 0.4118\ 0.2567]$ and η that solves $e(R + Q/\eta) = c$ is $\eta = 0.5994$ per kB.

To obtain C^* and C_*, we solve for g as the left eigenvector of $(R + Q/\eta)$ that corresponds to eigenvalue of c, that is, g satisfies $g(R + Q\eta) = cg$. Using that we get $g = [0.1746\ 0.7328\ 0.5533\ 0.3555]$ (although this is not unique, but notice that it appears in the numerator and denominator of both C^* and C_* and hence would be a nonissue). Next, notice that only in states $i = 1$ and $i = 2$, we have $r_i > c$. Thus, we have

$$C^* = \frac{g_1 + g_2 + g_3 + g_4}{\min(g_1/p_1, g_2/p_2)} = 0.6953$$

and

$$C_* = \frac{g_1 + g_2 + g_3 + g_4}{\max(g_1/p_1, g_2/p_2)} = 0.6561.$$

In Table 10.2, we contrast the bounds obtained as

$$C_* e^{-\eta x} \le P\{X > x\} \le C^* e^{-\eta x},$$

against the effective bandwidth approximation result in Problem 106, the exact result in Problem 85 and the CDE approximation in Problem 109. ∎

After describing one special case of an SMP, namely, CTMC, next we present another special case, the general on-off source, which is an alternating renewal process.

TABLE 10.2

Comparing the Exact Results against Approximations and Bounds

Method	Result
Exact computation	$\lim_{t \to \infty} P\{X(t) > x\} = 0.6757 e^{-0.5994x} - 0.0079 e^{-1.3733x}$
Effective bandwidth approx.	$\lim_{t \to \infty} P\{X(t) > x\} \approx e^{-0.5994x}$
CDE approx.	$\lim_{t \to \infty} P\{X(t) > x\} \approx 0.5208 e^{-0.5994x}$
Bounds	$0.6561 e^{-0.5994x} \le \lim_{t \to \infty} P\{X(t) > x\} \le 0.6953 e^{-0.5994x}$

Problem 112

Consider a source modulated by a two-state (on and off) process that alternates between the on and off states. The random amount of time the process spends in the on state (called *on-times*) has CDF $U(\cdot)$ with mean τ_U and the corresponding *off-time* CDF is $D(\cdot)$ with mean τ_D. The successive on and off-times are independent and on-times are independent of off-times. Fluid is generated continuously at rate r during the on state and at rate 0 during the off state. The source inputs traffic into an infinite-capacity buffer. The output capacity of the buffer is a constant c. State the stability condition, and assuming it is true obtain bounds for the steady-state buffer content distribution.

Solution

The stability condition is

$$\frac{r\tau_U}{\tau_U + \tau_D} < c.$$

We assume that the preceding condition is true and $c < r$ (otherwise the buffer would be empty in steady state). Following the notation described for the SMP bounds, we obtain the following. Define

$$\Phi(v) = \begin{bmatrix} 0 & \tilde{D}(vc) \\ \tilde{U}(-v(r-c)) & 0 \end{bmatrix},$$

where $\tilde{U}(\cdot)$ and $\tilde{D}(\cdot)$ are the LSTs of $U(t)$ and $D(t)$, respectively. We assume that $e(\Phi(\eta)) = 1$ has a solution and it implies that

$$e(\Phi(\eta)) = \sqrt{\tilde{U}(-\eta(r-c))\,\tilde{D}(\eta c)} = 1$$

can be solved (otherwise we would have to use the *effective bandwidth of general on/off source*). Hence, η is the smallest real-positive solution to

$$\tilde{U}(-\eta(r-c))\,\tilde{D}(\eta c) = 1.$$

Also, Equation 10.6 without the superscript k reduces to

$$h = [1 \quad \tilde{D}(\eta c)].$$

Furthermore, from Equations 10.7 through 10.9, we get

$$H = \frac{(1 - \tilde{D}(\eta c))r}{c(r-c)},$$

$$\Psi_{min} = \begin{bmatrix} 0 & \inf_x\left\{\dfrac{(\tau_U+\tau_D)\int_x^\infty e^{-\eta c(y-x)}dD(y)}{1-D(x)}\right\} \\ \inf_x\left\{\tilde{D}(\eta c)\dfrac{(\tau_U+\tau_D)\int_x^\infty e^{\eta(r-c)(y-x)}dU(y)}{1-U(x)}\right\} & 0 \end{bmatrix}$$

and

$$\Psi_{max} = \begin{bmatrix} 0 & \sup_x\left\{\dfrac{(\tau_U+\tau_D)\int_x^\infty e^{-\eta c(y-x)}dD(y)}{1-D(x)}\right\} \\ \sup_x\left\{\tilde{D}(\eta c)\dfrac{(\tau_U+\tau_D)\int_x^\infty e^{\eta(r-c)(y-x)}dU(y)}{1-U(x)}\right\} & 0 \end{bmatrix}.$$

Thus, we can derive bounds for the steady-state buffer content distribution X as

$$C_* e^{-\eta x} \leq P\{X > x\} \leq C^* e^{-\eta x}, \tag{10.12}$$

where

$$C^* = \frac{\tilde{U}(-\eta(r-c))-1}{\tau_U+\tau_D} \frac{r}{c(r-c)\eta\inf_x\left\{\dfrac{\int_x^\infty e^{\eta(r-c)(y-x)}dU(y)}{1-U(x)}\right\}} \tag{10.13}$$

and

$$C_* = \frac{\tilde{U}(-\eta(r-c))-1}{\tau_U+\tau_D} \frac{r}{c(r-c)\eta\sup_x\left\{\dfrac{\int_x^\infty e^{\eta(r-c)(y-x)}dU(y)}{1-U(x)}\right\}}. \tag{10.14}$$

If $U(\cdot)$ is IFR/DFR, the supremums and the infimums in the preceding equations can be obtained from Table 10.1. ∎

Next we consider a special case of the earlier problem (namely, the Erlang on-off source) to explore and explain the general on-off source.

Problem 113

The Erlang on-off source is one with $Erlang(N_U, \alpha)$ on-time distribution, $Erlang(N_D, \beta)$ off-time distribution, and fluid is generated at rate r when the source is on. Assuming stability, obtain bounds for the steady-state buffer content distribution. For a numerical example with $r = 15$, $c = 10$, $\tau_U = 1/70$, and $\tau_D = 1/30$, illustrate the bounds.

Solution

Note that $\tau_U = N_U/\alpha$ and $\tau_D = N_D/\beta$. Assume that the condition of stability is satisfied and $r > c$. Thus, we have

$$
\Phi(v) = \left[\begin{array}{cc} 0 & \tilde{D}(vc) \\ \tilde{U}(-v(r-c)) & 0 \end{array} \right] = \left[\begin{array}{cc} 0 & \left(\frac{\beta}{\beta+vc}\right)^{N_D} \\ \left(\frac{\alpha}{\alpha-v(r-c)}\right)^{N_U} & 0 \end{array} \right].
$$

It is possible to show that $e(\Phi(v)) = 1$ always has a solution. Hence, η is obtained by solving

$$
e(\Phi(\eta)) = \sqrt{\tilde{U}(-\eta(r-c))\,\tilde{D}(\eta c)} = 1
$$

and

$$
h = \left[1 \quad \left(\frac{\beta}{\beta+\eta c}\right)^{N_D} \right].
$$

Using the fact that the *Erlang* random variable has an increasing hazard rate function, we see that

$$
\Psi_{min} = \left[\begin{array}{cc} 0 & (N_U/\alpha + N_D/\beta)\left(\frac{\beta}{\beta+\eta c}\right)^{N_D} \\ \left(\frac{\beta}{\beta+\eta c}\right)^{N_D} (N_U/\alpha + N_D/\beta)\frac{\alpha}{\alpha-\eta(r-c)} & 0 \end{array} \right]
$$

and

$$
\Psi_{max} = \left[\begin{array}{cc} 0 & (N_U/\alpha + N_D/\beta)\frac{\beta}{\beta+\eta c} \\ \left(\frac{\beta}{\beta+\eta c}\right)^{N_D} (N_U/\alpha + N_D/\beta)\left(\frac{\alpha}{\alpha-\eta(r-c)}\right)^{N_U} & 0 \end{array} \right].
$$

Thus, we have the steady-state buffer content distribution bounded as

$$
C_* e^{-\eta x} \le P\{X > x\} \le C^* e^{-\eta x}, \tag{10.15}
$$

where

$$C^* = \frac{\left(\frac{\alpha}{\alpha-\eta(r-c)}\right)^{N_U} - 1}{\tau_U + \tau_D} \frac{r}{c(r-c)\eta\left\{\frac{\alpha}{\alpha-\eta(r-c)}\right\}}$$

(10.16)

and

$$C_* = \frac{\left(\frac{\alpha}{\alpha-\eta(r-c)}\right)^{N_U} - 1}{\tau_U + \tau_D} \frac{r}{c(r-c)\eta\left\{\left(\frac{\alpha}{\alpha-\eta(r-c)}\right)^{N_U}\right\}}.$$

(10.17)

Next consider the numerical example of an Erlang on-off source with on-time distribution $Erlang(N_U, \alpha)$ and off-time distribution $Erlang(N_D, \beta)$ with $r = 15$, $c = 10$, $\tau_U = 1/70$, and $\tau_D = 1/30$. We keep the means constant (i.e., τ_U and τ_D are held constant) but decrease the variances by increasing N_U and N_D. In Figure 10.11, we illustrate for four pairs of (N_U, N_D) (namely, $(1,1), (4,3), (9,8)$, and $(16,14)$) the logarithm of the upper and lower bounds on the limiting distribution of the buffer-content process. From the figure we notice that as the variance decreases, the bounds move further apart. Also note that C^* increases with decrease in variance and C_* decreases with decrease in variance. Since η increases with decrease in variance, the tail of the limiting distribution rapidly approaches zero. ∎

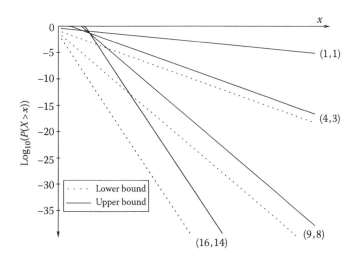

FIGURE 10.11
Logarithm of the upper and lower bounds as a function of x. (From Gautam, N. et al., *Prob. Eng. Inform. Sci.*, 13, 429, 1999. With permission.)

Remark 24

For the exponential on-off source, which is a special case of the Erlang on-off source, we get $C^* = C_* = r\beta/[c(\alpha + \beta)]$. Hence, the upper and lower bounds are equal resulting in

$$P\{X > x\} = \frac{r\beta}{c(\alpha + \beta)}e^{-\eta x},$$

where

$$\eta = \frac{c\alpha + c\beta - \beta r}{c(r - c)}$$

and this matches the result in Equation 9.22.   ∎

Next we consider a few extensions, especially in terms of multiple sources and networks. We begin by considering K CTMC sources inputing fluid into a buffer.

Problem 114

There are K independent fluid sources that input traffic into an infinite capacity buffer. Each source k is modulated by a CTMC $\{Z_k(t), t \geq 0\}$ with infinitesimal generator Q_k on state space $\{1, 2, ..., \ell_k\}$. Also,

$$p^k Q_k = 0 \quad \text{and} \quad \sum_{i=1}^{\ell_k} p_i^k = 1.$$

Fluid is generated at rate $r_k(Z_k(t))$ by source k at time t. Let R_k be the corresponding rate matrix. Fluid is removed from the buffer by a channel with constant capacity c. Let $X(t)$ be the amount of fluid in the buffer at time t. Obtain bounds for $P\{X(t) > x\}$ as $t \to \infty$ assuming that the buffer is stable.

Solution

For $k = 1, 2, ..., K$, let the effective bandwidth of source k be

$$eb_k(v) = e\left(R_k + \frac{Q_k}{v}\right),$$

where $e(M)$ is the largest real eigenvalue of the matrix M. We solve

$$\sum_{k=1}^{K} eb_k(\eta) = c$$

to obtain η. The left eigenvectors g^k are obtained by solving

$$g^k \left(R_k + \frac{Q_k}{\eta} \right) = eb_k(\eta)g^k.$$

We assume that the condition of stability

$$\sum_{k=1}^{K} \sum_{l=1}^{\ell_k} r_k(l)p_l^k < c$$

is satisfied. Hence as $t \to \infty$, $X(t) \to X$. Thus, the steady-state distribution of the buffer content is bounded as

$$C_* e^{-\eta B} \leq P\{X > B\} \leq C^* e^{-\eta B},$$

where

$$C^* = \frac{\prod_{k=1}^{K} \sum_{l=1}^{\ell_k} h_l^k}{\min_{i_1,\dots,i_k:\sum r_k(i_k)>c} \prod_{k=1}^{K} \left\{ h_{i_k}^k / p_{i_k}^k \right\}}$$

and

$$C_* = \frac{\prod_{k=1}^{K} \sum_{l=1}^{\ell_k} h_l^k}{\max_{i_1,\dots,i_k:\sum r_k(i_k)>c} \prod_{k=1}^{K} \left\{ h_{i_k}^k / p_{i_k}^k \right\}}. \qquad \blacksquare$$

In the previous problem, the denominators for the C_* and C^* expressions need to be computed carefully if one wants to avoid searching through the entire space of $\prod_{k=1}^{K} \ell_k$ terms. Of course, if the problem structure is special (such as identical sources) this would be straightforward. We will see one such example in a subsequent problem. Now we will consider another extension, namely, a tandem network.

Problem 115

An exponential on-off source with on-time parameter α, off-time parameter β, and rate r (fluid generation rate when on) generates traffic into an infinite-capacity buffer with output capacity c_1. The output from the buffer acts as an input to another infinite-capacity buffer whose output capacity is c_2. Assume for stability and nontriviality that

$$\frac{r\beta}{(\alpha + \beta)} < c_2 < c_1 < r.$$

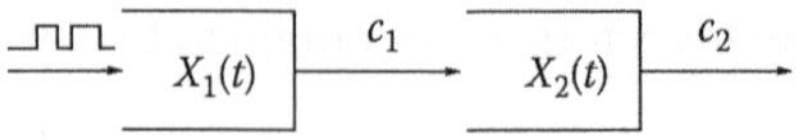

FIGURE 10.12
Exponential on-off input to buffers in tandem. (From Gautam, N. et al., *Prob. Eng. Inform. Sci.*, 13, 429, 1999. With permission.)

Obtain bounds for the buffer-content processes of the respective buffers $\{X_1(t), t \geq 0\}$ and $\{X_2(t), t \geq 0\}$. See Figure 10.12 for an illustration of the model.

Solution

The effective bandwidth of the exponential on-off source is

$$eb(v) = \frac{rv - \alpha - \beta + \sqrt{(rv - \alpha - \beta)^2 + 4\beta rv}}{2v}. \tag{10.18}$$

Since the system is stable, as $t \to \infty$, $X_1(t) \to X_1$ and $X_2(t) \to X_2$. From Remark 24, the expression for $P(X_1 > x)$ resulted in the upper and lower bounds being equal. In particular for this system, we have

$$P\{X_1 > x\} = \frac{r\beta}{c_1(\alpha + \beta)} e^{-\eta_1 x}, \tag{10.19}$$

where

$$\eta_1 = \frac{c_1\alpha + c_1\beta - \beta r}{c_1(r - c_1)}.$$

To obtain bounds for the distribution of X_2, we first model the input to the second buffer as a general on-off source with on-time distribution $U(t)$ (with mean $r/(c_1(\alpha + \beta) - r\beta)$), off-time distribution $D(t)$ (with mean $1/\beta$), and rate c_1. Clearly, the off-time corresponds to the consecutive time when buffer-1 remains empty (idle period) and the on time corresponds to the busy period (i.e., the consecutive time the buffer is nonempty). During the busy period of buffer-1, fluid flows from the first buffer to the second buffer at rate c_1, and no fluid flows between the buffers during the idle period. Also, the idle period is exponentially distributed with parameter β since once buffer-1 becomes empty (this can happen only when the original source if off), it would take $\exp(\beta)$ time for the source to switch to on. On the basis of this we have

$$D(t) = 1 - e^{-\beta t}.$$

Recall Problem 94 where we derived the LST of the busy period distribution $U(\cdot)$ as

$$\tilde{U}(w) = \begin{cases} \frac{w+\beta+c_1 s_0(w)}{\beta} & \text{if } w \geq w^* \\ \infty & \text{otherwise,} \end{cases}$$

where $w^* = (2\sqrt{c_1 \alpha \beta (r - c_1)} - r\beta - c_1\alpha - c_1\beta)/r$, $s_0(w) = (-b - \sqrt{b^2 + 4w(w + \alpha + \beta)c_1(r - c_1)})/(2c_1(r - c_1))$, and $b = (r - 2c_1)w + (r - c_1)\beta - c_1\alpha$. The LST of the distribution $D(\cdot)$ is

$$\tilde{D}(w) = \begin{cases} \frac{\beta}{\beta+w} & \text{if } w > -\beta \\ \infty & \text{otherwise.} \end{cases}$$

For this general on-off "pseudo" source that inputs traffic into the second buffer, we can compute its effective bandwidth, $eb_2(v)$, as

$$eb_2(v) = \begin{cases} eb_1(v) & \text{if } 0 \leq v \leq v^* \\ (eb_1(v^*) - c_1)\frac{v^*}{v} + c_1 & \text{if } v > v^*, \end{cases}$$

where

$$v^* = \frac{\beta}{r}\left(\sqrt{\frac{c_1\alpha}{\beta(r - c_1)}} - 1\right) + \frac{\alpha}{r}\left(1 - \sqrt{\frac{\beta(r - c_1)}{c_1\alpha}}\right) \qquad (10.20)$$

and $eb_1(v)$ is from Equation 10.18. Note that η_2 is obtained by solving

$$eb_2(\eta_2) = c_2.$$

If $\eta_2 \leq v^*$, we have

$$\Phi(\eta_2) = \begin{bmatrix} 0 & \tilde{D}(\eta_2 c_2) \\ \tilde{U}(-\eta_2(c_1 - c_2)) & 0 \end{bmatrix},$$

and $e(\Phi(\eta_2)) = 1$ (note that if $\eta_2 > v^*$, then $e(\Phi(\eta_2)) = 1$ has no solutions). Then we obtain h_2 by solving

$$[1 \ h_2]\,\Phi(\eta_2) = [1 \ h_2]$$

as $h_2 = \tilde{D}(\eta_2 c_2)$. If $\eta_2 > v^*$, we use the same h_2 since $\tilde{D}(\cdot)$ only gradually goes to infinite and the condition is mainly because of $\tilde{U}(\cdot)$. The situation is similar to the one in Figure 10.5(b). Hence, it would not cause any concerns. With this we proceed to obtain the bounds for the distribution of X_2.

Intuitively, a random variable with the distribution $U(t)$ is a DFR random variable (since $U(t)$ represents the busy period distribution). The intuition can be verified (after a lot of algebra) using the expression for $U(t)$ in Gautam et al. [39]. The steady-state distribution of the buffer-content process $\{X_2(t), t \geq 0\}$ is bounded as

$$C_{2*}e^{-\eta_2 x} \leq P(X_2 > x) \leq C_2^* e^{-\eta_2 x},$$

where

$$C_2^* = \frac{\frac{\tilde{D}(\eta_2 c_2)-1}{-\eta_2 c_2} + \frac{\tilde{U}(-\eta_2(c_1-c_2))-1}{\eta_2(c_1-c_2)}h_2}{\frac{h_2 c_1(\alpha+\beta)}{\beta(c_1(\alpha+\beta)-r\beta)} \lim_{x\to\infty}\left\{\frac{\int_x^\infty e^{\eta_2(c_1-c_2)y}dU(y)}{\int_x^\infty e^{\eta_2(c_1-c_2)x}dU(y)}\right\}}, \qquad (10.21)$$

$$C_{2*} = \frac{\frac{\tilde{D}(\eta_2 c_2)-1}{-\eta_2 c_2} + \frac{\tilde{U}(-\eta_2(c_1-c_2))-1}{\eta_2(c_1-c_2)}h_2}{\frac{h_2 c_1(\alpha+\beta)}{\beta(c_1(\alpha+\beta)-r\beta)}\tilde{U}(-\eta_2(c_1-c_2))}. \qquad (10.22)$$

∎

As a final extension, we consider a combination of multiple sources and a tandem buffer.

Problem 116

Consider the tandem buffers model in Figure 10.13. Input to the first buffer is from N independent and identical exponential on-off sources with on-time parameter α, off-time parameter β, and rate r. The output from buffer-1 is directly fed into buffer-2. The output capacities of buffer-1 and buffer-2 are c_1 and c_2, respectively. Assuming stability, obtain bounds on the limiting distributions of the contents of the two buffers.

Solution

We first obtain bounds on the contents of buffer-1 and then of buffer-2.

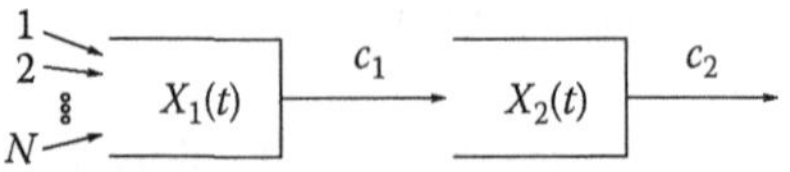

FIGURE 10.13
Tandem buffers model with multiple sources. (From Gautam, N. et al., *Prob. Eng. Inform. Sci.*, 13, 429, 1999. With permission.)

Buffer-1: Let $Z_1(t)$ be the number of sources that are in the on state at time t. Clearly, $\{Z_1(t), t \geq 0\}$ is an SMP (more specifically, a CTMC). Assume

$$\frac{Nr\beta}{\alpha + \beta} < c_1 < Nr$$

for stability (ensured by the first inequality) and nontriviality (the second inequality ensures that buffer-1 is not always empty). We can show that $\Phi(\delta)$ is given by

$$\phi_{ij}(\delta) = \begin{cases} \frac{i\alpha}{i\alpha + (N-i)\beta - (ir - c_1)\delta} & \text{if } j = i - 1 \\ \frac{(N-i)\beta}{i\alpha + (N-i)\beta - (ir - c_1)\delta} & \text{if } j = i + 1 \\ 0 & \text{otherwise}, \end{cases}$$

and $e(\Phi(\delta)) = 1$ always has solutions. Using the expression for $eb(v)$ in Equation 10.18 and solving for η_1 in $N\,eb(\eta_1) = c_1$, we get

$$\eta_1 = \frac{N(c_1\alpha + c_1\beta - N\beta r)}{c_1(Nr - c_1)}.$$

The eigenvectors are obtained by solving

$$h = h\Phi(\eta_1).$$

The limiting distribution of the buffer-content process $\{X_1(t) t \geq 0\}$ is given by

$$C_{1*}e^{-\eta_1 x} \leq P\{X > x\} \leq C_1^* e^{-\eta_1 x},$$

where

$$C_1^* = \frac{\sum_{i=0}^N \frac{h_i}{\eta_1(ir - c_1)}\left(\sum_{j=0}^N (\phi_{ij}(\eta_1)) - 1\right)}{\min_{i:ir > c_1} \frac{h_i}{p_i}\frac{1}{i\alpha + (N-i)\beta - \eta_1(ir - c_1)}},$$

$$C_{1*} = \frac{\sum_{i=0}^N \frac{h_i}{\eta_1(ir - c_1)}\left(\sum_{j=0}^N (\phi_{ij}(\eta_1)) - 1\right)}{\max_{i:ir > c_1} \frac{h_i}{p_i}\frac{1}{i\alpha + (N-i)\beta - \eta_1(ir - c_1)}},$$

and

$$p_i = \frac{a_i \tau_i}{\sum_{m=0}^N a_m \tau_m} = \frac{N!}{i!(N-i)!}\frac{\alpha^{N-i}\beta^i}{(\alpha+\beta)^N}.$$

Buffer-2: Let $\ell = \lceil c_1/r \rceil$. Define

$$Z_2(t) = \begin{cases} Z_1(t) & \text{if } X_1(t) = 0 \\ \ell & \text{if } X_1(t) > 0, \end{cases} \tag{10.23}$$

where $Z_1(t)$ is the number of sources on at time t. Let $R_1(t)$ be the output rate from the first buffer at time t. We assume that

$$\frac{Nr\beta}{\alpha + \beta} < c_2 < c_1.$$

We can see that the $\{Z_2(t), t \geq 0\}$ process is an SMP on state space $\{0, 1, \dots, \ell\}$ with kernel

$$G(t) = \big[G_{ij}(t)\big]$$

derived in Problem 98 (see Chapter 9).

From Kulkarni and Gautam [70], we have the effective bandwidth of the output from buffer-1, $eb_2(v)$, given by

$$eb_2(v) = \begin{cases} N\, eb_1(v) & \text{if } 0 \leq v \leq v^* \\ (N\, eb_1(v^*) - c_1)\frac{v^*}{v} + c_1 & \text{if } v > v^*, \end{cases}$$

where $eb_1(v)$ is from Equation 10.18 and

$$v^* = \frac{\beta}{r}\left(\sqrt{\frac{c_1\alpha}{\beta(Nr - c_1)}} - 1\right) + \frac{\alpha}{r}\left(1 - \sqrt{\frac{\beta(Nr - c_1)}{c_1\alpha}}\right).$$

Hence solving

$$eb_2(\eta_2) = c_2,$$

we get

$$\eta_2 = \min\left\{\frac{N(c_2\alpha + c_2\beta - N\beta r)}{c_2(Nr - c_2)}, \frac{h(v^*) - c_1 v^*}{c_2 - c_1}\right\},$$

where

$$h(v^*) = \frac{\left(rv^* - \alpha - \beta + \sqrt{(rv^* - \alpha - \beta)^2 + 4\beta r v^*}\right)N}{2}. \tag{10.24}$$

If $\eta_2 \leq v^*$, then

$$
\Phi_{ij}(\eta_2) = \begin{cases} \tilde{G}_{ij}(-\eta_2(ir - c_2)) & \text{if } 0 \leq i \leq \ell - 1, \\ \tilde{G}_{ij}(-\eta_2(c_1 - c_2)) & \text{if } i = \ell \end{cases}
$$

and $e(\Phi(\eta_2)) = 1$. We obtain h by solving

$$
h\Phi(\eta_2) = h.
$$

It can be shown that the random variables associated with the distribution $G_{\ell j}(x)/G_{\ell j}(\infty)$ have a decreasing failure rate. Hence, $\Psi_{min}(\ell, j)$ and $\Psi_{max}(\ell, j)$ occur at $x = \infty$ and $x = 0$, respectively. Thus, we can find bounds for the steady-state distribution of the buffer-content process $\{X_2(t), t \geq 0\}$ as follows. On the basis of Equations 10.7 through 10.9, removing k since $K = 1$, we can write

$$
H = \sum_{i=0}^{\ell} \frac{h_i}{\eta_2(\min(ir, c_1) - c_2)} \left(\sum_{j=0}^{\ell} (\phi_{ij}(\eta_2)) - 1 \right),
$$

and obtain $\Psi_{min}(i, j)$ as well as $\Psi_{max}(i, j)$ using Table 10.1.

Then the limiting distribution of the buffer content process is

$$
C_{2*}e^{-\eta_2 x} \leq \lim_{t \to \infty} P(X_2(t) > x) \leq C_2^* e^{-\eta_2 x}, \quad \forall x \geq 0,
$$

where

$$
C_2^* = \frac{H}{\min_{i,j:ir > c_2, j = i \pm 1} \Psi_{min}(i, j)}, \quad C_{2*} = \frac{H}{\max_{i,j:ir > c_2, j = i \pm 1} \Psi_{max}(i, j)}.
$$

In Figure 10.14, we illustrate the upper and lower bounds on the limiting distribution of the buffer-content process

$$
\lim_{t \to \infty} P\{X_2(t) > x\} = P\{X_2 > x\}
$$

for a numerical example with $\alpha = 1$, $\beta = 0.3$, $r = 1$, $c_1 = 13.22$, $c_2 = 10.71$, and $N = 16$. ∎

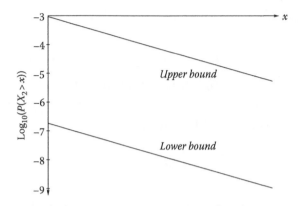

FIGURE 10.14
The upper and lower bounds as a function of x. (From Gautam, N. et al., *Prob. Eng. Inform. Sci.*, 13, 429, 1999. With permission.)

10.3 Multiclass Fluid Queues

In this section, we consider extending the analysis discussed so far to multiclass queues. This is akin to the extension to multiple classes seen in the discrete case in Chapter 5. In particular, we consider a system as described in Figure 10.15. There are N input buffers, one for each class of traffic. The input to buffer j ($j = 1, \ldots, N$) is from the K_j sources of class j. The ith source of class j is driven by an independent random environment process $\{Z_{ij}(t), t \geq 0\}$ for $i = 1, 2, \ldots, K_j$ and $j = 1, \ldots, N$. At time t, source i of class j generates fluid

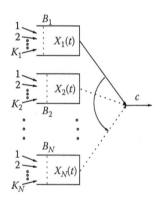

FIGURE 10.15
Multiclass fluid system. (From Kulkarni, V.G. and Gautam, N., *Queueing Syst. Theory Appl.*, 27, 79, 1997. With permission.)

at rate $r_{ij}(Z_{ij}(t))$ into buffer j. All the classes of fluids are emptied by a single channel of constant capacity c. At this time we do not specify the service scheduling policy for emptying the N buffers.

For example, we will consider policies such as: timed round-robin (polling) policy, where the scheduler serves the N buffers in a round-robin fashion; static priority service policy, where there is a priority order for each class and only when all higher priority buffers are empty this class would be served; generalized processor sharing (GPS) policy, where a fraction of channel capacity c is offered to all buffers simultaneously; and threshold policies, where both the buffer to serve and the fractions of capacity to be assigned depend on the amount of fluid in the buffers (using threshold values or switching curves). Notice that the buffers do not necessarily have a constant output capacity. However, all the results we have seen thus far have had constant output capacity. We will subsequently use a fictitious *compensating source* to address this.

However, we first describe the main objective, which is to analyze the buffer content levels in steady state. For that, let $X_j(t)$ be the amount of fluid in buffer j at time t. Assume that all N buffers are of infinite capacity. Assume that we can use the source characteristics to obtain the effective bandwidth of source i of class j as $eb_{ij}(v)$ for $i = 1, 2, \ldots, K_j$ and $j = 1, \ldots, N$. We use that for the performance analysis. The quality-of-service (QoS) criterion is mainly based on tail distribution of the buffer contents. In particular, for a given set of buffer levels $B_1, \ldots, B_N$, the probability of exceeding those levels must be less than $\epsilon_1, \ldots, \epsilon_N$, respectively. This for $j = 1, \ldots, N$,

$$\lim_{t \to \infty} P\{X_j(t) > B_j\} = P\{X_j > B_j\} < \epsilon_j.$$

Note that the preceding QoS can indirectly be used for bounds on delay as well.

The analysis in this section can be used not only in obtaining the tail probabilities but also for admission control. For that we assume that all sources of a particular class are stochastically identical. We assume that sources arrive at buffers, spend a random sojourn time generating traffic according to the respective environment process, and then depart from the buffers. We assume that the number of sources is slowly varying compared to sources changing states as well as buffer contents. In particular, we assume that steady state is attained well before the number of sources of each class changes. For such a system, our objective is to determine the feasible region $\mathcal{K}$ given by

$$\mathcal{K} = \{(K_1, \ldots, K_N) : \forall j \in [1, \ldots, N], P(X_j > B_j) < \epsilon_j\}.$$

We can use that for controlling admission in the following manner. If we have already admitted K_j sources of class j (for any $j \in [1, \ldots, N]$) and a new $K_j + 1$st source arrives, we can admit the new source if admitting it would result in the QoS criterion being satisfied for all sources. Otherwise, the source is rejected. This can easily be accomplished by maintaining a precomputed look-up table of the feasible region $\mathcal{K}$.

For the rest of this section, we will consider performance analysis for various service scheduling policies (such as timed round robin, static priority, generalized processor sharing, and threshold based). We will obtain admissible regions wherever appropriate and solve the admission control problem. However, prior to this we first describe how to analyze a buffer whose output capacity is varying as this is a recurring theme for all policies. In fact, the goal of the next section is to describe a unified approach to address time-varying output capacities so that they could be used subsequently in performance analysis.

10.3.1 Tackling Varying Output Capacity: Compensating Source

The analysis we saw earlier in this chapter for steady-state buffer content distribution's bounds and approximation assumed that the output channel capacity of any buffer is a constant. However, when we analyze multiclass fluids where each class has a buffer, the resulting channel capacity for any buffer is not a constant over time. Thus, to perform even the simplest approximation using the effective bandwidth analysis (which is common to all the approximations and bounds), we need the output channel capacity to be a constant over time. We consider the approach of using a fictitious compensating source. It can be used effectively to analyze time-varying output capacities when they are independent of the input environment process as well as buffer content process. We describe the model next with notation that will be used only in this section.

Consider a single buffer with infinite room for fluid. Fluid is generated into this buffer by a source modulated by a discrete environment process $\{Z(t), t \geq 0\}$. When the environment process is in state i, fluid enters the buffer at rate $r(i)$. The output capacity is governed by an independent discrete stochastic process $\{Y(t), t \geq 0\}$. When $Y(t) = j$ the buffer is offered a capacity $c(j)$. We assume that $\{Z(t), t \geq 0\}$ and $\{Y(t), t \geq 0\}$ are independent stochastic processes. Let $c = \max_j \{c(j)\}$ be the maximum possible channel capacity. To analyze the contents of such a buffer, we denote $\hat{X}(t)$ as the amount of fluid in the buffer (i.e., buffer content) at time t. The dynamics of the buffer content process $\{\hat{X}(t), t \geq 0\}$ is described by

$$\frac{d\hat{X}(t)}{dt} = \begin{cases} r(Z(t)) - c(Y(t)) & \text{if } \hat{X}(t) > 0 \\ \{r(Z(t)) - c(Y(t))\}^+ & \text{if } \hat{X}(t) = 0 \end{cases}$$

where $\{x\}^+ = \max(x, 0)$.

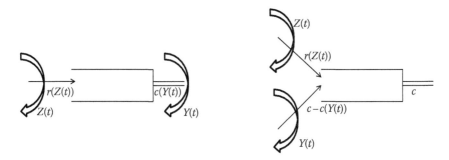

FIGURE 10.16
Original system (left) and equivalent fictitious system (right).

Now consider a fictitious infinite-sized buffer with constant output channel capacity c (recall that $c = \max_j\{c(j)\}$ based on the varying capacity $c(j)$ of the original buffer). There are two independent sources for this fictitious buffer. One source is same as that of the original buffer, that is, one that generates fluid at rate $r(Z(t))$ when the environment process is in state $Z(t)$ at time t. The other is what we call the *compensating source*. In particular, the compensating source generates fluid into this fictitious buffer at rate $c - c(Y(t))$ when the environment is in state $Y(t)$. Let $X(t)$ be the amount of fluid in this fictitious buffer at time t. The key realization to make is that if $X(0) = \hat{X}(0)$, then $X(t) = \hat{X}(t)$ for all $t \geq 0$. In fact, the dynamics of $\{X(t), t \geq 0\}$ is identical to that of $\{\hat{X}(t), t \geq 0\}$. However, notice that we know how to compute the steady-state distribution of $X(t)$; hence we can use that to obtain the steady-state distribution of $\hat{X}(t)$. The independence of $\{Z(t), t \geq 0\}$ and $\{Y(t), t \geq 0\}$ is mainly for the analysis, and the earlier result would hold even if they are dependent. Figure 10.16 summarizes what we described in the preceding text. We will use that extensively in the following sections.

10.3.2 Timed Round Robin (Polling)

Consider a multiclass fluid system as described in Figure 10.15. There are N input buffers, one for each class of traffic. The input to buffer j (for $j = 1, \ldots, N$) is from the K_j sources of class j. The ith source of class j is driven by an independent random environment process $\{Z_{ij}(t), t \geq 0\}$ for $i = 1, 2, \ldots, K_j$ and $j = 1, \ldots, N$. At time t, source i of class j generates fluid at rate $r_{ij}(Z_{ij}(t))$ into buffer j. All classes of fluids are emptied by a single channel of constant capacity c using a *timed round robin* or *polling* policy. A scheduler allocates the entire output capacity c to each of the N buffers in a cyclic fashion. In each cycle, buffer j gets the entire capacity for an interval of length τ_j. Note that during this interval, buffer j could be empty for some length of time. Hence, the scheduler is not work conserving. Let t_{so} be the total switch-over time during an entire cycle.

Assume that t_{so} does not change with time. The *cycle time* T is defined as the amount of time the scheduler takes to complete a cycle, and is given by

$$T = t_{so} + \sum_{j=1}^{N} \tau_j.$$

Using the earlier setting, our aim is to describe an approach to derive bounds and approximations for the steady-state buffer content distribution $P(X_j > x)$, for all $j \in [1, \ldots, N]$. We mainly concentrate on the effective bandwidth approximation and upper bounds (assuming the input sources are SMPs). Although we could consider the CDE approximation, since it is a relatively straightforward extension to the effective bandwidth approximation, we do not use it here (but it has been included in exercise problems at the end of the chapter). Nonetheless, for all approaches there are two commonalities—determining the stability conditions and doing an effective bandwidth analysis. This is done first followed by the QOS, numerical problems, and admission control.

It is crucial to realize that since $\tau_1, \tau_2, \ldots, \tau_N$ and t_{so} are known constants, for a given buffer (say j) the buffer contents and its dynamics do not depend on the parameters of any other buffer (i.e., any $i \neq j$). Therefore, it is convenient to analyze each buffer separately and hence we consider some $j \in [1, \ldots, N]$ and describe the analysis. Buffer j can be modeled as a single-buffer-fluid model with variable output capacity and input from K_j different sources, such that source i of class j is modulated by an environmental process $\{Z_{ij}(t), t \geq 0\}$. At time t, the input rate is $r_{ij}(Z_{ij}(t))$. The output capacity alternates between c (for τ_j units of time) and 0 (for $T - \tau_j$ units of time).

The condition for the stability of buffer j is

$$\sum_{i=1}^{K_j} E[r_{ij}(Z_{ij}(\infty))] < c\frac{\tau_j}{T}.$$

If the stability condition is satisfied, then $X_j(t) \to X_j$ as $t \to \infty$. Next we describe methods such as effective bandwidth approximation and SMP bounds to compute $P(X_j > x)$. To utilize these techniques, one needs to first transform the system into an appropriate one with a constant output channel capacity. Consider a single-buffer-fluid model for buffer j with a constant output channel capacity c whose input is generated by the original K_j sources and a fictitious compensating source.

The compensating source is such that it stays "on" for a deterministic amount of time $T - \tau_j$ and "off" for a deterministic amount of time τ_j. When the compensating source is on, it generates fluid at rate c and when

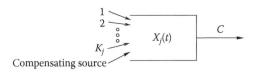

FIGURE 10.17
Transforming buffer j to a constant output capacity one using a compensating source. (From Gautam, N. and Kulkarni, V.G., *Queueing Syst. Theory Appl.*, 36, 351, 2000. With permission.)

it is off, it generates fluid at rate 0. Note that the compensating source is independent of the original K_j sources. As we saw in Section 10.3.1, the dynamics of the buffer-content process (of buffer j) remain unchanged for this transformed single-buffer-fluid model with $K_j + 1$ input sources (including the compensating source) and constant output capacity c (see Figure 10.17).

Using the results in Section 10.1.2, the effective bandwidth of the jth compensating source described earlier is given by

$$eb_j^s(v) = \frac{c(T - \tau_j)}{T}.$$

Note that the effective bandwidth of this deterministic source is indeed its mean traffic generation rate. Using the effective bandwidth approximation,

$$P(X_j > x) \approx e^{-\eta_j x},$$

where η_j is obtained by solving

$$\sum_{i=1}^{K_j} eb_{ij}(\eta_j) + c\frac{(T - \tau_j)}{T} = c.$$

Thus, based on the effective bandwidth approximation, the QoS criteria for all the classes of traffic are satisfied if for all $j = 1, 2, \ldots, N$,

$$e^{-B_j \eta_j} < \epsilon_j.$$

Similarly, it is also possible to check if the QoS criteria is satisfied using the upper SMP bound as $C_j^* e^{-B_j \eta_j} < \epsilon_j$ since

$$P(X_j > B_j) \leq C_j^* e^{-\eta_j B_j} < \epsilon_j.$$

Although the preceding results assume we know $eb_{ij}(v)$ and also how to compute C_j^*, next we present a specific example to clarify these aspects. Also, for the sake of obtaining closed-form algebraic expressions, we consider a rather simplistic set of sources.

Problem 117

Consider a multiclass fluid queueing system with N buffers, one for each class. For all $j = 1, \ldots, N$, the input to buffer j are K_j independent and identical alternating on-off sources that stay on for an exponential amount of time with parameter α_j and off for an exponential amount of time with parameter β_j. When a source is on, it generates traffic continuously at rate r_j into buffer j, and when it is off, it does not generate any traffic. The scheduler serves buffer j for a deterministic time τ_j at a maximum rate c and stops serving the buffer for a deterministic time $T - \tau_j$. Using effective bandwidth approximation and bounds, obtain expressions for $P(X_j > B_j)$. Then for the following numerical values $\alpha_j = 3$, $\beta_j = 0.2$, $r_j = 3.4$, $B_j = 30$, $\tau_j/T = 3/13$, $K_j = 10$, $c = 15.3$, and T varies from 0.01 to 0.40 while τ_j/T is fixed, graph using the two methods of approximate expressions for the fraction of fluid lost assuming that the size of the buffer is B_j.

Solution

The effective bandwidth of all the K_j sources combined is

$$K_j eb_j(v) = K_j \frac{r_j v - \alpha_j - \beta_j + \sqrt{(r_j v - \alpha_j - \beta_j)^2 + 4\beta_j r_j v}}{2v}.$$

Thus, solving for η_j in

$$K_j eb_j(\eta_j) = \frac{c\tau_j}{T},$$

we get

$$\eta_j = \frac{c\tau_j(\alpha_j + \beta_j) - r_j K_j \beta_j T}{\frac{c\tau_j}{K_j T}(r_j T K_j - c\tau_j)}.$$

Also, for the upper bound $C_j^* e^{-\eta_j B_j}$, we can obtain

$$
C_j^* = \frac{\left[\frac{r_j T K_j}{\alpha_j c \tau_j}\right]^{K_j} \left(\exp\left(\eta_j c \frac{\tau_j(T-\tau_j)}{T}\right) - 1\right)}{\min_{1 \le i \le K_j}\left\{\left(\frac{\alpha_j + \beta_j}{\alpha_j \beta_j}\right)^{K_j} \left(\frac{T K_j \beta_j}{T K_j \beta_j + \eta_j c \tau_j}\right)^{K_j - i}\right\} \eta_j c \tau_j} \left(\frac{T}{T - \tau_j}\right)
$$

$$
= \frac{\left[\frac{r_j T K_j}{\alpha_j c \tau_j}\right]^{K_j} \left(\exp\left(\eta_j c \frac{\tau_j(T-\tau_j)}{T}\right) - 1\right)}{\left(\frac{\alpha_j + \beta_j}{\alpha_j \beta_j}\right)^{K_j} \left(\frac{T K_j \beta_j}{T K_j \beta_j + \eta_j c \tau_j}\right)^{K_j - 1} \eta_j c \tau_j} \left(\frac{T}{T - \tau_j}\right)
$$

Recall that the effective bandwidth approximation yields $P(X_j > B_j) \approx e^{-B_j \eta_j}$ and the SMP bounds result in $P(X_j > B_j) \le C_j^* e^{-\eta_j B_j}$. Thus, for the numerical values listed in the problem, if we let B_j be the size of the buffer, then the fraction of fluids lost can be approximated as $P(X_j > B_j)$. With that understanding, the loss probability estimates using the effective-bandwidth technique is

$$
\text{loss(ebw)} = e^{-\eta_j B_j}
$$

and that using the SMP bound is

$$
\text{loss(smp)} = C_j^* e^{-\eta_j B_j}.
$$

Figure 10.18 shows the results for loss(ebw) and loss(smp), given the numerical values in the problem by varying T from 0.01 to 0.40 while keeping τ_j/T fixed.

Intuitively, we expect the loss probability to increase with T since an increase in T would increase the time the server does not serve the buffer. The SMP bounds estimate, loss(smp), increases with T and hence confirms our intuition. The effective-bandwidth estimate, loss(ebw), does not change with T. For small T, since loss(smp) < loss(ebw), we can conclude that the effective-bandwidth technique produces a conservative result. For large T, the estimate of the loss probability is smaller using the effective-bandwidth technique than the SMP bounds technique. This indicates that there may be a risk in using the effective-bandwidth technique as it could result in the QoS criteria not being satisfied. ∎

It is worthwhile pointing out that besides the fraction of fluid lost, we could also obtain other metrics. This can be used to compare the effective bandwidth method against the SMP bounds. In particular, say we continued with the same set of numerical values: $\alpha_j = 3$, $\beta_j = 0.2$, $r_j = 3.4$, $B_j = 30$,

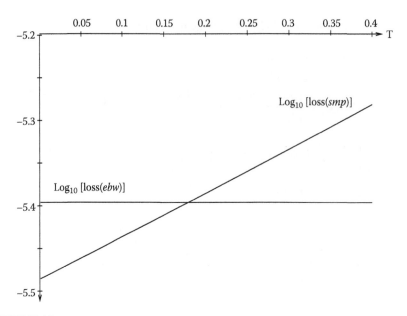

FIGURE 10.18
Estimates of the logarithms of loss probability. (From Gautam, N. and Kulkarni, V.G., *Queueing Syst. Theory Appl.*, 36, 351, 2000. With permission.)

$\tau_j/T = 3/13$, and $c = 15.3$, and if we are interested in the maximum number of class j sources that would ensure $P(X > B_j)$ to be less that ϵ_j, then the estimate of the maximum number of sources using the effective-bandwidth technique is

$$K_{j,\text{max}}^{ebw} = \left\lfloor \frac{1}{eb_j(-\log(\epsilon_j)/B_j)} \frac{c\tau_j}{T} \right\rfloor.$$

On the other hand, using the upper bound for an SMP we choose the largest integer $K_{j,\text{max}}^{smp}$ that satisfies

$$C_j^* e^{-\eta_j B_j} < \epsilon_j.$$

Figure 10.19 shows the results for $K_{j,\text{max}}^{ebw}$ and $K_{j,\text{max}}^{smp}$ when $\epsilon_j = 10^{-5}$ and T varies from 0.01 to 10.00 while τ_j/T is fixed. As T increases, we expect fewer sources to be allowable into the buffer so that long bursts of traffic can be avoided when the server is not serving. From the figure, $K_{j,\text{max}}^{smp}$ clearly conforms to our intuition. For large T, we may end up admitting more sources if we used the effective-bandwidth technique and hence the QoS criterion may not be satisfied.

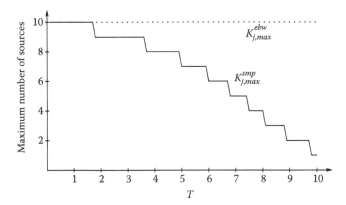

FIGURE 10.19
Estimate of the maximum number of sources. (From Gautam, N. and Kulkarni, V.G., *Queueing Syst. Theory Appl.*, 36, 351, 2000. With permission.)

Next, instead of estimating the maximum number of sources, say we are given the number of sources and want to know the minimum bandwidth c that would ensure that $P(X > B_j)$ is less than ϵ_j. We continue with the same set of numerical values: $\alpha_j = 3$, $\beta_j = 0.2$, $r_j = 3.4$, $B_j = 30$, and $\tau_j/T = 3/13$, with $K_j = 10$. The estimate of the smallest bandwidth required, c, using the effective-bandwidth technique is

$$c_{min}^{ebw} = K_j eb_j \left(-\frac{\log(\epsilon_j)}{B_j} \right).$$

The loss probability estimate using the SMP bounds decreases with increase in c. Therefore, we perform a search using the bisection method to pick a c between the mean and peak input rates that satisfies

$$C_j^* e^{-\eta_j B_j} = \epsilon_j,$$

and we denote the c value obtained as c_{min}^{smp} since it is the smallest output capacity that would result in satisfying the QoS criterion

$$C_j^* e^{-\eta_j B_j} < \epsilon_j.$$

Figure 10.20 shows the results for c_{min}^{ebw} and c_{min}^{smp} when $\epsilon_j = 10^{-5}$ and T varies from 0.01 to 0.40 while τ_j/T is fixed. Intuitively, the bandwidth required should increase with T so that all the buffer contents are drained out when the server is serving the buffer. The c_{min}^{smp} obtained using the SMP bounds technique is consistent with our intuition. On the other hand, c_{min}^{ebw} does not

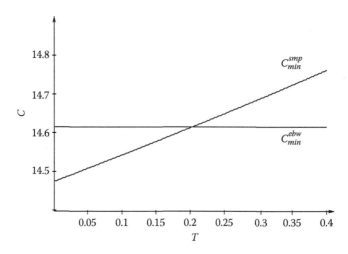

FIGURE 10.20
Estimates of the required bandwidth. (From Gautam, N. and Kulkarni, V.G., *Queueing Syst. Theory Appl.*, 36, 351, 2000. With permission.)

vary with T. Therefore, on using the effective-bandwidth technique one faces the risk of the QoS criteria not being satisfied.

In the next problem that deals with admission control for a two-class polling system with timed round-robin policy, we only consider SMP bounds.

Problem 118

Consider a multiclass fluid queueing system with $N = 2$. For $j = 1, 2$, class j fluid enters into buffer j from K_j exponential on-off sources with mean on-time $1/\alpha_j$ and mean off-time $1/\beta_j$. Fluid is generated by each source at rate r_j when the source is in the on-state. Fluid is emptied by a channel with capacity c that serves buffer j for τ_j time and has a total switch-over time of t_{so} per cycle. State an algorithm to determine the feasible region for this timed round-robin policy $\mathcal{K}^{trr}$ so that if $(K_1, K_2) \in \mathcal{K}^{trr}$, then $P(X_j > B_j) < \epsilon_j$ for $j = 1$ and $j = 2$. Graph the feasible region for the following numerical values:

$$\alpha_1 = 1, \quad \beta_1 = 0.3, \quad r_1 = 1.0, \quad \epsilon_1 = 10^{-6}, \quad B_1 = 8, \quad T = 1.22, \quad t_{so} = 0.02,$$
$$\alpha_2 = 1, \quad \beta_2 = 0.2, \quad r_2 = 1.23, \quad \epsilon_2 = 10^{-9}, \quad B_2 = 10, \quad \text{and} \quad c = 13.22.$$

To begin with, assume that the cycle time T and the switch-over time t_{so} are fixed known constants. However, the values τ_1 and τ_2 vary and are appropriately chosen such that $\tau_1 + \tau_2 + t_{so} = T$. Subsequently, consider the case where T is varied so that it is under different orders of magnitude compared to t_{so}.

Solution

An algorithm to compute the feasible region:

1. Set $\mathcal{K} = \emptyset$.

2. Let $\tau_1 = T$ and $\tau_2 = 0$. (The scheduler always serves only buffer-1, and hence there are no switch-over times and no compensating source.)

3. Obtain the maximum number of admissible class-1 sources K_1^{max} as the maximum value of K_1 such that

$$C_1^* e^{-\eta_1 B_1} < \epsilon_1,$$

where

$$C_1^* = \frac{[r_1 K_1 / (\alpha_1 c)]^{K_1}}{(\alpha_1 + \beta_1) / (\alpha_1 \beta_1)^{K_1} (K_1 \beta_1 / (K_1 \beta_1 + \eta_1 c))^{K_1 - 1}} \qquad (10.25)$$

and

$$\eta_1 = \frac{c(\alpha_1 + \beta_1) - r_1 K_1 \beta_1}{c / K_1 (r_1 K_1 - c)}.$$

4. $\mathcal{K} = \mathcal{K} \cup \{(0,0), (1,0), \ldots, (K_1^{max}, 0)\}$.

5. Let $\tau_2 = T$ and $\tau_1 = 0$. (The scheduler always serves only buffer-2, and hence there are no switch-over times and no compensating source.)

6. Obtain the maximum number of admissible class-2 sources K_2^{max} as the maximum value of K_2 such that

$$C_2^* e^{-\eta_2 B_2} < \epsilon_2,$$

where

$$C_2^* = \frac{[r_2 K_2 / (\alpha_2 c)]^{K_2}}{((\alpha_2 + \beta_2) / (\alpha_2 \beta_2))^{K_2} (K_2 \beta_2 / (K_2 \beta_2 + \eta_2 c))^{K_2 - 1}} \qquad (10.26)$$

and

$$\eta_2 = \frac{c(\alpha_2 + \beta_2) - r_2 K_2 \beta_2}{c / K_2 (r_2 K_2 - c)}.$$

7. $\mathcal{K} = \mathcal{K} \cup \{(0,1), (0,2), \ldots, (0, K_2^{max})\}$.

8. Set $K_1 = 1$.
9. While $K_1 < K_1^{max}$:
    (i) Compute the minimum required τ_1 ($\leq T - t_{so}$) such that the loss probability is less than ϵ_1.
    (ii) Compute the available $\tau_2 (= T - t_{so} - \tau_1)$.
    (iii) Given τ_2, compute the maximum possible K_2 value by minimizing over the set $\mathcal{A}^2$ for $K_2 + 1$ sources.
    (iv) $\mathcal{K} = \mathcal{K} \cup \{(K_1, 1), (K_1, 2), \ldots, (K_1, K_2)\}$.
    (v) $K_1 = K_1 + 1$.
10. Return $\mathcal{K}^{trr} = \mathcal{K}$.

Using the preceding algorithm, we plot the admissible region $\mathcal{K}^{trr}$ in Figure 10.21 for the following numerical values given in the problem. Note that there is a steep fall in the admissible region from $(0, 20)$ to $(1, 12)$. This is due to using Equation 10.26 for the K_2 sources and

$$
C_2^* = \frac{\left[\frac{r_2 TK_2}{\alpha_2 c \tau_2} \right]^{K_2} \left(\exp\left(\eta_{2c} \frac{\tau_2(T - \tau_2)}{T} \right) - 1 \right)}{\left(\frac{\alpha_2 + \beta_2}{\alpha_2 \beta_2} \right)^{K_2} \left(\frac{TK_2 \beta_2}{TK_2 \beta_2 + \eta_{2c} \tau_2} \right)^{K_2 - 1} \eta_{2c} \tau_2} \left(\frac{T}{T - \tau_2} \right)
$$

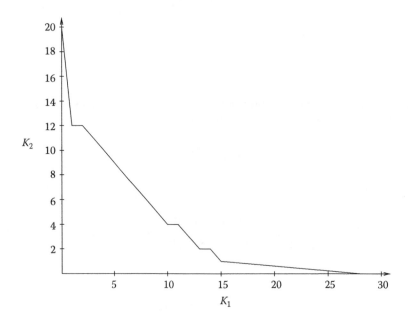

FIGURE 10.21
Admissible region $\mathcal{K}^{trr}$. (From Gautam, N. and Kulkarni, V.G., *Queueing Syst. Theory Appl.*, 36, 351, 2000. With permission.)

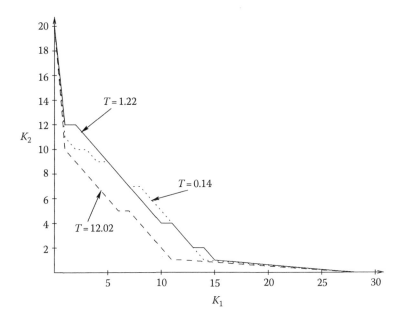

FIGURE 10.22
$\mathcal{K}^{trr}$ as a function of T. (From Gautam, N. and Kulkarni, V.G., *Queueing Syst. Theory Appl.*, 36, 351, 2000. With permission.)

for the $K_2 + 1$ sources (including the compensating source), respectively, for the two points $(0, 20)$ and $(1, 12)$. Note that the correct choice of τ_1 and τ_2 depend upon K_1 and $K_2 \in \mathcal{K}^{trr}$. Note that from Step 9 in the preceding algorithm, τ_1 (and hence τ_2) depends only on K_1. However, there could be other choices of (τ_1, τ_2) for a given (K_1, K_2).

Next we discuss the effect of varying T, the cycle time. Intuitively, for $T \gg t_{so}$, an increase in T would result in a smaller admissible region and a decrease in T would result in a bigger admissible region. We confirm our intuition by observing the results in Figure 10.22 (using the numerical values in the problem statement) for the cases $T = 1.22$ and $T = 12.02$. When T is approximately of the same order of magnitude as t_{so}, a significant fraction of the server off-time is the switch-over time. Hence, it is not clear how the admissible region would change with T in this case. From Figure 10.22, we can see that for the cases $T = 0.14$ and $T = 1.22$, one region is not the subset of the other. Hence, we conclude that it is not straightforward to obtain an optimal value of T such that the feasible region is maximized. Note that if $t_{so} = 0$, then the optimal T is such that $T \to 0$. ∎

That said, we move on to the next service scheduling policy, the static priority policy.

10.3.3 Static Priority Service Policy

Consider a multiclass fluid system as described in Figure 10.15. There are N input buffers, one for each class of traffic. The input to buffer j (for $j = 1, \ldots, N$) is from the K_j sources of class j. The ith source of class j is driven by an independent random environment process $\{Z_{ij}(t), t \geq 0\}$ for $i = 1, 2, \ldots, K_j$ and $j = 1, \ldots, N$. At time t, source i of class j generates fluid at rate $r_{ij}(Z_{ij}(t))$ into buffer j. All the classes of fluids are emptied by a single channel of constant capacity c using a *static priority* policy. Under this service policy, traffic of class j has higher service priority over traffic of class i, if $i > j$ for all i and j in $[1, \ldots, N]$. In other words, class-1 is given the highest priority and class-N the lowest priority. Two things make fluid priority queues different from their discrete counterparts: (i) there is no notion of preemption in fluids, and (ii) more than one class can be served at a time in fluids. Also, notice that unlike the timed round-robin policy we saw in the previous section, there is no switch-over time.

However, the scheduler serves traffic of class j only if there is no fluid of higher priority in the buffers (i.e., buffers $1, \ldots, j-1$ are empty). Thus, all the available channel capacity (a maximum of c) is assigned for the class-1 fluid, and the leftover channel capacity (if any) that class-1 does not need, to class-2 fluid. Any leftover channel capacity that class-1 and class-2 do not need is assigned to class-3 fluid, and so on. We will use "class" and "priority" interchangeably, that is, class-j is also priority-j or jth priority. Under such a policy, our objective is to derive the steady-state distribution of the contents of buffer j (for $j = 1, \ldots, N$). Let $X_j(t)$ be the amount of fluid in buffer j at time t. We will state the stability condition subsequently, but assuming it is satisfied, as $t \to \infty$, $X_j(t) \to X_j$. Under stability we would like to derive expressions for $P(X_j > x)$. We could use that to determine if $P(X_j > B_j) < \epsilon_j$ for some B_j and ϵ_j, which could be used in admission control decisions.

Before discussing methods to obtain $P(X_j > x)$ (for $j = 1, \ldots, N$), we describe the stability of the system. Notice that the scheduler is work-conserving and fluid is always removed from the system at rate c when any of the buffers is nonempty. Hence, if the steady-state arrival rate of fluid is smaller than the output capacity, the system must be stable. Thus, the stability condition can be stated as

$$\lim_{t \to \infty} \sum_{j=1}^{N} \sum_{i=1}^{K_j} E\{r_{ij}(Z_{ij}(t))\} < c. \tag{10.27}$$

As stated earlier in this section, we do assume that we can compute the effective bandwidth of all the sources with the effective bandwidth of the ith source of class-j being $eb_{ij}(v)$. With that we explain our analysis, first for buffer-1 and then for all the other buffers.

Buffer-1: Since priority-1 fluid gets uninterrupted service, the analysis is identical to the case where there is no other class traffic. Hence, we apply the results obtained for the single class traffic case earlier in this chapter. In particular, we could use effective bandwidth analysis, CDE approximation or SMP bounds and write $P(X_1 > x)$ as $e^{-\eta_1 x}$, $L_1 e^{-\eta_1 x}$, or $C_1^* e^{-\eta_1 x}$, respectively, where η_1 is the unique solution to

$$\sum_{i=1}^{K_1} eb_{i1}(\eta_1) = c. \tag{10.28}$$

The constants L_1 and C_1^* can be obtained using the CDE approximation (Section 10.2.2) and SMP bounds (Section 10.2.3), respectively. Thereby, we could use $P(X_1 > x)$ to determine if the QoS criteria $P\{X_1 > B_1\} \leq \epsilon_1$ is satisfied.

Buffer-j ($1 < j \leq N$): The capacity available to buffer j is 0 when at least one of the buffers $1, \ldots, j-1$ is nonempty and it is $\left[c - \sum_{k=1}^{j-1} \sum_{i=1}^{K_k} r_{ik}(Z_{ik}(t)) \right]^+$ if all the buffers $1, \ldots, j-1$ are empty. Let $R_{j-1}(t)$ be the sum of the output rates of the buffers $1, \ldots, j-1$ at time t with $R_0(t) = 0$. Therefore

$$R_{j-1}(t) = \begin{cases} c & \text{if } \sum_{k=1}^{j-1} X_k(t) > 0 \\ \min[c, \sum_{i=1}^{K_k} \sum_{k=1}^{j-1} r_{ik}(Z_{ik}(t))] & \text{if } \sum_{k=1}^{j-1} X_k(t) = 0. \end{cases} \tag{10.29}$$

Thus, the (time varying) channel capacity available for buffer j is $c - R_{j-1}(t)$ at time t. Any sample path of the buffer content process $\{X_j(t), t \geq 0\}$ remains unchanged if we transform the model for buffer j into one that gets served at a constant capacity c and an additional compensating source producing fluid at rate $R_{j-1}(t)$ at time t. Note that the compensating source j is independent of the K_j sources of priority j.

A critical observation to make is that the compensating source is indeed the output from a buffer whose input is the aggregated input of all $1, \ldots, j-1$ priority traffic and constant output capacity c. This observation is made in Elwalid and Mitra [32] and is immensely useful in the analysis. Consider the transformed model for the case $N = 2$ (a 2-priority model for ease of explanation) depicted in Figure 10.23. The sample paths of the buffer content processes $\{X_1(t), t \geq 0\}$ and $\{X_2(t), t \geq 0\}$ in this model are identical to those in the original system. Similarly, in the case of N priorities, such a tandem model is used.

Thus, buffer j can be equivalently modeled as one that is served at a constant rate c, but has an additional compensating source as described earlier. Let the effective bandwidth of this compensating source (which is the output traffic from a fictitious buffer with input corresponding to all $j-1$ higher

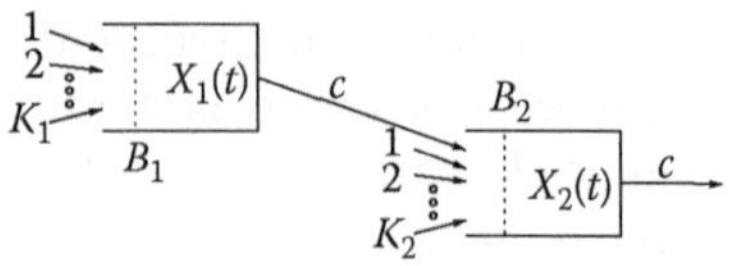

FIGURE 10.23
Equivalent $N=2$ priority system. (From Gautam, N. and Kulkarni, V.G., *Queueing Syst. Theory Appl.*, 36, 351, 2000. With permission.)

priority sources) be $eb_o^j(v)$, which is given by

$$eb_o^1(v) = 0 \quad \text{for all } v,$$

$$eb_o^j(v) = \begin{cases} \sum_{i=1}^{K_{j-1}} eb_{ij-1}(v) + eb_o^{j-1}(v) & \text{if } 0 \le v \le v_j^* \\ c - \frac{v_j^*}{v}\{c - \sum_{i=1}^{K_{j-1}} eb_{ij-1}\left(v_j^*\right) - eb_o^{j-1}\left(v_j^*\right)\} & \text{if } v > v_j^*, \end{cases}$$

$$(10.30)$$

where v_j^* is obtained by solving for v in the equation

$$\frac{d}{dv}\left[v\left(\sum_{i=1}^{K_{j-1}} eb_{ij-1}(v) + eb_o^{j-1}(v)\right)\right] = c. \qquad (10.31)$$

Combining the preceding results, we get η_j as the unique solution to

$$\sum_{i=1}^{K_j} eb_{ij}(\eta_j) + eb_o^j(\eta_j) = c, \quad \forall j = 1, 2, \ldots, N,$$

where $eb_o^1(\cdot) = 0$ and $eb_o^j(\cdot)$ is as in Equation 10.30. Note that the compensating source j is independent of the K_j sources of priority j. If it is possible to characterize the output process as a tractable stochastic process, we could use CDE approximation (Section 10.2.2) or SMP bounds (Section 10.2.3) to derive an expression for $P(X_j > x)$. Otherwise, we could always use the effective bandwidth approximation $P(X_j > x) \approx e^{-\eta_j x}$, for $j = 2, \ldots, N$. Thereby, we could use $P(X_j > x)$ to determine if the QoS criteria $P\{X_j > B_j\} \le \epsilon_j$ is satisfied. We will use the SMP bounds to obtain an approximation for $P(X_j > x)$ in the example we present next.

Problem 119

Consider two classes of traffic. The K_j class-j sources, for $j = 1, 2$, are independent and identical on-off sources with exponential on and off times, on-time parameter α_j, off-time parameter β_j, and on-time traffic generation rate r_j. Each class has its own buffer and class-1 is always given higher priority for a channel of capacity c. What is the condition of stability? Assuming that it is satisfied, use SMP bounds to derive an upper bound on $P(X_j > x)$ for $j = 1, 2$.

Solution

The mean arrival rate of traffic from each of the class-j sources is $r_j \beta_j / (\alpha_j + \beta_j)$. Hence, the stability condition is

$$\frac{K_1 r_1 \beta_1}{\alpha_1 + \beta_1} + \frac{K_2 r_2 \beta_2}{\alpha_2 + \beta_2} < c.$$

Assuming this condition is satisfied, we use SMP bounds to derive upper bounds of the type $P(X_j > x) \le C_j^* e^{-\eta_j x}$. For that we write down the effective bandwidth of each class-j source as $eb_j(v)$ given by

$$eb_j(v) = \frac{r_j v - \alpha_j - \beta_j + \sqrt{(r_j v - \alpha_j - \beta_j)^2 + 4\beta_j r_j v}}{2v}.$$

For the analysis, we first consider buffer-1. If $K_1 \le c/r_1$, then $P\{X_1 > x\} = 0$, since buffer-1 will always be empty. Now for the case $K_1 > c/r_1$, let η_1 be the solution to $K_1 eb_1(\eta_1) = c$. Then the steady-state distribution of the buffer-content process is bounded as

$$C_{*1} e^{-\eta_1 x} \le P\{X_1 > x\} \le C_1^* e^{-\eta_1 x},$$

where

$$\eta_1 = \frac{K_1(c\alpha_1 + c\beta_1 - K_1 \beta_1 r_1)}{c(K_1 r_1 - c)}, \tag{10.32}$$

$$C_1^* = \frac{\left(\frac{K_1 r_1}{K_1 r_1 - c} \frac{\alpha_1}{\alpha_1 + \beta_1}\right)^{K_1}}{\left(\frac{c\alpha_1}{\beta_1(K_1 r_1 - c)}\right)^{\lceil c/r_1 \rceil}},$$

and

$$C_{*1} = \left(\frac{K_1 r_1 \beta_1}{c(\alpha_1 + \beta_1)}\right)^{K_1}.$$

Thus, an upper bound on $P(X_1 > x)$ is $C_1^* e^{-\eta_1 x}$.

Now we consider buffer-2. We first model the K_2 exponential on-off sources as a single $(K_2 + 1)$-state SMP with the states denoting the number of priority-2 sources that are on and then derive expressions for H^1, $\Psi^1_{max}(i,j)$, and $\Psi^1_{min}(i,j)$ as defined in Equations 10.7 through 10.9. At buffer-2, besides the $(K_2 + 1)$-state SMP, we also have a compensating source that basically is the output from buffer-1. Recall Problem 116 where we derived the output from buffer-1 as an SMP (although in that problem there was no other source for buffer-2). Say we call the corresponding expressions H^2, $\Psi^2_{max}(i,j)$, and $\Psi^2_{min}(i,j)$ for the SMP model of the output from buffer-1. Therefore, we can analyze the input to buffer-2 as traffic from two sources (output from buffer-1 and the (K_2+1)-state SMP), each modulated by an SMP.

We begin by obtaining η_2. Using the effective bandwidth of the output from a buffer, we can show that η_2 solves either

$$K_1 \, eb_1(\eta_2) + K_2 \, eb_2(\eta_2) = c \quad \text{and} \quad \eta_2 \le v^*$$

or

$$\frac{v^*}{\eta_2} K_1 \, eb_1(v^*) + K_2 \, eb_2(\eta_2) = \frac{cv^*}{\eta_2} \quad \text{and} \quad \eta_2 > v^*,$$

where

$$v^* = \frac{\beta_1}{r_1}\left(\sqrt{\frac{c\alpha_1}{\beta_1(K_1 r_1 - c)}} - 1\right) + \frac{\alpha_1}{r_1}\left(1 - \sqrt{\frac{\beta_1(K_1 r_1 - c)}{c\alpha_1}}\right).$$

Using the preceding expression for η_2, we define

$$c^1 = K_2 \, eb_2(\eta_2)$$

and

$$c^2 = c - K_2 \, eb_2(\eta_2).$$

The (K_2+1)-state SMP
For $i = 0, 1, \ldots, K_2$ and $j = 0, 1, \ldots, K_2$, we define the following:

$$G^1_{i,j}(x) = \begin{cases} \frac{i\alpha_2}{i\alpha_2 + (K_2 - i)\beta_2}\left(1 - \exp\{-(i\alpha_2 + (K_2 - i)\beta_2)x\}\right) & \text{if } j = i - 1 \\ \frac{(K_2 - i)\beta_2}{i\alpha_2 + (K_2 - i)\beta_2}\left(1 - \exp\{-(i\alpha_2 + (K_2 - i)\beta_2)x\}\right) & \text{if } j = i + 1 \\ 0 & \text{otherwise,} \end{cases}$$

$$G_i^1(x) = 1 - \exp\{-(i\alpha_2 + (K_2 - i)\beta_2)x\},$$

$$\tau_i^1 = \frac{1}{i\alpha_2 + (K_2 - i)\beta_2},$$

$$P_{ij}^1 = G_{i,j}^1(\infty),$$

and

$$p_i^1 = \frac{a_i^1 \tau_i^1}{\sum_{m=0}^{K_2} a_m^1 \tau_m^1} = \frac{K_2!}{i!(K_2 - i)!} \frac{\alpha_2^{K_2-i} \beta_2^i}{(\alpha_2 + \beta_2)^{K_2}}.$$

Then, $\Phi^1(\eta_2)$ is given by

$$\phi_{i,j}^1(\eta_2) = \begin{cases} \dfrac{i\alpha_2}{i\alpha_2 + (K_2-i)\beta_2 - (ir_2 - c^1)\eta_2} & \text{if } j = i - 1 \\[2mm] \dfrac{(K_2-i)\beta_2}{i\alpha_2 + (K_2-i)\beta_2 - (ir_2 - c^1)\eta_2} & \text{if } j = i + 1 \\[2mm] 0 & \text{otherwise.} \end{cases}$$

Also, the eigenvectors are obtained by solving

$$h^1 = h^1 \Phi^1(\eta_2).$$

Therefore,

$$H^1 = \sum_{i=0}^{K_2} \frac{h_i^1}{\eta_2(ir_2 - c^1)} \left(\sum_{j=0}^{K_2} \left(\phi_{ij}^1(\eta_2) \right) - 1 \right)$$

and

$$\Psi_{max}^1(i,j) = \Psi_{min}^1(i,j)$$

$$= \frac{h_i^1 e^{-\eta_2(ir_2-c^1)x} \int_x^\infty e^{\eta_2(ir_2-c^1)y} dG_{ij}^1(y)}{\frac{p_i^1}{\tau_i^1} \int_x^\infty dG_{ij}^1(y)}$$

$$= \frac{h_i^1}{p_i^1} \frac{1}{i\alpha_2 + (K_2 - i)\beta_2 - \eta_2(ir_2 - c^1)}.$$

The output from buffer-1
We only consider the case when $K_1 > c/r_1$ (refer to Gautam and Kulkarni [40] for $K_1 \le c/r_1$). Let $M = \lceil c/r_1 \rceil$. Then the output from buffer-1 can

be modeled as an SMP on state space $\{0, 1, 2, \ldots, M\}$ (this directly follows Problem 116 albeit somewhat different notation). For $i = 0, 1, \ldots, M - 1$ and $j = 0, 1, \ldots, M$, let

$$G_{i,j}^2(t) = \begin{cases} \frac{i\alpha_1}{i\alpha_1 + (K_1 - i)\beta_1} \left(1 - \exp\{-(i\alpha_1 + (K_1 - i)\beta_1)t\}\right) & \text{if } j = i - 1 \\ \frac{(K_1 - i)\beta_1}{i\alpha_1 + (K_1 - i)\beta_1} \left(1 - \exp\{-(i\alpha_1 + (K_1 - i)\beta_1)t\}\right) & \text{if } j = i + 1 \\ 0 & \text{otherwise.} \end{cases}$$

Let

$$T = \min\{t > 0 : X_1(t) = 0\}.$$

Then for $j = 0, 1, \ldots, M - 1$, we have

$$G_{M,j}^2(t) = P\{T \le t, \bar{N}(T) = j | X_1(0) = 0, \bar{N}(0) = M\},$$

where $\bar{N}(t)$ denotes the number of priority-1 sources on at time t (note that $G_{M,M}^2(t) = 0$). We need $G^2(\infty) = [G_{i,j}^2(\infty)]$ in our analysis. We have for $i = 0, 1, \ldots, M - 1$ and $j = 0, 1, \ldots, M$,

$$G_{i,j}^2(\infty) = \begin{cases} \frac{i\alpha_1}{i\alpha_1 + (K_1 - i)\beta_1} & \text{if } j = i - 1 \\ \frac{(K_1 - i)\beta_1}{i\alpha_1 + (K_1 - i)\beta_1} & \text{if } j = i + 1 \\ 0 & \text{otherwise,} \end{cases}$$

$$G_{M,j}^2(\infty) = \tilde{G}_{M,j}^2(0),$$

where $\tilde{G}_{M,j}^2(s)$ is the LST of $G_{M,j}^2(t)$ that we have shown in Problem 116.

We also need the expression for the sojourn time τ_i^2 in state i, for $i = 0, 1, \ldots, M$. We have

$$\tau_i^2 = \begin{cases} \frac{1}{i\alpha_1 + (K_1 - i)\beta_1} & \text{if } i = 0, 1, \ldots, M - 1 \\ \sum_{j=0}^{M-1} \tilde{G}_{M,j}^{2\prime}(0) & \text{if } i = M \end{cases}$$

Then we have for $i = 0, 1, \ldots, M$

$$p_i^2 = \frac{a_i^2 \tau_i^2}{\sum_{k=0}^{M} a_k^2 \tau_k^2},$$

where

$$a^2 = a^2 \, G^2(\infty).$$

Define

$$\overline{\phi}_{ij}^2(\eta_2, m) = \begin{cases} \tilde{G}_{ij}^2(-\eta_2(ir_1 - c^2)) & \text{if } 0 \le i \le M - 1, \\ m\tilde{G}_{ij}^2(-\eta_2(c - c^2)) & \text{if } i = M. \end{cases}$$

Solve for m such that the Perron–Frobenius eigenvalue of $\overline{\Phi}^2(\eta_2, m)$ is 1. Hence, we obtain h^2 from

$$h^2 \overline{\Phi}^2(\eta_2, m) = h^2.$$

It can be shown that random variables with distribution $G_{Mj}^2(x)/G_{Mj}^2(\infty)$ have a decreasing failure rate. Hence, $\Psi_{min}^2(M, j)$ and $\Psi_{max}^2(M, j)$ occur at $x = \infty$ and $x = 0$, respectively. Thus, we have for $(i, j) \in \{0, 1, \dots, M\}$,

$$H^2 = \sum_{i=0}^{M} \frac{h_i^2}{\eta_2(ir_1 - c^2)} \left(\sum_{j=0}^{M} \left(\overline{\phi}_{ij}^2(\eta_2, m) \right) - 1 \right),$$

$$\Psi_{min}^2(i, j) = \inf_x \left\{ \frac{h_i^2 e^{-\eta_2(ir_1 - c^2)x} \int_x^\infty e^{\eta_2(ir_1 - c^2)y} dG_{ij}^2(y)}{\frac{p_i^2}{i\alpha_1 + (K_2 - i)\beta_1 - \eta_2(ir_1 - c^2)} \int_x^\infty dG_{ij}^2(y)} \right\},$$

and

$$\Psi_{max}^2(i, j) = \sup_x \left\{ \frac{h_i^2 e^{-\eta_2(ir_1 - c^2)x} \int_x^\infty e^{\eta_2(ir_1 - c^2)y} dG_{ij}^2(y)}{\frac{p_i^2}{i\alpha_1 + (K_2 - i)\beta_1 - \eta_2(ir_1 - c^2)} \int_x^\infty dG_{ij}^2(y)} \right\}.$$

Therefore, using the expressions for H^1, $\Psi_{max}^1(i, j)$, $\Psi_{min}^1(i, j)$, H^2, $\Psi_{max}^2(i, j)$, and $\Psi_{min}^2(i, j)$, we have

$$C_2^* = \frac{H^1 H^2}{\begin{subarray}{c} \min(i_1, j_1), (i_2, j_2): \min\{i_1 r_1, c\} + i_2 r_2 > c, \\ p_{i_1 j_1} > 0, p_{i_2 j_2} > 0 \, \Psi_{min}^1(i_1, j_1) \, \Psi_{min}^2(i_2, j_2) \end{subarray}}$$

and

$$C_{*2} = \frac{H^1 H^2}{\max(i_1,j_1),(i_2,j_2):\,\min\{i_1 r_1, c\} + i_2 r_2 > c,\ p_{i_1 j_1} > 0, p_{i_2 j_2} > 0\ \Psi_{max}^1(i_1,j_1)\ \Psi_{max}^2(i_2,j_2)}.$$

Thus, an upper bound on $P(X_2 > x)$ is $C_2^* e^{-\eta_{2^x}}$.   ∎

Having obtained an upper bound on the buffer content distribution using the SMP bounds in a rather tedious manner, it is natural to ask what if we had used one of the other techniques such as effective bandwidth approximation or CDE approximation. This will be the focus of the next problem, in particular with an admission control setting.

Problem 120

Consider the $N=2$ class system described in Problem 119, where all sources of a class are IID exponential on-off sources. Obtain admissible region $\mathcal{K}$ using effective bandwidth approximation, CDE approximation, as well as SMP bounds so that if $(K_1, K_2) \in \mathcal{K}$, then $P(X_1 > B_1) > \epsilon_1$ and $P(X_2 > B_2) > \epsilon_2$. Compare the approaches for the following numerical values:

$$\alpha_1 = 1.0, \quad \beta_1 = 0.2, \quad r_1 = 1.0, \quad \epsilon_1 = 10^{-9}, \quad B_1 = 10,$$

$$\alpha_2 = 1.0, \quad \beta_2 = 0.2, \quad r_2 = 1.23, \quad \epsilon_2 = 10^{-6}, \quad B_2 = 10, \quad and \quad c = 13.2.$$

Solution

Using the three different methodologies (effective bandwidth, CDE, and SMP bounds), we can obtain different admissible regions based on the expressions used for approximating $P(X_j > x)$ for $j=1,2$. We first describe them and provide notation:

- *Effective bandwidth approximation:* We obtain the admissible region $\mathcal{K}_{ebw}$ by using $P(X_j > x) \approx e^{-\eta_j x}$ for $j=1,2$, where η_j is described in Problem 119. Thus, we have

$$\mathcal{K}_{ebw} = \{(K_1, K_2) : e^{-\eta_1 x} \le \epsilon_1, e^{-\eta_2 x} \le \epsilon_2\}.$$

   Understandably, this is an approximation but one that is somewhat easy to compute.
- *Naive effective bandwidth approximation:* In fact, if we want an approximation that is even easier to compute, then we can get rid of the effective bandwidth of the output and just use the effective bandwidth of the input sources instead. Here, η_1 remains unchanged

(from the previous case). However, we define η as the solution to $K_1 e b_1(\eta) + K_2 e b_2(\eta) = c$ and this would be different from η_2 if $\eta_2 > v^*$. Using η_1 and η, we obtain the admissible region $\mathcal{N}$ by using $P(X_1 > x) \approx e^{-\eta_1 x}$ and $P(X_2 > x) \approx e^{-\eta x}$ for $j = 1, 2$, where η_j is described in Problem 119. Thus, we have

$$\mathcal{N} = \{(K_1, K_2) : e^{-\eta_1 x} \le \epsilon_1, e^{-\eta x} \le \epsilon_2\}.$$

We will always have $\mathcal{N} \subset \mathcal{K}_{ebw}$.

- *CDE approximation:* We obtain the admissible region $\mathcal{K}_{cde}^{(1)}$, by using $P(X_j > x) \approx L_j^{(1)} e^{-\eta_j x}$ for $j = 1, 2$, where η_j is described in Problem 119 and $L_j^{(1)}$ is obtained using the technique in Section 10.2.2. The full details are available in Kulkarni and Gautam [70]. Thus, we have

$$\mathcal{K}_{cde}^{(1)} = \{(K_1, K_2) : L_1^{(1)} e^{-\eta_1 x} \le \epsilon_1, L_1^{(1)} e^{-\eta_2 x} \le \epsilon_2\}.$$

Although this is an approximation, it is not clear whether it is conservative or not.

- *Bufferless CDE approximation:* We obtain the admissible region $\mathcal{K}_{cde}^{(2)}$, by using $P(X_j > x) \approx L_j^{(2)} e^{-\eta_j x}$ for $j = 1, 2$, where η_j is described in Problem 119; however, $L_j^{(2)}$ is obtained based on the meaning of L in the CDE approximation in Section 10.2.2. In particular, $L_j^{(2)}$ is the fraction of fluid lost in a bufferless system (i.e., if the capacity of the buffer j was zero). The details are again available in Kulkarni and Gautam [70]. Thus, we have

$$\mathcal{K}_{cde}^{(2)} = \{(K_1, K_2) : L_1^{(2)} e^{-\eta_1 x} \le \epsilon_1, L_1^{(2)} e^{-\eta_2 x} \le \epsilon_2\}.$$

One of the difficulties in using this approach in general is that it is not easy to determine the fraction of fluid lost in a bufferless system.

- *SMP bounds:* We obtain the admissible region $\mathcal{K}_{smp}$ by taking advantage of the fact that $P(X_j > x) \le C_j^* e^{-\eta_j x}$ for $j = 1, 2$, where η_j and C_j^* are described in Problem 119. Thus, we have

$$\mathcal{K}_{smp} = \{(K_1, K_2) : C_1^* e^{-\eta_1 x} \le \epsilon_1, C_2^* e^{-\eta_2 x} \le \epsilon_2\}.$$

This approach is guaranteed to be conservative, that is, the QoS constraint $P(X_j > B_j) < \epsilon_j$ for $j = 1, 2$ will be surely satisfied. However, as we saw in Problem 119, the analysis is cumbersome.

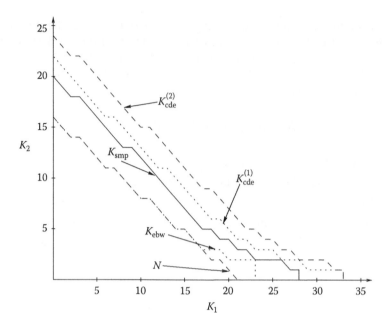

FIGURE 10.24
Regions $\mathcal{N}$, $\mathcal{K}_{ebw}$, $\mathcal{K}_{smp}$, $\mathcal{K}_{cde}^{(1)}$, $\mathcal{K}_{cde}^{(2)}$. (From Gautam, N. and Kulkarni, V.G., *Queueing Syst. Theory Appl.*, 36, 351, 2000. With permission.)

Next, we compare the region $\mathcal{K}_{smp}$ with the regions obtained using the CDE approximation, $\mathcal{K}_{cde}^{(1)}$ and $\mathcal{K}_{cde}^{(2)}$, as well as the regions obtained by effective-bandwidth approximation $\mathcal{K}_{ebw}$ and $\mathcal{N}$. We represent the regions under consideration in Figure 10.24 using the numerical values stated in the problem.

The region obtained by the SMP bounds, $\mathcal{K}_{smp}$, is conservative. Therefore, if an admissible region has points in $\mathcal{K}_{smp}$, then those points are guaranteed to satisfy the QoS criteria. Thus, the effective-bandwidth approximation produces overly conservative results for these parameter values. It is crucial to point out that although the effective bandwidth produces conservative results usually, it is not guaranteed to be conservative, unlike the results from SMP bounds. But in general, on one hand the effective-bandwidth approximation is computationally easy, on the other hand, it could either be too conservative (and hence leading to underutilization of resources) or be nonconservative (and hence unclear about meeting the QoS criteria). The CDE approximation, although computationally slower than the effective-bandwidth approximation, is typically faster than the SMP bounds technique. However, there are examples where we can show that the CDE approximation produces regions $\mathcal{K}_{cde}^{(1)}$ and $\mathcal{K}_{cde}^{(2)}$ with points (K_1, K_2) that would actually result in the QoS criteria not being satisfied. Using SMP bounds is computationally intensive. However, the computation can be done off line and the feasible region can be stored in a table. The computation needs to be repeated only when the input parameters change. ∎

We conclude this section by answering a natural question: How do the feasible region under timed round-robin compare against that of the static priority policy. Understandably, the policies are very different, but do they behave differently? We study that in the next problem, albeit using the same method, namely, the SMP bounds.

Problem 121

Consider a multiclass fluid queueing system with $N=2$ and IID exponential on-off sources for each class. For $j=1,2$, class j fluid enters into buffer j from K_j exponential on-off sources with mean on-time $1/\alpha_j$ and mean off-time $1/\beta_j$. Fluid is generated by each source at rate r_j when the source is in the on-state. Fluid is emptied by a channel with capacity c. Use the following numerical values:

$$\alpha_1 = 1, \quad \beta_1 = 0.3, \quad r_1 = 1.0, \quad \epsilon_1 = 10^{-6}, \quad B_1 = 8,$$
$$\alpha_2 = 1, \quad \beta_2 = 0.2, \quad r_2 = 1.23, \quad \epsilon_2 = 10^{-9}, \quad B_2 = 10, \quad and \quad c = 13.22.$$

Consider two policies: (1) timed round robin with $t_{so} = 0.02$ and $T = c(B_1 + B_2) + t_{so}$; and (2) static priority. Compare the two policies by viewing the admissible regions.

Solution

In Figure 10.25, we compare the two policies, timed round-robin and static priority, by viewing their respective admissible regions (using SMP bounds, hence the region corresponds to $\mathcal{K}_{smp}$ in the previous problem and $\mathcal{K}^{trr}$ in Section 10.3.2) for two-class exponential on-off sources with parameters given in the problem.

From the figure, we see the timed round-robin policy results in a smaller admissible region. This is because unlike the static priority service policy, the timed round-robin policy is not a work-conserving service discipline. In particular, there is time switching between buffers as well as time spent in buffers (recall that τ_1 and τ_2 are always spent) even if there is no traffic. However, static priority service policy does not achieve fairness among the classes of traffic. Therefore, it may not be an appropriate policy to use at all times. ∎

There are many other policies one could consider besides timed round-robin and static priority. We briefly describe three of them in the following section.

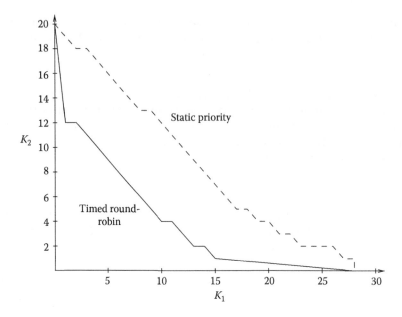

FIGURE 10.25
Timed round-robin versus static priority. (From Gautam, N. and Kulkarni, V.G., *Queueing Syst. Theory Appl.*, 36, 351, 2000. With permission.)

10.3.4 Other Policies

We consider three policies in this section. The first is called generalized processor sharing. It is fair like timed round-robin policy and work conserving like static priority. However, implementing it is a little tricky. Recall the setting described in Figure 10.15. There are N input buffers, one for each class of traffic. The input to buffer j $(j = 1, \ldots, N)$ is from the K_j sources of class j. The ith source of class j is driven by an independent random environment process $\{Z_{ij}(t), t \geq 0\}$ for $i = 1, 2, \ldots, K_j$ and $j = 1, \ldots, N$. At time t, source i of class j generates fluid at rate $r_{ij}(Z_{ij}(t))$ into buffer j. All the classes of fluids are emptied by a single channel of constant capacity c.

All classes of fluids are served using a *generalized processor sharing* (GPS) service scheduling policy, which is described in the following manner. Consider preassigned numbers $\phi_1, \phi_2, \ldots, \phi_N$ for each of the N classes such that

- $\phi_1 + \phi_2 + \cdots + \phi_N = 1$
- If all the input buffers have nonzero fluid, the scheduler allocates output capacity c in the ratio $\phi_1 : \phi_2 : \cdots : \phi_N$ to each of the N buffers
- If some buffers are empty, the scheduler allocates just enough capacity to those buffers (equal to the rates of traffic entering) so that

the buffers continue to be empty and the scheduler allocates the remaining output capacity c in the ratio of ϕ_j's of the remaining buffers

The GPS policy is in some sense the limiting timed round-robin policy, where $t_{so} = 0$ and for all j, $\tau_j \to 0$ such that $\tau_j / T \to \phi_j$. The only exception to timed round-robin is when a buffer is empty and here we assume that the system is work conserving. So empty buffers are served only for a fraction of their slot. The discrete version of the GPS is called the packetized general processor sharing (PGPS) or weighted fair queueing, which is well-studied in the literature. The quality-of-service aspects, effective bandwidths, and admission control for the GPS and PGPS have been addressed in detail in de Veciana et al. [24] and [22]. We recapitulate those results for GPS in the next problem for the case $N > 2$, although the $N = 2$ case should be referred to those articles.

Problem 122

What is the condition of stability? Assuming a stable system, let X_j be amount of fluid in buffer j in steady state for all $j \in [1, \ldots, N]$. Using effective bandwidth analysis, obtain an approximation for $P(X_j > B_j)$ for all $j \in [1, \ldots, N]$.

Solution

Notice that the scheduler is work conserving, in other words it is impossible that there is fluid in at least one buffer and the scheduler is draining at a rate lower than c. Thus, the condition for stability is

$$\sum_{j=1}^{N} \sum_{i=1}^{K_j} E[r_{ij}(Z_{ij}(\infty))] < c.$$

Using the effective bandwidth analysis is a little tricky since the compensating source is not easy to characterize except for some very special cases. For some $j \in [1, \ldots, N]$, take buffer j. It is guaranteed a minimum bandwidth of $\phi_j c$ at all times. However, the remaining $(1 - \phi_j)c$ (or greater, if buffer j is empty) is shared among all the sources in ratios according to the GPS scheme.

This is a little tricky to capture using a compensating source. Hence, we consider a fictitious compensating source that is essentially the output from a fictitious buffer with capacity $(1 - \phi_j)c$ and input being all the sources from all the classes except j. Thus, when the fictitious buffer is nonempty, buffer j gets exactly $\phi_j c$ capacity; however, when the fictitious buffer is empty, all unutilized capacity is used by buffer j. It is not difficult to check that this compensating source is rather conservative, that is, in the real setting, a lesser amount of fluid flows from the compensating source. In the special case when $N=2$ and $K_1 = K_2 = 1$ on-off source with on rates of class j source being larger than $\phi_j c$, this fictitious compensating source is identical to the real compensating source. One could certainly develop other types of compensating sources. The key idea here is a conservative one where unless all the other buffers are empty, the remaining capacity is not allocated to buffer j.

For such a compensating source, we solve for η_j as the unique solution to

$$\sum_{i=1}^{K_j} eb_{ij}(\eta_j) + \min\left((1 - \phi_j)c, \sum_{k \neq j}\sum_{i=1}^{K_k} eb_{ik}(\eta_j)\right) = c.$$

In fact, instead of the preceding expression one could have been more strict and written down the output effective bandwidth from the fictitious buffer. Thereby, we can obtain an approximation for the probability that there is more than B_j amount of fluid in buffer j as $P(X_j > B_j) \approx e^{-B_j\eta_j}$. Also for all $j = 1, 2, \ldots, N$,

$$e^{-B_j\eta_j} < \epsilon_j$$

would ensure some QoS guarantees for buffer j. ∎

That said, we move on to the next set of policies. Both are based on thresholds of buffer contents. They leverage upon results from this chapter as well as Chapter 9.

Problem 123

Consider a fluid queueing system with two infinite-sized buffers as shown in Figure 10.26. For $j = 1, 2$, fluid enters buffer j according to an alternating on-off process such that for an exponentially distributed time (with mean $1/\alpha_j$) fluid enters continuously at rate r_j and then no fluid enters for another

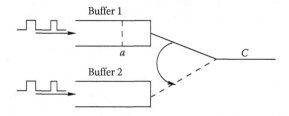

FIGURE 10.26
Two-buffer system. (From Aggarwal, V. et al., *Perform. Eval.*, 59(1), 19, 2004. With permission.)

exponentially distributed time (with mean $1/\beta_j$). When the off-time ends, another on-time starts, and so on. Let $X_j(t)$ be the amount of fluid in buffer j (for $j=1,2$) at time t. A scheduler alternates between buffers-1 and buffer-2 while draining out fluid continuously at rate c. Assume that $r_1 > c$ and $r_2 > c$. The policy adopted by the scheduler is as follows: as soon as buffer-1 becomes empty (i.e., $X_1(t)=0$), the scheduler switches from buffer-1 to buffer-2. When the buffer contents in buffer-1 reaches a (i.e., $X_1(t)=a$), the scheduler switches back from buffer-2 to buffer-1. We denote 0 and a as the thresholds for buffer-1. What is the stability condition? Assuming stability, derive an expression using SMP bounds for the steady-state distribution of the contents of buffer-2.

Solution

Note that the scheduler's policy is dependent only on buffer-1. That means even if buffer-2 is empty (i.e., $X_2(t)=0$), as long as buffer-1 has less than a (i.e., $X_1(t) < a$), the scheduler does not switch back to buffer-1. Also it is relatively straightforward to model the dynamics of buffer-1 and obtain the state probability $P(X_1 > x)$ for $x > a$ assuming the buffer is stable (and as $t \to \infty$, $X_1(t) \to X_1$). This analysis is described in Chapter 9 (Problem 96). Here we only consider bounds for $P(X_2 > x)$ assuming the system is stable. In fact, the stability condition (for limiting distributions of the buffer contents $X_1(t)$ and $X_2(t)$ to exist) is

$$\frac{r_1\beta_1}{\alpha_1 + \beta_1} + \frac{r_2\beta_2}{\alpha_2 + \beta_2} < c.$$

We now focus our attention on studying the buffer content process of buffer-2. In particular, our aim is to derive the limiting buffer-content distribution as our main performance measure. If we consider buffer-2 in isolation, its input is from an exponential on-off source but the output capacity alternates between c (for T_2 time, which is the time for the contents in buffer-1 to go from 0 to a) and 0 (for T_1 time, which is the time for the contents in buffer-1 to go from a to 0). To use the SMP bounds technique, which assumes that the output channel capacity is a constant, we first transform our model into an appropriate one with a constant output channel capacity.

Consider a single-buffer fluid model for buffer-2 with a constant output channel capacity c whose input is generated by the original exponential on-off source and a fictitious compensating source. The compensating source is such that it stays on for T_1 time units and off for T_2 time units. When the compensating source is on, it generates fluid at rate c, and it generates fluid at rate 0 when it is off. Note that the compensating source is independent of the original source. Clearly, the dynamics of the buffer-content process (of buffer-2) remain unchanged for this transformed single-buffer-fluid model with two input sources (including the compensating source) and constant output capacity c.

Notice that the on-time of the compensating source corresponds to the first passage time for buffer-1 contents to go from a to 0. The LST of this first passage time $\tilde{O}_1(\cdot)$ can be obtained from Problem 94 as

$$\tilde{O}_1(w) = \begin{cases} \frac{w+\beta_1+cs_0(w)}{\beta_1}e^{a\,s_0(w)} & \text{if } w \geq w^* \\ \infty & \text{otherwise,} \end{cases}$$

where

$$w^* = \frac{\left(2\sqrt{c\alpha_1\beta_1(r_1-c)} - r_1\beta_1 - c\alpha_1 - c\beta_1\right)}{r_1},$$

$$s_0(w) = \frac{-b - \sqrt{b^2 + 4w(w+\alpha_1+\beta_1)c(r_1-c)}}{2c(r_1-c)},$$

and $b = (r_1 - 2c)w + (r_1 - c)\beta_1 - c\alpha_1$. The mean on-time $E[T_1]$ can be computed as $E[T_1] = -d\tilde{O}_1(w)/dw$ at $w = 0$. Hence we have

$$E[T_1] = \frac{r_1 + a(\alpha_1 + \beta_1)}{c\alpha_1 + c\beta_1 - r_1\beta_1}.$$

Also, the off-time of the compensating source corresponds to the contents of buffer-1 to go from 0 to a for the first time. The LST of this first passage time $\tilde{O}_2(\cdot)$ can be obtained from Problem 95 as

$$\tilde{O}_2(s) = \frac{\beta_1}{\beta_1 + s}e^{-a\frac{\alpha_1 s+\beta_1 s+s^2}{r_1 s+r_1\beta_1}}.$$

Hence, the mean off-time $E[T_2]$ can be derived as $E[T_2] = -d\tilde{O}_2(s)/ds$ at $s = 0$ and is given by

$$E[T_2] = \frac{r_1 + a(\alpha_1 + \beta_1)}{r_1\beta_1}.$$

A detailed derivation of $\tilde{O}_1(\cdot)$ and $\tilde{O}_2(\cdot)$ is described in Aggarwal et al. [3].

Using this compensating source model and the SMP bounds analysis, we can derive the limiting distribution of the buffer contents of buffer-2. We first obtain the effective bandwidth of source 2 (the original source into buffer-2) using Equation 10.18 as

$$eb_2(v) = \frac{r_2 v - \alpha_2 - \beta_2 + \sqrt{(r_2 v - \alpha_2 - \beta_2)^2 + 4\beta_2 r_2 v}}{2v}.$$

Using the effective bandwidth of a general on-off source described in Section 10.1.2, we can derive the effective bandwidth of the compensating source ($eb_0(v)$) as the unique solution to

$$\tilde{O}_1(v\,eb_0(v) - cv)\,\tilde{O}_2(v\,eb_0(v)) = 1.$$

Then we can obtain η as the solution to

$$eb_0(\eta) + eb_2(\eta) = c. \tag{10.33}$$

Thus, the limiting distribution of the contents of buffer-2 is bounded as

$$LB \le \lim_{t \to \infty} P\{X_2(t) > x\} \le UB,$$

where

$$LB = \frac{r_2 \beta_2}{eb_2(\eta)(\alpha_2 + \beta_2)} \frac{1 - \tilde{O}_2(\eta eb_0(\eta))}{E(T_1) + E(T_2)} \frac{c}{eb_0(\eta)(c - eb_0(\eta))\eta} e^{-\eta x},$$

$$UB = \frac{r_2 \beta_2}{eb_2(\eta)(\alpha_2 + \beta_2)} \frac{1/\tilde{O}_2(\eta eb_0(\eta)) - 1}{E(T_1) + E(T_2)} \frac{c}{eb_0(\eta)(c - eb_0(\eta))\eta} e^{-\eta x},$$

and η is the solution to Equation 10.33. ∎

Policies of the type described in the preceding problem typically arise when there is a cost for switching from one buffer to the other. This is called a hysteretic policy (akin to hysteresis in magnetism). Usually when this switching cost is zero, the resulting policy is typically a threshold policy where one would always serve buffer-1 when the fluid is more than threshold a and serve buffer-2 otherwise. In contrast, the hysteretic policy avoids frequently switching between the buffers (due to the switching cost) and has two values (here they are a and 0) so that the switching occurs at those values from one to the other and back. We will consider a numerical problem in the exercises that also aims to optimize a. An interesting observation to make is that when $a = 0$, the policy reduces to static priority. In the next problem, we will consider a threshold policy as opposed to a hysteretic policy.

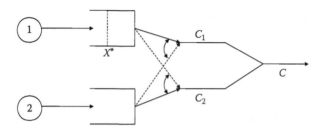

FIGURE 10.27
Two-buffer system. (From Mahabhashyam, S. et al., *Oper. Res.*, 56(3), 728, 2008. With permission.)

Problem 124

Consider a two-buffer fluid flow system illustrated in Figure 10.27. For $j = 1$, 2, class j fluid enters buffer j according to an alternating on-off process so that fluid enters continuously at rate r_j for an exponentially distributed time (on-times) with mean $1/\alpha_j$ and then no fluid enters (off-times) for another exponential time with mean $1/\beta_j$. The on and off times continue alternating one after the other. The buffers can hold an infinite amount of fluid; however, the contents of only one buffer is observed, buffer-1. There are two schedulers that drain fluid from the two buffers. Scheduler-1 has a capacity of c_1 and scheduler-2 has a capacity c_2, which are the maximum rates the respective schedulers can drain fluid. Let $X_j(t)$ be the amount of fluid in buffer j (for $j = 1, 2$) at time t. Fluid is drained from the two buffers in the following fashion. When $X_1(t)$ is nonzero, scheduler-1 serves buffer-1 and when $X_1(t) = 0$, scheduler-1 serves buffer-2. Also, if $X_1(t)$ is less than a threshold x^*, scheduler-2 removes fluid from buffer-2, otherwise it drains out buffer-1. Assuming stability, derive bounds for the steady-state fluid level in buffer-2.

Solution

Notice that when $X_1(t) = 0$, both schedulers serve buffer-2 and when $0 < X_1(t) < x^*$ scheduler-1 serves buffer-1 and scheduler-2 serves buffer-2, whereas, when $X_1(t) \geq x^*$, both schedulers serve buffer-1. Since only $X_1(t)$ is observed, the buffer-emptying scheme depends only on it. If $C_j(t)$ is the capacity available for buffer j at time t, then $C_1(t) = 0$ when $X_1(t) = 0$, $C_1(t) = c_1$ when $0 < X_1(t) < x^*$, and $C_1(t) = c_1 + c_2$ whenever $X_1(t) \geq x^*$. Capacity available for buffer-2 at any time t is $C_2(t) = c_1 + c_2 - C_1(t)$. The stability condition for the two-buffer system in Figure 10.27 is given by:

$$\frac{r_1 \beta_1}{\alpha_1 + \beta_1} + \frac{r_2 \beta_2}{\alpha_2 + \beta_2} < c_1 + c_2.$$

We assume that both r_1 and r_2 are greater than $c_1 + c_2$.

It is possible to obtain the state probability $P(X_1 > x)$ for $x > x^*$ assuming the buffer is stable (and as $t \to \infty$, $X_1(t) \to X_1$). This analysis is described in Problem 97 (see Chapter 9). Here we only consider bounds for $P(X_2 > x)$, where X_2 is the amount of fluid in buffer-2 in steady state. For buffer-2, the output capacity is not only variable but also inherently dependent on contents of buffer-1. The input for buffer-2 is from an exponential on-off source but the output capacity varies from zero to $(c_1 + c_2)$ depending on the buffer content in buffer-1. The variation of output capacity over time (say $\hat{O}(t)$) with respect to content of buffer-1 is as follows:

$$\hat{O}(t) = \begin{cases} 0 & when & X_1(t) \geq x^*, \\ c_2 & when & 0 < X_1(t) < x^*, \\ c_1 + c_2 & when & X_1(t) = 0. \end{cases}$$

Consider the queueing system (as depicted in Figure 10.28), where there are two input streams and a server with a constant output capacity $c_1 + c_2$. The first input stream is a *compensating source*, where fluid enters the queue at rate $c_1 + c_2 - \hat{O}(t)$ at time t. The second input stream is identical to source-2, where fluid enters according to an exponential on-off process with rates r_2 when on and 0 when off. The environment process that drives traffic generation for the compensating source can be modeled as a four-state SMP. Let $Z_1(t)$ be the environment process denoting the on-off source for buffer-1. If source-1 is on at time t, $Z_1(t) = 1$ and if source-1 is off at time t, $Z_1(t) = 0$. Consider the Markov regenerative sequence $\{(Y_n, S_n), n \geq 0\}$ where S_n is the nth regenerative epoch, corresponding to $X_1(t)$ equaling either x^* or 0, and Y_n is the state immediately following the nth Markov regenerative epoch such that

$$Y_n = \begin{cases} 1 & if & X_1(S_n) = 0 \text{ and } Z_1(S_n) = 0, \\ 2 & if & X_1(S_n) = 0 \text{ and } Z_1(S_n) = 1, \\ 3 & if & X_1(S_n) = x^* \text{ and } Z_1(S_n) = 0, \\ 4 & if & X_1(S_n) = x^* \text{ and } Z_1(S_n) = 1. \end{cases}$$

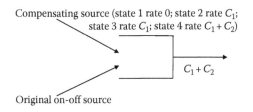

Compensating source (state 1 rate 0; state 2 rate C_1; state 3 rate C_1; state 4 rate $C_1 + C_2$)

$C_1 + C_2$

Original on-off source

FIGURE 10.28
Buffer-2 with compensating source. (From Mahabhashyam, S. et al., *Oper. Res.*, 56(3), 728, 2008. With permission.)

The compensating source is governed by an underlying environment process, which is an SMP with four states $\{1,2,3,4\}$. Fluid is generated at rates $0, c_1, c_1,$ and $c_1 + c_2$ when the SMP is, respectively, in states 1, 2, 3, and 4. Let $r(i)$ be the rate of fluid generation when SMP is in state i. Therefore, $r(1) = 0, r(2) = c_1, r(3) = c_1,$ and $r(4) = c_1 + c_2$. The kernel of SMP $G(t) = [G_{ij}(t)]$ will be computed first, which is given by

$$G_{ij}(t) = P\{Y_1 = j, S_1 \leq t \mid Y_o = i\}.$$

Since $X_1(t) = 0$ in State 1 and $X_1(t) > x^*$ in State 4, there cannot be a direct transition from State 1 to State 4 and vice versa without actually going through State 2 or 3 (where $0 < X_1(t) < x^*$). Therefore, $G_{14}(t) = G_{41}(t) = 0$. Also, clearly, $G_{ii}(t) = 0$ for $i = 1,$ 2, 3, 4. In addition, according to the definitions of States 2 and 3, $G_{23}(t) = G_{32}(t) = G_{42}(t) = G_{13}(t) = 0$.

The kernel matrix $G(t) = [G_{ij}(t)]$ of the SMP is given by

$$G(t) = \begin{bmatrix} 0 & G_{12}(t) & 0 & 0 \\ G_{21}(t) & 0 & 0 & G_{24}(t) \\ G_{31}(t) & 0 & 0 & G_{34}(t) \\ 0 & 0 & G_{43}(t) & 0 \end{bmatrix}.$$

The expressions $G_{12}(t),$ $G_{21}(t),$ $G_{24}(t),$ $G_{31}(t),$ $G_{34}(t),$ and $G_{43}(t)$ need to be obtained. Two of them are relatively straightforward to obtain, namely, $G_{12}(t)$ and $G_{43}(t)$. First consider $G_{12}(t)$. This is the probability that Y_n changes from 1 to 2 before time t, which is the same as the probability of the source-1 going from off to on. Hence $G_{12}(t)$ is given by

$$G_{12}(t) = 1 - e^{-\beta_1 t}.$$

Next consider $G_{43}(t)$. This is the probability that the buffer-2 content goes up from x^* and reaches x^* in time t. This is identical to the probability that the buffer content starts at zero, goes up, and comes back to zero within time t, that is, equivalent to the busy period distribution. The LST of $G_{43}(t)$ can be obtained by substituting appropriate terms in the busy period distribution of Problem 94. Hence,

$$\tilde{G}_{43}(w) = \begin{cases} \dfrac{w + \beta_1 + cs_0(w)}{\beta_1} & if \quad w > w^* \\ \infty & otherwise \end{cases}$$

where

$$s_0(w) = \dfrac{-b - \sqrt{b^2 + 4w(w + \alpha_1 + \beta_1)c(r_1 - c)}}{2c(r_1 - c)},$$

and $b = (r_1 - 2c)w + (r_1 - c)\beta_1 - c\alpha_1$, $w^* = (2\sqrt{c\alpha_1 \beta_1 (r_1 - c)} - r_1\beta_1 - c\alpha_1 + c\beta_1)/r_1$, and $c = c_1 + c_2$.

To obtain expressions for the remaining terms in the kernel of the SMP, namely, $G_{21}(t)$, $G_{24}(t)$, $G_{31}(t)$, and $G_{34}(t)$, turn to Problem 93. Using a subscript of 1 for α, β, r, and c, it is relatively straightforward to see that

$$G_{21}(t) = H_{21}(0, t)$$

$$G_{24}(t) = H_{22}(0, t)$$

$$G_{31}(t) = H_{11}(x^*, t)$$

$$G_{34}(t) = H_{12}(x^*, t)$$

with the RHS expressions corresponding to the notation in Problem 93 whose LSTs have already been derived there. Hence we have

$$\tilde{G}_{31}(w) = a_{11}(w)e^{S_1(w)x^*}\psi_1(w) + a_{21}(w)e^{S_2(w)x^*}\psi_2(w),$$

$$\tilde{G}_{21}(w) = a_{11}(w) + a_{21}(w),$$

$$\tilde{G}_{34}(w) = a_{12}(w)e^{S_1(w)x^*}\psi_1(w) + a_{22}(w)e^{S_2(w)x^*}\psi_2(w),$$

$$\tilde{G}_{24}(w) = a_{12}(w) + a_{22}(w),$$

where

$$S_1(w) = \frac{-\hat{b} - \sqrt{\hat{b}^2 + 4w(w + \alpha_1 + \beta_1)c_1(r_1 - c_1)}}{2c_1(r_1 - c_1)},$$

$$S_2(w) = \frac{-\hat{b} + \sqrt{\hat{b}^2 + 4w(w + \alpha_1 + \beta_1)c_1(r_1 - c_1)}}{2c_1(r_1 - c_1)},$$

with $\hat{b} = (r_1 - 2c_1)w + (r_1 - c_1)\beta_1 - c_1\alpha_1$ and for $i = 1, 2$,

$$\psi_i(w) = \frac{\beta_1}{w + \beta_1 + S_i(w)c_1}$$

and finally,

$$a_{11}(w) = \frac{e^{S_2(w)x^*}}{\delta(w)},$$

$$a_{12}(w) = \frac{-\psi_2(w)}{\delta(w)},$$

$$a_{21}(w) = \frac{-e^{S_1(w)x^*}}{\delta(w)},$$

$$a_{22}(w) = \frac{\psi_1(w)}{\delta(w)},$$

with $\delta(w) = e^{S_2(w)x^*}\psi_1(w) - e^{S_1(w)x^*}\psi_2(w)$.

Now that we have characterized the compensating source as an SMP, next we consider that and the original source-2 and obtain the effective bandwidths. The effective bandwidth of the compensating source can be computed as follows. For a given v such that $v > 0$, define the matrix $\chi(v, u)$ such that

$$\chi(v,u) = \begin{bmatrix} 0 & \tilde{G}_{12}(vu) & 0 & 0 \\ \tilde{G}_{21}(vu - c_1v) & 0 & 0 & \tilde{G}_{24}(vu - c_1v) \\ \tilde{G}_{31}(vu - c_1v) & 0 & 0 & \tilde{G}_{34}(vu - c_1v) \\ 0 & 0 & \tilde{G}_{43}(vu - c_1v - c_2v) & 0 \end{bmatrix}.$$

Let $e(A)$ be the Perron–Frobenius eigenvalue of a square matrix A. The effective bandwidth of the compensating source for a particular positive value of v (i.e., $eb_c(v)$) is given by the unique solution to

$$e(\chi(v, eb_c(v))) = 1.$$

Having obtained the effective bandwidth of the compensating source, the next step is to obtain that for the other source. Since the original source-2 has a CTMC as the environment process, its effective bandwidth (i.e., $eb_2(v)$) is

$$eb_2(v) = \frac{r_2v - \alpha_2 - \beta_2 + \sqrt{(r_2v - \alpha_2 - \beta_2)^2 + 4\beta_2 r_2 v}}{2v}.$$

Using the effective bandwidths of the original source 2 ($eb_2(v)$) and the compensating source ($eb_c(v)$), η can be obtained as the unique solution to

$$eb_2(\eta) + eb_c(\eta) = c_1 + c_2.$$

For notational convenience, use the following:

$$\gamma_2 = eb_2(\eta),$$

$$\gamma_c = eb_c(\eta).$$

Define $\Phi(\eta) = \chi(\eta, eb_c(\eta))$ such that $\phi_{ij}(\eta)$ is the ijth element of $\Phi(\eta)$. Let h be the left eigenvector of $\Phi(\eta)$ corresponding to the eigenvalue of 1, that is,

$$h = h\Phi(\eta).$$

Also note that $h = [h_1\ h_2\ h_3\ h_4]$. Next, using η, h, γ_2, γ_c, and other parameters defined earlier, bounds for $P(X_2 > x)$ can be obtained as

$$K_* e^{-\eta x} \le P(X_2 > x) \le K^* e^{-\eta x},$$

where

$$K^* = \frac{\left\{\frac{r_2\beta_2}{\gamma_2(\alpha_2+\beta_2)}\right\} \sum_{i=1}^{4} \left\{\frac{h_i}{\eta(r(i)-\gamma_c)}\left(\left[\sum_{j=1}^{4}\phi_{ij}(\eta)\right]-1\right)\right\}}{\min\,(i,j):p_{ij} > 0\ \Psi_{min}(i,j),}$$

and

$$K_* = \frac{\left\{\frac{r_2\beta_2}{\gamma_2(\alpha_2+\beta_2)}\right\} \sum_{i=1}^{4} \left\{\frac{h_i}{\eta(r(i)-\gamma_c)}\left(\left[\sum_{j=1}^{4}\phi_{ij}(\eta)\right]-1\right)\right\}}{\max\,(i,j):p_{ij} > 0\ \Psi_{max}(i,j)}$$

with $\Psi_{min}(i,j)$ and $\Psi_{max}(i,j)$ derived using the values in Table 10.3. ∎

Similarly, many other multiclass fluid queues can be analyzed using compensating sources. Also, there are other ways to evaluate things like

TABLE 10.3

Table of Ψ_{max} and Ψ_{min} Values

	IFR & $r(i) > \gamma_c$	IFR & $r(i) \le \gamma_c$	DFR & $r(i) > \gamma_c$	DFR & $r(i) \le \gamma_c$
$\Psi_{max}(i,j)$	$\dfrac{\phi_{ij}(-\eta(r(i)-\gamma_c))\tau_i h_i}{p_{ij}p_i}$	$\dfrac{\tau_i h_i \lambda_{ij}(\infty)}{p_i(\lambda_{ij}(\infty)-\eta(r(i)-\gamma_c))}$	$\dfrac{\tau_i h_i \lambda_{ij}(\infty)}{p_i(\lambda_{ij}(\infty)-\eta(r(i)-\gamma_c))}$	$\dfrac{\tilde{\phi}_{ij}(-\eta(r(i)-\gamma_c))\tau_i h_i}{p_{ij}p_i}$
$\Psi_{min}(i,j)$	$\dfrac{\tau_i h_i \lambda_{ij}(\infty)}{p_i(\lambda_{ij}(\infty)-\eta(r(i)-\gamma_c))}$	$\dfrac{\tilde{\phi}_{ij}(-\eta(r(i)-\gamma_c))\tau_i h_i}{p_{ij}p_i}$	$\dfrac{\tilde{\phi}_{ij}(-\eta(r(i)-\gamma_c))\tau_i h_i}{p_{ij}p_i}$	$\dfrac{\tau_i h_i \lambda_{ij}(\infty)}{p_i(\lambda_{ij}(\infty)-\eta(r(i)-\gamma_c))}$

Source: Mahabhashyam, S. et al., *Oper. Res.*, 56(3), 728, 2008. With permission.

joint distributions (we mainly considered marginal distributions of buffer contents). With that thought we conclude this chapter.

Reference Notes

The main focus of this chapter was to determine approximations and bounds for steady-state fluid levels in infinite-sized buffers. However, we did not present the underlying theory of large deviations that enabled this. Interested readers can refer to Shwartz and Weiss [98] as well as Ganesh et al. [38] for an excellent treatment of large deviations. The crucial point is that the tail events are extremely rare and, in fact, only analytical models can be used to estimate their probabilities suitably. There are simulation techniques too but they are typically based on a change of measure argument following the Radon–Nikodym theorem. Details regarding change of measures can be found in textbooks such as by Ethier and Kurtz [33]. In fact, the bounds described in this chapter are based on some exponential change of measure arguments in Ethier and Kurtz [33].

The common theme in this chapter is the concept of effective bandwidths (also called effective capacity). The theoretical underpinnings for effective bandwidth is based on large deviations and we briefly touched upon the Gärtner–Ellis condition. Further details on the Gärtner–Ellis conditions can be found in Kesidis et al. [61] and the references therein. An excellent tutorial on effective bandwidths is Kelly [60] and it takes a somewhat different approach defining it at time t whereas what we present is based on letting $t \to \infty$. Our results are mainly based on Elwalid and Mitra [30], Kesidis et al. [61], and Kulkarni [68] that show how to compute effective bandwidths of several types of traffic flows. Further, Chang and Thomas [16], Chang and Zajic [17], and de Veciana et al. [23] explain effective bandwidth computations for outputs from queues and extend the results to networks.

Once we know how to compute effective bandwidths, they can be used for approximating buffer content distributions. Recall from Section 9.2.3 that buffer content distributions can be obtained only when the sources are CTMCs (based on Anick et al. [5], Elwalid and Mitra [28, 29], and Kulkarni [69]). The effective bandwidth approximation lends itself well for computing the tail distributions. The results presented in this chapter (Section 10.2.1) are summarized from Elwalid and Mitra [30], Kesidis et al. [61], Krishnan et al. [65], and Kulkarni [68]. These results were fine-tuned by Elwalid et al. [31, 32] by considering Chernoff bounds (hence the CDE approximation in Section 10.2.2). Although effective bandwidth and CDE approximations are mainly for the tail probabilities, exponential bounds on the buffer content analysis (called SMP bounds because these require the sources to be

SMPs) described in Section 10.2.3 are based on Palmowski and Rolski [87, 88] and Gautam et al. [39].

These results can be suitably extended to multiclass fluid queues, where each class of fluid has a dedicated buffer. Perhaps the most well-studied policy is the priority rule that gained popularity to aid differentiated services and is based on its discrete counterpart. For a comprehensive study on effective bandwidths with priorities, see Berger and Whitt [9, 10], and Gautam and Kulkarni [70]. Policies such as generalized processor sharing are considered in de Veciana et al. [22, 24]. The results on timed round-robin policy and its comparison with static priority are based on Gautam and Kulkarni [40]. The analysis on threshold-based policies is based on Mahabhashyam et al. [76] and Aggarwal et al. [3].

Exercises

10.1 Consider a fluid source driven by a three-state CTMC environment process $\{Z(t), t \geq 0\}$ with generator matrix and rate matrix given by

$$Q = \begin{bmatrix} -\beta & \beta & 0 \\ \gamma & -\gamma - \delta & \delta \\ 0 & \alpha & -\alpha \end{bmatrix} \quad \text{and} \quad R = \begin{bmatrix} 0 & 0 & 0 \\ 0 & r_1 & 0 \\ 0 & 0 & r_2 \end{bmatrix}.$$

Derive the effective bandwidth of this source in terms of parameter v by computing the largest real positive eigenvalue of $Q/v + R$.

10.2 Consider two independent on-off sources that input fluid into a buffer. The on and off times for source i (for $i = 1, 2$) are exponentially distributed with parameters α_i and β_i, respectively. Also when source i is on, fluid is generated at rate r_i.

(a) Write down the effective bandwidth of each of the sources, let us call them $eb_1(v)$ and $eb_2(v)$.

(b) Now you can think of the input sources as a single four-state CTMC environment process $\{Z(t), t \geq 0\}$ with generator matrix and rate matrix given by

$$Q = \begin{bmatrix} -\beta_1 - \beta_2 & \beta_2 & \beta_1 & 0 \\ \alpha_2 & -\alpha_2 - \beta_1 & 0 & \beta_1 \\ \alpha_1 & 0 & -\alpha_1 - \beta_2 & \beta_2 \\ 0 & \alpha_1 & \alpha_2 & -\alpha_1 - \alpha_2 \end{bmatrix}$$

and

$$R = \begin{bmatrix} 0 & 0 & 0 & 0 \\ 0 & r_2 & 0 & 0 \\ 0 & 0 & r_1 & 0 \\ 0 & 0 & 0 & r_1 + r_2 \end{bmatrix}.$$

Note that the four states correspond to: both sources on, source-1 off and 2 on, source-2 off and source-1 on, and both sources off. Compute the effective bandwidth of this source, call it $eb(v)$.

(c) Show that the algebraic expression for $eb(v)$ is identical to the effective bandwidth of the net input to the buffer $eb_1(v) + eb_2(v)$.

10.3 Consider a single buffer fluid model with input from an on-off source with hyperexponential on-time CDF (for $x \geq 0$)

$$U(x) = 0.5(1 - e^{-2x}) + 0.3(1 - e^{-4x}) + 0.2(1 - e^{-x})$$

and Erlang off-time CDF (for $x \geq 0$)

$$D(x) = 1 - e^{-2x} - 2xe^{-2x}.$$

When the source is on, fluid enters the buffer at rate $r = 3$ Mbps and when the source is off, no fluid enters. The output channel capacity $c = 2$ Mbps.

(a) Compute $\tilde{U}(s)$ and $\tilde{D}(s)$, the LSTs of $U(x)$ and $D(x)$, respectively.

(b) The tail probability of the limiting buffer contents $P\{X > x\}$ for very large x can be obtained using effective bandwidths as

$$P\{X > x\} \approx e^{-\eta x}.$$

Show that η can be computed as the smallest real positive solution to

$$\tilde{U}(-\eta(r - c)) \, \tilde{D}(\eta c) = 1.$$

(c) Write a computer program and obtain the value of η numerically.

(d) For $x = 4$ Mb, compute the probability $P\{X > x\}$ by approximating it as

$$P\{X > x\} \approx e^{-\eta x}.$$

10.4 Consider a single buffer system with K independent and identical on-off sources with on-times and off-times distributed exponentially with parameters α and β, respectively. Each source produces fluid at rate r when it is on and at rate 0 when it is off. The buffer has output capacity c. Assume that for stability $Kr\beta/(\alpha+\beta) < c$. Show that

$$
v^* = \begin{cases} \frac{\beta}{r}\left(\sqrt{\frac{c\alpha}{\beta(Kr-c)}} - 1\right) + \frac{\alpha}{r}\left(1 - \sqrt{\frac{\beta(Kr-c)}{c\alpha}}\right) & \text{if } \frac{Kr\beta}{\alpha+\beta} < c < Kr \\ \infty & \text{if } c \geq Kr. \end{cases}
$$

Obtain the effective bandwidth of the output from the buffer in terms of v.

10.5 Consider a series network of three nodes in tandem, with each node having an infinite-sized buffer. Say you know the effective bandwidth of the source inputing traffic to buffer-1 (i.e., $eb_1(v)$). The traffic is processed at node 1 at a maximum capacity of c_1. This traffic flows into node 2. It gets processed at a maximum capacity of c_2 and then flows into node 3. At node 3, the processor uses a capacity c_3 to process traffic, which then exits the system. Using effective bandwidth approximations explain how you would compute the probability that none of buffers 1, 2, and 3 have more than B_1, B_2, and B_3 amount of fluids in them.

10.6 *Effective bandwidths for discrete time fluid sources*: Fluid arrives from a source at discrete time slots $1, 2, \ldots$ such that during a time slot, a random amount of fluid (distributed exponentially with mean $1/\lambda$) is generated by the source. Say Y_i is the amount of fluid generated in the ith slot, then $P(Y_i \leq y) = 1 - e^{-\lambda y}$. Define A_n as the total amount of fluid generated from slot 1 to n by the source. In fact, $A_n = Y_1 + Y_2 + \cdots + Y_n$. The effective bandwidth definition for discrete time sources is given by

$$
eb(v) = \lim_{n\to\infty} \frac{1}{vn} \log E\{\exp(vA_n)\}.
$$

Compute $eb(v)$ for the preceding source in terms of λ and v.

10.7 A multirate on-off source alternates between on and off states but when it is on, it can generate fluid at different rates. The off-times are according to an Erlang distribution with mean 6 and standard deviation $\sqrt{6}$. At the end of an off period the source switches to on-state-i with probability $i/6$ for $i = 1, 2, 3$. In on-state-i, fluid is generated by the source at rate $12/i$ for $i = 1, 2, 3$. The sojourn times in the ith on state is according to an Erlang distribution with mean

1 and standard deviation $0.5/\sqrt{i}$. Graph the effective bandwidth of this source $eb(v)$ versus v for $v \in [0,2]$.

10.8 Recall the in-tree network in Figure 10.10 considered in Problem 108. It is desired that the probability of exceeding buffer level of $b = 14\,kB$ must be less that 0.000001 in all seven buffers. Using effective bandwidth approximation, design the smallest output capacity c_j for $j = 1, \ldots, 7$ to achieve such a quality of service. Use the same numerical values of $\alpha = 5$, $\beta = 1$, and $r = 6$.

10.9 Consider Problem 110 and obtain bounds for $P(X > x)$ using SMP bounds. Compare the results against those based on CDE approximation described in Problem 110.

10.10 Solve Problem 117 using CDE approximation. In particular, using CDE approximation, obtain expressions for $P(X_j > B_j)$ for all $j = 1, \ldots, N$. Then graph the fraction of fluid lost assuming that the size of the buffer is B_j. Compare against SMP bounds and effective bandwidth approximation results presented in Problem 117.

10.11 Consider a static priority policy to empty fluids from three buffers with buffer-1 given the highest priority. Into buffer i, fluid enters from a general on-off source with on-time CDF $p_i(1 - e^{-3t}) + (1 - p_i)(1 - e^{-4t})$ and off-time CDF $1 - e^{-t} - te^{-t}$ for $i = 1, 2, 3$. Also, $p_1 = 0.5$, $p_2 = 0.4$, and $p_3 = 0.2$. Traffic is generated at the rate of 2 per unit time when any source is on. The output capacity $c = 1.8$. Using effective bandwidth approximations determine expressions for the probability that each of the buffers would exceed a level x in steady state.

10.12 For the setting in Problem 123, assume that there is a cost of C_s to switch from one buffer to another. What is the optimal value of a that would minimize the long-run average cost per unit time subject to satisfying the constraints that $P(X_j > B_j) < \epsilon_j$ for $j = 1, 2$. Use UB to ensure the constraint is satisfied. Illustrate the optimal solution for the following numerical values: $\beta_1 = 2$, $\alpha_1 = 8$, $r_1 = 2.645$, $\beta_2 = 3$, $\alpha_2 = 9$, $r_2 = 1.87$, $c = 1.06$, and $C_s = 100$. Also, $B_1 = 2.5$, $B_2 = 8$, $\epsilon_1 = 0.001$, and $\epsilon_2 = 0.01$.

Appendix A: Random Variables

This appendix chapter acts as a refresher to the topic of random variables typically covered in an elementary course on probability as well as describes some minor extensions. In addition, the contents of this chapter can be used as a reference when the corresponding results are stated in various portions of this textbook. Further, this chapter acts as a point of clarification for notation used in this book. For detailed explanation of the derivation of some of the results, the readers are encouraged to consult an elementary probability and stochastic process book. We begin the chapter by characterizing random variables.

A.1 Distribution and Moments

Any random variable X defined on a subset of the real line can be characterized by its cumulative distribution function (CDF), which we denote as $F_X(x)$ and defined as

$$F_X(x) = P\{X \leq x\}$$

for all $x \in (-\infty, \infty)$. In this book, we drop X from $F_X(x)$ and just call the CDF as $F(x)$ especially where there is only one random variable being considered. There are two basic types of random variables: discrete and continuous. Discrete random variables are defined on some discrete points on the real line. However, continuous random variables are defined on a set of open intervals on a real line but not on any discrete points. Of course, there is a class of random variables called mixture or hybrid random variables that are a combination of discrete and continuous random variables. The CDF for these random variables have jumps (or discontinuities) at the discrete points. Let $\mathcal{D}$ be the set of discrete points for a mixture random variable X, and for all $x \in \mathcal{D}$, let $p_x = p\{X = x\}$, be the magnitude of jumps.

We proceed with a generic mixture random variable with the understanding that both continuous and discrete random variables are special cases corresponding to $\mathcal{D} = \emptyset$ and $\sum_{x \in \mathcal{D}} p_x = 1$, respectively. Thus, let X be a mixture random variable with CDF $F(x)$ and discrete-point set $\mathcal{D}$. The expected value of X is defined as

$$E[X] = \int_{-\infty}^{\infty} x dF(x) + \sum_{x \in \mathcal{D}} x p_x$$

where the integral is Riemann type (however, it is indeed derived using the Lebesgue integral). We present an example to illustrate.

Problem 125

Let X be a random variable that denotes Internet packet size for TCP transmissions (in bytes). On the basis of empirical evidence, say the CDF of X is modeled as

$$F(x) = \begin{cases} 0 & \text{if } x < 40 \\ a\sqrt{x} + b & \text{if } 40 < x < 576 \\ 0.0001x + 0.6424 & \text{if } 576 < x < 1500 \\ 1 & \text{if } x > 1500 \end{cases}$$

where
$$a = 0.3/24 - \sqrt{40}$$
$$b = 0.25 - a\sqrt{40}$$

Notice that there are jumps or discontinuities in the CDF at 40, 576, and 1500 bytes. Hence, $\mathcal{D} = \{40, 576, 1500\}$ with p_x given by 0.25, 0.15, and 0.2076 for $x = 40$, $x = 576$, and $x = 1500$, respectively. Compute $E[X]$.

Solution
Using the definition

$$E[X] = \int_{-\infty}^{\infty} x dF(x) + \sum_{x \in \mathcal{D}} x p_x$$

with $\mathcal{D} = \{40, 576, 1500\}$ and p_x given in the problem statement, we have

$$E[X] = \int_{-\infty}^{40} x dF(x) + \int_{40}^{576} x dF(x) + \int_{576}^{1500} x dF(x)$$

$$+ \int_{1500}^{\infty} x dF(x) + 40 p_{40} + 576 p_{576} + 1500 p_{1500}$$

$$= 0 + \int_{40}^{576} 0.5 a \sqrt{x} dx + \int_{576}^{1500} 0.0001 x dx + 0 + 40$$

$$\times 0.25 + 576 \times 0.15 + 1500 \times 0.2076$$

$$= 580.4901 \text{ bytes.} \qquad \blacksquare$$

Further, the rth moment of a mixed random variable X is

$$E[X^r] = \int_{-\infty}^{\infty} x^r dF(x) + \sum_{x \in \mathcal{D}} x^r p_x$$

with $r = 1$ corresponding to the standard expected value $E[X]$. The variance of X can be computed as $V[X] = E[X^2] - \{E[X]\}^2$. That said, for the remainder of this chapter, we will only focus on discrete or continuous random variables but not on mixed ones. They are sometimes more elegantly handled individually. We present that in the following two sections.

A.1.1 Discrete Random Variables

A discrete random variable X is also characterized by its probability mass function (PMF) $p(x) = P\{X = x\}$ for all values of x for which X is defined, say $\mathcal{D}$. The PMF satisfies two important properties: $0 \le p(x) \le 1$ for all $x \in \mathcal{D}$ and $\sum_{x \in \mathcal{D}} p(x) = 1$. Sometimes we write the PMF of X as $p_X(x)$; however, we drop the X if it is the only random variable under consideration. We can compute the rth moment as

$$E[X^r] = \sum_{x \in \mathcal{D}} x^r p(x).$$

Next, we describe some commonly used discrete random variables, their PMFs, means, and variances.

1. *Discrete uniform distribution*
   - Description: If a random variable X assumes the values x_1, x_2, ..., x_k, with equal probabilities, then X is a discrete uniform random variable.
   - PMF:

$$p(x) = \frac{1}{k}, \quad x = x_1, x_2, \ldots, x_k.$$

   - Mean:

$$E[X] = \frac{\sum_{i=1}^{k} x_i}{k}.$$

   - Variance:

$$V[X] = \frac{\sum_{i=1}^{k} (x_i - E[X])^2}{k}.$$

2. *Bernoulli distribution*
   - Description: A Bernoulli trial can result in a success with probability p and a failure with probability q with $q = 1 - p$. Then the random variable X, which takes on 0 if the trial is a failure and 1 if the trial is a success, is called the Bernoulli random variable with parameter p.
   - PMF:

$$p(x) = px + (1 - p)(1 - x), \quad x = 0, 1.$$

   - Mean:

$$E[X] = p.$$

   - Variance:

$$V[X] = p(1 - p).$$

3. *Binomial distribution*
   - Description: A Bernoulli trial can result in a success with probability p and a failure with probability q with $q = 1 - p$. Then the random variable X, the number of successes in n independent Bernoulli trials is called the binomial random variable with parameters n and p.
   - PMF:

$$p(x) = \binom{n}{x} p^x q^{n-x}, \quad x = 0, 1, 2, \ldots, n.$$

   - Mean:

$$E[X] = np.$$

   - Variance:

$$V[X] = npq.$$

4. *Geometric distribution*
   - Description: A Bernoulli trial can result in a success with probability p and a failure with probability q with $q = 1 - p$. Then the random variable X, denoting the number of Bernoulli trials until a success is obtained is the geometric random variable with parameter p.

- PMF:

$$p(x) = pq^{x-1}, \quad x = 1, 2, \ldots .$$

- Mean:

$$E[X] = \frac{1}{p}.$$

- Variance:

$$V[X] = \frac{1-p}{p^2}.$$

5. *Negative binomial distribution*
   - Description: A Bernoulli trial can result in a success with probability p and a failure with probability q with $q = 1 - p$. Then the random variable X, denoting the number of the Bernoulli trial on which the kth success occurs is the negative binomial random variable with parameters k and p.
   - PMF:

$$p(x) = \binom{x-1}{k-1} p^k q^{x-k}, \quad x = k, k+1, k+2, \ldots .$$

   - Mean:

$$E[X] = \frac{k}{p}.$$

   - Variance:

$$V[X] = \frac{k(1-p)}{p^2}.$$

6. *Hypergeometric distribution*
   - Description: A random sample of size n is selected *without replacement* from N items. Of the N items, k may be classified as successes and $N - k$ are classified as failures. The number of successes, X, in this random sample of size n is a hypergeometric random variable with parameters N, n, and k.
   - PMF:

$$p(x) = \frac{\binom{k}{x}\binom{N-k}{n-x}}{\binom{N}{n}}, \quad x = 0, 1, 2, \ldots, n.$$

- Mean:

$$E[X] = \frac{nk}{N}.$$

- Variance:

$$V[X] = \frac{N-n}{N-1} n \frac{k}{N} \left(1 - \frac{k}{N}\right).$$

7. *Poisson distribution*

- Description: A Poisson random variable X with parameter λ, if its PMF is given by

$$p(x) = \frac{e^{-\lambda}(\lambda)^x}{x!}, \quad x = 0, 1, 2, \ldots$$

- Mean:

$$E[X] = \lambda.$$

- Variance:

$$V[X] = \lambda.$$

8. *Zipf distribution*

- Description: A random variable X with Zipf distribution taking on values $1, 2, \ldots, n$ has a PMF of the form

$$p(x) = \frac{1/x^s}{\sum_{i=1}^{n} 1/i^s}, \quad x = 1, 2, \ldots, n$$

where the parameter s is called the exponent.
- Mean:

$$E[X] = \frac{\sum_{i=1}^{n} 1/i^{s-1}}{\sum_{i=1}^{n} 1/i^s}.$$

- Variance:

$$V[X] = \frac{\sum_{i=1}^{n} 1/i^{s-2}}{\sum_{i=1}^{n} 1/i^s} - \{E[X]\}^2.$$

It is worthwhile for the reader to verify that the PMF properties are satisfied as well as verify $E[X]$ and $V[X]$ expressions for the various discrete random variables. Note that unlike the other distributions, Poisson and Zipf distributions descriptions are not based out of a random experiment. With these few examples, we move to continuous distributions.

A.1.2 Continuous Random Variables

A continuous random variable X is also characterized by its probability density function (PDF) $f(x) = dF(x)/dx$. The PDF satisfies two important properties: $f(x) \geq 0$ for all x and $\int_{-\infty}^{\infty} f(x) = 1$. Generally, we write the PDF of X as $f_X(x)$; however, we drop the X when it is the only random variable under consideration. To compute probabilities of events such as X being between x_1 and x_2 (such that $-\infty \leq x_1 < x_2 \leq \infty$), we have

$$P\{x_1 < X < x_2\} = \int_{x_1}^{x_2} f(x)dx = F(x_2) - F(x_1).$$

Also, we can compute the rth moment as

$$E[X^r] = \int_{-\infty}^{\infty} x^r f(x)dx.$$

Next, we describe some commonly used continuous random variables, their PDFs, CDFs, means, and variances.

1. *Exponential distribution (with parameter λ)*
   - PDF:

$$f(x) = \begin{cases} \lambda e^{-\lambda x}, & x > 0 \\ 0 & \text{elsewhere} \end{cases}$$

   - CDF:

$$F(x) = \begin{cases} 1 - e^{-\lambda x}, & x > 0 \\ 0 & \text{elsewhere} \end{cases}$$

   - Mean:

$$E[X] = \frac{1}{\lambda}$$

   - Variance:

$$V[X] = \frac{1}{\lambda^2}$$

2. *Erlang distribution (with parameters λ and k)*
   - PDF:

$$f(x) = \begin{cases} \lambda \frac{(\lambda x)^{k-1}}{(k-1)!} e^{-\lambda x} & x > 0 \\ 0 & \text{elsewhere} \end{cases}$$

   - CDF:

$$F(x) = \begin{cases} 1 - \sum_{r=0}^{k-1} e^{-\lambda x} \frac{(\lambda x)^r}{r!} & x > 0 \\ 0 & \text{elsewhere} \end{cases}$$

   - Mean:

$$E[X] = \frac{k}{\lambda}$$

   - Variance:

$$V[X] = \frac{k}{\lambda^2}$$

3. *Gamma distribution (with parameters α and β)*
   - PDF:

$$f(x) = \begin{cases} \frac{1}{\beta^\alpha \Gamma(\alpha)} x^{\alpha-1} e^{-x/\beta} & x > 0 \\ 0 & \text{elsewhere} \end{cases}$$

   where $\Gamma(\alpha) = \int_0^\infty x^{\alpha-1} e^{-x} dx$. If α is an integer, then $\Gamma(\alpha) = (\alpha - 1)!$.
   - CDF: There is no closed-form expression for the CDF in the generic case (exception is when α is an integer). For numerical values, use tables or software packages.

- Mean:

$$E[X] = \alpha\beta$$

- Variance:

$$V[X] = \alpha\beta^2$$

4. *Continuous uniform distribution (with parameters a and b)*
   - PDF:

$$f(x) = \begin{cases} \frac{1}{b-a}, & a \leq x \leq b \\ 0 & \text{elsewhere} \end{cases}$$

   - CDF:

$$F(x) = \begin{cases} 0 & x < a \\ \frac{x-a}{b-a}, & a \leq x \leq b \\ 1 & x > b \end{cases}$$

   - Mean:

$$E[X] = \frac{a+b}{2}$$

   - Variance:

$$V[X] = \frac{(b-a)^2}{12}$$

5. *Weibull distribution (with parameters α and β)*
   - PDF:

$$f(x) = \begin{cases} \alpha\beta x^{\beta-1} e^{-\alpha x^\beta} & x > 0 \\ 0 & \text{elsewhere} \end{cases}$$

   - CDF:

$$F(x) = \begin{cases} 1 - e^{-\alpha x^\beta} & x > 0 \\ 0 & \text{elsewhere} \end{cases}$$

   - Mean:

$$E[X] = \alpha^{-1/\beta} \Gamma\left(1 + \frac{1}{\beta}\right)$$

The function $\Gamma(\cdot)$ is described alongside the gamma distribution.
- Variance:

$$V[X] = \alpha^{-2/\beta} \left\{ \Gamma\left(1 + \frac{2}{\beta}\right) - \left[\Gamma\left(1 + \frac{1}{\beta}\right)\right]^2 \right\}$$

6. *Normal distribution (with parameters μ and σ^2)*
   - PDF:

$$f(x) = \frac{1}{\sqrt{2\pi}\sigma} e^{-(1/2)[(x-\mu)/\sigma]^2} \quad -\infty < x < \infty$$

   - CDF: There is no closed-form expression for the CDF. For numerical values, use tables or software packages.
   - Mean:

$$E[X] = \mu$$

   - Variance:

$$V[X] = \sigma^2$$

7. *Chi-squared distribution (with parameter v)*
   - Special case of the gamma distribution where $\alpha = v/2$ and $\beta = 2$, where v is a positive integer known as the degrees of freedom.
   - PDF:

$$f(x) = \begin{cases} \frac{1}{2^{v/2}\Gamma(v/2)} x^{v/2-1} e^{-x/2} & x > 0 \\ 0 & \text{elsewhere} \end{cases}$$

   where $\Gamma(a) = \int_0^\infty x^{a-1} e^{-x} dx$.
   - CDF: There is no closed-form expression for the CDF. For numerical values (except in special cases), use tables or software packages.
   - Mean:

$$E[X] = v$$

   - Variance:

$$V[X] = 2v$$

8. *Log-normal distribution (with parameters μ and σ^2)*
    - A continuous random variable X has a log-normal distribution if the random variable $Y = \ln(X)$ has a normal distribution with mean μ and standard deviation σ.
    - PDF:

$$f(x) = \begin{cases} \dfrac{1}{\sqrt{2\pi}x\sigma}e^{-(1/2)[(\ln(x)-\mu)/\sigma]^2} & x > 0 \\ 0 & \text{elsewhere} \end{cases}$$

    - CDF: There is no closed-form expression for the CDF. For numerical values, use tables or software packages.
    - Mean:

$$E[X] = e^{\mu + \sigma^2/2}$$

    - Variance:

$$V[X] = e^{2\mu + \sigma^2}(e^{\sigma^2} - 1)$$

9. *Beta distribution (with parameters α and β)*
    - PDF:

$$f(x) = \begin{cases} \dfrac{\Gamma(\alpha+\beta)}{\Gamma(\alpha)\Gamma(\beta)}x^{\alpha-1}(1-x)^{\beta-1} & 0 < x < 1 \\ 0 & \text{elsewhere} \end{cases}$$

    where $\Gamma(a) = \int_0^\infty x^{a-1}e^{-x}dx$.
    - CDF: There is no closed-form expression for the CDF. For numerical values, use tables or software packages.
    - Mean:

$$E[X] = \frac{\alpha}{\alpha + \beta}$$

    - Variance:

$$V[X] = \frac{\alpha\beta}{(\alpha + \beta)^2(\alpha + \beta + 1)}$$

10. *Pareto distribution (with parameters K and β)*
    - PDF:

$$f(x) = \begin{cases} \dfrac{\beta K^\beta}{x^{\beta+1}} & x \geq K > 0 \\ 0 & \text{elsewhere.} \end{cases}$$

- CDF:

$$F(x) = \begin{cases} 1 - \left(\frac{K}{x}\right)^{\beta} & x \geq K > 0 \\ 0 & \text{elsewhere.} \end{cases}$$

- Mean:

$$E[X] = \frac{K\beta}{\beta - 1} \quad \text{if} \beta > 1 \text{ while } E[X] = \infty \text{ if } \beta \leq 1.$$

- Variance:

$$V[X] = \frac{K^2\beta}{(\beta - 1)^2(\beta - 2)} \quad \text{if } \beta > 2 \text{ while } V[X] = \infty \text{ if } \beta \leq 2.$$

It is worthwhile to verify for each of the distributions that the PDF properties are satisfied. Also, compute $E[X]$ and $V[X]$ using the definitions and verify that expressions given for the various distributions. Before wrapping up, we briefly mention another metric that is frequently used in queueing called coefficient of variation (COV).

A.1.3 Coefficient of Variation

COV is a concept that is simple and gives an excellent platform to compare the variability in positive-valued random quantities. It is a dimensionless quantity that is a normalized measure of the variability. However, there are many things one has to be careful about and often misunderstood. We describe them next.

- COV is only defined for positive-valued random variables X and it is equal to

$$COV[X] = \frac{\sqrt{V[X]}}{E[X]}.$$

For example, the notion of COV does not exist for a normal random variable, say with mean 0 and variance 1.
- Exponential is *not* the only distribution with COV of 1. For example, a Pareto random variable with $\beta = 1 + \sqrt{2}$ and any $K > 0$ has a COV of 1. It is incorrect to use $M/G/1$ results for a $G/G/1$ queue with COV of arrivals equal to 1. The results will match only when the interarrival times are exponential.
- The relationship between COV and the hazard rate function of a positive-valued continuous random variable needs to be stated very carefully. First, let us define the hazard (or failure) rate function $h(x)$

of a positive-valued continuous random variable X with PDF $f(x)$ and CDF $F(x)$ as

$$h(x) = \frac{f(x)}{1 - F(x)}$$

for all x where $f(x) > 0$. Several references (e.g., Tijms [102] on page 438 and Wierman et al. [107] in Lemma 1) state that for a positive-valued continuous random variable X with hazard rate function $h(x)$, if $h(x)$ is increasing with x, then $COV[X] \leq 1$ and $h(x)$ is decreasing with x then $COV[X] \geq 1$. The preceding result is extremely useful and intuitive, but one has to be careful to interpret it and use it. For example, if one considers a Pareto random variable, $h(x)$ is decreasing for all $x \geq K$ but if $\beta > \sqrt{2} + 1$, then the COV is less than 1. Although it appears to be contradicting the preceding result, if one defines $h(x)$ for all $x \geq 0$, not just $x \geq K$ where $f(x) > 0$ for Pareto, then $h(x)$ is not decreasing for all $x > 0$ and the preceding result is valid. Also, it is crucial to realize that the result goes only in one direction (i.e., if $h(x)$ is increasing or decreasing for all $x \geq 0$, then COV would be <1 or >1). Knowing the COV does not reveal the monotonicity of the hazard rate function.

In addition, COV is different from "covariance," which is defined as $E[XY] - E[X]E[Y]$ for random variables X and Y. To avoid that confusion, in most instances in this book, we use squared coefficient of variation (SCOV) which is the square of COV, that is, the ratio of variance to square of the mean of a positive-valued random variable. As we conclude this section, it is worthwhile noting that nonnegative random variables are what we use predominantly in queueing. Thus, we give special considerations for nonnegative random variables. We just saw one such consideration, namely, COV. In the following section, we will describe generating functions (GFs) and transforms for discrete and continuous random variables, respectively.

A.2 Generating Functions and Transforms

Oftentimes while analyzing some queueing systems, it is not easy to obtain the distributions of random variables (such as CDF, PDF, or PMF). In those cases, GFs or transforms can be obtained, depending on whether the random variables are discrete or continuous, respectively. We next define GFs and subsequently two types of transforms, namely, Laplace–Stieltjes transform (LST) and Laplace transform (LT).

A.2.1 Generating Functions

Let X be a nonnegative integer-valued (discrete) random variable with PMF $p_X(x) = P\{X = x\} = p_x$ for $x = 0, 1, 2, \ldots$ (with the understanding that for the finite case where $X \leq a$, the expressions of p_x for $x > a$ will be zero). The GF (for any complex z with $|z| \leq 1$) is defined by

$$\Phi(z) = p_0 + p_1 z + p_2 z^2 + \cdots$$

$$= \sum_{j=0}^{\infty} p_j z^j$$

$$= E\left[z^X\right].$$

For example, if $X \sim \text{Binomial}(n, p)$, then $\Phi(z) = (pz + 1 - p)^n$. Also, if $X \sim \text{Geometric}(p)$, then $\Phi(z) = pz/[1 - (1 - p)z]$. Further, if $X \sim \text{Poisson}(\lambda)$, then $\Phi(z) = e^{-\lambda(1-z)}$. It would be a worthwhile exercise to verify them. However, what is crucial to note is that we usually do not know the PMF $p_X(x)$ but we are able to compute $\Phi(z)$. The question to ask is if we can go from $\Phi(z)$ to $p_X(x)$, or at least $E[X]$ and $V[X]$.

To do that, the following are some of the important properties:

$$\Phi(1) = 1$$

$$\Phi'(1) = E[X]$$

$$\Phi''(1) = E[X(X - 1)]$$

$$= E\left[X^2\right] - E[X]$$

$$= V[X] + \{E[X]\}^2 - E[X]$$

where $\Phi'(z)$ and $\Phi''(z)$ are the first and second derivatives of $\Phi(z)$ with respect to z. Also, in many instances, when we obtain $\Phi(z)$, there would be an unknown parameter that can be resolved by using the result $\Phi(1) = 1$.

A few other properties of a GF are described as follows:

1. GF uniquely identifies a distribution function. That means if you are given a GF, one can uniquely determine the PMF for the random variable.

2. The PMF can be derived from the GF using

$$P\{X = k\} = \frac{1}{k!}\frac{d^k}{dz^k}\Phi(z)|_{z=0}.$$

3. Moments of the random variables can be derived using

$$E[X_r] = \frac{d^r}{dz^r}\Phi(z)|_{z=1}$$

where X_r is the rth factorial power of X and written as $X_r = X(X-1)\ldots(X-r+1)$.

4. When the limits exist, the following hold:

$$\lim_{k\to\infty} P\{X = k\} = \lim_{z\to 1}(1-z)\Phi(z),$$

$$\lim_{k\to\infty} \frac{1}{k}\sum_{i=0}^{k} P\{X = i\} = \lim_{z\to 1}(1-z)\Phi(z).$$

See Chapter 2 for several examples of GFs used in contexts of queues. Next, we move to continuous random variables, which can also be described using GFs; however, for the purposes of this book, we mainly use transforms.

A.2.2 Laplace–Stieltjes Transforms

Consider a nonnegative-valued random variable X. The LST of X is given by

$$\tilde{F}_X(s) = E[e^{-sX}].$$

Therefore, if X is continuous, mathematically, the LST can be written (and computed) as

$$\tilde{F}_X(s) = \int_0^\infty e^{-sx}dF_X(x) = \int_0^\infty e^{-sx}f_X(x)dx$$

where $f_X(x)$ is the PDF of the random variable X. However, if X has a mixture of discrete and continuous parts, then $E[e^{-sX}]$ computation must be suitably adjusted as described in Section A.1. For the remainder of this section, we will assume X is continuous without any discrete parts.

Some examples of continuous random variables where LST of their CDF can be computed are as follows: If $X \sim \exp(\lambda)$, then $\tilde{F}_X(s) = \lambda/(\lambda+s)$; if $X \sim Erlang(\lambda,k)$, then $\tilde{F}_X(s) = (\lambda/(\lambda+s))^k$; if $X \sim Unif(0,1)$, then $\tilde{F}_X(s) =$

$1 - e^{-s}/s$. Similar to the PDF and CDF, for the LSTs too we drop the X and say $\tilde{F}(s)$ as the LST.

A few properties of LSTs are described as follows:

1. LST uniquely identifies a distribution function. That means if one is given an LST, one can uniquely determine the CDF for the random variable. However, obtaining the CDF or PDF from the LST is not easy except under three circumstances: (1) when only numerical inversion is needed (for details see Abate and Whitt [2] based on which there are several software packages available), (2) when the LST can be written as partial fractions of the form given in one of the earlier examples, or (3) when the corresponding LT and (converting LST to LT will be discussed in the following section) can be inverted.

2. Moments of the random variables can be directly computed using

$$E[X^r] = (-1)^r \frac{d^r}{ds^r} \tilde{F}(s)|_{s=0}.$$

3. The following properties of LSTs can be extended to any function $F(x)$ defined for $x \geq 0$ (not just CDFs):

   a. Let $F, G,$ and H be functions with nonnegative domain and range. Further, for scalars a and b, let $H(x) = aF(x) + bG(x)$. Then $\tilde{H}(s) = a\tilde{F}(s) + b\tilde{G}(s)$.

   b. Let $F(x)$, $G(x)$, and $H(x)$ be functions of x with nonnegative domain and range such that $F(0) = G(0) = H(0) = 0$. In addition, assume that $F(x)$, $G(x)$, and $H(x)$ either grow slower than e^{sx} or are bounded. Letting

   $$H(x) = \int_0^x F(x-u)dG(u) = \int_0^x G(x-u)dF(u),$$

   we have $\tilde{H}(s) = \tilde{F}(s)\tilde{G}(s)$.

   c. When the limits exist, the following hold:

   $$\lim_{t \to \infty} F(t) = \lim_{s \to 0} \tilde{F}(s),$$

   $$\lim_{t \to 0} F(t) = \lim_{s \to \infty} \tilde{F}(s),$$

   $$\lim_{t \to \infty} \frac{F(t)}{t} = \lim_{s \to 0} s\tilde{F}(s).$$

LSTs are used throughout this book starting with Chapter 2. Examples of their use can be found there. The main purpose is that in many instances the

CDF $F(x)$ of a random variable X cannot be directly obtained but it may be possible to obtain the LST $\tilde{F}(s)$. Using the LST sometimes we can obtain $F(x)$, however, we can always easily obtain $E[X]$ and $V[X]$ by taking derivatives. Next, we describe a closely related transform.

A.2.3 Laplace Transforms

Since LSTs can be interpreted as the expected value of a function of a random variable, they have been widely used in applied probability. However, in other fields, the LT is more widespread than LST. For any positive-valued Riemann integrable function $F(x)$ (not necessarily a CDF, although it can be) the LT is

$$F^*(s) = \int_0^\infty e^{-sx}F(x)dx.$$

We present some examples. As a trivial example, if $F(x)=1$, then $F^*(s)=1/s$. Further, if $F(x)=x^{n-1}e^{-ax}/(n-1)!$, then $F^*(s)=1/(s+a)^n$. Special cases such as $n=1$, $a=0$, and $n=2$. can be used in this relation to get LTs of x, x^n, e^{-ax}, etc. As a final example, if $F(x)=\sin(ax)$, then $F^*(s)=a/(s^2+a^2)$. There are several websites and books that describe LTs for many other functions (unlike LSTs which are not as widespread). For that reason, it may be a good idea to convert LSTs to LTs and then invert them. The relation between the LT and LST for a function $F(x)$ defined for $x \geq 0$ is

$$\tilde{F}(s) = sF^*(s)$$

provided $F(0)=0$ and $F(x)$ either grows slower than e^{sx} or is bounded.

A few properties of LTs are described:

1. Let F, G, and H be positive-valued functions. Also, for scalars a and b, $H(x)=aF(x)+bG(x)$. Then $H^*(s)=aF^*(s)+bG^*(s)$.
2. Let $f(x)=dF(x)/dx$ (it is not necessary that $F(x)$ is a CDF and $f(x)$ is a PDF). An extremely useful relation especially while solving differential equations is

$$f^*(s) = sF^*(s) - F(0).$$

3. Let $R(x) = \int_0^x F(u)du$, then

$$R^*(s) = \frac{F^*(s)}{s}.$$

4. Let $T(x) = x^n F(x)$, then

$$T^*(s) = (-1)^n \frac{d^n}{ds^n} F^*(s).$$

A.3 Conditional Random Variables

In many instances, we can obtain the distribution of a random variable or its moments when we know distributions of other random variables that govern its realization. In fact that is the reason sometimes we know the GF or LST of a random variable as we saw in the previous section. The question we seek to answer in this section is if we know the distribution of certain random variables, could we obtain that of others'. For example, if we know the distribution of interarrival times and service times as well as the steady-state number in a queueing system, could we obtain the steady-state distribution of the sojourn times? We have seen earlier in this book that is possible, however, what enabled that is the study of conditional random variables.

In a typical probability book, one usually encounters the topic of conditional random variables along with multiple random variables by studying their joint, conditional, and marginal distributions. Interested readers are encouraged to review appropriate texts in that regard. Here we restrict ourselves to situations more commonly found in queueing systems. Also, for the most part we only concentrate on the bivariate case, that is, when we have two random variables. We describe that next.

A.3.1 Obtaining Probabilities

Consider two random variables X and Y. Say we know the distribution of only X and we are interested in obtaining the distribution of Y. Here we also assume that if we are given X, we can characterize Y. In other words, we can obtain $P\{Y \leq y | X = x\}$. Thus, if we know $P\{Y \leq y | X = x\}$ for all x and y as well as the PMF $p_X(x)$ or PDF $f_X(x)$ of X depending on whether X is discrete or continuous, respectively, we can obtain the CDF of Y as

$$F_Y(y) = P\{Y \leq y\} = \begin{cases} \displaystyle\sum_x P\{Y \leq y | X = x\} p_X(x) & \text{if } X \text{ is discrete,} \\ \displaystyle\int_{-\infty}^{\infty} P\{Y \leq y | X = x\} f_X(x) dx & \text{if } X \text{ is continuous.} \end{cases}$$

(A.1)

This result is essentially the law of total probability that is typically dealt in an elementary probability book or course. We illustrate this expression using some examples. In many texts one would find examples where X and Y are either both discrete or both continuous and we encourage the readers to refer to them. However, we present examples where one of X or Y is discrete and the other is continuous.

Problem 126

A call center receives calls from three classes of customers. The service times are class-dependent: for class-1 calls they are exponentially distributed with mean 3 min; for class-2 calls they are according to an Erlang distribution with mean 4 min and standard deviation 2 min; and for class-3 calls they are according to a uniform distribution between 2 and 5 min. Compute the distribution of the service time of an arbitrary caller if we know that the probability the caller is of class i is $i/6$ for $i=1,2,3$.

Solution

Let Y be a continuous time random variable denoting the service time in minutes of the arbitrary caller and X be a discrete random variable denoting the class of that caller. From the problem statement, we know that $P(X=1)=1/6$, $P(X=2)=1/3$, and $P(X=3)=1/2$. We also can write down (based on the problem description) that for any $y \geq 0$,

$$P\{Y \leq y | X = 1\} = 1 - e^{-y/3}$$

$$P\{Y \leq y | X = 2\} = 1 - \left(1 + y + \frac{y^2}{2} + \frac{y^3}{6}\right) e^{-y}$$

$$P\{Y \leq y | X = 3\} = \min\left(\frac{y-2}{3}, 1\right) I(y > 2)$$

where $I(y > 2)$ is an indicator function that is 1 if $y > 2$ and 0 if $y \leq 2$. Thus, the CDF of Y based on Equation A.1 is given by

$$F_Y(y) = \frac{(1 - e^{-y/3})}{6} + \frac{\left[1 - \left(1 + y + \frac{y^2}{2} + \frac{y^3}{6}\right) e^{-y}\right]}{3}$$
$$+ \min\left(\frac{y-2}{3}, 1\right) \frac{I(y > 2)}{2}$$

for all $y \geq 0$. ∎

Having described an example where Y is continuous and X is discrete, next we consider a case where Y is discrete while X is continuous. When Y is discrete, we can convert CDF Equation A.1 to the PMF equation as

$$p_Y(y) = P\{Y = y\} = \begin{cases} \displaystyle\sum_x P\{Y = y | X = x\} p_X(x) & \text{if } X \text{ is discrete,} \\ \displaystyle\int_{-\infty}^{\infty} P\{Y = y | X = x\} f_X(x) dx & \text{if } X \text{ is continuous.} \end{cases}$$

With that understanding we describe the next problem.

Problem 127

The price of an airline ticket on a given day is modeled as a continuous random variable X, which is according to a Pareto distribution with parameters K and β (where β is an integer greater than 1 in this problem). The demand for leisure tickets during a single day follows a Poisson distribution with parameter C/X. What is the probability that the demand for leisure tickets on a given day is r?

Solution

Let Y be a random variable that denotes the demand for leisure tickets on a given day. We want $P\{Y = r\}$. To compute that, we use the fact that we know

$$P\{Y = r | X = x\} = e^{-C/x} \frac{(C/x)^r}{r!}.$$

Also, for $x \geq K$, we have

$$f_X(x) = \frac{\beta K^\beta}{x^{\beta+1}}$$

and $f_X(x) = 0$ when $X < K$. Thus, we have

$$P\{Y = r\} = \int_{-\infty}^{\infty} P\{Y = y | X = x\} f_X(x) dx$$

$$= \int_K^{\infty} e^{-C/x} \frac{(C/x)^r}{r!} \frac{\beta K^\beta}{x^{\beta+1}} dx$$

$$= \int_0^{1/K} e^{-Ct} \frac{(Ct)^r}{r!} \beta K^\beta t^{\beta-1} dt$$

$$= \frac{(r+\beta-1)!}{r!} \beta \left(\frac{K}{C}\right)^\beta \int_0^{1/K} e^{-Ct} C \frac{(Ct)^{r+\beta-1}}{(r+\beta-1)!} dt$$

$$= \frac{(r+\beta-1)!}{r!} \beta \left(\frac{K}{C}\right)^\beta \left(1 - \sum_{j=0}^{r+\beta-1} e^{-C/K} \frac{(C/K)^j}{j!}\right)$$

where the step in the middle is by making a transformation of variables letting $1/x = t$. ∎

A.3.2 Obtaining Expected Values

Here we extend the results for conditional probabilities to conditional expectation. Once again, let X and Y be random variables such that we know the distribution of X and we are interested in $E[Y]$ or $E[g(Y)]$ for some function $g(\cdot)$. We assume that it is possible to obtain $E[g(Y)|X=x]$. In fact we typically write the conditional expected values a little differently. Notice that $E[g(Y)|X]$ is a random variable and

$$E[g(Y)] = E[E[g(Y)|X]]$$

which in practice we write as

$$E[g(Y)] = \begin{cases} \displaystyle\sum_x E[g(Y)|X=x]p_X(x) & \text{if } X \text{ is discrete,} \\[2ex] \displaystyle\int_{-\infty}^{\infty} E[g(Y)|X=x]f_X(x)dx & \text{if } X \text{ is continuous.} \end{cases}$$

Thus, we can easily obtain moments of Y, LST of Y, etc. We illustrate that via a few examples.

Problem 128

The probability that a part produced by a machine is non-defective is p. By conditioning on the outcome of the first part type (defective or not), compute the expected number of parts produced till a non-defective one is obtained.

Solution

Let X be the outcome of the first part produced with $X=0$ denoting a defective part and $X=1$ denoting a non-defective part. Also, let Y be the number of parts produced till a non-defective one is obtained. The question asks to compute $E[Y]$. Although this problem can be solved by realizing that Y is a geometrically distributed random variable with probability of success p, the question specifically asks to condition on the outcome of the first part type.

Clearly, we have $E[Y|X=1]=1$ since if the first part is non-defective, then $Y=1$. However, $E[Y|X=0]=1+E[Y]$ since if the first part is defective, then it would take one plus, however, many parts to produce to get a non-defective part. Thus, we have

$$E[Y] = E[Y|X = 0]P\{X = 0\} + E[Y|X = 1]P\{X = 1\}$$

$$= (1 + E[Y])(1 - p) + 1p$$

and by solving for $E[Y]$ we get $E[Y]=1/p$. This is consistent with the expected value of a geometric random variable with probability of success p.   ∎

Problem 129

The average bus ride for Michelle from school to home takes b minutes; and it takes her on average w minutes to walk from school to home. One day when Michelle reached her bus stop to go home she found out that it would take a random time for the next bus to arrive and that random time is according to an exponential distribution with mean $1/\lambda$ minutes. Michelle decides to wait for a maximum of t minutes at the bus stop and then walk home if the bus does not arrive within t minutes. What is the expected time for Michelle to reach home from the time she arrived at the bus stop? If Michelle would like to minimize this, what should her optimal time t be?

Solution

Let X be the time the bus would arrive after Michelle gets to the bus stop and Y be the time she would reach home from the time she arrived at the bus stop. It is known that X is exponentially distributed with parameter λ. Also,

$$E[Y|X = x] = \begin{cases} x + b & \text{if } x \le t \\ t + w & \text{if } x > t \end{cases}$$

since $x \le t$ implies Michelle would ride the bus and vice versa. Thus, by unconditioning, we have

$$E[Y] = \int_0^\infty E[Y|X = x]\lambda e^{-\lambda x}dx$$

$$= \int_0^t (x + b)\lambda e^{-\lambda x}dx + \int_t^\infty (t + w)\lambda e^{-\lambda x}dx$$

$$= \frac{1}{\lambda}\left(1 - e^{-\lambda t} - \lambda t e^{-\lambda t}\right) + b\left(1 - e^{-\lambda t}\right) + (t + w)e^{-\lambda t}$$

which by rearranging terms we get $E[Y] = b + (1/\lambda) - (b + (1/\lambda) - w)e^{-\lambda t}$. To find the optimal t (call it t^*) that would minimize $E[Y]$ we find that if $b + (1/\lambda) > w$, then $t^* = 0$, otherwise $t^* = \infty$. Thus, from Michelle's standpoint if $b + (1/\lambda)$ is greater than w, she should not wait for the bus. However, if $b + (1/\lambda)$ is less than w, she should wait for the bus and not walk. ∎

Problem 130

The orders received for grain by a farmer add up to X tons, where X is an exponential random variable with mean $1/\beta$ tons. Every ton of grain sold brings a profit of p, and every ton that is not sold is destroyed at a loss of l. Let T be the tons of grains produced by the farmer, which is according to an Erlang distribution with parameters α and k. Any portion of orders that are not satisfied are lost without any penalty cost. What is the expected net profit for the farmer?

Solution

Let R be the net profit for the farmer, which is a function of X. The expected net profit conditioned on T is

$$E[R|T] = \int_0^T [px - l(T - x)]\beta e^{-\beta x}dx + \int_T^\infty pT\beta e^{-\beta x}dx$$

$$= \frac{p+l}{\beta}\left(1 - e^{-\beta T} - \beta Te^{-\beta T}\right) - lT\left(1 - e^{-\beta T}\right) + pTe^{-\beta T}$$

$$= \frac{p+l}{\beta}\left(1 - e^{-\beta T}\right) - lT.$$

To compute $E[R]$, we take the expectation of the expression to get

$$E[R] = E[E[R|T]] = \frac{p+l}{\beta}\left(1 - E[e^{-\beta T}]\right) - lE[T].$$

Since T is according to an Erlang distribution with parameters α and k, we have $E[T] = k/\alpha$ and $E[e^{-\beta T}] = (\alpha/(\alpha+\beta))^k$ by realizing that it is the LST at β. Thus, we have

$$E[R] = \frac{p+l}{\beta}\left[1 - \left(\frac{\alpha}{\alpha+\beta}\right)^k\right] - \frac{lk}{\alpha}.$$

∎

Notice that from the last two problems, we obtain some intriguing results mainly due to the exponential distribution properties. In that light it is worthwhile to delve into exponential distribution further, which is the objective of the following section.

A.4 Exponential Distribution

If we were to name one continuous random variable used most often in this book, it would be the exponential random variable. In fact, a large number of random variables such as Erlang, hyperexponential, Coxian, hypoexponential, and phase-type are indeed defined as a collection of exponential random variables. For that reason, it is worthwhile describing the exponential distribution's properties in one location so that they could be referred to if needed.

A.4.1 Characteristics

Before describing the properties of exponential random variables, we first recapitulate their characteristics so that they are in one location for easy reference. A nonnegative continuous random variable X is distributed exponentially with parameter λ if any of the following can be shown to hold:

1. The CDF of X is $1 - e^{-\lambda x}$ for all $x \geq 0$.
2. The PDF of X is $\lambda e^{-\lambda x}$ for all $x \geq 0$.
3. The LST of the CDF of X, $\tilde{F}(s)$, is $\lambda/(\lambda + s)$.

In other words, the three are equivalent. In fact in many instances, showing the LST form is simpler than showing the CDF or PDF. Now, if X is an exponentially distributed random variable with parameter λ, then we symbolically state that as $X \sim \exp(\lambda)$.

Another useful result to remember is that $P\{X > x\} = e^{-\lambda x}$ for $x \geq 0$. In fact the hazard rate function (defined as $f_X(x)/(1 - F_X(x))$ for any nonnegative random variable X) of the exponential random variable with parameter λ is indeed λ. Further, in terms of moments, the expected value of X is $E(X) = 1/\lambda$. Also, the variance of X is $V[X] = 1/\lambda^2$. Thus, the COV is 1 for the exponential random variable. Next, we describe some useful properties.

A.4.2 Properties

The following is a list of useful properties of the exponential random variable. They are presented without derivation. Interested readers are encouraged to refer to standard texts on probability and stochastic processes such as Kulkarni [67].

- *Memoryless property*: If $X \sim \exp(\alpha)$ and Y is any random variable, then

$$P(X - Y > t | X > Y) = P(X > t) = e^{-\alpha t}$$

and

$$P(X \leq t + Y | X > Y) = P(X \leq t) = 1 - e^{-\alpha t}.$$

This is called memoryless property because it is like as though the amount Y that X has "lived" has been forgotten and the remainder after that Y is still according to $\exp(\alpha)$. The constant hazard rate is also an artifact of this.
- *Minimum of independent exponentials*: If $X_1 \sim \exp(\alpha_1)$, $X_2 \sim \exp(\alpha_2)$, ..., $X_n \sim \exp(\alpha_n)$ and independent, then $\min(X_1, X_2, \ldots, X_n) \sim \exp(\alpha_1 + \alpha_2 + \cdots + \alpha_n)$ with mean $1/(\alpha_1 + \alpha_2 + \cdots + \alpha_n)$. In other words, the minimum of n independent exponential random variables is exponentially distributed with parameter equal to the sum of the parameters of the n independent variables.
- *Smallest of independent exponentials*: Let $X_1 \sim \exp(\alpha_1)$, $X_2 \sim \exp(\alpha_2)$, ..., $X_n \sim \exp(\alpha_n)$ and independent. Define $N = i$ if $X_i = \min(X_1, X_2, \ldots, X_n)$, then $P\{N = j\} = \alpha_j/(\alpha_1 + \alpha_2 + \cdots + \alpha_n)$ for all i and j between 1 and n. For example, when $n = 2$, we have $P(X_1 < X_2) = \alpha_1/(\alpha_1 + \alpha_2)$.
- *Sum of fixed number of independent identically distributed (IID) exponentials*: For $i = 1, \ldots, n$, let $X_i \sim \exp(\alpha)$ and independent, that is, $X_1, \ldots, X_n$ are IID exponential random variables. Define the sum of those n IID random variables as $Z = X_1 + \cdots + X_n$. Then Z is according to an Erlang distribution with parameters α and n.
- *Sum of geometric number of IID exponentials*: Let $X_1, X_2, \ldots,$ be a sequence of IID exponential random variables with parameter α. Define the sum of N of those IID random variables as $Z = X_1 + \cdots + X_N$, where N is a geometric random variable with parameter p (i.e., $P\{N = i\} = (1 - p)^{i-1} p$ for $i = 1, 2, \ldots$). Then Z is according to an exponential distribution with parameter $p\alpha$.
- *Sum of fixed number of independent exponentials*: Let $X_1 \sim \exp(\alpha_1)$, $X_2 \sim \exp(\alpha_2)$, ..., $X_n \sim \exp(\alpha_n)$ and independent. Define the sum of those n IID random variables as $Z = X_1 + \cdots + X_n$. Then the LST of the distribution of Z is

$$\tilde{F}_Z(s) = E[e^{-sZ}] = \prod_{i=1}^{n} \left(\frac{\alpha_i}{\alpha_i + s} \right).$$

The LST can be inverted and CDF obtained in two cases: (1) when all α_i values are equal, which result in the Erlang distribution,

(2) when all α_i values are different, which result in the hypoexponential distribution.

This leads us to study properties of collections of IID random variables, which is the focus of the following section.

A.5 Collection of IID Random Variables

We begin by stating two fundamental results for collection of random variables. For that, let $a_1, a_2, \ldots, a_n$ be n constants and $Z_1, Z_2, \ldots, Z_n$ be n random variables each with finite mean and variance. Then

$$E[a_1 Z_1 + a_2 Z_2 + \cdots + a_n Z_n] = a_1 E[Z_1] + a_2 E[Z_2] + \cdots + a_n E[Z_n]$$

with no other restrictions on the random variables. However,

$$V[a_1 Z_1 + a_2 Z_2 + \cdots + a_n Z_n] = a_1^2 V[Z_1] + a_2^2 V[Z_2] + \cdots + a_n^2 V[Z_n]$$

provided the random variables $Z_1, Z_2, \ldots, Z_n$ are pairwise independent.

Next, we consider the title of this section, that is, a collection of IID random variables $X_1, X_2, \ldots,$ with $E[X_i] = \tau$ and $V[X_i] = \sigma^2$ for all $i = 1, 2, \ldots$ (note that the random variables are independent). Now, define

$$\overline{X}_n = \frac{X_1 + X_2 + \cdots + X_n}{n}$$

for any n. Then based on the results in the previous paragraph, we have

$$E\left[\overline{X}_n\right] = \tau \quad \text{and} \quad V\left[\overline{X}_n\right] = \frac{\sigma^2}{n}.$$

In fact we also have $E[X_1 + X_2 + \cdots + X_n] = n\tau$ and $V[X_1 + \cdots + X_n] = n\sigma^2$.

Now, we ask the question what if we let $n \to \infty$ in this case? That is described in two quintessential convergence results in the theory of probability. According to the *strong law of large numbers*,

$$\overline{X}_n \to \tau$$

almost surely as $n \to \infty$. Also, if we define

$$T_n = \frac{\sqrt{n}\left(\overline{X}_n - \tau\right)}{\sigma}$$

then as $n \to \infty$, T_n converges in distribution to a normally distributed random variable with mean 0 and variance 1 according to the *central limit theorem*. Both strong law of large numbers and central limit theorem can be extended beyond the IID setting, but we do not present them here. Interested readers are referred to several websites and books that have these results.

Next, we consider two processes that track events as they occur with IID interevent times and study the process of counting events. We first present the Poisson process and generalize it to renewal processes subsequently.

A.5.1 Poisson Process

Although Poisson processes can be approached from a stationary and independent increments standpoint, we take a more elementary approach based on IID exponential random variables. Let X_1, X_2,..., be a sequence of IID random variables exponentially distributed with parameter λ. In particular, we assume that events occur one by one and the time between events is exponentially distributed. Let S_n be the time of the nth event with

$$S_n = X_1 + \cdots + X_n$$

being the sum of the first n exponentially distributed random variables. In addition, we assume that $S_0 = 0$ for the sake of notation. From the properties of exponential distributions, we know that S_n is according to an Erlang distribution with parameters λ and n.

Now, define $N(t)$ as the number of events in time $(0, t]$ so that the interevent times are IID exponentially with parameter λ. In particular,

$$N(t) = \max\{n \geq 0 : S_n \leq t\}$$

so that $\{N(t), t \geq 0\}$ is a counting process that increases by one every time an event occurs. Using the fact that the events $\{N(t) \geq n\}$ and $\{S_n \leq t\}$ are identical, we can derive

$$P\{N(t) = k\} = e^{-\lambda t} \frac{(\lambda t)^k}{k!}$$

for all $k = 0, 1, 2, \ldots$ and any $t \geq 0$. Thus, $N(t)$ is according to a Poisson distribution with $E[N(t)] = \lambda t$ and $V[N(t)] = \lambda t$. Hence, the counting process $\{N(t), t \geq 0\}$ is called a Poisson process with parameter λ, which we denote as $PP(\lambda)$. Next, we describe some properties most of which are derived from the fact that the Poisson process has stationary and independent increments (i.e., $N(t+s) - N(s)$ is independent of s as well as $N(t+u) - N(t)$ and $N(s+v) - N(s)$ are independent if $(t, t + u]$ and $(s, s + v]$ do not overlap):

- *Stationarity property*: If $\{N(t), t \geq 0\}$ is a $PP(\lambda)$, then

$$P\{N(t+s) - N(s) = k\} = e^{-\lambda t} \frac{(\lambda t)^k}{k!}$$

  for any $t > 0$ and $s > 0$. Notice that it is identical to $P\{N(t) = k\}$.
- *Independent increments*: If $\{N(t), t \geq 0\}$ is a $PP(\lambda)$, $0 \leq t_1 \leq t_2 \leq \cdots \leq t_n$ are fixed real numbers, and $0 \leq k_1 \leq k_2 \leq \cdots \leq k_n$ are fixed integers, then

$$P\{N(t_1) = k_1, N(t_2) = k_2, \ldots, N(t_n) = k_n\}$$

$$= P\{N(t_1) = k_1\} P\{N(t_2) - N(t_1) = k_2 - k_1\}$$

$$\ldots P\{N(t_n) - N(t_{n-1}) = k_n - k_{n-1}\}$$

$$= e^{-\lambda t_n} \frac{(\lambda t_1)^{k_1}}{k_1!} \frac{(\lambda(t_2 - t_1))^{k_2 - k_1}}{(k_2 - k_1)!} \cdots \frac{(\lambda(t_n - t_{n-1}))^{k_n - k_{n-1}}}{(k_n - k_{n-1})!}$$

  since nonoverlapping intervals have independent increments.
- *Covariance of overlapping intervals*: If $\{N(t), t \geq 0\}$ is a $PP(\lambda)$, and t, s, u, and v are fixed real numbers such that $0 \leq t \leq s \leq t + u \leq s + v$, then

$$Cov[N(t+u) - N(t), N(s+v) - N(s)] = \lambda(t + u - s)$$

  which is equal to the variance of the overlapping portion $(s, u + t]$.
- *Conditional event times*: If $\{N(t), t \geq 0\}$ is a $PP(\lambda)$, and S_1, S_2, $\ldots$, are event times, then for any nonnegative integers n and k,

$$E[S_k | N(t) = n] = \begin{cases} \frac{kt}{n+1} & \text{if } k \leq n, \\ t + \frac{k-n}{\lambda} & \text{if } k > n. \end{cases}$$

- *Superposition of Poisson processes*: For $i = 1, 2, \ldots, r$, let $\{N_i(t), t \geq 0\}$ be a $PP(\lambda_i)$. If $N(t) = N_1(t) + N_2(t) + \cdots + N_r(t)$, then $\{N(t), t \geq 0\}$ is a $PP(\lambda)$ with $\lambda = \lambda_1 + \lambda_2 + \cdots + \lambda_r$. In other words, by superimposing r independent Poisson processes we get a Poisson process with mean event rate equal to the sum of the mean event rates of the superimposed processes.
- *Splitting of Poisson processes*: Consider a $PP(\lambda)$, $\{N(t), t \geq 0\}$ where each event can be classified into r categories with probability p_1, p_2, $\ldots$, p_r of an event being of type $1, 2, \ldots, r$, respectively (such that $p_1 + p_2 + \cdots + p_r = 1$). Under such a Bernoulli split let $N_i(t)$ be the number of events of type i that occurred in time $(0, t]$ (note that

$N(t) = N_1(t) + N_2(t) + \cdots + N_r(t))$, then $\{N_i(t), t \geq 0\}$ is a $PP(p_i\lambda)$ for $i = 1, 2, \ldots, r$.

Having described Poisson processes and their properties, we briefly describe two extensions, namely nonhomogeneous Poisson processes (NPP) and compound Poisson processes (CPP) next. In a $PP(\lambda)$, the events take place at rate λ at an average for all $t \geq 0$, whereas in an NPP, the instantaneous average rate at which events take place is $\lambda(t)$. Define $\Lambda(t)$ such that

$$\Lambda(t) = \int_0^t \lambda(u) du.$$

Then we have the following probability

$$P\{N(t+s) - N(s) = k\} = \exp\{-[\Lambda(t+s) - \Lambda(s)]\} \frac{[\Lambda(t+s) - \Lambda(s)]^k}{k!}$$

where similar to the regular Poisson process, $N(u)$ is the number of events that occur in time $(0, u)$. Also, $E[N(t)] = \Lambda(t)$.

The concept of batch or bulk events can be modeled using CPP, which essentially is the same as a regular Poisson process with the exception that with every event, the counting process need not increase by one. Let $\{N(t), t \geq 0\}$ be a $PP(\lambda)$. Let $\{Z_n, n \geq 1\}$ be a sequence of IID random variables that is also independent of $\{N(t), t \geq 0\}$. Define

$$Z(t) = \sum_{n=1}^{N(t)} Z_n$$

for $t \geq 0$. The process $\{Z(t), t \geq 0\}$ is called a CPP. Let Z_n be an integer value random variable (then so is $Z(t)$). Let

$$p_k(t) = P\{Z(t) = k\}.$$

Then for $0 \leq t_1 \leq t_2 \ldots \leq t_n$, we have

$$P\{Z(t_1) = k_1, Z(t_2) = k_2, \ldots, Z(t_n) = k_n\}$$

$$= p_{k_1}(t_1) p_{k_2 - k_1}(t_2 - t_1) \ldots p_{k_n - k_{n-1}}(t_n - t_{n-1}).$$

Let $\{Z(t), t \geq 0\}$ be a CPP. Then

$$E[Z(t)] = \lambda t E[Z_1],$$

$$Var[Z(t)] = \lambda t E[Z_1^2].$$

A.5.2 Renewal Process

Having described Poisson processes, the next natural question to ask is: what if the interevent times are not exponentially distributed? That is the focus of this section. When the interevent times are generally distributed, the resulting counting process is called a renewal process. We provide a formal definition. Let Y_n denote the time between the nth and $(n-1)$st events, that is, interevent time. We assume that Y_n for all $n \geq 0$ are IID random variables with CDF $G(\cdot)$. Let S_n be the time of the nth event, therefore $S_n = Y_1 + \cdots + Y_n$, with $S_0 = 0$. Let $N(t)$ be the number of events till time t, therefore $N(t) = \max\{n : S_n \leq t\}$. Then $\{N(t), t \geq 0\}$ is a stochastic process called the renewal process.

Denote the probability

$$p_k(t) = P\{N(t) = k\}$$

Then using the LST of $p_k(t)$ one can show that

$$\tilde{p}_k(s) = \int_0^\infty e^{-st} dp_k(t) = (\tilde{G}(s))^k (1 - \tilde{G}(s))$$

where $\tilde{G}(s) = E[e^{-sY_n}]$, the LST of the CDF of Y_n. For example, if $Y_n \sim \exp(\lambda)$; $G(y) = P\{Y_n \leq y\} = 1 - e^{-\lambda y}$ for $y \geq 0$. Also, $\tilde{G}(s) = \lambda/(\lambda + s)$. In this case, $\{N(t), t \geq 0\}$ process is indeed a Poisson process with parameter λ. One can get by inverting the LST the following:

$$p_k(t) = P\{N(t) = k\} = e^{-\lambda t} \frac{(\lambda t)^k}{k!}.$$

As another example, let $Y_n \sim Erlang(m, \lambda)$; $G(y) = P\{Y_n \leq y\} = 1 - \sum_{r=0}^{m-1} e^{-\lambda y} (\lambda y)^r / r!$ for $y \geq 0$. Also, $\tilde{G}(s) = \lambda^m / (\lambda + s)^m$. Clearly,

$$\tilde{p}_k(s) = \left(\frac{\lambda}{\lambda + s}\right)^{mk} - \left(\frac{\lambda}{\lambda + s}\right)^{m(k+1)}.$$

In general, using the LST and inverting it to get the distribution of $N(t)$ is tricky. However, there are several results that can be derived without needing to invert. We present them here. For the remainder of this section, we assume that $E[Y_n] = \tau$ and $V[Y_n] = \sigma$, such that both τ and σ are finite. The main results (many of them being asymptotic) are as follows:

- As $t \to \infty$, the random variable $N(t)/t \to \tau$ almost surely.
- As $t \to \infty$, the random variable $N(t)$ converges in distribution to a normally distributed random variable with mean t/τ and variance $\sigma^2 t/\tau^3$.
- Let $M(t) = E[N(t)]$ with LST $\tilde{M}(s)$, then we have

$$\tilde{M}(s) = \frac{\tilde{G}(s)}{1 - \tilde{G}(s)},$$

$$\lim_{t \to \infty} \frac{M(t)}{t} = \frac{1}{\tau},$$

$$E[S_{N(t)+1}] = \tau(1 + M(t)),$$

$$\lim_{t \to \infty} \left[M(t) - \frac{t}{\tau} \right] = \frac{\sigma^2 - \tau^2}{2\tau^2}.$$

The next set of results have their roots in reliability theory, which would explain the terminology. We define the following variables: $A(t) = t - S_{N(t)}$, which is the time since the previous event, in reliability this would be the age; $B(t) = S_{N(t)+1} - t$, which is the time the next event would occur, in reliability this is the remaining life; and $C(t) = A(t) + B(t)$, which is the time between the previous and the next events, that is, in reliability this would be total life. It is possible to derive:

$$\lim_{t \to \infty} P\{B(t) \le x\} = \frac{1}{\tau} \int_0^x [1 - G(u)]du,$$

$$\lim_{t \to \infty} E[B(t)] = \frac{\tau^2 + \sigma^2}{2\tau},$$

$$\lim_{t \to \infty} E[A(t)] = \frac{\tau^2 + \sigma^2}{2\tau}.$$

Therefore, in the limit as $t \to \infty$, $B(t)$ as well as $A(t)$ are according to what is known as the *equilibrium distribution* of the interrenewal times. Interestingly, as $t \to \infty$, $E[C(t)] \to (\tau^2 + \sigma^2)/\tau$, and not τ which is what one would expect. This is called *inspection paradox*. In fact $(\tau^2 + \sigma^2)/\tau > \tau$.

Reference Notes

The contents of this chapter is a result of teaching various courses on probability and stochastic processes both at the undergraduate level and at the graduate level. There are several excellent books on the topics covered in this chapter. For example, the elementary material on probability, random variables, and expectations can be found in Ross [93]. However, the notations used in this chapter and a majority of results are directly from Kulkarni [67]. That would be a wonderful resource to look into the proofs and derivations for some of the results in this chapter. The notable exceptions not found in either of those texts are as follows: the discussion on mixture distributions, coefficient of variation (COV), some special distributions, as well as numerical inversion of LSTs and LTs. A description of relevant references are provided for those topics. Finally, for a rigorous treatment of probability, yet not abstract, an excellent source is Resnick [90], which also nicely explains topics such as law of large numbers and central limit theorem.

Exercises

A.1 Let X be a continuous random variable with PDF

$$f_X(x) = \begin{cases} \frac{1}{\pi} x \sin x & \text{if } 0 < x < \pi \\ 0 & \text{otherwise.} \end{cases}$$

Prove that

$$E(X^{n+1}) + (n+1)(n+2)E(X^{n-1}) = \pi^{n+1}.$$

A.2 The gamma function (which appears in gamma distribution and some other distributions as well) is given for $x > 0$ as

$$\Gamma(x) = \int_0^\infty t^{x-1} e^{-t} dt.$$

Using the PDF of the normal distribution, show that $\Gamma(1/2) = \sqrt{\pi}$.

A.3 The conditional variance of X, given Y, is defined by

$$Var(X|Y) = E(\{X - E(X|Y)\}^2 | Y).$$

Prove the conditional variance formula:

$$Var(X) = E(Var(X|Y)) + Var(E(X|Y)).$$

A.4 Suppose given $a, b > 0$, and let X, Y be two random variables with values in $\mathbb{Z}_+$ (set of nonnegative integers) and $\mathbb{R}_+$ (set of nonnegative real numbers), respectively. The joint distribution of X and Y is characterized by

$$P[X = n, Y \leq y] = b \int_0^y \frac{(at)^n}{n!} \exp(-(a+b)t)dt.$$

For $n \in \mathbb{Z}_+$, compute $E(Y|X = n)$. Also compute $E(X|Y)$.

A.5 The number of claims received at an insurance company during a week is a random variable with mean μ_1 and variance σ_1^2. The amount paid in *each* claim is a random variable with mean μ_2 and variance σ_2^2. Find the mean and variance of the amount of money paid by the insurance company each week. What independence assumptions are you making? Are these assumptions reasonable?

A.6 Bus A will arrive at a station at a random time between 10:00 a.m. and 11:00 a.m. tomorrow according to a beta distribution with parameters α and β. Therefore, if X denotes the number of hours after 10:00 a.m. bus A would arrive, then the PDF of X is

$$f_X(x) = \begin{cases} \frac{\Gamma(\alpha+\beta)}{\Gamma(\alpha)\Gamma(\beta)} x^{\alpha-1}(1-x)^{\beta-1} & 0 < x < 1 \\ 0 & \text{elsewhere} \end{cases}$$

where $\Gamma(a) = \int_0^\infty x^{a-1}e^{-x}dx$. Also, $E[X] = \alpha/(\alpha+\beta)$. Bus B will arrive at the same station at a random time uniformly distributed between the arrival time of bus A and 11:00 a.m. Find the expected value of the arrival time of bus B.

A.7 Two parts (call them A and B) are manufactured in parallel on two machines (call them machine 1 and machine 2). The processing time for part A on machine 1 is distributed exponentially with parameter α. Similarly, part B takes an $\exp(\beta)$ amount of time to process on machine 2. If the processing starts at the same time on both machines, what is the expected time to complete processing of both parts (i.e., the expected time for both machines to become idle)? **Hint:** Let X_A and X_B be random variables denoting the time to process jobs A and B on machines 1 and 2, respectively. Then define $Z = \max(X_A, X_B)$. Compute $E[Z]$.

A.8 The amount of time it takes to spot a defect in a cast is distributed exponentially with mean 15 min.

   a. What is the probability of spotting a defect within 10 min?
   b. If I have been unsuccessful in spotting a defect for the past 14 min, what is the expected time from now on that I will spot it?
   c. If there are four different students working on four similar casts, when do you expect the first among the four students to spot a defect?
   d. Continuing with the previous question, what is the probability that in 30 min exactly two students will spot defects and two will not?

A.9 A barber shop has three servers with expected service times of 20, 15, and 10 min. The servers have been busy for 5, 7, and 6 min, respectively with their current customers. What is the expected remaining time until the next service completion?

A.10 There are B light bulbs and they are all switched on simultaneously. The lifetime of each light bulb is exponentially distributed with mean θ hours. Assume $S_0 = 0$ and for $i \geq 1$, S_i denotes the time of ith failure (i.e., the ith time a light bulb dies). Define the counting process $\{N(t), t \geq 0\}$ such that $N(t) = \max\{n \geq 0 : S_n \leq t\}$ for $t \geq 0$. Answer the following questions assuming that $2 < B < \infty$:

i. $\{N(t), t \geq 0\}$ is a Poisson process. TRUE OR FALSE?

ii. What is $E[S_2 - S_1]$?

A.11 Let $\{N(t), t \geq 0\}$ be a renewal process with interrenewal times having the following density:

$$g(t) = \frac{\lambda^3 t^2 e^{-\lambda t}}{2} \quad t \geq 0.$$

Compute $P\{N(t) = k\}$ for $k = 0, 1, 2$.

A.12 Let $\{Y_n, n \geq 1\}$ be a sequence of IID $\exp(\mu)$ random variables and $\{Z_n, n \geq 1\}$ be a sequence of IID $\exp(\lambda)$ random variables. Define random variable X_n such that $X_n = Y_n + Z_n$ with probability α and $X_n = Y_n$ with probability $1 - \alpha$. The random variable X_n is said to be according to Coxian-2 distribution. Assume $\lambda \neq \mu$ and let $S_0 = 0$. Define $S_k = X_1 + \cdots + X_k$ for all $k \geq 1$. Argue that $\{S_n, n \geq 0\}$ is a renewal sequence. Then obtain an expression for $\tilde{p}_k(s)$ and the average interrenewal time τ.

Appendix B: Stochastic Processes

In this appendix chapter, we will review stochastic processes by describing the ones used in this book. For that, we will walk through modeling, analysis, and performance evaluation for stochastic systems. Essentially, a probabilistic model of a system evolving randomly in time can be modeled using stochastic process. In particular, queueing systems can be modeled using discrete-time Markov chains (DTMC), continuous-time Markov chains (CTMC), Markov regenerative processes (MRGPs), and Brownian motions. Hence we will review them here in that order.

B.1 Discrete-Time Markov Chains

Consider a system that is observed at discrete points in time. These time points can be equally spaced such as at the beginning of a day or an hour. They can also be randomly spaced, such as immediately after a customer departs from a queue or just before a customer joins a queue. The key feature is that observations are made at discrete points in time and not continuously. For such a system, let X_n be its state when observed for the nth time. Although X_n could be a vector, for the rest of this section we treat it as though it is a scalar with the understanding that it is possible to map any vector uniquely using an appropriate scalar notation. However, in many practical situations, it is wiser to use a vector notation. We will see examples of that subsequently.

Next we consider S, the state space for the system, which is the set of all possible values X_n can take for any n. Recall that the system under consideration evolves randomly from observation to observation but every time the system is in one of the states in S. The collection of random quantities $\{X_n, n \geq 0\}$ forms an ordered sequence. Such a sequence is called a stochastic process. For a stochastic process to be a DTMC, we require the system to satisfy the Markov property: given the current state, the future is independent of the past. For a DTMC, this is mathematically written as

$$P\{X_{n+1} = j | X_n = i, X_{n-1} = i_{n-1}, \ldots, X_0 = i_0\} = P\{X_{n+1} = j | X_n = i\}$$

for any $i \in S$, $j \in S$, and $i_k \in S$ for $k = 0, \ldots, n-1$.

Besides the Markov property, another property that is useful is time-homogeneity, which states that the probability of going from state i to state j does not depend on time. For a DTMC, this is mathematically written as

$$P\{X_{n+1} = j | X_n = i\} = P\{X_1 = j | X_0 = i\}$$

for any $i \in S$, $j \in S$, and $n \geq 0$. Although it is not necessary for a DTMC to be time-homogeneous, it is extremely convenient to analyze when it is so. In that case, we define p_{ij} as

$$p_{ij} = P\{X_{n+1} = j | X_n = i\}$$

for any $i \in S, j \in S$, and $n \geq 0$. Clearly, p_{ij} is the probability of going from state i to state j from one observation to the next. The matrix of p_{ij} values is called the transition probability matrix

$$P = [p_{ij}]$$

for all $i \in S$ and $j \in S$.

B.1.1 Modeling a System as a DTMC

There are six steps involved in modeling a system as a DTMC. They are as follows:

1. Define X_n, the state of the system at time n or nth observation (this should be done carefully so that Markov and time-homogeneity properties are satisfied)
2. Write down the state space S, which is a set of all possible values X_n can take
3. Verify if the Markov property is satisfied
4. Verify if the time-homogeneity property is satisfied
5. Construct the transition probability matrix P

$$P = [p_{ij}] = [P\{X_1 = j | X_0 = i\}]$$

   (note that for the transition probability matrix, each row sums to one)
6. Draw a transition diagram, that is, draw a directed network by drawing the node set (the state space S) and the arcs (i, j) if $p_{ij} > 0$, with arc cost p_{ij} for all $i \in S$ and $j \in S$

Next we present a few examples to illustrate the modeling process.

Problem 131

Packets arriving at a router are classified into two types: real-time (RT) and non-real-time (NR) packets. An RT packet follows another RT packet with probability 0.7 (therefore, the probability of an NR packet following an RT packet is 0.3). Similarly, an NR packet follows another NR packet with probability 0.6 (therefore, the probability of an RT packet following an NR packet is 0.4). Model the type of packets arriving at a router as a DTMC.

Solution

Let X_n denote the type of the nth packet (RT or NR) arriving at the router. Clearly, X_n can take only one of two values, RT or NR. Thus the state space is $S = \{RT, NR\}$. From the problem description to predict the type of the next packet, we only need to know the type of the current packet but nothing about the history. Also, the transition probabilities are time-invariant. Therefore, Markov and time-homogeneity properties are satisfied.

Now we can write down the transition probability matrix as follows:

$$P = \begin{array}{c} \\ RT \\ NR \end{array} \begin{array}{c} \overset{\displaystyle RT \quad NR}{\left[\begin{array}{cc} 0.7 & 0.3 \\ 0.4 & 0.6 \end{array} \right]} \end{array}.$$

Thus the probability of the next packet being NR given the current is NR is 0.7, which is the northwest corner of the P matrix. Notice the rows adding to one. We can also draw the transition diagram as described in Figure B.1. Thus the system is modeled as a DTMC. ∎

This is perhaps one of the simplest examples of a DTMC with two states. Next we state a slightly bigger example.

Problem 132

Consider three cell-phone companies A, B, and C. Every time a sale is announced, a thrifty graduate student switches from one company to another. If the student is with company A before a sale, he switches to B or C with probability 0.4 or 0.3, respectively. Likewise if he is with B, he

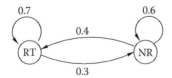

FIGURE B.1
Transition diagram for the RT/NR problem.

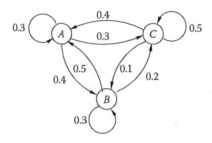

FIGURE B.2
Transition diagram for the cell-phone switching problem.

switches to A or C with probabilities 0.5 or 0.2, respectively. Finally, if he is with C, he switches to A or B with probability 0.4 or 0.1, respectively. Model the cell-phone company this student uses as a DTMC.

Solution

Let X_n denote the cell-phone company with which this particular student is just before the nth sale announcement. Since we are restricted to the three companies, the state space is $S = \{A, B, C\}$. Based on the problem description, notice that Markov and time-homogeneity properties are satisfied. Clearly, the transition probability matrix is

$$
P = \begin{array}{c c} & \begin{array}{c c c} A & B & C \end{array} \\ \begin{array}{c} A \\ B \\ C \end{array} & \left[\begin{array}{c c c} 0.3 & 0.4 & 0.3 \\ 0.5 & 0.3 & 0.2 \\ 0.4 & 0.1 & 0.5 \end{array} \right] \end{array}.
$$

Also, the transition diagram can easily be drawn as shown in Figure B.2. Thus the system is modeled as a DTMC. ∎

Next we present a case where the state space has infinite elements.

Problem 133

Consider a time division multiplexer from which packets are transmitted at times 0, 1, 2, etc. Packets arriving between time n and $n + 1$ have to wait until time $n + 1$ to be transmitted. However, at most one packet can be transmitted at a time. Let Y_n be the number of packets that arrive during time n to $n + 1$. Assume that $a_i = P\{Y_n = i\}$. Model the number of packets awaiting transmission as a DTMC.

Solution

Let X_n be the number of packets awaiting transmission just before time n (i.e., just before an opportunity to transmit). Clearly, $S = \{0, 1, 2, \ldots\}$. Based on the problem description we have $X_{n+1} = Y_n + \max(X_n - 1, 0)$ since immediately after time n there would be either one transmission if $X_n > 0$ or zero transmission if $X_n = 0$ (thus $\max(X_n - 1, 0)$ packets immediately after time n). Clearly, if X_n is known we can characterize X_{n+1} without knowing any history. Thus Markov property is satisfied. Also, Y_n is not dependent on n, hence the system is time-homogeneous. Therefore, we can obtain the transition probabilities p_{ij} as (for $i > 0$)

$$p_{ij} = P\{X_{n+1} = j | X_n = i\} = a_{j-i+1}.$$

Similarly,

$$p_{0j} = P\{X_{n+1} = j | X_n = 0\} = P\{Y_n = j | X_n = 0\} = a_j.$$

Therefore, the transition probability matrix $P = [p_{ij}]$ is

$$P = \begin{bmatrix} a_0 & a_1 & a_2 & a_3 & \cdots \\ a_0 & a_1 & a_2 & a_3 & \cdots \\ 0 & a_0 & a_1 & a_2 & \cdots \\ 0 & 0 & a_0 & a_1 & \cdots \\ 0 & 0 & 0 & a_0 & \cdots \\ \vdots & \vdots & \vdots & \vdots & \ddots \end{bmatrix}. \qquad \blacksquare$$

Notice in this example that we did not provide the transition diagram. This is fairly typical since there is a one-to-one correspondence between the transition probability matrix and the transition diagram. See the exercises for more example problems as well as Chapter 4.

B.1.2 Transient Analysis of DTMCs

Having described how to model a system as a DTMC, we take the next step, namely, analysis of DTMCs. We begin by discussing transient analysis in this section. For that, consider a DTMC $\{X_n, n \geq 0\}$ with state space S and transition probability matrix P. To answer the question about what is the probability that after n observations the DTMC will be in state j given it is in state i is the focus of this transient analysis. Note that we have so far only considered the case $n = 1$, which can be immediately obtained from the P matrix. The P matrix is also known as a one-step transition probability matrix. With that introduction, we define the probability of going from state i to state j in n steps as the n-step transition probability. It is defined and computed as

$$P\{X_n = j | X_0 = i\} = [P^n]_{ij}.$$

In other words, by raising the one-step transition probability matrix to the nth power, one gets the n-step transition probability matrix.

As an example, consider the DTMC in Problem 132 with

$$S = \{A, B, C\}$$

and

$$P = \begin{bmatrix} 0.3 & 0.4 & 0.3 \\ 0.5 & 0.3 & 0.2 \\ 0.4 & 0.1 & 0.5 \end{bmatrix}.$$

If at time 0, the student is with cell-phone company B, then after the third sale (i.e., $n = 3$) to obtain the probability that this student is with company C, it can be computed as follows. Firstly,

$$P^3 = \begin{bmatrix} 0.3860 & 0.2770 & 0.3370 \\ 0.3930 & 0.2760 & 0.3310 \\ 0.3870 & 0.2590 & 0.3540 \end{bmatrix}.$$

Thus we have

$$P\{X_3 = C | X_0 = B\} = [P^3]_{BC} = 0.3310$$

which is the element corresponding to row B and column C (second row and third column).

Continuing with this example, notice that as $n \to \infty$,

$$P^n \to \begin{bmatrix} 0.3882 & 0.2706 & 0.3412 \\ 0.3882 & 0.2706 & 0.3412 \\ 0.3882 & 0.2706 & 0.3412 \end{bmatrix}.$$

It appears as though in the long run, the state of the DTMC is independent of its initial state. In other words, irrespective of which phone company the graduate student started with, he/she would eventually be with A, B, or C with probability 0.3882, 0.2706, and 0.3412, respectively. Notice how the P^∞ matrix has all identical rows. This is the focus of the following section.

B.1.3 Steady-State Analysis of DTMCs

Consider a DTMC $\{X_n, n \geq 0\}$ with state space S and transition probability matrix P. As we saw in the previous example, as n approaches infinity, the matrix P^n for "well-behaved" DTMCs has identical rows (or the columns have the same elements). This will always be true for irreducible, aperiodic, and positive recurrent DTMCs, which we called earlier as "well-behaved."

The reader is encouraged to refer to any standard stochastic processes book for a definition but informally we provide the following definition. An irreducible DTMC is one for which it is possible to go from any state to any other state in one or more steps (i.e., there is a path from every node to every other node in the transition diagram). A periodic DTMC is one for which P^n as $n \to \infty$ does not exist. For example, a periodic DTMC with period-2 would have P^n for very large n toggle between two limits (likewise if the period is 3, it would toggle between three values and so on). An aperiodic DTMC is one that is not periodic. Finally, a positive recurrent DTMC is one where the probability of revisiting any state is one and the time of revisit is finite.

Define the long-run (or steady-state) probability of a "well-behaved" (i.e., irreducible, aperiodic, and positive recurrent) DTMC being in state j as

$$\pi_j = \lim_{n \to \infty} P\{X_n = j\}.$$

We are interested in π_j because it not only describes the probability of being in state j in steady state, but also represents the fraction of time the DTMC visited state j in the long run. That is due to what is called the ergodic theorem. As it turns out, it is also called as the stationary probability because if one started in state j with probability π_j, then at every observation, the DTMC would be in state j with probability π_j, that is, the system behaves like as though it is stationary.

One way to compute π_j is to raise P to a very large power. However, the method will not work for symbolic matrices, infinite state-space DTMCs as well as DTMCs with large dimensional state spaces. The easiest method to compute π_j is by obtaining the unique solution to the system of equations

$$\pi = \pi P$$

$$\sum_{j \in S} \pi_j = 1$$

where π is a row of π_i values for all $i \in S$, that is, $\pi = [\pi_i]$. As an example, consider Problem 131. The previous equations reduce to

$$(\pi_{RT} \ \pi_{NR}) = (\pi_{RT} \ \pi_{NR}) \begin{bmatrix} 0.7 & 0.3 \\ 0.4 & 0.6 \end{bmatrix}$$

$$\pi_{RT} + \pi_{NR} = 1.$$

Thus we have $(\pi_{RT} \ \pi_{NR}) = (4/7 \ 3/7)$. Therefore, in the long run, four-sevenths of the packets will be real-time and three-sevenths non-real-time.

Such an analysis is also very useful to describe the performance of systems in steady state. We use a terminology of cost; however, it is not necessary that the cost has a financial connotation. With that understanding, say when the system (DTMC) enters state i, it incurs a cost $c(i)$ on an average. Then the long-run average cost per unit time (or per observation or per slot) is

$$\sum_{i \in S} c(i) \pi_i$$

which can be computed once the steady-state probabilities π_i for all $i \in S$ are known.

As an example, consider Problem 133 describing a time-division multiplexer. Let π_j be the limiting probability that there are j packets in the multiplexer just before the nth attempted transmission (we assume that π_j for all $j \in S$ can be calculated). Let us answer the question: what is the average multiplexing delay for a packet that arrives immediately after an attempted transmission in the long run? If there are i packets in the multiplexer when this packet arrives, then this packet faces a latency (or delay) of $(i+1)\tau$ units of time. Note that τ is the time between successive multiplexing attempts. Therefore, the average multiplexing delay is

$$\pi_0 \tau + \tau \sum_{i=1}^{\infty} i \pi_i.$$

Also, notice that the throughput of the multiplexer is $(1 - \pi_0)/\tau$. In fact we can answer many such questions, for example, what is the probability that the delay for an arriving packet is greater than 6τ? If the buffer can store a maximum of 12 messages, what is the packet loss probability? and so on.

B.2 Continuous-Time Markov Chains

Having described DTMCs, the next natural stochastic process to consider is the CTMC. Like the DTMC, here too the states are discrete-valued. However, the system is observed continuously, hence the name CTMC. To explain further, consider a stochastic system which changes as events occur over time. The time between events must be exponentially distributed. For such a system let $X(t)$ be the state of the system at time t and this could be a scalar or a vector. Typically, when an event occurs, the state changes. Since the time between events are exponentially distributed, due to its memoryless property, we can predict $X(t+s)$ given $X(t)$ without knowing how we got to $X(t)$. Thus Markov property is satisfied and hence this is a Markov chain.

It is crucial to note that by modeling the state $X(t)$ appropriately, other distributions such as Erlang, hyperexponential, hypoexponential, coxian, and phase type can also be used by incorporating the phase of the distribution in the state. Also, for ease of analysis we assume that the system is time-homogeneous, that is, the parameters of the exponential distributions are not functions of time. We can write down the Markov and time-homogeneity properties mathematically as follows:

$$P\{X(t+s) = j | X(s) = i, X(u) \forall u \in [0,s)\} = P\{X(t+s) = j | X(s) = i\}$$
$$= P\{X(t) = j | X(0) = i\} = p_{ij}(t)$$

for any $i \in S, j \in S$, and $s \geq 0$, where S is the state space. Thereby the stochastic process $\{X(t), t \geq 0\}$ would be a CTMC.

A CTMC is typically characterized by its so-called infinitesimal generator matrix Q with rows and columns corresponding the current state and the *next* state in S. In other words, we keep track of epochs when the system state changes, that is, $X(t)$ changes with the understanding that between epochs the state remains a constant. Thus in some sense if we considered the epochs as observation times we indeed would have a DTMC. Next, to describe an element q_{ij} of the Q matrix, it is the rate at which an event that would take the system from state i to state j would occur. In other words, the triggering event that drives the system from state i to state j happens after $\exp(q_{ij})$ time. However, it is crucial to realize that j does not have to happen, another event may have occurred prior to that. With that description, next we describe how to model a system as a CTMC and state a few examples to clarify the earlier description.

B.2.1 Modeling a System as a CTMC

There are five steps involved in modeling a system as a CTMC. They are as follows:

1. Define $X(t)$, the state of the system at time t (this must be selected appropriately so that Markov and time-homogeneity properties are satisfied)

2. Write down the state space S, which is a set of all possible values $X(t)$ can take

3. Verify if Markov and time-homogeneity properties are satisfied (this is straightforward if the interevent times are exponentially distributed with time-invariant parameters)

4. Construct the generator matrix $Q = [q_{ij}]$ as follows:
   a. For $i \neq j$, q_{ij} is the rate of transitioning from state i to state j (this means that if no other event occurs then it would take an exponential amount of time with mean $1/q_{ij}$ to go from state i to state j;

also, if there are multiple events that can take the CTMC from i to j, then the rate q_{ij} is the sum of the rates of all the events)

b. For $i=j$, $q_{ij} = -\sum_{j \in S, j \neq i} q_{ij}$ (the rows in Q sum to zero and all diagonal elements in Q are negative)

5. Draw the rate diagram, that is, draw a directed network by drawing the node set (the state space S) and the arcs (i,j) if $q_{ij} > 0$, with arc cost q_{ij}

Next, we present a few examples to illustrate the modeling process.

Problem 134

Consider a machine that toggles between two states, up and down. The machine stays up for an exponential amount of time with mean $1/\alpha$ hours and then goes down. Then the machine stays down for an exponential amount of time with mean $1/\beta$ hours before it gets back up. Model the machine states using a CTMC.

Solution

Let $X(t)$ be the state of the machine at time t. Therefore, if $X(t)=0$, then the machine is down at time t. Also, if $X(t)=1$, the machine is up at time t. Clearly, we have the state space as $S=\{0,1\}$. The generator matrix in terms of $\{0,1\}$ by $\{0,1\}$ is

$$Q = \begin{matrix} & 0 & 1 \\ \begin{matrix} 0 \\ 1 \end{matrix} & \begin{bmatrix} -\beta & \beta \\ \alpha & -\alpha \end{bmatrix} \end{matrix}.$$

The rate diagram is provided in Figure B.3. Hence the system is modeled a CTMC. ∎

Problem 135

Consider a telephone switch that can handle at most N calls simultaneously. Assume that calls arrive according to $PP(\lambda)$ to the switch. Any call arriving

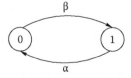

FIGURE B.3
Rate diagram for up/down machine.

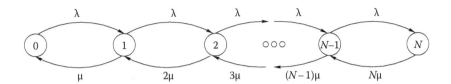

FIGURE B.4
Rate diagram for telephone switch.

when there are N other calls in progress receives a busy signal (and hence rejected). Each accepted call lasts for an exponential amount of time with mean $1/\mu$ amount of time (this is the duration of a phone call, also called hold times). Model the number of ongoing calls at any time as a CTMC.

Solution

Let $X(t)$ be the number of ongoing calls in the switch at time t. Clearly, there could be anywhere between 0 and N calls. Hence we have the state space as $S = \{0, 1, \ldots, N\}$. In many problem instances including this one, it is easier to draw the rate diagram and use it for analysis. In that light, the rate diagram is illustrated in Figure B.4. To explain that, consider some i such that $0 < i < N$. When $X(t) = i$, one of two events can occur: either a new call could arrive (this happens after $\exp(\lambda)$ time) or an existing call could complete (this happens after $\exp(i\mu)$ time). Of course if $X(t) = 0$, the only event that could occur is a new call arrival. Likewise if $X(t) = N$, the only event of significance is a call completing. Notice how memoryless property and minimum of exponentials property of exponential random variables are used in the description.

Then the generator matrix is

$$Q = \begin{bmatrix} -\lambda & \lambda & 0 & 0 & \cdots & 0 & 0 \\ \mu & -(\lambda + \mu) & \lambda & 0 & \cdots & 0 & 0 \\ 0 & 2\mu & -(\lambda + 2\mu) & \lambda & \cdots & 0 & 0 \\ \vdots & \vdots & \vdots & \vdots & \ddots & \vdots & \vdots \\ 0 & 0 & 0 & 0 & \cdots & N\mu & -N\mu \end{bmatrix}.$$

Notice how easy is the transition between the rate diagram and the generator matrix. ∎

Problem 136

Consider a system where messages arrive according to $PP(\lambda)$. As soon as a message arrives, it attempts transmission. The message transmission times are exponentially distributed with mean $1/\mu$ units of time. If no other message tries to transmit during the transmission time of this message, the transmission is successful. If any other message tries to transmit during this transmission, a collision results and all transmissions are terminated instantly. All messages involved in a collision are called backlogged and are forced to retransmit. All backlogged messages wait for an exponential amount of time (with mean $1/\theta$) before starting retransmission. Model the system called "unslotted Aloha" as a CTMC.

Solution

Let $X(t)$ denote the number of backlogged messages at time t and $Y(t)$ be a binary variable that denotes whether or not a message is under transmission at time t. Then we model the stochastic process $\{(X(t), Y(t)), t \geq 0\}$ as a CTMC. Notice that the state of the system is a two-tuple vector and the state space is

$$S = \{(0,0), (0,1), (1,0), (1,1), (2,0), (2,1), (3,0), (3,1), \ldots\}.$$

Say the state of the system at time t is (i,j) for some $(i,j) \in S$. If $j=1$, then one of three events can change the state of the system: a new arrival at rate λ would take the system to $(i+2, 0)$ due to a collision; a retransmission attempt at rate $i\theta$ would take the system to $(i+1, 0)$ due to a collision; and a transmission completion at rate μ would take the system to $(i, 0)$. However, if $j=0$, then one of two events can change the state of the system, a new arrival at rate λ would take the system to $(i+1, 1)$, and a retransmission at rate $i\theta$ would take the system to $(i-1, 1)$. Based on that we can show that the rate diagram would be as described in Figure B.5.   ∎

Notice in this example that we did not provide the Q matrix since it can easily be inferred from the rate diagram. See the exercises for more example problems as well as Chapters 2 and 3. Next, we move on to some analysis of CTMCs. Much like DTMCs, here too, we first present transient analysis and then move on to steady-state analysis.

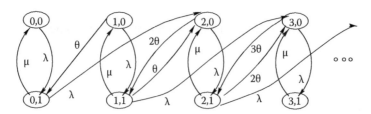

FIGURE B.5
Rate diagram for unslotted Aloha.

B.2.2 Transient Analysis of CTMCs

Consider a CTMC $\{X(t), t \geq 0\}$ with state space S and generator matrix Q. The probability of going from state i to state j in time t (possibly using multiple transitions) is

$$p_{ij}(t) = P\{X(t) = j | X(0) = i\}$$

for any $i \in S$ and $j \in S$. This is the same $p_{ij}(t)$ described for the Markov and time-homogeneity properties. The matrix $P(t) = [p_{ij}(t)]$ satisfies the following matrix differential equation:

$$\frac{dP(t)}{dt} = P(t)Q = QP(t)$$

with initial condition $P(0) = I$ and boundary condition $\sum_{j \in S} p_{ij}(t) = 1$ for every $i \in S$ and any $t \geq 0$. The solution to the differential equation can be written as

$$P(t) = \exp(Qt)$$

where the exponential of a square matrix A is defined as

$$\exp(A) = I + A + \frac{A^2}{2!} + \frac{A^3}{3!} + \cdots$$

which in the scalar special case would reduce to the usual exponential. It is crucial to notice that the solution works only if the CTMC has finite number of states. There are efficient ways of computing it especially when the entries of Q are numerical (and not symbolic). However, there are other techniques to use when Q is symbolic or if there are infinite elements in S.

As an example, consider a four-state CTMC $\{X(t), t \geq 0\}$ with $S = \{1, 2, 3, 4\}$ and

$$Q = \begin{bmatrix} -5 & 1 & 2 & 2 \\ 0 & -2 & 1 & 1 \\ 1 & 3 & -5 & 1 \\ 2 & 0 & 0 & -2 \end{bmatrix}.$$

Since $P(t) = \exp(Qt)$, we can compute at $t = 1$

$$P(1) = \exp(Q) = \begin{bmatrix} 0.1837 & 0.2938 & 0.1356 & 0.3869 \\ 0.1570 & 0.3522 & 0.1343 & 0.3565 \\ 0.1656 & 0.3328 & 0.1378 & 0.3638 \\ 0.2106 & 0.2270 & 0.1257 & 0.4367 \end{bmatrix}.$$

Therefore, if the CTMC is in state 1 at time $t=0$, then there is a probability of 0.2938 that it would be in state 2 at time $t=1$. Likewise $p_{34}(1)=0.3638$, but the CTMC could have jumped several states between $t=0$ when it was in state 3 and $t=1$ when it will be in state 4. As another example, $P\{X(15.7)=2|X(14.7)=3\}=0.3328$ due to time homogeneity.

Now, let us see what happens at $t=10$. Using $P(t)=\exp(Qt)$, we can compute at $t=10$

$$P(10) = \exp(10Q) = \begin{bmatrix} 0.1842 & 0.2895 & 0.1316 & 0.3947 \\ 0.1842 & 0.2895 & 0.1316 & 0.3947 \\ 0.1842 & 0.2895 & 0.1316 & 0.3947 \\ 0.1842 & 0.2895 & 0.1316 & 0.3947 \end{bmatrix}.$$

Notice how the rows are identical, that is, the columns have the same elements each. In other words, irrespective of which state 1 is currently, eventually with probability 0.1842 the system will be in state 1. This is similar to the steady-state behavior we saw with DTMCs. Next, we describe steady-state analysis of CTMCs.

B.2.3 Steady-State Analysis of CTMCs

Consider a CTMC $\{X(t), t \geq 0\}$ with state space S and generator matrix Q. As we saw in the previous section as t approaches infinity, the matrix $P(t)$ defined there has identical rows (or the columns have the same elements). However, that happened only for "well-behaved" CTMCs. It appears as though in the long run, the state of such a CTMC is independent of its initial state. This will always be true for irreducible and positive recurrent CTMCs (also called ergodic). For a definition of irreducible and positive recurrent CTMCs, we turn to DTMCs. In particular, recall the embedded DTMC that corresponds to the CTMC state, which is observed every time the system changes state. The CTMC is irreducible and positive recurrent if the embedded DTMC is irreducible and positive recurrent, respectively. We describe steady-state analysis only for such CTMCs.

Define the long-run (or steady-state) probability of the ergodic CTMC being in state j as

$$p_j = \lim_{t \to \infty} P\{X(t) = j\}.$$

We are interested in p_j because it not only describes the probability of being in a particular state in steady state, but also represents the fraction of times the CTMC spent in state j in the long run. Without loss of generality, assuming $S=\{0,1,\ldots,j,\ldots\}$, the easiest method to compute p_j is by obtaining the unique solution to the system of equations (called *balance equations*)

$$pQ = \bar{0}$$

$$\sum_{j \in S} p_j = 1$$

where
  p is the vector $(p_0, p_1, \ldots, p_j, \ldots)$
  $\bar{0}$ is a row vector of zeros

Also, when the system (CTMC) enters state i, it incurs a cost $c(i)$ per unit time on average. Again, the cost does not necessarily mean "dollar" cost but any other performance measures as well. Then the long-run average cost incurred per unit time is

$$\sum_{i \in S} c(i) p_i.$$

We illustrate these concepts using some examples.
  Consider the CTMC in Problem 134. The balance equations are

$$(p_0 \ p_1) \begin{bmatrix} -\beta & \beta \\ \alpha & -\alpha \end{bmatrix} = (0 \ 0)$$

$$p_0 + p_1 = 1.$$

Thus $(p_0 \ p_1) = (\alpha/(\alpha + \beta) \ \beta/(\alpha + \beta))$. Further, when the machine is up, it produces products at a rate of ρ per second and no product is produced when the machine is down. Then the long-run average production rate is $0 \times p_0 + \rho \times p_1 = \rho\beta/(\alpha + \beta)$.
  Next, consider the telephone switch system in Problem 135. Let $(p_0, p_1, \ldots, p_N)$ be the solution to $pQ = 0$ and $\sum p_i = 1$. For this system, the probability that an arriving call in steady state receives a busy signal (or is rejected) is p_N. Also, the long-run average rate of call rejection is λp_N and the long-run average switch utilization is $\sum_{i=0}^{N} i p_i$.
  Finally, for Problem 136, let $(p_{00}, p_{01}, p_{10}, p_{11}, p_{20}, p_{21}, \ldots)$ be the solution to $pQ = 0$ and $\sum_{(i,j) \in S} p_{ij} = 1$. Then the average number of backlogged messages in steady state is $\sum_{i=0}^{\infty} i(p_{i0} + p_{i1})$ and the long-run system throughput is $\sum_{i=0}^{\infty} p_{i1} \mu^2/(\mu + i\theta + \lambda)$.

B.3 Semi-Markov Process and Markov Regenerative Processes

In this section, we extend DTMCs and CTMCs to more general types of discrete-state stochastic processes. Interestingly, in some sense it also extends renewal processes. Although we do not present all the interesting results, the reader is encouraged to consider texts on stochastic processes in this regard. All we present here are things useful in the analysis of queues, and in particular we concentrate on steady-state performance measures.

B.3.1 Markov Renewal Sequence

Consider a stochastic process $\{Z(t), t \geq 0\}$ that models a system whose state at time t is a discrete value $Z(t)$. Say the state of the system is observed at discrete times $S_1, S_2, \ldots$, so that S_n is the time of the nth observation of the stochastic process $\{Z(t), t \geq 0\}$. Then the observation made at time S_n is called Y_n so that typically Y_n is a function of $Z(S_n)$ and S_n. For example, if an $M/G/1$ queue is observed at every departure, then $Z(t)$ could be the number in the system at time t, S_n is the time of nth departure and $Y_n = Z(S_n+)$, that is, the number in the system immediately after a departure. Usually if the nth observation changes the state of the stochastic process $\{Z(t), t \geq 0\}$, then typically Y_n is $Z(S_n+)$ or $Z(S_n-)$, that is, just before or just after the observation. However, if the observation does not change the system state, one usually uses $Y_n = Z(S_n)$. Define $\mathcal{J}$ as the set of all possible values Y_n can take for any n.

Given $S_0 = 0$ and $S_0 \leq S_1 \leq S_2 \leq S_3 \leq \ldots$, and $Y_n \in \mathcal{J}$ for all $n \geq 0$, the bivariate sequence $\{(Y_n, S_n), n \geq 0\}$ is called a *Markov renewal sequence* (MRS) if it satisfies the following:

$$P\{Y_{n+1} = j, S_{n+1} - S_n \leq x | Y_n = i, S_n, S_{n-1}, \ldots, Y_{n-1}, Y_{n-1}, \ldots\}$$

$$= P\{Y_{n+1} = j, S_{n+1} - S_n \leq x | Y_n = i\}$$

$$= P\{Y_1 = j, S_1 \leq x | Y_0 = i\}$$

for any $i \in \mathcal{J}$, $j \in \mathcal{J}$, and $x \geq 0$. The two equations here are similar to the Markov and time-homogeneity properties respectively. Now, for any $i \in \mathcal{J}$, $j \in \mathcal{J}$, and $x \geq 0$ define

$$G_{ij}(x) = P\{Y_1 = j, S_1 \leq x | Y_0 = i\}$$

such that the matrix

$$G(x) = [G_{ij}(x)]$$

is called the *kernel* of the MRS. We explain using an example.

Problem 137

Consider a $G/M/1$ queue with independent identically distribute (IID) inter-arrival times continuously distributed with common CDF $A(\cdot)$ and $\exp(\mu)$ service times. Let $Z(t)$ be the number of customers in the system at time t, S_n be the time of the nth arrival into the system with $S_0 = 0$, and $Y_n = Z(S_n-)$ be the number of customers in the system just before the nth arrival. For any $i \geq 0$ and $j \geq 0$ obtain $G_{ij}(x)$ for the MRS $\{(Y_n, S_n), n \geq 0\}$.

Solution

Clearly, we have $S_0 = 0$ and $S_0 \leq S_1 \leq S_2 \leq S_3 \leq \ldots$, since arrivals occur one by one and the $n + 1$st arrival occurs after the nth. Then define $\mathcal{J} = \{0, 1, 2, \ldots\}$. Then for any $i \in \mathcal{J}$ and any $j \in \mathcal{J}$, we have

$$G_{ij}(x) = P\{Y_1 = j, S_1 \leq x | Y_0 = i\}$$

$$= \begin{cases} \int_0^x e^{-\mu y} \frac{(\mu y)^{i+1-j}}{(i+1-j)!} dA(y) & \text{if } 0 < j \leq i+1, \\ A(x) - \sum_{i+1 \geq k > 0} G_{ik}(x) & \text{if } j = 0, \\ 0 & \text{otherwise.} \end{cases}$$

This result is due to the fact that when $i + 1 \geq j > 0$, $G_{ij}(x)$ is the probability of having exactly $i + 1 - j$ service completions in time S_1 and $S_1 \leq x$ (where S_1 is an interarrival time). Likewise if $j = 0$, then $G_{i0}(x)$ is the probability that the $i+1$st service completion occurs before S_1 and $S_1 \leq x$. Finally, if there are i customers just before time S_n, then just before time S_{n+1} it is not possible to have more than $i + 1$ customers in the system, so j must be less than or equal to $i + 1$ and thus $G_{ij}(x) = 0$ if $j > i + 1$. ∎

We will present other examples in the subsequent sections on semi-Markov processes (SMPs) and MRGPs. However, it is worthwhile restating three examples we have seen earlier. First of all, the usual renewal sequence $\{S_n, n \geq 0\}$ where S_n is the time of the nth event is a special case of an MRS where one could think of all Y_n values being identical. Hence the kernel reduces to a scalar $G(x)$, which is the CDF of the interrenewal times. Interestingly, as a second example, a DTMC $\{Y_n, n \geq 0\}$ with state space S and transition probability matrix P can be modeled as an MRS where $S_n = n$ since the system is observed at every unit of time. The third example is when $\{Z(t), t \geq 0\}$ is a CTMC with state space S and generator matrix Q. Then, S_n is the time of the nth transition of the CTMC and $Y_n = Z(S_n+)$, the state immediately after the nth transition. Then for any $i \in S$ and $j \in S$, $G_{ij}(x) = P\{Y_1 = j | Y_0 = i\} P\{S_1 \leq x | Y_0 = i\} = (1 - e^{q_{ii}x}) q_{ij}/(-q_{ii})$ since S_1 is independent of Y_1 when Y_0 is given.

There are some properties MRSs satisfy that are important to address. Say we are given an MRS $\{(Y_n, S_n), n \geq 0\}$ with kernel $G(x)$. Then the stochastic process $\{Y_n, n \geq 0\}$ is a DTMC with transition probability matrix $P = G(\infty)$. In fact in many analysis situations we may only have the LST of the kernel $\tilde{G}(s)$, then one can easily obtain $G(\infty)$ as $\tilde{G}(0)$ using one of the LST properties. Then, it is also crucial to notice that if we know the initial distribution $a = [P\{Y_0 = i\}]$, then the MRS is completely characterized by a and $G(x)$. With this, we move onto two stochastic processes that are driven by MRSs.

B.3.2 Semi-Markov Process

Recall the stochastic process $\{Z(t), t \geq 0\}$ on a countable state space and the MRS $\{(Y_n, S_n), n \geq 0\}$ described in the previous section. The SMP is a special type of $\{Z(t), t \geq 0\}$ where $Z(t)$ changes only at times S_n. In other words, $Z(t)$ versus t is piecewise constant and the only transitions occur at $S_0, S_1, S_2,$ ..., and hence $Y_n = Z(S_n+)$ is also equal to $Z(S_{n+1}-)$. A CTMC is a special type of SMP. A few other examples of SMPs can be found in Chapters 9 and 10. Toward the end of this section, we present another example. We first describe SMP analysis.

Since we do not use SMP transient analysis in this text, we do not present that here. Instead we go straight to steady-state behavior. Like we saw in DTMCs and CTMCs, here too for the steady-state analysis we require the SMP to be irreducible, aperiodic, and positive recurrent. For that, all we require is that the DTMC $\{Y_n, n \geq 0\}$ be irreducible, aperiodic, and positive recurrent, while the time between transitions $S_n - S_{n-1}$ have finite moments. Note that the SMP's aperiodicity is not an issue if the time between events are not all discrete. With that description in place, consider such an irreducible, aperiodic, and positive recurrent SMP $\{Z(t), t \geq 0\}$ on state space $\{1, 2, \ldots, \ell\}$. As before, S_n denotes the time of the nth jump epoch in the SMP with $S_0 = 0$. Recall Y_n as the state of the SMP immediately after the nth jump, that is,

$$Y_n = Z(S_n+).$$

Let

$$G_{ij}(x) = P\{S_1 \leq x; Y_1 = j | Y_0 = i\}.$$

The kernel of the SMP (which is the same as that of the MRS) is

$$G(x) = [G_{ij}(x)]_{i,j=1,\ldots,\ell} .$$

For the DTMC $\{Y_n, n \geq 0\}$, let the transition probability matrix be

$$P = G(\infty).$$

Let the conditional CDF

$$G_i(x) = P\{S_1 \le x | Y_0 = i\} = \sum_{j=1}^{\ell} G_{ij}(x)$$

and the expected time the SMP spends in state i continuously before transitioning out be

$$\tau_i = E(S_1 | Y_0 = i)$$

which can be computed from $G_i(x)$. Let

$$\pi_i = \lim_{n \to \infty} P\{Y_n = i\}$$

be the stationary distribution of the DTMC $\{Y_n, n \ge 0\}$. It is given by the unique nonnegative solution to

$$[\pi_1 \ \pi_2 \ \dots \ \pi_\ell] = [\pi_1 \ \pi_2 \ \dots \ \pi_\ell]P \quad \text{and} \quad \sum_{i=1}^{\ell} \pi_i = 1.$$

The stationary distribution of the SMP is given by

$$p_i = \lim_{t \to \infty} P\{Z(t) = i\} = \frac{\pi_i \tau_i}{\sum_{m=1}^{\ell} \pi_m \tau_m}.$$

Next, we present an example to illustrate SMP modeling and analysis.

Problem 138

Consider a system with two components, A and B. The lifetime of the system has a CDF $F_0(x)$ and mean μ_0. When the system fails, with probability q it is a component A failure and with probability $(1 - q)$ it is a component B failure. As soon as a component fails, it gets repaired and the repair time for components A and B have CDF $F_A(x)$ and $F_B(x)$, respectively, as well as means μ_A and μ_B, respectively. Model the system as an SMP and obtain the steady-state probability that the system is up with components A and component B are under repair.

Solution

Let $Z(t)$ be the state of the system with state space $\{0, A, B\}$ such that $Z(t) = 0$ implies the system is up and running at time t, whereas if $Z(t)$ is A or B, then at time t the system is down with component A or B, respectively, under repair. Let S_n denote the epoch when $Z(t)$ changes for the nth time and it corresponds to either the system going down or a repair completing. Also, Y_n is the state of the system immediately after time S_n. Then the process $\{Z(t), t \geq 0\}$ is an SMP with kernel (such that the states are in the order $0, A, B$)

$$G(x) = \begin{bmatrix} 0 & qF_0(x) & (1-q)F_0(x) \\ F_A(x) & 0 & 0 \\ F_B(x) & 0 & 0 \end{bmatrix}.$$

Having modeled the SMP, next we perform steady-state analysis. Notice that the DTMC $\{Y_n, n \geq 0\}$ is irreducible and positive recurrent, with transition probability matrix

$$P = G(\infty) = \begin{bmatrix} 0 & q & 1-q \\ 1 & 0 & 0 \\ 1 & 0 & 0 \end{bmatrix}.$$

Also, the conditional CDF $G_i(x) = P\{S_1 \leq x | Y_0 = i\} = \sum_{j=1}^{\ell} G_{ij}(x)$ for $i = 0$, A, B is $F_i(x)$ and the expected time the SMP spends in state i before transitioning out is μ_i (as described in the problem).

Then, we can compute (π_0, π_A, π_B), the stationary distribution of the DTMC $\{Y_n, n \geq 0\}$ by solving

$$[\pi_0 \ \pi_A \ \pi_B] = [\pi_0 \ \pi_A \ \pi_B]P \quad \text{and} \quad \pi_0 + \pi_A + \pi_B = 1.$$

Hence we get

$$[\pi_0 \ \pi_A \ \pi_B] = 0.5[1 \ q \ (1-q)].$$

The stationary distribution of the SMP for $i = 0, A, B$ can be computed using

$$p_i = \lim_{t \to \infty} P\{Z(t) = i\} = \frac{\pi_i \mu_i}{\pi_0 \mu_0 + \pi_A \mu_A + \pi_B \mu_B}.$$

Thus we have

$$[p_0 \ p_A \ p_B] = \frac{1}{\mu_0 + q\mu_A + (1-q)\mu_B}[\mu_0 \ q\mu_A \ (1-q)\mu_B]. \qquad \blacksquare$$

B.3.3 Markov Regenerative Processes

We briefly present an extension to SMPs, namely, MRGP. Essentially in the SMP, we were restricted by having $Z(t)$ stay constant between epochs that occur at times S_n for all $n \geq 0$. Here we generalize that to allow for $Z(t)$ to change between epochs. However, this stochastic process $\{Z(t), t \geq 0\}$ is precisely the one we considered in Section B.3.1. Although when we described MRS we did not pay attention to the $\{Z(t), t \geq 0\}$ process. We do that here. Indeed that process $\{Z(t), t \geq 0\}$ is an MRGP. However, since we considered MRGPs only very briefly in this book (see $G/M/1$ queues in Chapter 4), we do not delve deep into this concept and interested readers are suggested texts such as Kulkarni [67].

As described earlier, two immediate examples of MRGPs are as follows:

- $G/M/1$ queue with $Z(t)$ as the number in the system at time t, S_n as the time of the nth arrival, and $Y_n = Z(S_n-)$
- $M/G/1$ queue with $Z(t)$ as the number in the system at time t, S_n as the time of the nth departure, and $Y_n = Z(S_n+)$

For the steady-state analysis, consider an MRGP $\{Z(t), t \geq 0\}$ on state space S. For any $i \in S$ and $j \in S$, let

$$G_{ij}(x) = P\{S_1 \leq x; Y_1 = j | Y_0 = i\}.$$

The kernel of the MRGP (which is the same as that of the MRS) is

$$G(x) = [G_{ij}(x)].$$

For the DTMC $\{Y_n, n \geq 0\}$, let the transition probability matrix be

$$P = G(\infty).$$

Let the conditional CDF

$$G_i(x) = P\{S_1 \leq x | Y_0 = i\} = \sum_{j \in S} G_{ij}(x)$$

and

$$\tau_i = E(S_1 | Y_0 = i)$$

which can be computed from $G_i(x)$. Also, let

$$\pi_i = \lim_{n \to \infty} P\{Y_n = i\}$$

be the stationary distribution of the DTMC $\{Y_n, n \geq 0\}$ written in vector form as $\pi = [\pi_i]$ for all $i \in S$. It is given by the unique nonnegative solution to

$$\pi = \pi P \quad \text{and} \quad \sum_{i \in S} \pi_i = 1.$$

Define α_{kj} as the expected time spent in state j from time 0 to S_1, given that $Y_0 = k$, for all $k \in S$ and $j \in S$. This could be tricky to compute in some instances. The stationary distribution of the MRGP for any $j \in S$ is given by

$$p_j = \lim_{t \to \infty} P\{Z(t) = j\} = \frac{\sum_{k \in S} \pi_k \alpha_{kj}}{\sum_{k \in S} \pi_k \tau_k}.$$

Of course all this assumes that the stationary distribution exists, which only requires that the DTMC $\{Y_n, n \geq 0\}$ is irreducible and positive recurrent (assuming that the epochs S_n occur continuously over time). With that understanding, we move on to the final type of stochastic processes in this chapter and the only one where the states are not countable (note that other stochastic processes such as Ornstein–Uhlenbeck process and Gaussian process are not considered although they have been used in chapters of this book).

B.4 Brownian Motion

We begin this section by modeling a Brownian motion as a limit of a discrete random walk as done in Ross [91]. Consider a symmetric random walk: in each time unit, say one is equally likely to take a unit step either to the left or to the right. Take smaller and smaller steps at smaller and smaller time intervals. Say at each Δt time units we take a step of size Δx either to the left or to the right with equal probabilities. Let $X_i = 1$ if the ith step of length Δx is to the right, and $X_i = -1$ if the ith step of length Δx is to the left. We assume X_i are independent and

$$P\{X_i = 1\} = P\{X_i = -1\} = \frac{1}{2}.$$

Define $X(t)$ as the position at time t. Clearly

$$X(t) = X(0) + \Delta x(X_1 + X_2 + \cdots + X_{\lfloor t/\Delta t \rfloor}). \tag{B.1}$$

We let $X(0) = 0$, otherwise we will just look at $X(t) - X(0)$. Note that $E[X_i] = 0$ and $Var[X_i] = 1$. From Equation B.1, we have

$$E[X(t)] = 0,$$

$$Var[X(t)] = (\Delta x)^2 \left\lfloor \frac{t}{\Delta t} \right\rfloor. \tag{B.2}$$

Now let Δx and Δt become zero. However, we must be careful to ensure that the limit exists for Equation B.2. Therefore, we must have

$$(\Delta x)^2 = \sigma^2 \Delta t,$$

for some positive constant σ. Taking the limit as $\Delta t \to 0$, we have

$$E[X(t)] = 0,$$

and

$$Var[X(t)] \to \sigma^2 t.$$

From central limit theorem we have: $X(t)$ is normally distributed with mean 0 and variance $\sigma^2 t$. With that we now formally define a Brownian motion.

B.4.1 Definition of Brownian Motion

A stochastic process $\{X(t), t \geq 0\}$ satisfying the following properties is called a Brownian motion:

- $X(t)$ has independent increments; that is, for every pair of disjoint time intervals (s, t) and (u, v), $s < t \leq u < v$, the increments $\{X(t) - X(s)\}$ and $\{X(v) - X(u)\}$ are independent random variables. Therefore, the Brownian motion is a Markov process.
- Every increment $\{X(t) - X(s)\}$ is normally distributed with mean 0 and variance $\sigma^2(t - s)$.

Note that if $X(t)$ is Gaussian (i.e., normal)

$$P\{X(t) \leq x | X(s) = x_0\} = P\{X(t) - X(s) \leq x - x_0\} = \Phi(\alpha)$$

where
$$\alpha = (x - x_0)/(\sigma\sqrt{t - s})$$
$$\Phi(y) = \int_{-\infty}^{y} 1/\sqrt{2\pi} e^{-u^2/2} du$$

We assume that σ is a constant independent of t (hence time homogeneity) and x. When $\sigma = 1$, the process is called standard Brownian motion. Since any Brownian motion $X(t)$ can be converted into a standard Brownian motion $X(t)/\sigma$, we will use $\sigma = 1$ for the most part in this chapter (unless stated otherwise).

Next, we consider an important extension, namely, the geometric Brownian motion. Let $\{X(t), t \geq 0\}$ be a Brownian motion. Define $Y(t)$

such that

$$Y(t) = e^{X(t)}.$$

The process $\{Y(t), t \geq 0\}$ is called a *geometric Brownian motion*. Using the fact that the moment generating function of a normal random variable $X(t)$ with mean 0 and variance t is

$$E[e^{sX(t)}] = e^{ts^2/2},$$

we have

$$E[Y(t)] = E[e^{X(t)}] = e^{t/2},$$

$$Var[Y(t)] = E[Y^2(t)] - (E[Y(t)])^2 = e^{2t} - e^t.$$

As a rule of thumb, if absolute changes are IID, we use Brownian motion to model the process. However, if percentage changes are IID, we use geometric Brownian motion to model the process. For example, the percentage change of asset price in finance are IID and hence we model asset price using a geometric Brownian motion.

Another important extension is Brownian motion reflected at the origin. Let $\{X(t), t \geq 0\}$ be a Brownian motion. Define $Z(t)$ such that

$$Z(t) = |X(t)|.$$

The process $\{Z(t), t \geq 0\}$ is called *Brownian motion reflected at the origin*. The CDF of $Z(t)$ can be obtained for $y > 0$ as

$$P\{Z(t) \leq z\} = P\{X(t) \leq z\} - P\{X(t) \leq -z\}$$

$$= 2P\{X(t) \leq z\} - 1$$

$$= \frac{2}{\sqrt{2\pi t}} \int_{-\infty}^{z} e^{-x^2/2t} dx - 1.$$

Further, we have

$$E[Z(t)] = \sqrt{\frac{2t}{\pi}},$$

$$Var[Z(t)] = \left(1 - \frac{2}{\pi}\right) t.$$

As a final extension, we consider Brownian motion with drift. Let $B(t)$ be a standard Brownian motion. Define $X(t)$ such that

$$X(t) = B(t) + \mu t.$$

Then the process $\{X(t), t \geq 0\}$ is a Brownian motion with drift coefficient μ such that

- $X(0) = 0$
- $\{X(t), t \geq 0\}$ has stationary and independent increments
- $X(t)$ is normally distributed with mean μt and variance t

Note that the variance would be $\sigma^2 t$ if $B(t)$ was not a "standard" Brownian motion.

B.4.2 Analysis of Brownian Motion

Consider a process $\{X(t), t \geq 0\}$, which is a Brownian motion with drift coefficient μ and variance of $X(t)$ is $\sigma^2 t$. Let the CDF $F(t, x; x_0)$ be defined as follows:

$$F(t, x; x_0) = P\{X(t) \leq x | X(0) = x_0\}.$$

It can be shown that the CDF satisfies the following diffusion equation:

$$\frac{\partial}{\partial t} F(t, x; x_0) = -\mu \frac{\partial}{\partial x} F(t, x; x_0) + \frac{\sigma^2}{2} \frac{\partial^2}{\partial x^2} F(t, x; x_0). \tag{B.3}$$

Initial condition: $X(0) = x_0$ implies

$$F(0, x; x_0) = \begin{cases} 0 & \text{if } x < x_0 \\ 1 & \text{if } x \geq x_0. \end{cases}$$

If there is a reflecting barrier placed on the x-axis, the boundary condition is

$$F(t, 0; x_0) = 0 \quad \text{for all } t > 0.$$

Equation B.3 is also called forward Kolmogorov equation or Fokker–Planck equation. Next, we provide a solution to this system by presenting as a problem.

Problem 139

Consider the Brownian motion $\{X(t), t \geq 0\}$ with drift coefficient μ and variance of $X(t)$ is $\sigma^2 t$. Assume there is a reflecting barrier placed on the x-axis. Solve the PDE and obtain steady-state probabilities.

Solution

The solution to the PDE (Equation B.3) is

$$F(t, x; x_0) = \phi\left(\frac{x - x_0 - \mu t}{\sigma\sqrt{t}}\right) - e^{-2x\mu/\sigma^2}\phi\left(\frac{-x - x_0 - \mu t}{\sigma\sqrt{t}}\right).$$

In steady state as $t \to \infty$, let $F(t, x; x_0) \to F(x)$. Then $F(x)$ satisfies (letting $t \to \infty$ in Equation B.3)

$$0 = -\mu\frac{dF(x)}{dx} + \frac{\sigma^2}{2}\frac{d^2F(x)}{dx^2}.$$

The solution is

$$F(x) = 1 - e^{2x\mu/\sigma^2},$$

which is the CDF of an exponential random variable with parameter $-2\mu/\sigma^2$. ∎

B.4.3 Itô's Calculus

For a sufficiently smooth function $f(x)$, recall Taylor's expansion:

$$f(x + dx) = f(x) + f'(x)dx + \frac{1}{2}f''(x)(dx)^2 + o((dx)^2). \qquad \text{(B.4)}$$

By letting $df(x) = f(x + dx) - f(x)$, we can write

$$df(x) = f'(x)dx + \frac{1}{2}f''(x)(dx)^2 + o((dx)^2).$$

Dividing both sides by dx and letting $dx \to 0$, we get the familiar derivative formula:

$$\lim_{dx \to 0}\frac{f(x + dx) - f(x)}{dx} = f'(x).$$

For ordinary calculus, we do not need the second-derivative term in Equation B.4, as it is associated with higher-order infinitesimal $(dx)^2$. So, we can simply write

$$df(x) = f'(x)dx.$$

When Brownian motion is involved, things turn out a little differently. Let B_t be a standard Brownian motion (the same as $B(t)$ but to avoid too many parentheses in our formulae we use B_t). Consider a function $f(X_t)$, where

$$X_t = \mu t + \sigma B_t.$$

From Equation B.4, we have

$$df(X_t) = f'(X_t)dX_t + \frac{1}{2}f''(X_t)(dX_t)^2 + o((dX_t)^2).$$

Recall the fundamental relation

$$(dB_t)^2 = dt$$

when $dt \rightarrow 0$. Hence we have

$$(dX_t)^2 = (\mu dt + \sigma dB_t)^2$$

$$= \mu^2(dt)^2 + 2\mu\sigma(dt)(dB_t) + \sigma^2(dB_t)^2$$

$$= \sigma^2 dt,$$

where on the last line we have ignored terms of higher order than dt. Substituting in the Taylor's series expansion, and omitting higher-order terms, we have

$$df(X_t) = f'(X_t)dX_t + \frac{1}{2}f''(X_t)\sigma^2 dt. \tag{B.5}$$

In other words, the second-derivative term can no longer be ignored when Brownian motion is involved. This is the essence of Itô's calculus.

One of the main applications of this analysis is in finance. In particular, we show an example to obtain the probability distribution of asset price at time t. Let S_t denote asset price at time t. For a given time T, we would like to obtain the distribution of the random variable S_T. The dynamics of S_t is modeled as a geometric Brownian motion

$$\frac{dS_t}{S_t} = \mu dt + \sigma dB_t. \tag{B.6}$$

Now consider $\ln(S_t)$. We would like to obtain $d(\ln(S_t))$. Note that for $f(x) = \ln(x)$, we have $f'(x) = 1/x$ and $f''(x) = -1/x^2$. Therefore, we have (using

Equation B.6)

$$d(\ln(S_t)) = \frac{dS_t}{S_t} - \frac{1}{2}\frac{(dS_t)^2}{S_t^2}$$

$$= \mu dt + \sigma dB_t - \frac{1}{2}(\mu dt + \sigma dB_t)^2$$

$$= \mu dt + \sigma dB_t - \frac{1}{2}\sigma^2 dt.$$

Define ν such that $\nu = \mu - \sigma^2/2$. Taking integral from 0 to T, on both sides of the equation, we have

$$\ln(S_T) - \ln(S_0) = \nu T + \sigma B_T.$$

Rewriting in terms of S_T, we have

$$S_T = S_0 e^{\nu T + \sigma B_T}.$$

Therefore, S_T follows a log-normal distribution, or $\ln(S_T)$ follows a normal distribution:

$$\ln(S_T) \sim Normal\left(\ln(S_0) + \left(\mu - \frac{\sigma^2}{2}\right)T, \sigma^2 T\right).$$

It is important to note that while the geometric Brownian motion in Equation B.6 characterizes the dynamics of the asset price over time, this log-normal distribution only specifies the distribution of the asset price at a single time point T.

Reference Notes

Like the previous chapter, this chapter is also mainly a result of teaching various courses on stochastic processes especially at the graduate level. The definitions, presentations, and notations for the first part of this chapter (DTMC, CTMC, MRS, SMP, and MRGP) are heavily influenced by Kulkarni [67]. Another excellent resource for those topics is Ross [92]. For the Brownian part, the materials presented are based out of Ross [91] and Medhi [80]. Several topics such as diffusion processes, Ornstein–Uhlenbeck process, Gaussian process, and martingales have been left out. Some of these such as martingales can be found in both Ross [91] and Resnick [90]. Also, the topic of stochastic process limits (i.e., also not considered here) can be found in Whitt [105].

Exercises

B.1 A discrete-time polling system consists of a single communication channel serving N buffers in a cyclic order starting with buffer-1. At time $t=0$, the channel polls buffer-1. If it has any packets to transmit, the channel transmits exactly one and then moves to buffer-2 at time $t=1$. The same process repeats at each buffer until at time $t=N-1$ the channel polls buffer N. Then at time $t=N$, the channel polls buffer-1 and the cycle repeats. Now consider buffer-1. Let Y_t be the number of packets it receives during the interval $(t, t+1]$. Assume that $Y_t=1$ with probability p and $Y_t=0$ with probability $1-p$. Let X_n be the number of packets available for transmission at buffer-1 when it is polled for the nth time. Model $\{X_n, n \geq 1\}$ as a DTMC.

B.2 Consider a DTMC with transition probability matrix

$$P = \begin{pmatrix} p_0 & p_1 & p_2 & p_3 & \cdots \\ 1 & 0 & 0 & 0 & \cdots \\ 0 & 1 & 0 & 0 & \cdots \\ 0 & 0 & 1 & 0 & \cdots \\ \vdots & \vdots & \vdots & \vdots & \ddots \end{pmatrix}$$

where $p_j > 0$ for all j and $\sum_{j=0}^{\infty} p_j = 1$. Let $M = \sum_{j=0}^{\infty} (j\, p_j)$ and assume $M < \infty$. Show that $\pi_0 = 1/(1+M)$ and hence find the stationary probability distribution $\pi = (\pi_0\ \pi_1\ \pi_2\ \ldots)$.

B.3 Conrad is a student who goes out to eat lunch everyday. He eats either at a Chinese food place (C), at an Italian food place (I), or at a Burger place (B). The place Conrad chooses to go for lunch on day n can be modeled as a DTMC with state space $S = \{C, I, B\}$ and transition matrix

$$P = \begin{bmatrix} 0.3 & 0.3 & 0.4 \\ 1 & 0 & 0 \\ 0.5 & 0.5 & 0 \end{bmatrix}.$$

That means if Conrad went to the Chinese food place (C) yesterday, he will choose to go to the Burger place (B) today with probability 0.4, and also there is a 30% chance he will go to the Italian food place (*I*). Is the DTMC irreducible, aperiodic, and positive recurrent?

(a) After several days (assume steady state is reached) if you ask Conrad what he had for lunch, what is the probability that he would say "burger"?

(b) If on day $n = 1000$ Conrad went to the Chinese food place, what is the probability that on day $n = 1001$ he will go to the Italian food place?

(c) If all the students at Conrad's university follow the earlier Markov chain, in the long run, what fraction of the students going to lunch eat at the Italian food place?

(d) Continuing with the previous question, if the average cost per lunch is $1.90 in the Burger place, $5.70 in the Italian food place, and $3.80 in the Chinese food place, what is the average revenue per day per customer from all the three eating places together.

(e) How much does Conrad spend on lunch (on the average) everyday?

B.4 Consider a colony of amoebae whose lifetimes are independent exp(μ) random variables. During its lifetime, each amoeba produces offsprings (i.e., splits into two) in a $PP(\lambda)$ fashion. All amoebae behave independently. As soon as the last amoeba in the colony dies, a lab technician introduces a new amoeba after an exp(λ_0) time. Let $X(t)$ be the number of amoebae in the colony at time t. Model $\{X(t), t \geq 0\}$ as a CTMC.

B.5 Say that $\{X(t), t \geq 0\}$ is a CTMC with state space $S = \{0, 1, 2\}$ and infinitesimal generator matrix:

$$Q = \begin{bmatrix} - & 2 & 3 \\ 1 & -1 & - \\ - & 0 & -2 \end{bmatrix}.$$

Fill up the blanks in Q. If the CTMC is in state 0 at time 0, what is the probability that the CTMC would be in state 2 at time $t = 1$? Given that $p_0 = 2/9$, compute p_1 and p_2, where p_0, p_1, and p_2 form the limiting distribution of the CTMC. If the cost incurred per unit time in state i is $18i$, what is the long-run average cost incurred by the system?

B.6 There are two identical photocopy machines in an office. The up times of each machine is exponentially distributed with a mean up time of $1/\mu$ days. The repair times of each machine is exponentially distributed with mean $1/\lambda$ days. Let $X(t)$ be the number of machines that are up at time t. Model the $\{X(t), t \geq 0\}$ process as a CTMC under two conditions: one repair person and two repair persons. Would it better to employ one or two repair persons for this system? If a machine is up, it generates a revenue of $r per unit time. However, each repair person charges $c per unit time.

B.7 Consider a three-state SMP $\{Z(t), t \geq 0\}$ with state space $\{1, 2, 3\}$. The elements of the kernel of this SMP is given as follows: $G_{12}(t) = 1 - e^{-t} - te^{-t}$, $G_{21}(t) = 0.4(1 - e^{-0.5t}) + 0.3(1 - e^{-0.2t})$, $G_{23}(t) = 0.2(1 - e^{-0.5t}) + 0.1(1 - e^{-0.2t})$, $G_{32}(t) = 1 - 2e^{-t} + e^{-2t}$, and $G_{11}(t) = G_{13}(t) = G_{22}(t) = G_{31}(t) = G_{33}(t) = 0$. Obtain the probability that the SMP is in state i in steady state for $i = 1, 2, 3$.

B.8 Let $\{X(t), t \geq 0\}$ be a standard Brownian motion. Define $\{Z(t), t \geq 0\}$ such that $Z(t) = |X(t)|$, the Brownian motion reflected at the origin. Using the CDF of $Z(t)$, derive expressions for $E[Z(t)]$ and $Var[Z(t)]$.

B.9 For any constant k show that $\{Y(t), t \geq 0\}$ is a martingale if $Y(t) = \exp\{kB(t) - k^2 t/2\}$, where $B(t)$ is a standard Brownian motion. For that, all you need to show is the following is satisfied:

$$E[Y(t)|Y(u), 0 \leq u \leq s] = Y(s).$$

B.10 Let $\{B(t), t \geq 0\}$ be a standard Brownian motion and $X(t) = B(t) + \mu t$ for some constant μ. Compute $Cov(X(t), X(s))$.

References

1. S. Aalto, U. Ayesta, and R. Righter. On the Gittins index in the M/G/1 queue. *Queueing Systems*, 63(1–4), 437–458, 2009.
2. J. Abate and W. Whitt. Numerical inversion of Laplace transforms of probability distributions. *ORSA Journal on Computing*, 7, 36–43, 1995.
3. V. Aggarwal, N. Gautam, S.R.T. Kumara, and M. Greaves. Stochastic fluid-flow models for determining optimal switching thresholds with an application to agent task scheduling. *Performance Evaluation*, 59(1), 19–46, 2004.
4. S. Ahn and V. Ramaswami. Efficient algorithms for transient analysis of stochastic fluid flow models. *Journal of Applied Probability*, 42(2), 531–549, 2005.
5. D. Anick, D. Mitra, and M.M. Sondhi. Stochastic theory of a data handling system with multiple sources. *Bell System Technical Journal*, 61, 1871–1894, 1982.
6. L. Arnold. *Stochastic Differential Equations: Theory and Applications*, Krieger Publishing Company, Melbourne, FL, 1992.
7. F. Baccelli and P. Bremaud. *Elements of Queuing Theory: Palm Martingale Calculus and Stochastic Recurrences*, 2nd edn., Springer, Berlin, Germany, 2003.
8. F. Baskett, K.M. Chandy, R.R. Muntz, and F. Palacios. Open, closed and mixed networks of queues with different classes of customers. *Journal of the ACM*, 22, 248–260, 1975.
9. A.W. Berger and W. Whitt. Effective bandwidths with priorities. *IEEE/ACM Transactions on Networking*, 6(4), 447–460, August 1998.
10. A.W. Berger and W. Whitt. Extending the effective bandwidth concept to network with priority classes. *IEEE Communications Magazine*, 36, 78–84, August 1998.
11. G.R. Bitran and S. Dasu. Analysis of the $\Sigma PH_i/PH/1$ queue. *Operations Research*, 42(1), 159–174, 1994.
12. G. Bolch, S. Greiner, H. de Meer, and K.S. Trivedi. *Queueing Networks and Markov Chains*, 1st edn., John Wiley & Sons Inc., New York, 1998.
13. M. Bramson. Stability of queueing networks. *Probability Surveys*, 5, 169–345, 2008.
14. P.J. Burke. The output of a queuing system. *Operations Research*, 4(6), 699–704, 1956.
15. J.A. Buzacott and J.G. Shanthikumar. *Stochastic Models of Manufacturing Systems*, Prentice-Hall, New York, 1992.
16. C.S. Chang and J.A. Thomas. Effective bandwidth in high-speed digital networks. *IEEE Journal on Selected Areas in Communications*, 13(6), 1091–1100, 1995.
17. C.S. Chang and T. Zajic. Effective bandwidths of departure processes from queues with time varying capacities. In: *Fourteenth Annual Joint Conference of the IEEE Computer and Communication Societies*, Boston, MA, pp. 1001–1009, 1995.
18. X. Chao, M. Miyazawa, and M. Pinedo. *Queueing Networks: Customers, Signals, and Product Form Solutions*, John Wiley & Sons, New York, 1999.

19. H. Chen and D.D. Yao. *Fundamentals of Queueing Networks*, Springer-Verlag, New York, 2001.

20. N. Chrukuri, G. Kandiraju, N. Gautam, and A. Sivasubramaniam. Analytical model and performance analysis of a network interface card. *International Journal of Modelling and Simulation*, 24(3), 179–189, 2004.

21. G.L. Curry and N. Gautam. Characterizing the departure process from a two server Markovian queue: A non-renewal approach. In: *Winter Simulation Conference*, Miami, FL, pp. 2075–2082, 2008.

22. G. de Veciana and G. Kesidis. Bandwidth allocation for multiple qualities of service using generalized processor sharing. In: *IEEE Global Telecommunications (GLOBECOM-94)*, San Francisco, CA, pp. 1550–1554, 1994.

23. G. de Veciana, C. Courcoubetis, and J. Walrand. Decoupling bandwidths for networks: A decomposition approach to resource management. In: *Proceedings of Fourteenth Annual Joint Conference of the IEEE Computer and Communications Societies, 1994 (INFOCOM-94)*, Toronto, Ontario, Canada, pp. 466–473, 1994.

24. G. de Veciana, G. Kesidis, and J. Walrand. Resource management in wide-area ATM networks using effective bandwidths. *IEEE Journal on Selected Areas in Communications*, 13(6), 1081–1090, 1995.

25. J.G. Dai. Stability of fluid and stochastic processing networks. MaPhySto Miscellanea Publication, No. 9, 1999.

26. R.L. Disney and P.C. Kiessler. *Traffic Processes in Queueing Networks: A Markov Renewal Approach*, Johns Hopkins University Press, Baltimore, MD, 1987.

27. S.G. Eick, W.A. Massey, and W. Whitt. The physics of the $M_t/G/1$ queue. *Operations Research*, 41(4), 731–742, 1993.

28. A.I. Elwalid and D. Mitra. Analysis and design of rate-based congestion control of high speed networks. Part I: Stochastic fluid models, access regulation. *Queueing Systems: Theory and Applications*, 9, 29–64, 1991.

29. A.I. Elwalid and D. Mitra. Fluid models for the analysis and design of statistical multiplexing with loss priorities on multiple classes of bursty traffic. *IEEE Transactions on Communications*, 42(11), 2989–3002, 1992.

30. A.I. Elwalid and D. Mitra. Effective bandwidth of general Markovian traffic sources and admission control of high-speed networks. *IEEE/ACM Transactions on Networking*, 1(3), 329–343, June 1993.

31. A.I. Elwalid, D. Heyman, T.V. Lakshman, D. Mitra, and A. Weiss. Fundamental bounds and approximations for ATM multiplexers with applications to video teleconferencing. *IEEE Journal on Selected Areas in Communications*, 13(6), 1004–1016, 1995.

32. A.I. Elwalid and D. Mitra. Analysis, approximations and admission control of a multi-service multiplexing system with priorities. In: *Proceedings of Fourteenth Annual Joint Conference of the IEEE Computer and Communications Societies, 1995 (INFOCOM-95)*, Boston, MA, pp. 463–472, 1995.

33. S.N. Ethier and T.G. Kurtz. *Markov Processes: Characterization and Convergence*, 1st edn., John Wiley & Sons, Inc., New York, 1986.

34. M. Fackrell. Fitting with matrix-exponential distributions. *Stochastic Models*, 21, 377–400, 2005.

35. R.M. Feldman and C. Valdez-Flores. *Applied Probability and Stochastic Processes*, PWS Publishing Company, Boston, MA, 1995.

36. A. Feldmann and W. Whitt. Fitting mixtures of exponentials to long-tail distributions to analyze network performance models. *Performance Evaluation*, 31, 245–279, 1998.

37. M. Ferguson and Y. Aminetzah. Exact results for nonsymmetric token ring systems. *IEEE Transactions on Communications*, 33(3), 223–231, 1985.

38. A. Ganesh, N. O'Connell, and D. Wischik. Big queues. In: *Series: Lecture Notes in Mathematics*, Vol. 1838, Springer-Verlag, Berlin, Germany, 2004.

39. N. Gautam, V.G. Kulkarni, Z. Palmowski, and T. Rolski. Bounds for fluid models driven by semi-Markov inputs. *Probability in the Engineering and Informational Sciences*, 13, 429–475, 1999.

40. N. Gautam and V.G. Kulkarni. Applications of SMP bounds to multiclass traffic in high-speed networks. *Queueing Systems: Theory and Applications*, 36, 351–379, 2000.

41. N. Gautam. Buffered and unbuffered leaky bucket policing: Guaranteeing QoS, design and admission control. *Telecommunication Systems*, 21(1), 35–63, 2002.

42. N. Gautam. Pricing issues in web hosting services. *Journal of Revenue and Pricing Management*, 4(1), 7–23, 2005.

43. N. Gautam. Quality of service metrics. In: *Frontiers in Distributed Sensor Networks*, S.S. Iyengar and R.R. Brooks (eds.), Chapman & Hall/CRC Press, Boca Raton, FL, pp. 613–628, 2004.

44. N. Gautam. Queueing theory. In: *Operations Research and Management Science Handbook*, A. Ravindran (ed.), CRC Press, Taylor & Francis Group, Boca Raton, FL, pp. 9.1–9.37, 2007.

45. E. Gelenbe and I. Mitrani. *Analysis and Synthesis of Computer Systems*, Academic Press, London, U.K., 1980.

46. P.W. Glynn. Diffusion approximations. In: *Handbooks in Operations Research and Management Science Volume 2, Stochastic Models*, D.P. Heyman and M.J. Sobel (eds.), North-Holland, Amsterdam, the Netherlands, pp. 145–198, 1990.

47. W.J. Gordon and G.F. Newell. Closed queueing systems with exponential servers. *Operations Research*, 15(2), 254–265, 1967.

48. T.C. Green and S. Stidham. Sample-path conservation laws, with applications to scheduling queues and fluid systems. *Queueing Systems: Theory and Applications*, 36, 175–199, 2000.

49. D. Gross and C. M. Harris. *Fundamentals of Queueing Theory*, 3rd edn., John Wiley & Sons Inc., New York, 1998.

50. S. Halfin and W. Whitt. Heavy-traffic limits for queues with many exponential servers. *Operations Research*, 29, 567–588, 1981.

51. M. Harchol-Balter. Queueing disciplines. In: *Wiley Encyclopedia of Operations Research and Management Science*, John Wiley & Sons, New York, 2011.

52. J.M. Harrison. *Brownian Motion and Stochastic Flow Systems*, John Wiley & Sons Inc., New York, 1985.

53. D.P. Heyman and M.J. Sobel. *Stochastic Models in Operations Research, Volume I, Stochastic Processes and Operating Characteristics*, McGraw-Hill, New York, 1982.

54. M. Hlynka. Queueing theory page. http://web2.uwindsor.ca/math/hlynka/queue.html

55. M. Hlynka. List of queueing theory software. http://web2.uwindsor.ca/math/hlynka/qsoft.html

56. J.R. Jackson. Networks of waiting line. *Operations Research*, 5, 518–521, 1957.

57. M. Kamath. Rapid analysis of queueing systems software. http://www.okstate.edu/cocim/raqs/

58. O. Kella. Stability and non-product form of stochastic fluid networks with Levy inputs. *The Annals of Applied Probability*, 6(1), 186–199, 1996.

59. F.P. Kelly. *Reversibility and Stochastic Networks*, John Wiley & Sons Inc., Chichester, U.K., 1994.

60. F.P. Kelly. Notes on effective bandwidths. In: *Stochastic Networks: Theory and Applications*. F.P. Kelly, S. Zachary, and I.B. Ziedins (eds.), Oxford University Press, Oxford, UK, pp. 141–168, 1996.

61. G. Kesidis, J. Walrand, and C.S. Chang. Effective bandwidths for multiclass Markov fluids and other ATM sources. *IEEE/ACM Transactions on Networking*, 1(4), 424–428, 1993.

62. J.P. Kharoufeh and N. Gautam. Deriving link travel time distributions via stochastic speed processes. *Transportation Science*, 38(1), 97–106, 2004.

63. L. Kleinrock. *Queueing Systems*, Vol. 2, John Wiley & Sons Inc., New York, 1976.

64. H. Kobayashi. Application of the diffusion approximation to queueing networks. Part I: Equilibrium queue distribution. *Journal of the Association for Computing Machinery*, 21(2), 316–328, 1974.

65. K.R. Krishnan, A.L. Neidhardt, and A. Erramilli. Scaling analysis in traffic management of self-similar processes. In: *Proceedings of the 15th International Teletraffic Congress*, Washington, DC, pp. 1087–1096, 1997.

66. V.G. Kulkarni and T. Rolski. Fluid model driven by an Ornstein–Ühlenbeck process. *Probability in Engineering and Informational Sciences*, 8, 403–417, 1994.

67. V.G. Kulkarni. *Modeling and Analysis of Stochastic Systems*. Texts in Statistical Science Series, Chapman & Hall, Ltd., London, U.K., 1995.

68. V.G. Kulkarni. Effective bandwidths for Markov regenerative sources. *Queueing Systems: Theory and Applications*, 24, 137–153, 1996.

69. V.G. Kulkarni. Fluid models for single buffer systems. In: *Frontiers in Queueing*, Probab. Stochastics Ser., CRC Press, Boca Raton, FL, pp. 321–338, 1997.

70. V.G. Kulkarni and N. Gautam. Admission control of multi-class traffic with service priorities in high-speed networks. *Queueing Systems: Theory and Applications*, 27, 79–97, 1997.

71. T.G. Kurtz. Strong approximation theorems for density dependent Markov chains. *Stochastic Processes and Their Applications*, 6(3), 223–240, 1978.

72. R.C. Larson. Perspectives on queues: Social justice and the psychology of queueing. *Operations Research*, 35(6), 895–905, 1987.

73. G. Latouche and V. Ramaswami. *Introduction to Matrix Analytic Methods in Stochastic Modeling. ASA-SIAM Series on Statistics and Applied Probability*, Society for Industrial and Applied Mathematics (SIAM), Philadelphia, PA, 1999.

74. W.E. Leland, M.S. Taqqu, W. Willinger, and D.V. Wilson. On the self-similar nature of Ethernet traffic (Extended Version). *IEEE/ACM Transactions on Networking*, 2(1), 1–15, 1994.

75. S. Mahabhashyam and N. Gautam. On queues with Markov modulated service rates. *Queueing Systems: Theory and Applications*, 51(1-2), 89–113, 2005.

76. S. Mahabhashyam, N. Gautam, and S.R.T. Kumara. Resource sharing queueing systems with fluid-flow traffic. *Operations Research*, 56(3), 728–744, 2008.

77. A. Mandelbaum, W.A. Massey, and M.I. Reiman. Strong approximations for Markovian service networks. *Queueing Systems*, 30(1–2), 149–201, 1998.

78. A. Mandelbaum, W.A. Massey, M.I. Reiman, A. Stolyar, and B. Rider. Queue lengths and waiting times for multiserver queues with abandonment and retrials. *Telecommunication Systems*, 21(2–4), 149–171, 2002.

79. W.A. Massey and W. Whitt. Uniform acceleration expansions for Markov chains with time-varying rates. *The Annals of Applied Probability*, 8(4), 1130–1155, 1998.

80. J. Medhi. *Stochastic Models in Queueing Theory*, Elsevier Science, Boston, MA, 2003.

81. D.A. Menasce and V.A.F. Almeida. *Scaling for E-Business: Technologies, Models, Performance, and Capacity Planning*, Prentice Hall, Upper Saddle River, NJ, 2000.

82. S.P. Meyn. *Control Techniques for Complex Networks*, Cambridge University Press, New York, 2009.

83. M. Moses, S. Seshadri, and M. Yakirevich. HOM Software. http://www.stern.nyu.edu/HOM

84. A. Narayanan and V.G. Kulkarni. First passage times in fluid models with an application to two priority fluid systems. *Proceedings of the IEEE International Computer Performance and Dependability Symposium*, Urbana-Champaign, IL, 1996.

85. M.F. Neuts. *Matrix-Geometric Solutions in Stochastic Models—An Algorithmic Approach*, The Johns Hopkins University Press, Baltimore, MD, 1981.

86. T. Osogami and M. Harchol-Balter. Closed form solutions for mapping general distributions to quasi-minimal PH distributions. *Performance Evaluation*, 63, 524–552, 2006.

87. Z. Palmowski and T. Rolski. A note on martingale inequalities for fluid models. *Statistic and Probability Letter*, 31(1), 13–21, 1996.

88. Z. Palmowski and T. Rolski. The superposition of alternating on-off flows and a fluid model. Report no. 82, Mathematical Institute, Wroclaw University, 1996.

89. N.U. Prabhu. *Foundations of Queueing Theory*, Kluwer Academic Publishers, Boston, MA, 1997.

90. S.I. Resnick. *A Probability Path*, Birkhauser, Boston, MA, 1998.

91. S.M. Ross. *Stochastic Processes*, John Wiley & Sons Inc., New York, 1996.

92. S.M. Ross. *Introduction to Probability Models*, 8th edn., Academic Press, New York 2003.

93. S.M. Ross. *A First Course in Probability*, 8th edn., Pearson Prentice Hall, Upper Saddle River, NJ, 2010.

94. D. Sarkar and W.I. Zangwill. Expected waiting time for nonsymmetric cyclic queueing systems–Exact results and applications. *Management Science*, 35(12), 1463–1474, 1989.

95. L.E. Schrage and L.W. Miller. The queue M/G/1 with the shortest remaining processing time discipline. *Operations Research*, 14, 670–684, 1966.

96. R. Serfozo. *Introduction to Stochastic Networks*, Springer-Verlag, New York, 1999.

97. L.D. Servi. Fast algorithmic solutions to multi-dimensional birth-death processes with applications to telecommunication systems. In: *Performance Evaluation and Planning Methods for the Next Generation Internet*, A. Girard, B. Sanso, and F. Vazquez-Abad (eds.), Springer, New York, pp. 269–295, 2005.

98. A. Shwartz and A. Weiss. *Large Deviations for Performance Analysis*, Chapman & Hall, New York, 1995.

99. W.J. Stewart. *Introduction to the Numerical Solution of Markov Chains*, Princeton University Press, Princeton, NJ, 1994.

100. S. Stidham Jr. *Optimal Design of Queueing Systems,* CRC Press, Boca Raton, FL, 2009.
101. H. Takagi. Queueing analysis of polling models. *ACM Computing Surveys,* 20(1), 5–28, 1988.
102. H.C. Tijms. *A First Course in Stochastic Models,* John Wiley & Sons Inc., Bognor Regis, West Sussex, England, 2003.
103. W. Whitt. The queueing network analyzer. *The Bell System Technical Journal,* 62(9), 2779–2815, 1983.
104. W. Whitt. Departures from a queue with many busy servers. *Operations Research,* 9(4), 534–544, 1984.
105. W. Whitt. *Stochastic-Process Limits,* Springer, New York, 2002.
106. W. Whitt. Efficiency-driven heavy-traffic approximations for many-server queues with abandonments. *Management Science,* 50(10), 1449–1461, 2004.
107. A. Wierman, N. Bansal, and M. Harchol-Balter. A note comparing response times in the M/G/1/FB and M/G/1/PS queues. *Operations Research Letters,* 32, 73–76, 2003.
108. R.W. Wolff. *Stochastic Modeling and the Theory of Queues,* Prentice Hall, Englewood Cliffs, NJ, 1989.

Index

A

Acyclic queueing networks,
  Poisson flows
  automobile service station
    analysis and recommendation,
      321–324
    customer satisfaction, 320
    lucrative service packages, 319
    system description and model,
      320–321
  cyclic networks, 312
  departure processes
    Burke's theorem, 314
    $M/G/\infty$ and $M/G/s/s$ queues,
      314–315
    $M/M/1$ queue, 313–314
  feed-forward networks, 312
  node i, 312–313
  superpositioning and splitting
    Bernoulli splitting, 315–316
    exponential random
      variables, 315
    $G/G/m$ queue, 319
    minimum of exponentials, 315
    net customer-arrival process, 316
    nodes average number of
      customers, 317–318
    sojourn time distribution, 318
ALMGF, *see* Asymptotic logarithmic
      moment generating function
      (ALMGF)
Arrivals see time averages (ASTA)
    arrival point probabilities, 346
    arrival theorem, 348
    birth and death process, 349–352
    infinitesimal time interval, 347
    steady-state probabilities, 346–348
ASTA, *see* Arrivals see time averages
      (ASTA)
Asymptotic logarithmic moment
      generating function (ALMGF)
    alternating renewal process, 602–605
    CTMCs, 593–598

inflow characteristics, 590–593
SMPs, 598–601

B

Bagel queue, 332
Bay-Gull Bagels problem, 331,
      443–444
Bernoulli splitting, 315–316, 393–394
Bottleneck approximation
    definition, 400
    HDMA station, 423–424
    large C, 400–402
    small C, 405–406
Bounds and approximations
    analysis techniques, 188–189
    call centers staffing and
        work-assignment
      findings and adjustments, 200–201
      recommendation, major revamp,
        198–200
      TravHelp calls, 196–198
    general single server queueing
        system $(G/G/1)$, 189–192
    multiserver queues $(M/G/s, G/M/s,$
        and $G/G/s)$, 194–195
    performance measures, 188
Bounds, buffer content distribution
    CDE approximation, 627
    decreasing failure rate (DFR) random
        variable, 630
    description, 611
    Erlang on-off source, 638–640
    exact results $vs.$ approximations and
        bounds, 635
    expected amount of time, 632
    exponential random variable, 633
    four-state CTMC, 634
    increasing failure rate (IFR) random
        variable, 630
    k independent fluid sources, 640–641
    Laplace Stieltjes transform (LST), 628
    limiting distribution, 629
    ℓ-state irreducible CTMC, 631

on-off "pseudo" source, 643
on-times and off-times, 636–637
SMP source characteristics, 628
stability and nontriviality, 641–642
steady-state buffer content
    distribution, 634
tandem buffers model, 644–648
Braess' paradox, 336–337
Brownian motion
  analysis, 749–750
  central limit theorem, 747
  drift, 749
  geometric, 747–748
  Itô's calculus
    asset price, 751–752
    standard Brownian motion, 751
    Taylor's expansion, 750
  reflection
    description, 377–378, 441
    $G/G/1$ queue, 379–390
    origin, 748
  standard, 747
  stochastic process and properties, 747
  symmetric random walk, 746–747
Bulk arrival queues ($M^{[X]}/M/1$)
  L'Hospital's rule, 72–73
  Little's law, 73–74
  rate diagram, 71
  server processes, 70
Burke's theorem, 314

C

CDF, *see* Cumulative distribution
    function (CDF)
Central limit theorem, diffusion limit,
    484–488
Chernoff dominant eigenvalue (CDE)
    approximation
  description, 610–611
  input sources, 622–623
  ℓ-state irreducible CTMC, 623–625
  on-off source, 625–627
  stationary probability vector, 623
Closed Jackson networks
  arrivals see time averages
    arrival point probabilities, 346
    arrival theorem, 348
    birth and death process, 349–352

    infinitesimal time interval, 347
    steady-state probabilities, 346–348
  attributes, 339–340
  product-form solution
    balance equation, 340–342
    joint probability distribution, 340,
      342, 344–345
    marginal distribution, 343
    marginal probability vectors, 345
    N-dimensional stochastic process,
      340
    normalizing constant, 343–344
    routing probability matrix, 343
    steady-state probability, 340
  single-server closed Jackson
    networks
    database server system, 355–357
    mean performance measures, 353
    mean sojourn time, 353–354
    steady-state measures, 353
    three-tier architecture, e-business
      websites, 354–355
Closed queueing networks
  bottleneck approximation, large C,
    400–402
  MVA approximation, small C,
    402–406
Coefficient of variation (COV), 702–703
Compound Poisson processes (CPP),
    719–720
Conditional random variables
  expected values
    description, 711
    net profit, farmer, 713–714
    non-defective parts, 711–712
    optimal time, 712–713
  probability determination
    call center and service time,
      709–710
    law of total probability, 708–709
    leisure tickets, demand, 710–711
Continuous random variables
  distribution
    beta, 701
    chi-squared, 700
    Erlang, 698
    exponential, 697
    gamma, 698–699
    log-normal, 701

normal, 700
Pareto, 701–702
uniform, 699
Weibull, 699–700
moment computation, 697
Continuous-time Markov chains
(CTMCs)
buffer content analysis, 530–534
description, 732
environment process
definition of *h(v)*, 593–594
description, 593
eb(v) vs. v, 597
expression for *g(t)*, 595
mean rate and peak rate, 597
on-off source, 596
ergodic, 525–526
infinitesimal generator matrix, 733
notations, lists, 527
on-off source, 541–544, 577–578
QDB, 118
steady-state analysis
average cost and production rate,
739
balance equations, 738–739
long-run probability, 738
steady-state results and performance
evaluation, 534–537
system modeling
steps, 733–734
telephone switch, 734–735
unslotted Aloha, 735–736
up/down machine, 734
time-homogeneity properties, 733
transient analysis
four-state process, 737–738
matrix differential equation, 737
COV, *see* Coefficient of variation (COV)
CPP, *see* Compound Poisson processes
(CPP)
CTMCs, *see* Continuous-time Markov
chains (CTMCs)
Cumulative distribution function (CDF)
LST, 552
random variable T_1, 574–575
time τU and time τD calculation,
522–523
travel time, 550

D

Decreasing failure rate (DFR) random
variable, 630
Diffusion approximations
abandonments, 495–500
description, 484
$G/G/1$ queue, 483–484
limit and functional central limit
theorem, 484–488
multiserver queues, 488–495
time-varying parameters, 507–511
Discrete queue
fluid limits, 447
and fluid queues, 517
M-policy, 573
stability, 473, 474
stochastic, 467, 483
workload process, 517–518
Discrete random variables
distribution
Bernoulli, 694
binomial, 694
geometric, 694–695
hypergeometric, 695–696
negative binomial, 695
Poisson, 696
uniform, 693
Zipf, 696–697
moment computation, 693
Discrete-time Markov chains (DTMCs)
balance equations, 151–152
$G/M/1$ queue
arrival point probabilities, 177
balance equations, 170–173
bivariate stochastic process, 175
interarrival time, 168–169, 177–178
irreducible DTMC, 173
Lebesgue integral, 169
limiting distribution, 170, 178–180
limiting probability, 170, 172
lower Hessenberg matrix, 170
Markov property, 168
MRGP theory, 175–176
nondecreasing convex function,
172
Poisson process, 175–176
service times, 168
sojourn time distribution, 173–174

stability condition, 172–173,
    178–180
steady-state probabilities, 177
stochastic process, 168–169
transition probability matrix, 169
inter-observation times, 152
Markov property, 725
$M/G/1$ queue
    average sojourn time, 160–161
    balance equations, 156–157
    busy period distribution, 165–168
    customer departure, 154
    customers arrival probability,
        154–155
    exponential distributions, 154
    FCFS, 153
    generating functions, 157–159
    Lebesgue integral, 155
    L'Hospital's rule, 159, 161
    limiting distribution, 156
    Little's Law, 161
    long-run time-averaged number
        of entities, 160–161
    LST, 158, 161–164
    moments of steady-state sojourn
        time, 165
    Poisson process, 153
    Pollaczek-Khintchine equation,
        161
    Riemann integral, 155–156
    service time, 153
    stability condition, 159
    traffic intensity, 159
    transition probability matrix, 155
    upper Hessenberg matrix, 155
one-step transition probability
    matrix, 152
state space, 152
steady-state analysis
    average multiplexing delay, 732
    ergodic theorem, 731
    long-run average cost, 731–732
    "well-behaved", 730–731
steady-state probabilities, 152–153
system and state space, 725
system modeling
    cell-phone switching problem,
        727–728
    RT and NR packets, 727

steps, 726
time division multiplexer and
    packets, 728–729
time-homogeneity, 726
transient analysis
    n-step transition probability,
        729–730
    student and cell-phone company,
        730
    transition probability matrix, 726
Donsker's theorem, 487
DTMCs, *see* Discrete-time Markov
    chains (DTMCs)

E

Effective bandwidths
    ALMGF
        computation, 593–605
        inflow characteristics, 590–593
        output from queue, 607–609
        tail asymptotics, 611–622
Effective capacity, *see* Effective
    bandwidths
Emergency ward planning
    classification, patients, 267
    CTMC, 268
    experience, 269–271
    LST, 269
    30-minute ER commitment, 269
    Poisson process, 267
    recommendations, 271
    true emergencies, 271
    two-server systems, 267–268
    UTH, 266–267, 271
    waiting time, 267, 268
Environmental processes
    alternating renewal process
        on-off source, 602–604
        Q matrix, 604
        rate matrix, 605
        SMP approach, 602
    CTMCs, 522, 593–598
    SMPs, 523–525, 598–601
Erlang loss formula, 60–61
Erlang on-off source, 638–640
Exact computation method, 610
Exhaustive polling
    cycle time, 274–275

definition, 272–273
$M/G/1$ queue, server vacations, 275
Poisson process, 275
station time, 276
steady-state covariance, 276–277
Exponential interarrival and
    service times
  CTMC, 45
  network graph technique, 45
  solving balance equations, *see* Solving
    CTMC balance equations

F

FCFS, *see* First come first served (FCFS)
FCLT, *see* Functional central limit
    theorem (FCLT)
Finite-state Markov chains
  energy conservation, data centers
    concerns, 137
    CTMC, 140
    hysteretic policy, 139, 141
    powering on and off, 137
    servers processing speed, 137–138
  steady-state probabilities
    direct computation, 134–135
    eigenvalues and eigenvectors, 133
    finite-state approximation, 136
    transient analysis, 135–136
First come first served (FCFS) policy
  busy period, 292
  fair policy, 294–295
  and LCFS, 20–21
  multiclass $M/G/1$
    arrival process, 255
    performance measures, 255–256
    Poisson processes, 256
    Pollaczek–Khintchine formula,
      255
  open queueing networks, *see* Open
    queueing networks
  pseudo-conservation law, 282
  single class $G/G/1$ queue, 248
  steady-state expected workload,
    263–264
Fluid-flow queues
  computation, effective bandwidth
    and ALMGF
    CTMC source, 593–598

general On/Off source, 601–605
  SMP source, 598–601
  exercises, 687–690
  fluid removal, 589
  inflow characteristics
    ALMGF and effective
      bandwidths, 591–592
    $eb(v)$ vs. v, 592
    $h(v)$ vs. v, 591
    r^{mean} and r^{peak}, 592
    total amount of fluid generation,
      590
    of traffic flow, 590
    workload process, 591
  multiclass, *see* Multiclass fluid queues
  output effective bandwidth,
    607–609
  performance analysis, single queue
    bounds for buffer content
      distribution, 611, 627–648
    CDE approximation, 610–611,
      622–627
    effective bandwidths, tail
      asymptotics, 611–622
    exact computation, 610
    output capacity, buffer, 609
  single buffer, single environment
    process and output capacity,
    589–590
  traffic superposition, multiple
    sources, 605–607
Fluid models
  analyze, 447–448
  description, 512
  diffusion approximations
    abandonments, 495–500
    description, 484
    $G/G/1$ queue, 483–484
    limit and functional central limit
      theorem, 484–488
    multiserver queues, 488–495
  functional strong law, 450–454
  hydrodynamic limits, 447
  multiclass queueing networks, 512
  numerous books and monographs,
    512
  Rybko–Stolyar–Kumar–Seidman
    network, 513
  single queue, server, 448–450

stability analysis
  determination, 454
  multiclass queueing networks,
    virtual stations, 454–467
  queueing network stable, 475–483
  service rate, 454
  stable discrete network, 467–475
  time-varying parameters
    additional situations, 501
    diffusion approximation, 507–511
    $M_t/M/s_t$ queue, 500–501
    Poisson processes, 501–502
    uniform acceleration, 502–507
  Young Myong Ko section, 512–513
Fluid queues
  applications, 520
  deterministic
    functional strong law, 450–454
    single queue, server, 448–450
  vs. discrete, 516–518
Fokker–Planck equation, 385–386
FSLLN, *see* Functional strong law of
    large numbers (FSLLN)
Functional central limit theorem
    (FCLT)
  average arrival rate, 484
  fluid limit, 484–485
  $G/G/1$ queue with interarrival times,
    485–488
Functional strong law of large numbers
    (FSLLN)
  concept, 447
  fluid limit
    deterministic system, 451
    fluid queue, 453–454
    scaling, 452–453
    SLLN, 450–451
    stochastic process tracking,
      451–452
  strong approximation, 502–503

G

Gärtner–Ellis conditions, 611, 686
Gaussian process, characterization, 508
Generalized processor sharing (GPS)
    policy, 649, 674–676
General queueing networks
  analyze, 377

Bay-Gull Bagels problem, 443–444
closed queueing networks,
    algorithms
  bottleneck approximation, large C,
    400–402
  MVA approximation, small C,
    402–406
  description, 441
  Jackson networks, 377
  open queueing networks, *see* Open
      queueing networks
  reflected Brownian motion, 377–378
  Rybko–Stolyar–Kumar–Seidman
      network, 445–446
  sensor network, 444
  seven-node single-server queueing
      network, 442–443
  single-server and single-class
    Bernoulli splitting, 393–394
    flow through a Queue, 392–393
    $G/G/1$ Queue, 379–390
    superposition, 390–392
  single-server stations, 442
General single server queueing system
    $(G/G/1)$
  algorithm, 192
  empirical approximations, 192
  heavy-traffic queues, 192
  IFR/DFR, 190–191
  Pareto distribution, 193
  performance metrics, 189
  Pollaczek-Khintchine formula, 191
  positive-valued random variables,
    190–191
  random variables properties, 190
  traffic intensity, 189
Generating functions (GFs)
  definition, 704
  properties, 704–705
  queueing systems, 703
GFs, *see* Generating functions (GFs)
$G/G/s$ queues, *see* Matrix geometric
    methods (MGM)

H

Head-of-the-line (HL) discipline, 249
Host direct memory access (HDMA),
    421–424

I

Increasing failure rate (IFR) random variable, 630
Independent identically distributed (IID) random variables
central limit theorem, 716–717
finite mean and variance, 716
Poisson process
description, 717
NPP and CPP, 719–720
number of events, 717
properties, 717–719
renewal process
definition, 720
equilibrium distribution and inspection paradox, 721
LST, 720–721
reliability theory, 721
strong law of large numbers, 716–717
Itô's calculus
asset price, 751–752
standard Brownian motion, 751
Taylor's expansion, 750

J

Jackson networks, 377, 378, 395–396, 419, 428

K

Kelly networks, 361
Kolmogorov equation, 385–386

L

Laplace–Stieltjes transform (LST), 588
application, 706–707
computation, 705–706
properties, 706
renewal process, 720–721
Laplace transform (LT)
description, 707
properties, 707–708
Last come first served (LCFS) policy
and FCFS, 20–21
LCFS-PR, 247, 295, 367
nonpreemptive, 249, 283, 294, 305
L'Hospital's rule, 72–73, 159, 161

Little's Law, 12–14, 73–74, 161, 403, 427, 432–433, 499–500
flow system, 12
multiclass queues
renewal and arrival process, 244
scheduling policy, 245
service times, 245
traffic intensity, 246
Loss networks, 367–369
LST, *see* Laplace–Stieltjes transform (LST)
LT, *see* Laplace transform (LT)

M

Markov modulated fluid source (MMFS)
buffer content analysis, 530–534
buffer size, 540–541
CTMC on–off source, 541–547
environment CTMC, 538–540
five-state CTMC, 547–550
steady-state analysis and performance evaluation, 534–537
terminology and notation, 525–530
Markov regenerative processes (MRGPs)
SMP, 744–745
stationary distribution, 745–746
steady-state analysis, 745
Markov renewal sequence (MRS)
description, 740
$G/M/1$ queue, 741–742
kernel matrix, 740
Matrix geometric methods (MGM)
aggregated phase-type queue ($\Sigma PHi/PH/s$)
arrival point probabilities, 209
irreducible infinitesimal generator, 208
Kronecker product, 205–207
Kronecker sum, 206–208
multidimensional stochastic process, 206–207
performance measures, 209
phase-type arrivals superposition, 205
sojourn time distribution, 209

state-space explosion, 205
steady-state probabilities, 208
CTMC, 122–123
phase-type processes
exponential distributions, 202
fitting, 204
generalized Coxian distribution,
204
hyperexponential distribution,
204–205
hypoexponential distribution,
203–205
infinitesimal generator matrix, 202
moment-matching methods, 205
over-parameterization, 203, 205
phase-type distributions, 201–202
random variable, 202–203
square matrix, 203
semiconductor wafer fab application
Coxian distribution, 210
decomposition approach, 209
steady-state probabilities, 210–213
superpositioned process, 210
two-phase exponential processes,
210
use, 122
Mean value analysis (MVA)
approximation, small C, 402–406
$G/G/1$ queues, 187–188
$M/G/1$ queue
asymptotic values, 181
Poisson process, 181
Pollaczek–Khintchine formula,
183
X_n and X_{n+1}, 181–183
random variables properties, 180
renewal arrivals and general service
queues approximation
analytical closed-form
expressions, 184
interarrival time distribution, 184
notation, 184
performance measures, 185
SCOV, 184
sojourn time, 185–187
stochastic process, 180
MGM, *see* Matrix geometric methods
(MGM)
$M/G/1$ queue

DTMCs
average sojourn time, 160–161
balance equations, 156–157
busy period distribution, 165–168
customer departure, 154
customers arrival probability,
154–155
exponential distributions, 154
FCFS, 153
generating functions, 157–159
Lebesgue integral, 155
L'Hospital's rule, 159, 161
limiting distribution, 156
Little's law, 161
long-run time-averaged number
of entities, 160–161
LST, 158, 161–164
moments of steady-state sojourn
time, 165
Poisson process, 153
Pollaczek–Khintchine equation,
161
Riemann integral, 155–156
service time, 153
stability condition, 159
traffic intensity, 159
transition probability matrix, 155
upper Hessenberg matrix, 155
MVA
asymptotic values, 181
Poisson process, 181
Pollaczek-Khintchine
formula, 183
X_n and X_{n+1}, 181–183
processor sharing
balance equation, 225–227
context-switches, 221
density function, 222
fundamental calculus, 224
multidimensional stochastic
process, 222
multiprogramming limit, 221
partial differential equation, 222
performance measures, 222
sojourn time, 227
stability condition, 221–222
Taylor-series expansion, 223–225
time-quantum, 221

$M/G/\infty$ queue, modeling systems with ample servers
  busy period distribution, 216–218
  cumulative distribution function, 213
  interdeparture times distribution, 218–219
  nonhomogeneous Bernoulli splitting, 215
  performance measures, 214
  Poisson process, 213, 215
  Poisson random variable, 216
  regenerative process with regeneration epochs, 214
  split process, 215–216
  steady-state analysis, 214
  transient analysis, 214, 219–221
$M/G/s/s$ queue, telephone switch application
  density function, 228–229
  Erlang loss formula, 233
  holding times, 228
  IID random variables, 227
  joint probability, 228
  Markov process, 234
  multidimensional stochastic process, 228
  Poisson process, 227
  product-form solutions, 234
  reversible process, 234
  sojourn time distribution, 233
  steady-state balance equation, 230–232
  steady-state probability, 233
  Taylor-series expansion, 229–230
MMFS, *see* Markov Modulated Fluid Source (MMFS)
Monte-Carlo simulation methods, 549
M-policy, 573
MRGPs, *see* Markov regenerative processes (MRGPs)
MRS, *see* Markov renewal sequence (MRS)
Multiclass fluid queues
  admissible regions, 650
  buffer-2 with compensating source, 681–686
  compensating source, 649
  generalized processor sharing (GPS) policy, 649, 674–676

static priority service policy, 662–674
  structure, 648–649
  tackling varying output capacity, 650–651
  timed round-robin (polling) policy, 649, 651–661
  two infinite-sized buffer system, 676–681
Multiclass networks
  aggregate service rate, 366
  BCMP network, 364
  conditions, 364–365
  FCFS, 367
  steady-state probability, 365–366
  stochastic process, 365
  total arrival rate, 365
Multiclass queueing networks, virtual stations
  empty system, 465–467
  interarrival times/service times, 455
  number of jobs, 460–463
  numerical values, 463–465
  reentrant line, 455, 456–460
  simulations, 454
Multiclass queues
  computer communications, 243–244
  description, 241–242
  Little's law, 244–246
  optimal service-scheduling policies, 293–303
  partially completed service
    expected workload, waiting area, 252–253
    HL discipline, 249
    policies, 249
    server, expected workload, 250–252
    stochastic system, 249–250
    waiting and service areas, 250
  policy evaluation, *see* Policy evaluation, multiclass queues
  production, 243
  road transportation, 242–243
  work-conserving disciplines
    busy period distribution, 248
    policies, 247–248
    service times, 246, 248–249
    single-server queue, 246
    traffic intensity, 248
    workload, 246–247

Multidimensional Markov processes,
        221
Multiprogramming limit (MPL), 221
Multiserver and finite capacity queue
        model, 48–49
Multiserver queues
    abandonments
        balking, 495
        diffusion approximations,
            regimes, 495–496
        Ornstein–Uhlenbeck diffusion
            process, 497–500
        scaling, 496
        single-stage queueing system, 495
    diffusion approximation
        deterministic fluid model, 489
        fix β, increase λ and s, 492–495
        fix ρ, increase λ and s, 491–492
        fix s, increase λ, 490–491
        goal, 488–489
        $M/M/s$ queue, 489–490
MVA, *see* Mean value analysis (MVA)

N

Network interface card (NIC), cluster
        computing, 421–424
Nonhomogeneous Poisson processes
        (NPP), 500, 719
Non-real-time (NR) packets, 727
NPP, *see* Nonhomogeneous Poisson
        processes (NPP)
Numerical techniques and
        approximations
    finite-state Markov chains, *see*
        Finite-state Markov chains
    multidimensional birth and death
        chains
        arrival rates, Q matrix, 110–112
        CTMC, 93–94
        Diag(Q), 104–107
        inverses, 108–109
        matrices, 108–110
        optimal customer routing, 114–117
        PASTA, 113–114
        Poisson process, 103–104
        Servi algorithm, 103
        threshold policies, optimal
            control, 94–103

multidimensional Markov chains
        MGM, 122–124
        QBD process, 118–122
        variable service rate queues,
            124–132

O

Open Jackson networks
    Bagel queue, 332
    Bay-Gull Bagels schematic
        representation, 331
    Braess' paradox, 336–337
    cashier queue, 333
    categories, 324
    with deterministic routing
        conditions, 360–361
        Kelly networks, 361
        multiagent software system, 360
        product-form expression, 362
        route-node incidence matrix, 361
        steady-state probability, 361–362
    drinks queue, 332
    eat-in queue, 333
    Erlang distribution, 335
    external arrival rate vector, 332
    flow conservation and stability,
        325–326
    multiserver system, 335
    parallel system, 334–335
    performance measures, 330
    Poisson process, 324
    product-form solution
        balance equations, 327–329
        joint probability distribution,
            329–330
        marginal distribution, 330
        $M/M/s$ queue, 327
        N-dimensional stochastic process,
            326
        steady-state correlation, 326
        steady-state probability, 326–327
    routing probabilities, 332
    serial system, 333–334
    smoothies queue, 332
    stability condition, 332
    with state-dependent arrivals and
        service
        marginal distribution, 359

mean sojourn times, 360
normalizing constant, 359
service rate, 358
steady-state probability, 358–359
Whittle network, 360
waiting time computation, 338
Open queueing networks
decomposition algorithm
feed-forward network, 396
Jackson networks, 395–396
mean and standard deviation,
service times, 397
nodes *i*, 396–397
reflected Brownian motions,
394–395
routing probabilities, 397, 398–399
multiclass and multiserver, FCFS
flow across multiple classes,
411–413
flow through multiple servers, 411
network interface card, 421–424
preliminaries, 407–409
QNA algorithm, 414–421
service discipline, 406–407
single $G/G/m$ queue, 410–411
superposition and splitting of
flows, 413–414
multiclass and single-server,
priorities
approaches, 435–436
exponential case, 425–431
general case, 431–435
low and high-priority jobs,
424–425
MVA-based algorithm, 436–437
SRBM, 435
state-space-collapse-based
algorithm, 437–441
Optimal service-scheduling policies
multiclass queues
Gittins index policy, 300–302
nonpreemptive policy, 302–303
SRPT, 300
stochastic dynamic program, 303
setting and classification
classification, 293–294
description, 293
fair policy, 294–295
objective function, 295–296

single class queues
FCFS, 297
Gittins index policy, 296–297
IFR and DFR, service times,
298–300
SRPT, 296

P

Pareto distribution, 193
Poisson arrivals *see* time process
(PASTA), 113–114
Poisson distribution, 155
Policy evaluation, multiclass queues
description, 253–254
emergency ward planning
classification, patients, 267
CTMC, 268
experience, 269–271
LST, 269
30-minute ER commitment, 269
Poisson process, 267
recommendations, 271
true emergencies, 271
two-server systems, 267–268
UTH, 266–267, 271
waiting time, 267, 268
FCFS, 254–256
$G/G/1$ queue, 254
knowledge, service times
description, 282–283
mean waiting time, 283
Pollaczek–Khintchine formula,
283
processor sharing and preemptive
LCFS, 284
PSJF, 285–287
SPTF, 284–285
SRPT, 287–293
web servers, 283
nonpreemptive priority, $M/G/1$
Cμ rule, 260
definition, 257
Little's Law, 259
mapping, classes and priorities,
256–257
optimal priority assignment,
260–262
performance measures, 260

Poisson process, 259
    random variables, 257–258
    steady-state probability, 258
    waiting time, 259
online orders, fast-food restaurants,
        254
polling models, *see* Polling models
preemptive resume priority, $M/G/1$
    average workload, 264–265
    modification, 262–263
    performance measures, 265–266
    Pollaczek–Khintchine formula, 263
    sojourn time, 263
    steady-state expected workload,
        263–264
    total service time, 263
Pollaczek–Khintchine equation, 161, 183,
        191, 255, 283, 387–389
Polling models
    exhaustive polling, *see* Exhaustive
        polling
    gated policy
        cycle time, 277–278
        definition, 273
        expression, average waiting time,
            278–280
        service time, 278
        token rings, 277
    limited service
        definition, 273
        long-run fraction, 281
        mean cycle time, 281
        pseudo-conservation law, 282
        random time, 281
        transmission, packets, 280
        waiting time, symmetric case, 282
    regenerative process, 273
    round-robin scheduling, 272
    single server, 272, 273
Preemptive shortest job first (PSJF)
        policy
    description, 285–286
    expression, sojourn time,
        286–287
    and SRPT, 294
Product-form networks
    generalizations, 357–358
    loss networks, 367–369
    multiclass networks

aggregate service rate, 366
    BCMP network, 364
    conditions, 364–365
    FCFS, 367
    steady-state probability, 365–366
    stochastic process, 365
    total arrival rate, 365
open Jackson-like networks with
        deterministic routing
    conditions, 360–361
    Kelly networks, 361
    multiagent software system, 360
    product-form expression, 362
    route-node incidence matrix, 361
    steady-state probability, 361–362
open Jackson networks with
        state-dependent arrivals and
        service
    marginal distribution, 359
    mean sojourn times, 360
    normalizing constant, 359
    service rate, 358
    steady-state probability, 358–359
    Whittle network, 360
quasi-reversibility, 358
steady-state distributions, 357
Psychology, queueing
    checkout lines, 37–38
    customer behavior, 40
    customer satisfaction, 36–37
    elevator problem, 37
    human nature, 39
    QoS, 40–41

Q

QNA, *see* Queueing network analyzer
        (QNA)
Quality-and efficiency-driven (QED),
        495–496
Quasi-birth-death (QBD) processes
    CTMC, 118
    MGM, *see* Matrix geometric methods
        (MGM)
    structure, 121
Queueing fundamentals and notations
    arrival process, 19
    discrete queues, 18–19
    FCFS, 20–21

$G/G/s$ queue special case, 35–36
Kendall notation, 19–20
modeling, queueing system, 26–29
relationship, system metrics
  $G/G/s/K$ queue, 34–35
  regenerative cycle, 30
  renewal process, 29–30
single-station queueing system, 18
terminology
  variables, 22
  workload and number sample
    path, 22–23
Queueing network analyzer (QNA),
  414–421
Queueing networks
acyclic queueing networks, *see*
    Acyclic queueing networks,
    Poisson flows
closed Jackson networks, *see* Closed
    Jackson networks
open Jackson networks, *see* Open
    Jackson networks
product-form networks, *see*
    Product-form networks
stability analysis, fluid models
  determination, 454
  multiclass queueing networks,
    virtual stations, 454–467
  queueing network stable, 475–483
  service rate, 454
  stable discrete network, 467–475
Queues analysis
applications
  computer, communication, and
    information systems, 2–3
  production systems, 3–4
  service systems, 4
modeling, 5–6
scope and methods, 7–8

R

Random variables
conditional, *see* Conditional random
    variables
distribution and moments
  continuous random variables, *see*
    Continuous random variables
  COV, 702–703

description, 691
discrete random variables, *see*
    Discrete random variables
expected value, mixture random
    variable, 691–692
TCP transmission, 692–693
exponential distribution
  characteristics, 714
  properties, 714–716
GFs
  definition, 704
  properties, 704–705
  queueing systems, 703
IID, collection, *see* Independent
    identically distributed (IID)
    random variables
LST
  application, 706–707
  computation, 705–706
  properties, 706
LT
  description, 707
  properties, 707–708
Real-time (RT) packets, 727
Retrial queue
  description, 65–66
  Ethernet cable, 66–70
  rate diagram, 66–67
Rybko–Stolyar–Kumar–Seidman
    network, 445–446, 513

S

Semi-Markov decision process (SMDP),
  97
Semi-Markov process (SMP)
  conditional CDF and expected time,
    743
  defined, $e(\Lambda(u,v))$, 599
  eb(v) vs. *v*, 600–601
  environmental processes, 523–525
  kernel and transition probability
    matrix, 742
  LST of the kernel, 600
  MRGP, 598
  stationary distribution, 743
  steady-state analysis, 742
  system modeling and steady-state
    probability, 743–744

Semi-martingale reflected Brownian
   motion (SRBM), 435
Servi algorithm, 103, 107–110, 112, 116,
   117
Shortest processing time first (SPTF)
   policy, 284–285
Shortest remaining processing time
   (SRPT) policy
  busy period
    calculation, 290–293
    definition, 288
    equilibrium random variable,
     288–289
    service times, 289–290
  preemptive priority, 287
  service times, 289–290
  sojourn time, 287–288
Single buffer fluid model
  buffer content analysis, 530–534
  environment process
    characterization, 522–525
  preliminaries, performance analysis,
    521–522
  steady-state and performance
    evaluation, 534–537
  terminology and notation, 525–530
Single queue, performance analysis
  bounds for buffer content
    distribution, 611, 627–648
  CDE approximation, 610–611,
    622–627
  effective bandwidths, tail
    asymptotics, 611–622
  exact computation, 610
  output capacity, buffer, 609
Single-server closed Jackson networks
  database server system, 355–357
  mean performance measures, 353
  mean sojourn time, 353–354
  steady-state measures, 353
  three-tier architecture, e-business
    websites, 354–355
SLLN, *see* Strong law of large numbers
   (SLLN)
SMDP, *see* Semi-Markov decision
   process (SMDP)
SMP, *see* Semi-Markov process (SMP)
Solving CTMC balance equations
  arc cuts

Erlang loss formula, 60–61
$M/M/1/K$ queue, 58–60
$M/M/s$ queue, 57–58
multiserver and finite capacity
   queue model, 48–49
node flow balance, 46–47
notations, 55–56
steady-state analysis, 49–55
steady-state probabilities, 47–48
bandwidth-sensitive traffic
  arc cut, 84–85
  class i connection, 81–82
  independent reversible processes,
   83
  optimal policy, 82
generating functions
  CTMC, 61–62, 75
  $M/M/1$, 63–65
  $M^{[X]}/M/1$, 70–74
  number of requests, 77–78
  polynomials, 76
  pricing, performance and
   availability, 74
  retrial queue, 65–70
  server breakdown, 74
  server up-and downtimes, 77
  single server queue, 74–75
  steady-state probabilities, 62–63
reversibility
  processes, 79–80
  properties, 80–81
Squared coefficient of variation (SCOV),
   184
SRBM, *see* Semi-martingale reflected
   Brownian motion (SRBM)
Static priority service policy
  admissible region K, 670–671
  buffer-1, 663
  buffer-j, 663–664
  bufferless CDE approximation, 671
  CDE approximation, 671
  $(K2 + 1)$-state SMP, 666–667
  mean arrival rate of traffic, 664–665
  multiclass fluid queueing system,
   673–674
  output from buffer-1, 668–669
  Perron–Frobenius eigenvalue, 669
  region K_{smp}, 672
  scheduler, 662

SMP bounds, 671
steady-state arrival rate, 662
vs. timed round-robin, 674
Stochastic fluid-flow queues
applications, 518–520
channel capacity, 588
concerning level to nominal,
563–566
continuous state space processes,
584–585
decades, 584
discrete *vs.* fluid queues, 516–518
environment process
characterization, 522–525
exponential on–off source, 566–573
finite-sized buffer, 585
infinite-sized buffer, 585
Laplace–Stieltjes transform
(LST), 588
leaky bucket, 586
MMFS, single buffer, *see* Markov
modulated fluid source
(MMFS)
nominal to excessive, 560–563
output capacity, 576–580
output valve/sink, 573–576
PDE and boundary conditions,
551–555
positive/negative drifts, 580–584
preliminaries, performance analysis,
521–522
token pool, 586–587
water flow, reservoir, 555–559
Stochastic processes
Brownian motion
analysis, 749–750
central limit theorem, 747
definition, 747–749
Itô's calculus, 750–752
symmetric random walk,
746–747
CTMCs, *see* Continuous-time Markov
chains (CTMCs)
DTMCs, *see* Discrete-time Markov
chains (DTMCs)
MRGPs
SMP, 744–745
stationary distribution, 745–746
steady-state analysis, 745

MRS
description, 740
$G/M/1$ queue, 741–742
kernel matrix, 740
SMP
conditional CDF and expected
time, 743
kernel and transition probability
matrix, 742
stationary distribution, 743
steady-state analysis, 742
system modeling and steady-state
probability, 743–744
Strong law of large numbers (SLLN),
450–451
Systems analysis
asymptotically stationary and
ergodic flow systems, 11–12
flow systems
and entities, 8–9
Poisson process, 14–15
single-product inventory system,
16–17
limiting averages, 10–11
Little's Law, 12–14
stability and flow conservation, 10

T

Tackling varying output capacity,
650–651
Tail asymptotics, effective bandwidths
computation
buffer-content process, 617–618
CTMC on-off source, 612–613
feed-forward in-tree network,
618–619
four-state CTMC, 613–615
mean fluid-flow rate, 619–620
"multiple" independent sources,
617
network structure, 620
stable $M/M/1$ queue, 615–617
described, 611
Gärtner–Ellis conditions, 611
Threshold policies, optimal control
CTMC, 103
monotonic switching curve, 98
queues control, 102

scheduler's options, arrivals, 95–96
single server queues, 95
SMDP, 97
structure, 96
Timed round-robin (polling) policy
   admissible region K^{trr}, 660–661
   cycle time T, defined, 652
   effective bandwidth approximation,
     653–654
   estimation, maximum number of
     sources, 657
   estimation, required bandwidth, 658
   feasible region, 658–660
   logarithms of loss probability, 656
   QoS criteria, 653
   scheduler, 651–652
   SMP bounds, 654–655
   stability of buffer j, 652–653
Traffic superposition, multiple sources
   net fluid input rate, 606
   single infinite-sized buffer, 605–606
TravHelp calls, 196–198
Two-buffer fluid flow system
   buffer-2 with compensating source,
     681
   general on-off source, 678–679
   hysteretic policy, 679
   kernel of SMP, 682–683
   LST, 678
   multiclass fluid queues, 685
   nth Markov regenerative epoch, 681
   Perron–Frobenius eigenvalue, 684

schedulers, 680
scheduler's policy, 677
structure, 676–677

U

University Town Hospital (UTH)
   30-Minute ER Commitment, 266
   recommendations, 271
   upper management, 266–267
   urgent care centers, 266
UTH, *see* University Town Hospital
   (UTH)

V

Variable service rate queues
   CTMC, 125–126
   elastic traffic, 129–130
   iteration, 131–132
   processing rate, 124
   QBD process, 127
   single server queue, 125
   web servers, 128–129
Virtual interface architecture (VIA), 422

W

Whittle network, 360
Workload process
   definition, 516
   depleted rate c, 517
   discrete and fluid queues, 517
   fluid queue, 518

9 781439 806586

An environmentally friendly book printed and bound in England by www.printondemand-worldwide.com

PEFC Certified

This product is
from sustainably
managed forests
and controlled
sources

PEFC

PEFC/16-33-415   www.pefc.org

This book is made of chain-of-custody materials; FSC materials for the cover and PEFC materials for the text pages.

#0128 - 161115 - C0 - 234/156/43 [45] - CB - 9781439806586